Membranes for Membrane Reactors

Membranes for Membrane Reactors

Preparation, Optimization and Selection

Edited by

ANGELO BASILE

Institute on Membrane Technology, CNR, c/o University of Calabria, Rende, CS, Italy

FAUSTO GALLUCCI

Chemical Process Intensification, Faculty of Chemical Engineering and Chemistry, Eindhoven University of Technology, Eindhoven, The Netherlands

WILEY

A John Wiley and Sons, Ltd, Publication

This edition first published 2011

Registered office
John Wiley & Sons Ltd, The Atrium, Southern Gate, Chichester, West Sussex, PO19 8SQ, United Kingdom

For details of our global editorial offices, for customer services and for information about how to apply for permission to reuse the copyright material in this book please see our website at www.wiley.com.

Library of Congress Cataloging-in-Publication Data

Membranes for membrane reactors : preparation, optimization, and selection /
[edited by] Angelo Basile, Fausto Gallucci.
p. cm.
Includes bibliographical references and index.
ISBN 978-0-470-74652-3 (hardback)
1. Membrane reactors. I. Basile, Angelo (Angelo Bruno) II. Gallucci, Fausto.
TP248.25.M45M46 2011
660′.2832–dc22

2010044313

A catalogue record for this book is available from the British Library.

Print ISBN: 9780470746523
ePDF ISBN 9780470977552
oBook ISBN: 9780470977569
ePub ISBN: 9780470977576

Set in 9.5/11.5 pt Times Roman by Thomson Digital, Noida, India.
Printed and bound in Singapore by Markono Print Media Pte Ltd.

Contents

Contributors

J. Arias-Pardilla, *Centro de Electroquímica y Materiales Inteligentes, Universidad Politécnica de Cartagena, Cartagena, Spain*

A. Basile, *Institute on Membrane Technology, CNR, c/o University of Calabria, Rende, CS, Italy*

F. Basile, *Università di Bologna, Bologna, Italy*

N. L. Basov, *Topchiev Institute of Petrochemical Synthesis, Moscow, Russia*

P. Benito, *Università di Bologna, Bologna, Italy*

A. Berenguer-Murcia, *Universidad de Alicante, Departamento de Química Inorgánica, Alicante, Spain*

M. Broglia, *ERSE S.p.A., Milano, Italy*

R. S. Brusa, *Dipartimento di Fisica, Università di Trento, Trento, Italy*

M. Bryjak, *Department of Polymer and Carbon Materials, Wroclaw University of Technology, Wroclaw, Poland*

M. G. Buonomenna, *Department of Material and Chemical Engineering, University of Calabria and Consortium INSTM ,Rende, CS, Italy*

G. Capannelli, *Department of Chemistry and Industrial Chemistry, University of Genoa, Genoa, Italy*

D. Cazorla-Amorós, *Universidad de Alicante, Departamento de Química Inorgánica, Alicante, Spain*

R. Checchetto, *Dipartimento di Fisica, Università di Trento, TN, Italy*

S.-H. Choi, *Institute on Membrane Technology, ITM-CNR, c/o University of Calabri, Rende, CS, Italy; and Green Chemistry and Environmental Biotechnology, University of Science and Technology, Daejeon, Korea*

P. Daraei, *Membrane Research Center, Department of Chemical Engineering, Razi University, Kermanshah, Iran*

E. Drioli, *Institute on Membrane Technology/Department of Material and Chemical Engineering, ITM-CNR, c/o University of Calabria, Rende CS Italy*

S. D. Dubrovensky, *St. Petersburg State Institute of Technology, St. Petersburg, Russia*

M. M. Ermiloa, *Topchiev Institute of Petrochemical Synthesis, Moscow, Russia*

C. Evangelisti, *Department of Chemistry and Industrial Chemistry, University of Pisa, Pisa, Italy*

G. Fornasari, *Università di Bologna, Bologna, Italy*

F. Galiano, *Institute on Membrane Technology, ITM-CNR, c/o University of Calabria, Rende CS, Italy*

F. Gallucci, *Chemical Process Intensification, Faculty of Chemical Engineering and Chemistry, Eindhoven University of Technology, Eindhoven, The Netherlands*

J. Galuszka, *Natural Resources Canada, CanmetENERGY, Ontario, Canada*

I. Gancarz, *Department of Polymer and Carbon Materials, Wroclaw University of Technology, Wroclaw, Poland*

M. A. Gibson, *CSIRO Materials Science and Engineering, Clayton, Victoria, Australia*

T. Giddings, *Natural Resources Canada, CanmetENERGY, Ontario, Canada*

L. Guo, *State Key Laboratory of Multiphase Flow in Power Engineering, Xi'an Jiaotong University, Xi'an, PR China*

F. I. Hai, Environmental Engineering, The University of Wollongong, New South Wales, Australia

K. Haraya, *Research Institute for Innovation in Sustainable Chemistry, National Institute of Advanced Industrial Science and Technology, Tsukuba, Japan*

A. Inoue, *Institute for Materials Research, Tohoku University, Sendai, Japan.*

B. F. K. Kingsbury, *Department of Chemical Engineering, Imperial College London, London, UK*

S. Kurungot, *Physical and Materials Chemistry Division, National Chemical Laboratory, Pune, India*

V.I. Lebedeva, *Topchiev Institute of Petrochemical Synthesis, Russian Academy of Sciences, Moscow, Russia*

K. Li, *Department of Chemical Engineering, Imperial College London, London, UK*

M. Li, *State Key Laboratory of Multiphase Flow in Power Engineering, Xi'an Jiaotong University, Xi'an, PR China*

A. Linares-Solano, *Universidad de Alicante, Departamento de Química Inorgánica, Alicante, Spain*

S. S. Madaeni, *Membrane Research Center, Department of Chemical Engineering, Razi University, Kermanshah, Iran*

A. A. Malkov, *St. Petersburg State Institute of Technology, St. Petersburg, Russia*

A. A. Malygin, *St. Petersburg State Institute of Technology, St. Petersburg, Russia*

M. Menéndez, *Aragón Institute of Engineering Research, University of Zaragoza, Zaragoza, Spain*

S. V. Mikhaylovskiy, *St. Petersburg State Institute of Technology, St. Petersburg, Russia*

A. Miotello, *Dipartimento di Fisica, Università di Trento, Trento, Italy*

M. Monti, *Università di Bologna, Bologna, Italy*

E. Morallón, *Universidad de Alicante, Departamento de Química Física and Instituto Universitario de Materiales, Alicante, Spain*

A. P. Nemudry, *Institute of Solid State Chemistry and Mechanochemistry, SB RAS, Novosibirsk, Russia*

R. O'Hayre, *Metallurgical and Materials Engineering, Colorado School of Mines, Golden, Colorado, USA*

N. V. Orekhova, *Topchiev Institute of Petrochemical Synthesis, Moscow, Russia*

N. Panziera, *Department of Chemistry and Industrial Chemistry, University of Pisa, Pisa, Italy*

I. V. Petrova, *Topchiev Institute of Petrochemical Synthesis, Russian Academy of Sciences, Moscow, Russia*

J. W. Phair, *Division of Fuel Cells and Solid State Chemistry, Risø National Laboratory for Sustainable Energy, The Technical University of Denmark, Roskilde, Denmark*

P. Pinacci, *ERSE S.p.A., Milano, Italy*

E. Pitzalis, *CNR, Institute of Chemical and Physical Processes, Pisa, Italy*

V. I. Roldughin, *Frumkin Institute of Physical Chemistry and Electrochemistry, Russian Academy of Sciences, Moscow, Russia*

V. A. Sadykov, *Boreskov Institute of Catalysis, SB RAS, Novosibirsk, Russia*

E. Scavetta, *Università di Bologna, Bologna, Italy*

R. Sennen Brusa, *Dipartimento di Fisica, Università di Trento, Trento, Italy*

H. Susanto, *Lehrstuhl für Technische Chemie II, Universität Duisburg-Essen, Essen, Germany; and Department of Chemical Engineering, Universitas Diponegoro, Semarang, Indonesia*

C. Téllez, *Aragón Institute of Engineering Research, University of Zaragoza, Zaragoza, Spain*

G. F. Tereshchenko, *Topchiev Institute of Petrochemical Synthesis, Russian Academy of Sciences, Moscow, Russia*

M. Tonelli, *Università di Bologna, Bologna, Italy*

J. Tong, *Metallurgical and Materials Engineering, Colorado School of Mines, Golden, Colorado, USA*

S. Tosti, *ENEA, Dipartimento FPN, CR ENEA Frascati, Frascati (RM), Italy*

M. Ulbricht, *Lehrstuhl für Technische Chemie II, Universität Duisburg-Essen, Essen, Germany*

T. Uragami, *Faculty of Chemistry, Materials and Bioengineering Kansai University Suita, Osaka, Japan*

A. Vaccari, *Università di Bologna, Bologna, Italy*

F. Varela-Gandía, *Universidad de Alicante, Departamento de Química Inorgánica, Alicante, Spain*

G. Vitulli, *Advanced Catalysts Srl, Pisa, Italy*

V. V. Volkov, *Topchiev Institute of Petrochemical Synthesis, Russian Academy of Sciences, Moscow, Russia*

Z. Wu, *Department of Chemical Engineering, Imperial College London, London, UK*

T. Yamaguchi, *Chemical Resources Laboratory, Tokyo Institute of Technology, Yokohama, Japan*

S.Yamaura, *Institute for Materials Research, Tohoku University, Sendai, Japan*

M. Yoshimune, *Research Institute for Innovation in Sustainable Chemistry, National Institute of Advanced Industrial Science and Technology, Tsukuba, Japan*

V. V. Zyryanov, *Institute of Solid State Chemistry and Mechanochemistry, SB RAS, Novosibirsk, Russia*

Glossary

Nomenclature

A	membrane area (m^2)
b	Langmuir adsorption parameter (kPa^{-1})
b_o	Langmuir pre-exponential parameter (kPa^{-1})
Bi	Biot number ($\frac{h_W \cdot d_t/2}{\lambda_{er}}$)
$\tilde{c}_i$	dimensionless $\boldsymbol{i}$ component concentration
c_i	$\boldsymbol{i}$ component concentration ($\frac{kmol}{m^3}$)
$c_{p,i}$	specific heat of the component $\boldsymbol{i}$ ($\frac{kJ}{kmol \times K}$ or $\frac{kJ}{kg \times K}$)
$c_{p,\text{mix}}$	gas mixture specific heat ($\frac{kJ}{kg \times K}$)
Da	Damköhler number
D_e	effective diffusivity in the solid particle ($\frac{m^2}{h}$)
D_{ez}	effective axial mass diffusivity ($\frac{m^2}{h}$)
D_{er}	effective radial mass diffusivity ($\frac{m^2}{h}$)
d_p	equivalent particle diameter (m)
d_t	internal tubular reactor diameter (m)
E_a	apparent activation energy ($\frac{kJ}{kmol}$)
f	friction factor
F_{sweep}	sweeping gas flow rate ($\frac{kmol}{h}$)
F_{TOT}	inlet total gas mixture flow rate ($\frac{kmol}{h}$)
G	Gibbs free energy ($\frac{kJ}{mol}$)
h_f	heat transport coefficient between gas and solid phase ($\frac{kJ}{h \times m^2 \times K}$)
$h_{i,perm}$	i component enthalpy in permeation zone ($\frac{kJ}{mol}$)
$h_{i,reac}$	i component enthalpy in reaction zone ($\frac{kJ}{mol}$)
$(-\Delta H_j)$	heat of reaction, j ($\frac{kJ}{mol}$)
h_{perm}	forced convection heat transport coefficient in the permeation zone ($\frac{kJ}{h \times m^2 \times K}$)
h_W	heat transport coefficient near the tube wall ($\frac{kJ}{h \times m^2 \times K}$)
$h_{W,f}$	wall heat transport coefficient for the fluid phase ($\frac{kJ}{h \times m^2 \times K}$)
$h_{W,s}$	wall heat transport coefficient for the solid phase ($\frac{kJ}{h \times m^2 \times K}$)
Ji	i component flux through the membrane ($\frac{kmol}{h \times m^2}$)
K_i	adsorption equilibrium constant of the component $\boldsymbol{i}$ (kPa^{-1})
k_j	rate constant of reaction j (depends on reaction)
$K_{eq,j}$	equilibrium constant of reaction j
k_{mem}	membrane thermal conductivity ($\frac{kJ}{h \times m \times K}$)
k_{met}	tube wall conductivity ($\frac{kJ}{h \times m \times K}$)
L	reactor length (m)

M_i	molecular weight of the component $\boldsymbol{i}$ ($\frac{\text{kg}}{\text{kmol}}$)
M_m	average molecular weight of the gas mixture ($\frac{\text{kg}}{\text{kmol}}$)
p	pressure (kPa)
$\tilde{p}$	dimensionless pressure in the reaction zone
Δp	pressure drop along the reactor (kPa)
p_0	inlet pressure (kPa)
P_i	permeability coefficient ($\frac{\text{kmol}\times\text{m}}{\text{h}\times\text{m}^2\times\text{kPa}^n}$)
P_i^0	permeability pre-exponential factor ($\frac{\text{kmol}\times\text{m}}{\text{h}\times\text{m}^2\times\text{kPa}^n}$)
Pe_a	axial Peclèt number
Pe_{cr}	critical Peclèt number
Pe_{mr}	radial mass Peclèt number ($\frac{u_z d_p}{D_{er}}$)
p_i	i component partial pressure (kPa)
$p_{i,p}$	i component partial pressure on the permeate (downstream) side (kPa)
$p_{i,r}$	i component partial pressure on the retentate (upstream) side (kPa)
$p_{i,s}$	i component partial pressure on the catalyst surface (kPa)
Pr	Prandtl number ($\frac{c_p \mu_g}{k_g}$)
Pr_{perm}	Prandtl number in the permeation (downstream) zone
q_m	heat flux through the membrane ($\frac{\text{kW}}{\text{m}^2}$)
q_r	heat flux through the walls ($\frac{\text{kW}}{\text{m}^2}$)
r	radial coordinate (m)
$\tilde{r}$	radial dimensionless coordinate
R	universal gas constant ($\frac{\text{kJ}}{\text{kmol}\times\text{K}}$)
Re	Reynolds number
Re_p	Reynolds number referred to catalyst particle diameter ($\frac{G d_p}{\mu_g}$)
Re_{perm}	Reynolds number in the permeation (downstream) zone
$r_{i,j}$	reaction rate of component i in reaction j ($\frac{\text{kmol}}{\text{h}\times\text{m}^2}$ or $\frac{\text{kmol}}{\text{h}\times\text{kg}_{\text{cat}}}$)
Sc	Schmidt coefficient ($\frac{\mu}{\rho_g D}$)
s_s	specific surface area of metal
St	Stanton number
S/C	steam to carbon ratio
T	temperature (K)
T_{bi}	normal boiling temperature of component i (K)
T_{in}	inlet temperature (K)
T_m	membrane temperature (K)
T_W	wall temperature (K)
U	overall heat transport coefficient ($\frac{\text{kJ}}{\text{h}\times\text{m}^2\times\text{K}}$)
u_s	gas superficial velocity ($\frac{\text{m}}{\text{s}}$)
V_p	pellet volume (m^3)
$V_{reactor}$	reactor volume (m^3)
w	halfwidth of diffraction peaks
W/F	apparent residence time ($\frac{\text{kg}_{\text{cat}}\text{s}}{\text{mol}}$)
X_i	i component conversion
y_i	molar fraction of component $\boldsymbol{i}$
Y_i	overall yield to species $\boldsymbol{i}$
z	axial coordinate (m)
$\tilde{z}$	axial dimensionless coordinate

Greek symbols

α	perm-selectivity
δ	membrane thickness (m)
ε	void fraction of packing
η	effectiveness factor
η_j	effectiveness factor of reaction *j*
θ	angular position of X-ray reflection
λ	wavelength (m)
λ_e	effective thermal conductivity in the solid particle $(\frac{kJ}{h \times m \times K})$
λ_{ea}	effective axial thermal conductivity $(\frac{kJ}{h \times m \times K})$
λ_{er}	effective radial thermal conductivity $(\frac{kJ}{h \times m \times K})$
λ_g	gas phase thermal conductivity $(\frac{kJ}{h \times m \times K})$
λ_i	thermal conductivity of the component $\boldsymbol{i}$ $(\frac{kJ}{h \times m \times K})$
λ_S	thermal conductivity of packing material $(\frac{kJ}{h \times m \times K})$
ξ	radial coordinate inside the particle (m)
μ_g	gas mixture viscosity $(\frac{kJ}{h \times m})$
μ_i	viscosity of the component $\boldsymbol{i}$ $(\frac{kJ}{h \times m})$
ρ_B	catalytic bed density $(\frac{kg}{m^3})$
ρ_g	gas density $(\frac{kg}{m^3})$
ρ_m	weight concentration of metals in the solution $(g\,l^{-1})$
ρ_s	catalyst density $(\frac{kg}{m^3})$
$\hat{\phi}_i$	fugacity coefficient of species *i* in the gas mixture (kPa)
Λ	mean free path (m)
Γ	thermodynamic correction factor (–)

Subscripts or superscripts

app	apparent
eff	effective parameter
g	gas translational
i	component in mixture
j	reaction number
s	surface
sat	saturation
SM	Stefan–Maxwell

Acronyms

AR	melamine-formaldehyde resin
CA	cellulose acetate
CM	catalytic membrane
CMC	catalytic membrane contactors
CMR	catalytic membrane reactor

DMF	N,N-dimethylformamide
DO	dissolved oxygen
EC	ethyl cellulose
MC-CN	deposition of Pd onto outer surface of membrane in continuous regimes
MC-CS	deposition of Pd onto outer surface of membrane in consecutive regimes
MR	membrane reactor
NMP	N-methyl-2-pyrrolidone
PAA	polyacrilic acid
PAI	poly(amide imide)
PAN	polyacrylonitrile
pCM	polymeric catalytic membrane
PDMS	poly(dimethylsiloxane)
PEBA	poly(ether-b-amide)
PEI	polyetherimide
PEMFC	polymer electrolyte membrane fuel cell
PES	polyether sulfone
PP	polypropylene
ppb	parts per billion
ppm	parts per million
PPO	poly(2,6-dimethyl-1,4phenylene oxide)
PSF	polysulfone
PVDF	polyvinylidene
PVP	poly(vinylpyrrolidone)
THF	tetrahydrofurane
TR	traditional reactor (reactor without membranes)
WHSV	weight hourly space velocity

Introduction – A Review of Membrane Reactors

Fausto Gallucci[1], *Angelo Basile*[2] *and Faisal Ibney Hai*[3]

[1]*Chemical Process Intensification, Faculty of Chemical Engineering and Chemistry, Eindhoven University of Technology, Eindhoven, The Netherlands*
[2]*Institute on Membrane Technology, CNR, c/o University of Calabria, Rende, CS, Italy*
[3]*Environmental Engineering, The University of Wollongong, New South Wales, Australia*

1 Introduction

In recent decades, membrane catalysis has been studied by several research groups, and significant progress in this field is summarised in several review articles [7,143,146,154,194,195,202].

Considering a IUPAC definition [131], a membrane reactor (MR) is a device for simultaneously performing a reaction (steam reforming, dry reforming, autothermal reforming, etc.) and a membrane-based separation in the same physical device. Therefore, the membrane not only plays the role of a separator, but also takes place in the reaction itself. The term membrane bioreactor (MBR), however, refers to the coupling of biological treatment with membrane separation in contrast to the sequential application of membrane separation downstream of classical biotreatment [117,237]. This introduction comprises a review of both MR (Sections 2–5) and MBR (Section 6).

2 Membranes for Membrane Reactors

The membranes can be classified according to their nature, geometry and separation regime. In particular, they can be classified into organic, inorganic and organic/inorganic hybrids.

The choice of membrane type to be used in MRs depends on parameters such as the productivity, separation selectivity, membrane life time, mechanical and chemical integrity at the operating conditions and, particularly, the cost.

The discovery of new membrane materials was the key factor for increasing the application of the membrane in the catalysis field. The significant progress in this area is reflected in an

Membranes for Membrane Reactors: Preparation, Optimization and Selection, Edited by Angelo Basile and Fausto Gallucci

increasing number of scientific publications, which have grown exponentially over the past few years, as shown by McLeary *et al.* [154].

Generally, the membranes can be even classified into homogenous or heterogeneous, symmetric or asymmetric in structure, solid or liquid; they can possess a positive or negative charge as well as they can be neutral or bipolar. In all cases, a driving force as a gradient of pressure, concentration, etc., is applied in order to induce the permeation through the membrane.

Thus, the membranes can be categorised according to their nature, geometry and separation regime [125].

The first classification is by their nature, which distinguishes the membranes into biological and synthetic ones, which differ completely for functionality and structure. *Biological membranes* are easy to manufacture, but present many disadvantages such as limited operating temperature (below 100 °C), limited pH range, drawbacks related to the clean up, susceptibility to microbial attack due to their natural origin [248].

Synthetic membranes can be subdivided into organic (polymeric) and inorganic (ceramic, metal) ones. *Polymeric membranes* commonly operate between 100 and 300 °C [34], *inorganic ones* above 250 °C. Moreover, inorganic membranes show both wide tolerance to pH and high resistance to chemical degradation. Referring to the organic membranes, it can be said that the majority of the industrial membrane processes are made from natural or synthetic polymers. Natural polymers include wool, rubber (polyisoprene) and cellulose, whereas synthetic polymers include polyamide, polystyrene and polytetrafluoroethylene (Teflon).

In the viewpoint of the morphology and/or membrane structure, the inorganic membranes can be even subdivided into porous and metallic. In particular, as indicated by IUPAC [131] definition, *porous membranes* can be classified according to their pore diameter: microporous ($dp < 2\,nm$), mesoporous ($2\,nm < dp < 50\,nm$) and macroporous ($dp > 50\,nm$).

Metallic membranes can be categorised into supported and unsupported ones. Supported dense membranes offer many advantages unmatched by the porous ceramic membranes. In particular, many efforts were devoted to develop dense metallic layers deposited on a porous support (alumina, silica, carbon, zeolite) for separating hydrogen with a noncomplete perm-selectivity, but lowering the costs related to the dense metallic membranes. In fact, the kind of membranes based on palladium and its alloy is used for gas separation and in MR field for producing pure H_2 [143] and presents as main drawback the high cost.

2.1 Polymeric Membranes

Basically, all polymers can be used as membrane material but, owing to a relevant difference in terms of their chemical and physical properties, only a limited number of them is practically utilised. In fact, the choice of a given polymer as a membrane material is not arbitrary, but based on specific properties, originating from structural factors. Ozdemir *et al.* [169] gives an overview of the commercial polymers used as membranes as well as of other polymers having high potentially for application as a membrane material. However, many industrial processes involve operations at high temperatures. In this case, polymeric membranes are not suitable and, therefore, inorganic ones are preferred.

2.2 Inorganic Membranes

Inorganic membranes are commonly constituted by different materials as ceramic, carbon, silica, zeolite, oxides (alumina, titania, zirconia) as well as palladium, silver and so forth, and their alloys.

Table 1 *Advantages and disadvantages of inorganic membranes with respect to polymeric membranes*

Advantages	Disadvantages
Long-term stability at high temperatures	High capital cost
Resistance to harsh environments (chemical degradation, pH, etc.)	Embrittlement phenomenon (in the case of dense Pd membranes)
Resistance to high pressure drops	Low membrane surface per module volume
Inertness to microbiological degradation	Difficulty of achieving high selectivities in large scale microporous membranes
Easy cleanability after fouling	Generally low permeability of the highly hydrogen selective (dense) membranes at medium temperatures
Easy catalytic activation	Difficult membrane to module sealing at high temperature

They can operate at elevated temperatures. In fact, they are stable at temperatures ranging from 300 to 800 °C and in some cases (ceramic membranes) usable over 1000 °C [234]. They present also high resistance to chemical degradation. As previously said, the inorganic membranes present a high cost as main drawback.

Table 1 sketches the most important advantages and disadvantages of inorganic membranes with respect to polymeric ones.

So, although inorganic membranes are more expensive than the polymeric ones, they possess advantages such as resistance towards solvents, a well defined stable pore structure (in the case of porous inorganic membranes), high mechanical stability and elevated resistance at high operating temperatures.

2.2.1 Metal Membranes

Conventionally, dense metal membranes are used for hydrogen separation from gas mixtures and in MR area. Palladium and its alloys are the dominant materials for preparing this kind of membranes due to its high solubility and permeability of hydrogen. Unfortunately, owing to the low availability of palladium in the nature, it results to be very expensive. Recently, supported thin metallic membranes are realised by coating a thin layer of palladium (showing thickness ranging from submicron to few microns) on a ceramic support. In this case, the advantages include reduced material costs, improved resistance to mechanical strength and higher permeating flux.

Otherwise, dense membranes selectively permeable only to hydrogen based on tantalum, vanadium, nickel and titanium are considered valid and less expensive alternative with respect to the palladium and its alloy.

A problem associated with metal membranes is the surface poisoning, which can be more significant for thin metal membranes. The influence of poisons such as H_2S or CO on Pd-based membranes is a serious problem. These gases are adsorbed on the palladium surface blocking available dissociation sites for hydrogen. The effect of small amounts of H_2S may be minimised by operating at higher temperature or by using a protective layer of platinum. CO can easily desorb at operating temperatures above 300 °C [5].

2.2.2 Ceramic Membranes

These membranes are made from aluminium, titanium or silica oxides. They show as advantages of being chemically inert and stable at high temperatures. This stability makes ceramic microfiltration and ultrafiltration membranes particularly suitable for food, biotechnology and pharmaceutical applications in which membranes require repeated steam sterilisation and chemical cleaning. Ceramic membranes have been also proposed for gas separation as well as for application in MRs.

However, some problems remain to be solved: difficulties in proper sealing of the membranes in modules operating at high temperature, extremely high sensitivity of membranes to temperature gradient leading to membrane cracking, chemical instability of some perovskite-type materials.

2.2.3 Carbon Membranes

Carbon molecular sieve (CMS) membranes have been identified as very promising candidates for gas separation, both in terms of separation properties and stability. CMS are porous solids containing constricted apertures that approach the molecular dimensions of diffusing gas molecules. As such, molecules with only slight differences in size can be effectively separated through molecular sieving [84].

CMS membranes can be obtained by pyrolysis of many thermosetting polymers such as poly (vinylidene chloride) or PVDC, poly(furfural alcohol) or PFA, cellulose triacetate, polyacrylonitrile or PAN and phenol formaldehyde and carbon membranes can be divided into two categories: supported and unsupported.

2.2.4 Zeolite Membranes

Zeolites are microporous crystalline alumina silicate with an uniform pore size. Zeolites are used as catalysts or adsorbents in the form of micron- or submicron-sized crystallites embedded in millimetre-sized granules.

One of the main drawbacks related to these membranes is represented by their relatively low gas fluxes compared to other inorganic membranes. Moreover, another important problem is represented by the zeolites thermal effect. The zeolite layer can exhibit negative thermal expansion, that is, in the high temperature region the zeolite layer shrinks, but the support continuously expands, resulting in thermal stress problems for the attachment of the zeolite layer to the support as well as for the connection of the individual microcrystals within the zeolite layer [35].

2.3 Membrane Housing

Concerning the applications of both organic and inorganic membranes, several configurations are conventionally used for the membrane housing. Generally, a modular configuration (parallel, in series and so on) may be combined for producing the desired effect. Membrane housing provides support and protection against operating pressures. Plate and frame, spiral wound, tubular and hollow fibre systems are the most common membrane housing configurations. The advantages and disadvantages of the different membrane elements are listed in Table 2.

Table 2 *Advantages and disadvantages of different kinds of membrane housing*

Figure	Membrane module	Advantages	Disadvantages
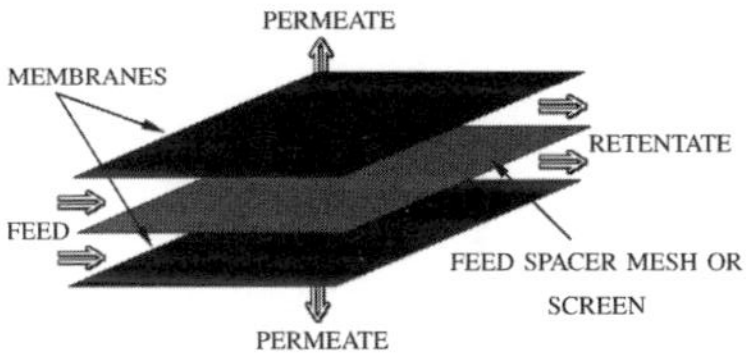	Flat sheet/plate and frame	• Moderate membrane surface/volume ratio • Well-developed equipment • Easy membrane replacement	• Susceptible to plugging at flow stagnation points • Difficult to clean • Expensive
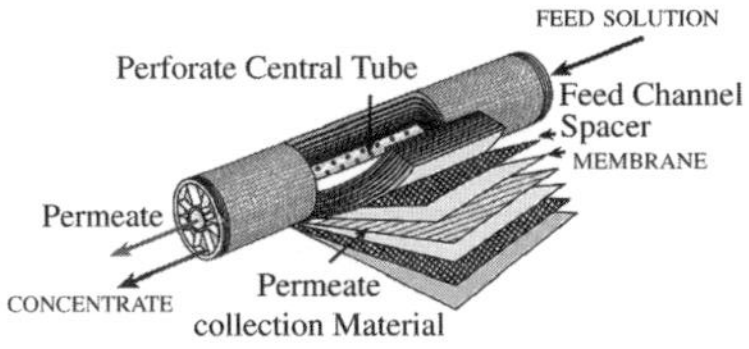	Spiral wound	• Compact • Good membrane surface/volume ratio • Minimum energy consumption • Low capital/operating cost	• Not suitable for very viscous fluid • Difficult to clean • Faulty membrane – change whole module
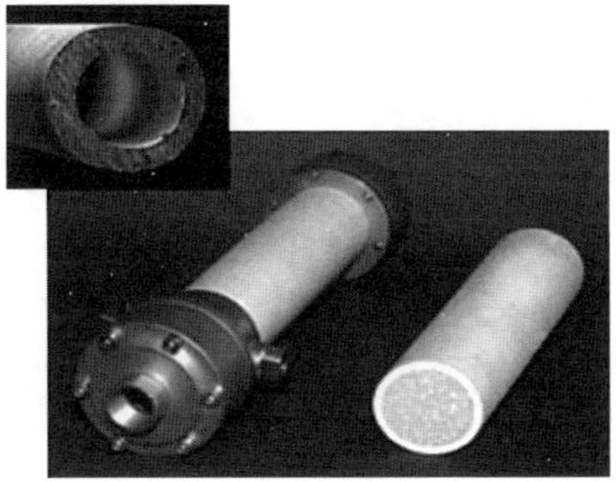	Hollow fibre	• Compact • Excellent membrane surface/volume ratio • Low energy consumption	• Susceptible to end-face fouling • Susceptible to plugging by particulates • Single fibre damage – replace entire module

(continued)

Table 2 (*Continued*)

Figure	Membrane module	Advantages	Disadvantages
	Tubular	• Easy to clean • Feed stream with particulate matter can be put through membrane • Good hydrodynamic control • Individual tubes can be replaced	• Relative high volume required for unit membrane area • High energy consumption • Relatively expensive

2.4 Membrane Separation Regime

Mass transport through porous and dense membranes occurs with different mechanisms. In porous membranes, molecular transport occurs depending on the membrane properties. In particular, macroporous materials, such as α–alumina, provide no separating function and are mainly used to create controlled dosing of a reactant or to support a dense or mesoporous separation layer. Transport through mesoporous membranes, such as Vycor glass or γ–alumina, is governed by Knudsen diffusion. These membranes are used as composite membranes with macroporous support materials. Microporous membranes, such as carbon molecular sieves, porous silica and zeolites, offer higher separation factors due to their molecular sieving effect.

2.4.1 Porous Membrane

The different transport mechanisms in porous membranes are presented below:

1. *Poiseuille (viscous) mechanism* (Figure 1) – This mechanism occurs when the average pore diameter is bigger than the average free path of fluid molecules. In this case, no separation takes place (Saracco 1994).
2. *Knudsen mechanism* (Figure 2) – When the average pore diameter is similar to the average free path of fluid molecules, Knudsen mechanism takes place. In this case, the flux of the component through the membrane is calculated by means of this equation [195]:

$$J_i = \frac{G}{\sqrt{2 \cdot M_i \cdot R \cdot T}} \cdot \frac{\Delta p_i}{\delta} \tag{1}$$

3. *Surface diffusion* (Figure 3) – This mechanism is achieved when one of the permeating molecules is adsorbed on the pore wall. This type of mechanism can reduce the effective pore dimensions obstructing the transfer of different molecular species [121].
4. *Capillary condensation* (Figure 4) – When one of the component condenses within the pores due to capillary forces, this type of mechanism takes place. Generally, the capillary condensation favours the transfer of relatively large molecules [136].
5. *Multi-layer diffusion* (Figure 5) – When the molecule–surface interactions are strong multi-layer diffusion occurs. This mechanism is like to an intermediate flow regime between surface diffusion and capillary condensation [229].
6. *Molecular sieving* (Figure 6) – This takes place when pore diameters are very small, allowing the permeation of only the smaller molecules.

2.4.2 Dense Metallic Membranes

In dense metallic membranes, molecular transport occurs through a solution–diffusion mechanism. In particular, in a dense palladium-based membrane, hydrogen atoms interact with

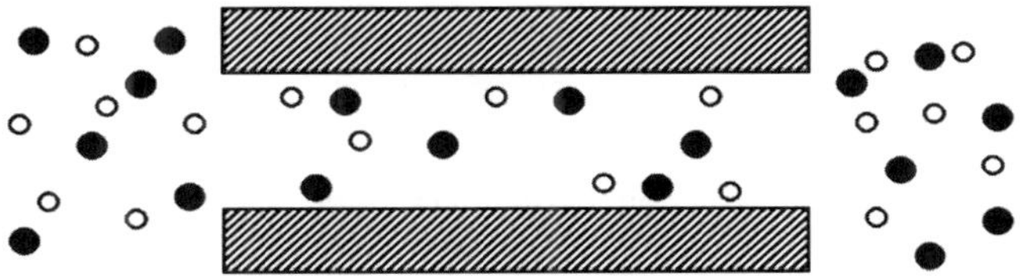

Figure 1 *Poiseuille mechanism*

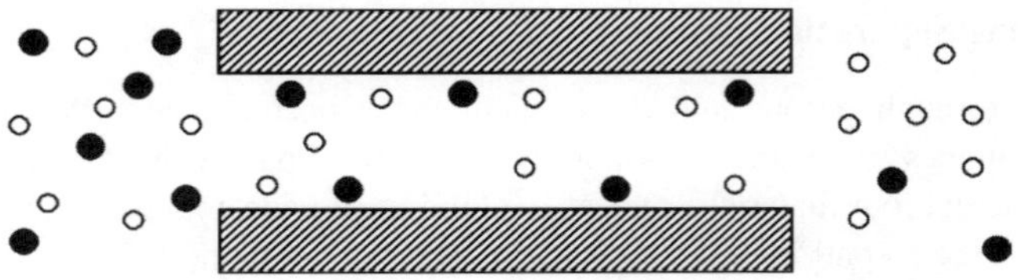

Figure 2 *Knudsen mechanism*

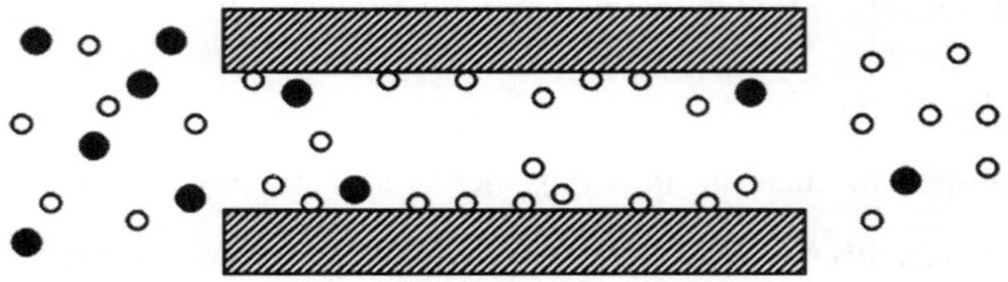

Figure 3 *Surface diffusion*

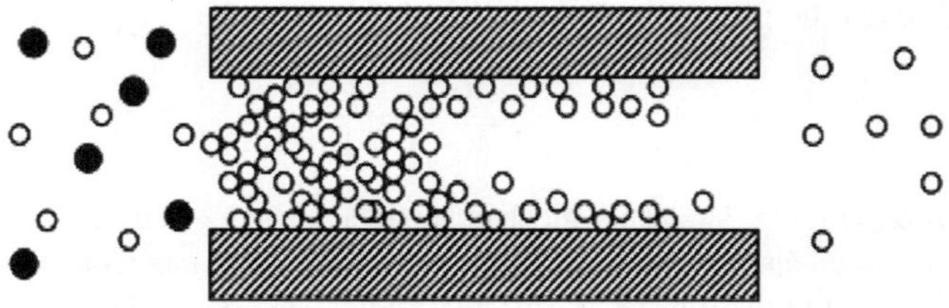

Figure 4 *Capillary condensation*

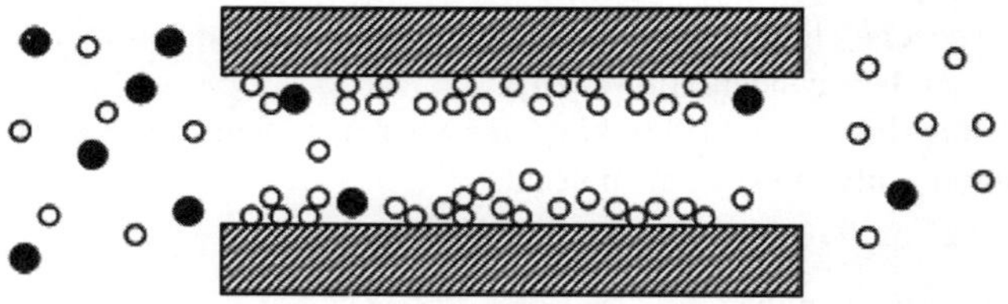

Figure 5 *Multi-layer diffusion*

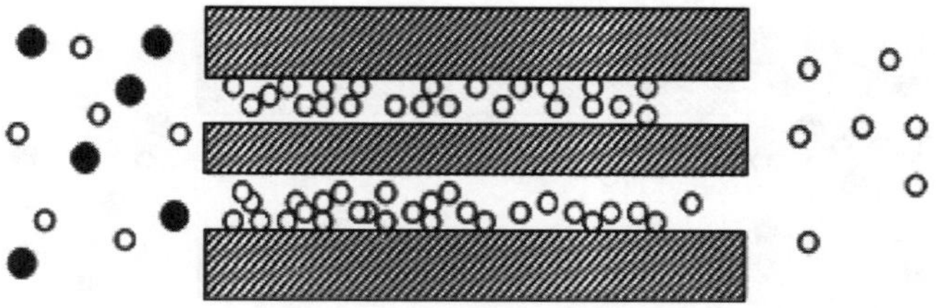

Figure 6 *Molecular sieving*

palladium metal. Hydrogen permeation through the membrane is a complex process with several stages:

- dissociation of molecular hydrogen at the gas/metal interface,
- adsorption of the atomic hydrogen on membrane surface;
- dissolution of atomic hydrogen into the palladium matrix;
- diffusion of atomic hydrogen toward the opposite side;
- recombination of atomic hydrogen to form hydrogen molecules at the gas/metal interface;
- desorbtion of hydrogen molecules.

At a fixed temperature, the hydrogen permeation flux through a dense palladium membrane can be expressed by means of this relation:

$$J_{H2} = Pe_{H2} \cdot (p^n \mathrm{H2, ret} - p^n \mathrm{H2, perm})/\delta \quad (2)$$

where J_{H2} is the hydrogen flux through the membrane, Pe_{H2} is the hydrogen permeability, δ is the membrane thickness, $p_{H2,ret}$ and $p_{H2,perm}$ are the hydrogen partial pressures at the retentate and permeate sides, respectively, and n (in the range 0.5–1.0) is the dependence factor of the hydrogen flux on the hydrogen partial pressure.

When the pressure is relatively low [202], $n = 0.5$ and Equation (2) becomes the Sieverts–Fick law:

$$J_{H2,Sieverts} = Pe_{H2} \cdot (p^{0.5}{}_{H2,ret} - p^{0.5}{}_{H2,perm})/\delta \quad (3)$$

The thickness of a dense palladium membrane is very important because it represents a compromise between two factors. On one hand, a thinner membrane offers less flow resistance and, hence, a higher permeability. On the other hand, practical fabrication technology limits the thickness of the membrane with respect to mechanical integrity and strength.

Moreover, palladium alloys are preferred over pure palladium for two reasons. Firstly, the hydrogen permeability of some palladium alloys is higher than those of pure palladium. Secondly, pure palladium can become brittle after different thermal and hydrogenation cycles. The choice of alloying other different metals to the palladium has been studied, for example, by Hwang *et al.* [113]. The authors found that the palladium alloyed show different hydrogen fluxes depending on the metal content (Figure 7).

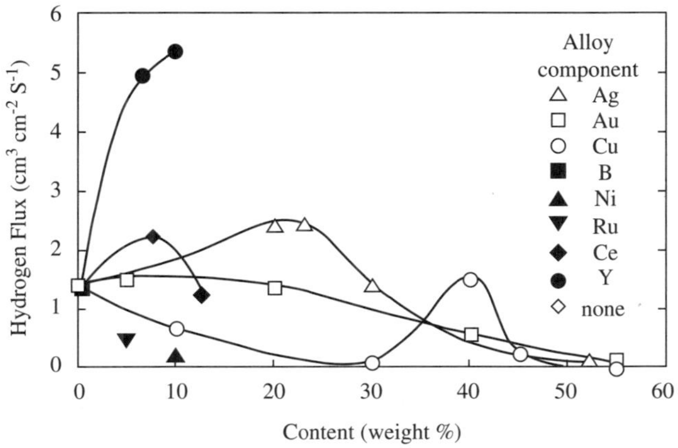

Figure 7 *Hydrogen flux through palladium alloy membranes against metal content [202]*

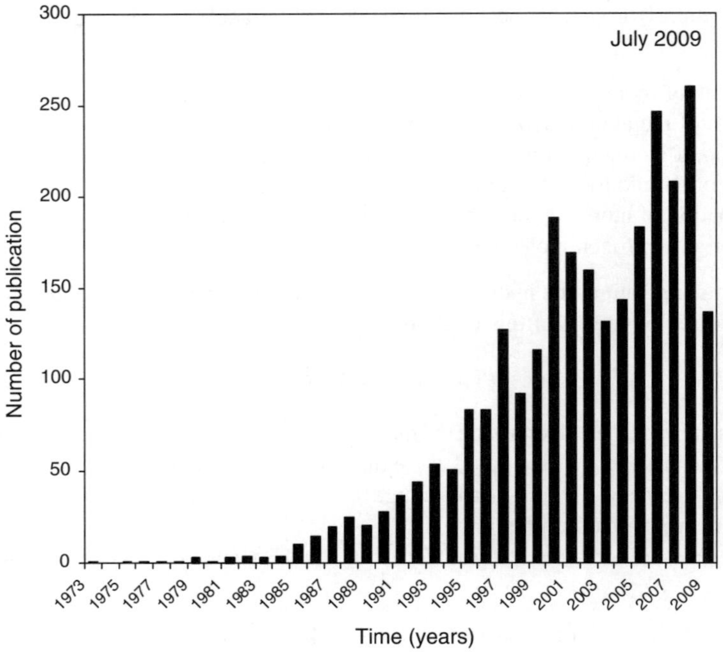

Figure 8 *Number of publications versus time*

3 Salient Features of Membrane Reactors

As already said, a membrane reactor combines the chemical reaction and gas separation. The significant progress in the field of MRs is reflected in the increasing number of publications, as shown in Figure 8.

Many heterogeneous gas–solid catalytic processes of industrial relevance (conventionally carried out using fixed, fluidised or trickle bed reactors) involve the combination of operations at high temperatures and in chemically harsh ambient. For these two factors, inorganic membranes are favourite over polymeric materials.

A MR can show flat (Figure 9) or tubular (Figure 10) geometry. In tubular MR, the density of packed bed could be improved using multichannel tubular monoliths and depositing the catalyst inside the pores.

Generally, the MRs can be also subdivided as reported below (and shown in Figure 10):

- catalytic membrane reactors (CMR);
- packed bed membrane reactors (PBMR);
- catalytic nonperm-selective membrane reactors (CNMR),
- nonperm-selective membrane reactors (NMR);
- reactant-selective packed bed reactors (RSPBR).

3.1 Applications of Membrane Reactors

Membrane reactors are mainly used to carry out the reactions limited by the equilibrium conversion such as water gas shift and so on. In fact, in a MR the separation capability of a

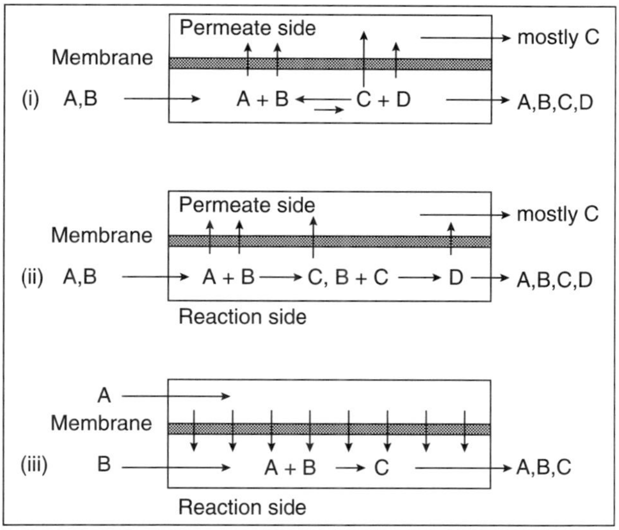

Figure 9 *Flat membrane reactor: (i) removal of product for a limited reaction thermodynamically, (ii) removal of favourite product, (iii) controlled addition of reactants*

membrane is utilised to improve the performance of a catalytic system. Usually, there are two main generic approaches: selective product separation (extractor) and selective reactant addition (distributor), as shown in Figure 11 [119]. The first MR type facilitates the in–situ removal of one of the products (Figure 11a). For example, for steam reforming reactions, H_2 yield and CO_2 product selectivity in TRs are limited by thermodynamics. By selective removal of H_2 from the reaction side, the thermodynamic equilibrium restrictions can be overcome. Due to the shift effect, both high H_2 yields and high CO_2 selectivities can be achieved. Moreover, this effect allows operation at milder reaction conditions in terms of temperature and pressure [257].

The second kind of MR uses the membrane to control the contact within reactants (Figure 11b). Both a perm-selective and a nonperm-selective membranes can be used to feed distributively one of the reactants. For partial oxidation reactions in TRs, O_2 rich feed results in low product selectivity and high reactant conversions. On the contrary, low oxygen content feed results in high product selectivity but lower conversions. Using a membrane for distributive feeding of O_2 along the axial coordinate of the catalytic bed, both high reactant conversions and high product selectivities can be combined [30,58,119,196]. An additional advantage of this approach is that the reactant (hydrocarbon) and O_2 feeds are not premixed and, hence, the possibility of realising mixtures as well as the flame back firing into the feed lines are greatly reduced. Moreover, the feed distribution can represent a promising approach for fast reactions.

3.2 Advantages of the Membrane Reactors

With respect to TRs, a MR permits the improvement of the performances in terms of reaction conversion, products selectivity, and so on. In fact, by means of the so called *shift effect*, the thermodynamic equilibrium restrictions can be overcome. At least, MRs behaviour could be the same of a TR working at the same MRs operating conditions.

Keizer *et al.* [124] studied the performances of several MRs using different kind of membranes. As reported in Figure 12, they represent the dependence of the cyclohexane

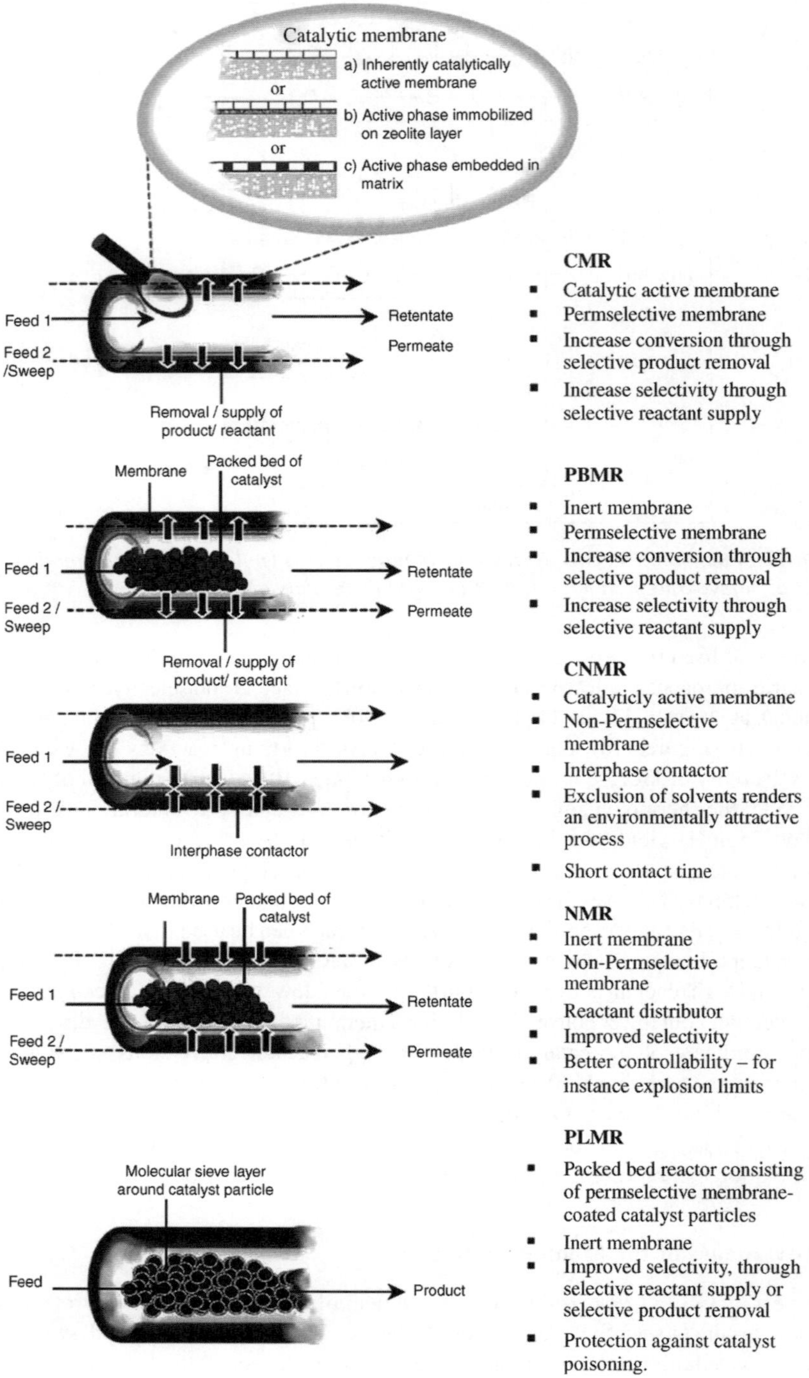

Figure 10 *Classification of MRs based on the function and position of membrane [154]. Reprinted from Microporous and Mesoporous Materials, McLeary, E. E., Jansen, J. C., Kapteijn, F., Zeolite based films, membranes and membrane reactors: Progress and prospects. Vol. 90. Copyright (2006) with permission from Elsevier*

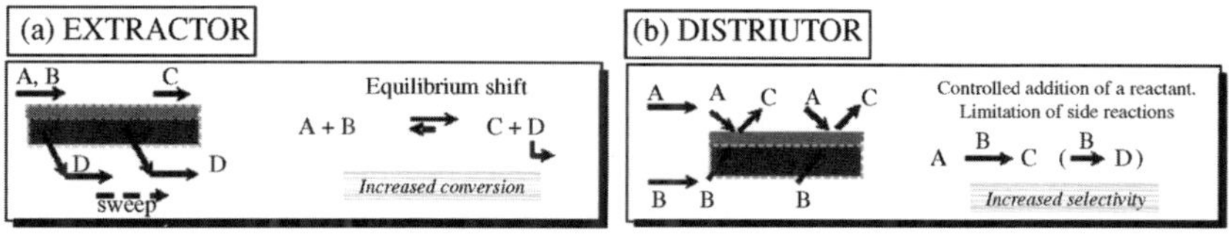

Figure 11 *Two approaches in membrane reactors [257]*

conversion as a function of the parameter *H*, defined as permeation to reaction ratio and considering the Damköhler number (Da) equal to 1. The line with $H = 0$ represents a TR, while other lines correspond to a different type of MRs. In particular, lines 1 and 2 refer to MRs governed by Knudsen transport mechanisms, lines 3 and 4 refer to microporous MRs and lines 5 and 6 refer to dense ones. Two regions can be distinguished. The first one corresponds to low permeation to reaction rate ratios. In this region, microporous MRs show the same behaviors of dense and mesoporous ones. However, the performance of each MR types in terms of conversion is better than the TRs ones.

At higher *H* values, the difference in the MRs properties are visible. MRs with a finite separation factor show an optimum permeability/reaction rate region. Above optimum the reactant loss due to permeation induces a detrimental effect on the conversion. The higher the separation the higher the conversion in this optimum region. MRs with infinite separation factors for hydrogen do not show this conversion drawback since no loss of reactants occurs. Thus, they maintain the conversion at a high value.

As shown, this kind of membranes represents an important issue concerning the MR performances in terms of conversion, hydrogen selectivity and so on.

Thus, the main advantage of using MRs is represented by the combination of reaction and hydrogen separation, leading to a reduction of capital cost and better reactor performances. Moreover, they allow also controlling additions of reactants and coupling of reactions [195].

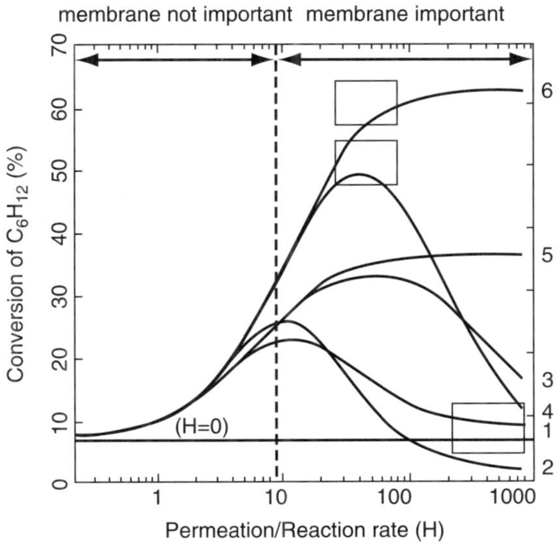

Figure 12 *Conversion of cyclohexane versus H [124]*

4 Hydrogen Production by Membrane Reactors

The world economy is mainly based on the exploitation of fossil fuels (oil, coal, methane) [160], according to data provided by International Energy Agency reported in Figure 13. In particular, the primary energy source is oil, but owing to the decrease of its reserves and to the increase of the environmental pollution due to emissions of CO_2 and other greenhouse gases (in the world, more than 75.0% of CO_2 emissions comes from burning of fossil fuels and, in the past 70 years, more than 30.0% of CO_2 increment as volume percentage was registered in the atmosphere [152]), it is strongly necessary to develop new technologies as well as to exploit renewable materials as alternative to the derived fossil fuels.

For example, fuel cells have been identified as one of the most promising technologies for the future clean energy industry [210]. They can be applied to large-scale stationary systems for distributed power generation as well as for small-scale portable power supplying devices for microelectronic equipment and auxiliary power units in vehicles [241]. Compared to other types of fuel cells, PEMFCs generate more power for a given volume or weight of fuel cell. This high-power density characteristic makes them compact and lightweight. PEMFCs are fed by pure hydrogen and only few ppm of CO (<10) may be tolerated by the anodic Pt catalysts. For this reason, it is strictly necessary to use a pure or at least CO-free hydrogen stream for feeding a PEM fuel cell.

Industrially, hydrogen is produced in fixed bed reactors by means of reforming reactions of fossil fuels such as natural gas, gasoline and so on. Nevertheless, as previously mentioned, in order to solve the problems related to the environmental pollution, it is necessary the exploitation of renewable materials. Therefore, hydrogen could be produced using *clean* fuels [96].

The steam reforming reaction is conventionally carried out in fixed bed reactors and produces a stream containing hydrogen with other byproduct gases like mainly CO, CH_4 and CO_2. Therefore, in the viewpoint of feeding a PEMFC, hydrogen needs to be purified by means of the following processes: water gas shift (WGS) reaction, pressure swing adsorption and/or Pd

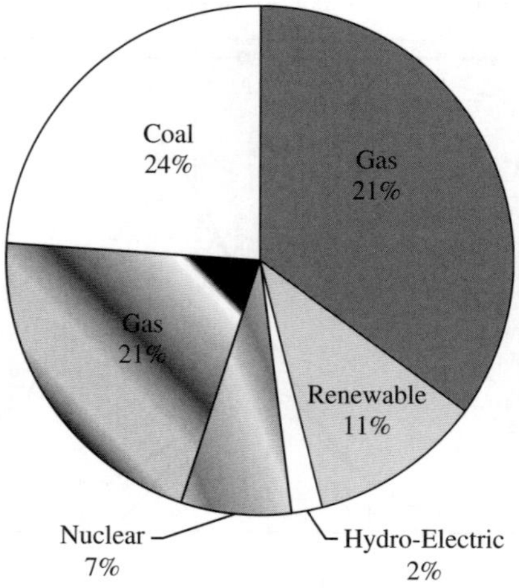

Figure 13 *Energy sources used in the world provided by International Energy Agency (http://www.eia.doe.gov/pub/international/iealf/tablee1p.xls)*

membrane separation, etc. Otherwise, it could be economically more advantageous to use a hydrogen perm-selective MR, able to both carry out the reaction and remove pure hydrogen in the same device [19,42,61,153,223,232]. In particular, with respect to the traditional reactors (TRs), MRs are able to:

- combine chemical reaction and hydrogen separation in only one system reducing the capital costs;
- enhance the conversion of equilibrium limited reactions;
- achieve higher conversions than TRs, operating at the same MR conditions, or the same conversion, but operating at milder conditions;
- improve yield and selectivity.

Moreover, as previously said, the most useful membranes offering a complete hydrogen perm-selectivity are the dense palladium-based ones [146]. The transport mechanism related to hydrogen permeation through a dense Pd-based membrane is solution/diffusion [240]. Generally, when a dense Pd membrane is exposed to a hydrogen stream at low temperatures ($< 300\,°C$), the embrittlement phenomenon takes place owing to the various typologies of expansion of the reticular constants in the Pd-H systems. A possible solution is represented by alloying palladium with elements, such as silver or copper, in order to obtain Pd-H phases with increased reticular step and able of anticipating the reticular expansion from hydrogen [108].

Since the 1960s, hydrogen production by MRs has been mainly studied using dense Pd-based membranes and microporous silica membranes. Pd membranes for H_2 production overcome all the other candidate materials due to of the very high solubility of H_2 in pure Pd (Figure 14) and for their infinite perm-selectivity to H_2.

In particular, Pd adsorbs 600 times its volume of H_2 at room temperature [120]. For this characteristics, Pd or Pd alloys on metallic or ceramic supports have been widely studied [6,71,126,140,173,189,203,221,224,228,238,259].

Therefore, as stated in the first part of this section, it is very interesting to investigate the production of hydrogen by means reforming reaction of renewable sources, using the innovations connected to the MRs. However, in the following a small overview is presented on hydrogen production based on the classic processes such as methane steam reforming, methane dry reforming and partial oxidation of methane as well as water gas shift reaction coupled with the use of membrane reactors.

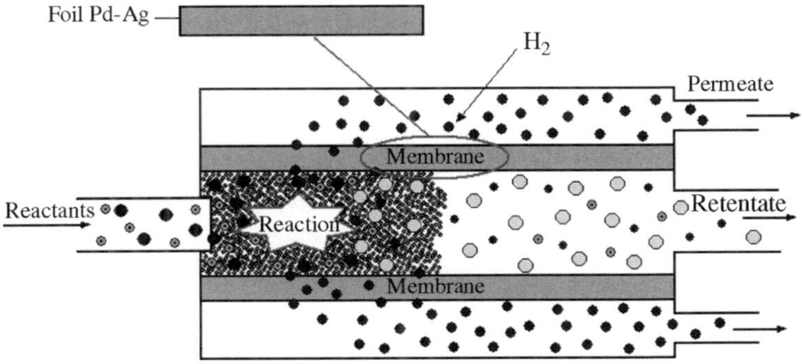

Figure 14 *Conceptual scheme of a dense Pd-Ag MR*

4.1 Methane Steam Reforming

Conventionally, hydrogen is produced by exploiting methane as a derived fossil fuel in reforming reactions. Currently, 80.0–85.0% of the worldwide hydrogen supply is produced by a methane steam reforming (SRM) reaction in fixed bed reactors [205]:

$$CH_4 + 2H_2O \rightleftarrows 4H_2 + CO_2 \quad \Delta H^{\circ}_{298k} = 165.0\,\text{kJ mol}^{-1} \tag{4}$$

Alternatively, methane could be renewably obtained via biogas generated by the fermentation of organic matter including manure, wastewater sludge, municipal solid waste (including landfills) or any other biodegradable feedstock, under anaerobic conditions. The composition of biogas varies depending on the origin of the anaerobic digestion process. Advanced waste treatment technologies can produce biogas with 55.0–75.0% of CH_4 using *in situ* purification techniques [187].

However, most part of the specialised literature on SRM area is devoted to study the optimal reaction conditions and the most adequate catalyst usable during the reaction in TRs.

In recent decades, the alternative technology of the membrane reactors has been applied to SRM reaction in order to produce hydrogen with the advantages previously reported in this work. In particular, recent reviews regarding the state of the art on the hydrogen production via SRM reaction performed by MRs have been published [11,188]. Moreover, different scientific papers deal on various MRs (based mainly on dense palladium and its alloy membranes) for hydrogen production by SRM reaction [40,41,88,98,222,226].

4.2 Dry Reforming of Methane

Another approach for hydrogen production in MRs is the dry reforming of methane:

$$CH_4 + CO_2 = 2CO + 2H_2 \quad \Delta H^{\circ}_{298k} = +247.0\,\text{kJ mol}^{-1} \tag{5}$$

In particular, methane dry reforming reaction could reduce the amount of greenhouse gases present in the atmosphere. An important limitation for making the methane dry reforming a commercially viable reaction using TRs is due to thermodynamics, which limits the conversion. Nevertheless, in a MR, methane (and carbon dioxide) conversion can be increased though the reaction products (or preferentially only hydrogen) are selectively removed from the reaction side.

Gallucci *et al.* [89] performed the dry reforming reaction in both TR and MR with the aim of consuming carbon dioxide and producing hydrogen. Moreover, by using the dense Pd membrane reactor, the carbon deposition on the catalyst is drastically reduced and a CO-free hydrogen stream is produced. At 450 °C, the maximum CO_2 conversion obtained in the MR was around 20.0% versus 14.0% achieved in the TR.

Haag *et al.* [98] studied the methane dry reforming reaction in a composite MR, where the membrane was constituted of a thin, catalytically inactive nickel layer, deposited by electroless plating on asymmetric porous alumina with acceptable hydrogen perm-selectivity at high temperature. Ferreira-Aparicio *et al.* [76] analysed the applicability of mesoporous ceramic filters in a MR to carry out the dry reforming of methane with carbon dioxide.

4.3 Partial Oxidation of Methane

Both steam reforming and dry reforming of methane are endothermic reactions. On the contrary, the partial oxidation of methane (POM) is an exothermic reaction, in which the main drawback in

TRs is represented by the thermodynamics. For example, the pressure increase gives a decrease in equilibrium methane conversions:

$$CH_4 + 1/2O_2 = CO + 2H_2 \quad \Delta H^{\circ}_{298k} = -36.0\,\text{kJ mol}^{-1} \tag{6}$$

Therefore, a MR allows these thermodynamic limitations to be overcome, reaching a high methane conversion at low temperature with respect to a TR.

By using a dense Pd-based MR with respect to a TR exercised at same conditions, Basile *et al.* [14,15] stated that:

- the methane conversion is remarkably higher in MRs than in the TRs, at a fixed temperature.
- the Pd-based MR shows the highest methane conversion (96.0% at 550 °C and 1.2 bar).
- the MR methane conversions exceed the thermodynamic equilibrium conversion.

Yin *et al.* [255] used a tubular MR for correlating air separation with catalytic POM. The MR consisted of three annular layers: a porous and thin cathodic layer, a dense and thin mixed conducting layer and a porous, thick anodic layer. At 850 °C, high methane conversion (>90.0%), CO selectivity (>90.0%) and hydrogen selectivity (>80.0%) were obtained as best result.

Cheng *et al.* [43] using a MR equipped with a Pd-based membrane for carrying out the POM reaction, obtained as best result 97.0% of hydrogen purity, 85.0% of methane conversion and 98.0% of oxygen conversion.

4.4 Water Gas Shift Reaction Performed in Membrane Reactors

Conventionally, the WGS reaction is limited in terms of thermodynamic constrains. As a consequence, the interest of scientists seems quite justified in searching for alternatives to TRs [157]. In different scientific works, the WGS reaction carried out in MRs was analysed while paying attention to the influence of different parameters such as reaction temperature and pressure as well as sweep gas flow rate and feed molar ratio. In particular, two opposite effects on the MR system occur when increasing the reaction temperature. A temperature increase induces a positive effect in terms of higher hydrogen permeability through the membrane, enhancing the hydrogen permeating flux from the reaction to the permeate side, resulting in a shift towards the reaction products with a consequent increase of CO conversion. On the contrary, since the WGS reaction is exothermic, at higher temperature a detrimental effect on the equilibrium CO conversion is produced.

4.5 Outlines on Reforming Reactions of Renewable Sources in Membrane Reactors

Clean and renewable sources can be produced for example by biomass, which mainly presents the following advantages:

- It is a renewable source.
- It is widely available.
- It can be processed and converted into liquid fuel (biofuel).

Moreover, using biomass energy, the carbon dioxide atmospheric levels are not increased because of the cycles of regrowth for plants and trees; the use of biomass can also decrease the amount of methane, emitted from the decay of organic matter;

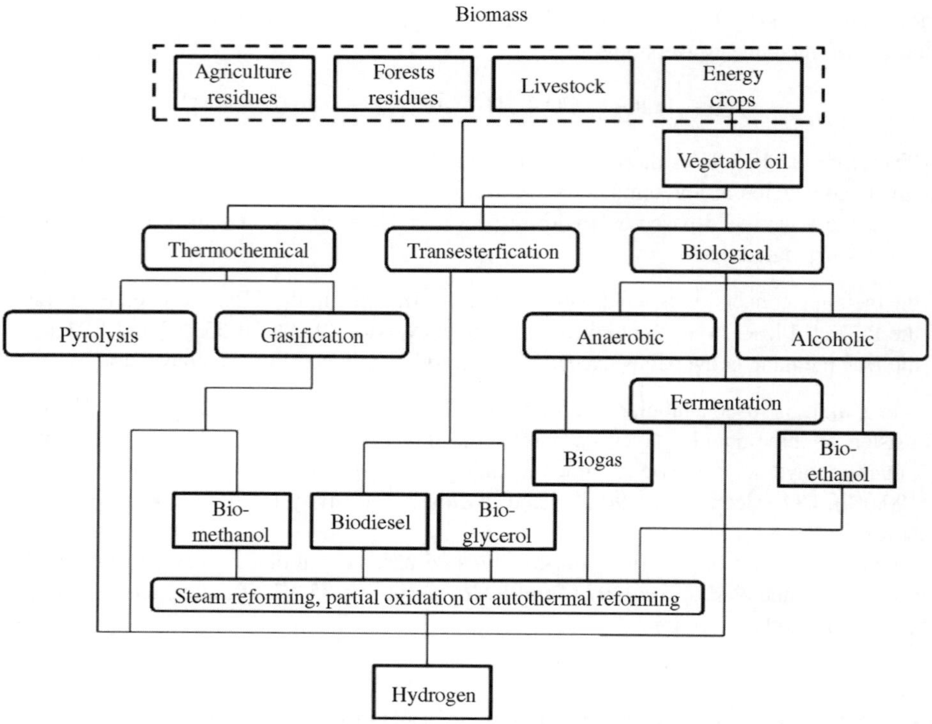

Figure 15 *Selected hydrogen production technologies from various biomass [250]*

An outline of production methods of the biosources is shown in Figure 15, whereas a list of the main biofuels is reported below:

1. bioethanol: ethanol produced from biomass and/or the biodegradable fraction of waste;
2. biomethanol: methanol produced from biomass;
3. biodiesel: a methyl ester produced from vegetable or animal oil;
4. bioglycerol: glycerol produced as byproduct of biodiesel production;
5. biogas: a fuel gas produced from biomass and/or the biodegradable waste that can be treated in a purification plant in order to achieve a quality similar to the natural gas.

The biosources shown in Figure 15 can be converted in hydrogen via reforming reactions (autothermal reforming, steam reforming, partial oxidative steam reforming). Therefore, in the following sections, a summary of scientific studies made since the 2000s on steam reforming reactions of biosources performed in MRs is given. In particular, a small overview on the membrane type, the operative conditions, and performances in terms of hydrogen recovery and reaction conversion obtained performing the steam reforming reaction of different biosources in MRs is reported in Table 3.

The steam reforming is an endothermic reaction, which is generally carried out in TRs at high temperatures ($>600\,^{\circ}$C) and pressures (>10 bar). Vice versa, as illustrated in Table 3, the MRs reaction temperatures commonly range between 250 and 600 °C and the pressure varies between 1 and 8 bar. Moreover, Table 3 illustrates also the MR ability to obtain almost complete conversion and a pure or, at least, CO-free hydrogen stream to be fed for example to a PEMFC.

Table 3 *Operative conditions and performance of MRs used to carry out the steam reforming reaction of biosources*

Authors, reference	Membrane	Catalyst	Temperature (°C)	Pressure (bar)	Hydrogen recovery (%)	Conversion (%)
Bioethanol steam reforming						
Gernot *et al.* [93]	Composite Pd-based	BIOSTAR	600			100
Ethanol steam reforming						
Basile *et al.* [17,19] and Gallucci *et al.* [87]	Dense Pd-Ag	$Ru\text{-}Al_2O_3$	400	1.3		99
Methanol steam reforming						
Wieland *et al.* [247]	$Pd_{75}Ag_{25}$		300	25	96	
	$Pd_{60}Cu_{40}$				78	
	Pd/V/Pd				-	
Basile *et al.* [16]	Dense Pd-Ag	$Cu/ZnO/Al_2O_3$	250	1.3		80
Basile *et al.* [18,20]	MR1	$Ru\text{-}Al_2O_3$	350	1.3		30
	MR2		600			100
	MR3		350			45
			550			65
			350			87
			450			100
Basile *et al.* [21]	Dense Pd-Ag	$CuOAl_2O_3ZnOMgO$	300	1.3		100
Damle [61]	Pd-Ag	WGS catalyst	450	5.2	15	76
				5.2	45	73
			550	7.9	50	72
				11.4	53	71
			600	11.4	60	75

(continued)

Table 3 **(*Continued*)**

Authors, reference	Membrane	Catalyst	Temperature (°C)	Pressure (bar)	Hydrogen recovery (%)	Conversion (%)
Bioethanol steam reforming						
Bioglycerol steam reforming						
Iulianelli *et al.* [115]	Dense Pd-Ag	Co/Al_2O_3	400	4.0	60	94
Dimethyl ether steam reforming						
Park *et al.* [171]	Composite membrane	Cu/Al_2O_3	350			50
			450			100
Acetic acid steam reforming						
Basile *et al.* [21]	Dense Pd-Ag (MR1)	$Ni\text{-}Al_2O_3$	400	2.5	32	100
			450		36	100
	Dense Pd-Ag (MR2)	$Ni\text{-}Al_2O_3$/ $Ru\text{-}Al_2O_3$	400		26	92
			450		32	98
Iulianelli *et al.* [115]	Dense Pd-Ag	$Ni\text{-}Al_2O_3$	400	4.0	70	

5 Other Examples of Membrane Reactors

Ultrapure hydrogen production is surely the field in which membrane reactors are being applied, because of the possibility of combining the separation and reaction in one compact reactor, resulting in both higher conversion than traditional systems and pure hydrogen production (if dense hydrogen selective membranes are used). However, membrane reactors can be used in different other applications. In this second part of the review, the recent developments in the application of membrane reactors for different reaction systems, including membrane bioreactors, are discussed.

5.1 Zeolite Membrane Reactors

Among the different inorganic membrane reactors, zeolite membrane reactors gained increasing interest during the last twenty years, as demonstrated by the growing number of scientific publications and patents presented in literature (some of them discussed below).

Zeolites present a crystalline and ordered structure along with a narrow pore distribution. Zeolites are hydrated aluminosilicates, with an open crystalline structure constituted by tetrahedral SiO_4 and AlO_4^- units linked by oxygen atoms. They are structurally unique since they have cavities or pores with molecular dimension as a part of their crystalline structure as indicated by Meier [155] and Weitkamp [244]. Around 50 zeolites have been found in nature and more than 1500 types of zeolite have been synthesised. The Structure Commission of the International Zeolite Association (IZA) is in charge to approve zeolite structures, which are classified using a three-letter code, included in the *Atlas of Zeolite Structure Types*. When a zeolite is arranged as a layer and it performs as a diffusion barrier we have a zeolite membrane. The quality and then the mass transport characteristics of the zeolite membrane mainly depend on the zeolite type and synthesis, presence of a support and obviously the involved specie along with the operating conditions.

Table 4 reports the main investigators on zeolite membrane reactors. Zeolite-based membrane reactors have been used for different applications, such as xylene isomerisation [64,218,258], ethanol esterification [62], hydrolysis of olive oil [204], methanol production [85] and various others.

In particular, Tarditi *et al.* [218] synthesised a membrane made of ZSM-5 films supported on porous SS tubes to be used for separation of xylene isomers. This separation is quite important for

Table 4 *Major investigators on zeolite membrane reactors (Scopus has been used for the search)*

Investigator	Institution	Number of papers
Yeung, K.L. (King Lun)	Hong Kong University of Science and Technology	15
Dalmon, J.A. (Jean Alain)	Institut de Recherches sur la Catalyse et l'Environnement de Lyon	10
Jansen, J.C. (Jacobus)	Delft University of Technology	9
Coronas, J. (Joaqún)	Universidad de Zaragoza	9
Menendez, M. (Miguel)	Universidad de Zaragoza	7
Santamaria, J. (Jesus)	Universidad de Zaragoza	7
Gora, L. (Leszek)	Delft University of Technology	7

refinery industries. In fact, the most valuable *p*-xylene should be separated from the other isomers. Generally the isomers are separated by distillation of *m*-xylene and successive crystallisation of *o*-xylene, a quite energy intensive separation route. The use of the MFI zeolite membranes for xylene separation appears as a good alternative to the conventional route. Their results indicate that ZSM-5 membranes can be used for increasing the *p*-xylene yield. Based on the permeation characteristic found for ZSM-5 membrane, Deshayes *et al.* [64] formulated a model for xylene isomerisation in the membrane reactor. With optimised kinetics, an industrial scale reactor was simulated by taking into consideration practical restrictions on the pressure drop and on the effective diameters of the membrane tubes which were kept within physical and constructive feasibility. Within these boundaries, the authors were able to optimise their reactor confirming that a ZSM-5 membrane reactor can give 12% increase in *p*-xylene production with respect a conventional reactor. Recently, Zhang *et al.* [258], performed an extensive study on the effects of operating conditions and membrane stability. The use of zeolite membrane reactors (mordenite and zeolite A membranes) was studied by de la Iglesia *et al.* [62] for the esterification of ethanol to ethyl acetate with simultaneous water removal. Tubular membrane reactor configuration has been used where catalyst was packed inside the membrane tubes. Both membranes used were able to shift the equilibrium reaction due to product removal during the reaction. The possibility of removing water and methanol via a zeolite membrane during methanol synthesis was studied by Gallucci *et al.* [85]. A zeolite A membrane was used in a packed bed membrane reactor where a commercial catalyst was used for carbon dioxide hydrogenation. The experimental results show a good performance of the membrane reactor with respect to the traditional reactor: at the same experimental conditions, CO_2 conversion for the membrane reactor was higher than that related to the traditional reactor. Zeolite membranes can be also used in a Fischer–Tropsch reaction system for water removal as indicated, for example, by Rohde *et al.* [190].

5.2 Fluidised Bed Membrane Reactor

Fluidised bed membrane reactors are being studied for different applications and by different research groups as indicated in Table 5.

The integration of membranes (dense or porous, generally non catalytic) inside a fluidised bed reactor, allows to combine the benefits of both separation through membrane and benefits derived from fluidisation regime. It is well known that packed bed membrane reactors suffer from the same disadvantages of packed bed reactors; that is to say: Relatively high pressure drop, possible mass transfer limitations owing to the relatively large particle size to be used, radial temperature

Table 5 *Major investigators on fluidised bed membrane reactors (Scopus has been used for the search)*

Investigator	Institution	Number of papers
Elnashaie, S.S.E.H. (Said)	Misr University for Science and Technology, Egypt	23
Grace, J.R. (John)	The University of British Columbia, Canada	15
Kuipers, J.A.M.	University of Twente, The Netherlands	12
Abashar, M.E.E. (Mohamed)	King Saud University, Saudi Arabia	6
Rahimpour, M.R. (Mohammad)	Shiraz University, Iran	5
Yan, Y. (Yibin)	Auburn University, USA	5

and concentration profiles, difficulties in reaction heat removal or heat supply, low specific membrane surface area per reactor volume.

On the other hand, as summarised in the review presented by Deshmukh [67], the main advantages of the fluidised bed membrane reactors are:

- Negligible pressure drop; no internal mass and heat transfer limitations because of the small particle sizes that can be employed.
- Isothermal operation.
- Flexibility in membrane and heat transfer surface area and arrangement of the membrane bundles.
- Improved fluidisation behavior as a result of:
  - ✓ compartmentalisation, that is, reduced axial gas back mixing.
  - ✓ Reduced average bubble size due to enhanced bubble breakage, resulting in improved bubble to emulsion mass transfer.

Some disadvantages are of course foreseen such as:

- Difficulties in reactor construction and membrane sealing at the wall.
- Erosion of reactor internals and catalyst attrition.

The last disadvantage can be really critical if high selective thin layer membrane is used inside the fluidised bed. Any erosion on the membrane surface can result in a decreased perm-selectivity and a decrease in overall membrane reactor performance. For this reason, membranes to be used in fluidised membrane reactors should be protected by erosion, perhaps by using a porous media between the membrane layer and the fluidised bed. Fluidised bed membrane reactors for pure hydrogen production are studied by different research groups (for example, [1,89,90,150,183]). In this case, as discussed in the first part of the review, Pd-based membranes are inserted in fluidised bed reactors where reforming of hydrocarbons takes place. On the other hand, fluidised bed membrane reactors have also been proposed for different applications. In particular, Deshmukh *et al.* [65,66] developed a membrane-assisted fluidised bed reactor for the partial oxidation of methanol. At first the authors performed cold experiments in order to study gas phase back mixing (via tracer injection technique) and bubble to emulsion phase mass transfer (using ultrasound experiments). A scheme of the tracer injection set up is reported in Figure 16. With this technique, the authors demonstrated that effective compartmentalisation of the fluidised bed is realised especially in the case of gas permeation through horizontal membranes inserted in the fluidised bed. Based on this study, the same authors [66] built a small scale experimental set up for partial oxidation of methanol inside a fluidised bed membrane reactor. The effects of different operating conditions on the methanol conversion to formaldehyde have been evaluated and the results compared with a phenomenological model for the fluidised bed membrane reactor. The experimental set up is shown in Figure 17.

The authors demonstrated that distributive feeding of oxygen in a fluidised bed membrane reactor produces an increased overall formaldehyde yield and throughput without pronounced conversion of formaldehyde to carbon monoxide.

Prof. Rahimpour's group proposed the application of fluidised bed membrane reactors for different reaction systems. In particular, an example of a fluidised bed applied to a Fischer–Tropsch reaction system can be found in Ref. [184]. In this work, Pd-based membranes are inserted in a fluidised bed in a two reactors plant as indicated in Figure 18. The addition of hydrogen through the Pd-based membranes inside the fluidised bed keeps the H_2/CO ration to an optimal value (or close to it), resulting in a better overall performance. On the other hand, the use of fluidised bed membrane reactors also solves some drawbacks of packed-bed reactors already discussed such as high pressure drop, heat transfer problem and internal mass transfer limitations.

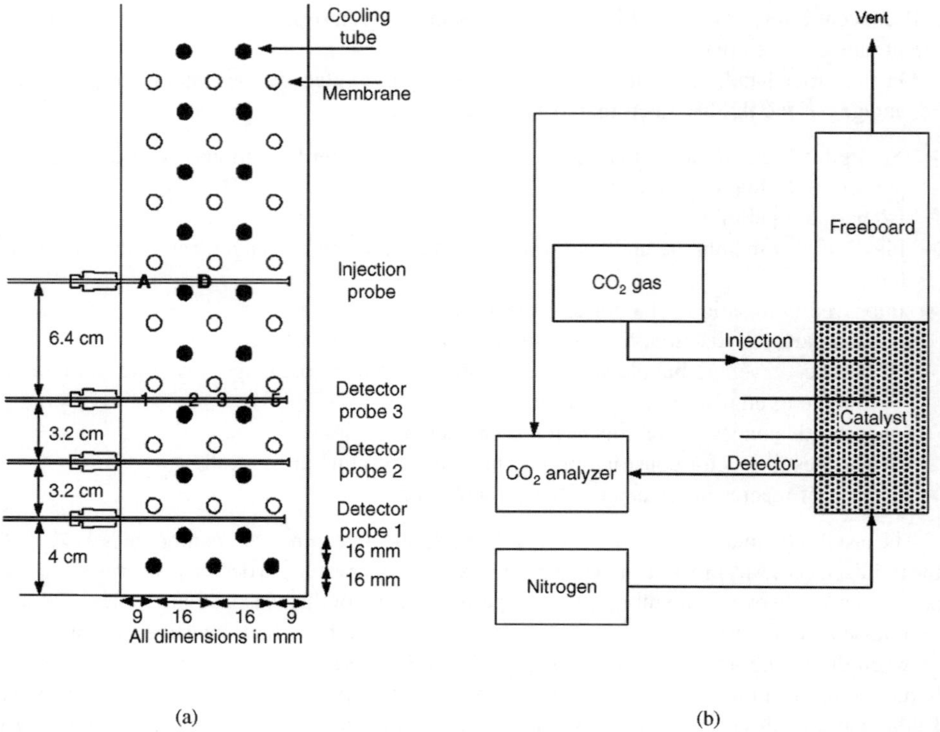

Figure 16 *Schematic of the experimental setup to measure gas back-mixing in the membrane-assisted fluidised bed reactor by steady-state tracer experiments: (a) detailed side view of the reactor, (b) simplified flow-sheet. Reprinted from Industrial & Engineering Chemistry Research, Deshmukh, S. A. R. K., et al., Development of a membrane-assisted fluidized bed reactor. 1. Gas phase back-mixing and bubble-to-emulsion phase mass transfer using tracer injection and ultrasound experiments. Vol. 44, 5955–5965. Copyright (2005) with permission from American Chemical Society*

5.3 Perovskite Membrane Reactors

The first report of oxygen permeation through perovskite-based materials was probably Teraoka and coworkers [219], who studied the oxygen flux through 10 mm disk-shaped perovskite-based material. More than 20 years after this interesting report, no industrial applications of perovskite membrane exist yet. This is mainly due to difficulties in membrane/module sealing at high temperature (high temperature needed for achieving a reasonable oxygen flux), to problems in membrane stability, and to lack of membrane modules with high surface area per volume. The last problem is addressed by using hollow fibre membrane reactor configurations as discussed in the following section. In this section, some examples of perovskite membranes used in membrane reactors are presented. The main investigators on perovskite membrane reactors are reported in Table 6.

A recent example of perovskite membrane reactor has been presented by Sun and coworkers [213] in order to oxidise the ammonia to NO (for nitric acid production). Actually, 80% of the ammonia is used for fertilisers production, and a big part is first converted to nitric acid through a high temperature oxidation on platinum–rhodium alloy catalyst. This reaction is well

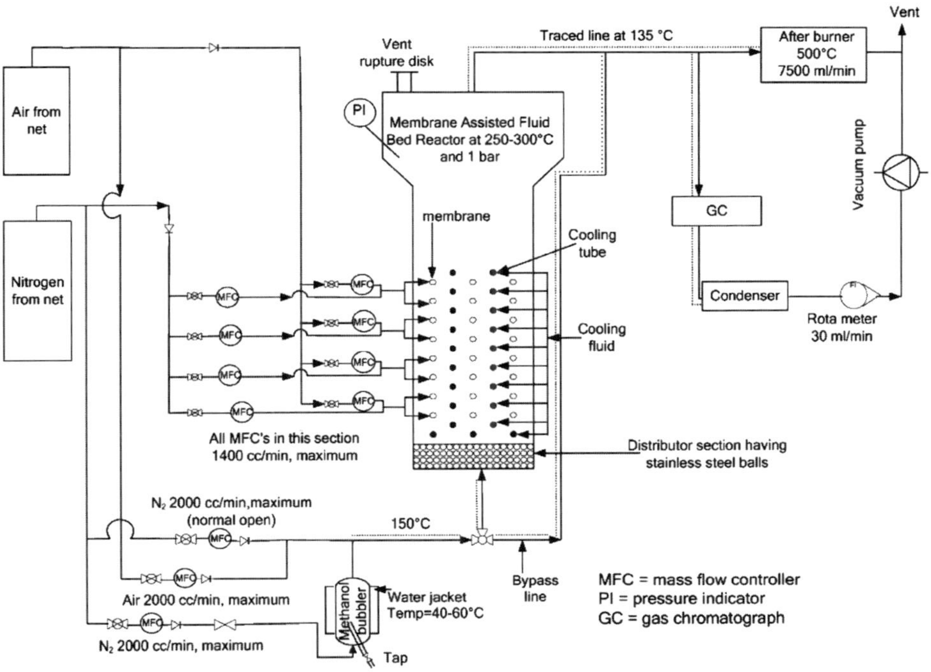

Figure 17 *Experimental setup for partial oxidation of methanol to formaldehyde in a fluidised bed membrane reactor. Reprinted from Industrial & Engineering Chemistry Research, Deshmukh, S. A. R. K., et al., Development of a membrane-assisted fluidized bed reactor. 2. Experimental demonstration and modeling for the partial oxidation of methanol. Vol. 44, 5966–5976. Copyright (2005) with permission from American Chemical Society*

known since years and also well optimised in terms of catalyst. However, still some technological problems have to be faced. In particular, this operation is quite cost intensive also due to catalyst loss as oxides. This problem is being studied by exploring some other catalysts such as Cr_2O_3 or Co_3O_4. However, N_2O emissions from these plants are the greatest among chemical industries, which means that costly N_2O capture systems are required.

Table 6 *Major investigators on perovskite membrane reactors (Scopus has been used for the search)*

Investigator	Institution	Number of papers
Yang, W. (Weishen)	Dalian Institute of Chemical Physics Chinese Academy of Sciences, China	18
Wang, H. (Hai)	Ministry of Education China, School of Chemistry and Chemical Eng., China	18
Caro, J. (Jürgen)	University of Hannover, Germany	14
Kharton, V.V. (Vladislav)	Consejo Superior de Investigaciones Cientficas, Spain	12
Yaremchenko, A.A. (Aleksey)	Centre for Research in Ceramics and Composite Materials, Portugal	10
Schiestel, T. (Thomas)	Fraunhofer-Institut für Grenzflächen und Bioverfahrenstechnik, Germany	9

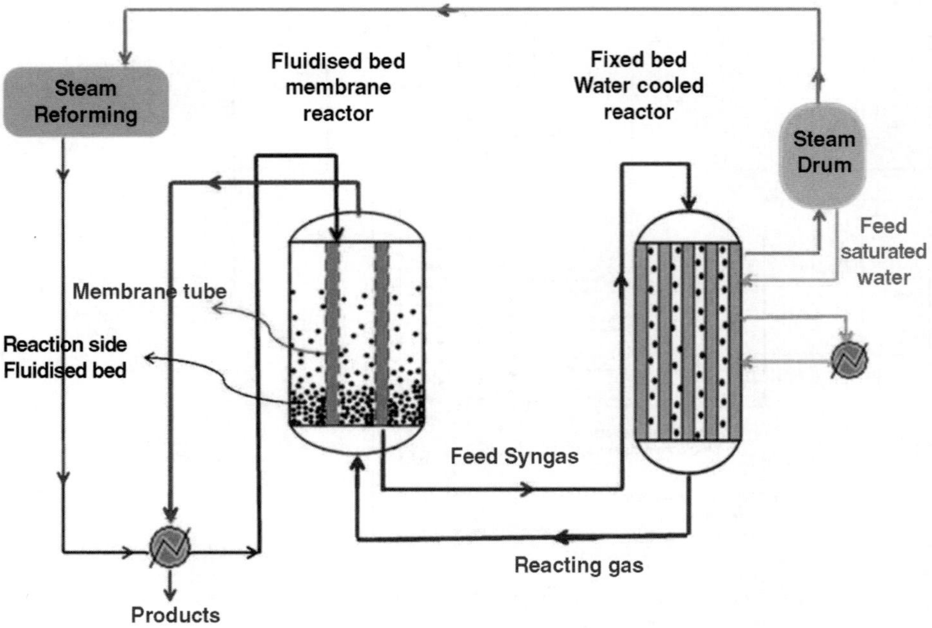

Figure 18 *Schematic diagram of a fluidised bed membrane dual-type Fischer–Tropsch reactor. Reprinted from Fuel Processing Technology, Rahimpour, M. R., Elekaei, H., A comparative study of combination of Fischer–Tropsch synthesis reactors with hydrogen-permselective membrane in GTL technology. Vol. 90, 747–761. Copyright (2009) with permission from Elsevier*

Sun *et al.* [213] show the application of a perovskite membrane reactor to carry out the separation of oxygen and the reaction in one unit. The scheme of the reactor is depicted in Figure 19.

Air is fed from one side of the membrane, and oxygen is dissociated on the membrane surface into O_2^-. The O_2^- ion diffuses through the membrane due to the difference in oxygen partial pressure between the two membrane surfaces. On the other membrane surface, ammonia is fed and selectively reacts with the oxygen diffusing through the membrane to form NO.

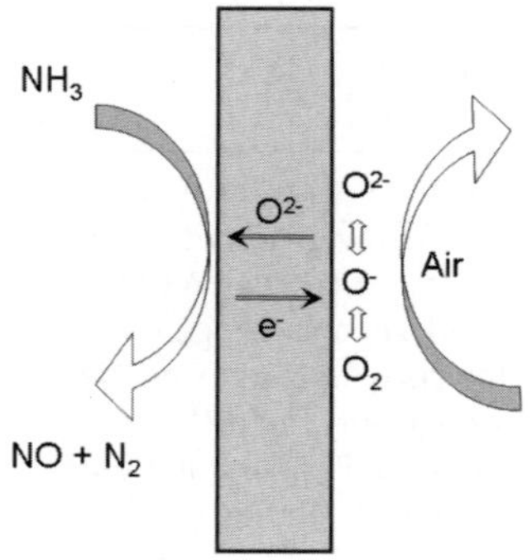

Figure 19 *Scheme of the perovskite membrane reactor for ammonia oxidation; after [213]*

The membrane proposed and studied by Sun is a $Ba_{0.5}Sr_{0.5}Co_{0.8}Fe_{0.2}O_{3-\delta}$ perovskite in form of discs with a thickness of 1.4 mm. The membrane is housed in a reactor consisting in two quartz tubes. Most of the articles related to perovskite membranes in membrane reactor deals with reactions involving natural gas or hydrocarbons. The following paragraphs discuss examples of membrane reactors for oxidative coupling of methane [168], partial oxidation of methane [141] and methane reforming [253].

Another system where perovskites can be used as membrane reactors is the methane partial oxidation, an interesting route for producing syngas from methane. The reaction is discussed as a promising route for hydrogen production (in dense Pd-based membrane reactors) in Section 4.3:

$$CH_4 + 0.5O_2 \Leftrightarrow CO + 2H_2 \tag{7}$$

When focusing on syngas production the reaction leads to a syngas with a ratio $H_2/CO = 2$, which makes the reaction system more interesting than the typical methane steam reforming. In fact, according to Kleinert [129], the syngas ratio = 2 is the optimum for different postprocessing systems, while the methane steam reforming gives a syngas ratio = 3.

The problem of POM is that pure oxygen is needed in order to produce syngas, resulting in quite expensive air separation units and makes the operation quite risky because the direct contact between methane and pure oxygen at high temperature may result in explosions. Air can not be used because nitrogen would contaminate the syngas and would also react to produce NOx. For this reason, oxygen permeating membranes are quite attractive for this reaction system too.

Li and coworkers [141] studied POM in a $BaCe_{0.1}Co_{0.4}Fe_{0.5}O_{3-\delta}$ membrane reactor by using a LiLaNi-based catalyst. Another interesting work has been presented by Yaremchenko *et al.* [253], who studied the effect of perovskite-like tubular membrane for reforming of methane. The aim is to demonstrate the possibility to carry out the POM reaction with oxygen addition through tubular membranes (thus with higher area/volume than the flat membrane). The promising results will drive the research towards an optimisation of the membrane material as well as a better reactor design (such as hollow fibre membrane reactor) in order to optimise the oxygen flux inside the reactor.

5.4 Hollow Fibre Membrane Reactors

An interesting membrane reactor configuration is the hollow fibre membrane reactor, which allows achieving much higher membrane area/reactor volume than the other membrane reactors configurations. The membrane area available is an important parameter for all the membrane systems. However, it becomes really important when membranes with low permeation fluxes are used. A good example is the use of polymeric membranes in gas separation. It is quite evident that the driving force for industrial exploiting of polymeric membrane systems, for example in natural gas treatment as well as dialysis applications, was the availability of hollow fibre membranes and membrane modules.

Following this example, many other membrane applications are looking for hollow fibre availability. For example, in case of perovskite membranes the membrane flux is generally quite low and the hollow fibre configuration is quite interesting. A pioneer on ceramic hollow fibre membrane is surely Prof. Li, who contributes to this book with a chapter on hollow fibre membrane preparation. The main investigators of hollow fibre membrane reactors are summarised in Table 7. In the following some examples of hollow fibre perovskite membrane applications will be discussed. So far, laboratory studies on hollow fibre ceramic membrane applications use a single ceramic membrane in a configuration tube in tube or at best few hollow fibres in tubes in shell configuration. The typical tube in tube configuration is reported in Figure 20.

Table 7 Major investigators on hollow fibre membrane reactors (Scopus has been used for the search)

Investigator	Institution	Number of papers
Fane, A.G. (Anthony)	Nanyang Technological University, Singapore	5
Tan, X. (Xiaoyao)	Shandong University of Technology, China	4
Haginaka, J. (Jun)	Mukogawa Women's University, Japan	4
Li, K. (Kang)	Imperial College London, UK	4
Caro, J. (Jürgen)	University of Hannover, Germany	4
Schiestel, T. (Thomas)	Fraunhofer-Institut für Grenzflächen und Bioverfahrenstechnik, Germany	4

The applications of hollow fibre membrane reactors are in principle the same applications in which distributed oxygen feeding can be beneficial for the reaction system. A good example, often studied is the oxidative coupling of methane (OCM). This reaction system is known to be the direct route for transforming methane into C2 products. This route is surely more economical interesting than the indirect route in which methane is first converted into syngas and then the Fischer–Tropsh process is used to convert syngas into higher hydrocarbons. OCM has been then studied by using dense ceramic membranes, and C2 selectivity up to 95% has been obtained. However, due to the low oxygen flux and to the low membrane area per volume the total yield was not higher than 10%.

To overcome this drawback, a hollow fibre membrane reactor has been proposed by Tan and Li [216,217]. They prepared a $La_{0.6}Sr_{0.4}Co_{0.2}Fe_{0.8}O_{3-R}$ (LSCF) hollow fibre membrane by phase inversion/sintering technique. For the details of the techniques please see the following chapters. A tubes in shell configuration (with five hollow fibres in a ceramic shell) has been used for this research. To reduce the stresses on the membranes due to the difference in thermal expansion between membranes and shell material, the authors used some rubber tubes at the ends of the membrane. Long membranes were used so that just the central part of the reactor was used in a furnace, while the extremities were relatively cold. This is a good practice in order to avoid high temperature sealing problems.The OCM reaction was carried out in this membrane reactor, and the results suggest that C2 yield depends on both reaction temperature and oxygen flux through membranes. Promising 15.3% C2 yield, even though with low selectivity 44% ca, suggests that the development of OCM in perovskite hollow fibre membrane reactors is an interesting field to be explored.

Hollow fibre membrane reactors have also been studied for different reaction system. Kleinert and coworkers [129] applied this kind of reactor to partial oxidation of methane.

As already discussed above, the problem of POM is that pure oxygen is needed in order to produce syngas, resulting in quite expensive air separation units. By using perovskite-like hollow fibre membrane reactor, a higher membrane area for the air separation is available.

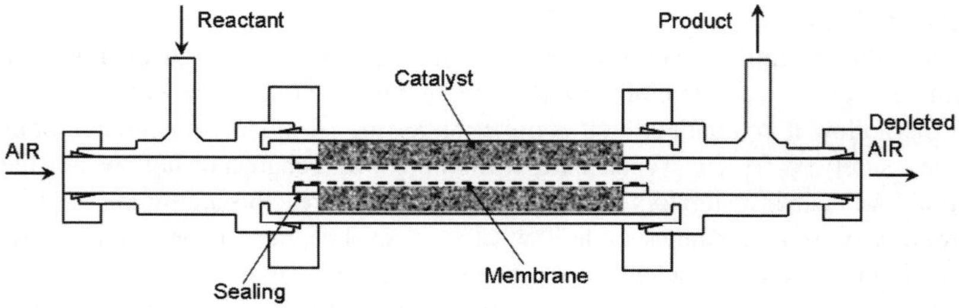

Figure 20 Schematic diagram of a tube in tube hollow fibre membrane reactor for laboratory tests

The perovskite membranes used by Kleinert *et al.* [129] were produced from Ba(Co, Fe, Zr) O_{3-d} (BCFZ) powder via phase inversion spinning technique. A tube in tube configuration has been used while the catalyst was packed in the shell side of the reactor.

In their paper the authors show that the membrane was able to give quite interesting results with a methane conversion of 82% and a Co selectivity of 83%. Moreover the membrane was quite stable under the reactive conditions investigatedA comparison between the performance of perovskites in hollow fibre configuration and disk geometry as been carried out by Caro and coworkers [33]. The work is mainly based on structural study and oxygen permeation. A quite stable syngas production over 120 h of stream has been obtained, with a 80% methane conversion and 82.5% CO selectivity, confirming the possibility of carrying out the POM reaction in hollow fibre membrane reactors.

A perovskite hollow fibre membrane reactor has been applied by Wang and coworkers [239] for the oxidative dehydrogenation of ethane to ethylene. The direct oxidation of ethane to ethylene is an alternative route to the typical thermal steam cracking routes. However, as already discussed for OCM, the co-feeding of ethane and oxygen in a fixed bed reactor is not an option, being the deep oxidation to CO_2 a thermodynamically favoured reaction.

5.5 Catalytic Membrane Reactors

A direct survey of the main investigators on catalytic membrane reactors is quite complicated because various authors erroneously call catalytic membrane reactor a reactor in which a catalyst is somehow packed inside the reactor. Indeed, this kind of reactor should be called packed bed membrane reactor. A catalytic membrane reactor is a special reactor where the membrane acts as separation layer and as catalyst as well. The membrane can be either self catalytic [72], or can be made catalytic by coating the surface of a dense membrane [22], or by depositing the catalyst material inside the pores of the membrane [78], or by casting a solution containing the polymeric material and the catalytic material [63].

Both experimental and theoretical studies have been presented on catalytic membrane reactors. A very active group in modeling polymeric catalytic membrane reactors is the group of Mendes who modeled different reaction systems in polymeric CMR with quite detailed models (for an example, see [157]).

Concerning the experimental works, both polymeric and inorganic catalytic membrane reactors have been used. Fritsch [78] produced porous polymeric membranes with high fluxes with the casting machine available at GKSS (Germany). The authors followed two different routes for producing the catalytic membranes as previously indicated. Both a catalyst containing casting solution and the pore filling catalyst material have been used.

The membranes were used for the hydrogenation of sunflower oils to edible oils. The approach proposed is quite interesting since with the catalytic membrane with high fluxes the authors are able to both overcome the problem of catalyst separation from the edible oil (catalyst normally used are either expensive or toxic) and the problem related to high pressure drops induced by high viscous oils.

Bobrov and coworkers [29] produced a catalytic membrane by depositing a catalytic layer on gas separation inorganic membrane by using the chemical vapour deposition technique. This is a quite standard procedure to produce catalytic membranes, as indicated in the following chapters. The membrane produced was used for propane dehydrogenation demonstrating that probably the catalytic membranes are much more suitable for this reaction than the dense selective hydrogen permeating membranes (more difficult to produce and less stable).

Table 8 *Major investigators on photocatalytic membrane reactors (Scopus has been used for the search)*

Investigator	Institution	Number of papers
Morawski, A.W. (Antoni)	Polish Academy of Sciences, Poland	11
Palmisano, L. (Leonardo)	University of Palermo, Italy	8
Molinari, R. (Raffaele)	University of Calabria, Italy	8
Tomaszewska, M. (Maria)	Polytechnic University Szczecin, Poland	5
Toyoda, M. (Masahiro)	Kyoto University, Japan	5
Thiruvenkatachari, R.	University of Technology Sydney, Australia	4

5.6 Photocatalytic Membrane Reactors

An interesting new system to be taken into account is the photocatalytic membrane reactor system where the photocatalysis is somehow improved by membrane separation. The field is studied by various scientists and a list of the main investigators is reported in Table 8.

The photocatalytic membrane reactor can be built in two different ways. What we strictly would call a photocatalytic membrane reactor is a reactor in which the membrane is placed in contact with the reactants and on which the light is irradiated via an internal or external light source. A typical scheme can be for example the one reported in Figure 21. For similar schemes, see [47,50,112,227]. A second way to work with photocatalytic membrane reactor is to separate the reaction system and the membrane separation system (ultafiltration, or other) in two different steps (Figure 22). For examples using this scheme, see [9,159,161,162].

The membrane often serves as separator of the suspended photocatalyst particles from the treated media [112]. In other cases the photocatalyst can be impregnated into the membrane media which also acts as support or the membrane itself can be photocatalytic [227]. Moreover, the membrane can act as separator of the reaction products [159]. Typical applications of photocatalytic membrane reactors are the photodegradation of water pollutants [162], Photoreaction to obtain more valuable products [159] and photooxidation of pollutant vapour compounds [227].

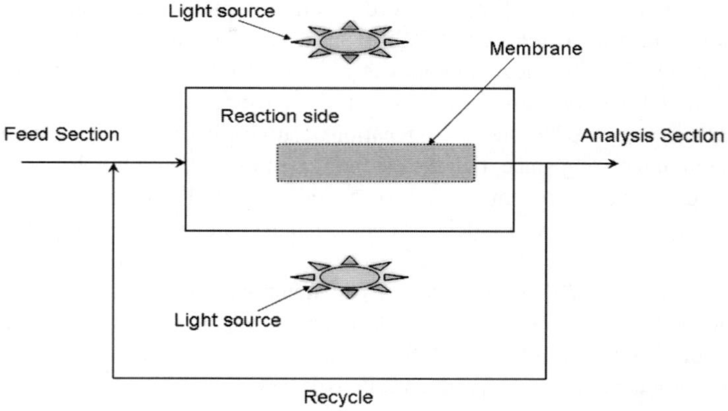

Figure 21 *Typical integrated photocatalytic membrane reactor*

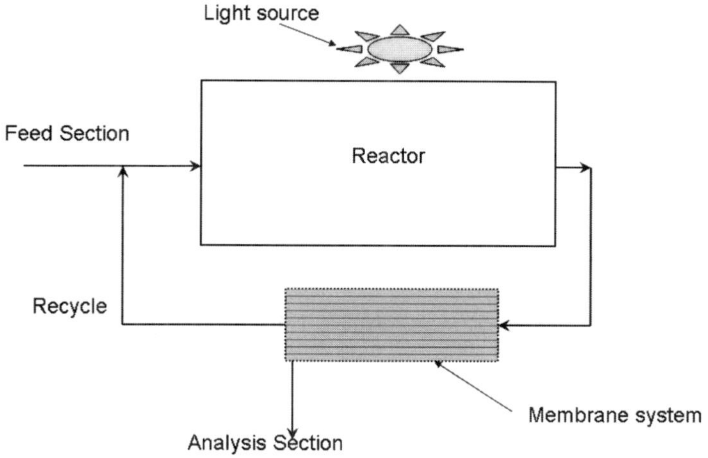

Figure 22 *Typical photocatalytic reactor coupled with membrane separation system*

Where the membrane is used as external separation system, the problem reduces to a study of membrane filtration. In this case, often commercial membrane filtration system can be easily used. Different is the case in which a membrane is photocatalytic or it is supporting the catalyst. In this case, as indicated by Tsuru [227], the membrane needs to be prepared with a tailored amount of catalyst, with particular attention to the membrane pore size distribution and membrane photocatalytic activity towards the reaction of interest.

6 Membrane Bioreactor

Membrane separation in a membrane bioreactor (MBR) combines clarification and filtration of a conventional activated sludge (CAS) process into a simplified, single step process. Membranes are seldom used by themselves to filter untreated wastewater, since fouling prevents the establishment of steady-state conditions and because water recovery is too low [118,201]. However, when used in conjunction with the biological process, biological process converts dissolved organic matter into suspended biomass, reducing membrane fouling and allowing recovery to be increased. On the other hand, the membrane filtration process introduced into bioreactors not only replaces the settling unit for solid–liquid separation but also forms an absolute barrier to solids and bacteria and retain them in the process tank, giving rise to several advantages (see Section 6.4) over the CAS.

6.1 A Brief History of the MBR Technology Development

The progress of membrane manufacturing technology and its applications led to the replacement of tertiary treatment steps by microfiltration or ultrafiltration. Parallel to this development, microfiltration or ultrafiltration was used for solid/liquid separation in the biological treatment process. The original process was introduced by Dorr–Olivier, Inc., who combined the use of an activated sludge bioreactor with a crossflow membrane filtration loop [206]. By pumping the mixed liquor at a high pressure into the membrane unit, the permeate passes through the membrane and the concentrate is returned to the bioreactor. The idea of replacing the settling tank of the CAS process was attractive; but it was difficult to justify the use of such a process because of the high cost of membranes, low economic value of the product (tertiary effluent) and the potential rapid loss of performance due to fouling. The first generation MBRs only found applications in niche areas with special needs like isolated trailer parks or ski resorts, for example. The breakthrough for the MBR, however, came in 1989 by submerging the membranes in the reactor itself and withdrawing the treated water through membranes [251]. In this development, membranes were suspended in the reactor above the air diffusers. The diffusers provided the oxygen necessary for treatment to take place and scour the surface of the membrane to remove deposited solids [44,45,122,236].

Two broad trends have emerged today, namely submerged MBRs and sidestream MBRs. Submerged technologies tend to be more cost effective for larger scale lower strength applications, and sidestream technologies are favoured for smaller scale higher strength applications. The sidestream MBR envelope has been extended in recent years by the development of the air lift concept, which bridges the gap between submerged and crossflow sidestream MBR, and may have the potential to challenge submerged systems in larger scale applications [175]. Figure 23 presents simplified schematics of MBR formats.

Along with aerobic MBRs, anaerobic MBRs (AnMBR) were also developed. Anaerobic biological treatment systems can offer a number of advantages over their anaerobic counterparts [24]. The operational costs associated with anaerobic systems are typically lower than with aerobic systems, and anaerobic systems also generate less waste sludge. In addition, the energy associated with the biogas produced during anaerobic biological treatment can be recovered. On the other hand, it may be possible to overcome some of the treatment limitations of anaerobic systems by coupling with membrane separation. For instance, in case of low-strength wastewater treatment, the biomass growth yield and growth rate are relatively low, resulting in a low overall net biomass production. The net biomass production must exceed the net biomass loss to the effluent for a biological treatment system to function properly. However, in conventional anaerobic biological treatment systems, the net biomass loss to the effluent is governed by the relatively poor settling characteristics of the biomass. In AnMBR it is possible to maintain an adequate biomass concentration. Again, although anaerobic biological treatment systems can effectively remove the bulk of the organic contaminants present in a wastewater, they are typically not effective at removing residual levels of soluble and colloidal organic contaminants. In AnMBR residual organics can be retained in the system independently of the hydraulic throughput, enabling these contaminants to be hydrolysed and biodegraded [212].

The first test of the concept of using membrane filtration with anaerobic treatment of wastewater appears to have been reported by Grethlein [97]. The first commercially available AnMBR was developed by Dorr–Oliver in the early 1980s for high-strength whey processing wastewater treatment. That process, however, was not applied at full scale, possibly due to high membrane costs [214]. The Ministry of International Trade and Industry (MITI), Japan launched a six-year research and development project named *Aqua-Renaissance '90* in 1985 with the particular objective of developing energy-saving and smaller footprint water treatment processes utilising side stream anaerobic MBR to produce reusable water from industrial wastewater and sewage. However, it was difficult to significantly reduce the energy consumption by adopting the side stream operation using a big recirculation pump [252]. The ADUF (anaerobic digestion ultrafiltration) process developed in South Africa in 1987 for industrial wastewater treatment [192] is perhaps the only example of current AnMBR utilisation. However, recently there has been a resurgence of interest in submerged AnMBRs [117]. This chapter will focus on aerobic

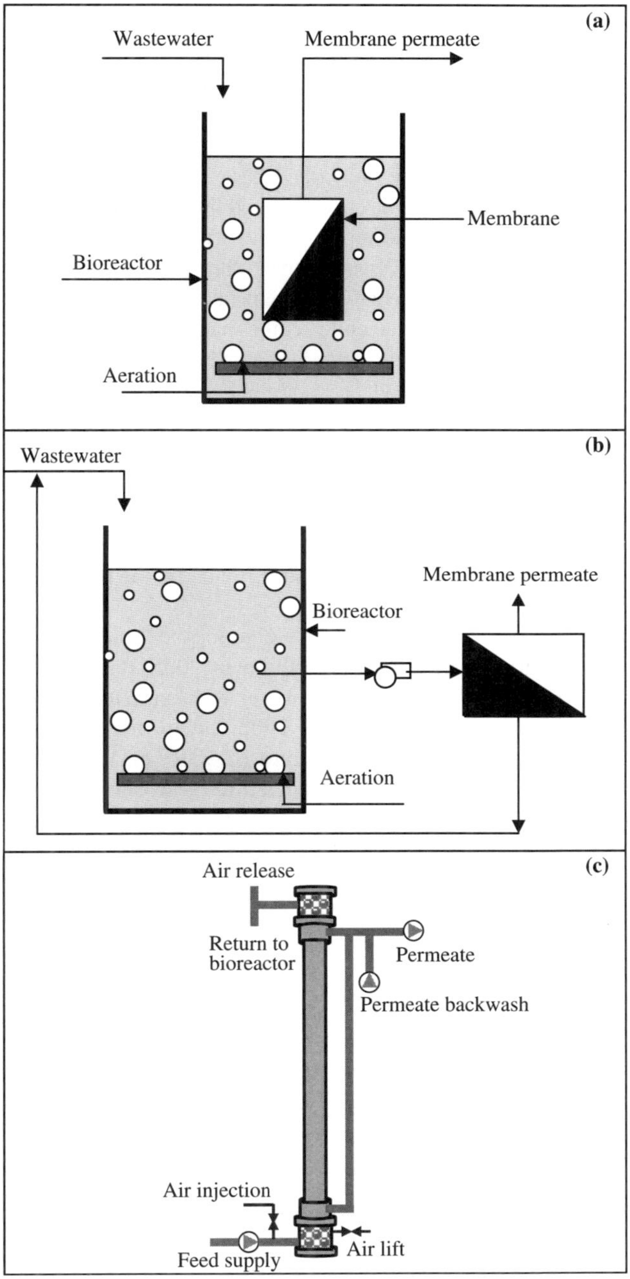

Figure 23 *Simplified representation of MBR formats: (a) submerged MBR, (b) sidestream MBR, (c) airlift MBR*

MBRs. Further details on AnMBR can be derived from the comprehensive reviews by Liao [142] and Berube [24].

6.2 Market Value and Drivers

MBR costs have declined sharply since the early 1990s, falling typically by a factor of 10 in 15 years. As MBR technology has become accepted, and the scale of installations has increased, there has been a steady downward trend in membrane prices, which is still continuing.

Driven principally by legislation and water stress, the global MBR market is expected to display double-digit growth [104]. Legislation is the primary driving force in Northern Europe and parts of the United States and Canada, while in other parts of the world, notably, China, India, Australia and the Middle East, water stress is the dominant issue. Legislation has always been one of the strongest drivers for advanced water treatment solutions, such as MBR systems, in developed regions such as Europe and North America. It is now becoming an influential factor in the Indian and Chinese markets as well. Another strong driver for the expansion of the MBR market has been the increasing level of water stress, particularly in the arid regions of Australia and the Pacific Rim, the Middle East, parts of South Africa and North America and Southern Europe. Besides these two key market drivers, increased funding and incentives allied with decreasing costs and a growing confidence in the performance of the technology are contributing to the expansion of MBR market.

The global market for membrane bioreactor technology doubled over a five-year period from 2000 to reach a market value of \$217 million in 2005, this from a value of around \$10 million in 1995 (Figure 24). The global market was estimated to increase from \$217 million in 2005 to \$296 million by the end of 2008 and further to \$488 million in 2013, corresponding to a compound annual growth rate (CAGR) of 10.5% [104]. Most MBRs use aerobic microbial processes, which are expected to be worth \$291 million in 2008 and \$483 million in 2013, a CAGR of 10.7%. Municipal and domestic wastewater treatment

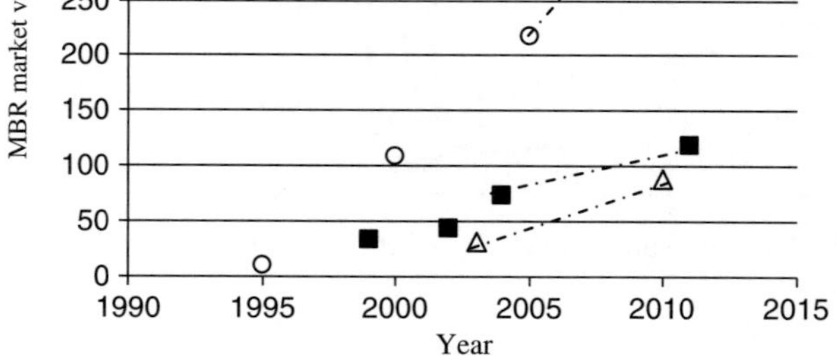

Figure 24 *Estimated MBR market value. (Dotted lines indicate projected values.) Data sources: [79–83,104]*

generated $96.6 million in 2005. Values of $142 million and $249 million were projected for 2008 and 2013, respectively, with a CAGR of 11.9% in this segment. Growth rates of MBR systems are not the same for all regions and are not increasing from the same base. Common to all regions, however, is the fact that sales of the technology are growing faster than the gross domestic products of countries installing the systems and more rapidly than the industries that use them.

In Europe, the total MBR market for industrial and municipal users was estimated to have been worth 25.3 million in 1999 and 32.8 million in 2002 [79]. In 2004, the European MBR market was valued at $57 million [83]. Market projections for the future indicated that the 2004 figure is expected to rise annually by 6.7%, with the European MBR market more than doubling its size by 2011 [83]. Market analysis conducted by Frost [80] revealed that revenues in the United States and Canadian membrane bioreactor (MBR) markets totalled US$32.2 million in 2003, and are projected to reach $89.0 million in 2010. In fact, the United States and Canadian MBR market achieved a revenue totalling over $750 million from membrane-based water purification, desalination and waste treatment in 2003, and it was projected to reach $1.3 billion in 2010 [80–82]. According to some analysts, the MBR market in the United States is growing at a significantly faster rate than other sectors of the US water industry, such that within some subsectors, such as the filtration market, technologies like membrane filters or ultraviolet radiation are growing at rates in excess of 15% [152].

The Asia Pacific region has emerged as a strong market for MBR systems, following the established markets in North America and Europe, and could become one of the most promising and largest regional markets by 2010.

6.3 Commercially Available MF/UF Membranes for MBR

This section examines the commercial MF/UF membrane systems for water treatment in general and the membranes for MBR in particular.

6.3.1 Membrane Geometry

The configurations of membranes used for MBR are based on either planar or cylindrical geometry. Principal membrane configurations currently employed in practice are:

1. *Hollow fibre module:* In the hollow fibre (HF) module, either large amounts of hollow fibres make a bundle or single fibres are mounted horizontally (Figure 25), and the ends of the fibres are sealed in epoxy block connected with the outside of the housing. The water can flow from inside to the outside of the membrane or from outside to the inside. These membranes can work under pressure and under vacuum.
2. *Plate and frame membrane modules:* Plate and frame membrane modules comprise of flat sheet (FS) membranes with separators and/or support membranes. The pieces of these sheets are clamped onto a plate. The water flows across the membrane and permeate gets collected through pipes emerging from the interior of the membrane module in a process that operates under vacuum.

   HF and FS modules are mostly immersed directly in mixed liquor with permeate drawn through the membranes using vacuum pumps.
3. *Tubular membranes:* Typically, tubular membranes are encased in pressure vessels and mixed liquor is pumped into them. These are predominantly used for sidestream configurations.

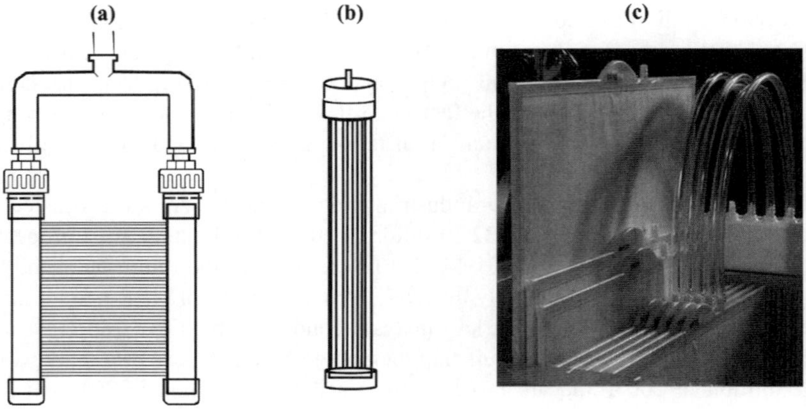

Figure 25 Submerged membrane modules: (a) horizontally mounted single hollow fibres, (b) hollow fibre bundles, (c) flat sheet membrane module

6.3.2 Mode of Operation: Inside-Out Versus Outside-In Flow

Inside-out flow and outside-in flow refers to the direction of feed water passing through the membrane, as well as the orientation of the feed water in relation to the membrane surface. For example, in an outside-in system, the feed water surrounds the membrane, and the filtrate is collected from inside of the hollow tube fibres (lumen). The outside-in scheme has the advantage of a larger membrane surface area, which allows for a slightly higher flow than the inside-out model while still maintaining the same flux rate and solids concentration. An inside-out system places the feed water inside the fibres, and the filtrate is collected on the outside of the membranes. The feed water enters the fibres at one end of the membrane element, and the discharge passes through the element and exits the fibres on the other side. The filtrate is collected inside the element on the outside of the fibres. The submerged hollow fibre and flat sheet membranes follow outside-in and sidestream tubular membranes follow inside-out operation.

6.3.3 Membrane Materials and Material Properties

Membranes can be manufactured in a wide variety of materials. These materials differ in their performance characteristics including mechanical strength, fouling resistance, hydrophobicity, hydrophillicity, and chemical tolerance.

Membrane materials can be classified as either hydrophilic or hydrophobic. Hydrophilic materials readily adsorb water. The surface chemistry allows these materials to be wetted forming a water film or coating on their surface. Hydrophobic membrane materials have little or no tendency to adsorb water. Water tends to *bead* on hydrophobic surfaces into discrete droplets.

The hydrophilic and hydrophobic properties of a membrane material are related to the surface tension of the material. The fundamental importance of surface tension comparison is that liquids having lower surface tension values will generally spread on materials of higher surface tension values. Figure 26 summarises surface tension values of some polymeric membrane materials. The higher the surface tension value of the material, the more hydrophilic the material.

The degree of hydrophilicity or hydrophobicity influences the wettability and applied pressure requirements for water flow through the membrane. Hydrophilic membranes require less

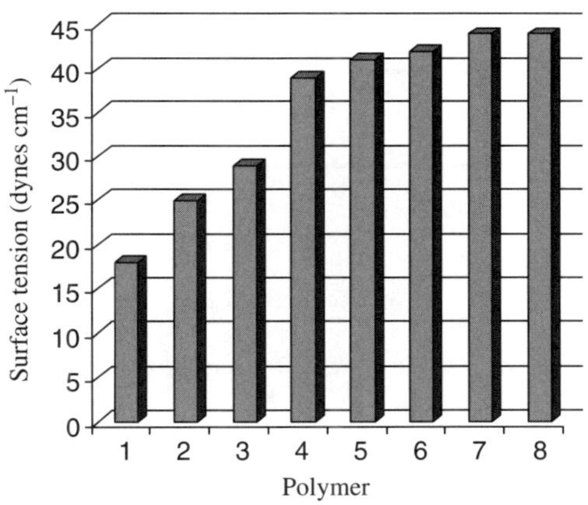

Figure 26 *Surface tension of common polymers for membranes. Polymers: 1. polytetrafluoroethylene, 2. polyvinylidene fluoride, 3. polypropylene, 4. polyvinyl chloride, 5. polysulfone, 6. polycarbonate, 7. polyacrylonitrile, 8. cellulose*

operating pressure than hydrophobic membranes. Particles that foul in aqueous media tend to be hydrophobic. Hydrophobic particles tend to cluster or group together to form colloidal particles because this lowers the interfacial free energy (surface tension) due to surface area exposure. General tendency will favour particle attachment to any material less hydrophilic than water because less exposure of hydrophobic particles can be achieved by attachment of the particles to the membrane surface. Accordingly, hydrophilic membranes tend to exhibit greater fouling resistance than hydrophobic membranes.

6.3.4 Features of Commercial MBR Technologies

As stated earlier, the first generation MBRs in wastewater treatment used a sidestream format, in which feed was pumped from the bioreactor through an external membrane system. This approach was suitable for the early stage small scale applications for difficult to treat feeds. An alternative format was developed in the 1990s using modules submerged in the bioreactor tank, or in an adjoining compartment. This was much more cost effective for treating larger scale flows with more easily treatable wastewater.

The submerged format is available with modules either in a flat sheet configuration or as hollow fibres or capillary membranes. Originally, the favoured concept was to submerge the modules directly into the bioreactor for simplicity. However, in order to gain better control of the balance between the biological and filtration treatment capacity it is now more common to use the membrane in an external membrane tank [32]. The external arrangement allows the size and design of the membrane tank to be optimised independently, with the practical advantages for operation and maintenance. Almost all submerged MBRs are either vertically oriented PVDF HF modules of outside diameter predominantly between 2.0 and 2.8 mm, or FS rectangular membranes 1.0–1.2 m in depth with a membrane separation between 6 and 10 mm, the exceptions being:

- the Huber VRM rotating membrane module,
- the Mitsubishi Rayon horizontal *Sterapore* membrane, which is a horizontally oriented polyethylene (PE),
- the Koch–Puron membrane (polyethersulfone, PES), the Polymem (polysulfone, PS) and the Ultraflo (polyacrylonitrile, PAN).

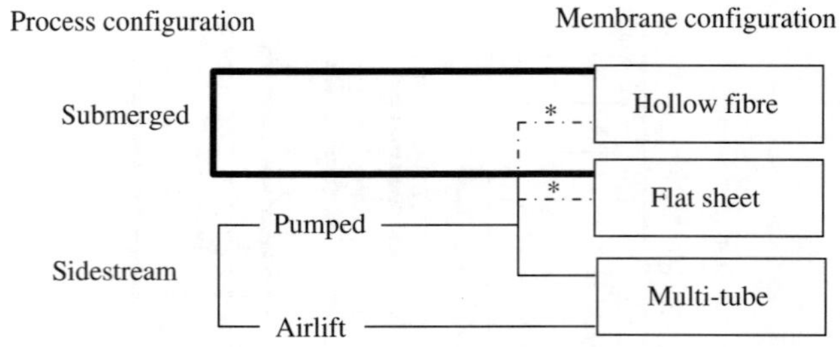

Figure 27 *Features of commercial MBR technologies (asterisk and dotted line indicates rare example)*

The principal differences between the most common products therefore arises from:

- the membrane and panel substrate materials, in the case of the FS submerged MBRs,
- the use of reinforcement (for HF modules).

The sidestream approaches are also divided into two formats – the long established traditional method of cross flow, now used only for the most difficult feeds, and the newer concept of air lift, which uses air to recirculate the feed and thereby significantly reduces energy demand. With a few exceptions, both the sidestream formats use mainly tubular membranes. Figure 27 shows the process and membrane configurations of commercial MBR technologies.

Flat sheet systems have the advantage of relatively low manufacturing cost compared to hollow fibre systems. However, packing density tends to be significantly lower than a hollow fibre system (for example by a factor of 2.5–3.0). Therefore, flat sheet systems tend to have a cost advantage for small to medium scale systems, whereas hollow fibre becomes more attractive at large scale due to the footprint advantage [175]. The comparison is made more complicated, however, since aeration costs for hollow fibre systems are often lower. This means that the most cost effective solution for total treatment costs at medium scale is closely contested, and both approaches are found across the size range due to site specific circumstances which could favour either solution. Air lift technology has a similar power cost to the submerged technology.

Figure 28 depicts the membrane system design principles followed by different membrane companies for submerged MBR.

6.4 Advantages of MBR over CAS

The combination of activated sludge with membrane separation in the MBR results in efficiencies of footprint, effluent quality and residuals production that cannot be attained when these same processes are operated in sequence. The advantages offered by the MBR process over CAS processes are widely recognised, and of these the ones most often cited are [117]:

- Production of high quality, clarified and largely disinfected permeate product in a single stage. The membrane has an effective pore size $<0.1\ \mu m$ – significantly smaller than the pathogenic bacteria and viruses in the sludge.
- Independent control of solids and hydraulic retention time (SRT and HRT, respectively). In a CAS separation of solids is achieved by sedimentation, which then relies on growth of the mixed liquor solid particles (of flocs) to a sufficient size ($>50\ \mu m$) to allow their removal by settlement. This then demands an appropriately long HRT for growth. In an MBR the particles need only be larger than the membrane pore size.

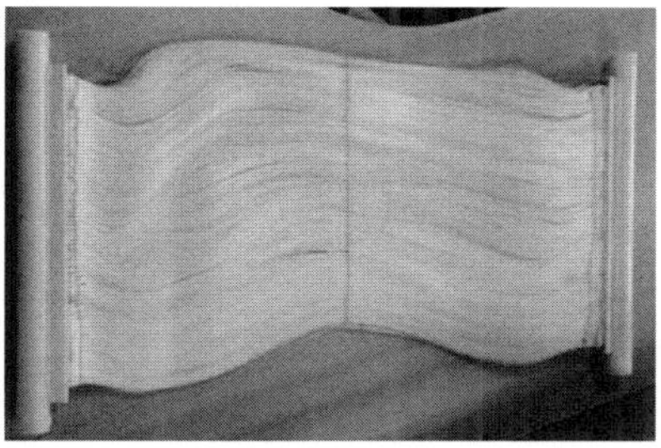

(a) Horizontally mounted single hollow fibres (Mitsubishi Rayon)

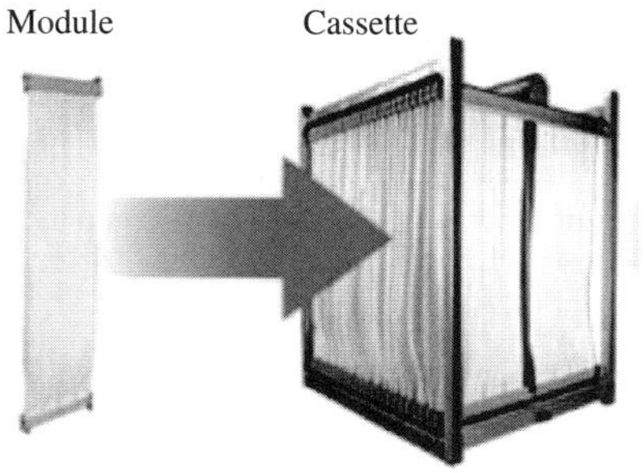

(b)Vertically mounted reinforced hollow fibres (Zenon)

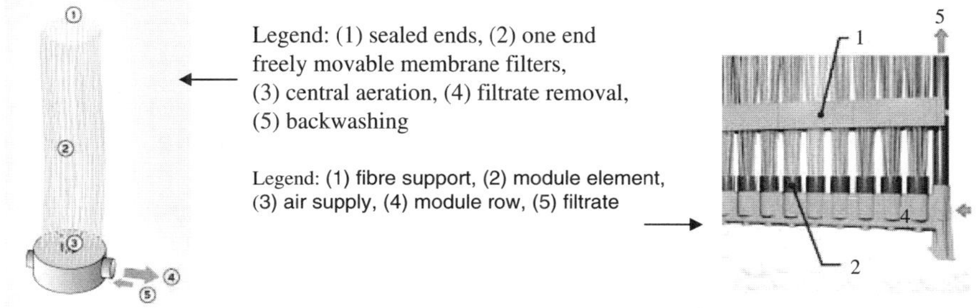

(c) One end free vertically mounted, braid-supported hollow fibres (Puron)

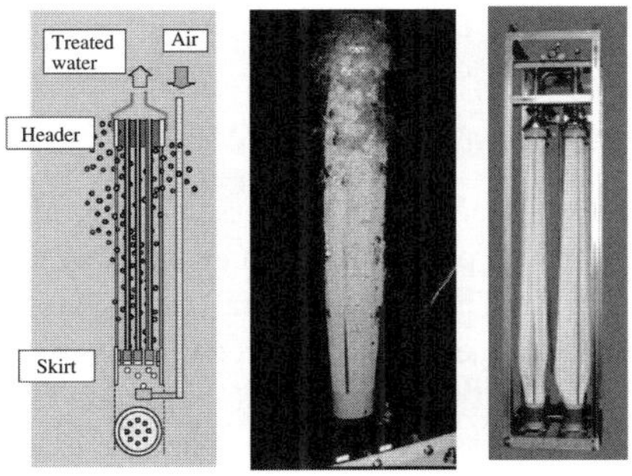

(d) Densely packed hollow fibre bundles with unique aeration arrangement (Asahi kasei)

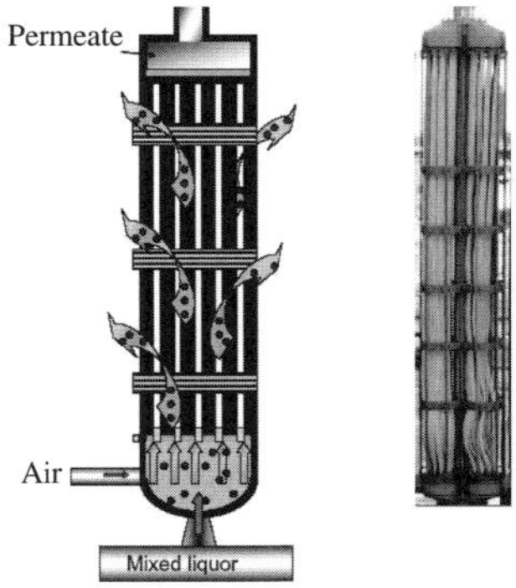

(e) Vertically mounted hollow fibre bundles with unique two-phase jet cleaning arrangement (Memcor)

Figure 28 *Membrane system design principles*

- Operation at higher mixed liquor suspended solids (MLSS) concentrations, which reduces the required reactor size and promotes the development of specific nitrifying bacteria, thereby enhancing ammonia removal.
- Reduced sludge production, which results from operation at long SRTs because the longer the solids are retained in the biotank the lower the waste solids (sludge) production.

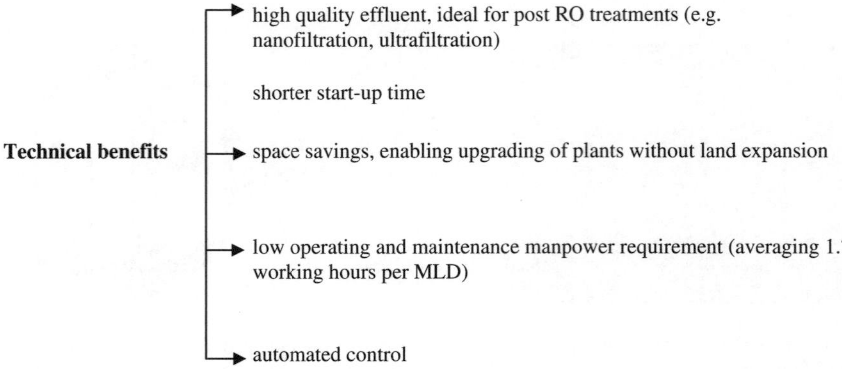

Figure 29 *Comparative technical benefits of MBR over CAS*

Of these advantages, it is the intensity of the process (i.e., the smaller size of the plant compared to conventional treatment) and the superior quality of the treated product water that are generally most important in practical wastewater treatment applications. The MBR system is particularly attractive when applied in situations where long biological solids retention times are necessary and physical retention and subsequent hydrolysis are critical to achieving biological degradation of pollutants [39]. Figure 29 depicts the technical benefits of MBR.

6.5 Organics and Nutrients Removal in MBR

6.5.1 Removal of Organic Matter and Suspended Solids

Several investigations on treatment efficiencies of MBR and CAS processes operating under comparable conditions have shown significantly improved performance of an MBR in terms of COD, NH_3-N and SS removals [44,45,106,165,251]. There are several factors that may contribute to the lower organic carbon content of MBR effluents as compared to CAS processes. Côté [59,60] attributed the improved COD removal to the avoidance of biomass washout problems commonly encountered in activated sludge process, as well as to complete particulate retention by the membrane. Since typical sludge concentrations for immersed MBRs are between 15 and 25 g l^{-1}, elimination of organic matter and turbidity is almost independent of SRT, and average removals normally achieved for COD and SS are over 90 and nearly 100%, respectively [60]. The robustness of MBR treatment regarding turbidity and organic matter removals was confirmed in several studies [74]. It is assumed that there is an upper limit for organic loading rate in an MBR under which degradation performance is independent of biomass concentration and organic loading rate [191].

MBR produces quality effluent suitable for reuse applications or as a high quality feed water source for reverse osmosis treatment. Indicative output quality include SS < 1 mg l^{-1}, turbidity < 0.2 NTU and up to 4 log removal of virus (depending on the membrane nominal pore size). In addition, it provides a barrier to certain chlorine-resistant pathogens such as *Cryptosporidium* and *Giardia*. In comparison to the conventional activated sludge process which typically achieves 95%, COD removal can be increased to 96–99% in MBRs [211].

6.5.2 Nutrient Removal

The MBR process can be configured in many different ways depending on project-specific nutrient removal objectives. Anoxic zones before or after the aerobic treatment may be used for denitrification, depending on the effluent nitrate and total nitrogen requirements. Anaerobic zones may be used to achieve enhanced biological phosphorus removal in any of its possible configurations [181].

Because of the low growth rate and poor cell yield of nitrifying bacteria, nitrification is generally a rate-limiting step in biological nitrogen removal performance. The key requirement for nitrification to occur is that the net rate of accumulation of biomass (and hence the net rate of withdrawal of biomass from the system) is less than the growth rate of nitrifying bacteria [12]. Long SRTs applied in MBR prevent nitrifying bacteria from being washed out from the bioreactor, improving the nitrification capability of the activated sludge. Moreover, nitrifiers are less endangered by faster-growing heterotrophic bacteria. Many studies have proven that MBR can operate as a high rate nitrifying technology that can be applied in the nitrification of wastewater containing a high concentration of ammonia nitrogen [91].

On the other hand, the denitrification process requires anoxic conditions. To enhance denitrification, usually an anoxic tank is added upstream from the aerated tank. Anoxic conditions can also be introduced by operating MBR in an intermittent aeration mode. In the intermittently aerated MBR, ammonium is nitrified mostly to nitrate and most phosphates are removed during the aerobic period (aeration), where the accumulated nitrate is completely denitrified during the anoxic period (non-aeration), and phosphorus (P) is taken up. The net P removal is achieved by wasting sludge after the aerobic period when the biomass contains a high level of polyphosphates (polyP) [52].

In WWTPs, biological P removal is accomplished by introduction of an anaerobic phase in the wastewater treatment line ahead of the aerobic phase and recycling of sludge through the anaerobic and aerobic phase [207]. Exposing mixed liquor to an anaerobic/aerobic sequence selects phosphate-accumulating microorganisms (PAOs) due to a competition between PAOs and other aerobic organisms. This competition mechanism is based on a complete anaerobic uptake of the lower fatty acids by the polyp organisms (i.e., PAOs), which assures that in the aerobic phase, no fatty acids are left. The polyP organisms use the stored internal substrate during aerobic conditions while other aerobic organisms are lacking substrate. This process is usually referred to as the enhanced biological phosphorus removal (EBPR) process. EBPR process can be established in MBR treatment unit by operating it in intermittent aeration mode. Moreover, phosphorus removal will be significantly improved in an MBR by a physical retention of PAOs, whose size is typically larger than 0.5 μm. Since an MF membrane (0.2 μm) acts as a physical barrier to retain the PAOs in the reactor, sufficient biomass is provided for the EBPR mechanism to take place [25,230].

Intermittently aerated MBR can achieve nitrogen and phosphorus removal by a simultaneous nitrification and denitrification, P uptake and P release in the same reactor in accordance with time cycle of aeration and non-aeration. Furthermore, there is an increasing interest for the application of MBR as a technology for phosphorus recycling, since the P content of sludge is expected to increase when prolonging SRT [138].

6.6 Recalcitrant Industrial Wastewater Treatment by MBR

High organic loadings and hardly biodegradable compounds are the major characteristics of industrial waste streams that render alternative treatment techniques such as the MBR desirable [130]. The efficiency in the removal of the organic load depends on the type of industrial process that has been implemented and consequently on the quantity of nonbiodegradable compounds.

MBRs in recent years have been proved to be effective and economically feasible for treatment of various kinds of high strength, refractory, and/or toxic wastewaters. The mechanisms of refractory chemical oxygen demand (COD) removal, however, still remain to be well documented. This section addresses this question by providing a brief review of the studies dealing with recalcitrant industrial wastewater treatment by MBR and offers unique insights into this matter.

6.6.1 Micropollutants

Several pharmaceuticals, ingredients of personal care products and so-called endocrine disrupting compounds (EDCs) are detected in surface waters all over the world. Most of those compounds are of anthropogenic origin and wastewater treatment plant (WWTP) effluents are important point discharges for the presence of endocrine disrupting compounds and residuals of pharmaceuticals in rivers, streams and surface waters [27,56,178].

Eight pharmaceuticals, two polycyclic musk fragrances and nine endocrine disrupting chemicals were analysed by Clara [55] in several WWTPs and a pilot scale MBR. As in conventional WWTP also the removal potential of MBRs was found dependent on the SRT. Weiss [242] found for half of the studied compounds, such as benzotriazole, 5-tolyltriazole (5-TTri), benzothiazole-2-sulfonate and 1,6-naphthalene disulfonate (1,6-NDSA), removal by MBR was significantly better than in CAS, while no improvement was recorded for the other half (1,5-NDSA, 1,3-NDSA, 4-TTri and naphthalene-1-sulfonate). Reif [186] reported high removal of nonsorbing but biodegradable compounds like the anti-inflammatories ibuprofen and naproxen, and moderate removal of sorbing but recalcitrant compounds; while they observed very limited removal of nonsorbing recalcitrant compounds like carbamazepine or diclofenac. Bernhard [23] observed that nondegradable micropollutants, such as EDTA and carbamazepine were not eliminated at all during WW treatment by any technique. However, the MBR showed significantly better removals compared to CAS for the investigated non-adsorbing and poorly biodegradable, persistent polar pollutants (P3), such as diclofenac, mecoprop and sulfophenylcarboxylates. Schroder [200] reported > 98.0 and 97.8% removals for 4-NP and BPA, respectively, in MBR and > 98.0 and 91.6% removals in CAS. Mass balance proved biodegradation as the main elimination mechanism for 4-NP and BPA in both processes.

Chen [40] showed that MBR could remove bisphenol-A (BPA) little more effectively than conventional activated sludge (CAS) under conditions of equal sludge loadings ranging from 0.046 to 10.2 $g\,kg^{-1}\,day^{-1}$. However, MBR could bear much higher volume loadings than CAS and still achieve the same BPA removal efficiencies. The results suggested that biodegradation dominated the BPA removal process. In a study by Snyder [208] microfiltration and ultrafiltration only were found to reject very few target compounds. An MBR, on the other hand, was effective for reducing the concentration of many EDC/PPCPs, while several other compounds remained unaffected. The removal was likely related to biodegradability of the individual compound. Lyko [149] demonstrated that the ultrafiltration membrane used in an MBR was able to partly remove the macromolecular DOC of the wastewater, while micropollutants tended to adsorb and associate with these removed macromolecules. In a study by Spring [209], a pilot scale MBR was more effective at removing cholesterol, coprostanol, stigmastanol, estrogenic species (E1, EE2), and BPA to low ng/L levels than a full-scale CAS plant receiving the same wastewater. In a study by Kimura [128], compared with CAS, MBRs exhibited much better removal regarding ketoprofen and naproxen. Removal efficiencies of the PhACs were found to be dependent on their molecular structure such as number of aromatic rings or inclusion of chlorine. Urase [231] obtained higher removal of acidic pharmaceutical substances in the case of lower pH operation because of the increased tendency of adsorption to the sludge particles. Kim [127] found MBR system to be efficient for hormones (e.g., estriol, testosterone, androstenedione) and certain pharmaceuticals (e.g., acetaminophen, ibuprofen, caffeine) with approximately 99% removal, but not efficient for erythromycin, TCEP, trimethoprim, naproxen, diclofenac and carbamazepine. Radjenovic [182] confirmed enhanced elimination of several pharmaceutical residues poorly removed by the CAS treatment (e.g., mefenamic acid, indomethacin, diclofenac, propyphenazone, pravastatin, gemfibrozil), whereas the anti-epileptic drug carbamazepine and diuretic hydrochlorothiazide bypassed both the systems. Results in a study by Gobel [95] indicated that together with the high SRT, the low substrate loading may lead to an increased biodiversity of the active biomass in MBR, resulting in a broader range of micropollutant degradation pathways available.

In a review of the factors influencing the removal of organic micropollutants from wastewater, Cirja [54] concluded that hydrophobic compounds are removed from the liquid phase via adsorption, and possibly through biodegradation processes when the SRT is high enough. Owing to the compactness of MBR plant and the high organic load that can be applied, this process is promising concerning the removal of micropollutants, which are eliminated at high SRT and biomass concentration.

6.6.2 *Dye Wastewater*

Large amounts of dyes are annually produced and applied in many different industries, including the textile, cosmetic, paper, leather, pharmaceutical and food industries [145]. The release of coloured wastewater in the ecosystem is a remarkable source of esthetic pollution, eutrophication and perturbations in aquatic life. Virtually all the known physicochemical and biological techniques have been explored for decolorisation [100], none has emerged as a panacea [101].

While treating real textile dye wastewater, Yun [256] observed removal efficiencies of 54, 79, 99 and 100%, for colour, COD, BOD and SS, respectively, with the MBR process as compared to the corresponding values of 51, 70, 96 and 60% for a sequencing batch process. As compared to an extended aeration biological process, Malpei [151] obtained higher COD removal and colour abatement, besides much higher removal efficiency for suspended solids and microorganisms in MBR. While running a pilot MBR plant in parallel to one existing WWTP (activated sludge + clariflocculation + ozonation) for the treatment of textile wastewater, Lubello [147] observed higher COD and SS removal, similar colour removal and slightly lower surfactant removal in the MBR. Based on another study, Lubello [148] summarised the relative advantages of the MBR system compared to the conventional WWTP as follows: lower effluent COD value; very little standard deviation of the effluent quality; no problems due to filamentous microorganisms; complete nitrification, and heterotrophic bacteria growth without the need to add nutrients due to the biomass retention in the MBR. Brik [31] observed a distinct relationship between sludge growth and colour removal, suggesting dye removal predominantly by adsorption on sludge and proposed post treatment by nanofiltration for water reuse. In a study by Schoeberl [198], a combination of a membrane bioreactor and subsequent nanofiltration proved to meet all requirements for reuse.

Yun [256] observed an opposite tendency in treatment efficiency between COD and dye as a function of DO level and suggested that an alternative operation of aerobic and anoxic phases is indispensable to efficiently biodegrade both COD and dye molecules in dye wastewater treatment with MBR. Using an adapted biomass, Badani *et al.* [10] obtained 97% COD and 70% colour reduction in an MBR. Hai et al. [100,102] developed a submerged membrane fungi reactor for excellent colour and TOC removal.

6.6.3 Tannery Wastewater

The high complexity of the tannery wastewater matrix originates from a wide range of components such as: raw materials (skins) residues, excess dosage of reagents, tanning agents, dyes and surfactants [163]. The choice of the treatment process is strictly related to the presence of significant fractions of slowly hydrolysable and inhibiting compounds requiring technologies such as MBR, which increase the sludge age and consequently the capacity to degrade substrates usually not biodegraded in conventional wastewater treatment plants.

In a study by Munz [163], as compared to CAS, MBR showed a higher COD removal (+4%) and a more stable and complete nitrification while treating the same tannery wastewater. Artiga [8] reported about 86% COD removal from tannery wastewater in an MBR. Chung [51] reported 50% as the optimum ratio of the volume of anoxic denitrification tank to aerobic nitrification tank to treat tannery wastewater in MBR. It was also found that supplementation of phosphorus to maintain a COD : T-P ratio of 100 : 1 was needed to achieve the best performance.

6.6.4 Landfill Leachate

Landfill leachate can broadly be defined as the liquid produced from the decomposition of waste and infiltration of rainwater in the landfill [123]. Improvements in landfill engineering are aimed at reducing leachate production, collection and treatment prior to discharge [75]. Nitrification is generally readily achievable, with >95% removal of ammonia reported through the exclusive application of biological techniques to the treatment of both young and old leachates [3,107,245]. However, COD removal is considerably more challenging, with removal efficiency values from 20% to >90% reported according to leachate characteristics (origin and age), process type and process operational facets [70,92,220].

In view of the high strength of the landfill leachate, Visvanathan [235] utilised an aerobic thermophilic MBR and demonstrated an average COD removal of 62–79%, with a gradual increase in BOD supplementation. Laitinen [134], while treating landfill leachate by a sequencing batch reactor (SBR) and a submerged MBR reported that MBR effluent was significantly better in quality and had lower variations. Alvarez-Vazquez [4] presented a critical assessment of data from existing plants where MBR-based treatment schemes appeared to achieve greater mean COD removal (80%) across all installations, for less biotreatable feed waters (BOD/COD = 0.03–0.16), than conventional systems which achieve COD removals of around 63% at feed water BOD/COD ratios of 0.21–0.3. Moreover, they do so at generally lower HRTs, and thus correspondingly higher loading rates (1–3 kg COD m^{-3} day^{-1} compared with < 0.25 kg COD m^{-3} day^{-1} for conventional treatment schemes) and so lower footprints.

6.6.5 Oil Contaminated Wastewater

The composition of the wastewater from petroleum refinery is a function of the processing units involved and generally contains: hydrocarbons (aliphatic or aromatic), phenolic compounds (phenol, methylphenol, dimethyphenol), sulphur, mercaptans, oil, solvents or chlorines [73]. Bienati [26] demonstrated that MBR was able to treat industrial oil contaminated wastewater with high removal efficiency (about 98%), under short HRT (about 10 h) and high biomass concentration. The results from the study conducted by Rahman [185], who employed a crossflow MBR showed a COD removal efficiency of more than 93% from refinery wastewater with limited influence of HRT. Trials in an MBR with a high activated sludge concentration of up to 48 g l^{-1} showed that oily wastewater also containing surfactants was biodegraded with high efficiency [199]. Moreover, Sutton *et al.* [215] designed and operated a cost-effective full-scale MBR system that can be expected to perform technically at a level equal to or better than a conventional oily wastewater treatment system.

Olive mill wastewater (OMW) is a foul smelling phenol-rich, acidic wastewater with high organic load (BOD up to 100 g l^{-1}, COD up to 200 g l^{-1}) [193]. Dhaouadia [69] observed complete phenols removal, but moderate total COD removal from OMW in an external ceramic MBR, indicating post treatment requirements. Conversely, while treating a high phenol-containing wastewater, Barrios-Martinez [13] confirmed a phenol-free permeate, along with 50 g day^{-1} phenol degradation in the MBR containing acclimatised biomass. While treating dissolved air flotation (DAF) pretreated pet food wastewater, Kurian [133] observed the following COD and oil/grease removal efficiencies: thermophillic MBR (75–98, 66–86%) and conventional MBR (94, 98%), respectively, and high concentrations of VFA, equivalent to 50–73% of total COD, were recorded in the thermophillic MBR effluent. In an MBR with submerged multichannel flat sheet ceramic membranes, Blocher [28], while working with degreasing solutions from surface refining processes, achieved retention and subsequent biodegradation of hydrocarbons, while the membrane allowed relatively high permeation of the surfactants for reuse.

6.6.6 Insight into Recalcitrant Compound Removal in MBR

Studies suggest that build up of heterotrophic bacteria having special degrading capacity [95,148] or easier acclimatisation [13,69] is possible in MBR due to the complete retention of biomass by membrane and application of longer SRT. Accordingly better and/or stable removal performance may be expected in MBR.

For moderate to considerably biodegradable compounds, the volumetric biodegradation rate in MBR may improve due to high biomass concentration [28]. However, that may not be the case for the hardly biodegradable or nonbiodegradable compounds [55]. As in conventional process the removal potential in MBRs also depends on the SRT. However, MBRs achieve a high SRT within a compact reactor. Conversely, other studies have reported that MBR is not superior for well degradable compounds that are already extensively degraded in conventional treatment, nor is it superior for recalcitrant compounds that are not amenable to biodegradation [23,242]. However, they agreed that for most compounds of intermediate removal in conventional treatment, the MBR is clearly superior.

The membrane filtration step provides an effluent with high quality in terms of suspended solid; hence the recalcitrant compounds adsorbed on suspended solid remains within the retentate phase where they may undergo biodegradation [231]. At times, significant retention of soluble organics on the cake layer over the membrane and subsequent biodegradation may occur [102,256].

6.7 Recent Advances in Membrane Bioreactors Design/Operation

Several interesting modifications to conventional design of MBR have been proposed in the literature in order to enhance removal performance and/or mitigate membrane fouling. These include bioaugmented MBR [94,102,180], settling-enhanced inclined plate MBR [249], integrated anoxic–aerobic MBR [36,37,100,103], compact Jet loop-type MBR [170,254], biofilm MBR [100,137,167], MBR followed by reverse osmosis (RO) or submerged nanofiltration MBR [48,49,179], forward osmosis MBR [2,57] and membrane distillation bioreactor [176,177].

6.8 Development Challenges

MBR technology is facing some research and development challenges (Figure 30). Among the challenges underscored by the experts, membrane fouling is one of the most serious problems that have retarded faster commercialisation of MBR technology. Two other issues closely related to membrane fouling and performance are: (i) a requirement for mechanical pretreatment and (ii) maintaining membrane integrity.

6.8.1 Membrane Fouling

Membrane fouling and its consequences in terms of plant maintenance and operating costs limit the widespread application of MBRs [135]. Membrane fouling can be defined as the undesirable deposition and accumulation of microorganisms, colloids, solutes, and cell debris within pores or on membrane surface [158]. It results from the interaction between the membrane material and the components of the activated sludge liquor, which include biological flocs formed by a large range of living microorganisms along with soluble and colloidal compounds. Thus it is not surprising that the fouling behavior in MBRs is more complicated than that in most membrane

Research and development challenges
– Membrane fouling
– Effective pretreatment requirements
– Membrane lifespan
– Cost
– Plant capacity (scale up)
– Exchangeability of membrane modules

Problems often encountered by plant operators
– membrane fouling during permeate backpulsing
– entrained air impacting suction pump operation
– bioreactor foaming
– inefficient aeration due to partial clogging of aerator holes
– no significant decrease of biosolid production
– scale build up on membrane and piping
– corrosion of concrete, hand rails and metallic components due to corrosive vapour produced during high-temperature NaOCl cleaning
– membrane delamination and breakage during cleanings
– odour from screening, compaction, drying beds and storage areas
– failure of control system

Figure 30 Technology bottlenecks

applications. The suspended biomass has no fixed composition and varies both with feed water composition and MBR operating conditions employed.

All the parameters involved in the design and operation of MBR processes have an influence on membrane fouling [135,158]. Three main categories of factors can be identified: membrane and module characteristics, feed and biomass parameters and operating conditions.

When the operating flux is below the critical flux, particles accumulation in the region of membranes can be effectively prevented. However, due to physicochemical solute–membrane material interactions, the membrane permeability will decrease over time, even when MBRs are operated in subcritical (below critical flux) conditions. Techniques used to remove fouling includes physical cleaning (permeate backwashing, air backwashing, relaxation, sonification etc.) and chemical cleaning. Certain other measures can be adopted to limit fouling. These include membrane modification (optimisation of membrane characteristics, membrane module design etc.), optimisation of operating conditions like aeration, SRT etc., and modification of biomass characteristics (developing aerobic granular sludge, addition of adsorbent/flux enhancers/coagulant etc.).

6.8.2 Pre-Treatment Requirement

Pre-treatment is one of the most critical factors for ensuring a stable and continuous MBR operation. Because of membrane sensitivity to the presence of foreign bodies, fine pre-screening of the feed (and sometimes of the mixed liquors) must occur. The type of sieve installed is very important regarding the total screening of hair and fibres [77,197]. The fact that even intensive long-term pilot plant trials can fail to suggest the effective scale up design of the sieve [156] adds to the problem.

6.8.3 Maintaining Membrane Integrity

A major problem facing MBR systems is the loss of membrane integrity, which leads to the permeate quality deterioration and ineffective backwashing. This may be caused by faulty installation, frequent and/or extended contact between membrane and cleaning solution causing delamination of the membrane, scoring and cleaving of the membrane resulting from the presence of abrasive or sharp-edged materials in the influent, and operating stress and strain occurring in the system due to fibre movement and membrane backwashing. A better understanding of the effect of membrane material, age and fouling on membrane integrity may be gained from hollow fibre tensile test reported in the literature [46].

6.9 Future Research

Future research in MBRs is likely to focus on reduction in energy demand through more frugal use of membrane aeration in immersed systems [117]. This will rely on a better understanding of membrane channel clogging and chemical cleaning. Lesjean [139] opines that academic research is addressing only some of the crucial issues. For instance, while many publications on fouling are being produced and some cost studies are conducted, no significant research efforts have addressed membrane lifespan, pretreatment and scale up issues. Yamamoto [252] contends that in addition to the alleviation of the technology bottlenecks, a radical shift from the conventional concept of 'organic wastewater treatment' to 'water/biomass production' is necessary. This can be materialised by developing next generation

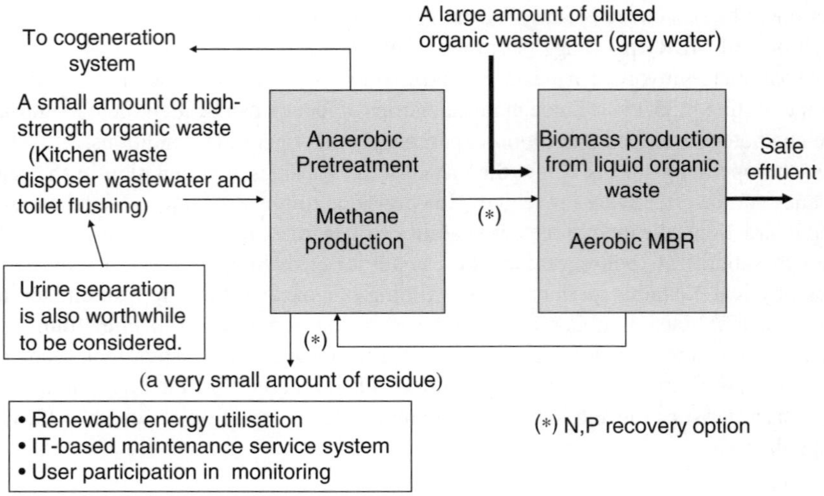

Figure 31 *Next generation MBR system: anaerobic combination for on site small scale advanced treatment. Reprinted from Book of Proceedings, Final MBR-Network Workshop, 31 March-1 April, 2009, Berlin*

MBRs where the membrane acts as a separator of water and biomass, and biomass is utilised for energy production (Figure 31).

7 Conclusion

In this review, membrane reactors and bioreactor have been introduced and discussed. A Membrane Reactor (MR) is a (multiphase) device in which reaction and separation (through a membrane) take place simultaneously. MR can be used either to increase the conversion (circumventing equilibrium limitations via the Le Chatelier principle) or to increase the selectivity (through distributive feeding of a reactant through the membrane). The core of the MR is the membrane, which can be either organic and inorganic, porous or dense. In the following chapters, the preparation methods of various membranes to be used in membrane reactors will be discussed. In this review the different membrane reactors (packed bed, fluidised bed, micro, hollow fibre) have been discussed for various processes (e.g. hydrogen production, syngas production etc.).

Membrane bioreactor (MBR) process is a hybrid system amalgamating membrane separation with biological treatment. Right from its inception, the MBR technology became a dependable option for high strength wastewater treatment. Over the past decade, submerged membrane bioreactor (MBR) processes have experienced unprecedented growth also in domestic and municipal wastewater treatment owing to several advantages including excellent effluent quality, low sludge production, small foot print, and flexibility in future expansion. The superiority of the MBR lies in the fact that it is compatible with a variety of post treatments. This paves the way for designing effective hybrid systems having the MBR as the core, for water reuse. New studies in this field are expected to tackle the development challenges illustrated here.

References

1. M.E.E. Abashar, S.S.E.H. Elnashaie, Feeding of oxygen along the height of a circulating fast fluidized bed membrane reactor for efficient production of hydrogen, *Chem. Eng. Res.Des.*, **85**, 1529–1538 (2007).
2. A. Achilli, T.Y. Cath, E.A. Marchand and A.E. Childress, The forward osmosis membrane bioreactor: A low fouling alternative to MBR processes, *Desalination*, **239**, 10–21 (2009).
3. W.Y. Ahn, M.S. Kang, S.K. Yim and K.H. Choi, Nitrification of leachate with submerged membrane bioreactor pilot scale, in *Membrane Technology in Environmental Management*, ed. by Yamamoto, K., Tokyo, Japan, 432–435 (1999).
4. H. Alvarez-Vazquez, B. Jefferson and S. Judd, Membrane bioreactors vs conventional biological treatment of landfill leachate: a brief review. *J Chem Technol Biotechnol.*, **79**, 1043–1049 (2004).
5. H. Amandusson, L.G. Ekedahl, H. Dannetun, The effect of CO and O_2 on hydrogen permeation through a palladium membrane, *Appl. Surf. Sci.*, **153**, 259–267 (2000).
6. H. Amandusson, L.G. Ekedahl, H. Dannetun, Hydrogen permeation through surface modified Pd and Pd-Ag membranes, *J. Membrane Sci.*, **193**, 35–47 (2001).
7. J.N. Armor, Catalysis with permselective inorganic membranes, *J. Membrane Sci.*, **147**, 217–233 (1998).
8. P. Artiga, E. Ficara, F. Malpei, J.M. Garrido and R. Mendez, Treatment of two industrial wastewaters in a submerged membrane bioreactor, *Desalination*, **179**, 161–169 (2005).
9. K. Azrague, E. Puech-Costes, P. Aimar, M.T. Maurette, F. Benoit-Marquie, Membrane photoreactor (MPR) for the mineralisation of organic pollutants from turbid effluents, *J. Membrane Sci.*, **258**, 71–77 (2005).
10. Z. Badani, H. Ait-Amar, A. Si-Salah, M. Brik and W. Fuchs, Treatment of textile waste water by membrane bioreactor and reuse, *Desalination*, **185**, 411–417 (2005).
11. L. Barelli, G. Bidini, F. Gallorini, S. Servili, Hydrogen production through sorption-enhanced steam methane reforming and membrane technology: A review, *Energy*, **33**, 554–570 (2008).
12. D. Barnes and P.J. Bliss, in *Biological control of nitrogen in wastewater treatment*, E. Spon and F.N. Spon (Eds), Spon Pr, New York, 1983.
13. A. Barrios-Martinez, E. Barbot, B. Marrot, P. Moulin and N. Roche, Degradation of synthetic phenol-containing wastewaters by MBR, *J Membrane Sci.*, **281**, 288–296 (2006).
14. A. Basile, L. Paturzo, An experimental study of multilayered composite palladium membrane reactors for partial oxidation of methane to syngas, *Catal. Today*, **67**, 55–64 (2001a).
15. A. Basile, L. Paturzo, F. Laganà, The partial oxidation of methane to syngas in a palladium membrane reactor: simulation and experimental studies, *Catal. Today*, **67**, 65–75 (2001b).
16. A. Basile, F. Gallucci, L. Paturzo, A dense Pd/Ag membrane reactor for methanol steam reforming: Experimental study, *Catal. Today*, **104**, 244–250 (2005).
17. A. Basile, F. Gallucci, A. Iulianelli, S. Tosti, E. Drioli, The pressure effect on ethanol steam reforming in membrane reactor: experimental study, *Desalination*, **200**, 671–672 (2006a).
18. A. Basile, S. Tosti, G. Capannelli, G. Vitulli, A. Iulianelli, F. Gallucci, E. Drioli, Co-current and counter-current modes for methanol steam reforming membrane reactor: Experimental study, *Catal. Today*, **118**, 237–245 (2006b).
19. A. Basile, Hydrogen production using Pd-based membrane reactors for fuel cells, *Top. Catal.*, **51**, 107–122, (2008a).
20. A. Basile, F. Gallucci, A. Iulianelli, S. Tosti, CO-free hydrogen production by ethanol steam reforming in a Pd-Ag membrane reactor, *Fuel Cells*, **1**, 62–68 (2008b).
21. A. Basile, F. Gallucci, A. Iulianelli, F. Borgognoni, S. Tosti, Acetic acid steam reforming in a Pd–Ag membrane reactor: The effect of the catalytic bed pattern, *J. Membrane Sci.*, **311**, 46–52 (2008d).

22. S. Bhatia, C.Y. Thien, A. R. Mohamed, Oxidative coupling of methane (OCM) in a catalytic membrane reactor and comparison of its performance with other catalytic reactors, *Chem. Eng. J.*, **148**, 525–532 (2009).
23. M. Bernhard, J. Muller, J. Knepper, Biodegradation of persistent polar pollutants in wastewater: Comparison of an optimised lab-scale membrane bioreactor and activated sludge treatment. *Water Res.*, **40**, 3419 –3428 (2006).
24. P.R. Bérubé, E.R. Hall and P.M. Sutton, Parameters Governing the Permeate Flux in an Anaerobic Membrane Bioreactor Treating Low Strength Municipal Wastewater: A Literature Review, *Water Environment Research*, **78**, 887–896 (2006).
25. J.J. Beun, J.J. Heijnen and M.C. M Van Loosdrecht, N-Removal in a granular sludge sequencing batch airlift reactor, *Biotechnol Bioeng*, **75**, 82–92 (2001).
26. B. Bienati, A. Bottino, G. Capannelli and A. Comite, Characterization and performance of different types of hollow fibre membranes in a laboratory-scale MBR for the treatment of industrial wastewater, *Desalination*, **231**, 133–140 (2008).
27. M.A. Blackburn and M.J. Waldock, Concentrations of alkylphenols in rivers and estuaries in England and Wales. *Water Res.*, **29**, 1623–1629 (1995).
28. C. Blocher, U. Bunse, B. SeBler, H. Chmiel and H.D. Janke, Continuous regeneration of degreasing solutions from electroplating operations using a membrane bioreactor. *Desalination*, **162**, 315–326 (2004).
29. V.S. Bobrov, N.G. Digurov, V.V. Skudin, Propane dehydrogenation using catalytic membrane, *J. Membrane Sci.*, **253**, 233–242 (2005).
30. R. Bredesen, K. Jordal, O. Bolland, High-temperature membranes in power generation with CO_2 capture, *Chem. Eng. Process.*, **43**, 1129–1158 (2004).
31. M. Brik, P. Schoeberl, B. Chamam, R. Braun and W. Fuchs, Advanced treatment of textile wastewater towards reuse using a membrane bioreactor, *Process Biochem.*, **41**, 1751–1757 (2006).
32. P.J. Brow, Ballyclare MBR WwTW: Cost and challenges of compliance with stringent standards. In: Proceedings of the 4th International Water Association Conference on Membranes for Water and Wastewater Treatment, 15–17 May, Harrogate, UK. CD ISBN number: 978-1-86194-127-5 2007.
33. J. Caro, H.H. Wang, C. Tablet, A. Kleinert, A. Feldhoff, T. Schiestel, M. Kilgus, P. Koelsch, S. Werth, Evaluation of perovskites in hollow fibre and disk geometry in catalytic membrane reactors and in oxygen separators, *Catal. Today*, **118**, 128–135 (2006).
34. Catalytica®, Catalytic membrane reactors: concepts and applications, Catalytica Study N. 4187 MR, (1988).
35. J. Cejka, H. Van Corna, A. Corma, F. Schuth, Introduction to zeolite science and practice, Studies in surface science and catalysis, **168**, Elsevier, Amsterdam, The Netherlands, 3rd revised edition, 2007.
36. S.R. Chae, S.T. Kang, Y. Watanabe and H.S. Shin, Development of an innovative vertical submerged membrane bioreactor (VSMBR) for simultaneous removal of organic matter and nutrients, *Water Research*, **40**, 2161–2167 (2006a).
37. S.R. Chae, Y.T. Ahn, S.T. Kang and H.S. Shin, Mitigated membrane fouling in a vertical submerged membrane bioreactor (VSMBR), *Journal of Membrane Science*, **280**, 572–581 (2006b).
38. J.S. Chang, C.Y. Chang, A.C. Chen, L. Erdei and S. Vigneswaran, Long-term operation of submerged membrane bioreactor for the treatment of high strength acrylonitrile-butadiene-styrene (ABS) wastewater: effect of hydraulic retention time, *Desalination*, **191**, 45–51 (2006).
39. T.K. Chen, J.N. Chen, C.H. Ni, G.T. Lin and C.Y. Chang, Application of a membrane bioreactor system for opto-electronic industrial wastewater treatment– a pilot study. *Water Science and Technology*, **48**, 195–202 (2003).

40. J. Chen, X. Huang and D. Lee, Bisphenol A removal by a membrane bioreactor, *Process Biochem.*, **43**, 451–456 (2008).
41. Y. Chen, Y. Wang, H. Xu, G. Xiong, Efficient production of hydrogen from natural gas steam reforming in palladium membrane reactor, *Appl. Catal. B: Env.*, **80**, 283–294 (2008).
42. Y.S. Cheng, M.A. Pena, J.L. Fierro, D.C.W. Hui, K.L. Yeung, Performance of alumina, zeolite, palladium, Pd-Ag alloy membranes for hydrogen separation from Towngas mixture, *J. Membrane Sci.*, **204**, 329–340 (2002).
43. Y.S. Cheng, M.A. Peña, K.L. Yeung, Hydrogen production from partial oxidation of methane in a membrane reactor, *J. Taiwan Inst. Chem. Eng.*, **40**, 281–288 (2009).
44. C. Chiemchaisri, K. Yamamoto and S. Vigneswaran, Household membrane bioreactor in domestic wastewater Treatment, *Water Sci. Technol.*, **27**, 171–178 (1993).
45. C. Chiemchaisri, Y.K. Wong, T. Urase and K. Yamamoto, Organic stabilisation and nitrogen removal in a membrane separation bioreactor for domestic wastewater treatment (1993). *Filtr. Sep.*, **30**, 247–252 (1993)
46. A.E. Childress, J. Daugherty, C. Chen, P. Le-Clech and G.L. Leslie, Hollow fiber membrane integrity: Mechanics and modeling, *Desalination*, **180**, 5–14 (2005).
47. S.S. Chin, T.M. Lim, K. Chiang, A.G. Fane, Factors affecting the performance of a low-pressure submerged membrane photocatalytic reactor, *Chem. Eng. J.*, **130**, 53–63 (2007).
48. J.H. Choi, S. Dockko, K. Fukushi and K. Yamamoto, A novel application of a submerged nanofiltration membrane bioreactor (NF MBR) for wastewater treatment, *Desalination*, **146**, 413–420 (2002).
49. J.H. Choi, K. Fukushi and K. Yamamoto, (2007) A submerged nanofiltration membrane bioreactor for domestic wastewater treatment: the performance of cellulose acetate nanofiltration membranes for long-term operation. *Separation and Purification Technology*, **52**, 470–477 (2007).
50. K.-H. Choo, R. Tao, M.-J. Kim, Use of a photocatalytic membrane reactor for the removal of natural organic matter in water: Effect of photoinduced desorption and ferrihydrite adsorption, *J. Membrane Sci.*, **322**, 368–374 (2008).
51. Y.J. Chung, H.N. Choi, S.E. Lee and J.B. Cho, Treatment of tannery wastewater with high nitrogen content using anoxic/oxic membrane bio-reactor (MBR). *J Environ Sci Health, Part A*, **39**, 1881–1890 (2004).
52. J. Chung, Y. Kim, D.J. Lee, H. Shim and J.O. Kim, Characteristics of denitrifying phosphate accumulating organisms in an anaerobic-intermittently aerobic process, *Environ Eng Sci*, **23**, 981–993 (2006).
53. N. Cicek, A review of membrane bioreactors and their potential application in the treatment of agricultural wastewater, *Can. Biosys. Eng.*, **45**, 6.37–6.49 (2003).
54. M. Cirja, P. Ivashechkin, A. Schaeffer, F. Philippe and X. Corvini, Factors affecting the removal of organic micropollutants from wastewater in conventional treatment plants (CTP) and membrane bioreactors (MBR). *Rev Environ Sci Biotechnol*, **7**, 61–78 (2008).
55. M. Clara, B. Strenn, O. Gans, E. Martinez, N. Kreuzinger and H. Kroiss, Removal of selected pharmaceuticals, fragrances and endocrine disrupting compounds in a membrane bioreactor and conventional wastewater treatment plants, *Water Res*, **39**, 4797–4807 (2005).
56. T. Colborn, J. Peterson Myers and D. Dummanoski, *Our Stolen Future*; Little, Brown and Co.: Boston, MA, USA, 1996.
57. E.R. Cornelissen, D. Harmsen, K.F. de Korte, C.J. Ruiken, J.J. Qin, H. O.o. H and L.P. Wessels (2008) Membrane fouling and process performance of forward osmosis membranes on activated sludge, *Journal of Membrane Science*, **319**, 158–168 (2008).
58. J. Coronas, J. Santamaria, Catalytic reactors based on porous ceramic membranes, *Catal. Today*, **51**, 377–389 (1999).
59. P. Côté, H. Buisson, C. Pound and G. Arakaki, Immersed membrane activated sludge for the reuse of municipal wastewater, *Desalination*, **113**, 189–196 (1997).

60. P. Côté, H. Buisson and M. Praderie, Immersed membranes activated sludge process applied to the treatment of municipal wastewater, *Water Sci Technol*, **38**, 437–442 (1998).
61. A.S. Damle, Hydrogen production by reforming of liquid hydrocarbons in a membrane reactor for portable power generation – Experimental studies, *J. Power Sour.*, **186**, 167–177 (2009).
62. O. de la Iglesia, R. Mallada, M. Menendez, J. Coronas, Continuous zeolite membrane reactor for esterification of ethanol and acetic acid, *Chem. Eng. J.*, **131**, 35–39 (2007).
63. K.C. de Souza Figueiredo, V.M. Martins Salim, C.P. Borges, Synthesis and characterization of a catalytic membrane for pervaporation-assisted esterification reactors, *Catal. Today*, **133/135**, 809–814 (2008).
64. A.L. Deshayes, E.E. Mirò, G.I. Horowitz, Xylene isomerization in a membrane reactor. Part II. Simulation of an industrial reactor, *Chem. Eng. J.*, **122**, 149–157 (2006).
65. S.A.R.K. Deshmukh, J.A. Laverman, A.H.G. Cents, M. Van Sint Annaland, J.A.M. Kuipers, Development of a membrane-assisted fluidized bed reactor. 1. Gas phase back-mixing and bubble-to-emulsion phase mass transfer using tracer injection and ultrasound experiments, *Ind. Eng. Chem. Res.*, **44**, 5955–5965 (2005a).
66. S.A.R.K. Deshmukh, J.A. Laverman, M. Van Sint Annaland, J.A.M. Kuipers, Development of a membrane-assisted fluidized bed reactor. 2. Experimental demonstration and modeling for the partial oxidation of methanol, *Ind. Eng. Chem. Res.*, **44**, 5966–5976 (2005b).
67. S.A.R.K. Deshmukh, S. Heinrich, L. Mörl, M. van Sint Annaland, J.A.M. Kuipers, Membrane assisted fluidized bed reactors: Potentials and hurdles, *Chem. Eng. Sci.*, **62**, 416–436 (2007a).
68. S.A.R.K. Deshmukh, M. van Sint Annaland, J.A.M. Kuipers, Gas back-mixing studies in membrane assisted bubbling fluidized beds, *Chem. Eng. Sci.*, **62**, 4095–4111 (2007b).
69. H. Dhaouadia and B. Marrot, Olive mill wastewater treatment in a membrane bioreactor: Process feasibility and performances, *Chem Eng J.*, **145**, 225–231 (2008).
70. A. Ding, Z. Zhang, J. Fu and L. Cheng, Biological control of leachate from municipal landfills, *Chemosphere*, **44**, 1–8 (2001).
71. R. Dittmeyer, V. Höllein, K. Daub, Membrane reactors for hydrogenation and dehydrogenation processes based on supported palladium, *J. Molecular Catal. A: Chemical*, **173**, 135–184 (2001).
72. X. Dong, C. Zhang, X. Chang, W. Jin, N. Xu, A Self-Catalytic Membrane Reactor Based on a Supported Mixed-Conducting Membrane, *AIChE J.*, **54**, (6), 1678–1680 (2008).
73. W.W. Eckenfelder, J.F. Malina and J.W. Patterson, *Toxicity Reduction: Evaluation and Control*; Technomic Publishing Co. Inc: Ford, DL, ISBN 0-87762-905-6, 1992.
74. X.J. Fan, Nitrification and mass balance with a membrane bioreactor for municipal wastewater treatment, *Water Sci Technol.*, **34**, 129–136 (1996).
75. G.J. Farquhar, Leachate production and characteristics, *Can. J. Civ. Eng.*, **16**, 317–325 (1989).
76. P. Ferreira-Aparicio, I. Rodrguez-Ramos, A. Guerrero-Ruiz, On the applicability of membrane technology to the catalyzed dry reforming of methane, *Appl. Catal. A: Gen.*, **237**, 239–252 (2002).
77. F.B. Frechen, W. Schier and M. Wett, Pre-treatment of municipal MBR applications in Germany – current status and treatment efficiency, *Water Practice and Technology*, **1**, doi: 10. 2166/WPT.2006057 (2006).
78. D. Fritsch, G. Bengtson, Development of catalytically reactive porous membranes for the selective hydrogenation of sunflower oil, *Catal. Today*, **118**, 121–127 (2006).
79. Frost and Sullivan, MBR: A buoyant reaction in Europe, Market Research Report, June (2003).
80. Frost and Sullivan, US advanced water treatment equipment markets, Market Research Report, June (2004a).
81. Frost and Sullivan, US and Canada membrane bioreactor markets, Market Research Report, June (2004b).
82. Frost and Sullivan, US desalination plant market, Market Research Report, January (2004c).
83. Frost and Sullivan, European report: introduction and executive summary, Market Research Report, August (2005).

84. A.B. Fuertes, T.A. Centeno, Preparation of supported asymmetric carbon molecular sieve membranes, *J. Membrane Sci.*, **144**, 105–111 (1998).
85. F. Gallucci, L. Paturzo, A. Basile, An experimental study of CO2 hydrogenation into methanol involving a zeolite membrane reactor, *Chem. Eng. Proc.*, **43**, 1029–1036 (2004).
86. F. Gallucci, S. Tosti, A. Basile,Synthesis, characterization and applications of palladium membranes, Eds. R. Malada, M. Menendez, Elsevier, cap.8, (2007a).
87. F. Gallucci, A. Basile, S. Tosti, A. Iulianelli, E. Drioli, Methanol and ethanol steam reforming in membrane reactors: An experimental study, *Int. J. Hydrogen Eng.*, **32**, 1201–1210 (2007b).
88. F. Gallucci, S. Tosti, A. Basile, Pd–Ag tubular membrane reactors for methane dry reforming: A reactive method for CO_2 consumption and H_2 production, *J. Membrane Sci.*, **317**, 96–105 (2008c).
89. F. Gallucci, M. Van Sint Annaland, J.A.M. Kuipers, Autothermal reforming of methane with integrated CO_2 capture in a novel fluidized bed membrane reactor. Part 1: Experimental demonstration, *Topics in Catal.*, **51**, 133–145 (2008a).
90. F. Gallucci, M. Van Sint Annaland, J.A.M. Kuipers, Autothermal reforming of methane with integrated CO_2 capture in a novel fluidized bed membrane reactor. Part 2: Comparison of reactor configurations, *Topics in Catal.*, **51**, 146–157 (2008b).
91. M. Gao, M. Yang, H. Li, Y. Wang and F. Pan, Nitrification and sludge characteristics in a submerged membrane bioreactor on synthetic inorganic wastewater, *Desalination*, **170**, 177–185 (2004).
92. H. Garcia, J.L. Rico and P. Garcia, Comparison of anaerobic treatment of leachates from an urban-solid-waste landfill at ambient temperature and 35 °C. *Bioresource Technol.*, 1996, **58**, 273–277 (1996).
93. E. Gernot, F. Aupretre, A Deschamps, F. Epron, P. Merecot, D. Duprez, C. Etievant, Production of hydrogen from bioethanol in catalytic membrane reactor, 16th Conférence Mondiale de l'Hydrogène Energie (WHEC16), 13–16 June 2006-Lyon (France), http://www.ceth.fr/download/presse/art_ceth_3.pdf.
94. W. Ghyoot, D. Springael, Q. Dong, S.V. Roy, G. Nuyts and L. Diels, Bioaugmentation with the clc-element carrying Pseudomonas putida BN210 in a membrane separation bioreactor, *Water Sci. Technol.*, **41**, 279–286 (2000).
95. A. Göbel, C.S. McArdell, A. Joss, H. Siegrist and W. Giger, Fate of sulfonamides, macrolides, and trimethoprim in different wastewater treatment technologies, *Sci. Total Environ.*, **372**, 361–371 (2007).
96. V. Goltsov, N. Veziroglu, From hydrogen economy to hydrogen civilization, *Int. J. Hydrogen Eng.*, **26**, 909–915 (2001).
97. H.E. Grethlein, Anaerobic digestion and membrane separation of domestic wastewater. *Journal of Water Pollution Control Federation*, **50**, 754–763 (1978).
98. S. Haag, M. Burgard, B. Ernst, Beneficial effects of the use of a nickel membrane reactor for the dry reforming of methane: Comparison with thermodynamic predictions, *J. Catal.*, **252**, 190–204 (2007).
99. F.I. Hai, K. Yamamoto, K. Fukushi and F. Nakajima, Fouling resistant compact hollow-fiber module with spacer for submerged membrane bioreactor treating high strength industrial wastewater, *Journal of Membrane Science* **317**, 34–42 (2008a).
100. F.I. Hai, K. Yamamoto and F. Nakajima and K. Fukushi, Removal of structurally different dyes in submerged membrane fungi reactor—Biosorption/PAC-adsorption, membrane retention and biodegradation, *J. Mem. Sci.*, **325**, 395–403 (2008b).
101. F.I. Hai, K. Yamamoto and K. Fukushi, Hybrid treatment systems for dye wastewater. *Crit. Rev. Environ. Sci. Technol.*, **37**, 315–377 (2007).
102. F.I. Hai, K. Yamamoto and K. Fukushi, Development of a submerged membrane fungi reactor for textile wastewater treatment, *Desalination*, **192**, 315–322 (2006a).

103. F.I. Hai, K. Yamamoto and K. Fukushi, Performance of newly developed hollow fiber module with spacer in integrated anaerobic–aerobic fungi reactor treating textile wastewater, *Desalination*, **199**, 305–307 (2006b).
104. S. Hanft, *Membrane Bioreactors in the Changing World Water Market*, Business Communications Company, Report C-240, June 2008.
105. O.J. Hao, H. Kim and P.C. Chiang, Decolorization of Wastewater. *Crit. Rev. Environ. Sci. Technol.*, **30**, 449–505 (2000).
106. R.D. Holbrook, K.A. Massie and J.T. Novak, A comparison of membrane bioreactor and conventional-activated-sludge mixed liquor and biosolids characteristics, *Water Environ Res*, **77**, 323–330 (2005).
107. N.J. Horan, H. Gohar and B. Hill, Application of a granular activated carbon—biological fluidised bed for the treatment of landfill leachates containing high concentrations of ammonia, *Water Sci Technol.*, **36**, 369–375 (1997).
108. K. Hou, R. Hughes, Preparation of thin and highly stable Pd/Ag composite membranes and simulative analysis of transfer resistance for hydrogen separation, *J. Membrane Sci.*, **214**, 43–55 (2003).
109. http://www.eia.doe.gov/pub/international/iealf/tablee1p.xls, *World Energy Intensity: Total Primary Energy Consumption per Dollar of Gross Domestic Product using Purchasing Power Parities, 1980–2004 (XLS)*, Energy Information Administration, US Department of Energy. 23 August 2006.
110. http://www.juniper.co.uk/Publications/mbt_report.html, Juniper, *MBT: A Guide for Decision Makers – Processes, Policies and Markets*, March 2005.
111. http://www.nfcrc.uci.edu/EnergyTutorial/pemfc.html, *Energy Tutorial*, National Fuel Cell Research Center, University of California, Irvine, March 2009.
112. X. Huang, Y. Meng, P. Liang, Y. Qian, Operational conditions of a membrane filtration reactor coupled with photocatalytic oxidation, *Sep. and Pur. Techn.*, **55**, 165–172 (2007).
113. S.T. Hwang, K. Kammermeyer, Techniques in chemistry: membranes in separation, Wiley Interscience, New York (1975).
114. A. Iulianelli, T. Longo, A. Basile, CO-free hydrogen production by steam reforming of acetic acid carried out in a Pd–Ag membrane reactor: The effect of co-current and counter-current mode, *Int. J. Hydrogen Eng.*, **33**, 4091–4096 (2008).
115. A. Iulianelli, T. Longo, S. Liguori, A. Basile, Production of hydrogen via glycerol steam reforming in a Pd-Ag membrane reactor over Co-Al_2O_3 catalyst, *Asia-Pac. J. Chem. Eng.*, in press, doi: 10.1002/apj.365.
116. O. Iyoha, R. Enick, R. Killmeyer, B. Howard, B. Morreale, M. Ciocco, Wall-catalyzed water-gas shift reaction in multi-tubular Pd and 80 wt%Pd–20 wt%Cu membrane reactors at 1173 K, *J. Membrane Sci.*, **298**, 14–23 (2007).
117. S. Judd, The status of membrane bioreactor technology, *TRENDS in Biotechnology*, **26**, 109–116 (2008).
118. S. Juddand B. Jefferson (eds.) *Membranes for Industrial Wastewater Recovery and Re-use*. Oxford: Elsevier Science Ltd., 2003.
119. A. Julbe, D. Farrusseng, C. Guizard, Porous ceramic membranes for catalytic reactors– overview and new ideas, *J. Membrane Sci.*, **181**, 3–20 (2001a).
120. A. Julbe, C. Guizard, Role of membranes and membrane reactors in the hydrogen supply of fuel cells, *Ann. Chim. Sci. Mat.*, **26**, 79–92 (2001b).
121. A. Kapoor, R. T. Yang, C. Wong, Surface diffusion, *Catal. Rev.*, **31**, 129–214 (1989).
122. E. Kayawake, Y. Narukami and M. Yamagata, Anaerobic digestion by a ceramic membrane enclosed reactor, *Journal of fermentation and bioengineering*, **71**, 122–125 (1991).
123. J.D. Keenan, R.L. Steiner and A.A. Fungaroli, Landfill leachate treatment. *J. Water Pollut. Control Fed.* **56**, 33–39 (1984).

124. K. Keizer, V.T. Zaspalis, R.S.A. De Lange, M.P. Harold, A.J. Burggraaf, Membrane reactors for partial oxidation and dehydrogenation reactions, *Membr. Proc. Sep. Pur.*, J.G. Crespo and Boddeker, eds., Netherlands, 415–429 (1994).
125. K.C. Khulbe, C.Y. Feng, T. Matsuura,Synthetic polymeric membranes, characterization by atomic force microscopy, Springer (2007).
126. E. Kikuchi, Membrane reactor application to hydrogen production, *Catal. Today*, **56**, 97–101 (2000).
127. S.D. Kim, J. Cho, I.S. Kim, B.J. Vanderford and S.A. Snyder, Occurrence and removal of pharmaceuticals and endocrine disruptors in South Korean surface, drinking, and waste waters. *Water Res.*, **41**, 1013–1021 (2007).
128. K. Kimura, H. Hara and Y. Watanabe, Removal of pharmaceutical compounds by submerged membrane bioreactors (MBRs). *Desalination*, **178**, 135–140 (2005).
129. A. Kleinert, A. Feldhoff, T. Schiestel, J. Caro, Novel hollow fibre membrane reactor for the partial oxidation of methane, *Catal. Today*, **118**, 44–51 (2006).
130. N. Klinkow, J. Oleksi-Frenzel and M. Jekel, Toxicity-directed fractionation of organic compounds in tannery wastewater with regard to their molecular weight and polarity. *Water Res.*, **32**, 2588–2592 (1998).
131. W.J. Koros, Y.H. Ma, T. Shimidzu, Terminology for membranes and membrane processes, *J. Membrane Sci.*, **120**, 149–159 (1996).
132. K.H. Krauth and K.F. Staab, Pressurized biomembrane rector for wastewater treatment. *Hydrotop*, **94**, 555–562 (1998).
133. R. Kurian, C. Acharya, G. Nakhla and A. Bassi, Conventional and thermophilic aerobic treatability of high strength oily pet food wastewater using membrane-coupled bioreactors. *Water Res.*, **39**, 4299–4308 (2005).
134. N. Laitinen, A. Luonsi and J. Vilen, Landfill leachate treatment with sequencing batch reactor and membrane bioreactor, *Desalination*, **191**, 86–91 (2006).
135. P. Le-Clech, V. Chen and T.A.G. Fane, Fouling in membrane bioreactors used in wastewater treatment, *Journal of Membrane Science*, **284**, 17–53 (2006).
136. K.H. Lee, S. T. Hwang, Transport of condensible vapors through a microporous vycor glass membrane, *J. Coll. Int. Sci.*, **110**, 544–555 (1986).
137. W.N. Lee, I.J. Kang and C.H. Lee, Factors affecting filtration characteristics in membrane-coupled moving bed biofilm reactor, *Water Research*, **40**, 1827–1835 (2006).
138. B. Lesjean, R. Gnirss, C. Adam, M. Kraume and F. Luck, Enhanced biological phosphorus removal process implemented in membrane bioreactors to improve phosphorous recovery and recycling, *Water Sci. Technol.*, **48**, 87–94 (2003).
139. B. Lesjean, S. Rosenberger, J.C. Schrotter and A. Recherche, Membrane-aided biological wastewater treatment – an overview of applied systems, *Membrane Technol.*, **2004**, 5–10 (2004).
140. Z.Y. Li, H. Maeda, K. Kusakabe, S. Morooka, H. Anzai, S. Akiyama, Preparation of palladium-silver alloy membranes for hydrogen separation by the spray pyrolysis method, *J. Membrane Sci.*, **78**, 247–254 (1993).
141. Q. Li, X. Zhu, Y. He, W. Yang, Partial oxidation of methane in $BaCe_{0.1}Co_{0.4}Fe_{0.5}O_{3-\delta}$ membrane reactor, *Catal. Today*, in press, doi: 10.1016/j.cattod.2009.03.019 (2009).
142. B.Q. Liao, J.T. Kraemer and D. Bagley, (2006) Anaerobic membrane bioreactors: applications and research directions. *Critical Reviews in Environmental Science and Technology*, **36**, 489–530 (2006).
143. Y.S. Lin, Microporous and dense inorganic membranes: current status and prospective, *Sep. Purif. Tech.*, **25**, 39–55 (2001).
144. C. Liu, X. Huang and H. Wang, Start-up of a membrane bioreactor bioaugmented with genetically engineered microorganism for enhanced treatment of atrazine containing wastewater, *Desalination*, **231**, 12–19 (2008).

145. N.D. Lourenço, J.M. Novais and H.M. Pinheiro, Effect of some operational parameters on textile dye biodegradation in a sequential batch reactor. *J. Biotechnol.*, **89**, 163–174 (2001).
146. G.Q. Lu, J.C. Diniz de Costa, M. Duke, S. Giessler, R. Socolow, R.H. Williams, T. Kreutz, Inorganic membranes for hydrogen production and purification: a critical review and perspective, *J. Coll. Interface Sci.*, **314**, 589–603 (2007).
147. B. Lubello and R. Gori, Membrane bio-reactor for textile wastewater treatment plant upgrading, *Water Sci. Technol.*, **52**, 91–98 (2005).
148. C. Lubello, S. Caffaz, L. Mangini, D. Santianni and C. Caretti, MBR pilot plant for textile wastewater treatment and reuse, *Water Sci Technol.*, **55**, 115–124 (2007).
149. S. Lyko, T. Wintgens and T. Melin, Estrogenic trace contaminants in wastewater — possibilities of membrane bioreactor technology. *Desalination*, **178**, 95–105 (2005).
150. A. Mahecha-Botero, T. Boyd, A. Gulamhusein, N. Comyn, C.J. Lim, J.R. Grace, Y. Shirasaki, I. Yasuda, Pure hydrogen generation in a fluidized-bed membrane reactor: Experimental findings, *Chem. Eng. Sci.*, **63**, 2752–2762 (2008).
151. F. Malpei, L. Bonomo and A. Rozzi, Feasibility study to upgrade a textile wastewater treatment plant by a hollow fibre membrane bioreactor for effluent reuse, *Water Sci Technol.*, **47**, 33–39 (2003).
152. G. Marbán, T. Valdés-Solís, Towards the hydrogen economy? *Int. J. Hydrogen Eng.*, **32**, 1625–1637 (2007).
153. Y. Matsumura, J. Tong, Methane steam reforming in hydrogen-permeable membrane reactor for pure hydrogen production, *Top. Catal.*, **51**, 123–132 (2008).
154. E.E. McLeary, J.C. Jansen, F. Kapteijn, Zeolite based films, membranes and membrane reactors: Progress and prospects, *Microp. and Mes. Mat.*, **90**, 198–220 (2006).
155. W.M. Meier, Zeolites and zeolite-like materials, *Pure and Appl. Chem.*, **58**, 1323–1328 (1986).
156. T. Melin, B. Jefferson, D. Bixio, C. Thoeye, W. De Wilde, J. De Koning, J. van der Graaf and T. Wintgens, *Membrane bioreactor technology for wastewater treatment and reuse Desalination*, **187**, 271–282 (2006).
157. D. Mendes, A. Mendes, L. M. Madeira, A. Iulianelli, J.M. Sousa, A. Basile, The water-gas shift reaction: from conventional catalytic systems to Pd-based membrane reactors – a review, *Asia-Pac. J. Chem. Eng.*, doi: 10.1002/apj.364 (2009).
158. F. Meng, S.R. Chae, A. Drews, M. Kraume, H.S. Shin and F. Yang F, Recent advances in membrane bioreactors (MBRs): membrane fouling and membrane material, *Water Res.*, **43**, 1489–1512 (2009).
159. R. Molinari, A. Caruso, T. Poerio, Direct benzene conversion to phenol in a hybrid photocatalytic membrane reactor, *Catal. Today*, **144**, 81–86 (2009).
160. P. Moriarty, D Honnery, Intermittent renewable energy: The only future source of hydrogen?, *Int. J. Hydrogen Eng.*, **32**, 1616–1624 (2007).
161. S. Mozia, A.W. Morawski, M. Toyoda, M. Inagaki, Effectiveness of photodecomposition of an azo dye on a novel anatase-phase TiO2 and two commercial photocatalysts in a photocatalytic membrane reactor (PMR), *Sep. and Pur. Technol.*, **63**, 386–391 (2008).
162. S. Mozia, A.W. Morawski, M. Toyoda, T. Tsumura, Effect of process parameters on photodegradation of Acid Yellow 36 in a hybrid photocatalysis–membrane distillation system, *Chem. Eng. J.*, **150**, 152–159 (2009).
163. G. Munz, M. Gualtiero, L. Salvadori, B. Claudia and L. Claudio, Process efficiency and microbial monitoring in MBR (membrane bioreactor) and CASP (conventional activated sludge process) treatment of tannery wastewater, *Bioresource Technol.*, **99**, 8559–8564 (2008).
164. M. Ni, D.Y.C. Leung, M.K.H. Leung, A review on reforming bio-ethanol for hydrogen production, *Int. J. Hydrogen Eng.*, **32**, 3238–3247 (2007).
165. H.Y. Ng and S.W. Hermanowicz, Membrane bioreactor operation at short solids retention times: performance and biomass characteristics, *Water Res*, **39**, 981–992 (2005).

166. C.A. Ng, D. Sun and A.G. Fane, Operation of membrane bioreactor with powdered activated carbon addition, *Separation Science and Technology*, **41**, 1447–1466 (2006).
167. H.H. Ngo, W. Guo and W. Xing, Evaluation of a novel sponge-submerged membrane bioreactor (SSMBR) for sustainable water reclamation, *Bioresource Technology*, **99**, 2429–2435 (2008).
168. L. Olivier, S. Haag, C. Mirodatos, A.C. van Veen, Oxidative coupling of methane using catalyst modified dense perovskite membrane reactors, *Catal. Today*, **142**, 34–41 (2009).
169. S.S. Ozdemir, M.G. Buonomenna, E. Drioli, Catalytic polymeric membranes: next term Preparation and application, *Appl. Catal. A: Gen.*, **307**, 167–183 (2006).
170. S. Park, K.M. Yeon and C.H. Lee, Hydrodynamics and microbial physiology affecting performance of a new MBR, membrane-coupled high performance compact reactor, *Desalination*, **172**, 181–188 (2005).
171. S.J. Park, D.W. Lee, C.Y. Yu, K.Y. Lee, K.H. Lee, Hydrogen production from a DME reforming-membrane reactor using stainless steel-supported Knudsen membranes with high permeability, *J. Membrane Sci.*, **318**, 123–128 (2008).
172. P. Parvasi, A. Khosravanipour Mostafazadeh, M.R. Rahimpour, Dynamic modeling and optimization of a novel methanol synthesis loop with hydrogen-permselective membrane reactor, *Int. J. Hydrogen Eng.*, **34**, 3717–3733 (2009).
173. L. Paturzo, A. Basile, E. Drioli, High temperature membrane reactors and integrated membrane operations, *Reviews in Chem. Eng.*, **18**, 511–551 (2002).
174. B. Pauwels, F. Ngwa, S. Deconinck and W. Verstraete, Effluent quality of a conventional activated sludge and a membrane bioreactor system treating hospital wastewater, *Environ Technol.*, **27**, 395–402 (2006).
175. G. Pearce, Introduction to membranes – MBRs: Manufacturers' comparison: part 1. *Filtration and Separation*, March 2008, 28–31 (2008).
176. J. Phattaranawik, A.G. Fane, A.C.S. Pasquier and W. Bing, A novel membrane bioreactor based on membrane distillation, *Desalination*, **223**, 386–395 (2008).
177. J. Phattaranawik, A.G. Fane, A.C.S. Pasquier, W. Bing and F.S. Wong, Experimental study and design of a submerged membrane distillation bioreactor, *Chemical Engineering and Technology*, **32**, 38–44 (2009).
178. C.E. C. Purdom, P.A. Hardiman, V.J. Bye, N.C. Eno, C.R. Tyler, J.P. Sumpter, Estrogenic effects of effluents from sewage-treatment works. *Chem Ecol.*, **8**, 275–285 1994.
179. J.J. Qin, K.A. Kekre, G.Tao, M.H. Oo, M.N. Wai, T.C. Lee, V. Bala and S. Harry (2006) New option of MBR-RO process for production of NEWater from domestic sewage. *Journal of Membrane Science*, **272**, 70–77 (2006).
180. Y.Y. Qu, J.T. Zhou, J. Wang, L.L. Xing, N. Jiang, M. Gou and M.S. Uddin, Population dynamics in bioaugmented membrane bioreactor for treatment of bromoamine acid wastewater. *Bioresource Technol.*, **100**, 244–248 (2009).
181. J. Radjenovic, M. Matosic, I. Mijatovic, M. Petrovic and D. Barcelo,Membrane Bioreactor (MBR) as an Advanced Wastewater Treatment Technology, *in the Handbook of Environmental Chemistry: Emerging Contaminants from Industrial and Municipal Waste*, Vol. 5, Part S/2, 37–101, Springer, Berlin (2008).
182. J. Radjenovic, M. Petrovic and D. Barcelo, Fate and distribution of pharmaceuticals in wastewater and sewage sludge of the conventional activated sludge (CAS) and advanced membrane bioreactor (MBR) treatment, *Water Res.*, **43**, 831–841 (2009).
183. M.R. Rahimpour, Enhancement of hydrogen production in a novel fluidized-bed membrane reactor for naphtha reforming, *Int. J. Hydrogen Eng.*, **34**, 2235–2251 (2009a).
184. M.R. Rahimpour, H. Elekaei, A comparative study of combination of Fischer–Tropsch synthesis reactors with hydrogen-permselective membrane in GTL technology, *Fuel Proc. Technol.*, **90**, 747–761 (2009b).

185. M.M. Rahman and M.H. Al-Malack, Performance of a crossflow membrane bioreactor (CF–MBR) when treating refinery wastewater, *Desalination*, **191**, 16–26 (2006).
186. R. Reif, S. Suárez, F. Omil and J.M. Lema, Fate of pharmaceuticals and cosmetic ingredients during the operation of a MBR treating sewage, *Desalination*, **221**, 511–517 (2008).
187. B.K. Richards, F.G. Herndon, W.J. Jewell, R.J. Cummings, T.E. White, In situ methane enrichment in methanogenic energy crop digesters, *Biomass and Bioenergy*, **6**, 275–282 (1994).
188. J.A. Ritter, A.D. Ebner, State-of-the-art adsorption and membrane separation processes for hydrogen production in the chemical and petrochemical industries, *Sep. Sci. Techn.*, **42**, 1123–1193 (2007).
189. F. Roa, J. D. Way, R. L. McCormik, S. N. Paglieri, Preparation and characterization of Pd-Cu composite membranes for hydrogen separation, *Chem. Eng. J.*, **93**, 11–22 (2003).
190. M.P. Rohde, G. Schaub, S. Khajavi, J.C. Jansen, F. Kapteijn, Fischer–Tropsch synthesis with in situ H_2O removal – Directions of membrane development, *Microp. and Mesop. Mat.*, **115**, 123–136 (2008).
191. S. Rosenberger, R. Witzig, W. Manz, U. Szewzyk and M. Kraume, Operation of different membrane bioreactors: experimental results and physiological state of the micro-organisms, *Water Sci Technol*, **41**, 269–277 (2000).
192. W.R. Ross, J.P. Barnard, N.K. Strohwald, C.J. Grobler and J. Sanetra, Practical application of the ADUF process to the full-scale treatment of maize-processing effluent. *Water Science and Technology*, **25**, 27–39 (1992).
193. L. Saez, J. Perez and J. Martinez, Low molecular weight phenolics attenuation during simulated treatment of wastewaters from olive oil mills in evaporation ponds. *Water Res.*, **26**, 1261–1266 (1992).
194. M.J.G. Sanchez, T.T. Tsotsis, *Catalytic Membranes and Membrane Reactors*, 1st edn, Wiley-VCH, Weinheim (2002).
195. G. Saracco, V. Specchia, Catalytic inorganic membrane reactors: present experience and future opportunities, *Catal. Rev. Sci. Eng.*, **36**, 305–384 (1994).
196. G. Saracco, H.W.J.P. Neomagus, G.F. Versteeg, W.P.M. van Swaaij, High-temperature membrane reactors: potential and problems, *Chem. Eng. Sci.*, **54**, 1997–2017 (1999).
197. W. Schier, F.B. Frechen and S. Fischer, Efficiency of mechanical pre-treatment on European MBR plants, *Desalination*, **236**, 85–93 (2009).
198. P. Schoeberl, M. Brik, R. Braun and W. Fuchs, Treatment and recycling of textile wastewater case study and development of a recycling concept, *Desalination*, **171**, 173–183 (2004).
199. W. Scholz and W. Fuchs, Treatment of oil contaminated wastewater in a membrane bioreactor. *Wat. Res.*, **34**, 3621–3629 (2000).
200. H.F. Schröder, The elimination of the endocrine disrupters 4-nonylphenol and bisphenol A during wastewater treatment – Comparison of conventional and membrane assisted biological wastewater treatment followed by an ozone treatment, *Water Pract. Tech.* 2006, **1**, doi: 10.2166/WPT.2006060, (2006).
201. G.A. Schrader, A. Zwijnenburg and M. Wessling, The effect of WWTP effluent zeta-potential on direct nanofiltration performance. *Journal of Membrane Science*, **266**, 80–93 (2005).
202. J. Shu, B.P.A. Grandjean A. Vanneste, S. Kaliaguine, Catalytic palladium-based membrane reactors: a review, *Can. J. Chem. Eng.*, **89**, 1036–1060 (1991).
203. A. Shu, A. Adnot, B.P.A. Grandjean, S. Kaliaguine, Structurally stable composite Pd-Ag alloy membranes: Introduction of a diffusion barrier, *Thin Solid Films*, **286**, 72–79 (1996).
204. A. Shukla, A. Kumar, Experimental Studies and Mass-Transfer Analysis of the Hydrolysis of Olive Oil in a Biphasic Zeolite-Membrane Reactor Using Chemically Immobilized Lipase, *Ind. Eng. Chem. Res.*, **43**, 2017–2029 (2004).
205. A.P. Simpson, A.E. Lutz, Energy analysis of hydrogen production via steam methane reforming, *Int. J. Hydrogen Eng.*, **32**, 4811–4820 (2007).

206. C.V. Smith, D. DiGregorio and R.M. Talcott, The use of ultrafiltration membranes for activated sludge separation. In: *Proceedings of the 24th Annual Purdue Industrial Waste Conference*, 6–8 May, Lafayette, Indiana, USA 1996.
207. G.J.F. Smolders, J. Van der Meij, M.C.M. Van Loosdrecht and J.J. Heijnen, A structured metabolic model for anaerobic and aerobic stoichiometry and kinetics of the biological phosphorus removal process, *Biotechnol Bioeng*, **47**, 277 (1995).
208. S.A. Snyder, S. Adham, A.M. Redding, F.S. Cannon, J. DeCarolis, J. Oppenheimer, E.C. Wert and Y. Yoon, Role of membranes and activated carbon in the removal of endocrine disruptors and pharmaceuticals, *Desalination*, **202**, 156–181 (2007).
209. A.J. Spring, D.M. Bagley, R.C. Andrews, S. Lemanik and P. Yang, Removal of endocrine disrupting compounds using a membrane bioreactor and disinfection, *J. Environ. Eng. Sci.*, **6**, 131–137 (2007).
210. B.A. Stambouli, E. Traversa, Fuel cells, an alternative to standard sources of energy, *Renew. Sust. Energy Rev.*, **6**, 295–304 (2002).
211. T. Stephenson, S. Judd, B. Jeferson and K. Brindle, *Membrane Bioreactors for Wastewater Treatment*. London: IWA Publishing, 2000.
212. D. C. Stuckey and A. Hu, The Submerged Anaerobic Membrane Bioreactor (SAMBR): An Intensification of Anaerobic Wastewater Treatment, *Proceedings of the International Water Association–Leading Edge Conference on Drinking Water and Wastewater Treatment Technologies*, Noordwijk/Amsterdam, Netherlands, May 26–28; International Water Association: London 2003.
213. S. Sun, M. Rebeilleau-Dassonneville, X. Zhu, W. Chu, W. Yang, Ammonia oxidation in Ba0.5Sr0.5Co0.8Fe0.2O3-d membrane reactor, *Catal. Today*, in press (2009).
214. P.M. Sutton, P.R. Li S.R. Korchin, Dorr-oliver's fixed film suspended growth anaerobic systems for industrial wastewater treatment and energy recovery. In: *Proceedings of the 37th Annual Purdue Industrial Waste Conference*, pp. 667–675, 11–13 May, Lafayette, Indiana, USA, 1983.
215. P.M. Sutton, P.N. Mishra and P.M. Crawford, Combining Biological and Physical Processes for Complete Treatment of Oily Wastewaters. *Int. Biodeter. Biodegrad.*, **33**, 3–21 (1994).
216. X. Tan, K. Li, Oxidative Coupling of Methane in a Perovskite Hollow-Fiber Membrane Reactor, *Ind. Eng. Chem. Res.*, **45**, 142–149 (2006).
217. X. Tan, Z. Pang, Z. Gu, S. Liu, Catalytic perovskite hollow fibre membrane reactors for methane oxidative coupling, *J. Membrane Sci.*, **302**, 109–114 (2007).
218. A.M. Tarditi, S. Irusta, E.A. Lombardo, Xylene isomerization in a membrane reactor. Part I: The synthesis of MFI membranes for the p-xylene separation, *Chem. Eng. J.*, **122**, 167–174 (2006).
219. Y. Teraoka, H.M. Zhang, S. Furukawa, N. Yamazoe, Oxygen permeation through perovskite-type oxides, *Chem. Lett.*, **11**, 1743–1746 (1985).
220. H. Timur, I. Ozturk, M. Altinbas, O. Arikan and B.S. Tuyluoglu, Anaerobic treatability of leachate: a comparative evaluation for three different reactor systems, *Water Res.*, **42**, 287–292 (2000).
221. J. Tong, R. Shirai, Y. Kashima, Y. Matsumura, Preparation of a pinhole-free Pd-Ag membrane on a porous metal support for pure hydrogen separation, *J. Membrane Sci.*, **260**, 84–89 (2005a).
222. J. Tong, Y. Matsumura, H. Suda, K. Haraya, Thin and dense Pd/CeO2/MPSS composite membrane for hydrogen separation and steam reforming of methane, *Sep. Pur. Techn.*, **46**, 1–10 (2005b).
223. S. Tosti, L. Bettinali, V. Violante, Rolled thin Pd and Pd–Ag membranes for hydrogen separation and production, *Int. J. Hydrogen Eng.*, **25**, 319–325 (2000).
224. S. Tosti, A. Adrover, A. Basile, V. Camilli, G. Chiappetta, V. Violante, Characterization of thin wall Pd-Ag rolled membranes, *Int. J. Hydrogen Eng.*, **28**, 105–112 (2003).
225. S. Tosti, L. Bettinali, Diffusion bonding of Pd-Ag membranes, *J. Materials Sci.*, **39**, 3041–3046 (2004).

226. T. Tsuru, H. Shintani, T. Yoshioka, M. Asaeda, A bimodal catalytic membrane having a hydrogen-permselective silica layer on a bimodal catalytic support: Preparation and application to the steam reforming of methane, *Appl. Catal. A: Gen.*, **302**, 78–85 (2006a).
227. T. Tsuru, T. Kan-no, T. Yoshioka, M. Asaeda, A photocatalytic membrane reactor for VOC decomposition using Pt-modified titanium oxide porous membranes, *J. Membrane Sci.*, **280**, 156–162 (2006b).
228. S. Uemiya, T. Matsuda, E. Kikuchi, Hydrogen permeable palladium–silver alloy membrane supported on porous ceramics, *J. Membrane Sci.*, **56**, 315–325 (1991).
229. R.J.R. Ulhorn, K. Keizer, A.J. Burggraaf, Gas transport and separation with ceramic membranes. Part I. Multilayer diffusion and capillary condensation, *J. Membrane Sci.*, **66**, 259–269 (1992).
230. Z. Ujang, M.R. Salim and S.L. Khor, The effect of aeration and non-aeration time on simultaneous organic, nitrogen and phosphorus removal using an intermittent aeration membrane bioreactor, *Water Sci Technol.*, **46**, 193–200 (2002).
231. T. Urase, C. Kagawa and T. Kikuta, Factors affecting removal of pharmaceutical substances and estrogens in membrane separation bioreactors, *Desalination*, **178**, 107–113 (2005).
232. G. Valenti, F. Macchi, Proposal of an innovative, high efficiency, large-scale hydrogen liquefier, *Int. J. Hydrogen Eng.*, **33**, 3116–3121 (2008).
233. C. Vanninia, G. Munz, G. Mori, C. Lubello, F. Vernia and P. Giulio, Sulphide oxidation to elemental sulphur in a membrane bioreactor: Performance and characterization of the selected microbial sulphur-oxidizing community, *Syst. Appl. Microbiol.*, **31**, 461–473 (2008).
234. H.M. Van Veen, M. Bracht, E. Hamoen, P.T. Alderliesten, Feasibility of the application of porous inorganic gas separation membranes in some large-scale chemical processes Fundamentals of inorganic membrane science and technology, edited A.J. Burggraaf, L. Cot, Elsevier, **14**, 641–681 (1996).
235. C. Visvanathan, M.K. Choudhary, M.T. Montalbo and V. Jegatheesan, Landfill leachate treatment using thermophilic membrane Bioreactor, *Desalination*, **204**, 8–16 (2007).
236. C. Visvanathan, Y. Byung-Soo, S. Muttamara and R. Maythanukhraw, Application of air backflushing technique in membrane bioreactor, *Water Science and Technology*, **36**, 259–266 (1997).
237. C. Visvanathan, R. Ben Aim and K. Parameshwaran, Membrane separation bioreactors for wastewater treatment, *Critical Reviews in Environmental Science and Technology*, **30**, 1–48 (2000).
238. D. Wang, J. Tong, H. Xu, Y. Matsamura, Preparation of palladium membrane over porous stainless steel tube modified with zirconium oxide, *Catal. Today*, **93–95**, 689–693 (2004).
239. H. Wang, C. Tablet, T. Schiestel, J. Caro, Hollow fiber membrane reactors for the oxidative activation of ethane, *Catal. Today*, **118**, 98–103 (2006).
240. T.L. Ward, T. Dao, Model of hydrogen permeation behavior in palladium membranes, *J. Membrane Sci.*, **153**, 211–231 (1999).
241. J. Wee, Applications of proton exchange membrane fuel cell systems, *Renew. Sust. Energy Rev.*, **11**, 1720–1738 (2007).
242. S. Weiss and T. Reemtsma, Membrane bioreactors for municipal wastewater treatment – A viable option to reduce the amount of polar pollutants discharged into surface waters? *Water Res.*, **42**, 3837–3847 (2008).
243. X.-H. Wen, C.-H. Xing and Y. Qian, Akinetic model for the prediction of sludge formation in a membrane bioreactor, *Process Biochem.*, **35**, 249–254 (1999).
244. J. Weitkamp, Zeolites and catalysis, *Solid State Ionics*, **131**, 175–188 (2000).
245. U. Welander, T. Henrysson and T. Welander, Biological nitrogen removal from municipal landfill leachate in a pilot scale suspended carrier biofilm process, *Water Res.*, **32**, 1564–1570 (1998).
246. B. Wichitsathian, S. Sindhuja, C. Visvanathan and K.H. Ahn, Landfill leachate treatment by yeast and bacteria based membrane bioreactors, *J. Environ. Sci. Health A*, **39**, 2391–2404 (2005).

247. S. Wieland, T. Melin, A. Lamm, Membrane reactors for hydrogen production, *Chem. Eng. Sci.*, **57**, 1571–1576 (2002).
248. Y. Xia, Y. Lu, K. Kamata, B. Gates, Y. Yin,Macroporous materials containing three-dimensionally periodic structures, Chemistry of Nanostructured Materials (Ed.: Yang, P.), World Scientific 69–100 (2003).
249. C.H. Xing, K. Yamamoto and K. Fukushi, Performance of an inclined-plate membrane bioreactor at zero excess sludge discharge, *Journal of Membrane Science*, **275**, 175–186 (2006).
250. J. Xuan, M.K.H. Leung, D.Y.C. Leung, M. Ni, A review of biomass-derived fuel processors for fuel cell systems, *Renew. Sust. Energy Rev.*, **13**, 1301–1313 (2009).
251. K. Yamamoto, H. Hiasa, M. Talat and T. Matsuo, Direct solid liquid separation using hollow fiber membranes in activated sludge aeration tank. *Water Science and Technology*, **21**, 43–54 (1989).
252. K. Yamamoto,Submerged MBR technology: An unfinished international adventure of 20 years. In: Lesjean B and Leiknes T (eds.) *Final MBR-Network Workshop,* pp.83–84. Berlin, Germany, March 31–April 1, 2009.
253. A.A. Yaremchenko, V.V. Kharton, A.A. Valente, F.M.M. Snijkers, J.F.C. Cooymans, J.J. Luyten, F.M.B. Marques, Performance of tubular SrFe(Al)O3–SrAl2O4 composite membranes in CO2- and CH4-containing atmospheres, *J. Membrane Sci.*, **319**, 141–148 (2008).
254. K.M. Yeon, J.S. Park, C.H. Lee and S.M. Kim, Membrane coupled high performance compact reactor: a new MBR system for advanced wastewater treatment, *Water Research*, **39**, 954–1961 (2005).
255. X. Yin, L. Hong, Z. Liu, Integrating air separation with partial oxidation of methane–A novel configuration of asymmetric tubular ceramic membrane reactor, *J. Membrane Sci.*, **311**, 89–97 (2008).
256. M.A. Yun, K.M. Yeon, J.S. Park, C.H. Lee, J. Chun,and D.J. Lim, Characterization of biofilm structure and its effect on membrane permeability in MBR for dye wastewater treatment, *Water Res.*, **40**, 45–52 (2006).
257. J. Zaman, A. Chakma, Inorganic membrane reactors, *J. Membrane Sci.*, **92**, 1–28 (1994).
258. C. Zhang, Z. Hong, X. Gu, Z. Zhong, W. Jin, N. Xu, Silicalite-1 zeolite membrane reactor packed with HZSM-5 catalyst for meta-xylene isomerization, *Ind. Eng. Chem. Res.*, **48**, 4293–4299 (2009).
259. Y. Zhang, J. Gwak, Y. Murakoshi, T. Ikeara, R. Maeda, C. Nishimura, Hydrogen permeation characteristics of thin Pd membrane prepared by microfabrication technology, *J. Membrane Sci.*, **277**, 203–209 (2006).
260. J.M. Zheng, J.M. Sousa, D. Mendes, L.M. Madeira, A. Mendes, Theoretical analysis of conversion enhancement in isothermal polymeric catalytic membrane reactors, *Catal. Today*, **118**, 228–236 (2006).

1

Microporous Carbon Membranes

Miki Yoshimune and Kenji Haraya

Research Institute for Innovation in Sustainable Chemistry, National Institute of Advanced Industrial Science and Technology (AIST), Tsukuba, Japan

1.1 Introduction

There is growing interest in the development of microporous inorganic membranes made of zeolites, silica, carbon, or similar materials, whose separation mechanisms are controlled mainly by the molecular sieving effect. Such inorganic membranes are capable of achieving excellent separation efficiencies and, unlike conventional polymeric membranes, can function at high temperatures or in harsh environments. Carbon membranes have the greatest potential among these inorganic membranes because of the relative ease with which they can be produced and their resulting low cost.

Figure 1.1 shows the general types of carbon membranes together with a classification of their gas transport mechanisms into various categories, such as molecular sieving, surface diffusion, Knudsen diffusion, and viscous flow (VS), together with the ranges of pore sizes that correspond to each particular mechanism. Microporous carbon membranes can be categorised into two types: (i) carbon molecular sieve (CMS) membranes (Figure 1.1a) and (ii) nanoporous carbon membranes (Figure 1.1b). CMS membranes, first prepared by Koresh and Soffer [1], have micropores with diameters of approximately 0.3–0.5 nm, and they are characterised by high selectivities in gas separations as a result of the selective permeation of smaller gas molecules. Nanoporous carbon membranes were designed by Rao and Sircar [2–4] as selective surface flow (SSF) membranes, and have larger micropores (0.5–0.7 nm) than CMS membranes.

Because separations using microporous carbon membranes have attracted consistently high levels of research interest, they are the subject of a number of excellent reviews and books [5–9]. This chapter presents an overview of recent researches on microporous carbon membranes and explores their possible applications in membrane reactors. Section 1.3 reviews and discusses the

Membranes for Membrane Reactors: Preparation, Optimization and Selection, Edited by Angelo Basile and Fausto Gallucci

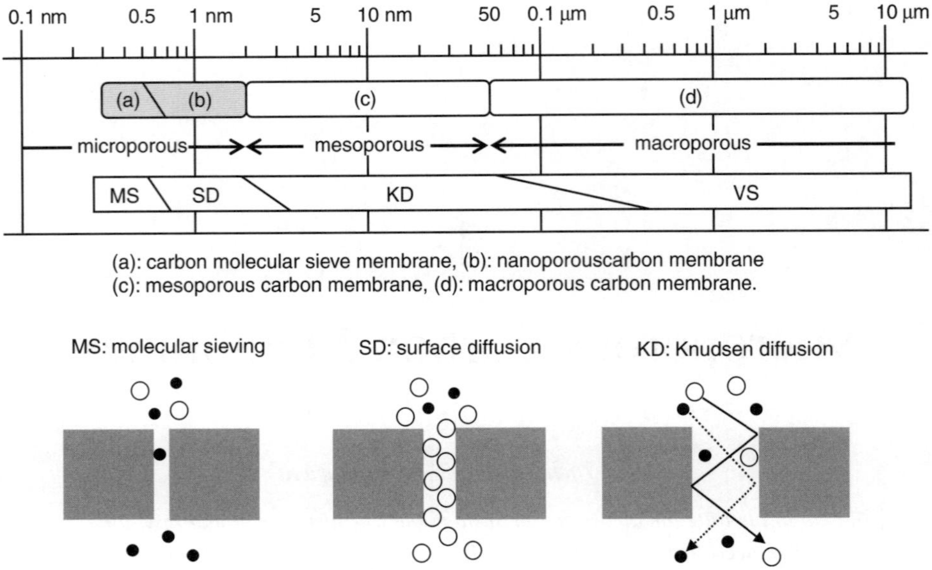

Figure 1.1 *Types of carbon membranes and transport mechanisms*

factors that control the preparation of high-performance microporous carbon membranes. Trends in mixed-matrix carbon membranes prepared from polymeric precursors that incorporate inorganic materials such as metals, metal oxides, or zeolites are discussed in Section 1.3.10. These incorporation methods can also be used to prepare catalytic membranes for use in membrane reactors; such membranes are discussed in Section 1.5.

1.2 Transport Mechanisms in Carbon Membranes

The microporous carbon membranes that are used for gas separation usually have a turbostratic structure [10] in which layer planes of graphite-like microcrystallites are randomly stacked. Figure 1.2 shows that there are lattice vacancies in the microcrystallites and that pores are formed from imperfections in the packing between microcrystalline regions.

The mechanism of gas transport through porous carbon membranes is essentially the same as that in other inorganic porous membranes. When the pore diameter (d_p) is greater than the mean free path of the gas molecule (λ), intermolecular collisions predominate and the transport of gas molecules through porous membranes under a pressure or a concentration gradient corresponds to viscous flow and is nonselective.

When d_p is smaller than λ, collisions between the gas molecules and the pore walls predominate so that the transport of gas molecules is controlled by the thermal mean velocity of the gas molecules ($v = \sqrt{8RT/\pi M}$). In the case of a capillary pore with a diameter of d_p, the diffusion of the gas can be described by Equation (1.1).

$$D_K = -\frac{1}{3} d_p \sqrt{\frac{8RT}{\pi M}} \tag{1.1}$$

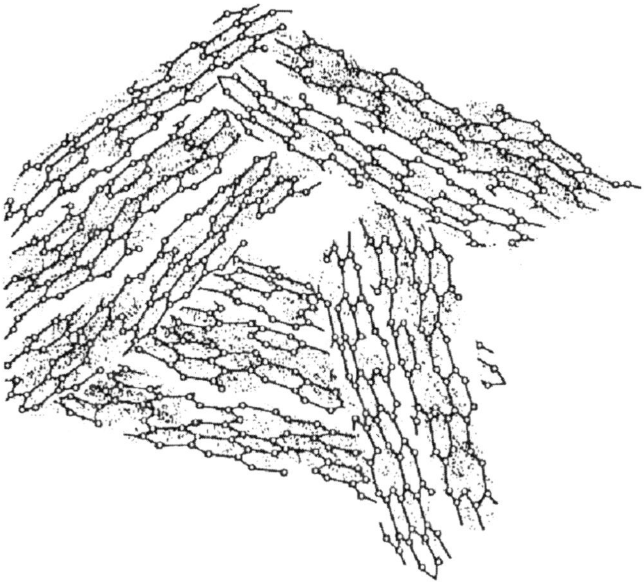

Figure 1.2 *Structure of turbostratic graphite. This article was published in Handbook of Carbon, Graphite, Diamond, and Fullerenes. Vol. 3, Pierson, H., Graphite Structure and Properties, 48, Copyright (1993) with permission from Elsevier*

Here, D_k is the Knudsen diffusion coefficient, R is the universal gas constant, T is the absolute temperature, and M is the molecular weight of the penetrant gas. On the basis of Knudsen diffusion, the selectivity (i.e., the ideal separation factor) of a gas pair A–B is given by the expression $\sqrt{M_B/M_A}$.

When the temperature is within the range where adsorption of gas molecules on the pore walls becomes important, transport of the gas molecules along the surface (surface diffusion) occurs in combination with Knudsen flow. The effects of surface diffusion increase with decreasing d_p and they produce selectivity in the flow as a result of selective adsorption. Selective surface flow (SSF) membranes, as named by Rao and Sircar [2–4], operate in this regime. SSF membranes can achieve high performances in separations of gas mixtures consisting of a readily adsorbed species and a component that is not readily adsorbed, such as mixtures of hydrocarbons with hydrogen. If penetrants are condensable, such as vapours, the condensates can completely fill the pores resulting in capillary condensation that blocks the permeation of noncondensable components. This mechanism has been observed in other inorganic porous membranes, but has not yet been reported in carbon membranes.

When d_p is of a similar size to that of a gas molecule (0.5 nm or less), selective transport as a result of a molecular sieving effect can be observed. Smaller molecules pass readily through the pores, whereas the passage of larger molecules is obstructed or highly restricted. Microporous carbon membranes in this regime are usually known as carbon molecular sieve (CMS) membranes. Typical examples of the permeances of various gases through a CMS membrane are plotted in Figure 1.3 as a function of the size of the gas molecule. This figure shows that the membrane is not only effective in separating mixtures of gases of different molecular sizes, such as H_2/CH_4, H_2/C_3H_8, He/N_2, or N_2/SF_6, but also in separating gases of similar molecular sizes, such as O_2/N_2, CO_2/CH_4, CO_2/N_2, or C_3H_6/C_3H_8.

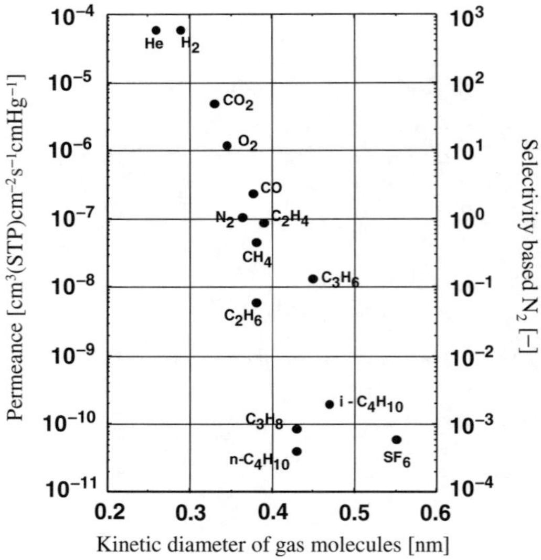

Figure 1.3 *Gas permeance and selectivity of a CMS membrane derived from a polyimide hollow fibre measured at 25 °C*

Because diffusion is an activated process in both CMS and polymeric membranes, the diffusion coefficient (D) can be expressed by an Arrhenius-type relationship:

$$D = D_0 \exp(-E_D/RT) \tag{1.2}$$

Here, E_D is the energy of activation required for a gas molecule to execute a diffusive jump from one cavity to another, and D_0 is the temperature-independent pre-exponential term. The diffusion selectivity of A–B gas molecules can be expressed as follows:

$$\frac{D_A}{D_B} = \frac{D_{0,A}}{D_{0,B}} \exp\left[\frac{-(E_{D,A} - E_{D,B})}{RT}\right] \tag{1.3}$$

The exponential term is an energetic selectivity. For gas molecules that differ in both size, and shape, complex configurational effects related to factors affecting D_0 for the components A and B can occur. These configurational selectivity contributions to the D_A/D_B ratio are often referred to as the entropic selectivity [11]. The excellent selectivity observed in CMS membranes is the result of a favourable contribution from this factor, which is generally lacking in conventional polymeric membranes.

1.3 Methods for the Preparation of Microporous Carbon Membranes

1.3.1 General Preparation and Characterisation

Microporous carbon membranes are generally formed by pyrolysing polymeric precursor membranes. Pyrolysis (or carbonisation) is the process whereby the precursor membrane is heated to a pyrolysis temperature in the range 500–1000 °C under a controlled atmosphere, such as a vacuum or an inert gas (N_2, He, or Ar), at a specific heating rate and then held at the pyrolysis

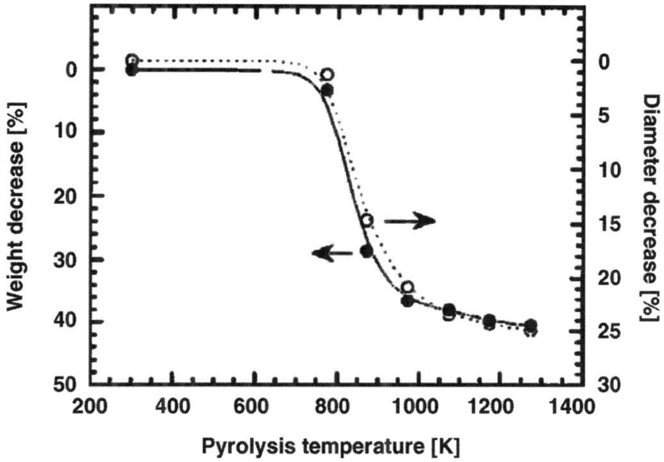

Figure 1.4 *Changes in the weight and diameter of Kapton polyimide films as a function of the pyrolysis temperature (○ = diameter; ● = weight). Reprinted from Journal of Physical Chemistry B, Suda, H., Haraya, K., Gas permeation through micropores of carbon molecular sieve membranes derived from kapton polyimide. Vol 101, 3988–3994. Copyright (1997) with permission from American Chemical Society*

temperature for a sufficiently long thermal soak time [8]. Gaseous decomposition products are evolved during the pyrolysis of the polymeric precursor, resulting in formation of micropores in the membrane; this is accompanied by a considerable loss in weight and dimensional shrinkage. Figure 1.4 shows a typical example in which a weight loss of up to 40% and shrinkage by up to 25% were observed during pyrolysis of circular films of a polyimide [12]. The pyrolysed membranes are sometimes post-treated by chemical vapour deposition (CVD) or by activation processes to improve their performance.

The greatest interest in the resulting carbon membranes is in evaluating the possibilities for their use as separation membranes. For this reason, the permeabilities of gases or vapours through the membranes are usually measured by using a permeation test apparatus. In some cases, pervaporation tests are also performed to test the separation performances for organic solutions such as water–ethanol or benzene–cyclohexane [6]. Gas permeability or permeance through flawless carbon membranes depends mainly on the size of the gas molecules, as shown in Figure 1.3, so that the relationship can be considered as an index of the pore size distribution.

The microstructures of carbon membranes are generally investigated by several analytical techniques; these include gas adsorption measurements, wide angle X-ray diffraction (WAXD), high-resolution transmission electron microscopy (TEM), scanning electron microscopy (SEM), thermogravimetric analysis (TGA), Fourier-transform infrared spectroscopy (FTIR), and X-ray photoelectron spectroscopy (XPS). Gas adsorption measurements using N_2, CO_2, or hydrocarbons as sorbing gases provide information on the pore size, the pore size distribution, and the specific surface area. WAXD is used to evaluate the degree of packing of the microporous carbon structures, whose interlayer distance (i.e., the average d spacing value) is calculated by using the Bragg equation, $d = n\lambda/2\sin\theta$, where d is the d-spacing, θ the diffraction angle, λ the wavelength of the X-ray radiation, and n is an integral number (1, 2, 3, etc). The average d spacing does not indicate the essential pore dimensions, but is believed to provide a measure of the length of the diffusion pathway for gas molecules through the carbon membranes. These

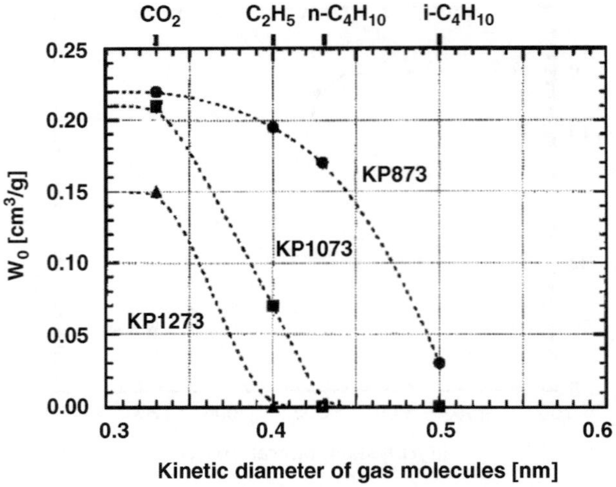

Figure 1.5 *Pore size distributions for Kapton CMS pyrolysed at various temperatures (873, 1073 and 1273 K). Limiting micropore volume (W_0) is plotted against the kinetic diameters of sorbate probe molecules. Reprinted from Journal of Physical Chemistry B, Suda, H., Haraya, K., Gas permeation through micropores of carbon molecular sieve membranes derived from kapton polyimide. Vol 101, 3988–3994. Copyright (1997) with permission from American Chemical Society*

properties relating to the pore structure are strongly dependent on the pyrolysis conditions and the nature of the polymeric precursors. The dependence of the pore size distribution on the pyrolysis temperatures is illustrated in Figure 1.5, where the average pore diameters decreased from about 0.45 to about 0.35 on increasing the pyrolysis temperature from 873 to 1273 K.

High-resolution TEM can give information on the pore structure in the form of a visual image. An example is shown in Figure 1.6, where the black regions represent the carbon matrix. These pictures show that the membrane pyrolysed at the higher temperature (1273 K) developed layer planes of graphite-like microcrystallites.

It is also important to measure other parameters related to the course of the pyrolysis reaction. TGA is used to determine the decomposition temperature, and it can also be used to study the effects of the atmosphere on weight losses from the membrane during pyrolysis. Simultaneous

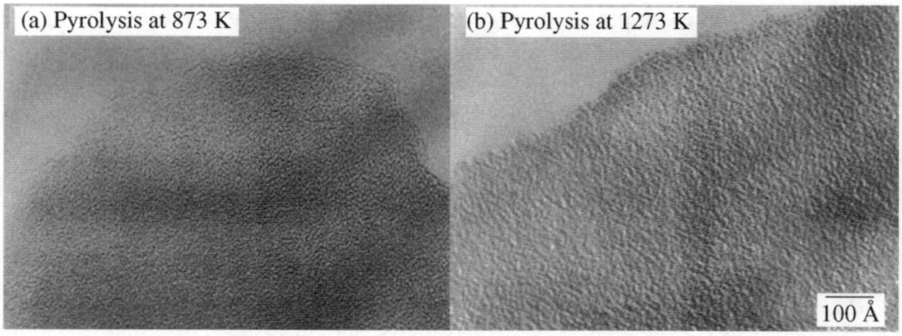

Figure 1.6 *High-resolution transmission electron micrographs of CMS membranes pyrolysed at 873 and 1273 K. Reprinted from Journal of Physical Chemistry B, Suda, H., Haraya, K., Gas permeation through micropores of carbon molecular sieve membranes derived from kapton polyimide. Vol 101, 3988–3994. Copyright (1997) with permission from American Chemical Society*

studies of off gases by means of mass spectrometry (MS) provide information on the chemical groups that are decomposed at high temperatures. FTIR, X-ray photoelectron spectroscopy (XPS), and elemental analyses of the precursor and of carbonised membranes pyrolysed at various temperatures are very helpful in providing an understanding of the changes in chemical structure that occur during pyrolysis.

1.3.2 Classification of Carbon Membranes

Carbon membranes can be grouped into two categories: (i) unsupported or freestanding carbon membranes and (ii) composite or supported carbon membranes. Unsupported membranes are generally produced as flat films, capillary tubes, or hollow fibres, whereas supported membranes are generally in the form of flat sheets or tubes. Many polymers, such as cellulose derivatives [1,13], polyacrylonitrile (PAN) [14], polyimides [15–64], phenolic resins [65–80], poly(furfuryl alcohol) (PFA) [81–92], poly(vinylidene chloride) [2,3,93–96], and poly(phenylene oxide) (PPO) [97–99], have been used as precursors for the production of carbon membranes. To prepare microporous carbon membranes that show a high performance in gas separations, it is important to select an appropriate polymeric precursor and to optimise the conditions for its pyrolysis. In the following sections, factors that control the preparation of carbon membranes are reviewed and discussed.

1.3.3 The Pyrolysis Process

In inert or vacuum atmospheres, the heat treatment of polymers can be separated into three processes: (i) annealing at 100–400 °C, (ii) intermediate heating at 400–500 °C, and (iii) pyrolysis to form carbon at 500–1000 °C [62]. The pyrolysis process is governed by several parameters such as the heating rate, the final pyrolysis temperature, the thermal soak time, and the pyrolysis atmosphere.

1.3.3.1 Pyrolysis Temperature

The pyrolysis temperature is generally chosen to be above the decomposition point of the polymer but below the graphitisation temperature (500–1000 °C). Koresh and Soffer [1] studied the effects of the carbonisation temperature by preparing membranes at 800 and 950 °C, and found that membranes pyrolysed at the higher temperature exhibited lower permeabilities but higher permselectivities. In a study on hollow fibre CMS membranes derived from 6FDA/BPDA–DAM polyimide precursors (abbreviations for dianhydrides and diamines of polyimides are listed in Table 1.1), Geiszler and Koros [19] found that increasing the final pyrolysis temperature from 500 to 800 °C decreased the permeability but increased the permselectivity. Suda and Haraya [12] reported that CMS membranes from pyrolysed Kapton (PMDA–ODA polyimide) showed a decrease in gas permeabilities but an increase in permselectivities on increasing the final pyrolysis temperature in the range 600–1000 °C. Similar trends in the relationship between permeabilities, permselectivities, and the pyrolysis temperature have been reported for P84 (BTDA–TDI/MDI) polyimide CMS membranes [35], Matrimid (BTDA–DAPI) polyimide CMS membranes [38], BTDA–ODA polyimide CMS membranes [52,55], and PFA CMS membranes [108]. As a general rule, an increase in the pyrolysis temperature reduces the permeability of CMS membranes but increases their selectivity [5]. High pyrolysis temperatures produce increased crystallinity, increased density, and lower average interplanar spacings in CMS membranes [22,30,38,91,92].

Table 1.1 *Abbreviations for dianhydrides and diamines of polyimide precursors*

Dianhydrides	
BPDA	3,3′,4,4′-Biphenyl tetracarboxylic acid dianhydride
6FDA	2,2-Bis(3,4-dicarboxyphenyl)-hexafluoro-propane dianhydride
BTDA	3,3′,4,4′-Benzophenone tetracarboxylic dianhydride
PMDA	Pyromellitic dianhydride
ODPA	4,4′-Oxydiphthalic dianhydride
NTDA	1,4,5,8-Naphthalene tetracarboxylic dianhydride
Diamines	
DAM	
TrMPD	2,4,6-Trimethyl-1,3-phenylene diamine
m-TMPD	
ODA	4,4′-Oxydianaline
DDBT	Dimethyl-3,7-diaminodiphenyl-thiophene-5,5′-dioxide
DABA	3,5-Diaminobenzoic acid
Durene	2,3,5,6-Tetramethyl-1,4-phenylene diamine
m-PDA	*meta*-Phenylenediamine
p-PDA	*para*-Phenylenediamine
DBA	1,3-Diamino benzoic acid
2,4-DAT	2,4-Diaminotoluene
TMMDA	Tetramethylmethylenedianaline
BDSA	4-4′-Diamino 2,2′-biphenyl disulfonic acid
BDSA	Benzidine-2,2′-disulfonic acid
BAHFDS	2,2-Bis[4-(4-aminophenoxy)phenyl] hexafluropropane disulfonic acid
ODADS	4,4′-Diaminodiphenyl ether-3,3′-disulfonic acid
BAPF	9,9′-Bis(4-aminophenyl)fluorene
p-intA	4,4′-Diaminodiphenylacetylene
DABZ	3,3′-Diaminobenzidine
DAI	5,7-Diamino-1,1,4,6-tetramethylindane
DAPI	5(6)-Amino-1-(4′-aminophenyl)-1,3-trimethylindane
TDI	Methylphenylenediamine
MDI	Methylenediamine

Several researchers have made detailed studies on the effects of the pyrolysis temperature and have reported exact results. Increasing the pyrolysis temperature from 500 °C results in an increase in the gas permeability of the carbonised membranes by one or two order of magnitude, with a maximum at around 650–750 °C. On further increasing the pyrolysis temperature, the resulting carbon membranes become less permeable. In some cases, maxima in the permselectivities are also observed. Hayashi *et al.* [46] reported that BPDA–ODA polyimide CMS membranes pyrolysed at 550–700 °C showed maximal permeabilities, whereas those pyrolysed at 800 °C exhibited peak He/N_2 selectivity. Kusuki *et al.* [42] found that a BPDA-aromatic diamine polyimide hollow fibre CMS membrane pyrolysed at 650 °C displayed maximal permeability to H_2, whereas one pyrolysed at 850 °C exhibited peak H_2/CH_4 selectivity. Centeno and Fuertes [69] observed a peak in the He permeability of a phenolic resin-based CMS membrane carbonised at around 700 °C. In studies on 6FDA/BPDA-DDBT copolyimide hollow fibre CMS membranes, Yoshino *et al.* [45] observed maximum permeabilities for membranes pyrolysed at 550 °C, whereas the peak permselectivities occurred at 650 °C. Yoshimune *et al.* [97,98] found that CMS membranes based on PPO and PPO derivatives pyrolysed at 650 °C exhibited maximal permeabilities, whereas peak permselectivities were observed at different pyrolysis temperatures.

These results suggest that pores appear at about 500 °C and enlarge as their numbers increase at temperatures up to 550–700 °C. Heating to a higher temperature causes the pores to shrink or disappear. This behavior certainly depends on the physical properties of the polymers, so that the suitable carbonisation temperature needs to be chosen individually for a selected polymer precursor to attain optimal performance in gas separations.

1.3.3.2 Thermal Soak Time

The thermal soak time can have various effects on the performance of the final membrane. Varying the thermal soak time, particularly at the final pyrolysis temperature, can be used to fine tune the permeation properties of a CMS membrane effectively. Kim *et al.* [52] showed that lengthening the thermal soak time for BPDA-ODA polyimide-based CMS membranes increased their selectivity but decreased their permeability. Yoshino *et al.* [45] reported that thermal soaking was effective in increasing the C_3H_6/C_3H_8 selectivity of 6FDA/BPDA-DDBT-based hollow fibre CMS membranes.

1.3.3.3 Heating Rate

The heating rate determines the rate of evolution of volatile components from a polymeric membrane during pyrolysis and consequently affects the nature of the pores that are formed in the resulting carbon membranes. Widely different heating rates have been used, ranging from 0.2 to 13.3 °C min^{-1}. Lower heating rates favor the formation of small pores and increase the crystallinity of the resulting carbon, thereby giving carbon membranes with a higher selectivity [12]. Higher heating rates can lead to the formation of pinholes, microscopic cracks, blisters, or distortions, which in extreme cases can render the membranes useless for gas separation [8]. From the practical standpoint, however, an optimal heating rate that is not too low should be chosen, because low heating rates increase the costs and time involved in producing carbon membranes.

1.3.3.4 Atmosphere

The pyrolysis is generally conducted in vacuum or under an atmosphere of an inert gas to prevent undesired burn off and chemical damage to the final carbonised membranes. Geizler and Koros [19] examined the pyrolysis of 6FAD/BPDA-DAM polyimide-based hollow fibre CMS membranes in vacuum and under an inert gas, and they concluded that vacuum pyrolysis produced more-selective but less-productive membranes than did pyrolysis in an inert atmosphere of He, Ar, or CO_2. Yoshino *et al.* [45] reported similar results for hollow fibre CMS membranes prepared from 6FDA/BPDA-DDBT. Su and Lua [31] examined the effects of the carbonisation atmosphere (Ar, He, N_2, or vacuum) on the membrane structure and transport properties of Kapton-derived CMS membranes pyrolysed at 600 and 800 °C. They found that carbonisation in vacuum at 600 and 800 °C gave membranes with low gas permeabilities but maximal ideal selectivities for O_2/N_2 of 9.37 and 17.76, respectively, whereas carbonisation at these temperatures under argon gave membranes with maximal selectivities for CO_2/CH_4 of 93.35 and 476.74, respectively.

1.3.4 Pretreatment

In some cases, pretreatments have been used to condition polymeric precursors before pyrolysis. The most common pretreatment is pre-oxidation, which allows the polymeric precursor to retain its form and structure during pyrolysis as a result of the formation of crosslinks in the polymer that increase its thermal stability. Kusuki *et al.* [42], Tanihara *et al.* [43], and Okamoto *et al.* [44] treated asymmetric hollow fibre membranes composed of a polyimide derived from BPDA and an aromatic diamine by oxidation in air at 400 °C for 30 min before carbonisation. They found that the pre-oxidation treatment was effective in preventing softening of the precursors during pyrolysis, which would otherwise have resulted in carbon membranes with a poor performance. Barsema *et al.* [34] pretreated P84 polyimide hollow fibre membranes by oxidation in air at 300 °C for 1 h before carbonisation. David and Ismail [18] showed that PAN hollow fibre membranes were thermally stabilised by heating at 250 °C in air or oxygen for 30 min. Yoshimune *et al.* [97,98] and Lee *et al.* [99] pretreated precursors of PPO or its derivatives by oxidation in air at 280 °C for 45 min to prevent melting during pyrolysis. Poly(phthalazinone ether sulfone ketone) (PPESK) precursors [100–102] have also been pre-oxidised in air at 400 °C for 30 min for the same reason.

Beside oxidation, chemical modifications have also been applied as pretreatments. Tin *et al.* [36,37] examined the effects of crosslinking modification and nonsolvent pretreatment of Matrimid and P81 precursors on the properties of the final CMS membranes.

1.3.5 Post-Treatment

The main purpose of post-treatments is to adjust the pore size distribution or to repair flaws in carbon membranes, thereby enhancing their performance. In general, oxidation processes are used for the former purpose and CVD treatments for the latter.

Koresh and Soffer [1] used an activation process in an oxidizing gas to improve the performance of a cellulose-derived carbon membrane by enlargement of its pores. Hayashi [49] and Kusakabe *et al.* [50] examined the oxidation of carbonised BPDA–ODA polyimide-based CMS membranes by treatment in O_2/N_2 mixtures at 300 °C. This treatment caused a broadening of the pore size distribution, resulting in a marked improvement in gas permeation accompanied by retention of O_2/N_2 permselectivity. Hayashi *et al.* [48] examined the post-treatment of BPDA–ODA polyimide CMS membranes by CVD through pyrolysis of propylene at 650 °C. The treated membrane showed increased permselectivities of 14 for O_2/N_2 and 73 for CO_2/N_2 at 35 °C.

Post-oxidation is also an effective method for increasing the permeability of CMS membranes to larger gas molecules such as hydrocarbons. Treatment of Kapton-based CMS membranes with water vapour at 400 °C increased their permeabilities by several orders of magnitude and gave a notably high selectivity of more than 100 for C_3H_6/C_3H_8 at 35 °C [23].

Fuertes and coworkers [70,71] produced supported CMS membranes by pyrolysis at 700 °C under vacuum of a novolac-type phenolic resin deposited on the inner face of a tubular ceramic ultrafiltration membrane. In some cases, pretreatment or post-treatment by aerial oxidation at 75–350 °C was examined. The separation performance for olefin/paraffin hydrocarbon mixtures was increased by pre-oxidation, post-oxidation, or CVD treatments of the carbonised membrane. Post-treatment by CVD followed by aerial oxidation is a key treatment in scaling up the area of a membrane without a loss of performance and will be discussed later in Section 1.4.

Fuertes [72] also used air oxidation at 300–475 °C of supported tubular carbon membranes to transform their gas-permeation properties into those of selective adsorption and surface diffusion carbon membranes.

1.3.6 Polymer Precursors

1.3.6.1 Cellulose Derivatives

Koresh and Soffer [1] prepared CMS membranes by carbonizing polymeric hollow fibres of an undisclosed composition. The precursor is likely to have been a hollow cellulose fibre, because

Carbon Membranes Ltd. (Israel) commercialised cellulose-derived CMS membranes on the basis of the research of Koresh and Soffer. Lagorsse *et al.* [13] performed a detailed characterisation of the CMS membranes produced by Carbon Membranes Ltd.

Otherwise, the only reports on cellulose-derived CMS membranes appear in several papers by Hägg and coworkers. These were metal-loaded CMS membranes and, as such, are discussed later in Section 1.3.10.

1.3.6.2 Polyimides

Polyimides are a precursor of choice for many researchers, probably because of their high glass-transition temperatures (T_g), ease of processability, and good separation performance as polymeric membranes [5].

6FDA/BPDA–DAM Polyimide: Jones and Koros [15] prepared CMS membranes by carbonizing commercially available asymmetric hollow fibre membranes of 6FDA/BPDA–DAM copolyimide. Carbonisation was carried out at two temperatures, 500 or 550 °C. Selectivities for O_2/N_2 in the range 11.0–14.0 and selectivities for CO_2/N_2 of about 55 were achieved. Exposure of the resulting membranes to volatile organic compounds (VOCs) at ambient temperatures resulted in losses in both permeability and selectivity [16], but the membranes could be partially regenerated by exposing them to propylene at a pressure of 1.03 MPa. Jones and Koros [17] also studied the effects of humidity on O_2/N_2 selectivity and permeability by using feeds with a relative humidity of between 23 and 85%. Some losses in performance occurred at all humidity levels, but these were reduced by rendering the surfaces of the membranes hydrophobic by coating them with thin layers of Teflon AF1600 or AF2400 [18].

In subsequent studies, Geizler and Koros [19] examined the effects of various pyrolysis atmospheres on the separation performance of asymmetric CMS hollow fibre membranes. Vacuum pyrolysis produced more-selective but less productive CMS membranes than did pyrolysis in an inert atmosphere.

Ghosal and Koros [11] prepared dense CMS films from 6FDA/BPDA–DAM polyimides and studied the changes in the intrinsic permeability and selectivity during the pyrolysis process.

Vu *et al.* [20] carbonised two types of polyimide hollow fibre membranes prepared from 6FDA/BPDA–DAM and from Matrimid 5218 (a polyimide from BTDA and DAPI). They investigated the performances of the two types of CMS membrane for the separation of CO_2/CH_4 at pressures of up to 6.89 MPa to confirm that the mechanical properties, permeability, and selectivity of the membranes were stable at high pressures. Vu *et al.* also examined the effects of condensable vapour impurities on the performance of the CMS hollow fibre membranes [21]. The experiments showed that the CMS membranes retained their CO_2/CH_4 selectivity and underwent a maximum 20% reduction in their CO_2 permeability. Furthermore, a simple in situ regeneration procedure involving moderate heating (70–90 °C) with dry N_2 purge gas resulted in almost complete recovery of permeability to CO_2 without any loss in CO_2/CH_4 selectivity.

Kapton (PMDA–ODA polyimide): Kapton is a polyimide obtained by curing the polyamic acid prepared by condensation of PMDA with ODA. Because both Kapton and its precursor polyamic acid are available commercially, they are frequently used as starting materials for CMS membranes. Suda and Haraya [12,22,23] prepared flat carbon membranes by carbonizing Kapton film. When the pyrolysis process was suitably controlled, the Kapton CMS membrane showed a high performance with a H_2/N_2 selectivity of 4700 and an O_2/N_2 selectivity of 36 at 35 °C. Because Kapton is almost insoluble in any solvent, the precursor PMDA–ODA polyamic acid has been used to prepare asymmetric capillary precursor membranes by a phase-inversion technique [24,25]. Ogawa and Nakano [26,27] also adapted this method to prepare CMS capillary membranes suitable for CO_2/CH_4 separation. Fuertes *et al.* [28] prepared asymmetric flat-type CMS membranes from PMDA–ODA polyamic acid membranes obtained by spin-coating the polymer onto a porous carbon disk with subsequent phase inversion.

Hatori *et al.* [29] obtained a Kapton CMS membrane by pyrolysis at 1000 °C; this membrane showed an ideal H_2/CO separation factor of 5900; this showed that it might be possible to reduce the CO content of hydrogen for use in fuel cells from 1% to 2 ppm. Recently, Lua and Su [30,31] have investigated the effects of the pyrolysis temperature and atmosphere on the final pore structure and gas-permeation properties of Kapton CMS membranes.

Matrimid and P84 polyimides: Matrimid (a polyimide prepared from BTDA and DAPI) and P84 (a polyimide prepared from BTDA and 80% TDI + 20% MDI) are commercially available polyimides that are sometimes used as precursors of CMS membranes. These polyimides can be conveniently cast into any form because they are soluble in various solvents. Fuertes *et al.* [32] prepared asymmetric flat CMS membranes from P84 and from Matrimid. Steel and Koros [33] produced dense freestanding CMS membranes from Matrimid and they examined the effects of the pyrolysis temperature on the ultramicropore distributions, which were related to the performances of the membrane in separating O_2/N_2, CO_2/CH_4, and C_3H_6/C_3H_8 mixtures. Vu *et al.* [20] prepared CMS hollow fibre membranes as described above. Barsema *et al.* [34] produced hollow fibre CMS membranes from P84 polyimide ultrafiltration membranes, but these exhibited no separation property for any gases. They also prepared CMS membranes containing nanoclusters of silver from a silver-containing P84 precursor, as discussed in Section 1.3.10. Tin *et al.* [35] prepared flat CMS membranes from P84 film. The membrane obtained by pyrolysis at 800 °C showed a CO_2 permeability of 500 Barrers [$1\ \text{Barrer} = 10^{-10}\ \text{cm}^3\ \text{(STP) cm cm}^{-2}\ \text{s}^{-1}\ \text{cmHg}^{-1} = 3.35 \times 10^{-16}\ \text{mol m m}^{-2}\ \text{s}^{-1}\ \text{Pa}^{-1}$] and a CO_2/CH_4 selectivity of 89, which was the highest efficiency among a group of CMS membranes derived from four commercially available polyimides [P84, Matrimid, Kapton, and Ultem (a polyether imide)].

Recently, Favvas *et al.* [39] produced hollow fibre CMS membranes by pyrolysis of hollow fibres of Matrimid under inert (N_2) or reactive ($CO_2 + H_2O$) atmospheres. The sizes of the pores were affected by the temperature of the carbonisation process, whereas the pore volume was affected by the environmental conditions for pyrolysis.

Polyether imides: Most polyimides used as precursors of CMS membranes are either very expensive commercial materials or are available only on a laboratory scale. In this context, one polyimide-based material that can be used economically is the commercially available polyether imide Ultem 1000. Fuertes *et al.* [40] used Ultem 1000 to prepare CMS membrane supported on a macroporous carbon substrate. Sedigh *et al.* [41] prepared tubular CMS membranes by pyrolysing polyether imide coated on the inside of a mesoporous tubular support.

BPDA–aromatic diamine polyimide: Kusuki and coworkers [42,43] prepared asymmetric CMS hollow fibre membranes from asymmetric hollow fibre membranes composed of a polyimide derived from BPDA and an unspecified aromatic diamine, produced by UBE Industries (Japan). They developed a pyrolysis method for the continuous preparation of hollow fibre CMS membranes. For a feed gas mixture of 50% H_2 in CH_4 at 80 °C, a CMS membrane carbonised at 700 °C showed a H_2 permeability of 1000 GPU [$1\ \text{GPU} = 10^{-6}\ \text{cm}^3\ \text{(STP) cm}^{-2}\ \text{s}^{-1}\ \text{cmHg}^{-1} = 3.35 \times 10^{-10}\ \text{mol m}^{-2}\ \text{s}^{-1}\ \text{Pa}^{-1}$] and a H_2/CH_4 selectivity of 132, whereas a membrane carbonised at 850 °C displayed a H_2 permeability of 180 GPU and H_2/CH_4 selectivity of 631. Tanihara *et al.* [43] examined the influence of trace levels of toluene vapour (7500 ppm) and concluded that these had little effect on the permeation properties of the CMS membrane.

Okamoto *et al.* [44] evaluated the olefin/paraffin separation properties of asymmetric CMS hollow fibre membranes that they prepared by pyrolysis of asymmetric hollow fibre membranes of a BPDA–DDBT/DABA copolyimide. At 100 °C, the permeabilities for C_3H_6 and C_4H_6 were

50 and 80 GPU, respectively, and the selectivities for C_3H_6/C_3H_8 and C_4H_6/C_4H_{10} were 13 and 50, respectively. Other asymmetric CMS hollow fibre membranes have been prepared from 6FDA/BPDA–DDBT copolyimide [45]; the CMS membranes pyrolysed at 540 °C for 1 h displayed the best performance in terms of C_3H_6 permeability (26 GPU) and selectivity (22) for a 50 : 50 C_3H_6/C_3H_8 mixture at 100 °C.

Laboratory-synthesised polyimides: Many laboratory-synthesised polyimides have also been used as precursors for carbonised membranes. Hayashi *et al.* [46,47] synthesised BPDA–ODA polyimide and then used it to produce CMS membranes supported on a porous alumina tube. The CMS membrane pyrolysed at 800 °C showed a permeability to CO_2 of 300 GPU and a selectivity to CO_2/CH_4 of 100 at 30 °C. Hayashi *et al.* also reported the possibility of separating olefins from paraffins by using a CMS membrane derived from BPDA–ODA polyimide carbonised at 700 °C. The CMS membrane exhibited permeabilities of approximately 30 GPU for C_2H_4 and 6 GPU for C_3H_6 at 100 °C. The selectivities were 4–5 for C_2H_4/C_2H_6 and 25–29 for C_3H_6/C_3H_8 systems.

Fuertes *et al.* [51] used a polyamic acid prepared from BPDA and *p*-PDA to prepare precursor membranes. The precursor membranes were converted into polyimide membranes that were subsequently pyrolysed to give flat supported CMS membranes.

Lee and coworkers [52–56] performed a series of studies on CMS membranes derived from BTDA–aromatic diamine polyimides. Kim *et al.* [52] synthesised BTDA–ODA polyimide and used it as a precursor for the production of flat CMS membranes. The membrane exhibited an attractive separation potential compared with CMS membranes derived from PMDA–ODA.

Other rigid polyimides: Kita *et al.* [57] synthesised a polypyrrolone from 6FDA and DABZ. The polymer had a ladder structure in the backbone chains that provided an enhanced permeability and maintained the permselectivity of gases by inhibiting both chain packing and intermolecular motions. The CMS membranes carbonised at 700 °C under nitrogen showed a better membrane performance than those prepared by carbonisation of polyimides.

Xiao *et al.* [58] synthesised four polyimides from BTDA, 6FDA, ODPA, and BPDA with DAI (BTDA–DAI, ODPA–DAI, BPDA–DAI, and 6FDA–DAI), and they investigated the effects of the chemical structure and physical properties of the rigid polyimides on the performance of the derived carbon membranes. At low pyrolysis temperatures, polyimides with a high fractional free volume (FFV) and a low thermal stability gave carbon membranes with bigger pores and higher gas permeabilities. Xiao *et al.* [59] also synthesised crosslinked copolyimides from 6FDA, durene, and *p*-intA, and they carbonised the resulting films at 800 °C under vacuum. Thermally induced crosslinking occurred through the acetylene groups present in the p-intA units, which are believed to form naphthalene structures by a Diels–Alder-type reaction. Carbon membranes derived from copolyimides with large numbers of internal acetylene units showed much better gas separation performances than did those derived from polyimides without internal acetylene units.

1.3.6.3 Phenolic Resins

Phenolic resins, which are very popular and inexpensive polymers, have also been used as precursors for preparing CMS membranes. Shusen *et al.* [65] produced free-standing asymmetric flat carbon membranes by carbonizing thermosetting phenol–formaldehyde resin films at 800–950 °C under a N_2 atmosphere, followed by oxidation on one side. The resulting membrane exhibited molecular sieve-like flow properties. Kita *et al.* [66] prepared tubular CMS membranes by subjecting a phenolic resin coated on the surface of a porous tube of α-alumina to pyrolysis at 600 °C under a N_2 atmosphere. This coating–carbonisation cycle was repeated four or five times. The resulting membranes showed an excellent separation performance for alkenes/alkanes and CO_2/N_2.

Zhou *et al.* [67,68] produced highly permeable CMS membranes from a phenolic resin, and they investigated the effects of the pyrolysis temperature, the dip-coating conditions, and the number of coating/pyrolysis cycles on the gas-permeation properties of the membranes. Membranes obtained under optimal preparation conditions exhibited an O_2 permeability of 30 GPU and an ideal O_2/N_2 separation factor of 12 at 35 °C.

Fuertes and coworkers [69–74] published a series of studies on CMS membranes made of phenolic resin. They produced CMS membranes consisting of a thin (2 μm) microporous carbon layer obtained by pyrolysis of a film of a novolac-type phenolic resin supported on a macroporous carbon disk substrate (pore size, 1 μm; porosity, 30%). The carbon membrane prepared by carbonisation at 700 °C showed high selectivities for the separation of permanent gases, such as the O_2/N_2 system (selectivity 10 at 25 °C).

Centeno *et al.* [74] investigated processing variables (the heat-treatment temperature, heating rate, soaking time, and atmosphere) for pyrolysing phenolic resin-based carbon membranes. The membranes that they obtained at temperatures of around 700 °C behaved as selective adsorption and surface-diffusion membranes and were highly effective for the recovery of hydrocarbons from hydrocarbon/N_2 mixtures. An increase in the carbonisation temperature to 800 °C caused a significant decrease in the gas permeability and led to a CMS membrane that displayed good capabilities for the separation of O_2/N_2, CO_2/CH_4, and olefin/paraffin mixtures. Heat treatment of phenolic resin films at temperatures of around 900–1000 °C produced CMS membranes that showed high permselectivities for mixtures of gases with molecular sizes smaller than 0.4 nm.

Zhang *et al.* [77] produced a CMS mixed-matrix carbon membrane. The coating solution was obtained by dispersing some CMS with a median particle size of 0.37 μm, prepared by high-temperature pyrolysis of walnut hulls, in an ethanol solution of a novolac phenol–formaldehyde resin (PFNR). A green porous tubular support of the same resin was dipped in the mixed-matrix coating solution and then carbonised at 800 °C. After carbonisation, the tube was activated in CO_2 for 20–60 min at the same temperature that was used for the carbonisation.

Blue Membranes GmbH (Germany) produces large-scale CMS membranes pyrolysed from supported flat sheet precursor of a mixed polymer of phenolic and epoxy resins. Lagorsse *et al.* [78] have examined the basic properties of this membrane. The procedure used for constructing modules containing these membranes is discussed in Section 1.5.

1.3.6.4 Poly(Furfuryl Alcohol)

Poly(furfuryl alcohol) (PFA) has been used extensively as a precursor for CMS membranes. Because PFA is a liquid at room temperature, all membranes derived from it are composite membranes supported by porous substrates. Chen and Yang [81] coated PFA onto a macroporous graphite disk support; carbonisation of the resulting assembly gave a PFA-CMS membrane. Sedigh *et al.* [82] also used PFA as a precursor for the preparation of supported CMS films. The membranes were tested by using single gases (H_2, CO_2, CO, CH_4, and Ar), binary mixtures of CO_2 and CH_4, and a quaternary mixture of CO_2, CO, H_2, and CH_4. Separation factors for CO_2/CH_4 in the range 34–37 were obtained for the binary and quaternary mixtures.

Wang *et al.* [83] used the vapour deposition polymerisation (VDP) technique to coat furfuryl alcohol (FA) onto γ-Al_2O_3/α-Al_2O_3 or glass/α-Al_2O_3 support tubes. The support tubes were pretreated with an acid catalyst and exposed to FA vapours at 90 °C. The tubes were then heated at 200 °C to crosslink the PFA polymer and carbonised at 600 °C. In comparison with certain PFA–CMS membranes prepared by dip-coating techniques, the membranes prepared by VDP had similar CO_2/CH_4 selectivities but lower CO_2 permeabilities.

Foley and coworkers [84–90] published a series of studies on PFA–CMS membranes formed on sintered stainless steel supports. Solutions of PFA in acetone (50–60 wt%) were coated by hand brushing [84] or spray coating [85] onto porous stainless steel disks (0.2 μm pore size). Shiflett and Foley [86] developed an ultrasonic spray-coating method using a 25 wt% solution of PFA in acetone to make the PFA precursor. The PFA–CMS membranes formed on sintered stainless steel tubes with a pore size of 0.2 μm exhibited high capabilities for the separation of O_2/N_2, He/N_2, and H_2/N_2. Shiflett and Foley [87] also explored a protocol involving high-temperature pyrolysis of the initial layers followed by lower-temperature pyrolysis of subsequent layers to give membranes with a higher permeation flux; in addition, they examined the modification of PFA–CMS membranes by using additives such as titanium dioxide, small-pore high-silica zeolite, or PEG. They also used a new automated ultrasonic spray system to prepare uniform PFA films with improved reproducibility for the production of PFA–CMS membranes [88].

Anderson *et al.* [91] have used a similar method for making PFA–CMS membranes supported on a porous stainless steel disk. Positron-annihilation lifetime spectrometry and wide-angle X-ray diffraction studies showed that the size of the micropores decreased and the porosity increased with increasing pyrolysis temperature. Studies on the performances of the resulting membrane supported these finding, in that significant increases in permeability, related to an increase in porosity, were observed on increasing the pyrolysis temperature.

1.3.6.5 Vinylidene Chloride Copolymers

Rao and Sircar [2,3] obtained carbon membranes by pyrolysis of a poly(vinylidene chloride) (PDVC)–acrylate terpolymer latex coated on a porous graphite support. The resulting membrane separated H_2/hydrocarbon mixtures by selective adsorption and surface diffusion of the larger component (the hydrocarbon). This membrane is referred to as a selective surface flow (SSF) membrane. The diameter of the pores in the membrane was found to be in the range 0.5–0.6 nm [3], which is larger than those of CMS membranes. The researchers extended their method to produce tubular membranes, and they demonstrated the possibility of an SSF membrane/pressure-swing adsorption hybrid process for the production of pure hydrogen [93–95].

Centeno and Fuertes [96] formed carbon membranes by pyrolysing poly(vinylidene chloride-co-vinyl chloride) (PVDC-PVC) films supported on porous carbon disks. These membranes showed molecular sieving properties, for example, a high permselectivity of 14 for the O_2/N_2 pair. Pre-oxidation in air at 200 °C for 6 h improved the permselectivity, but resulted in a decrease in gas permeability.

1.3.6.6 Novel Polymer Precursors in Recent Research

Yoshimune *et al.* [97] produced novel CMS membranes from poly(phenylene oxide) (PPO) and its derivatives. They synthesised PPO derivatives with functional groups such as SO_3H, CO_2H, Br, $SiMe_3$, or PPh_2 by one-step reactions, and they cast these derivatives into hollow fibre configurations with symmetrically dense structures. Gas permeabilities and selectivities for He, H_2, CO_2, O_2, and N_2 for the pyrolysed membranes were as high as those observed in polyimide-based CMS membranes. The highest performance was shown by trimethylsilyl ($SiMe_3$)-substituted PPO CMS membrane pyrolysed at 650 °C, for which the O_2 permeability was 125 Barrers and the O_2/N_2 selectivity was 10 at 25 °C. Yoshimune *et al.* [98] made a further detailed investigation of $SiMe_3$-substituted PPO CMS; they also investigated supported CMS membranes derived from dip-coated PPO on a porous ceramic tube with a pore size of 0.1 μm [99].

Zhang *et al.* [100–102] prepared CMS membranes from poly(phthalazinone ether sulfone ketone) (PPESK) as a novel polymeric precursor. The maximum permselectivities of these membranes for H_2/N_2, CO_2/N_2 and O_2/N_2 gas pairs were as high as 278.5, 213.8, and 27.5, respectively. For PPESK, oxidative stabilisation before carbonisation was beneficial for the preparation of carbon membranes with a good gas separation performance in that it helped to shift the pore size distribution to a smaller pore size and increased the maximum pore volume in the carbon matrix. The researchers also found that it was possible to tune the gas permeability and microstructure of the carbon membranes by varying the chemical structure of the precursors (i.e., the sulfone-to-ketone ratio) or the conditions for the preparation of the membrane [102].

Most recently, Chng *et al.* [103] explored carbon membranes derived from an interpenetrating network of poly(aryl ether ketone) and 2,6-bis-(4-azidobenzylidene)-4-methylcyclohexanone (PAEK/azide). Their concept involved designing a polymer precursor that consisted of a thermally stable part and a thermal labile part at the molecular level, so as to create CMS membranes with relatively large pores that were suitable for C_3H_6/C_3H_8 separation. PAEK/azide (80 : 20) pyrolysed at 550 °C exhibited the best C_3H_6/C_3H_8 separation performance of these polymers with a C_3H_6 permeability of 48 Barrers and a C_3H_6/C_3H_8 selectivity of 44. Their work is based on the preparation concept discussed in the next section, which is employing a precursor that involves molecular size poregen and pyrolysing it at an intermediate temperature.

1.3.7 Adjustments of Pore Structures

Although many CMS membranes show excellent selectivities for O_2/N_2 or C_3H_6/C_3H_8 pairs in comparison with polymer membranes, they have lower permeabilities to O_2 or C_3H_6, which reduces their attractiveness. Recently, several researches have investigated strategies for enhancing the permeability of CMS membranes while retaining their selectivitity. Pyrolysis at intermediate temperature has been examined for preventing losses in permeability associated with shrinkage or loss of micropores as a result of carbonisation at high temperatures. Changing the structure of the polymeric precursor by modifying its fractional free volume and blending of the polymeric precursors with a porogen are two techniques that are intended to raise the numbers of pores and increase the pore volume, resulting in an enhanced permeability.

1.3.7.1 Intermediate Structure

Barsema *et al.* [62] subjected films of Matrimid polyimide to various heat treatments at between 300 and 525 °C, and they investigated the intermediate structures that evolved at temperatures between the annealing temperature and the carbonisation temperature. The T_g of this polymer was 323 °C. The permeabilities to noncondensable gases (N_2 and O_2) were depressed by heat treatments at below the T_g of the polymer. A peak in permeability was observed for polymer treated at 350 °C. Above 350 °C, increasing formation of charge-transfer complexes and the resulting densification of the polymer structure led to a gradual decrease in the permeability. However, the permeability increased significantly again when the films were exposed to temperature above 475 °C, as a result of the onset of thermal decomposition.

Shao *et al.* [63] synthesised a 6FDA–durene polyimide ($T_g = 425$ °C) and then pyrolysed it by a heat-treatment procedure at temperatures from 50 to 800 °C under vacuum. The gas permeability increased with increasing treatment temperature. The maximum increase in the permeabilities to O_2, N_2, and CH_4 occurred at 475 °C. Carbon membranes pyrolysed at various heating rates (1 or 3 °C min^{-1}) showed differences in their transport performances at low pyrolysis temperatures. At a high temperature (800 °C), however, the resultant carbon membranes pyrolysed by the various protocols showed similar, but superior, gas-transport performances. Shao *et al.* [64] also observed that the casting solvent affected the morphologies and gas-transport properties of membranes made from a novel copolyimide of 6FDA with PMDA and TMMDA, as well as those of its derived carbon membranes. The differences between CMS membranes derived from precursors with different morphologies decreased as the pyrolysis temperature was increased. At low pyrolysis temperatures, the structure and separation performance of the CMS membranes were significantly affected by the decomposition temperature of the precursor; however, at higher pyrolysis temperatures, the factor that dominated the structure and performance of the CMS membranes was the pyrolysis temperature, because of the complete degradation of the polymeric precursor.

1.3.7.2 Polymers with Modified Fractional Free Volumes

Xiao *et al.* [38] examined the effects of brominating a Matrimid precursor before it was carbonised to produce carbon membranes. The lower thermal stability and higher FFV of brominated Matrimid resulted in higher gas permeabilities for carbon membranes pyrolysed at low pyrolysis temperatures, whereas the selectivity remained similar to those of membranes obtained by pyrolysis of the original Matrimid precursor under the same conditions.

Park *et al.* [55] synthesised polyimides from BTDA and a 9 : 1 mixture of ODA with *m*-PDA, 2,4-DAT, or *m*-TMPD. These comonomers contain no methyl substituent (*m*-PDA), one methyl substituent (2,4-DAT), or three methyl substituents (mTMPD). The resulting films were carbonised to produce flat dense CMS membranes. The introduction of methyl substituents on the rigid polyimide backbone increased the FFV of the polyimides, and the gas permeabilities typically increased with the FFV. However, the CMS membranes prepared by pyrolysis of each of these polyimides in an inert atmosphere at 600 and 800 °C showed similar gas-permeation behaviors.

1.3.7.3 Polymers Incorporating a Porogen

Okamoto and coworkers [60,61,67] succeeded in enhancing the gas permeability and selectivity of CMS membranes by introducing sulfonic acid groups into rigid structured precursor polymers. The sulfonic acid groups acted as pore-forming agents (porogens) by decomposing at temperatures below those required for carbonisation. It is noteworthy that the resulting CMS membranes were pyrolysed at intermediate temperatures of 450–500 °C.

Zhou *et al.* [67] synthesised a thermosetting phenolic resin with pendant sulfonic acid groups by treating a resol-type phenolic resin (PF) with a novolac-type sulfonated phenolic resin (SPF). During the pyrolysis of this thermosetting PF/SPF resin, large amounts of gaseous molecules of similar and small sizes, such as H_2O and SO_2, were evolved between 110 and 350 °C. Highly permeable CMS membranes were then obtained by pyrolysis of PF/SPF (45 : 55) precursor membranes dip-coated onto porous alumina tubes (average pore size: 0.14 μm, diameter: 2.3 mm, porosity: 40–48%). For example, a membrane pyrolysed at 500 °C for 1.5 h displayed the highest O_2 permeability (240 GPU) at 1 atm and 35 °C of any CMS membrane reported in the literature, but only showed a mediocre O_2/N_2 separation factor of 5.2.

Flexible pyrolytic membranes were prepared by Okamoto and coworkers [60,61] through pyrolysis of dense and flat membranes of sulfonated polyimides. The selection of a suitable pyrolysis temperature at which most of the sulfonic acid groups decomposed but no substantial cleavage of the polyimide backbone occurred was important. The membrane prepared by pyrolysis of a copolymer of NTDA with an 8 : 2 mixture of BDSA and BAPF at 450 °C exhibited a C_3H_6 permeability of 18 Barrers and a C_3H_6/C_3H_8 selectivity of 26 at 35 °C. The membrane prepared by pyrolysis of a polymer of NTDA with BAHFDS at 450 °C exhibited a C_3H_6 permeability of 29 Barrers and a C_3H_6/C_3H_8 selectivity of 29. The researchers concluded that this is a promising approach for the preparation of membrane materials for olefin/paraffin separations.

Kim *et al.* [53,54] evaluated the effects of blending a labile polymer [poly(2-vinylpyrrolidone)] (PVP)] with a BTDA–ODA polyimide on the gas-permeation performance of the final CMS membranes. The permeabilities of gases through the CMS membranes were enhanced by the introduction of the thermally labile PVP, and decreased as the final pyrolysis temperature was increased. A CMS membrane prepared from a blend containing high-molecular weight (50 000) PVP pyrolysed at 550 °C exhibited an enhancement in the O_2 permeability from 560 to 810 Barrers and a reduction in the O_2/N_2 selectivity from 10 to 7. Kim *et al.* [56] synthesised a series of copolyimides from BTDA and complex diamines of ODA and *m*-PDA (8 : 2), ODA, and 1,3-diaminobenzoic acid (DBA) (8 : 2), and ODA and DBA (5 : 5). In these polyimides, the molar ratios of DBA, which has a pendant carboxylic acid in the diamine moiety, were 0, 20, and 50% respectively. The decomposition of the pendant carboxylic acid side-groups during pyrolysis of the polymers resulted in large pore volumes in the resulting carbon matrix, which significantly affected the gas separation performance of the final CMS membranes. The gas permeabilities of the CMS membranes pyrolysed at 700 °C increased with increasing carboxylic acid group content. The CMS membranes derived from the polyimide containing 50 mol% diamines showed a maximum gas permeability for O_2 of 707 Barrers and an O_2/N_2 selectivity of 9 at 25 °C.

Zhang *et al.* [79] prepared CMS membranes by pyrolysing a blend of PFNR and poly(ethylene glycol) (PEG) spray coated onto a porous carbon membrane support. This confirmed that PEG is a significant porogen for the formation of micropores during pyrolysis of the blend and that the presence of the micropores increased the number of diffusion pathways available for transport of gas molecules through the CMS membrane.

Nishiyama *et al.* [80] produced microporous carbon membranes on an alumina support by the pyrolysis of cationic tertiary amine/anionic polymer composites. The precursor solutions contained a thermosetting resorcinol/formaldehyde (R/F) polymer and a cationic tertiary amine. Three types of cationic tertiary amine with different chain lengths were used: (i) tetramethylammonium bromide (TMAB), (ii) tetrapropylammonium bromide (TPAB), and (iii) cetyl(trimethyl)ammonium bromide (CTAB). A porous structure was produced by decomposition of the amine, and the resulting pores assisted the further gasification of the RF polymer at high temperatures. Studies on the permeation of pure gases of various molecular sizes showed that the pore sizes of the carbon membranes prepared by using TMAB, TPAB, and CTAB were 0.4, 0.5 and >0.55 nm, respectively.

1.3.8 Modification of Porous Substrates

The porous substrate itself is another important factor in relation to supported thin-film CMS membranes, and it needs to be flawless to give a good separation performance. Although repetitive coating or coating–pyrolysis cycles are useful for producing flawless thin layers, modifications of the porous substrate by controlling the pore size or modifying the nature of the surface are generally more effective. Foley and coworkers [89,90] modified porous stainless steel supports with silica nanoparticles that were slip-cast into the pores. In comparison with membranes prepared on unmodified porous stainless steel supports, PFA-based CMS membranes prepared on the modified support showed an improvement of about a two orders of magnitude in their oxygen permeability while retaining their selectivity. Recently, Song *et al.* [92] reported a novel PFA–CMS membrane supported on porous coal-based carbon tubes with an average pore diameter of 0.11 μm and a porosity of 40.3%. Membranes prepared by one-step coating followed by pyrolysis at 600–900 °C showed a good separation performance for gas pairs such as H_2/N_2, CO_2/N_2, O_2/N_2, and CO_2/CH_4; the highest permselectivities at 25 °C reached 465.0, 58.8, 13.2, and 160.5, respectively.

Wei *et al.* [75,76] explored a novel method for producing composite CMS membranes in which the coated layer and support were carbonised simultaneously. An alcoholic solution of PFNR containing a little tetrahexamethylenetetramine or hexamine was coated onto a green porous resin support of the same material and the assembly was subjected to pyrolysis. An advantage of this technique is that the membrane layer undergoes shrinkage at the same rate as the support during heating and carbonisation, which helps to prevent cracks from forming as a result of differences in the rates of shrinkage of the two parts.

1.3.9 Current Status

Data on O_2/N_2 separation performances taken from literature sources are plotted in Figure 1.7, which also shows the upper limits for polymer membranes [104]. Results for O_2/N_2 selectivity and O_2 permeability are sometimes used to benchmark the properties and characteristics of membranes as a function of the synthesis variables. It can be clearly seen that most CMS membranes have O_2/N_2 selectivities that are roughly three times higher than the upper limit for polymer membranes, when the selectivities are compared at the same O_2 permeability. Previous reports have shown that CMS membranes have considerable promise for commercial applications, not only in O_2/N_2

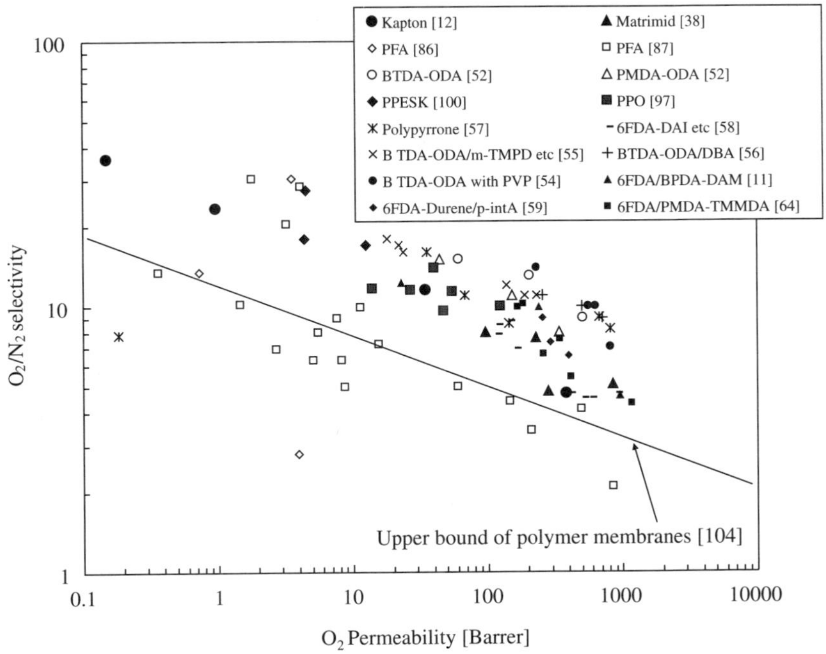

Figure 1.7 *Comparison of the oxygen permeabilities and permselectivities of various CMS membranes with the upper limit for polymer membranes*

separation, but also in hydrogen purification and the separation of CO_2/CH_4, CO_2/N_2, or C_3H_6/C_3H_8, which are processes where conventional polymer membranes do not perform adequately. New studies are continuing to extend the boundaries of the performance of CMS membranes, for example, by incorporation of activated nanoparticles, as discussed in the next section.

Although detailed descriptions of pervaporation are omitted from this chapter, it should be noted that CMS membranes have shown preferential permeability to water in water/ethanol mixtures with selectivities of 100 to 600 [6]. These values are not as high as those of zeolite membranes, but are worth taking into account in relation to further investigations with a view to developing practical applications.

1.3.10 Mixed-Matrix Carbon Membranes

In this chapter, a mixed-matrix carbon membrane is defined as the product of the pyrolysis of a mixed-matrix membrane comprising a continuous polymer matrix incorporating nanosized inorganic particles dispersed throughout the polymer [105,106]. Table 1.2 shows some representative examples of mixed-matrix carbon membranes that have been reported in the literature. Metal ions or particles, silica particles, zeolites, or carbons are normally used as inorganic particles to improve the properties of the prepared membranes. Such nanosized particles increase the polarity of membranes, form interlayer spaces, or have an affinity toward target gases; they can also function as catalysts in catalytic membrane reactors, as discussed in Section 1.5.

One method for preparing mixed-matrix carbon membranes involves the incorporation of metals into carbon membranes. Kim *et al.* [107] prepared carbon membranes containing alkali metal ions such as Li^+, Na^+, or K^+ by pyrolysis of metal-substituted sulfonated polyimides. Barsema *et al.* [108,109] found that the incorporation of nanosized Ag clusters into a carbon matrix resulted in an increase in selectivity to O_2 over N_2, accompanied by a substantial increase in the permeability. Figure 1.8 is a schematic representation of the structure of the functionalised carbon. The authors concluded that the main route of diffusion through the membrane was route (c), where the gas molecules pass through the free volume between the Ag nanoclusters and the carbon matrix. Yoda and coworkers [110,111] investigated carbon membranes containing dispersed Pt and Pd particles prepared through doping of polyimide films by supercritical impregnation. Hägg and coworkers [112–114] produced cellulose-derived carbon membranes and examined the effects of various additives, including oxides of Ca, Mg, Fe(III), and Si, and nitrates of Ag, Cu, and Fe(III). Yoshimune *et al.* [115] prepared hollow fibre carbon membranes derived from sulfonated poly (phenylene oxide) containing various metal ions (Na^+, Mg^{2+}, Al^{3+}, Ag^+, Cu^{2+}, or Fe^{3+}). Zhang *et al.* [116] developed carbon membranes filled with nanosized Ni particles by dispersing nanoparticulate nickel into the precursor solution for the phenolic resin.

There has been a recent growth in research interest on the use of silica and zeolites as additives for mixed-matrix carbon membranes. An excellent example is given by Park and Lee and coworkers [117–122], who prepared carbon membranes containing silica by pyrolysis of a poly (imide siloxane). Figure 1.9 shows the proposed model for the structure of the carbon–silica membranes. During pyrolysis, the imide domains are converted into a carbon-rich phase, and the siloxane domains are transformed into a silica-rich phase containing small carbon clusters. The gas separation properties of the resulting carbon–silica membranes were highly dependent on the siloxane content and the siloxane chain length in the imide siloxane precursor. In similar research performed by Yoshimune *et al.* [98], silica-containing carbon membranes were prepared from trimethylsilyl-substituted poly(phenylene oxide). Rajagopalan *et al.* [89] reported that the addition of silica nanoparticles to a PFA precursor solution improved the O_2 permeabilities of the resulting membranes within the same range of O_2/N_2 selectivities.

Table 1.2 *Representative examples of mixed-matrix carbon membranes*

Precursor	Additive	Configuration	Refs
Metal ions or particles			
Sulfonated polyimide	Li^+, Na^+, K^+	Flat film	[107]
P84 polyimide	Ag nanoclusters	Flat film	[108]
P84 polyimide + SPEEK[a]	Ag nanoclusters	Flat film	[109]
Kapton polyimide	Pt or Pd nanoparticles	Flat film	[110,111]
Cellulose	CaO, MgO, Fe_2O_3, SiO_2, Ag^+, Cu^{2+}, Fe^{3+}	Flat film	[112–114]
Sulfonated PPO	Na^+, Mg^{2+}, Al^{3+}, Ag^+, Cu^{2+}, Fe^{3+}	Hollow fibre	[115]
Phenolic resin	Ni nanoparticles	Supported tube	[116]
Silica particles			
Imide-siloxane block copolymer	Oligomeric organosiloxane	Flat film, supported tube	[117–121]
Imide-siloxane block copolymer + Al_2O_3	Oligomeric organosiloxane	Flat film	[122]
Trimethylsilylated PPO	Chloro(trimethyl)silane	Hollow fibre	[98]
PFA	SiO_2 nanoparticles	Supported film	[89]
Zeolites			
Matrimid polyimide	Zeolite KY	Flat film	[123]
Polyimide	ZSM-5	Flat film	[124]
Phenolic resin	NaA zeolite	Supported tube	[125,126]
Matrimid polyimide/PSF[b]	Zeolite beta	Hollow fibre	[127]
P84 polyimide/PES[c]	Zeolite beta	Hollow fibre	[128]
Polyimide/phenolic resin	Silicalite-1	Supported tube	[129]
Phenolic resin	NaA zeolite	Supported tube	[130]
PPESK	ZSM-5	Flat film	[131]
Polyimide	Zeolite T	Flat film	[132]
Carbons			
PFA	Carbon black	Supported film	[90]
PEI/PVP	CNT	Supported tube	[133]
Polyimide	CNT	Supported tube	[134]

[a]Sulfonated poly(ether ether ketone)
[b]Polysulfone
[c]Poly ether sulfone

In the case of zeolite-containing carbon membranes, various combinations of precursor polymers and zeolites have been reported; these are summarised in Table 1.2 [123–132]. Basically, the membranes were prepared by pyrolysis of precursor membranes consisting of nanosized zeolite particles dispersed in a polymer matrix. The resultant zeolite-containing carbon membranes showed some improvements in gas permeability or permselectivity. It is believed that the increase in gas permeability is due to the presence of ordered microchannels in the zeolite and of interfacial gaps between the zeolite and the carbon matrix in the membranes, as shown in Figure 1.8.

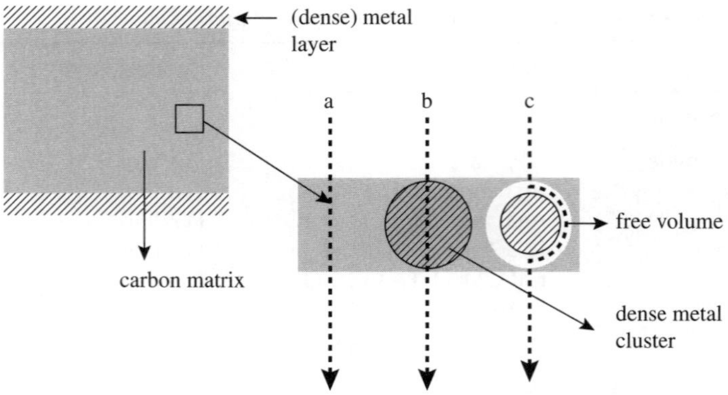

Figure 1.8 *Possible routes for diffusion through the functional part of an Ag-containing carbon membrane. Reprinted from Journal of Membrane Science, Barsema, J. N., et al., Functionalized carbon molecular sieve membranes containing Ag-nanoclusters. Vol 219, 47–57. Copyright (2003) with permission from Elsevier*

Other approaches are based on the addition of nanosized carbon particles, such as carbon black or carbon nanotubes (CNTs), to the carbon matrix (Table 1.2) [90,133,134]. These carbons are also effective filler materials, as discussed above, and they play roles in improving the transport properties of the membranes.

Figure 1.10 shows the H_2 permeability and selectivity of several of the mixed-matrix CMS membranes discussed above. In comparison to the line for the Kapton CMS membrane, which is shown here as the boundary for pure CMS membranes, some of mixed-matrix CMS membranes exhibit significantly superior performances. The techniques introduced here will help to improve the trade-off between the permeability and the selectivity of carbon membranes, as well as providing novel functionalities in membrane reactors.

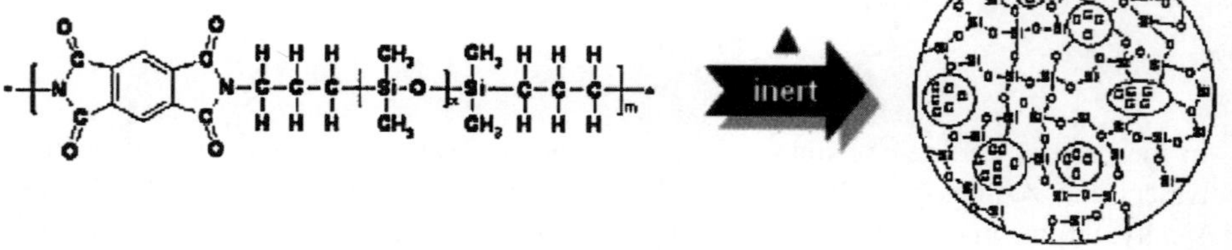

Figure 1.9 *An illustrative process for the conversion of (a) polyimide and (b) poly(imide siloxane) precursors during inert pyrolysis. Reprinted from Journal of Membrane Science, Park, H. B., et al., Pyrolytic carbon membranes containing silica derived from poly(imide siloxane): the effect of siloxane chain length on gas transport behavior and a study on the separation of mixed gases. Vol. 235 87–98. Copyright (2004) with permission from Elsevier*

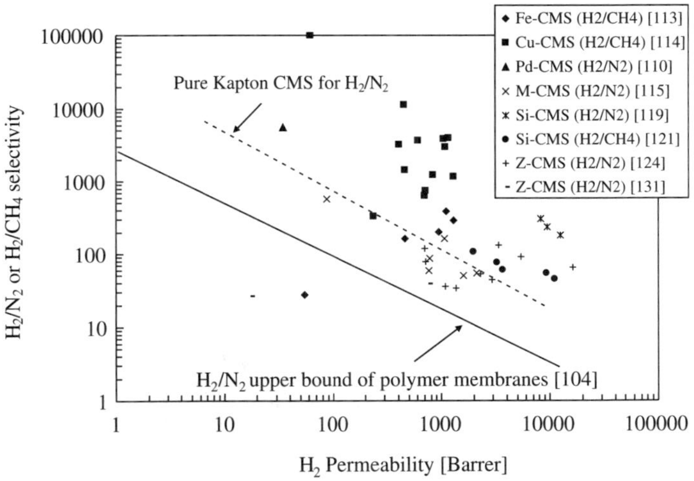

Figure 1.10 *Comparison of the hydrogen permeabilities and permselectivities of various mixed-matrix CMS membranes with the upper limits for polymer membranes*

1.4 Membrane Modules

As can be seen in above sections, there are many reports in the literature regarding research on CMS membranes that focused on preparing membranes with improved gas separation performance. In many of these research efforts, small or laboratory-scale membrane modules were constructed and their performance was tested. Several groups have, however, made attempts to improve methodology for construction of large-scale CMS membrane modules. The combination of the continuous pyrolysation apparatus invented by Kusuki *et al.* [42,43] and the preparation method of flexible pyrolytic membranes found by Okamoto *et al.* [60,61] will lead easy fabrication of large-scale membrane modules with CMS hollow fibres. The selection and modification of porous stainless steel for supported CMS membranes conducted by Foley *et al.* [84–90] also will be a strategy to construct a large module because welding can be used to seal between supports and module shells.

To our knowledge, only two large-scale CMS membrane modules have been reported in the literature. Lagorssee *et al.* [13,78] surveyed these two modules and characterised their membranes. The first one is a module containing CMS hollow fibre membranes that was commercialised by Carbon Membranes Ltd (Israel), a company that has since ceased to trade. Carbon Membranes Ltd prepared *primary* carbon membranes by pyrolysing dense hollow fibres of cellulose, and they used a posterior CVD treatment (using propylene as the source) on the bore side of the fibres to fill the existing pores and produce a thin CVD layer [5,13]. The *secondary* membranes were then oxidised (activated) to create a pore structure of tailored dimensions within the CVD layer. The resulting membranes had an asymmetric structure and improved selectivity and permeability. The highest packing density reported for these modules was $2000\,m^2\,m^{-3}$ (10 000 fibres, $4\,m^2$ per module).

Another large-scale module is manufactured by Blue Membranes GmbH (Germany), who have developed a new concept of CMS flat membrane with a honeycomb configuration [78]. This has a high packing density (up to 2500 $m^2\,m^{-3}$, 10 m^2 per module). The flat support is an industrial-grade paper modified with ceramic fibres. The support is coated with a polymer precursor layer by an imprinting technique. The precursor is a mixed polymer system composed of a phenolic resin (Phenodur PR 515) and an epoxy resin (Beckopox EP309). Figure 1.11

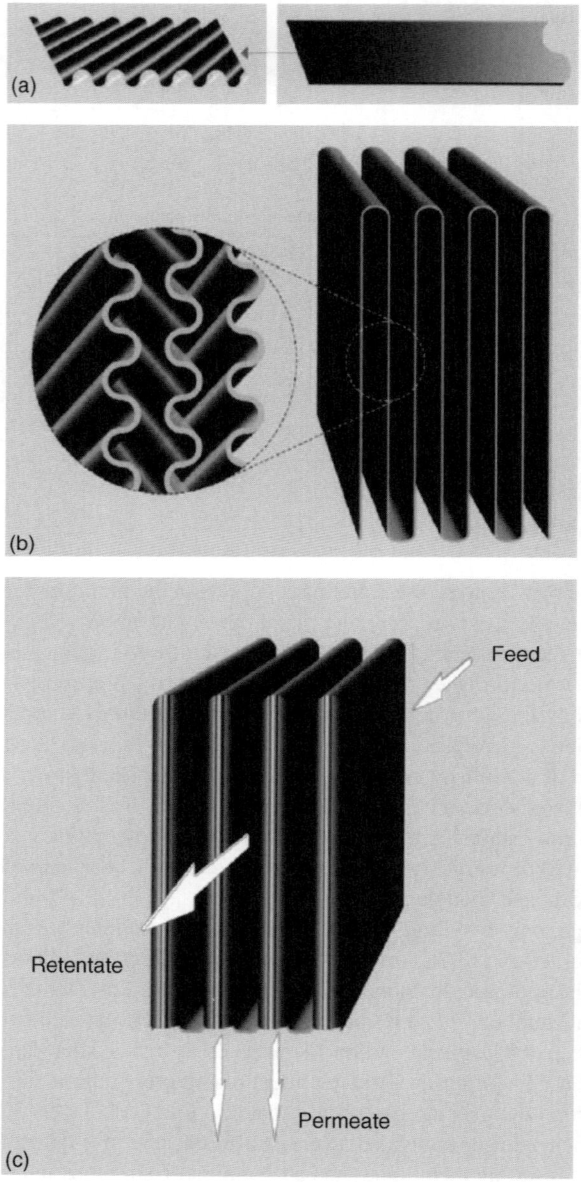

Figure 1.11 *Schematic of the procedure for producing honeycomb CMS membrane modules: (a) flat corrugated precursor; (b) pleated precursor; (c) membrane module. Reprinted from Carbon, Lagorsse, S., et al., Novel carbon molecular sieve honeycomb membrane module: configuration and membrane characterization. Vol. 43, 809–819. Copyright (2005) with permission from Elsevier*

illustrates the procedure for the production of the honeycomb CMS membrane module. First, a stamping process is used to form a corrugated precursor sheet (Figure 1.11a) with a pattern that is oriented diagonally to the length of the sheet. The sheet is then pleated (Figure 1.11b) and the pleated corrugated segments are overlapped to form flow channels. The permeate and feed channels of the module are made independent of one another by sealing at the edges. The module is subsequently pyrolysed under N_2 at 780 °C. Finally, the molecular sieving properties are tuned by a CVD process with a subsequent activation step. This CVD plus activation procedure is similar to the technique adopted by Carbon Membranes Ltd, and is the one of methods described by Feurtes and Menendez [71]. Post-treatment by CVD followed by activation is a possible strategy for adjusting the pore structure of large-scale CMS membrane modules.

1.5 Applications of Membranes in Membrane Reactor Processes

As a result of the molecular sieving effect, CMS membranes show excellent permselectivities for H_2 from mixtures with gases with larger molecules such as CH_4. Carbon membranes can therefore be considered as possible alternatives to Pd-based membranes for use in membrane reactors for the production of H_2. In comparison with Pd-based membranes, carbon membranes show clear advantages in terms of their chemical resistance to sulfur-containing species and their lower production costs. However, carbon membranes are very fragile and cannot be used in oxidizing atmospheres, so only a few applications of carbon membranes in membrane reactors have been reported [135–142]. Table 1.3 summarises recent advances in the use of carbon membrane reactors for H_2-related reactions.

The first experimental carbon membrane reactor was reported in 2000 by Itoh and Haraya [135]. The membrane reactor for dehydrogenation of cyclohexane consisted of a bundle of 20 hollow carbon fibres produced by pyrolysis of hollow polyimide fibres containing 0.5 wt% Pt/Al_2O_3 pellets as a catalyst. Figure 1.12 is a schematic showing the carbon membrane reactor that was developed in this study. Dehydrogenation of cyclohexane to benzene was carried out at 195 °C and atmospheric pressure. The carbon membrane reactor produced a conversion

Table 1.3 *Examples of carbon membrane reactors*

Precursor	Catalyst	Reaction	Temperature	Refs
Polyimide	Pt/Al_2O_3	Dehydrogenation of cyclohexane	195 °C	[135]
PFA	Pt	Hydrogenation of olefins	~175 °C	[136]
Phenolic resin	H_3PO_4	Hydration of propylene	130 °C	[137]
Cellulose	Cr_2O_3/Al_2O_3	Dehydrogenation of isobutane	~500 °C	[138,139]
Phenolic resin + PEG	$Cu/ZnO/Al_2O_3$	Methanol steam reforming	~250 °C	[140]
—[a]	$CuO/ZnO/Al_2O_3$	Water gas shift reaction	250 °C	[141]

[a]Not disclosed

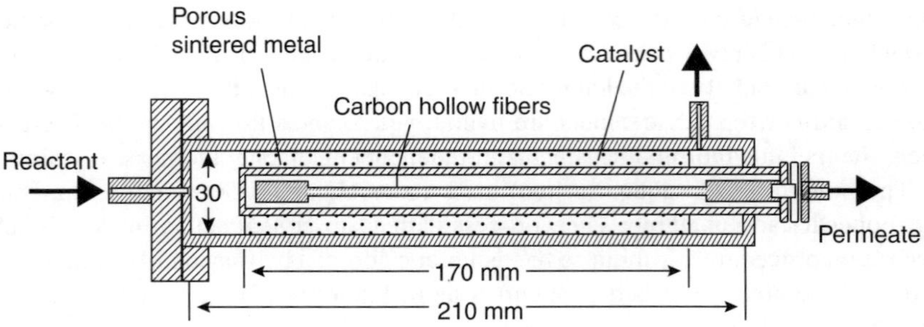

Figure 1.12 *Schematic of a carbon membrane reactor. Reprinted from Catalysis Today, Itoh, N., Haraya, K., A carbon membrane reactor. Vol. 56, 103–111. Copyright (2000) with permission from Elsevier*

that was somewhat better than the equilibrium conversion; this was supported by a mathematical model for a limited range of reaction conditions.

Strano *et al.* [136] synthesised a Pt-loaded nanoporous carbon membrane derived from PFA for the selective hydrogenation of olefins (propylene, 1-butene, isobutylene) to the corresponding alkanes (propane, butane, isobutane) in a catalytic membrane reactor. This nanoporous carbon membrane had a mean pore size of about 0.5 nm and was selective toward alkanes. The Pt metal catalyst, which had a calculated mean particle size of 7.1 nm, was highly dispersed within the membrane layer. The catalytic membranes showed a good selectivity for the hydrogenation of olefins because of the selective transport porosity of the membrane and because of shape-selective catalytic effects.

Lapkin *et al.* [137] developed a porous carbon membrane contactor for the hydration of propylene; the contactor uses phenolic resin-derived carbon membranes with an average pore diameter of 0.7 nm. A liquid-phase aqueous catalyst was fed continuously to the membrane reactor, and gaseous propylene was fed to the opposite side of the membrane. Because the carbon membrane was unaffected by the presence of strong acids such as phosphoric acid, stable operation of the membrane contactor was achieved in alcohol production. Among the advantages of this reactor are recovery of the catalyst and effective separation of the product from the reaction mixture.

Sheintuch and coworkers [138,139] tested a carbon membrane reactor for dehydrogenation of isobutane at high temperatures (450–500 °C) on chromia/alumina catalyst pellets. The membrane module, which consisted of 100 hollow carbon fibres fabricated by Carbon Membranes Ltd., had a hydrogen-to-isobutene permeability ratio in excess of 100. Although the results obtained were superior to those achieved with a corresponding fixed-bed reactor, the authors concluded that the improvement was due to sweeping nitrogen transport and dilution. Simulations of the behavior of the membrane reactor showed a poor agreement with the experiments in this case.

The use of carbon membranes in methanol steam reforming reactors was studied by Zhang *et al.* [140]. The carbon membrane was prepared from a novolac-type phenolic resin and PEG on a green support. Methanol steam reforming was performed at 200–250 °C with a $Cu/ZnO/Al_2O_3$ catalyst, and the carbon membrane reactor was compared with a conventional fixed-bed reactor. A higher conversion of methanol and a lower yield of carbon monoxide were achieved as a result of the enhanced potential of the carbon membrane.

Harale *et al.* [141] used a carbon membrane reactor in the water gas shift reaction for the production of H_2. They proposed a hybrid adsorbent membrane reactor (HAMR) system that combines the reaction and membrane separation steps with adsorption. $CuO/ZnO/Al_2O_3$ was used as the catalyst, and a layered double hydroxide was selected as the CO_2 adsorbent. The carbon membranes, which were 25.4 cm long and had an outside diameter of 0.57 cm, showed high hydrogen permeation fluxes at 250 °C; the methods used for their preparation were, however, not discussed. The experimental results agreed well with model predictions, and the HAMR system has the potential for providing improved yields of H_2 with reduced CO concentrations.

Recently, by using a one-dimensional mathematical model, Sá *et al.* [142] examined the potential advantages of a carbon membrane reactor (CMR) in comparison with a Pd-membrane reactor (Pd-MR) for the production of H_2 by steam reforming of methanol. The study focused on the analysis of the methanol conversion, the selectivity of the H_2/CO reaction, the CO concentration at the permeate side, and the H_2 recovery. It concluded that the CMR gives a higher H_2 recovery than the Pd-MR at high H_2 concentrations, but the Pd-MR is more advantageous at lower H_2 production rates.

For successful use in membrane reactors, carbon membranes require not only a high separation selectivity, but also a high permeability so that the rate of permeation is comparable to the rate of the catalytic reaction. The key challenges in this context are in reducing the thickness of the membrane without introducing defects, and in scaling up of techniques for the fabrication of carbon membranes with large surface areas. Furthermore, for commercial applications, the carbon membrane should be prepared in the form of a honeycomb or hollow fibre module to provide the additional benefits of a low drop in pressure and a high surface to volume ratio. It would also be advantageous to shift the thermodynamic equilibrium by improving the porous structure of the carbon membranes.

1.6 Final Remarks and Conclusions

Recent technological developments in membrane science have permitted the fabrication of a huge variety of microporous carbon membranes, as discussed in this chapter. Although only a few studies have been made on the efficacy of carbon membranes in membrane reactors, those that have been made suggest several applications of carbon membranes in membrane reactors, not only for H_2-related reactions, but also for other reactions involving gases such as H_2O, CO_2, and NH_3, because of the high selectivities and superior chemical resistances of carbon membranes. Recently, CMS membranes have been used as H_2O-selective membranes for the separation of H_2O–alcohol mixtures by pervaporation, and it would be interesting to apply them in membrane reactors for water-related reactions. Scaling up of carbon membranes is one of the most challenging hurdles to surmount for further advancement of their use in membrane reactors. It will be necessary to prepare high-quality carbon membranes with large surface areas in a reliable and cost-effectively manner, and to integrate these membranes into process modules with high-temperature seals. These challenges are daunting, but the use of new polymer precursors, new production methods, and new module-fabrication techniques offers considerable scope for overcoming them. The proper design of reactors with regard to heat and mass transport issues and separation processes is also a significant factor.

References

1. J.E. Koresh, A. Soffer, Molecular sieve carbon permselective membrane. Part 1. Presentation of a new device for gas mixture separation, *Sep. Sci. Technol.*, **18**, 723–734 (1983).
2. M.B. Rao, S. Sircar, Nanoporous carbon membranes for separation of gas mixtures by selective surface flow, *J. Membr. Sci.*, **85**, 253–264 (1993).
3. M.B. Rao, S. Sircar, Nanoporous carbon membrane for gas separation, *Gas Sep. Purif.*, **7**, 279–284 (1993).
4. M.B. Rao, S. Sircar, Performance and pore characterization of nanoporous carbon membrane for gas separation, *J. Membr. Sci.*, **110**, 109–118 (1996).
5. P.J. Williams, W.J. Koros, in: N.N. Li, A.G. Fane, W.S.W. Ho, T. Matsuura (eds), *Advanced Membrane Technology and Applications*, John Wiley & Sons, Inc., New Jersey, pp. 599–632 (2008).
6. H. Kita, in: Y. Yampolskii, I. Pinnau, B.D. Freeman (eds), *Material Science of Membranes for Gas and Vapor Separation*, John Wiley & Sons Ltd, Chichester, pp. 337–354 (2006).
7. A.F. Ismail, K. Li, in: R. Mallada, M. Menendez (eds), *Inorganic Membranes: Synthesis, Characterization and Applications*, Elseevier, Oxford, pp. 81–120 (2008).
8. S.M. Saufi, A.F. Ismail, Fabrication of carbon membranes for gas separation – a review, *Carbon*, **42**, 241–259 (2004).
9. A.F. Ismail, L.I.B. David, A review on the latest development of carbon membranes for gas separation, *J. Membr. Sci.*, **193**, 1–18 (2001).
10. H.O. Pierson, *Handbook of Carbon, Graphite, Diamond, and Fullerenes*. Noyes Publication, Park Ridge, N.J., (1993).
11. A.S. Ghosal, W.J. Koros, Air separation properties of flat sheet homogeneous pyrolytic carbon membranes, *J. Membr. Sci.*, **174**, 177–188 (2000).
12. H. Suda, K. Haraya, Gas permeation through micropores of carbon molecular sieve membranes derived from kapton polyimide, *J. Phys. Chem. B*, **101**, 3988–3994 (1997).
13. S. Lagorsse, F.D. Magalhaes, A. Mendez, Carbon molecular sieve membranes sorption, kinetic and structural characterization, *J. Membr. Sci.*, **241**, 275–287 (2004).
14. L.I.B. David, A.F. Ismail, Influence of the thermostabilization process and soak time during pyrolysis process on the polyacrylonitrile carbon membranes for O_2/N_2 separation, *J. Membr. Sci.*, **213**, 285–291 (2003).
15. C.W. Jones, W.J. Koros, Carbon molecular sieve gas separation membranes-I. Preparation and characterization based on polyimide precursors, *Carbon*, **32**, 1419–1425 (1994).
16. C.W. Jones, W.J. Koros, Carbon molecular sieve gas separation membranes-II. Regeneration following organic exposure, *Carbon*, **32**, 1427–1432 (1994).
17. C.W. Jones, W.J. Koros, Characterization of ultramicroporous carbon membranes with humidified feeds, *Ind. Eng. Chem. Res.*, **34**, 158–163 (1995).
18. C.W. Jones, W.J. Koros, Carbon composite membranes: a solution to adverse humidity effects, *Ind. Eng. Chem. Res.*, **34**, 164–167 (1995).
19. V.C. Geiszler, W.J. Koros, Effect of polyimide pyrolysis conditions on carbon molecular sieve membrane properties, *Ind. Eng. Chem. Res.*, **35**, 2999–3003 (1996).
20. D.Q. Vu, W.J. Koros, S.J. Miller, High pressure CO_2/CH_4 separation using carbon molecular sieve hollow fibre membranes, *Ind. Eng. Chem. Res.*, **41**, 367–380 (2002).
21. D.Q. Vu, W.J. Koros, S.J. Miller, Effect of condensable impurities in CO_2/CH_4 gas feeds on carbon molecular sieve hollow fibre membranes, *Ind. Eng. Chem. Res.*, **42**, 1064–1075 (2003).
22. H. Suda, K. Haraya, Molecular sieving effect of carbonized kapton polyimide membrane, *J. Chem. Soc. Chem. Commun.*, **1995**, 1179–1180 (1995).

23. H. Suda, K. Haraya, Alkene/alkene permselectivities of a carbon molecular sieve membrane, *Chem. Commun.*, **1997**, 93–94 (1997).
24. K. Haraya, H. Suda, H. Yanagishita, S. Matsuda, Asymmetric capillary membranes of a carbon molecular sieve, *J. Chem. Commun.*, **17**, 1781–1782 (1995).
25. J. Petersen, M. Matsuda, K. Haraya, Capillary carbon molecular sieve membranes derived from kapton for high temperature gas separation, *J. Membr. Sci.*, **131**, 85–94 (1997).
26. M. Ogawa, Y. Nakano, Gas permeation through carbonized hollow fibre membranes prepared by gel modification of polyamic acid, *J. Membr. Sci.*, **162**, 189–198 (1999).
27. M. Ogawa, Y. Nakano, Separation of CO_2/CH_4 mixture through carbonized membrane prepared by gel modification, *J. Membr. Sci.*, **173**, 123–132 (2000).
28. A.B. Fuertes, D.M. Nevskaia, T.A. Centeno, Carbon composite membranes from matrimid and kapton polyimide, *Microporous Mesoporous Mater.*, **33**, 115–125 (1999).
29. H. Hatori, H. Takagi, Y. Yamada, Gas separation properties of molecular sieving carbon membranes with nanopore channels, *Carbon*, **42**, 1169–1173 (2004).
30. A.C. Lua, J. Su, Effects of carbonization on pore evolution and gas permeation properties of carbon membranes from Kapton polyimide, *Carbon*, **44**, 2964–2972 (2006).
31. J. Su, A.C. Lua, Effects of carbonization atmosphere on the structural characteristics and transport properties of carbon membranes prepared from Kapton polyimide, *J. Membr. Sci.*, **305**, 263–270 (2007).
32. A.B. Fuertes, D.M. Nevskaia, T.A. Centeno, Carbon composite membranes from matrimid and kapton polyimides for gas separation, *Microporous Mesoporous Mater.*, **33**, 115–125 (1999).
33. K.M. Steel, W.J. Koros, Investigation of porosity of carbon materials and related effects on gas separation properties, *Carbon*, **41**, 253–266 (2003).
34. J.N. Barsema, N.F.A. van der Vegt, G.H. Koops, M. Wessling, Carbon molecular sieve membranes prepared from porous fibre precursor, *J. Membr. Sci.*, **205**, 239–246 (2002).
35. P.S. Tin, T.S. Chung, Y. Liu, R. Wang, Separation of CO_2/CH_4 through carbon molecular sieve membranes derived from P84 polyimide, *Carbon*, **42**, 3123–3131 (2004).
36. P.S. Tin, T.S. Chung, S. Kawi, M.D. Guiver, Novel approaches to fabricate carbon molecular sieve membranes based on chemical modified and solvent treated polyimides, *Microporous Mesoporous Mater.*, **73**, 151–160 (2004).
37. P.S. Tin, T.S. Chung, A.J. Hill, Advanced fabrication of carbon molecular sieve membranes by nonsolvent pretreatment of precursor polymers, *Ind. Eng. Chem. Res.*, **43**, 6476–6483 (2004).
38. Y. Xiao, Y. Dai, T.S. Chung, M.D. Guiver, Effects of brominating matrimid polyimide on the physical and gas transport properties of derived carbon membranes, *Macromolecules*, **38**, 10042–10049 (2005).
39. E.P. Favvas, G.C. Kapantaidakis, J.W. Nolan, A.C. Mitropoulos, N.K. Kanellopoulos, Preparation, characterization and gas permeation properties of carbon hollow fibre membranes based on Matrimid 5218 precursor, *J. Mat. Proc. Technol.*, **186**, 102–110 (2007).
40. A.B. Fuertes, T.A. Centeno, Carbon molecular sieve membranes from polyetherimide, *Microporous Mesoporous Mater.*, **26**, 23–26 (1998).
41. M.G. Sedigh, L. Xu, T.T. Tsotsis, M. Sahimi, Transport and morphological characteristics of polyetherimide-based carbon molecular sieve membranes, *Ind. Eng. Chem. Res.*, **38**, 3367–3380 (1999).
42. Y. Kusuki, H. Shimazaki, N. Tanihara, S. Nakanishi, T. Yoshinaga, Gas permeation properties and characterization of asymmetric carbon membranes prepared by pyrolyzing asymmetric polyimide hollow fibre membrane, *J. Membr. Sci.*, **134**, 245–253 (1997).
43. N. Tanihara, H. Shimazaki, Y. Hirayama, S. Nakanishi, T. Yoshinaga, Y. Kusuki, Gas permeation properties of asymmetric carbon hollow fibre membranes prepared from asymmetric hollow fibre, *J. Membr. Sci.*, **160**, 179–186 (1999).

44. K. Okamoto, S. Kawamura, M. Yoshino, H. Kita, Y. Hirayama, N. Tanihara, *et al.*, Olefin/paraffin separation through carbonized membranes derived from an asymmetric polyimide hollow fibre membrane, *Ind. Eng. Chem. Res.*, **38**, 4424–4432 (1999).
45. M. Yoshino, S. Nakamura, H. Kita, K. Okamoto, N. Tanihara, Y. Kusuki, Olefin/paraffin separation performance of carbonized membranes derived from an asymmetric hollow fibre membrane of 6FDA/BPDA-DDBT copolyimide, *J. Membr. Sci.*, **215**, 169–183 (2003).
46. J. Hayashi, M. Yamamoto, K. Kusakabe, S. Morooka, Simultaneous improvement of permeance and permselectivity of 3,3′,4,4′-biphenyltetracarboxylic dianhydride-4,4′-oxydianiline polyimide membrane by carbonization, *Ind. Eng. Chem. Res.*, **34**, 4364–4370 (1995).
47. J. Hayashi, H. Mizuta, M. Yamamoto, K. Kusakabe, S. Morooka, Separation of ethane/ethylene and propane/propylene system with a carbonized BPDA-pp′ODA polyimide, *Ind. Eng. Chem. Res.*, **35**, 4176–4181 (1996).
48. J. Hayashi, H. Mizuta, M. Yamamoto, K. Kusakabe, S. Morooka, Pore size control of carbonized BPDA-pp′ODA polyimide membrane by chemical vapor deposition of carbon, *J. Membr. Sci.*, **124**, 243–251 (1997).
49. J. Hayashi, M. Yamamoto, K. Kusakabe, S. Morooka, Effect of oxidation on gas permeation of carbon molecular sieving membranes based on BPDA-pp′ODA polyimide, *Ind. Eng. Chem. Res.*, **36**, 2134–2140 (1997).
50. K. Kusakabe, M. Yamamoto, S. Morooka, Gas permeation and micropore structure of carbon molecular sieving membranes modified by oxidation, *J. Membr. Sci.*, **149**, 59–67 (1998).
51. A.B. Fuertes, T.A. Centeno, Preparation of supported carbon molecular sieve membrane, *Carbon*, **37**, 679–684 (1999).
52. Y.K. Kim, H.B. Park, Y.M. Lee, Preparation and characterization of carbon molecular sieve membranes derived from BTDA-ODA polyimide and their separation properties, *J. Membr. Sci.*, **255**, 265–273 (2005).
53. Y.K. Kim, H.B. Park, Y.M. Lee, Carbon molecular sieve membranes derived from thermally labile polymer containing blend polymers and their gas separation properties, *J. Membr. Sci.*, **243**, 9–17 (2004).
54. Y.K. Kim, H.B. Park, Y.M. Lee, Gas separation properties of carbon molecular sieve membranes derived from polyimide/polyvinylpyrroldone blends: Effect of the molecular weight of polyvinylpyrrolidone, *J. Membr. Sci.*, **251**, 159–167 (2005).
55. H.B. Park, T.K. Kim, J.M. Lee, S.Y. Lee, Y.M. Lee, Relationship between chemical structure of aromatic polyimides and gas permeation properties of their carbon molecular sieve membranes, *J. Membr. Sci.*, **229**, 117–127 (2004).
56. Y.K. Kim, J.M. Lee, H.B. Park, Y.M. Lee, The gas separation properties of carbon molecular sieve membranes derived from polyimides having carboxylic acid groups, *J. Membr. Sci.*, **235**, 139–146 (2004).
57. H. Kita, M. Yoshino, K. Tanaka, K. Okamoto, Gas permselectivity of carbonized polypyrrolone membrane, *J. Chem. Soc. Chem. Commun.*, 1997, 1051–1052 (1997).
58. Y. Xiao, T.-S. Chung, M.L. Chng, S. Tamai, A. Yamaguchi, Structure and properties relationships for aromatic polyimides and their derived carbon membranes: experimental and simulation approaches, *J. Phys. Chem. B*, **109**, 18741–18748 (2005).
59. Y. Xiao, T.-S. Chung, H. M. Guana, M.D. Guiver, Synthesis, cross-linking and carbonization of co-polyimides containing internal acetylene units for gas separation, *J. Membr. Sci.*, **302**, 254–264 (2007).
60. W. Zhou, T. Watari, H. Kita, K. Okamoto, Gas permeation properties of flexible pyrolytic membranes from sulfonated polyimides, *Chem. Lett.*, **2002**, 534 (2002).

61. M. N. Islam, W. Zhou, T. Honda, K. Tanaka, H. Kita, K. Okamoto, Preparation and gas separation performance of flexible pyrolytic membranes by low-temperature pyrolysis of sulfonated polyimides, *J. Membr. Sci.*, **261**, 17–26 (2005).
62. J.N. Barsema, S.D. Klijnsta, N.F.A. van der Vegt, G. H. Koops, M. Wessling, Intermediate polymer to carbon gas separation membranes based for Matrimid PI, *J. Membr. Sci.*, **238**, 93–102 (2004).
63. L. Shao, T.S. Chung, K.P. Pramoda, The evolution of physiochemical and transport properties of 6FDA-durene toward carbon membranes; from polymer, intermediate to carbon, *Microporous Mesoporous Mater.*, **84**, 59–68 (2005).
64. L. Shao, T.S. Chung, G. Wensley, S.H. Goh, K.P. Pramoda, Casting solvent effects on morphologies, gas transport properties of a novel 6FDA/PMDA-TMMDA copolyimide membrane and its derived carbon membranes, *J. Membr. Sci.*, **244**, 77–87 (2004).
65. W. Shusen, Z. Meiyun, W. Zhizhong, Asymmetric molecular sieve carbon membranes, *J. Membr. Sci.*, **109**, 267–270 (1996).
66. H. Kita, H. Maeda, K. Tanaka, K. Okamoto, Carbon molecular sieve membrane prepared from phenolic resin, *Chem. Lett.*, **1997**, 179–180 (1997).
67. W. Zhou, M. Yoshino, H. Kita, K. Okamoto, Carbon molecular sieve membranes derived from phenolic resin with a pendant sulfonic acid group, *Ind. Eng. Chem. Res.*, **40**, 4801–4807 (2001).
68. W. Zhou, M. Yoshino, H. Kita, K. Okamoto, Preparation and gas permeation properties of carbon molecular sieve membranes based on sulfonated phenolic resin, *J. Membr. Sci.*, **217**, 55–67 (2003).
69. T.A. Centeno, A.B. Fuertes, Supported carbon molecular sieve membranes based on phenolic resin, *J. Membr. Sci.*, **160**, 201–211 (1999).
70. T.A. Centeno, A.B. Fuertes, Carbon molecular sieve membranes derived from a phenolic resin supported on porous ceramic tubes, *Sep. Purif. Tech.*, **25**, 379–384 (2001).
71. A.B. Fuertes, I. Menendez, Separation of hydrocarbon gas mixtures using phenolic resin-based carbon membranes, *Sep. Purif. Tech.*, **28**, 29–41 (2002).
72. A.B. Fuertes, Adsorption-selective carbon membrane for gas separation, *J. Membr. Sci.*, **177**, 9–16 (2000).
73. A.B. Fuertes, Effect of air oxidation on gas separation properties of adsorption-selective carbon membranes, *Carbon*, **39**, 697–706 (2001).
74. T.A. Centeno, J.L. Vilas, A.B. Fuertes, Effects of phenolic resin pyrolysis conditions on carbon membrane performance for gas separation, *J. Membr. Sci.*, **228**, 45–54 (2004).
75. W. Wei, H. Hu, L. You, G. Chen, Preparation of carbon molecular sieve membrane from phenol-formaldehyde novolac resin, *Carbon*, **40**, 465–467 (2002).
76. W. Wei, G. Qin, H. Hu, L. You, G. Chen, Preparation of supported carbon molecular sieve membrane from novolac phenol-formaldehyde resin, *J. Membr. Sci.*, **303**, 80–85 (2007).
77. X. Zhang, H. Hu, Y. Zhu, S. Zhu, Effect of carbon molecular sieve on phenol formaldehyde novolac resin based carbon membranes, *Sep. Purif. Technol.*, **52**, 261–265 (2006).
78. S. Lagorsse, A. Leite, F.D. Magalhaes, N. Bischofberger, J. Rathenow, A. Mendes, Nobel carbon molecular sieve honeycomb membrane module: configuration and membrane characterization, *Carbon*, **43**, 809–819 (2005).
79. X. Zhang, H. Hu, Y. Zhu, S. Zhu, Carbon molecular sieve membranes derived from phenol formaldehyde novolac resin blended with poly(ethylene glycol), *J. Membr. Sci.*, **289**, 86–91 (2007).
80. N. Nishiyama, Y.-R. Dong, T. Zheng, Y. Egashira, K. Ueyama, Tertiary amine-mediated synthesis of microporous carbon membranes, *J. Membr. Sci.*, **280**, 603–609 (2006).

81. Y.D. Chen, R.T. Yang, Preparation of carbon molecular sieve membrane and diffusion of binary mixtures in the membrane, *Ind. Eng. Chem. Res.*, **33**, 3146–3153 (1994).
82. M.G. Sedigh, W.J. Onstot, L. Xu, W.L. Peng, T.T. Tsotsis, M. Sahimi, Experiments and simulation of transport and separation of gas mixtures in carbon molecular sieves membranes, *J. Phys. Chem. A*, **102**, 8580–8589 (1998).
83. H. Wang, L. Zhang, G.R. Gavalas, Preparation of supported carbon membranes from furfuryl alcohol by vapor deposition polymerization, *J. Membr. Sci.*, **177**, 25–31 (2000).
84. M. Acharya, B.A. Raich, H.C. Foley, M.P. Harold, J.J. Lerou, Metal-supported carbogenic molecular sieve membranes: synthesis and applications, *Ind. Eng. Chem. Res.*, **36**, 2924–2930 (1997).
85. M. Acharya, H.C. Foley, Spray-coating of nanoporous carbon membranes for air separation, *J. Membr. Sci.*, **161**, 1–5 (1999).
86. M.B. Shiflett, H.C. Foley, Ultrasonic deposition of high-selectivity nanoporous carbon membranes, *Science*, **285**, 1902–1905 (1999).
87. M.B. Shiflett, H.C. Foley, On the preparation of supported nanoporous carbon membranes, *J. Membr. Sci.*, **179**, 275–282 (2000).
88. M.B. Shiflett, H.C. Foley, Reproducible production of nanoporous carbon membranes, *Carbon*, **39**, 1421–1446 (2001).
89. R. Rajagopalan, A. Merritt, A. Tseytlin, H.C. Foley, Modification of macroporous stainless steel supports with silica nanoparticles for size selective carbon membranes with improved flux, *Carbon*, **44**, 2051–2058 (2006).
90. A. Merritt, R. Rajagopalan, H.C. Foley, High performance nanoporous carbon membranes for air separation, *Carbon*, **45**, 1267–1278 (2007).
91. C.J. Anderson, S.J. Pas, G. Arora, S.E. Kentish, A.J. Hill, S.I. Sandler, G.W. Stevens, Effect of pyrolysis temperature and operating temperature on the performance of nanoporous carbon membranes, *J. Membr. Sci.*, **322**, 19–27 (2008).
92. C. Song, T. Wang, X. Wang, J. Qiu, Y. Cao, Preparation and gas separation properties of poly (furfuryl alcohol)-based C/CMS composite membranes, *Sep. Purif. Technol.*, **58**, 412–418 (2008).
93. M. Anand, M. Langsam, M.B. Rao, S. Sircar, Multicomponent gas separation by selective surface flow (SSF) and poly-trimethylpropyne (PTMSP) membranes, *J. Membr. Sci.*, **123**, 17–25 (1997).
94. S. Sircar, M.B. Rao, C.M.A. Thaeron, Selective surface flow membrane for gas separation, *Sep. Sci. Technol.*, **34** (10), 2081–2093 (1999).
95. S. Sircar, W.E. Waldron, M.B. Rao, M. Anand, Hydrogen production by hybrid SMR-PSA-SSF membrane system, *Sep. Purif. Technol.*, **17**, 11–20 (1999).
96. T.A. Centeno, A.B. Fuertes, Carbon molecular sieve gas separation membranes based on poly(vinylidene chloride-co-vinyl chloride), *Carbon*, **38**, 1067–1073 (2000).
97. M. Yoshimune, I. Fujiwara, H. Suda, K. Haraya, Novel carbon molecular sieve membranes derived from poly(phenylene oxide) and its derivatives for gas separation, *Chem. Lett.*, **34**, 958–959 (2005).
98. M. Yoshimune, I. Fujiwara, K. Haraya, Carbon molecular sieve membranes derived from trimethylsilyl substituted poly(phenylene oxide) for gas separation, *Carbon*, **45**, 553–560 (2007).
99. H.J. Lee, M. Yoshimune, H. Suda, K. Haraya, Gas permeation properties of poly(2,6-dimethyl-1,4-phenylene oxide) (PPO) derived carbon membranes prepared on a tubular ceramic support, *J. Membr. Sci.*, **279**, 372–379 (2006).
100. B. Zhang, T. Wang, S. Zhang, J. Qiu, X. Jian, Preparation and characterization of carbon membranes made from poly(phthalazinone ether sulfone ketone), *Carbon*, **44**, 2764–2769 (2006).

101. B. Zhang, T. Wang, S. Liu, S. Zhang, J. Qiu, Structure and morphology of microporous carbon membrane materials derived from poly(phthalazinone ether sulfone ketone), *Microporous Mesoporous Mater.*, **96**, 79–83 (2006).
102. T. Wang, B. Zhang J. Qiu, Y. Wu, S. Zhang, Y. Cao, Effects of sulfone/ketone in poly(phthalazinone ether sulfone ketone) on the gas permeation of their derived carbon membranes, *J. Membr. Sci.*, **330**, 319–325 (2009).
103. M.L. Chung, Y. Xiao, T.S. Chung, M. Toriida, S. Tamai, Enhanced propylene/propane separation by carbonaceous membrane derived from poly(aryl erher ketone)/2,6-bis(4-azidobenzylidene)-4-methyl-cyclohexanone interpenetrating network, *Carbon*, **47**, 1857–1866 (2009).
104. L.M. Robeson, The upper bound revisited, *J. Membr. Sci.*, **320**, 390–400 (2008).
105. M. Jia, K.V. Peinemann, R.D. Behling, Molecular sieving effect of the zeolite-filled silicone rubber membranes in gas permeation, *J. Membr. Sci.*, **57**, 289–292 (1991).
106. M. Moaddeb, W.J. Koros, Gas transport properties of thin polymeric membranes in the presence of silicon dioxide particles, *J. Membr. Sci.*, **125**, 143–163 (1997).
107. Y.K. Kim, H.B. Park, Y.M. Lee, Carbon molecular sieve membranes derived from metal-substituted sulfonated polyimide and their gas separation properties, *J. Membr. Sci.*, **226**, 145–158 (2003).
108. J.N. Barsema, J. Balster, V. Jordan, N.F.A. van der Vegt M. Wessling, Functionalized carbon molecular sieve membranes containing Ag-nanoclusters, *J. Membr. Sci.*, **219**, 47–57 (2003).
109. J.N. Barsema, N.F.A. van der Vegt, G.H. Koops, M. Wessling, Ag-functionalized carbon molecular-sieve membranes based on polyelectrolyte/polyimide blend precursors, *Adv. Funct. Mat.*, **15**, 69–75 (2005).
110. S. Yoda, A. Hasegawa, H. Suda, Y. Uchimaru, K. Haraya, T. Tsuji, K. Otake, Preparation of a platinum and palladium/polyimide nanocomposite film as a precursor of metal-doped carbon molecular sieve membrane via supercritical impregnation, *Chem. Mater.*, **16**, 2363–2368 (2004).
111. H. Suda, S. Yoda, A. Hasegawa, T. Tsuji, K. Otake, K. Haraya, Gas permeation properties of carbon molecular sieve membranes dispersed with palladium nano particles via supercritical CO_2 impregnation, *Desalination*, **193**, 211–214 (2006).
112. J.A. Lie, M.B. Hägg, Carbon membranes from cellulose and metal loaded cellulose, *Carbon*, **43**, 2600–2607 (2005).
113. J.A. Lie, M.B. Hägg, Carbon membranes from cellulose: synthesis, performance and regeneration, *J. Membr. Sci.*, **284**, 79–86 (2006).
114. D. Grainger, M.B. Hägg, Evaluation of cellulose-derived carbon molecular sieve membranes for hydrogen separation from light hydrocarbons, *J. Membr. Sci.*, **306**, 307–317 (2007).
115. M. Yoshimune, I. Fujiwara, H. Suda, K. Haraya, Gas transport properties of carbon molecular sieve membranes derived from metal containing sulfonated poly(phenylene oxide), *Desalination*, **193**, 66–72 (2006).
116. L. Zhang, X. Chen, C. Zeng, N. Xu, Preparation and gas separation of nano-sized nickel particle-filled carbon membranes, *J. Membr. Sci.*, **281**, 429–434 (2006).
117. H.B. Park, I.Y. Suh, Y.M. Lee, Novel pyrolytic carbon membranes containing silica: preparation and characterization, *Chem. Mater.*, **14**, 3034–3046 (2002).
118. H.B. Park, Y.M. Lee, Pyrolytic carbon-silica membrane: a promising membrane material for improved gas separation, *J. Membr. Sci.*, **213**, 263–272 (2003).
119. H.B. Park, C.H. Jung, Y.K. Kim, S.Y. Nam, S.Y. Lee, Y.M. Lee, Pyrolytic carbon membranes containing silica derived from poly(imide siloxane): the effect of siloxane chain length on gas transport behavior and a study on the separation of mixed gases, *J. Membr. Sci.*, **235**, 87–98 (2004).
120. H.B. Park, S.Y. Lee, Y.M. Lee, Pyrolytic carbon membranes containing silica: morphological approach on gas transport behavior, *J. Mol. Struct.*, **739**, 179–190 (2005).

121. H.B. Park, Y.M. Lee, Fabrication and characterization of nanoporous carbon/silica membranes, *Adv. Mater.*, **17**, 477–483 (2005).
122. S.H. Han, G.W. Kim, C.H. Jung, Y.M. Lee, Control of pore characteristics in carbon molecular sieve membranes (CMSM) using organic/inorganic hybrid materials, *Desalination*, **233**, 88–95 (2008).
123. P.S. Tin, T.S. Chung, L. Jiang, S. Kulprathipanja, Carbon-zeolite composite membranes for gas separation, *Carbon*, **43**, 2025–2027 (2005).
124. Q. Liu, T. Wang, C. Liang, B. Zhang, S. Liu, Y. Cao, and J. Qiu, Zeolite married to carbon: a new family of membrane materials with excellent gas separation performance, *Chem. Mater.*, **18**, 6283–6288 (2006).
125. Z. Zhou, J. Yang, Y. Zhang, L. Chang, W. Sun, J. Wang, NaA zeolite/carbon nanocomposite thin films with high permeance for CO_2/N_2 separation, *Sep. Purif. Technol.*, **55**, 392–395 (2007).
126. Z. Zhou, J. Yang, L. Chang, Y. Zhang, W. Sun, J. Wang, Novel preparation of NaA/carbon nanocomposite thin films with high permeance for CO_2/CH_4 separation, *Chin. Chem. Lett.*, **18**, 455–457 (2007).
127. L.Y. Jiang, T.S. Chung, R. Rajagopalan, Dual-layer hollow carbon fibre membranes for gas separation consisting of carbon and mixed matrix layers, *Carbon*, **45**, 166–172 (2007).
128. Y. Li, T.S. Chung, Exploratory development of dual-layer carbon-zeolite nanocomposite hollow fibre membranes with high performance for oxygen enrichment and natural gas separation, *Microporous Mesoporous Mater.*, **113**, 315–324 (2008).
129. C. Kong, J. Wang, J. Yang, J. Lu, A. Wang, Q. Zhao, Thin carbon-zeolite composite membrane prepared on ceramic tube filter by vacuum slip casting for oxygen/nitrogen separation, *Carbon*, **45**, 2848–2850 (2007).
130. C. Zeng, L. Zhang, X. Cheng, H. Wang, N. Xu, Preparation and gas permeation of nano-sized zeolite NaA-filled carbon membranes, *Sep. Purif. Technol.*, **63**, 628–633 (2008).
131. B. Zhang, T. Wang, Y. Wu, Q. Liu, S. Liu, S. Zhang, J. Qiu, Preparation and gas permeation of composite carbon membranes from poly(phthalazinone ether sulfone ketone), *Sep. Purif. Technol.*, **60**, 259–263 (2008).
132. Q. Liu, T. Wang, H. Guo, C. Liang, S. Liu, Z. Zhang, Y. Cao, D.S. Su, J. Qiu, Controlled synthesis of high performance carbon/zeolite T composite membrane materials for gas separation, *Microporous Mesoporous Mater.*, **120**, 460–466 (2009).
133. P.S. Rao, M.Y. Wey, H.H. Tseng, I.A. Kumar, T.H. Weng, A comparison of carbon/nanotube molecular sieve membranes with polymer blend carbon molecular sieve membranes for the gas permeation application, *Microporous Mesoporous Mater.*, **113**, 499–510 (2008).
134. H.H. Tseng, I.A. Kumar, T.H. Weng, C.Y. Lu, M.Y. Wey, Preparation and characterization of carbon molecular sieve membranes for gas separation – the effect of incorporated multi-wall carbon nanotubes, *Desalination*, **240**, 40–45 (2009).
135. N. Itoh, K. Haraya, A carbon membrane reactor, *Catal. Today*, **56**, 103–111 (2000).
136. M.S. Strano, H.C. Foley, Synthesis and characterization of catalytic nanoporous carbon membranes, *AIChE. J.*, **47**, 66–78 (2001).
137. A.A. Lapkin, S.R. Tennison, W.J. Thomas, A porous carbon membrane reactor for the homogeneous catalytic hydration of propene, *Chem. Eng. Sci.*, **57**, 2357–2369 (2002).
138. G. Sznejer, M. Sheintuch, Application of a carbon membrane reactor for dehydrogenation reactions, *Chem. Eng. Sci.*, **59**, 2013–2021 (2004).
139. M. Sheintuch, I. Efremenko, Analysis of a carbon membrane reactor: from atomistic simulations of single-file diffusion to reactor design, *Chem. Eng. Sci.*, **59**, 4739–4746 (2004).

140. X. Zhang, H. Hu, Y. Zhu, S. Zhu, Methanol steam reforming to hydrogen in a carbon membrane reactor system, *Ind. Eng. Chem. Res.*, **45**, 7997–8001 (2006).
141. A. Harale, H.T. Hwang, P.K.T. Liu, M. Sahimi, T.T. Tsotsis, Experimental studies of a hybrid adsorbent-membrane reactor (HAMR) system for hydrogen production, *Chem. Eng. Sci.*, **62**, 4126–4137 (2007).
142. S. Sá, H. Silva, J.M. Sousa, A. Mendes, Hydrogen production by methanol steam reforming in a membrane reactor: palladium vs carbon molecular sieve membranes, *J. Membr. Sci.*, **339**, 160–170 (2009).

2

Metallic Membranes by Wire Arc Spraying: Preparation, Characterisation and Applications

Sayed Siavash Madaeni and Parisa Daraei

Membrane Research Center, Department of Chemical Engineering, Razi University, Kermanshah, Iran

2.1 Introduction

Inorganic membranes are increasingly being utilised to separate gas mixtures or to improve catalytic reaction rates at high temperatures. Gas separation at relatively high temperatures is very attractive since many petrochemical processes could be enhanced by gas separation with membranes [1]. Among the first groups of dense inorganic membranes studied extensively in the past decade are metallic membranes [2]. These membranes can be supported by other metals, ceramics and polymers. Moreover they may include a support free thin separating layer.

Metal membranes appear to be suitable for water clarification due to their high treatment capability for removal of microorganisms and particulates [3]. Leiknes *et al.* investigated the feasibility and potential of using inorganic metal microfilter in a submerged membrane configuration with coagulation pretreatment for drinking water production [4].

Metallic membranes can be prepared using various methods. Almost all preparation techniques include specified temperature and pressure conditions, thermal treatments and various modifications on support or membrane [1, 5]. The methods such as electroplating, electroless plating, chemical vapour deposition (CVD), physical vapour deposition (PVD), wet powder spraying (WPS), rolling, vacuum evaporation and chemical deposition are used for preparing

Membranes for Membrane Reactors: Preparation, Optimization and Selection, Edited by Angelo Basile and Fausto Gallucci

the support and/or dense metallic membranes. Only the sintering technique leads to preparation of the porous metallic membrane.

Recently wire arc spraying which is a common technique for production of protective layers on industrial components has been applied to prepare porous stainless steel membrane for various applications including water purification [6]. In this technique stainless steel is melted and sprayed toward a solid substrate to form a porous coating. The deposited metal is immediately solidified and forms a porous metallic membrane.

2.2 Thermal Spraying

2.2.1 Definition and Types

Thermal spraying is a group of coating processes in which finely sprayed metallic or nonmetallic materials are deposited in a molten or semi-molten condition to form a layer on a substrate. The coating may be in the form of powder, ceramic rod, wire, or molten materials [7, 8].

In the early 1900s Schoop and his coworkers developed equipment and techniques for producing coatings using molten and powder metals. Several years later, in 1912, the first instrument for the spraying of solid metal in wire form was produced. This simple device was based on the principle that if a wire rod was fed into an intense flame, it would melt and, if the flame was surrounded by a stream of compressed gas, the molten metal would become atomised and readily propelled onto a surface to form a coating layer [7].

Nowadays, this process contains some various plasma spraying methods like powder, wire and molten metal flame spraying, detonation flame spraying, high velocity oxy/fuel flame spraying (HVOF) and different electrical techniques like nontransferred plasma arc spraying, radio frequency (RF) plasma spraying and wire arc spraying [7, 9]. Each process exhibits different cost, materials flexibility and coating performance capabilities. As a surfacing technology, thermal spray is compatible with most materials, is affordable, rapid and can produce relatively thick (0.01–0.1 inches, or 0.25–2.54 mm) protective coatings [10].

2.2.2 Applications

Thermal spray coatings have been in use for over 100 years. They are of high quality to protect aircraft engine components and biomedical prostheses. Many industrial components can be protected and have their lives extended or enhanced using thermal spray. Hard, wear-resistant, coatings are used in automotive engines, insulators are sprayed, chemical reactors are repaired against corrosion, pumps are restored, bridges are coated and aircraft bodies are protected [10]. Therefore, thermal spray is commonly applied for depositing a protective layer over the surfaces of industrial components for increasing the resistance against corrosion, abrasive wear, and high thermal shock [7, 10–12].

2.2.3 Wire Arc Spraying

Electric arc spraying is a thermal spray process in which an arc is formed between two consumable electrodes of a coating material. Compressed gas is used to atomise and propel the material to the substrate [7]. In the electric arc spraying, a spray of molten metallic droplets is created by the impingement of a fast moving gas upon the continuously melting tips of consumable wires fed into a DC arc formed between the wires [7, 13]. In fact, the molten end of the wires are dispersed and accelerated by a gas stream (air or inert gas) [7, 13, 14]. The temperature in the arc can reach 5000 °C. The particle velocity lies in the range 100–300 $m\,s^{-1}$. The process is simple and can be operated either manually or in an automated manner. Hence, this technique is an inexpensive process to produce a porous structure in a short time. It is possible to spray a wide range of metals, alloys and metal matrix composites (MMCs) [14, 15]. This process differs from other thermal spray methods. There is no external heat source as in other spray processes. Heating and melting occur when two electrically opposed charged wires, comprising the spray material, are fed together in such a manner that a controlled arc occurs at their intersection. Electric arc spraying has the advantage of not requiring the use of oxygen and a combustible gas. This has the ability to process metals at high spray rates; and is less expensive compared to either plasma or wire flame spraying [7]. The technique is versatile and reliable, may readily be automated with easy operation and portable instrumentation. In electric arc spray solely electrically conductive wires can be sprayed [14, 15].This is the major disadvantage of the process.

Practically wire arc spraying is carried out using a simple gun (Figure 2.1). Two guides direct the wires to an arcing point. Behind this point a nozzle directs a stream of high-pressure gas or air onto the arcing point where it atomises the molten metal and carries it to the substrate (Figure 2.2) [7]. Indeed, the gas stream is responsible for forming the initial droplets after removal of the molten material from the wire (primary break up), for atomising the large droplets (secondary break up), and for accelerating them toward the substrate [13, 15]. Typically, power settings of about 450 A can spray over 50 $kg\,h^{-1}$. Some units push the wire to the gun while others pull the wire into the arc. Controlling the operating parameters is carried out by volt and ampere meters and air regulators [7].

Figure 2.1 *Wire arc spraying gun*

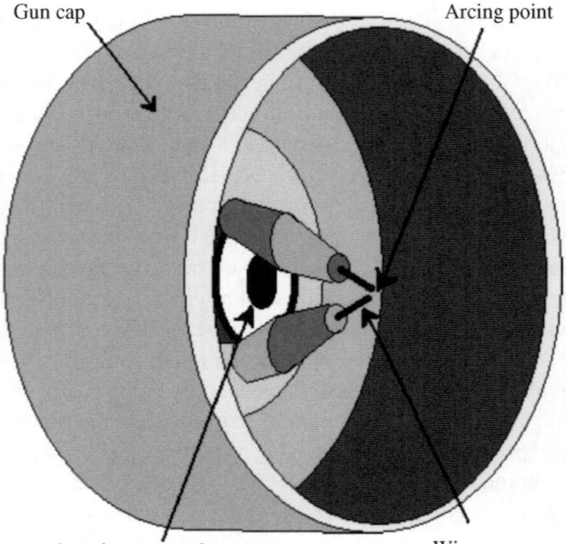

Figure 2.2 Wire arc spraying gun cap

2.3 Preparation of Membranes

2.3.1 Preparation of Inorganic Membranes Using Thermal Spraying

Thermal spray techniques include several types. These techniques have no broad application in membrane preparation. However, high velocity oxyfuel flame spraying (HVOF) has been applied more than the other techniques. For instance Hollein *et al.* [16] prepared the membrane using this method. They coated palladium on the surface of a layer with the fine pores. Palladium powder was injected into the nozzle of a spray gun using nitrogen as the carrier gas. It was partially molten in a fuel/air flame at temperatures between 2600 and 2900 °C, accelerated to high velocities (500–600 m s^{-1}), and propelled onto the surface of the substrate. The palladium powder had a grain size $\leq$45 μm. An ethylene/air mixture served as fuel gas. The substrates were cleaned with ethyl alcohol prior to the coating. The coating occurred at a normal air atmosphere. Neither protective gas nor vacuum were applied [16]. The HVOF technique exhibited superior performance using stainless steel substrates. This can tolerate relatively rough surfaces and is able to provide better adhesion of the Pd film to the support [16]. In a similar work, Dittmeyer *et al.* [17] prepared supported palladium membrane on porous stainless steel using HVOF for hydrogenation/dehydrogenation membrane reactor.

Thermal spray technology can be employed for preparing ceramic membranes to separate oxygen from gas stream [18]. Indeed, Kulikani *et al.* used plasma spraying to deposit yttria-stabilised zirconia (YSZ) coating on a porous support. Plasma spraying is a high temperature process. The feed stock powder was fed into a plasma arc, where it was melted and accelerated to high velocity, before it impinged onto a substrate [18]. The study demonstrated the potential of plasma spraying for producing porous (using air plasma spray process) and dense (using vacuum plasma spray process) membranes. A similar work was conducted by Huang and Dittmeyer [19].

2.3.2 Preparation of Metallic Membranes Using Wire Arc Spraying

Wire arc spraying has been used to prepare porous stainless steel membrane by Madaeni *et al.* for the first time [6]. Applying this technique for membrane preparation has not been previously tackled. The method requires a thermal resistant substrate with desirable surface properties. The cylindrical pieces of steel may be utilised as an appropriate substrate. For preparing the substrates, firstly, they should be scrapped by grinding paper to clean the substrate surface and removes impurities. Polishing the substrate led to the detachment of the sprayed layer before an appropriate thickness was reached as well as to bending of the metal layer. Coarsening the substrate surface is necessary for partially sticking of sprayed layer to the substrate. Coarsening can be carried out by sand blast apparatus. This should be performed homogeneously.

The substrates should be fixed perpendicularly for spraying. Rima 410 stainless steel wire with 1.5 mm diameter can be used as coating material [6]. The metal coatings could be prepared with various thicknesses (approximately 0.5–1.5 mm). Finally, the deposited metal layer on substrate surface may be detached carefully. The produced porous metallic layer is the stainless steel membrane (Figure 2.3).

2.3.3 Advantages and Disadvantages

The wire arc spraying procedure for preparation of porous metallic membranes possesses some advantages compared to the existing techniques. The low cost and rapid fabrication are the key advantages of the current technique. Other advantages include easy scale up, simplicity of producing membrane with various thicknesses and appropriate mechanical strength without the need for an additional support. Most of the available techniques use limited kinds of metals. For instance the laminable metals such as Ag and Pd may be employed in rolling method [15]. In the aforementioned procedure various types of conductive materials

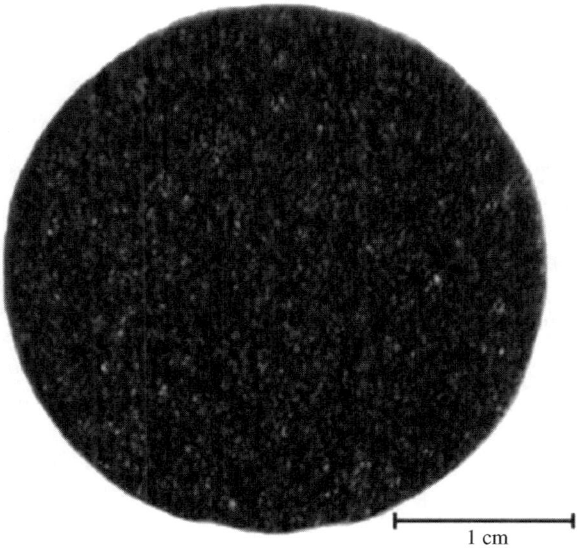

Figure 2.3 *Stainless steel membrane*

may be employed. Moreover both symmetric and asymmetric membranes may be obtained by simple techniques including manipulating the spraying conditions such as multidistance spraying.

The wire form materials can be used in thermal spraying which is extensively available. There is no need to micronise and pulverise the material similar to the required conditions in sintering [15]. A specific purity is required for most procedures including electroplating, electroless plating, PVD, CVD and rolling. Contrary to these requirements, wire arc spraying is compatible with numerous types of metals, metal composites and alloys [7, 8].

The current wire arc spraying technique results in a membrane with limited porosity. This leads to low water flux. Generally there are various procedures for porosity maximisation such as using cooling system after spraying, reducing the substrate temperature, etc. The appropriate types of metals or alloys, gun distance and atomiser air pressure may lead to higher porosity.

2.4 Characterisation of Prepared Metallic Membrane

2.4.1 Metallographic Tests

Optical microscopy image analysis may be applied to elucidate vital properties including porosity, metal oxide content, average pore size and pore size distribution for both surface and cross section of the prepared membranes. The microphotographs may be analysed using computer programs such as Olysia m3. The percentage of surface area for each existing phase can be estimated by this method. The arc spraying deposits include three obvious phase (pore, metal oxide, metal matrix) [11] as shown in Figure 2.4. The software converts the microphotographs to a threshold form (Figure 2.5) and then computes the area percentage of each phase. The same technique can be used for determining the pore size distribution and average pore diameter.

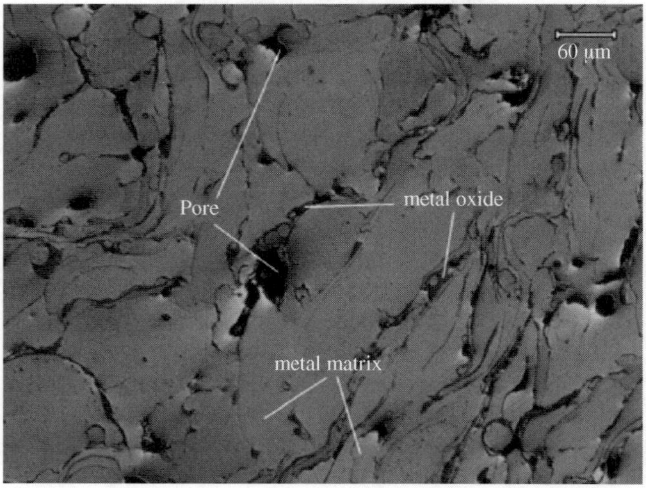

Figure 2.4 *Optical microscopy image of the cross section of the prepared membrane*

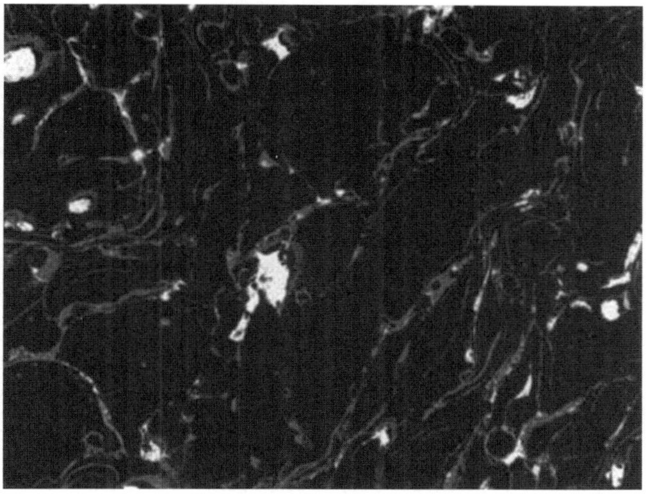

Figure 2.5 *Threshold converted form of Figure 2.4*

2.4.1.1 Porosity

During a thermal spraying process droplets are propelled and impacted onto a solid substrate. The molten or semi-molten metal droplets spread and ultimately solidify and form the splats [14, 21]. Accumulation of such splats provides a porous metal layer. In this process the extent of material flow and the rate of heat transfer control the properties of the prepared layer. The formation of pores can be elaborated based on three types of parameter. The parameters are:

1. Rate of mass and heat transfer in splatting process,
2. Substrate properties,
3. Interaction between splat and interface.

The first set of parameters is a function of temperature, velocity, viscosity and surface tension of the splat. The vital substrate properties are roughness, thermal conductivity and temperature. The third parameters include wettability and thermal conductivity of the interface [22, 23]. The connected pores increase the permeability of porous metallic membrane.

The unsuccessful connection of splats produces the pore in the deposited layer. This may be obtained by accumulation of irregular lamellae; reduction of splat temperature; gas bubbles which are trapped in molten metal during atomisation; gas solubility in molten metal and shrinkage of splats during solidification. Thermal and/or tensile quenching stresses enhance the pore formation by incomplete intersplat connections [18]. Vital parameters affecting the process are spraying variables such as gun distance or stand off distance, atomiser gas pressure and spraying atmosphere. In the normal application of spraying i.e. coating, the main purpose is reducing the number and extend of the pores. However with the goal of membrane preparation, the spraying conditions should be manipulated to establish the acceptable/highest porosity.

Few studies explained the effect of spray distance on coating porosity. Daengmool *et al.* [24] reported that the coating porosity is less dependent on spray distance. They found that, shorter distance produces larger splat size. In other words, the porosity is diminished at short gun distance. This is represented in schematic model of splat forming at short and long distances (Figure 2.6). Another study [6] on the effect of gun distance (Figure 2.7) indicates that the surface

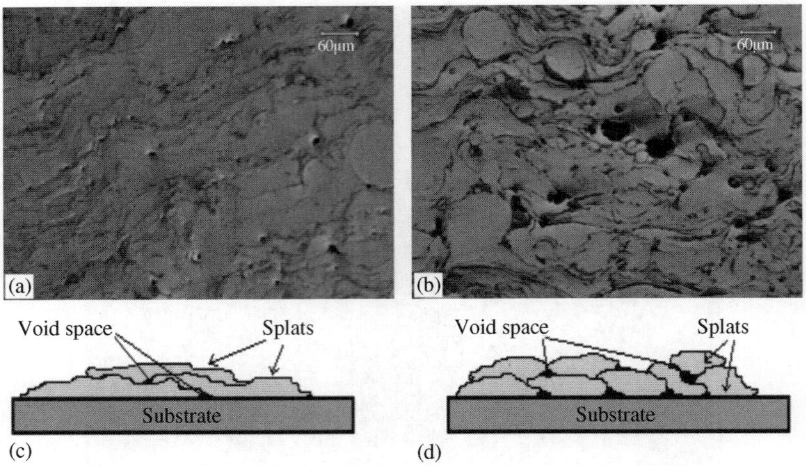

Figure 2.6 Optical microphotograph of cross section of sprayed layer at 15 and 40 cm (a, b), suggested schematic model for splat accumulation in two spraying distance (c, d). Reprinted from Journal of Membrane Science, Madaeni, S. S., Aalami-Aleagha, M. E., and Daraei, P., Preparation and characterization of metallic membrane using wire arc spraying. Vol. 320, 1–2. Copyright (2008) with permission from Elsevier

porosity of the prepared metal membrane is decreased at distances over 40 cm. This can be explained with the oxide content of the sprayed layer. Fe is the main element in stainless steel wire, and FeO is probably the greatest part of the formed metal oxides during spraying. The melting point of FeO is low compared to the steel and may remain liquid for longer time. This facilitates the deposition of splash droplets and moves them to the surface pores leading to lower porosity in surface [25]. Accordingly at longer gun distances, surface porosity is diminished whereas, the cross section porosity is steadily increased.

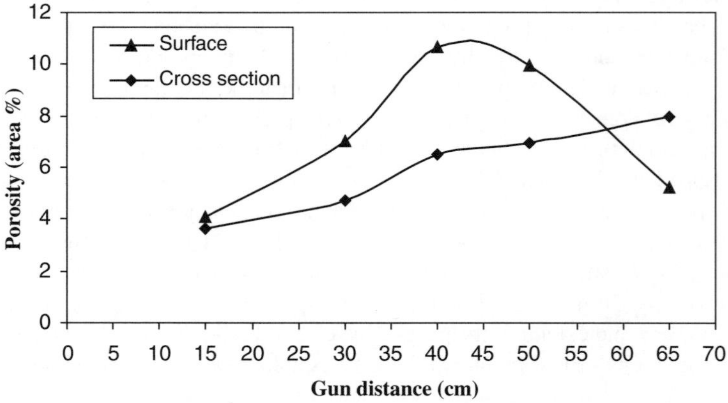

Figure 2.7 Effect of gun distance on membrane porosity in both surface and cross section. Reprinted from Journal of Membrane Science, Madaeni, S. S., Aalami-Aleagha, M. E., and Daraei, P., Preparation and characterization of metallic membrane using wire arc spraying. Vol. 320, 1–2. Copyright (2008) with permission from Elsevier

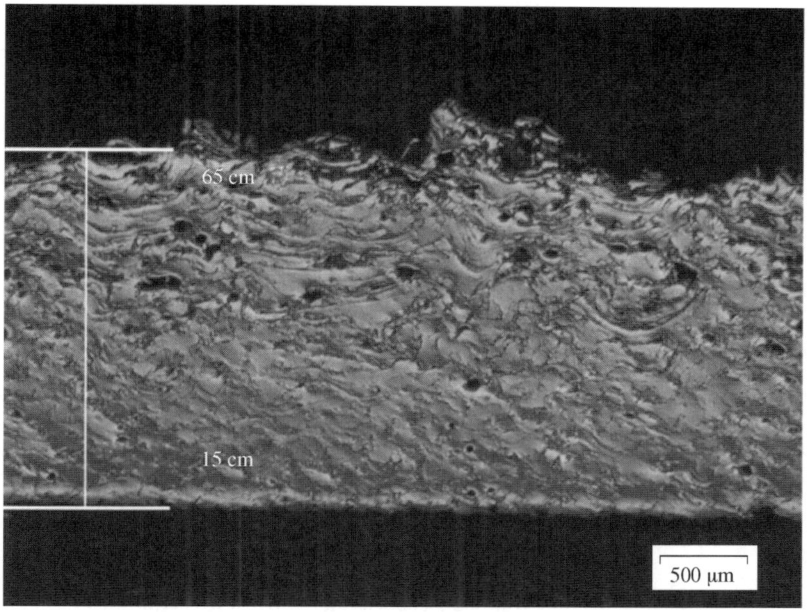

Figure 2.8 *Optical microscopy image of membrane cross section (prepared at multi distance condition). Reprinted from Journal of Membrane Science, Madaeni, S. S., Aalami-Aleagha, M. E., and Daraei, P., Preparation and characterization of metallic membrane using wire arc spraying. Vol. 320, 1–2. Copyright (2008) with permission from Elsevier*

Figure 2.8 indicates a special spraying condition which was started at 15 cm spray distance and increased to 65 cm during time. Decrease of the porosity and pore size is clearly presented in this micrograph of the cross section of multidistance spraying deposits. The image analysis for determining the porosity of the membranes may not be the optimum procedure. For confirmation of the image analysis outcomes, a simple procedure may be performed. In this technique, the porosity of the membrane can be measured using isopropanol intrusion into the membrane matrix. For this purpose, the specimen should be immersed in isopropanol for 2 days. Isopropanol is an appropriate diffusing fluid due to its low surface tension compared to e.g. water. The specimen may be dried with hot air and weighed as W_1. The membrane should be immersed in isopropanol again for 3 days. The wet specimen should be removed from alcohol and the surface be dried and weighed as W_2. Then the specimen can be completely dried by hot air for 30 min and weighed for the last time as W_3. The value of W_2–W_1 or W_2–W_3 is the weight of alcohol which has diffused in inner pores of metallic film. The third weighing was performed for certainty of W_1 value. The void spaces in the membrane structure or membrane porosity can be calculated using the weight and density of isopropanol and membrane. Porosity measurement using isopropanol indicates that the total porosity of sprayed layer is decreased after an increase at spray distance around 40 cm (Figure 2.9). This confirms the image analysis results indicating membrane porosity.

Atomiser air pressure is another parameter affecting the membrane porosity. An increase in atomiser air pressure decreases the porosity of the established layer. This may be attributed to enhancement of spraying intensity, which leads to a decline in the size of molten metal droplets.

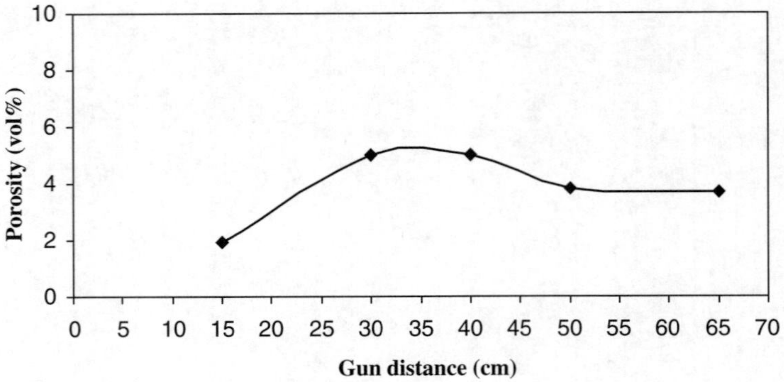

Figure 2.9 Porosity measurement using isopropanol. Reprinted from Journal of Membrane Science, Madaeni, S. S., Aalami-Aleagha, M. E., and Daraei, P., Preparation and characterization of metallic membrane using wire arc spraying. Vol. 320, 1–2. Copyright (2008) with permission from Elsevier

By decreasing the particle size, the trapped gas between the particles is diminished. This results in reduction of pore number in deposited layer.

2.4.1.2 Oxide Content

The hot molten metal droplets travel in air prior to the contact with the substrate and solidification. Accordingly formation of metal oxide in droplets is inevitable. Oxidation of the steel during the electric arc spraying occurs in three ways [25]:

1. Oxidation of primary droplets during atomisation and flight in air,
2. Oxidation of secondary droplets during splashing,
3. Oxidation of the top surface deposition.

The oxide content of the prepared membrane is a function of type and pressure of atomiser gas pressure and other spray conditions such as spray distance. Previous studies demonstrated an increase in oxide content with lengthening the gun distance [25, 26]. At longer gun distances, in flight droplets of melted alloy propel longer direction in air. They are more favourable for reaction with air oxygen due to the high temperature. Hence, the increasing trend in oxide content of coating is approximately predictable. Another study [6] showed that the oxide content of membranes in both surface and cross section was approximately constant after 45 cm gun distances. This may be related to the limitation of surface oxidation of metal droplets, that is, oxidation of metal droplets occurs until top surface of the hot droplets is in direct contact with surrounding air oxygen. After formation of an oxide layer on the top surface, no noticeable oxidation of remaining metal is carried out. Therefore, oxide content of sprayed coatings at distances longer than 45 cm, exhibit no considerable change.

Increasing the atomiser air pressure leads to great disintegration of the particles and high velocity in traveling to substrate surface. This decreases the residence time of particles in air and the time for oxide formation. Increasing the atomiser air pressure decreases the droplets temperature considerably. This results in lower oxide formation.

2.4.2 Performance

The filtration capability of the prepared membranes can be tested by purifying water and removal of some pollutant like ion and dye particles. The metallic membranes can be installed in dead end cell. The ion rejection can be calculated during a specified time at constant pressure. According to a previous research [6] this type of membrane can undergo under 7 bars of transmembrane pressure. Saline water may be prepared by addition of NaCl to distilled water and used as the feed. The conductivity of the feed (C_f) and permeate (C_p) can be measured for calculating the ion rejection (R) as follows:

$$R(\%) = \frac{C_f - C_p}{C_f} \times 100 \qquad (2.1)$$

In another set of experiments, the efficiency of metallic membranes for removing particles (blue indigo dye) was investigated. The blue indigo contains particles in the range of 0.5–3.0 μm. A suspension of the dye was used with certain turbidity value. For investigating the capability of membrane for preventing the passage of blue indigo dye, the turbidity of water was measured. The rejection of dye particles was calculated using Equation 2.1 with substituting the conductivities by dye concentration (turbidity value in mg l^{-1}).

The prepared membrane by wire arc spraying can efficiently remove ions and turbidity from water [6]. Figure 2.10 illustrates the capability of the prepared membrane for removing salt from water. Adsorption of anions is the most possible mechanism for retention of ions. The membrane prepared at 40 cm gun distance exhibited the highest porosity and interconnections between void spaces in the membrane matrix. On the other hand, the oxide content of this membrane was moderate. Therefore, the available metal surface area for adsorbing ions in this membrane was extensive leading to removal of ions. By contacting metal and water, a thin layer of metal oxide is generated on the metal surface. The thin film adsorbs water molecules. The available Cl^- results in an exchange between adsorbed H_2O and chloride ions, that is, ions are polarised in the vicinity of the metal surface. The negative charge of the chloride ions attracts the cations. Accordingly, ion permeation is diminished [6].

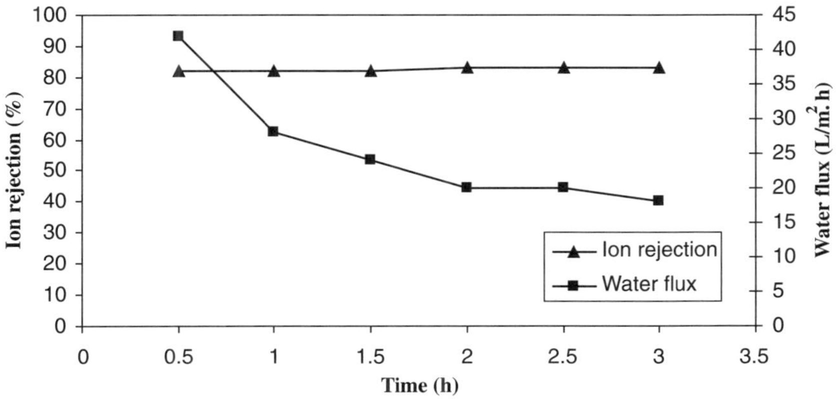

Figure 2.10 Ion rejection of prepared metallic membrane at 40 cm gun distance and 3.5 bar atomiser air pressure during 3 h

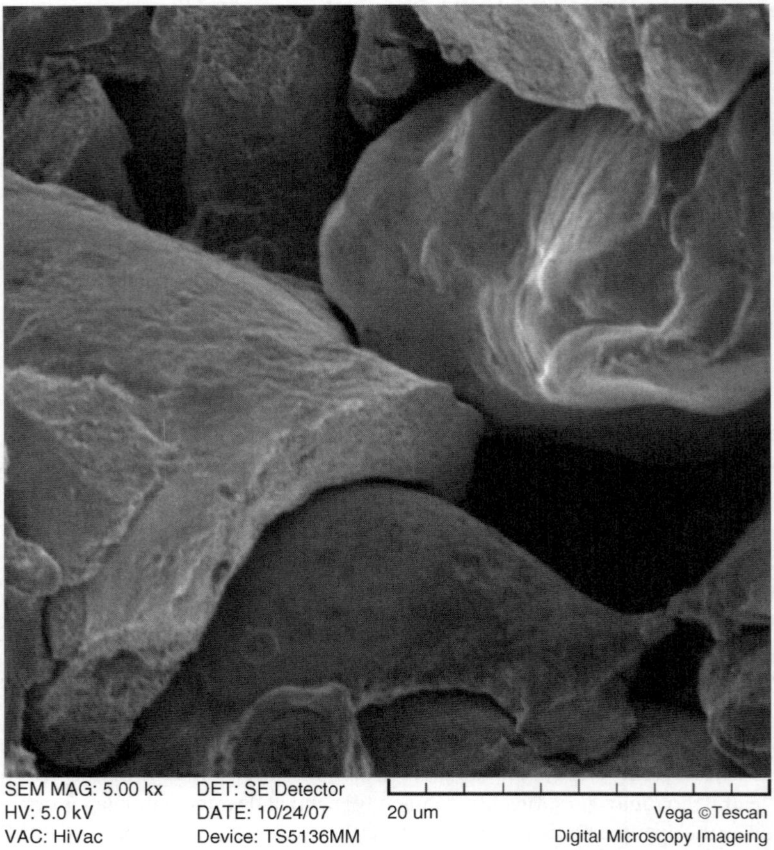

Figure 2.11 *SEM micrograph of cross section of virgin membrane. Reprinted from Journal of Membrane Science, Madaeni, S. S., Aalami-Aleagha, M. E., and Daraei, P., Preparation and characterization of metallic membrane using wire arc spraying. Vol. 320, 1–2. Copyright (2008) with permission from Elsevier*

Additionally, the removal of the dye particles was tested for purification capability of the prepared metallic membrane. The membrane was able to remove blue indigo dye particles and reduce the turbidity. The main mechanism for this removal is particle deposition in the membrane pores. Deposition of indigo particles is clearly observable by comparison the SEM micrographs of virgin (Figure 2.11) and used membrane (Figure 2.12) exhibiting deposited indigo particles.

2.5 Applications of Prepared Metallic Membrane

2.5.1 Water Treatment

Membranes are widely used in water purification and waste water treatment. This includes pressure-driven processes from microfiltration to reverse osmosis and combined membrane with

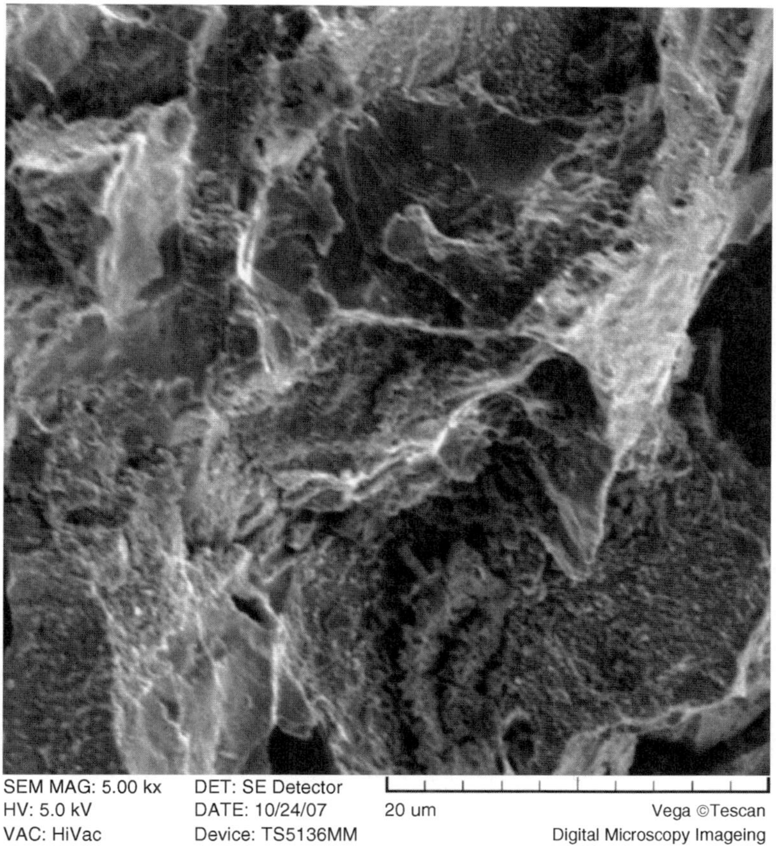

Figure 2.12 *SEM micrograph of cross section of membrane used for removal of indigo dye from water. Reprinted from Journal of Membrane Science, Madaeni, S. S., Aalami-Aleagha, M. E., and Daraei, P., Preparation and characterization of metallic membrane using wire arc spraying. Vol. 320, 1–2. Copyright (2008) with permission from Elsevier*

activated sludge or membrane bioreactors (MBR). The application of hybrid systems combining traditional water treatment with membrane processes is widely accepted in water purification [27]. Few examples of hybridisation are membrane with photoreactor for the purification of turbid waters and coupled system of membrane with powdered activated carbon to remove soluble organic contaminants by adsorption [28–30].

The ideal characteristics for employed membrane in water treatment are: high water flux, superior salt rejection, pronounced resistance to biological attack, appropriate resistance to fouling induced by colloidal and suspended material, inexpensiveness, simplicity to form thin layers, high chemical and physical strength, tolerating high pressures and high thermal stability (resistance against high temperatures without deforming or degradation). These characteristics indicate the capability of metallic membranes for application in water treatment.

2.5.2 Gas Purification

Gas separation at relatively high temperatures is very attractive since many petrochemical plants looking for gas separation at elevated operating temperature. Accordingly metallic membranes are increasingly being employed to separate gas mixtures at high temperatures. Metallic membranes including mesoporous, microporous and dense membranes are finding applications for this process. Extensive studies are conducted to apply palladium dense membranes for purifying hydrogen [1, 16, 17, 19].

The porous stainless steel membrane by wire arc spraying is able to remove 95% of hydrogen sulfide from gas mixture on the basis of the following reaction:

$$Fe + H_2S \rightarrow FeS + H_2$$

Accordingly, gas purification can be considered as an important field for application of metallic membranes. This is due to the compatible characteristics for severe reaction condition (high temperature and pressure) of so many gaseous reactions.

2.5.3 Membrane Reactors

The membrane prepared by wire arc spraying can be utilised as a catalyst for gaseous reactions. According to Figure 2.13 the prepared membrane by wire arc spraying exhibit a tortuous structure. This makes the membrane suitable for catalytic reactions and provides a porous bed for efficient contact among the reactive materials. On the other hand, the membrane may contain materials with catalytic activity (e.g. Ni, Zn etc.). Moreover most of the materials with appropriate catalytic activity such as titanium, palladium, aluminum and zinc can be sprayed by this technique to prepare catalytic membranes. Furthermore the prepared metallic membranes are able to absorb water and organic solvents (isopropanol) and simultaneously are gas permeable. This demonstrates the role of the prepared metallic membranes in membrane rector and capability for being catalytic diffuser. Additionally, the metallic membranes are thermal resistant. Therefore, they are compatible with hot gases or liquids and may be operated at high temperature reactions. The reactions are enforced to proceed by selective removal of one or more products from the reaction mixture in a membrane reactor.

2.6 Final Remarks and Conclusions

Wire arc spraying can be used to produce support free porous metallic membrane. This method of membrane preparation has an advantage of using so many electrical conductive materials. Improving the membrane properties may be achieved by changing spraying variables such as gun distance and atomiser air pressure. Stainless steel membranes by wire arc spraying are capable of removing ions and dye particles from water. This indicates ability of the prepared metallic membrane for water treatment. Furthermore, the membrane can reduce hydrogen sulfide content of gas mixtures. These membranes are potentially capable for considering as catalytic membrane due to their porosity and composition including metals and metal oxides. The high thermal stability of metallic membrane makes it propitious for high temperature reactions. Accordingly, wire arc spraying technique is a method for producing metallic membranes to be applied in membrane reactors.

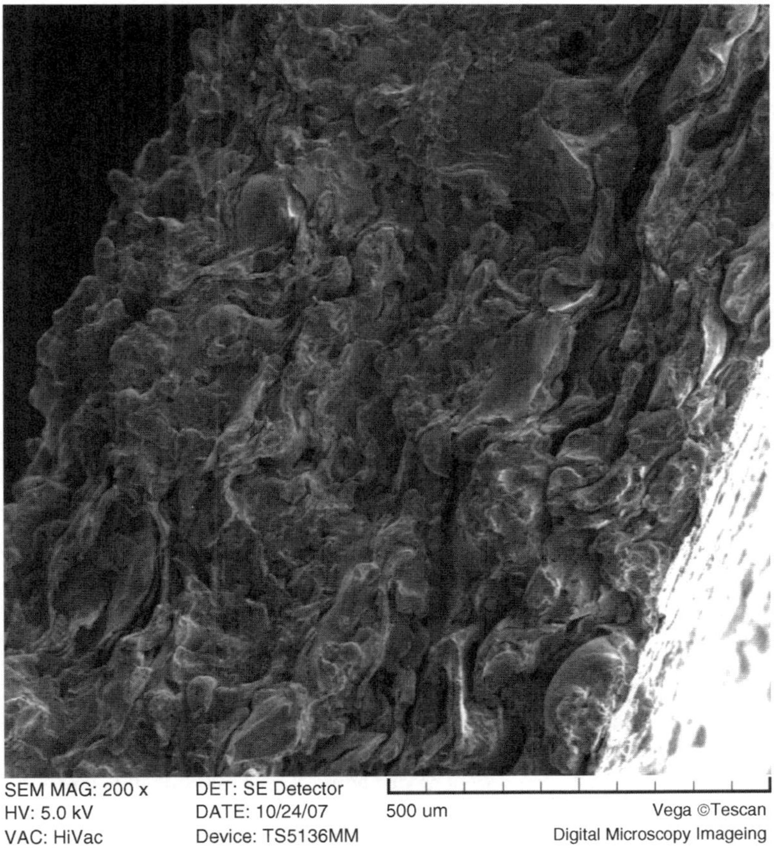

Figure 2.13 *SEM microphotograph of metallic membrane cross section*

References

1. A. Li, W. Liang and R. Hughes, Characterization and permeation of palladium/stainless steel composite membrane, *J. Membr. Sci.*, **149**, 259–268 (1998).
2. Y.S. Lin, Microporous and dense inorganic: current estates and prospective, *Sep. Purif. Technol.*, **25**, 39–55 (2001).
3. R.H. Kim, S. Lee and J.O. Kim, Application of a metal membrane for rainwater utilization: filtration characteristics and membrane fouling, *Desalination*, **177**, 121–132 (2004).
4. T. Leiknes, H. Odegaard and H. Myklebust, Removal of natural organic matter (NOM) in drinking water treatment by coagulation–microfiltration using metal membranes, *J. Membr. Sci.*, **242**, 47–55 (2004).
5. F. Roa, J.D. Way, R.L. McCormick and S.N. Paglieri, Preparation and characterization of Pd-Cu composite membranes for hydrogen separation, *Chem. Eng. J.*, **93**, 11–22 (2003).
6. S.S. Madaeni, M.E. Aalami-Aleagha and P. Daraei, Preparation and characterization of metallic membrane using wire arc spraying, *J. Membr. Sci.*, **320**, 541–548 (2008).

7. F.J. Hermanek, What is Thermal Spray, International Thermal Spray Association, www.thermal-spray.org (2000).
8. M.L. Berndt and C.C. Berndt, Thermal Spray Coatings, *J. Protective Coatings and Linings*, 1–3 (2003).
9. M. Yamada, Y. Kouzaki, T. Yasui and M. Fukumoto, Fabrication of iron nitride coatings by reactive RF plasma spraying, *Surf. Coat. Tech.*, **201**, 1745–1751 (2006)
10. J. Kim, H. Yang, K. Baik, B.G. Seong, C. Lee and S.Y. Hwang, Development and properties of nanostructured thermal spray coatings, *Current Applied Physics*, **6**, 1002–1006 (2006)
11. B.S. Schorr, K.J. Stein and A.R. Marder, Characterization of thermal spray coatings, *Mater. Charact.*, **42**, 93–100 (1999).
12. S. Sampath, X.Y. Jiang, J. Matejicek, L. Prchlik, A. Kulkarni and A. Vaidya, Role of thermal spray processing method on the microstructure, residual stress and properties of coatings: an integrated study for Ni – 5 wt% Al bond coats, *Mater. Sci. Eng.*, **A364**, 216–231 (2004).
13. A.P. Newbery, P.S. Grant and R.A. Neiser, The velocity and temperature of steel droplets during electric arc spraying, *Surf. Coat. Tech.*, **195**, 91–101 (2005).
14. I. Gedzevičius and A.V. Valiulis, Influence of the particles velocity on the arc spraying coating adhesion, *Mater. Sci.*, **9**, 334–337 (2003).
15. S. Nourouzi and A. Vardelle, Effect of nozzle daimeter and atomizing air pressure on characteristics of in-flight particles in wire arc spraying, *Iranian Journal of Materials Science and Engineering*, **2**, 28–34 (2005).
16. V. Höllein, M. Thornton, P. Quicker and R. Dittmeyer, Preparation and characterization of palladium composite membranes for hydrogen removal in hydrocarbon dehydrogenation membrane reactors, *Catalysis Today*, **67**, 33–42 (2001).
17. R. Dittmeyer, V. Höllein and K. Daub, Membrane reactors for hydrogenation and dehydrogenation processes based on supported palladium, *J. Mol. Catal. A – Chem.*, **173**, 135–184 (2001).
18. A.A. Kulkarni, S. Sampath, A. Goland, H. Herman, A. J. Allen, J. Ilavsky, W. Gong and S. Gopalan, Plasma spray coatings for producing next-generation supported membranes, *Top. Catal.*, **32**, 241–249 (2005).
19. Y. Huanga and R. Dittmeyer, Preparation of thin palladium membranes on a porous support with rough surface, *J. Membr. Sci.*, **302**, 160–170 (2007).
20. S.S. Tosti, L. Bettinala, S. Castelli, F. Sarto, S. Sdaglione and V. Violante, Sputtered, electroless, and rolled palladium–ceramic membranes, *J. Membr. Sci.*, **196**, 241–249 (2002).
21. M. Raessi, J. Mostaghimi and M. Bussmann, Effect of surface roughness on splat shapes in the plasma spray coating process, *Thin Solid Films*, **506/507** 133–135 (2006).
22. J.C. Fang, W.J. Xub and Z.Y. Zhaoa, Arc spray forming, *J. Mater. Process. Tech.*, **164/165** 1032–1037 (2005)
23. R.S.C. Paredes, S.C. Amico and A.S.C.M. d'Oliveira, The effect of roughness and pre-heating of the substrate on the morphology of aluminium coatings deposited by thermal spraying, *Surf. Coat. Tech.*, **200**, 3049–3055 (2006).
24. R. Daengmool, S. Wirojanupatump, S. Jiansirisomboon and A. Sopadang, Effect of spray parameters on stainless steel arc sprayed coating, *4th Thailand Material Science and Technology Conference, Thailand*, Article Code MP03 (2006).
25. A.P. Newbery and P.S. Grant, Oxidation during electric arc spray forming of steel, *J. Mater. Process. Tech.*, **178**, 259–269 (2006).
26. S. Deshpande, S. Sampath and H. Zhang, Mechanisms of oxidation and its role in microstructural evolution of metallic thermal spray coatings – Case study for Ni–Al, *Surf. Coat. Tech.*, **200**, 5395–5406 (2006).
27. K. Konieczny a, D. Sąkolb and M. Bodzeka, Efficiency of the hybrid coagulation–ultrafiltration water treatment process with the use of immersed hollow-fiber membranes, *Desalination*, **198**, 102–110 (2006).

28. M.A.S. Rodrigues, F.D.R. Amado, J.L.N. Xavier, K.F. Streit, A.M. Bernardes and J.Z. Ferreira, Application of photoelectrochemical–electrodialysis treatment for the recovery and reuse of water from tannery effluents, *Journal of Cleaner Production*, **16**, 605–611 (2008).
29. K. Azrague, P. Aimar, F. Benoit-Marquie and M.T. Maurette, A new combination of a membrane and a photocatalytic reactor for the depollution of turbid water, *Appl. Catal. B – Environ.*, **72**, 197–205 (2006).
30. H.K. Oha, S. Takizawaa, S. Ohgakia, H. Katayamaa, K. Ogumaa and M.J. Yub, Removal of organics and viruses using hybrid ceramic MF system without draining PAC, *Desalination*, **202**, 191–198 (2007).

3

Inorganic Hollow Fibre Membranes for Chemical Reaction

Benjamin F. K. Kingsbury, Zhentao Wu and K. Li

Department of Chemical Engineering, Imperial College London, London, UK

3.1 Introduction

The combination of high chemical, thermal and mechanical resistance has made ceramic membranes an attractive alternative to polymeric varieties. Due to the high surface area/volume ratios achieved by hollow fibre configurations, ceramic hollow fibre membrane performance may greatly exceed that of other membrane systems. The ability to operate at high temperatures and pressures and in corrosive environments allows ceramic membranes to be used in a variety of applications including filtration for corrosive fluids [1], high temperature membrane reactors [2–4], solid oxide fuel cells [5] and membrane contactors [6] as well as robust membrane supports [7]. In particular, membrane reactors combining reaction and separation within the same unit have many advantages over conventional reactor designs and have therefore attracted great interest over recent years. The membrane component of the reactor can either serve to uniformly disperse the reactants prior to their reaction or to selectively separate a product from a reaction mixture. In the latter case, the use of a membrane to separate a product from an equilibrium limited reaction leads to greater conversions and lower operating temperatures [8–10]. In addition, the different phases of the reaction mixture are separated and individual control of the feed and permeate phases can be achieved, leading to a greater level of control over the operating parameters of the reactor. Hollow fibre membrane configurations offer extremely high surface area to volume ratios and are therefore desirable in order to reduce module costs and to increase reactor efficiencies. However, production of ceramic hollow fibres can be costly and time consuming, especially if an

Membranes for Membrane Reactors: Preparation, Optimization and Selection, Edited by Angelo Basile and Fausto Gallucci

asymmetric structure is required, due to the multistep nature of the preparation process. By using a combine phase inversion and sintering technique, asymmetric hollow fibre membranes can be prepared in a single step which greatly reduces production costs and allows for the preparation of a wide range of asymmetric structures. For example, an asymmetric pore structure can be used as either a porous membrane and a matrix for catalyst deposition, or as a porous support for the coating of a gas separation layer and a matrix for catalyst deposition, in a highly compact multifunctional catalytic membrane reactor [11,12].

In this chapter, several aspects of the preparation procedure from raw materials to a functioning catalytic membrane reactor for the dehydrogenation of propane are discussed. This includes the many factors that affect hollow fibre morphology during the preparation process and how proper control of these parameters allows for the development of a membrane reactor which may be customised for a wide range of applications and operating conditions. In addition, different methods for the impregnation of catalyst within the asymmetric hollow fibre support structure and for the deposition of a palladium or palladium/silver hydrogen permselective membrane are introduced and the performance of the developed membrane reactor is compared to a more traditional reactor design.

3.2 Preparation of Inorganic Hollow Fibre Membranes

Despite the many beneficial properties of ceramic membranes such as their ability to operate at high temperature and under harsh operating conditions, in some cases the high production costs are prohibitive in their implementation. In many applications an asymmetric membrane structure is desired consisting of a thin separation layer and a support structure to provide mechanical strength and ideally the support structure will offer as little as possible resistance to the permeation of reactants and/or products.

Traditional methods of fibre preparation such as dry spinning a system of inorganic material and binder [13–15], wet spinning a solution and/or sol containing a suitable inorganic material, depositing fibres from the gas phase onto a substrate or the pyrolysis of polymers [16] yield only symmetric structures and often many additional stages are required to achieve asymmetry, which are both costly and time consuming. By using a combined phase inversion and sintering technique, not only can asymmetric ceramic hollow fibres be prepared in a single step but a wide range of morphologies can be achieved in which fibres possess multimodal pore size distributions which may vary over orders of magnitude within the same fibre.

Preparation of ceramic hollow fibre membranes using the combined phase inversion and sintering technique consists of 3 principal stages, namely the preparation of the ceramic/polymer/solvent spinning suspension, the spinning of the fibre precursor and the calcination of the precursors to yield the final product, a ceramic hollow fibre membrane. The properties of the final product are dependent on factors at each of the three stages. A ceramic suspension containing ceramic material, polymer binder and solvent as the principal components is prepared and extruded through a tube-orifice spinneret into a precipitation bath containing a non-solvent for the polymer. Simultaneously, a non-solvent for the polymer is supplied through the tube of the tube-orifice spinneret into the lumen of the nascent fibre. A flow diagram of the preparation process is shown in Figure 3.1, indicating the various stages involved in the formation of ceramic hollow fibres from starting materials using the combined phase inversion and sintering technique.

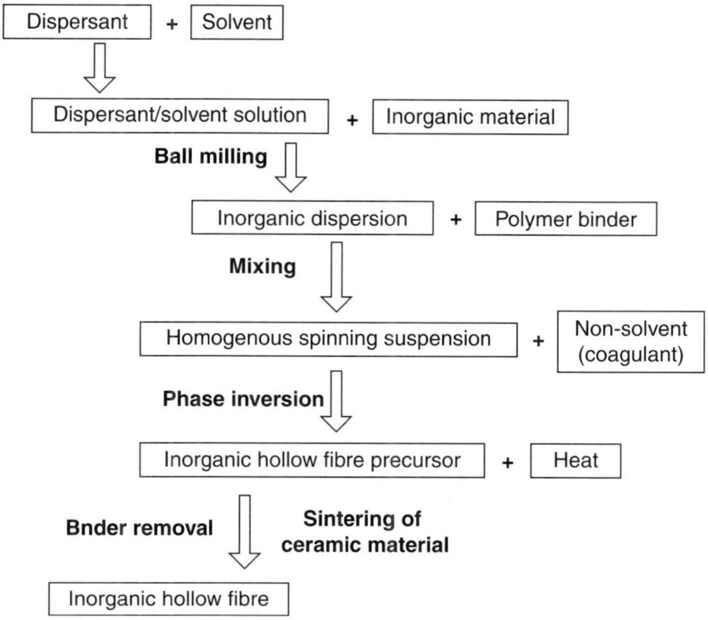

Figure 3.1 *Flow diagram showing the stages involved in the combined phase inversion and sintering technique for ceramic hollow fibre fabrication. Reprinted from Catalysis Today, Benjamin F.K. Kingsbury, Zhentao Wu and K. Li, A morphological study of ceramic hollow fibre membranes: A perspective on multifunctional catalytic membrane reactors, In Press, Corrected Proof. Copyright (2010) with permission from Elsevier*

This method is an adaption of the immersion induced phase inversion method first described by Loeb and Sourirajan [17] for asymmetric polymeric membrane formation. After the addition of a desired amount of ceramic powder to a polymer solution, if well mixed, the ceramic/polymer/solvent system can be seen as a suspension of polymer coated ceramic particles. Once immersed in a non-solvent for the polymer which is miscible with the solvent, solvent/non-solvent exchange takes place leading to precipitation of the polymer phase. Ceramic particles are immobilised once precipitation has taken place and the membrane macrostructure can be largely determined at this point by manipulating and adjusting the various parameters of the phase inversion process.

Great effort has been made to both control and understand the formation mechanisms for the wide range of structures observed in polymeric membrane formation [18–24]. However, due to the large differences between polymeric and ceramic systems, in particular the low polymer concentration, this information is of limited use during ceramic membrane preparation. In fact, only two morphologies have so far been observed in ceramic/polymer systems, that is: (i) finger-like voids and (ii) a sponge-like structure. The macrostructure of the fibre precursor designed during the phase inversion process is retained during sintering (calcination/heat treatment) and finger-like voids above a certain size are not usually eliminated, although at elevated sintering temperature sponge-like regions will densify and eventually become gas tight for some ceramic materials [25].

Hydrodynamically unstable viscous fingering is a well known phenomenon that occurs at the interface between fluids with different viscosities in the first moments of mixing and may be applied to explain the formation of finger-like voids in ceramic membrane precursors. When the suspension is in contact with non-solvent, a steep concentration gradient results in solvent/non-solvent exchange, a rapid increase in local viscosity and finally precipitation of the polymer phase. However, due to instabilities at the interface between the suspension and the non-solvent there is a tendency for viscous fingering to occur, initiating the formation of finger-like voids. If viscous fingering is initiated at the fibre surface then the entrances to the finger-like voids will form surface pores. The relative thickness of finger-like and sponge-like regions greatly affects the properties of the membrane such as membrane mechanical strength and permeation flux and, due to the versatility of ceramic hollow fibre membranes, it is essential that fibre morphology can be controlled so that it may be tailored to a specific application.

3.2.1 Preparation of the Suspension

A suspension containing ceramic particles, a polymer binder and a solvent is prepared. The ceramic material is dispersed in a solvent with the aid of a dispersant and is milled for a predetermined duration. To the ceramic dispersion, a polymer binder which is soluble in the solvent is added and milling is continued until a homogenous ceramic/polymer suspension is achieved, after which the suspension is degassed to remove air bubbles and dissolved gas. The properties and composition of the suspension determine not only the properties of the ceramic hollow fibre precursor but also greatly affect the properties of the fibre after calcination.

Al_2O_3/polymer ratio: A hollow fibre precursor formed through the combined phase inversion and sintering technique contains Al_2O_3, which is one of the most commonly employed ceramic membrane materials, as well as the polymer binder. During the sintering process, the polymer binder is removed and the Al_2O_3 hollow fibre is ultimately formed. Therefore, the Al_2O_3 particle size and content in the spinning suspension play an important role in determining the pore size and porosity of the final product. The effect of variation in the Al_2O_3/polymer ratio can be seen from the results shown in Figure 3.2. All the fibre samples were calcined at a temperature of 1600 °C and were prepared from spinning suspensions containing 1 μm alumina particles and polyethersulfone as the binder and it can be seen that both the pore size and the surface porosity decrease as the Al_2O_3/polyethersulfone ratio is increased. This indicates that in order to produce a denser membrane, a higher Al_2O_3 content in the spinning suspension must be maintained. In addition, the Al_2O_3 content in the spinning suspension plays an important role in determining the fibre mechanical strength. Figure 3.3 illustrates the effect of the Al_2O_3 content on the fibre mechanical strength and gas permeability. It can be seen that the three point bending strength (3P value) is greatly enhanced as the Al_2O_3/polyethersulfone ratio is increased. Compared to the sintering temperature, an increase of the Al_2O_3 powder content in the spinning suspension results in a much more obvious effect on the fibre mechanical strength. It therefore follows that in order to produce an Al_2O_3 hollow fibre membrane with higher mechanical strength, a higher Al_2O_3 content in the spinning suspension must be maintained. At an Al_2O_3/Polyethersulfone ratio of 7 or greater the reduction in gas permeability is tailed, indicating that the membrane is transformed to a much denser structure.

Particle size distribution: Pores in the sintered ceramic hollow fibre membranes are voids left between packed particles, having neither a regular shape nor a regular size. The particle size distribution in the spinning suspension has a marked effect on the pore size distribution of the calcined fibres. Figure 3.4 shows the effect of the weight percent of 0.01 μm particles in the spinning suspension on the pore size of the resultant hollow fibres, that is, 0.01 μm particles blended with 1 μm particles, $w_{0.01}$ over total Al_2O_3 particles, w_T ($w_{0.01}/w_T$,%). It can be seen that the maximum pore size decreases as the content of 0.01 μm Al_2O_3 particles is increased, however,

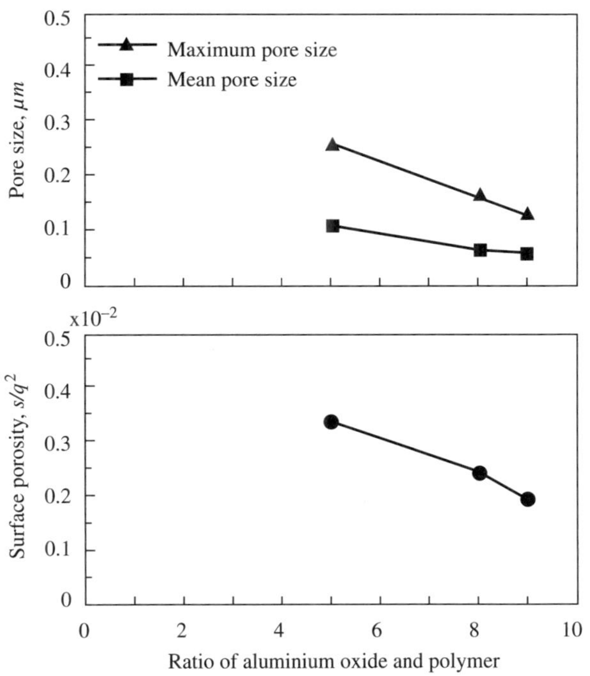

Figure 3.2 *Effect of Al_2O_3/polymer ratio on the membrane surface pore size for fibres prepare with 1 μm particles and sintered at 1600 °C. Reprinted from Journal of Membrane Science, Xiaoyao Tan, Shaomin Liu and K. Li, Preparation and characterization of inorganic hollow fiber membranes, 188(1) 87–95. Copyright (2001) with permission from Elsevier*

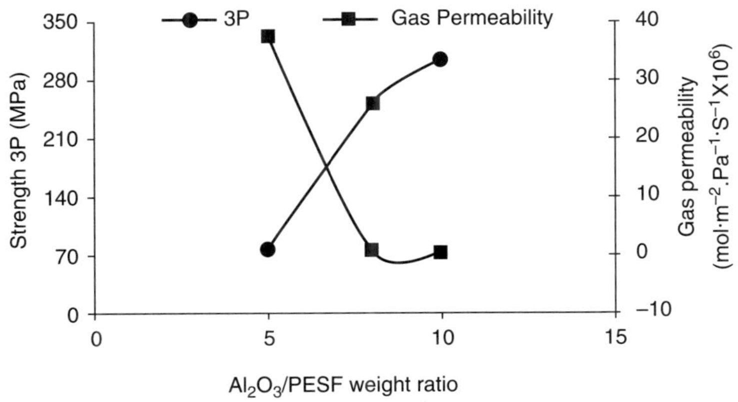

Figure 3.3 *Effect of Al_2O_3/polymer ratio on the mechanical strength and gas permeability of hollow fibre membranes prepared from 1.0 μm Al_2O_3 particles and calcined at 1550 °C. Reprinted from Ceramics International, Liu, K. Li and R. Hughes, Preparation of porous aluminium oxide (Al_2O_3) hollow fibre membranes by a combined phase-inversion and sintering method, 29 (8) 2003 875–881*

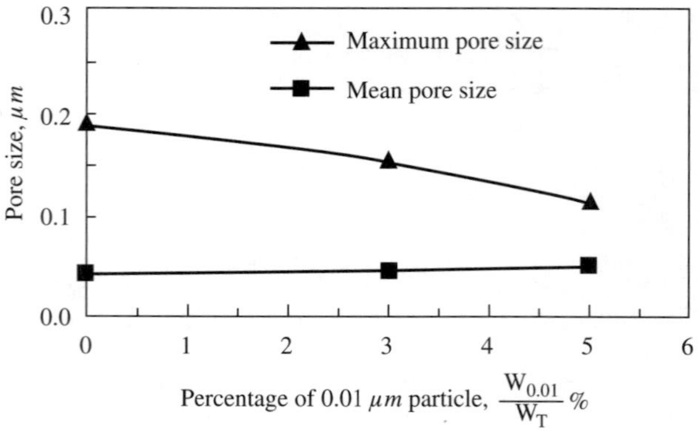

Figure 3.4 *Effect of the 0.01 μm Al_2O_3 particle content on membrane surface pore size for fibres prepared with an Al_2O_3/polymer ratio of 8 and calcined at 1550°C. Reprinted from Journal of Membrane Science, Xiaoyao Tan, Shaomin Liu and K. Li, Preparation and characterization of inorganic hollow fiber membranes, 188(1) 87–95. Copyright (2001) with permission from Elsevier*

the reverse is observed for the average pore size. This suggests that the pores of the hollow fibres prepared may be made more uniform in size by adding 0.01 μm particles to the spinning suspension. The effect of the weight percent of 0.01 μm particles on the porosities of the resultant hollow fibres is illustrated in Figure 3.5. It can be seen that both the surface and the volumetric porosities increase as the content of 0.01 μm particles is increased. This result is contrary to the analysis conducted by Lu [26], who showed theoretically that the porosity of a ceramic powder compact reduces when smaller particles are used alone for the preparation of the powder compact. However, a polymer binder is present in fibre precursors prepared using the combined phase inversion and sintering technique and in view of the complexity of this process, the relationship between the powder size and the structure of the final resultant membrane cannot be predicted adequately using the theoretical techniques employed by Lu.

3.2.2 Preparation of the Membrane Precursors

Both the micro and macrostructure of ceramic hollow fibres prepared using the combined phase inversion and sintering technique are strongly dependant on a great many parameters throughout the entire preparation process. As discussed previously, viscous fingering which occurs as the less viscous fluid (non-solvent) intrudes into the more viscous fluid (suspension film) is responsible for the asymmetric structures that can be achieved using this preparation method. As the extent and nature of the viscous fingering phenomenon is dependent on the relative viscosities of the two phases, by varying parameters such as the air-gap, the internal coagulant flow rate and the precipitation rate of the polymer phase during spinning, the suspension viscosity can be controlled throughout the spinning process and the morphology of the resulting ceramic hollow fibre precursor can be controlled in this way. Mechanical strength is an extremely important property of a ceramic hollow fibre and as well as being strongly influenced by properties of the suspension, such as the particle size distribution and the total solids loading, it is also very dependent on the morphology of the fibre with respect to the thickness of the sponge-like layer, which provides the bulk of the mechanical strength. Although mechanical strength can be increased by increasing the calcination temperature in the final stage of fibre preparation, this

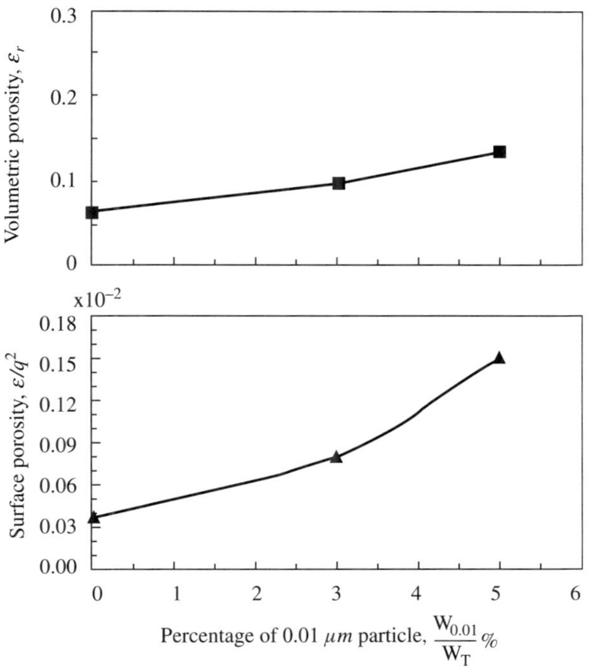

Figure 3.5 *Effect of the 0.01 μm Al_2O_3 particle content on membrane porosity for fibres prepared with an Al_2O_3/polymer ratio of 8 and calcined at 1550 °C. Reprinted from Journal of Membrane Science, Xiaoyao Tan, Shaomin Liu and K. Li, Preparation and characterization of inorganic hollow fiber membranes, 188 (1) 87–95. Copyright (2001) with permission from Elsevier*

may be accompanied by an unwanted reduction in porosity and an increase in pore size. Therefore, control of the thickness of the sponge-like region of the fibre and the effect that this change in thickness has on other fibre properties, such as the pore size distribution, is critical when preparing fibres using the combined phase inversion and sintering method.

Air-gap: Figure 3.6a–g shows precursor fibres spun with air-gaps of 0 cm, 2 cm and 15 cm, an internal coagulant flow rate of 10 ml min^{-1} and an extrusion velocity of approximately 3.5 cm s^{-1}. It can be seen from Figure 3.6a that the morphology of fibres spun directly into a non-solvent bath (0 m airgap) consists of finger-like voids originating from both the inner and outer fibre surfaces which extend almost to the centre of the fibre cross section. A central sponge-like region is present which provides the majority of the mechanical strength and separation characteristics. Maximum void length is approximately the same for voids originating from both the inner and outer surfaces and as in all the prepared fibres a void length distribution exists, some being only a few microns in length while others penetrate far into the fibre cross section. This structure may not be ideal for some of the principal applications of ceramic hollow fibre membranes such as solvent filtration which generally require the separation layer (packed pore mostly originated from the sponge-like region after heat treatment) to be at either the inner or outer edge. However, the above structure may be beneficial for the development of catalytic membrane reactors, as finger-like voids may serve as substrates for catalyst particle impregnation. For example, a multifunctional catalytic membrane reactor could be developed with different catalytic functions

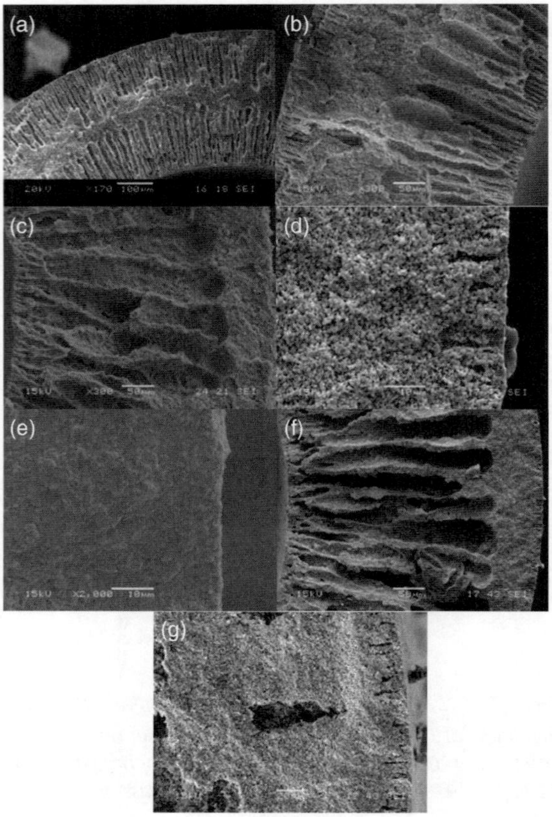

Figure 3.6 *Cross sectional images of precursor fibres (a–c) and sintered fibres (d–g): (a) 0 cm air gap, (b) 2 cm air gap, (c) 15 cm air gap, (d) outer edge (15 cm air gap) calcined at 1450 °C, (e) outer edge (15 cm air gap) calcined at 1600 °C, (f) 15 cm air gap calcined at 1450 °C, (g) an isolated void. Reprinted from Catalysis Today, Benjamin F.K. Kingsbury, Zhentao Wu and K. Li, A morphological study of ceramic hollow fibre membranes: A perspective on multifunctional catalytic membrane reactors, In Press, Corrected Proof. Copyright (2010) with permission from Elsevier*

at the inner and outer surfaces with the central region of the membrane determining the permeation characteristics. This could be achieved by depositing two different types of catalyst targeted at different reactions within the inner and outer finger-like voids respectively, as shown schematically in Figure 3.7. Compounds A and B are reacted in Reaction Zone 1 in the presence of Catalyst 1 to yield product C. The product mixture is then enriched in a particular component as it passes through the separation layer, after which it is reacted with compound D in Reaction Zone 2 in the presence of Catalyst 2 to yield product E. Not only does the separation layer serve to separate reactants from products but it also disperses the permeate phase uniformly in Reaction Zone 2. Such a scheme could be employed to combine reforming and watergas shift reactions within the same hollow fibre membrane reactor. For example, a mixture of methanol and water could be converted to hydrogen, carbon dioxide and carbon monoxide in Reaction Zone 1. The permeate mixture, enriched in hydrogen as a consequence of passing through the separation layer, may undergo a watergas shift reaction in Reaction Zone 2 with steam being fed into the

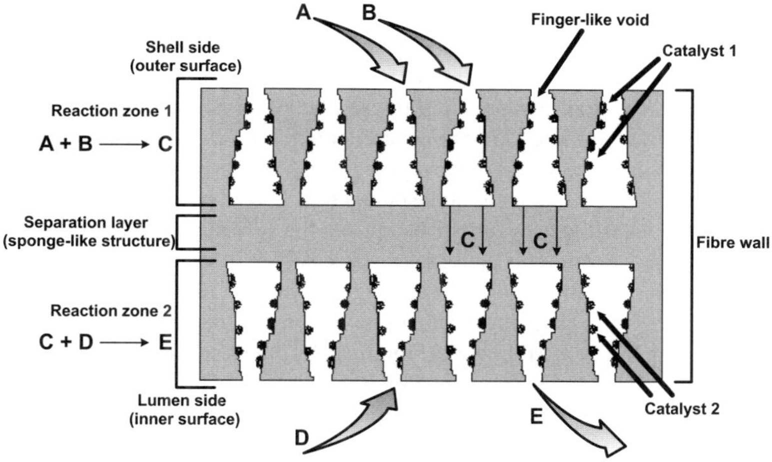

Figure 3.7 *Schematic diagram of a multifunctional inorganic hollow fibre membrane reactor performing two complementary reactions*

lumen side of the fibre, thus increasing the concentration of hydrogen further and reducing the concentration of carbon monoxide. This scheme has been shown to be preferable to the use of a separation layer with a high selectivity for hydrogen due to the low permeance associated with such a selective layer [27]. Greater reactor fluxes and efficiencies can be achieved if the permeance of the separation layer is increased at the expense of hydrogen selectivity and the water–gasshift reaction is employed to further increase the hydrogen concentration in the permeate stream. As with the combination of reforming and watergas shift reactions, an additional increase in reactor efficiency may result from the union of exothermic and endothermic reactions as the heat generated during the exothermic reaction is available to shift the equilibrium of the endothermic reaction towards the product side. In a variation of this scheme, reaction of A with B may yield two or more products which permeate through the sponge-like region and react together in Reaction Zone 2. In this case, additional reactants are not supplied to the lumen side of the fibre.

Figure 3.6b shows the fibre morphology resulting from a 2 cm air-gap. As shown, the finger-like voids extend from the inner surface across approximately 50% of the fibre cross section but void length at the outer surface has been greatly reduced. A sponge-like region occupying approximately 35% of the fibre cross section is present between the inner and outer finger-like voids. The size and number of voids at the outer edge are further reduced when the air-gap is increased to 15 cm as shown in Figure 3.6c. As can be seen, finger-like voids extend from the inner surface across approximately 80% of the fibre cross section with the remaining 20% consisting of a sponge-like region. Close examination of the outer edge of the fibre, as shown in Figure 3.6d at increased magnification, reveals the presence of small finger-like voids despite the increased air-gap. However, it is observed from Figure 3.6e that these small voids can be eliminated with high temperature treatment at 1600 °C and that the membrane macrostructure is retained during the calcination process, as seen for the fibre calcined at 1450 °C shown in Figure 3.6f.

The morphology of the fibre shown in Figure 3.6a is believed to result from hydrodynamically unstable viscous fingering occurring simultaneously and to a similar extent at both the inner and outer fibre surfaces. The situation is somewhat different when a 2 cm air-gap is present (Figure 3.6b). In this case simultaneous solvent evaporation and moisture (non-solvent) condensation causes a local viscosity increase in the outer region of the fibre prior to immersion. During the time the nascent fibre is exposed to the atmosphere (2 cm air gap), the viscosity of the outer region increases and finger-like voids originating at the inner surface penetrate into the fibre cross section. As the nascent fibre makes contact with the non-solvent bath the increased viscosity of the outer region, as a result of exposure to the atmosphere, inhibits the growth of finger-like voids at the outer edge. However, non-solvent influx from the precipitation bath does still occur, further increasing the suspension viscosity as it penetrates the nascent fibre cross section. This increases the viscosity in front of finger-like voids growing from the inner surface, limiting their length to approximately 50% of the fibre cross section. Therefore finger-like voids forming from the inner surface can be increased in length by increasing the air gap to 15 cm as shown in Figure 3.6c. Under these conditions the viscosity of the outer region is higher when the nascent fibre is immersed and non-solvent influx is reduced, while at the same time voids originating at the inner surface have more time to penetrate further towards the outer edge. The presence of small finger-like voids in the outer region shows that the suspension viscosity in this region is not above the threshold at which the finger-like voids can be totally suppressed.

It has already been shown that a reduction in air-gap gives rise to an increase in the thickness of the sponge-like region and consequently an increase in mechanical strength [28]. However, the appearance of isolated voids in the sponge-like structure, as shown in Figure 3.6g, has been observed which will lead to a reduction in mechanical strength. Even if the viscosity near the suspension/air interface has increased considerably before immersion, non-solvent will still diffuse into and through this region when the nascent fibre is immersed. As non-solvent diffuses through the outer sponge-like region of high viscosity and reaches a region of lower viscosity, viscous fingering takes place and isolated void formation is initiated. Isolated voids can be eliminated if the suspension viscosity in this region is above the viscous fingering threshold before non-solvent influx occurs and it has been shown that both the internal coagulant flow rate and the air-gap are important parameters in this respect.

Therefore, the effect of air-gap on the fibre morphology and pore size of fibres prepared with air-gaps of between 3–13 cm and an internal coagulant flow rate of 7 ml min^{-1} was investigated and the data is shown in Figure 3.8a–f and Figure 3.9, respectively. Both the relationship between the air-gap and the thickness of the sponge-like region as well as the presence of isolated voids can be clearly seen and, as described previously, an increase in air-gap leads to an increase in the length of the finger-like voids. In addition, due to the relatively low internal coagulant flow rate of 7 ml min^{-1}, isolated voids are also observed in the sponge-like region of the fibre. It can be seen from mercury intrusion data, Figure 3.9, that finger-like void entrance size varies greatly with the highest values of between 4–5 μm for fibres prepared with air-gaps of 3, 7, 9 and 13 and values of 0.4 and 1.3 μm for fibres prepared with air-gaps of 5 and 11 cm, respectively. Both isolated voids and a reduction in the size or elimination of the entrances to the finger-like voids are detrimental to the function of the asymmetric membrane or membrane support for many applications. In addition, it can be seen from the data in Figure 3.9 that at an internal coagulant flow rate of only 7 ml min^{-1} the pore size of the entrances to the finger-like voids is unpredictable and cannot be correlated with air-gap.

However, it can be seen from the SEM images of fibres prepared with air-gaps between 1–13 cm and with an internal coagulant flow rate of 12 ml min^{-1}, Figure 3.10a–f , that despite the progressive increase in the thickness of the sponge-like region as the air-gap is reduced, accounting for over 50% of the thickness of the fibre cross section in fibres prepared using an air-gap of only 1 cm, the almost total absence of isolated voids is apparent. In addition, despite

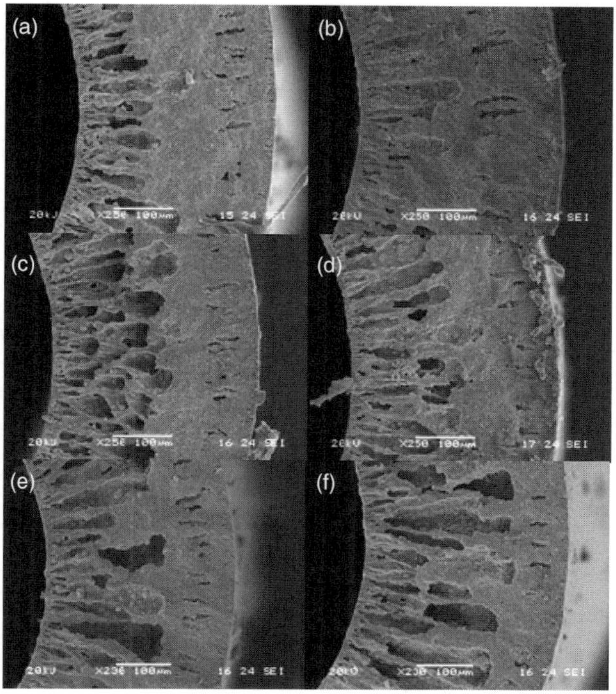

Figure 3.8 Cross-sectional SEM images of fibres prepared with air-gaps between 3 and 13 cm, an internal coagulant flow rate of 7 ml min^{-1} and calcined at 1450 °C: (a) 3 cm, (b) 5 cm, (c) 7 cm, (d) 9 cm, (e) 11 cm, (f) 13 cm. Reprinted from Catalysis Today, Benjamin F.K. Kingsbury, Zhentao Wu and K. Li, A morphological study of ceramic hollow fibre membranes: A perspective on multifunctional catalytic membrane reactors, In Press, Corrected Proof. Copyright (2010) with permission from Elsevier

the reduction in the length of the air-gap, the mercury intrusion data shown in Figure 3.11 demonstrates that the pore size of the entrances to the finger-like voids remains at between 7–9 μm. Therefore, the thickness of the sponge-like region and hence the mechanical strength of the fibre can be controlled by varying the air-gap only if the internal coagulant flow rate is sufficient to prevent the formation of isolated voids and to prevent a decrease in the size of the entrances to the finger-like voids.

Internal coagulant flow rate: Fibre morphology is also strongly influenced by the flow rate of the internal coagulant during the spinning process. SEM images and mercury intrusion data for fibres prepared using a 15 cm air-gap and calcined at 1450 °C with internal coagulant flow rates of 3, 6, 7, 8, 9, 12 and 15 ml min^{-1} are shown in Figure 3.12a–g and Figure 3.13 respectively. The progressive increase in the thickness of the sponge-like region as the internal coagulant flow rate is decreased is clear from the SEM images and the characteristic bimodal pore size distribution associated with fibres prepared in this way is evident from the mercury intrusion data, Figure 3.13. In all cases the peak representing the sponge-like region can be found at approximately 0.1–0.2 μm. However, the peak representing the pores at the inner surface formed by the entrances to the finger-like voids, resulting from the viscous fingering phenomenon, is present at 1.2 μm for a flow rate of 6 ml min^{-1}, at approximately 5.5 μm for flow rates of between

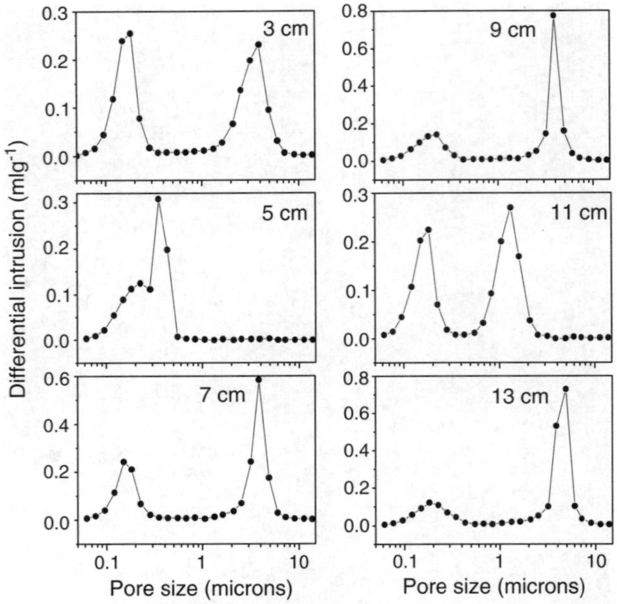

Figure 3.9 *Mercury intrusion data for fibres prepared with air-gaps between 3 and 13 cm, an internal coagulant flow rate of 7 ml min^{-1} and calcined at 1450 °C*

7–12 ml min^{-1} and at 9 μm at a flow rate of 15 ml/ min^{-1}. Notably absent is a second peak representing the pores formed by the entrances to the finger-like voids for the fibre prepared with an internal coagulant flow rate of only 3 ml min^{-1}. In this case, the driving force for the formation of finger-like voids is insufficient to overcome the viscosity of the suspension at the interface between the suspension film and the internal coagulant. Consequently, viscous fingering in not initiated at this interface but is instead initiated within the suspension film itself and a sponge-like region is formed at the inner fibre surface. Also notable in the fibre prepared with an internal coagulant flow rate of 3 ml min^{-1} are finger-like voids originating from the shell side of the fibre. A corresponding peak in the mercury intrusion data is also absent, indicating that the finger-like voids do not originate at the outer fibre surface. In this case, finger-like voids are isolated within the sponge-like structure and serve only to reduce the mechanical strength of the fibre. Therefore, although the thickness of the sponge-like region of the fibre can be increased but varying the internal coagulant flow rate, this is also associated with a reduction in the size of the entrances to the finger-like voids as the flow rate is reduced.

Non-solvent additive: The importance of the suspension viscosity on fibre morphology has been demonstrated by the addition of water as a non-solvent additive to increase the suspension viscosity [28]. Increasing the viscosity by other means, such as by varying the ratio of solvent/polymer/alumina or by the introduction of additives such as polyvinylpyrrolidone (PVP), may affect the membrane properties considerably and for this reason, water is a good choice as a viscosity enhancer. The addition of water to spinning suspensions, measured as a percentage of the total solvent content, caused an increase in viscosity as shown in Figure 3.14. The suspension viscosity value of 12.1 Pa.s at 0 wt% water increased to 16.7, 18.5 and 22 Pa.s at 2, 4 and 6 wt% water content respectively. Further addition of water resulted in large viscosity increases to 33.3

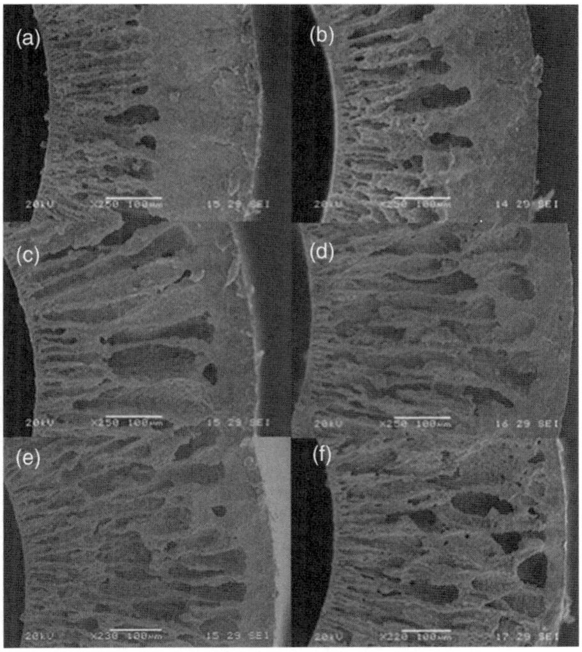

Figure 3.10 *Cross-sectional SEM images of fibres prepared with air-gaps between 1 and 13 cm, an internal coagulant flow rate of 12 ml min^{-1} and calcined at 1450 °C: (a) 1 cm, (b) 3 cm, (c) 7 cm, (d) 9 cm, (e) 11 cm, (f) 13 cm. Reprinted from Catalysis Today, Benjamin F.K. Kingsbury, Zhentao Wu and K. Li, A morphological study of ceramic hollow fibre membranes: A perspective on multifunctional catalytic membrane reactors, In Press, Corrected Proof. Copyright (2010) with permission from Elsevier*

and 45 Pa.s at 8 and 10 wt% respectively. The rate of viscosity increase as a function of water content is seen to increase at 6 wt% indicating that if all other parameters remain the same, fibre morphology may change abruptly at this point. The extrusion pressure was increased in line with viscosity in order to maintain a constant extrusion rate over the range of suspension compositions.

Figure 3.15a–j illustrates the effect of spinning suspension viscosity on fibre morphology for calcined fibres prepared with a 15 cm air-gap and an internal coagulant flow rate of 10 ml min^{-1}. It can be seen in Figure 3.15a that, at 0 wt% water content, finger-like voids extent across approximately 80% of the fibre cross section with the remaining 20% consisting of a sponge-like region. At higher magnification, Figure 3.15b shows the existence of small finger-like voids at the outer fibre edge which are present in fibres prepared with water concentrations below 6 wt%. Addition of 2 wt% water results in a large reduction in finger-like void length to approximately 40% of the cross section as shown in Figure 3.15c and in the appearance of inward facing isolated voids in the centre of the sponge-like region which can be seen in Figure 3.15d, e. Figure 3.15f shows the effect of the addition of 4 wt% water. Again, finger-like voids from the inner surface penetrate about 40% of the cross section but isolated void formation is reduced in this case. The total elimination of isolated voids resulting from the addition of 6 wt% water can be seen in Figure 3.15g and at higher magnification, Figure 3.15h

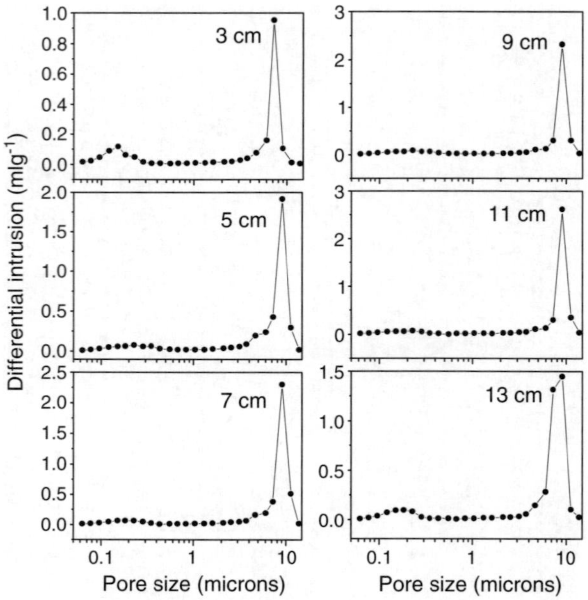

Figure 3.11 *Mercury intrusion data for fibres prepared with air-gaps between 3 and 13 cm, an internal coagulant flow rate of 12 ml min and calcined at 1450 °C*

shows the elimination of small finger-like voids at the outer edge. However, it is shown in Figure 3.15g that void penetration from the inner surface remained unchanged at 40%, while Figure 3.15i shows a reduction in void length to about 30% at 8 wt% water content. Total void elimination resulted with the addition 10 wt% water, that is, the symmetric membrane structure shown in Figure 3.15j.

The presence of water in the spinning suspension has two effects. Firstly, the initial suspension viscosity is increased and secondly, the suspension viscosity increases more rapidly when in contact with non-solvent, or as a result of solvent evaporation, as the polymer phase is closer to its precipitation point. An increase in suspension viscosity inhibits viscous fingering at both the inner and outer surfaces of the fibre and finger-like void length is reduced. Close examination of the outer fibre edges reveals small finger-like voids – shown at higher magnification in Figure 3.15b, d – resulting from non-solvent influx from the precipitation bath which, although small, are likely to diminish mechanical strength by reducing the integrity of the sponge-like structure. Despite the presence of a 15 cm air-gap, at water concentrations below 6 wt% the viscosity of the outer region is below the viscous fingering threshold when the fibre is immersed and small finger-like voids are seen in all fibres regardless of the internal coagulant flow rate. These small finger-like voids are eliminated if the viscosity of the outer region is above a critical value before the outer fibre surface is brought into contact with non-solvent. This threshold is exceeded for fibres prepared with at least 6 wt% water and a 15 cm air-gap. It can also be seen from Figure 3.15d, e that the inward facing isolated voids observed in fibres prepared with less than 6 wt% water content and an internal coagulant flow rate of 10 ml min^{-1} are formed by non-solvent influx from the shell side of the fibre. This is not the case for fibres prepared at 2 and 4 wt% water with an internal coagulant flow rate of 10 ml min^{-1}. However, by increasing the internal coagulant flow rate so as to maintain a constant ratio between flow rate and suspension viscosity,

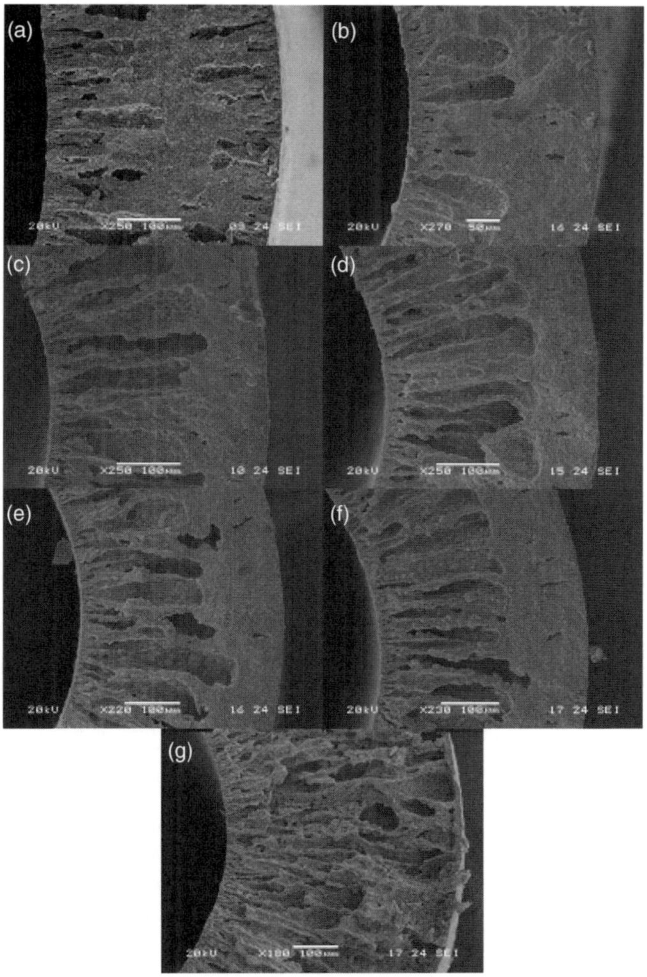

Figure 3.12 Cross-sectional SEM images of fibres prepared with varying flow rates and calcined at 1450 °C. Flow rate: (a) 3 ml min^{-1}, (b) 6 ml min^{-1}, (c) 7 ml min^{-1}, (d) 8 ml min^{-1}, (e) 9 ml min^{-1}, (f) 12 ml min^{-1} and (g) 15 ml min^{-1}. Reprinted from Catalysis Today, Benjamin F.K. Kingsbury, Zhentao Wu and K. Li, A morphological study of ceramic hollow fibre membranes: A perspective on multifunctional catalytic membrane reactors, In Press, Corrected Proof. Copyright (2010) with permission from Elsevier

isolated voids can be eliminated in all fibres prepared from suspensions with water as a non-solvent additive, as seen in Figure 3.16a–e . In addition, it is observed that small finger-like voids present at the outer fibre surface at a water concentration of 4 wt% (Figure 3.16f) are eliminated as the water concentration is increased to 6 wt% (Figure 3.16g). Under these circumstances, internal coagulant influx and finger-like void growth from the inner surface are greater. Consequently, the viscosity of the region in which isolated voids are observed is above the viscous fingering threshold before non-solvent influx from the precipitation bath occurs and a sponge-like structure

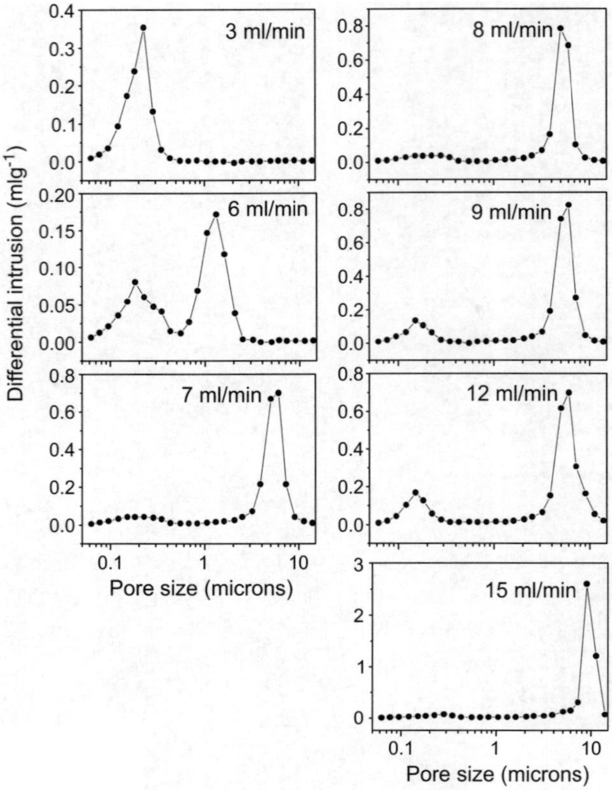

Figure 3.13 *Mercury intrusion data for fibres prepared with internal coagulant flow rates between 3 and 15 ml min^{-1} calcined at 1450 °C*

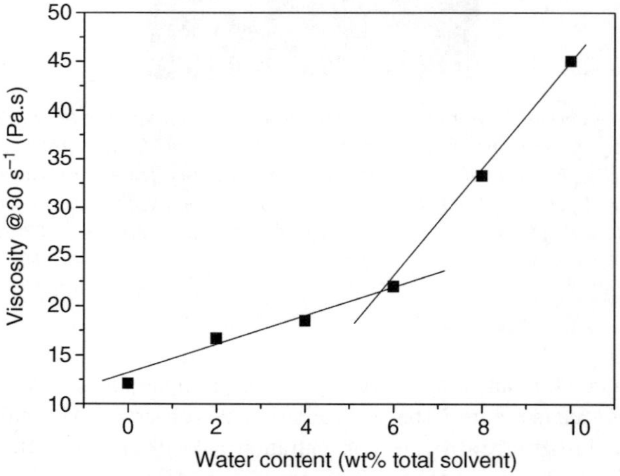

Figure 3.14 *Viscosity of spinning suspensions prepared with 0–10 wt% water (as a percentage of the total solvent content). Reprinted from Catalysis Today, Benjamin F.K. Kingsbury, Zhentao Wu and K. Li, A morphological study of ceramic hollow fibre membranes: A perspective on multifunctional catalytic membrane reactors, In Press, Corrected Proof. Copyright (2010) with permission from Elsevier*

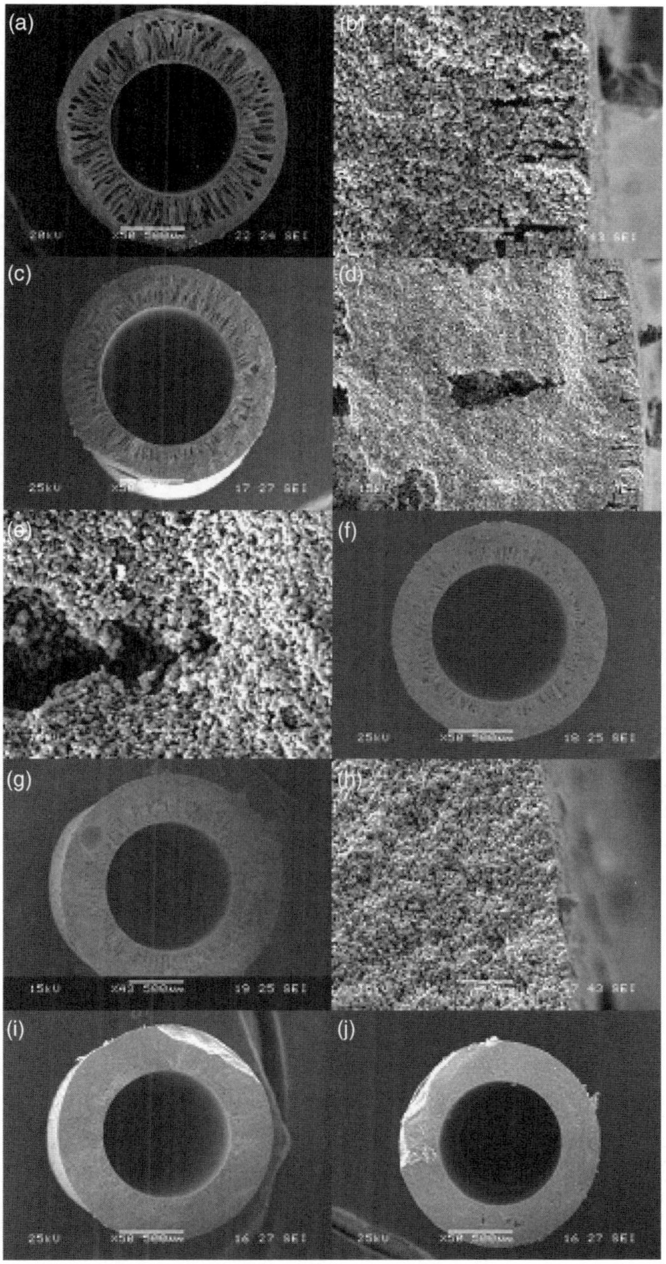

Figure 3.15 *Cross-sectional images of fibres made from spinning suspensions containing varying amounts of water: (a) 0 wt% water, (b) 4 wt% water, outer edge, (c) 2 wt% water, (d) 4 wt% water, isolated void, (e) 4 wt% water, enlargement of isolated void, (f) 4 wt% water, (g) 6 wt% water, (h) 6 wt% water, outer edge, (i) 8 wt% water, (j) 10 wt% water. Reprinted from Journal of membrane science, Benjamin F. K. Kingsbury and K Li, 328 (1-2) 134-140. A morphological study of ceramic hollow fibre membranes. Copyright (2009) with permission from Elsevier*

Figure 3.16 Cross-sectional images of fibres made from spinning suspensions containing varying amounts of water, prepared with a flow rate to suspension viscosity at 30 s^{-1} ratio of 0.83 and calcined at 1450 °C: (a) 2 wt% water, 13.8 ml min^{-1}, (b) 4 wt% water, 15.6 ml min^{-1}, (c) 6 wt% water, 18 ml min^{-1}, (d) 8 wt% water, 27.6 ml min^{-1}, (e) 10 wt% water, 37.2 ml min^{-1}, (f) 4 wt% water, 15.6 ml min^{-1} (outer edge), (g) 6 wt% water, 18 ml min^{-1} (outer edge). Reprinted from Journal of membrane science, Benjamin F. K. Kingsbury and K Li, 328 (1-2) 134-140. A morphological study of ceramic hollow fibre membranes. Copyright (2009) with permission from Elsevier

results. In addition it can be seen that, due to the increased internal coagulant flow rate, finger-like void structures are present in the fibre prepared with 10 wt% water, indicating that in order to produce a symmetric hollow fibre at an internal coagulant flow rate of 37.2 ml min^{-1} the water concentration must be increased.

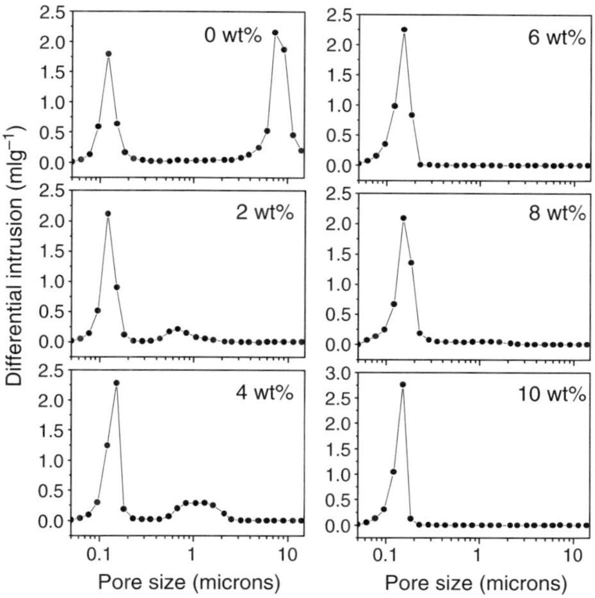

Figure 3.17 *Mercury intrusion data for fibres prepared with 0–10 wt% water (as a percentage of the total solvent content) calcined at 1200 °C*

As discussed, the relative ratios of the finger-like void and sponge-like regions in asymmetric hollow fibres, that is, the thickness of the sponge-like region of the fibre can be varied by adding a non-solvent additive to the spinning suspension. Figure 3.17 shows mercury intrusion data for fibres prepared from suspensions containing water as a non-solvent additive, calcined at 1200 °C and with an internal coagulant flow rate of 10 ml min^{-1}. The data clearly demonstrates the dramatic effect that the viscosity increase has on the pore size of the entrances to the finger-like voids. A peak at approximately 0.1–0.2 μm represents the sponge-like region of the fibre and varies little as the concentration of water is increased. This is not the case however for the pores formed by the entrances to the finger-like voids, which are dramatically reduced in size from 5 μm to approximately 1 μm even after the addition of only 2 and 4 wt% water. Further addition of water results in the elimination of this peak indicating that the entrances to the finger-like voids are no longer present at the inner fibre surface. Therefore, despite the presence of finger-like voids and the asymmetry of the hollow fibre cross section, the mercury intrusion data shows that the addition of water to the spinning suspension first results in a reduction in the size of the pores formed by the entrances to the finger-like voids and then in the formation of a sponge-like region at the inner surface of the fibre. This trend can also be clearly seen from the mercury intrusion data for fibres prepared with increased internal coagulant flow rates and calcined at 1200 °C, Figure 3.18. Again, as the concentration of water in the spinning suspension is increased, a dramatic reduction in the pore size of the entrances to the finger-like voids is observed, culminating in the elimination of finger-like void entrances at the inner fibre surface at a water concentration of 10 wt%. If the suspension viscosity is above a critical value, the viscous fingering phenomenon is not observed and finger-like void growth is prevented. In this case, due to the presence of non-solvent in the spinning suspension and the large availability of non-solvent near the surface of the suspension film, the viscosity in this region is prohibitively high and

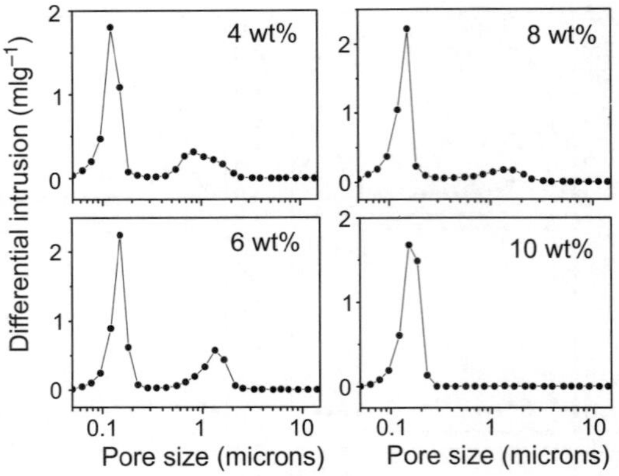

Figure 3.18 *Mercury intrusion data for fibres prepared with 0–10 wt% water (as a percentage of the total solvent content) with a flow rate to suspension viscosity at 30 s^{-1} ratio of 0.83 and calcined at 1200 °C*

viscous fingering does not occur. However, as non-solvent diffuses further into the suspension film, where the availability of non-solvent is reduced, the concentration of solvent is higher and the suspension viscosity is lower, viscous fingering then occurs and initiates the formation of finger-like voids in this region. Consequently, the morphology that results is that of a sponge-like region at the inner fibre surface and finger-like voids that originate from within the bulk of the fibre. It is not only the initial suspension viscosity that is important in determining the morphology but also the rate of the increase in viscosity once the phase inversion process has been initiated, which is determined by the precipitation rate of the polymer phase. This can be demonstrated by the addition of ethanol as a non-solvent additive to the spinning suspension. Although the solubility of polyethersulfone in ethanol is lower than in NMP but higher than in water, the suspension viscosity initially decreases as the concentration of ethanol is increased and is only seen to increase at higher ethanol concentrations, Figure 3.19. The effect of ethanol addition on the pore size of the entrances to the finger-like voids for fibres calcined at 1450 °C can be seen in Figure 3.20. The data shows that the pore size is reduced slightly to between 5 and 6 μm after the addition of 2 and 4 wt% ethanol, when compared to the data for the fibre prepared without the addition of a non-solvent additive, as seen in Figure 3.17. However, as the concentration of ethanol is increased to 6 and 10 wt% the pore size drops to approximately 1 μm despite the reduction in suspension viscosity to 8.5 Pa.s, demonstrating the importance of the polymer precipitation rate in determining the pore size of the entrances to the finger-like voids. When controlling the thickness of the sponge-like region by the addition of a non-solvent additive, the pore size of the entrances to the finger-like voids is substantially reduced and the finger-like voids may no longer be accessible above a certain non-solvent concentration, depending on the non-solvent in question.

3.2.3 Calcination

Previous studies have shown that densification of sponge-like regions occurs during calcination, causing a decrease in permeability at high Al_2O_3/polyethersulfone ratios [29], and may eventually result in a gas tight membrane at high calcination temperatures [28]. It can be clearly

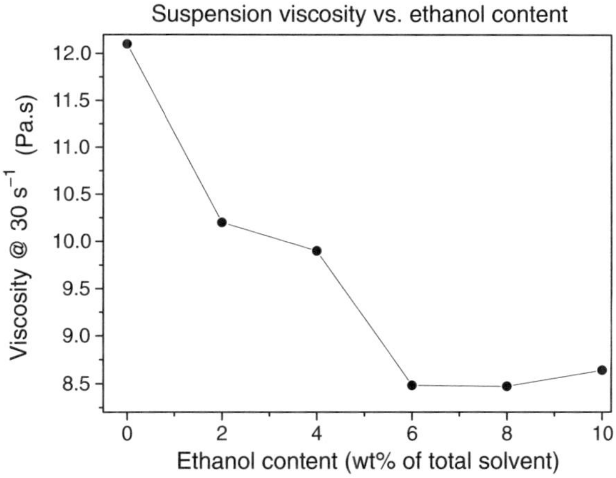

Figure 3.19 *Viscosity of spinning suspensions prepared with 0–10 wt% ethanol (as a percentage of the total solvent content). Reprinted from Catalysis Today, Benjamin F.K. Kingsbury, Zhentao Wu and K. Li, A morphological study of ceramic hollow fibre membranes: A perspective on multifunctional catalytic membrane reactors, In Press, Corrected Proof. Copyright (2010) with permission from Elsevier*

seen Figure 3.6d, e showing SEM images of the sponge-like region of a fibre calcined at 1450 and 1600 °C respectively, and prepared with an Al_2O_3/polymer ratio of approximately 10, that densification of this region has indeed occurred. This is in contrast to the finger-like void region which is retained even at elevated sintering temperature as shown by a comparison of a precursor fibre, Figure 3.6c and the same fibre calcined at 1450 °C, Figure 3.6g. It can also be seen from the gas permeation data shown in Figure 3.21 that as the calcination temperature is increased the gas permeation flux decreases, suffering a dramatic drop between 1400 and 1450 °C. At 1500 °C the permeation flux is barely detectable and is shown to be totally eliminated in fibres calcined at 1600 °C.

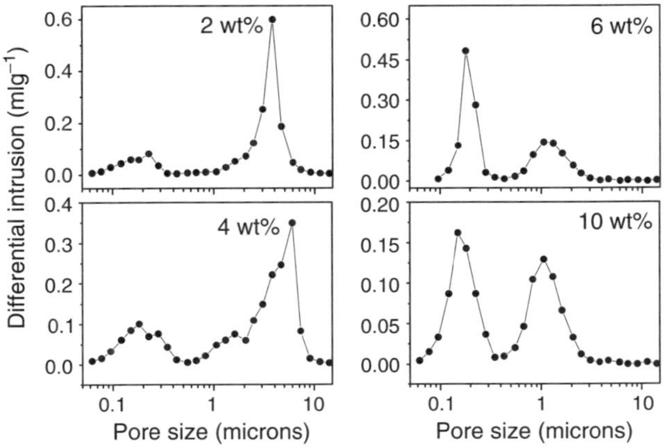

Figure 3.20 *Mercury intrusion data for fibres prepared with 0–10 wt% ethanol (as a percentage of the total solvent content), with an internal coagulant flow rate of 10 ml min^{-1} and calcined at 1450 °C*

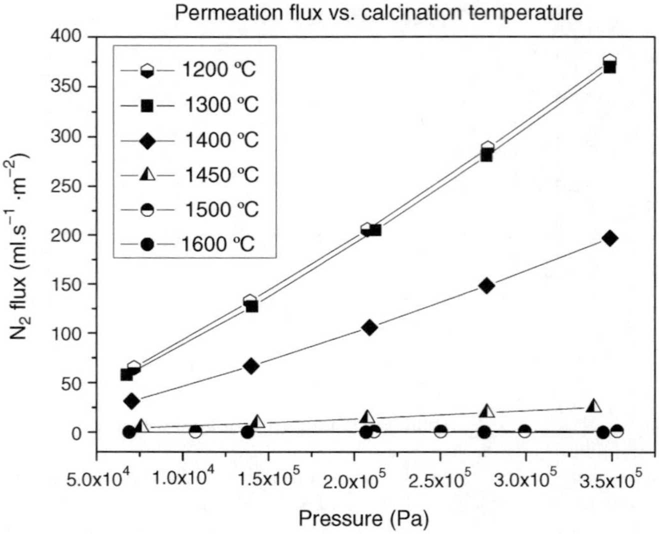

Figure 3.21 *Gas permeation data for a fibre prepared with a 15 cm air-gap, an internal coagulant flow rate of 10 ml min^{-1} and calcined at temperatures of between 1200 and 1600 °C. Reprinted from Journal of membrane science, Benjamin F. K. Kingsbury and K Li, 328 (1-2) 134–140. A morphological study of ceramic hollow fibre membranes. Copyright (2009) with permission from Elsevier*

In order to impart the fibre with sufficient mechanical strength for practical applications, a minimum sintering temperature is required, which depends on the fibre morphology. The effect of sintering temperature on mechanical strength can be seen from Figure 3.22 for asymmetric fibres prepared with a 15 cm air-gap, an internal coagulant flow rate of 10 ml min^{-1} and an Al_2O_3/polymer ratio of approximately 10. The mechanical strength of the fibre is a function of

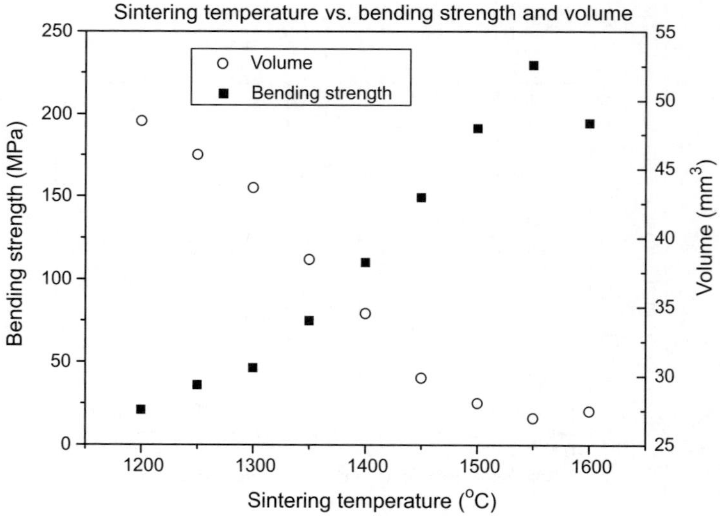

Figure 3.22 *Fibre volume and bending strength as a function of temperature*

both the porosity and the grain size of the ceramic material. As the sintering temperature is increased the material undergoes densification and the mechanical strength increases. However, at temperatures above 1550 °C densification is seen to tail off and continued sintering results in an increase in grain size, leading to a reduction in mechanical strength. As sintering temperature is a critical factor in determining the mechanical strength of the hollow fibre membrane or membrane support structure and a minimum sintering temperature is often required, it is important to determine the effect of sintering temperature on the pore size distribution. Pore size data generated using mercury intrusion porosimetry and a bubble point method can be seen in Figures 3.23 and 3.24, respectively. The data demonstrates the effect of sintering temperature on hollow fibres prepared using a 15 cm air-gap, an Al_2O_3/polymer ratio of approximately 10 and calcined at temperatures between 1200 and 1600 °C, at intervals of 100 °C. An additional interval at 1450 °C is included due to the large amount of morphological change that occurs at temperatures between 1400 and 1500 °C. In order to determine the pore size distribution of both through and dead end pores present in the sponge-like region of the fibre, as well as the pores at the inner fibre surface formed by the entrances to the finger-like voids, both mercury intrusion porosimetry and a bubble point method were employed. It can be seen from mercury intrusion data that a bimodal pore size distribution exists and that as the sintering temperature is increased from 1200 to 1600 °C the peak pore size of the sponge like region increases from 0.12 μm at 1200 °C to approximately 0.18 μm at 1400 °C. This increase is accompanied by a progressive reduction in pore volume which continues up to 1500 °C, after which the sponge-like region is seen to densify [28]. It can be seen from pore size data generated using a bubble point method, Figure 3.24, that the peak pore size for the sponge-like region is present between 0.14–0.16 μm

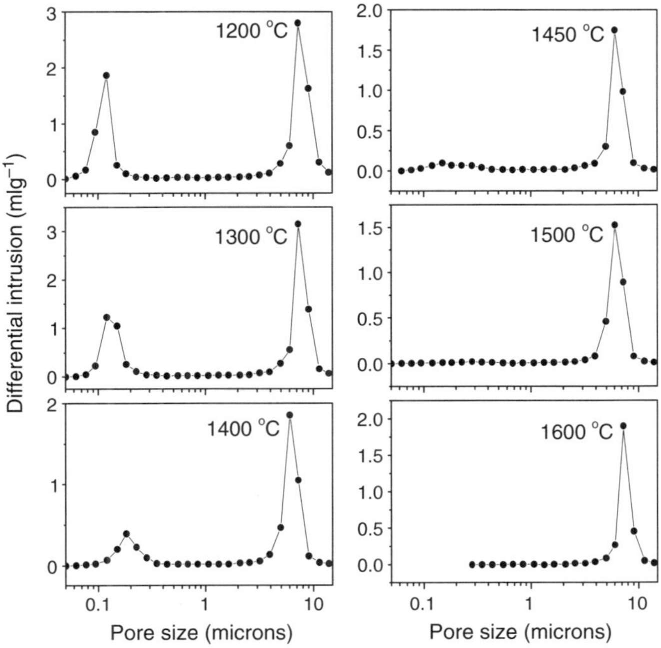

Figure 3.23 Mercury intrusion data of pore size versus sintering temperature for an asymmetric hollow fibre prepared with an air-gap of 15 cm and an internal coagulant flow rate of 10 ml min^{-1}

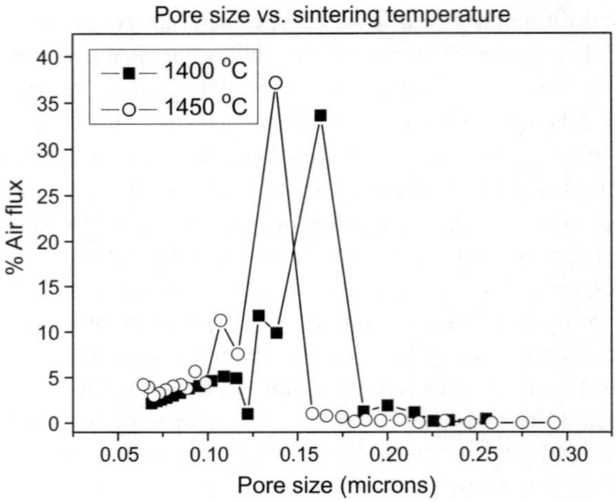

Figure 3.24 *Bubble point data of pore size versus sintering temperature for an asymmetric hollow fibre prepared with an air-gap of 15 cm and an internal coagulant flow rate of 10 ml min*$^{-1}$. *Reprinted from Catalysis Today, Benjamin F.K. Kingsbury, Zhentao Wu and K. Li, A morphological study of ceramic hollow fibre membranes: A perspective on multifunctional catalytic membrane reactors, In Press, Corrected Proof. Copyright (2010) with permission from Elsevier*

for fibres prepared under the same conditions and calcined at 1400 and 1450 °C. As only through pores are measured using this technique the pore size distribution of both through and dead end pores present in the sponge-like region of the fibre is shown to compare well, establishing that mercury intrusion porosimetry can be used to infer the pore size of both through and dead end pores. It can be seen from mercury intrusion data that a reduction in pore size is not observed for the pores at the inner fibre surface formed by the entrances to the finger-like voids, resulting from the viscous fingering phenomenon, represented by the peak between 6–7 μm. As the sintering temperature is increased, despite the elimination of porosity and the increase in pore size for the sponge-like region, the pore size of the pores at the inner surface formed by the entrances to the finger-like voids is first reduced slightly to approximately 6 μm and then returns to 7 μm at 1600 °C. Therefore, it has been demonstrated that for fibres prepared under these conditions, finger-like void entrances are open and accessible even at elevated temperature.

3.3 Coating of Pd/Ag Membranes

Pd or Pd/Ag membranes may be coated onto the outer surface of asymmetric alumina hollow fibre substrates by an electroless plating technique [30]. Prior to coating, the substrates are cleaned and activated by a conventional Pd-Sn activation procedure. The activation process consists of successive immersion of the substrates in a tin (II) chloride ($SnCl_2$) solution and a palladium chloride ($PdCl_2$) solution at room temperature. Deionised water and 0.1 M HCl are used to rinse the samples between the immersions. The activation process is repeated 6 times, after which the substrate surface turns brown. While the formation of a pure Pd membrane is achieved using an electroless plating technique followed by heat treatment at 450 °C for 4 h under hydrogen, Pd/Ag composite membranes are coated onto the outer surface of the substrates by a

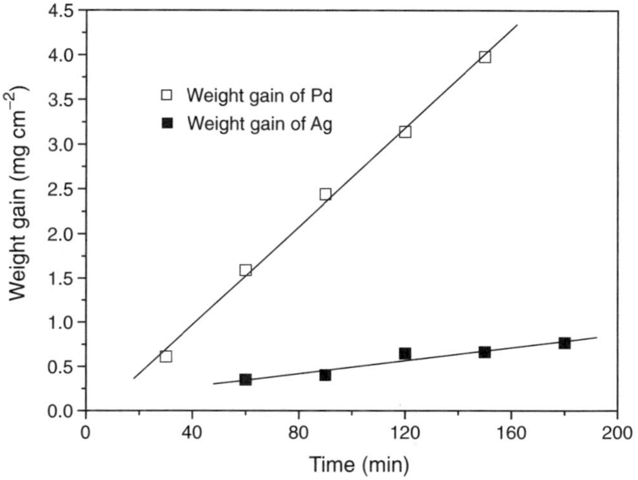

Figure 3.25 *Weight gain of Pd and Ag during electroless plating. Reprinted from Zhentao Wu, Irfan M. D. Hatim, Benjamin F.K. Kingsbury, Ejiro Gbenedio and K. Li, A novel inorganic hollow fiber membrane reactor for catalytic dehydrogenation of propane, AIChE Journal 55 (9) 2389–2398. Copyright (2009) with permission from John Wiley and Sons*

coating-diffusion electroless plating method [30] where the plating bath is refreshed every hour. For coating of a Pd/Ag membrane, prior to deposition of Ag, the surface of the Pd membrane must be reactivated so as to produce uniform Ag plating. The composition of the Pd/Ag membranes can be controlled by the plating rates of Pd and Ag as shown in Figure 3.25. The weight gain of Pd and Ag increases linearly with time during the electroless plating and the mechanism of the electroless plating (ELP) technique is based on the controlled autocatalytic reduction of the metastable metallic salt complexes on the target surfaces [31]. Nonlinear behaviour is observed when the ELP is carried out continuously without changing the ELP solution, in which case the number of the Pd nucleation sites and the concentration of the metallic salt complexes change with time. This is not the case if the ELP solution is refreshed every hour, as a result of which the Pd and Ag plating behaviour is linear with respect to time with the plating rate of Pd being approximately 6.5 times that of Ag. This plating profile can be further confirmed by preparing Pd/Ag membranes with different thicknesses, which are calculated based on the theoretical densities of Pd and Ag. The calculated membrane thicknesses agree well with those rom SEM characterisation.

Figure 3.26 shows the morphology of a Pd/Ag membrane with a thickness of around 8 μm. The reason for choosing a relatively thick membrane for SEM as well as the subsequent Energy Dispersive Spectroscopy (EDS) analysis is to highlight the microstructural change of the membranes during the electroless plating process. As can be seen from Figure 3.26a, b, a uniform Pd layer was first deposited on the activated substrate with the Pd grains tightly bound on the surface without observable defects. A less defined grain boundary on the surface of the Ag layer is shown in Figure 3.26c. A clear double-layer structure was obtained after the deposition of Ag as shown in Figure 3.26d. Figure 3.26e illustrates that the boundary between the Pd and Ag layers disappeared after heat treatment at 650 °C for 12 h under argon due to intermetallic diffusion between the layers. Although some small pores are present on the surface of the composite membrane after the heat treatment (Figure 3.26f), these pores are dead volumes as the

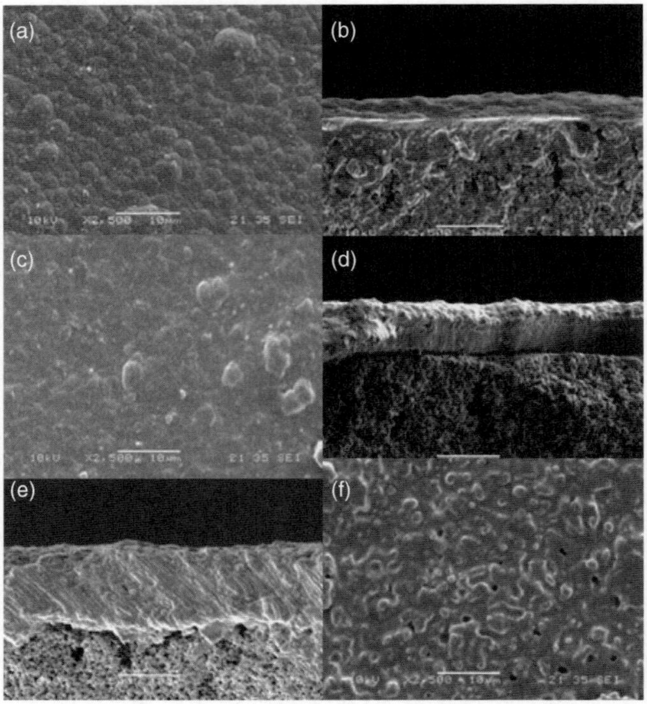

Figure 3.26 *SEM images of Pd/Ag composite membranes during coating and diffusion electroless plating: (a) surface of Pd layer, (b) cross section of Pd layer, (c) surface of Ag layer, (d) cross section of Pd and Ag layers before heat treatment, (e) cross section of Pd/Ag alloy after heat treatment, (f) surface of Pd/Ag alloy after heat treatment. Reprinted from Zhentao Wu, Irfan M.D. Hatim, Benjamin F.K. Kingsbury, Ejiro Gbenedio and K. Li, A novel inorganic hollow fiber membrane reactor for catalytic dehydrogenation of propane, AIChE Journal 55 (9) 2389–2398. Copyright (2009) with permission from John Wiley and Sons*

membrane as a whole is gas-tight with excellent H_2/N_2 selectivity. A comparison of hydrogen permeation data for both Pd and Pd/Ag membranes coated on identical α-alumina substrates can be found in Figures 3.27 and 3.28 respectively. It can be seen that in both cases hydrogen permeation flux increases with increasing temperature. The linear relationship between the hydrogen flux and the square root of the pressure difference indicates that the pressure exponent (n) is equal to 0.5. Therefore, bulk diffusion is the controlling step in hydrogen permeation [30]. The apparent activation energy (E_a) of hydrogen permeation through the Pd/Ag membrane, which was calculated according to the Arrhenius plot shown in Figure 3.28, was 5.4 kJ mol^{-1}, agreeing well with the reported values of 6.38 [32], 5.73 [33], and 5.7 kJ mol^{-1} [34]. However, higher E_a values of 9.8 [35], 11.36 [36], and 15.5 kJ mol^{-1} [37] have also been reported, indicating that the Ea value of Pd/Ag membranes is an indication of the net effects of a number of factors. Irrespective of the uniformity of the membrane, the Ea of the Pd/Ag membranes is closely related to the weight ratio between Pd and Ag. For example, Pd/Ag membranes with 20–23 wt% of Ag possess a relatively low Ea. Also, Ea is dependent on the controlling step of the hydrogen permeation. For instance, the Ea value for the hydrogen permeation process controlled by bulk diffusion is lower than that controlled by surface exchange. In contrast, the apparent activation

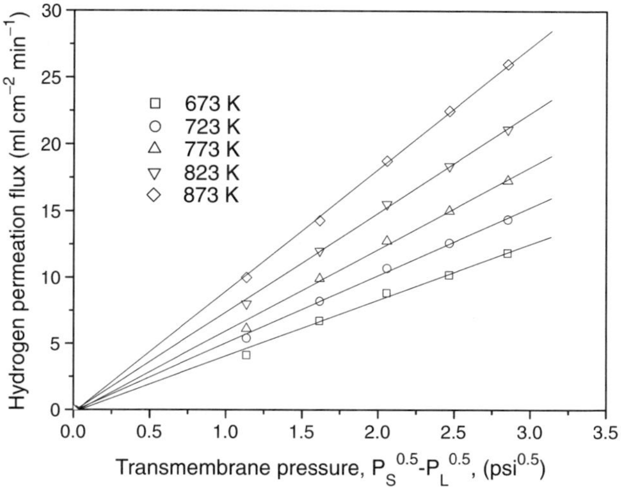

Figure 3.27 *Relationship between square root of pressure difference and hydrogen permeation flux through the alumina supported Pd composite hollow fibre membrane at different temperatures*

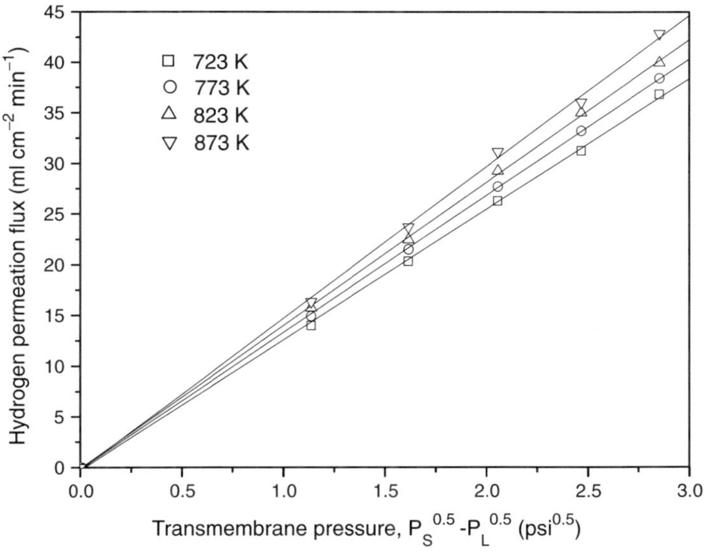

Figure 3.28 *Relationship between square root of pressure difference and hydrogen permeation flux through the composite alumina supported Pd/Ag composite hollow fibre membrane at different temperatures. Reprinted from Zhentao Wu, Irfan M.D. Hatim, Benjamin F.K. Kingsbury, Ejiro Gbenedio and K. Li, A novel inorganic hollow fiber membrane reactor for catalytic dehydrogenation of propane, AIChE Journal 55 (9) 2389–2398. Copyright (2009) with permission from John Wiley and Sons*

energy (E_a) of hydrogen permeation through the pure Pd membrane, which was calculated according to the Arrhenius plot shown in Figure 3.27, was approximately 19.1 kJ mol^{-1}, which is consistent with the values of 14.8 kJ mol^{-1} [38] and 21 kJ mol^{-1} [39] of other reported composite Pd/alumina hollow fibre membranes.

Energy Dispersive Spectroscopy (EDS) was also carried out to investigate the elemental distribution across the Pd/Ag membrane cross section before and after heat treatment. As shown in Figure 3.29a , a small amount of Pd is observed at approximately 2 μm beneath the substrate surface before heat treatment, which indicates that the Pd plating solution penetrated slightly into the substrate during the electroless plating process. In contrast to Figure 3.29a, significant change in the elemental distribution across the membrane occurs after heat treatment as shown in Figure 3.29b. The Tamman temperature (TM) is defined as the temperature at which considerable

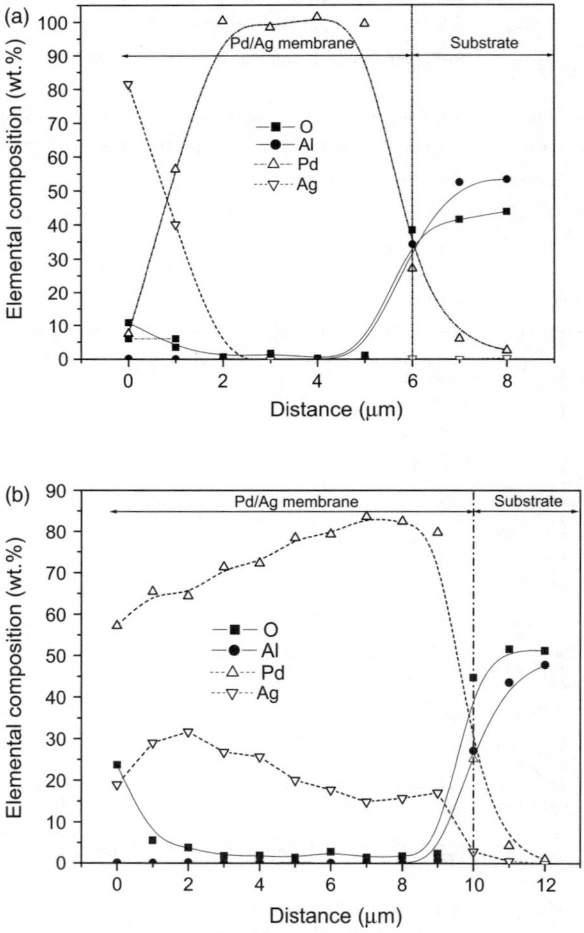

Figure 3.29 *EDS analysis of Pd/Ag membranes: (a) before heat treatment, (b) after heat treatment. Reprinted from Zhentao Wu, Irfan M.D. Hatim, Benjamin F.K. Kingsbury, Ejiro Gbenedio and K. Li, A novel inorganic hollow fiber membrane reactor for catalytic dehydrogenation of propane, AIChE Journal 55 (9) 2389–2398. Copyright (2009) with permission from John Wiley and Sons*

thermal vibration occurs in a lattice and is estimated at half the melting point of a material [40]. As the TM of Ag, which is about 344 °C, is much lower than that of Pd (641 °C), it is reasonable to assume that silver atoms migrate significantly into the palladium layer during heat treatment. As can be seen in Figure 3.29b, the concentration of Ag decreases slightly towards the substrate, while the concentration of Pd is increased. This indicates that a higher temperature and/or longer time would be required for a more uniform elemental distribution in such a thick membrane. Although a number of studies on metallic interdiffusion have been carried out to determine the diffusion coefficient of Ag in Pd, it is still difficult to accurately predict the alloying process. For example, the diffusion coefficients from several groups are quite different from each other [40]. One of the possible reasons is that, apart from the material nature of Pd and Ag, the alloying process can be determined by the membrane microstructure, which is closely related to the preparation processes and the surface properties of the substrates. Hou *et al.* [35] obtained homogeneous Pd/Ag membranes by annealing the membrane at 600 °C for 10 h, while Uemiya *et al.* [41] concluded that high temperatures of above 800 °C were necessary to obtain a homogeneous Pd/Ag alloy membrane. However, area scans performed randomly on the cross section of other thinner samples revealed an average Pd/Ag ratio of 78:22, which is considered to be an indication that the Pd/Ag membranes have a homogeneous distribution of Pd and Ag.

3.4 Catalyst Impregnation

Two methods of catalyst deposition have been used to impregnate the alumina hollow fibre substrate with catalyst: (1) submicron Pt (0.5 wt%)/γ-alumina catalyst particles are prepared and deposited into the substrate of a hollow fibre membrane reactor (iHFMR-I) before the deposition of a Pd/Ag hydrogen permselective membrane on the outer fibre surface. Prior to catalyst deposition, the catalyst particles are dispersed in an aqueous medium using Arlacel P135 as a dispersant. Only the catalyst particles suspended evenly in the aqueous medium are used for subsequent catalyst deposition and as a result, a very small amount of the catalyst is deposited uniformly in the finger-like voids of the asymmetric alumina substrates. As can be seen from Figure 3.30a, b, the surfaces of the finger-like voids in iHFMR-I are covered by a thin layer of submicron sized catalyst particles, while the sponge-like regions of the substrate remain unaffected. Figure 3.30b shows the finger-like void surface at high magnification and indicates that the catalyst particles are sparsely deposited on the surface of the finger-like voids. This agrees with the result that, after catalyst deposition, the average weight gain of the asymmetric alumina hollow fibre substrates (30 cm in length) is only approximately 0.74%. As a consequence of the limited weight gain associated with this method a second method of catalyst impregnation was investigated: (2) The asymmetric alumina hollow fibre substrates are functionalised by depositing SBA-15 mesoporous silica inside the finger-like voids using a sol-gel technique, followed by treatment with a solution containing Pt catalyst precursors. As can be seen in Figure 3.31a–d , SBA-15 is successfully deposited within the alumina hollow fibre substrates. Instead of fully blocking the finger-like voids, there is a gap between the deposited SBA-15 and the wall of the finger-like voids after sintering. In this case a pure Pd hydrogen permselective membrane was coated on the shell side of the fibre to form a catalytic hollow fibre membrane reactor (iHFMR-II).

A comparison of the surface area increase and weight gain for the two methods for catalyst impregnation can be seen in Figure 3.32. In comparison with method (1), in which the catalyst

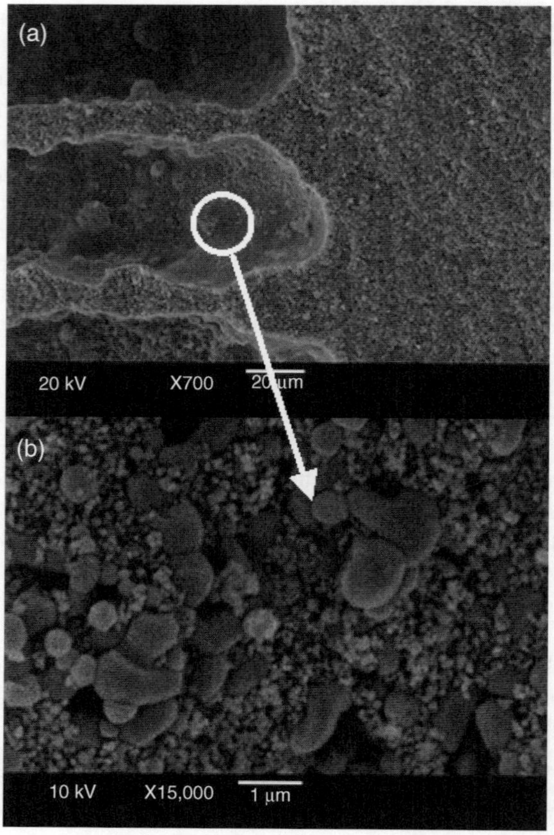

Figure 3.30 SEM images of catalyst deposition in the hollow fibre substrate: (a) catalyst particles deposited into the finger-like voids, (b) finger-like voids (high magnification). Reprinted from Zhentao Wu, Irfan M.D. Hatim, Benjamin F.K. Kingsbury, Ejiro Gbenedio and K. Li, A novel inorganic hollow fiber membrane reactor for catalytic dehydrogenation of propane, AIChE Journal 55 (9) 2389–2398. Copyright (2009) with permission from John Wiley and Sons

was first deposited on γ-alumina particles before impregnation into the finger-like voids of the hollow fibre support, method (2) is much more efficient and achieves greater loading of catalyst with a higher surface area, and as a result, yields a significantly higher catalyst surface area in the functionalised substrates. The increase in the surface area of the reaction zone (5 cm in length) in iHFMR-II, which is calculated based on the amount of SBA-15 deposited, is about 5.19 m^2, which is much higher than the value of 0.22 m^2 obtained in the previous work for iHFMR-I.

3.5 Application in Chemical Reaction

Due to the versatility of the combined phase inversion and sintering technique it is possible to prepare a wide range of ceramic hollow fibre morphologies for use in an extensive range of applications. One promising application in particular is the use of an asymmetric alumina hollow

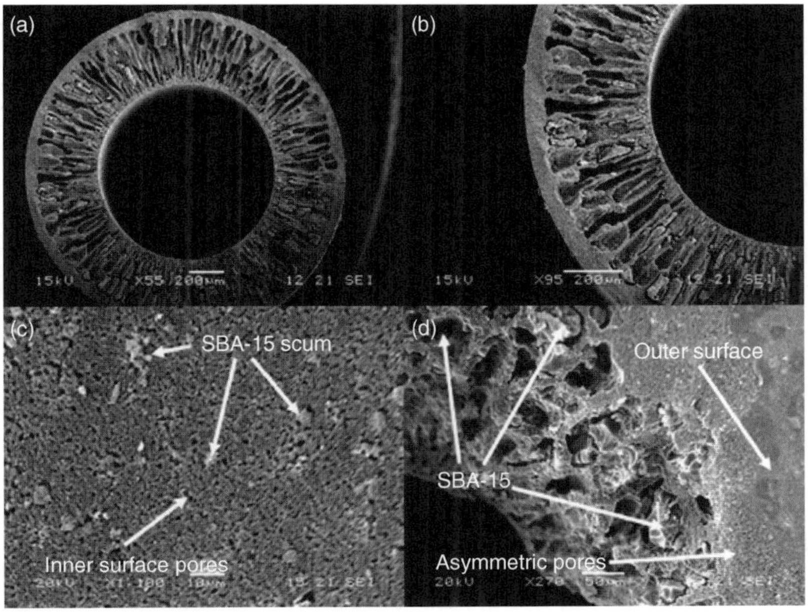

Figure 3.31 SEM images of the functionalised alumina hollow fibre substrate: (a) whole view, (b) cross section, (c) inner surface and (d) top view from outer surface. Reprinted from Catalysis Today (in press corrected proof) doi:10.1016/j.cattod.2010.04.044. Ejiro Gbenedio, Zhentao Wu, Irfan Hatim, Benjamin F.K. Kingsbury and K. Li, A multifunctional Pd/alumina hollow fibre membrane reactor for propane dehydrogenation. Copyright (2010) with permission from Elsevier

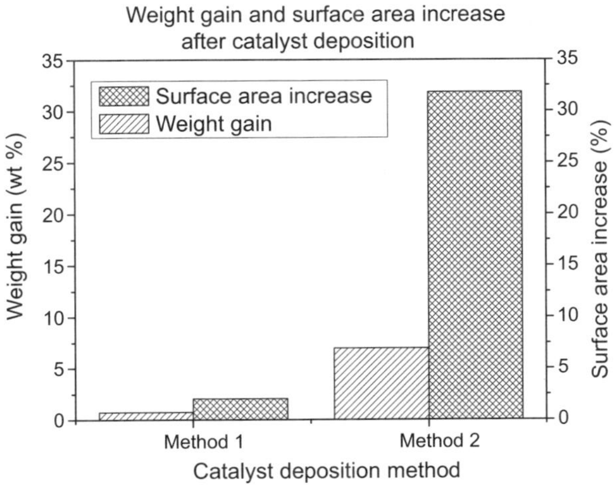

Figure 3.32 Comparison of weight gain and surface area increase for different catalyst impregnation methods. Reprinted from Catalysis Today, Benjamin F.K. Kingsbury, Zhentao Wu and K. Li, A morphological study of ceramic hollow fibre membranes: A perspective on multifunctional catalytic membrane reactors, In Press, Corrected Proof. Copyright (2010) with permission from Elsevier

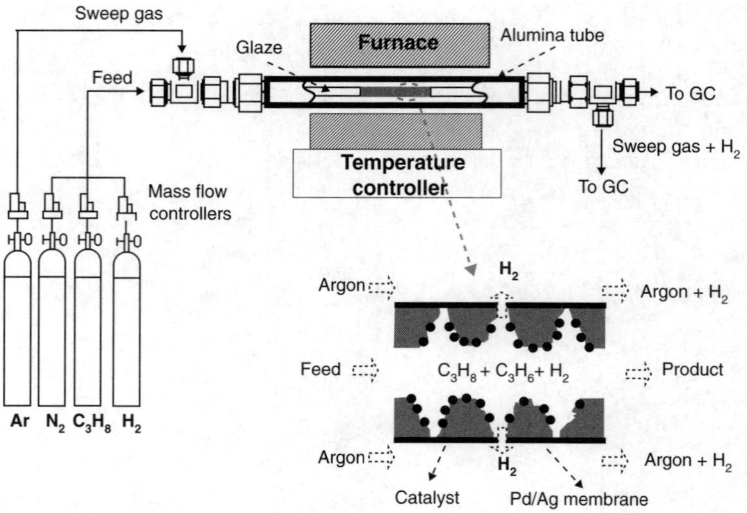

Figure 3.33 *Schematic diagram of the apparatus for the dehydrogenation of propane to propene. Reprinted from Zhentao Wu, Irfan M.D. Hatim, Benjamin F.K. Kingsbury, Ejiro Gbenedio and K. Li, A novel inorganic hollow fiber membrane reactor for catalytic dehydrogenation of propane, AIChE Journal 55 (9) 2389–2398. Copyright (2009) with permission from John Wiley and Sons*

fibre in a catalytic membrane reactor. In this case, finger-like voids generated during the fibre preparation process may be impregnated with catalyst while the sponge-like region of the fibre may serve as a separation layer or as a substrate for a separation layer such as a hydrogen permselective membrane. Inorganic hollow fibre membrane reactors (iHFMR-I and iHFMR-II) were designed, bases on this principal, for the dehydrogenation of propane to propene.

Performance of iHFMR-I: A schematic of the apparatus used for the dehydrogenation of propane to propene is shown in Figure 3.33 showing the membrane reactor with catalyst deposited within the finger-like voids of the asymmetric alumina hollow fibre. The performance of iHFMR-I is compared with a fixed bed reactor (FBR-I), in which 0.1 g of Pt (0.5 wt %)/γ-alumina catalyst is packed into the centre of a dense ceramic tube of 9 mm in diameter. As can be seen in Figure 3.34, the propane conversion in the FBR started at about 22% and then reduced to about 8% after 40 min, which is close to the equilibrium conversion of propane at 450 °C. In the mean time, the propene selectivity increased from about 34 to 80%. This trend agrees well with the work of Lobera [42], in which the coke content increased abruptly in approximately the first 10 min, causing rapid catalyst deactivation at the initial stage of the reaction. In the case of iHFMR-I, the initial propane conversion was measured to be about 37%, which is approximately 68% higher than that of the FBR which proves that the use of iHFMR-I substantially shifts the reaction to the product side and gives rise to an increase in propane conversion. As can be seen from Table 3.1, the STY of iHFMR-I is more than 6 times that of the FBR. It should be noted that this value can be further increased to approximately 14 when smaller ceramic substrates (outer diameter = 1.3 mm) with similar asymmetric structures are employed [43]. Moreover, the STY of iHFMR-I, in view of the amount of catalyst employed, is more than 60 times that of the FBR, indicating that iHFMR-I is more efficient for propene production.

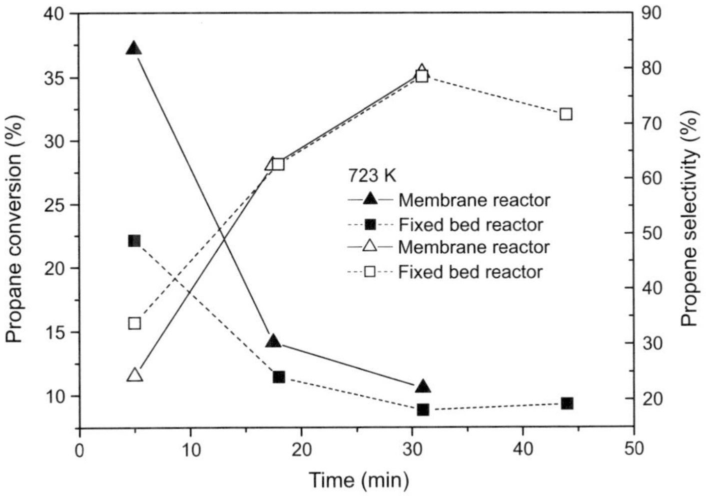

Figure 3.34 *Propane conversion (solid symbols) and propene selectivity (open symbols) of the inorganic hollow fibre membrane reactor (iHFMR-I) and fixed bed reactor (FBR-I). Reprinted from Zhentao Wu, Irfan M.D. Hatim, Benjamin F.K. Kingsbury, Ejiro Gbenedio and K. Li, A novel inorganic hollow fiber membrane reactor for catalytic dehydrogenation of propane, AIChE Journal 55 (9) 2389–2398. Copyright (2009) with permission from John Wiley and Sons*

Performance of iHFMR-II: For comparison, the catalytic activity of the Pt (1 wt%)/SBA-15 catalyst, prepared by a conventional wetness impregnation method, is investigated in a fixed bed reactor (FBR-II) using the same operating conditions as for iHFMR-II. Approximately 0.1 g of the catalyst is packed into the centre of a dense ceramic tube of 9 mm in diameter. Initial propane conversion values refer to the analysis of the reaction products after 5 minutes on stream. As can be seen in Figure 3.35, the propane conversion started at about 75.3% at 500 °C and then reduced to about 9% after about 120 min of operation, in agreement with the work of Lobera *et al.* [28], in which the coke content increased abruptly in approximately the first 10 min, causing rapid catalyst deactivation at the initial stage of the reaction. In the meantime, the propene selectivity increased sharply from about 2 to 70% after around 35 min on stream.

In contrast with the FBR, the initial propane conversion of iHFMR-II was lower and was measured at approximately 48.7%, as shown in Figure 3.36. It has been proved that the

Table 3.1 *Space–time yields for a fixed bed reactor (FBR-I) and iHFMR-I*

Space–time yield of iHFMR-I			Space–time yield of FBR-I		
(min)	(mol propzene m^{-3} h^{-1})	(g propene g^{-1} catalyst h^{-1})	(min)	(mol propene m^{-3} h^{-1})	(g propene g^{-1} catalyst h^{-1})
5.0	2488.60	12.82	5	398.62	0.21
17.5	2431.61	12.52	18	379.64	0.20
31.0	2298.63	11.84	31	366.99	0.19
			44	352.22	0.18

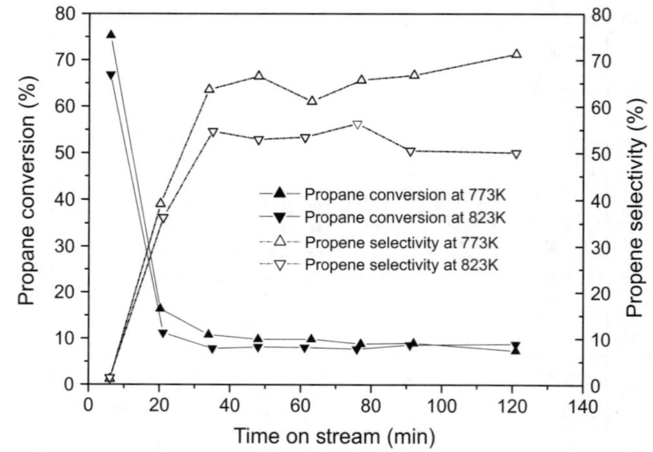

Figure 3.35 *The catalytic properties of Pt (1 wt%)/SBA-15 catalysts in the dehydrogenation of propane to propene in a fixed bed reactor (FBR-II). Reprinted from Catalysis Today (in press corrected proof) doi:10.1016/j.cattod.2010.04.044. Ejiro Gbenedio, Zhentao Wu, Irfan Hatim, Benjamin F.K. Kingsbury and K. Li, A multifunctional Pd/alumina hollow fibre membrane reactor for propane dehydrogenation. Copyright (2010) with permission from Elsevier*

use of Pd-based membranes to remove hydrogen as a product from a dehydrogenation reaction enhances the catalyst deactivation in membrane reactors as a consequence of faster coke formation [3]. As a result, the propane conversion values, which may be significantly higher in the first several minutes of reaction, may drop sharply before the product sample representing the initial propane conversion value was taken for analysis. As the reaction proceeds, the propane conversions of the iHFMR-II and FBR are comparable.

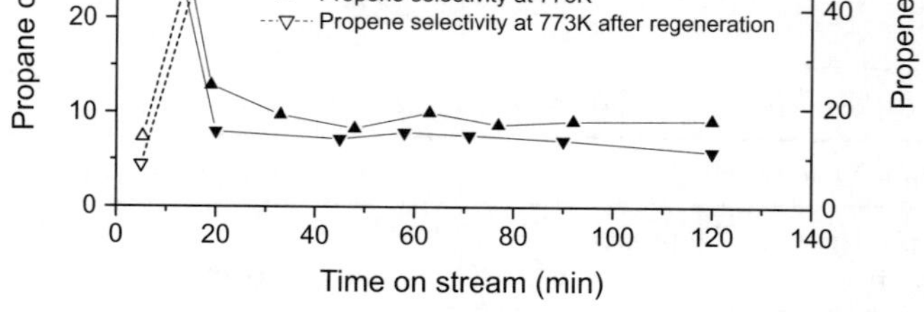

Figure 3.36 *Dehydrogenation of propane to propene using multifunctional iHFMR-II. Reprinted from Catalysis Today (in press corrected proof) doi:10.1016/j.cattod.2010.04.044. Ejiro Gbenedio, Zhentao Wu, Irfan Hatim, Benjamin F.K. Kingsbury and K. Li, A multifunctional Pd/alumina hollow fibre membrane reactor for propane dehydrogenation. Copyright (2010) with permission from Elsevier*

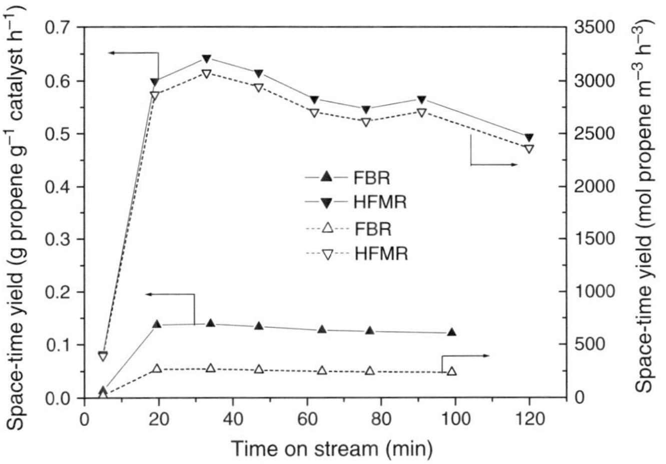

Figure 3.37 *Space–time yields (STY) for FBR-II and iHFMR-II. Reprinted from Catalysis Today (in press corrected proof) doi:10.1016/j.cattod.2010.04.044. Ejiro Gbenedio, Zhentao Wu, Irfan Hatim, Benjamin F.K. Kingsbury and K. Li, A multifunctional Pd/alumina hollow fibre membrane reactor for propane dehydrogenation. Copyright (2010) with permission from Elsevier*

It can be seen in Figure 3.37, that the volumetric space–time yield (STY) of iHFMR-II is more than 10 times that of the FBR. As with iHFMR-I it should be noted here that this value can be further increased to approximately 20 when smaller ceramic substrates (outer diameter = 1.3 mm) with similar asymmetric structures are employed [27]. Moreover, the STY of iHFMR-II, in view of the amount of catalyst employed, is more than 4 times that of the FBR, indicating that iHFMR-II is more efficient in propene production.

3.6 Final Remarks and Conclusions

Ceramic membranes may be used for a wide range of applications due to their stability and durability under harsh operating conditions. In particular, their ability to operate at high temperatures allow for their use in high temperature membrane reactors. By using a ceramic hollow fibre configuration the surface area to volume ratio of a membrane module can be greatly increased, leading to a reduction in production cost and an increase in efficiency. Although the combination of a high surface area to volume ratio and extreme durability is extremely desirable, the production cost of ceramic membranes is often prohibitive and alternative technologies are favoured for this reason. These high production costs are principally a result of the expensive and time consuming multistep process that is required for preparation of membranes with asymmetric structures. This is in contrast to the preparation of most asymmetric polymer membranes, in which the well known phase inversion process is employed to achieve asymmetry in a single step. Therefore, by using the a combined phase inversion and sintering technique, a wide range of asymmetric ceramic membrane structures can be prepared in a single step. In addition, the versatility of the technique with respect to the type of asymmetric structures that can be prepared is extremely useful when designing membranes for specific applications. This versatility has been demonstrated by the application of such an asymmetric structure in a catalytic membrane reactor for the dehydrogenation of propane to propene. In this novel design, a highly asymmetric membrane structure consisting of finger-like voids and a sponge-like structure is utilised as both a substrate for catalyst and as a substrate for a hydrogen permselective layer. By preparing the membrane reactor in this way, reactor space time yield is increased with respect to both the volume of the reactor and to the amount of catalyst used. The reduction in production cost combined with the versatility of the combined phase inversion and sintering method for the production asymmetric ceramic hollow fibre membranes allows for membrane technology to become more competitive, in particular for high temperature applications.

Acknowledgements

The authors gratefully acknowledge the research funding provided by EPSRC in the United Kingdom (grant No. EP/E000231/1). A project studentship provided by EPSRC to one of the authors, Benjamin F. K. Kingsbury is also gratefully acknowledged.

References

1. R. Weber, H. Chmiel, V. Mavrov, *Desalination*, **157**, 113 (2003).
2. J. Galuszka, R.N. Pandey, S. Ahmed, *Catalysis Today*, **46**, 83 (1998).
3. J.N. Keuler, L. Lorenzen, *Journal of Membrane Science*, **202**, 17 (2002).
4. H.W.J.P. Neomagus, G. Saracco, H.F.W. Wessel, G.F. Versteeg, *Chemical Engineering Journal*, **77**, 165 (2000).
5. C.C. Wei, K. Li, *Industrial and Engineering Chemistry Research*, **47**, 1506 (2008).
6. S. Koonaphapdeelert, K. Li, *Desalination*, **200**, 581 (2006).
7. A. Julbe, D. Farrusseng, C. Guizard, *Journal of Membrane Science*, **181**, 3 (2001).
8. J. Galuszka, R.N. Pandey, S. Ahmed, *Catalysis Today*, **46**, 83 (1998).
9. J.N. Armor, *Journal of Membrane Science*, **147**, 217 (1998).
10. H. Weyten, J. Luyten, K. Keizer, L. Willems, R. Leysen, *Catalysis Today*, **56**, 3 (2000).
11. E. Gbenedio, Z. Wu, I. Hatim, B.F.K. Kingsbury, K. Li, *Catalysis Today*, in press (2010).
12. Z. Wu, I. Hatim, B.F.K. Kingsbury, E. Gbenedio, K. Li, *AIChE Journal*, in press (2010).
13. R. Terpstra, *Key Engineering Materials*, **132**, 1770 (1997).
14. H. Brinkman, *American Ceramic Society Bulletin*, **78**, 51 (1999).
15. J. Smid, C.G. Avci, V. Günay, R.A. Terpstra, J.P.G.M. Van Eijk, *Journal of Membrane Science*, **112**, 85 (1996).
16. J. Koresh, *Separation Science and Technology*, **18**, 723 (1983).
17. S. Loeb, S. Sourirajan, *Advances in Chemistry Series*, **38**, 117 (1963).
18. Z.S. Li, *Journal of Polymer Science Part B: Polymer Physics*, **43**, 498 (2005).
19. S.A. McKelvey, W.J. Koros, *Journal of Membrane Science*, **112**, 29 (1996).
20. J. Barzin, B. Sadatnia, *Polymer*, **48**, 1620 (2007).
21. J. Ren, Z. Li, F.S. Wong, *Journal of Membrane Science*, **241**, 305 (2004).
22. K.W. Lee, B.K. Seo, S.T. Nam, M.J. Han, *Desalination*, **159**, 289 (2003).
23. M.R. Pekny, J. Zartman, W.B. Krantz, A.R. Greenberg, P. Todd, *Journal of Membrane Science*, **211**, 71 (2003).
24. M.A. Frommer, R.M. Messalem, *Industrial & Engineering Chemistry Product Research and Development*, **12**, 328 (1973).

25. K. Li, X. Tan, Y. Liu, *Journal of Membrane Science*, **272**, 1 (2006).
26. G.Q. Lu, *Journal of Materials Processing Technology*, **59**, 297 (1996).
27. D.W. Lee, S.J. Park, C.Y. Yu, S.K. Ihm, K.H. Lee, *Journal of Membrane Science*, **316**, 63 (2008).
28. B.F.K. Kingsbury, K. Li, *Journal of Membrane Science*, **328**, 134 (2009).
29. X. Tan, S. Liu, K. Li, *Journal of Membrane Science*, **188**, 87 (2001).
30. S.N. Paglieri, J.D. Way, *Separation and Purification Reviews*, **31**, 1 (2002).
31. L.Q. Wu, N. Xu, J. Shi, *Industrial and Engineering Chemistry Research*, **39**, 342 (2000).
32. Y. Yildirim, E. Gobina, R. Hughes, *Journal of Membrane Science*, **135**, 107 (1997).
33. H. Yoshida, *Journal of the Less Common Metals*, **89**, 429 (1983).
34. E. Kikuchi, S. Uemiya, *Gas Separation and Purification*, **5**, 261 (1991).
35. K. Hou, R. Hughes, *Journal of Membrane Science*, **214**, 43 (2003).
36. Y.S. Cheng, M.A. Peña, J.L. Fierro, D.C.W. Hui, K.L. Yeung, *Journal of Membrane Science*, **204**, 329 (2002).
37. S. Uemiya, N. Sato, H. Ando, Y. Kude, T. Matsuda, E. Kikuchi, *Journal of Membrane Science*, **56**, 303 (1991).
38. J. Tong, *Chemical Communications*, **2006**, 1142 (2006).
39. B.K. Nair, J. Choi, M.P. Harold, *Journal of Membrane Science*, **288**, 67 (2007).
40. J. Shu, A. Adnot, B.P.A. Grandjean, S. Kaliaguine, *Thin Solid Films*, **286**, 72 (1996).
41. S. Uemiya, T. Matsuda, E. Kikuchi, *Journal of Membrane Science*, **56**, 315 (1991).
42. M.P. Lobera, C. Téllez, J. Herguido, M. Menéndez, *Applied Catalysis A: General*, **349**, 156 (2008).
43. C.C. Wei, O.Y. Chen, Y. Liu, K. Li, *Journal of Membrane Science*, in press (2010).

4

Metallic Membranes Prepared by Cold Rolling and Diffusion Welding

Silvano Tosti

ENEA, Unità Tecnica Fusione, CR ENEA Frascati, Frascati (RM), Italy

4.1 Introduction

The hydrogen is known to interact with the metals surfaces and bulk: mainly these characteristics are considered for preparing the catalysts to be used for the hydrogenation/dehydrogenation reactions. Another important hydrogen/metal process which is extensively taken into consideration is the hydrogen permeation: in fact, by interacting with the surfaces of the metals and diffusing into their lattice the hydrogen atoms permeate selectively dense metal walls. Many applications of Pd-based membranes for hydrogen separation and production are reported [1–4]. Especially, both ceramic and metallic porous supports have been used for thin Pd or Pd alloy layers. Usually, a Pd-ceramic membrane consists of three parts: a Pd or Pd alloy layer, a ceramic porous support and an intermediate ceramic layer prepared in order to improve the metal/ceramic adhesion [5]. Nevertheless, as a main drawback the intermediate ceramic layer characterised by a small pore size could decrease the hydrogen permeance of the composite membrane. The metal porous supports are expected to reduce the mismatching between the Pd alloy layer and the support due to the different thermal expansion coefficients of the coupled materials, even though the intermetallic diffusion can occur at high temperature and poison the Pd alloy. In these cases, an intermediate ceramic layer of appropriate thermal expansion coefficient can be developed: for an example, the use of a middle layer (yittria–zirconia) has been proposed for covering 446 stainless steel (SS) porous supports with Pd alloy films of thickness less than 10 μm [6]. Really, the yittria–zirconia layer has a thermal expansion coefficient of 11 $\mu m\ m^{-1}\ K^{-1}$ which is equal to

Membranes for Membrane Reactors: Preparation, Optimization and Selection, Edited by Angelo Basile and Fausto Gallucci

the stainless steel one (11 μm m^{-1} K^{-1}) and close to the Pd alloys one (13.9 μm m^{-1} K^{-1}). Finally, another configuration of composite Pd-based membrane consists of a dense Pd or Pd alloy thin layer supported by a silicon carrier characterised by rectangular perforations [7].

Among the metals, the system Pd/H has been widely studied [2,8]: in fact, the hydrogen dissolves into Pd by forming an hydride which at low temperature presents the coexistence of two phases (α and β) characterised by different lattice parameters thus involving the embrittlement of the Pd operated under hydrogen at temperature below 300 °C. In order to avoid this drawback, Pd alloys have been introduced for manufacturing hydrogen gas separators. Especially, by adding 20–25 wt% silver the presence of the two hydride phases is reduced and, furthermore, the hydrogen permeation is maximised. Such a Pd alloy is commercially used to prepare permeator tubes of wall thickness 100–150 μm: however, the high cost of the precious metal limits the use of these membranes to niche applications.

The hydrogen permeation through dense metals is a mass transfer mechanism which consists of several steps. In a simplified model, the hydrogen by interacting with the metal surface (adsorption) enters the metal lattice where diffuses according to a Fick's law. In the opposite metal surface, by desorption the hydrogen atoms leaves the metal and goes back to the gaseous phase. The overall mass transfer phenomenon is ruled by the well known Sieverts' law [9,10]:

$$J = P\frac{\left(p_r^{0,5} - p_p^{0,5}\right)\mathrm{A}}{\delta} \tag{4.1}$$

where J is the hydrogen permeation flow rate (mol s^{-1}), P the hydrogen permeability through the metal (mol m^{-1} s^{-1} Pa$^{-0.5}$), p_r and p_p the hydrogen partial pressure at upstream and downstream side, respectively, A is the membrane area (m^2) and δ the membrane thickness (m).

The above expression shows that by reducing the metal thickness the hydrogen permeation flux proportionally increases; moreover, a reduction of the Pd alloy thickness proportionally reduces the cost of the membrane. As a result, a thickness reduction by a factor two involves a reduction of the cost per unit of hydrogen permeated by a factor 4. Therefore, the reduction of the metal thickness is a key aspect to be considered when Pd-based membranes are applied in the processes for producing hydrogen.

Several synthesis processes have been developed in order to produce Pd-based membranes consisting of thin metal layers coated over porous support: for an example, electroless, sputtering, CVD and other methods have been applied for covering ceramic or metal porous tubes with thin (from μm to tens of μm) Pd alloy films [9–15]. Accordingly to a general behaviour, such composite membranes present as higher hydrogen permeability as lower hydrogen permselectivity, and viceversa. Particularly, it is demonstrated that Pd-ceramic membranes cannot be completely selective and durable [14]. Therefore, when pure hydrogen has to be separated the use of dense Pd-based membranes has been considered: commercial Pd-Ag tubes of wall thickness 0.100–0.150 mm are available. The question to be solved concerns the minimum membrane thickness which withstands the mechanical stresses (due to the transmembrane pressure differences) and ensures the complete hydrogen selectivity (absence of membrane defects and its long life). Under the operating conditions of many processes to separate hydrogen (temperatures of 300–400 °C), Pd-Ag permeator tubes of wall thickness about 50 μm can withstand operative transmembrane pressures up to 200–300 kPa. Accordingly, a cold rolling and diffusion welding technique for producing Pd-Ag thin wall tubes has been developed.

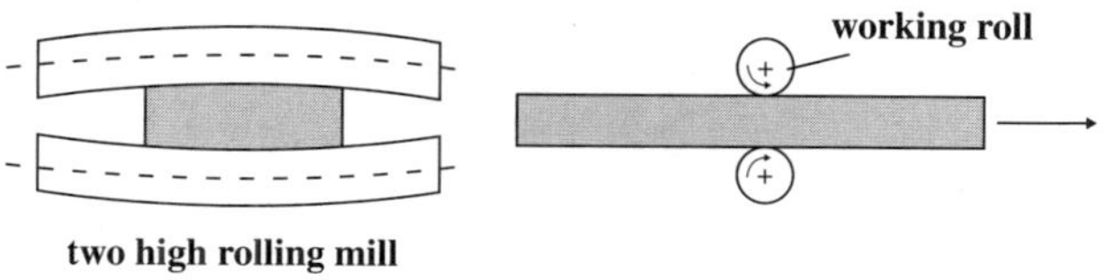

Figure 4.1 Rolls bending of a two high rolling mill: the worked metal foils are curved

4.2 Preparation Method

The production of Pd-Ag thin wall tubes consists of two main processes: the reduction of the metal foils thickness and their joining in order to form the tube.

4.2.1 Cold Rolling

The cold rolling is a practical metal working applied for reducing the thickness of metal sheets: several kinds of rolling mills are used according to the production requirements (hardness of the material, rolling speed, minimum thickness, etc.). Particularly, the minimum thickness which can be obtained for a given material depends on the diameter of the milling rolls: as smaller the diameter as thinner the worked metal sheet. However, the use of small diameter rolls involves their bending under operation: such behaviour produces curved metal foils, see Figure 4.1. Consequently, in order to both produce very thin and straight foils, a four-high rolling mill has been used: in such a device two larger support rolls constrains the working rolls of small diameter by avoiding their bending, see Figures 4.2 and 4.3.

The mechanical characteristics of the Pd-Ag alloys depend on the silver content: especially, the commercial alloy (23–25% wt. of silver) exhibits tensile strength values comparable to the steel one, see Figure 4.4 [16].

Upon rolling, the hardness of the commercial Pd-Ag alloy passes from about 100 HB (Brinell hardness) to 190 HB [16], see Figure 4.5: therefore, during the metal working (cold rolling) thermal treatments are necessary in order to anneal the material. These thermal treatments are carried out under controlled atmosphere for avoiding the oxidation at high temperature of the Pd

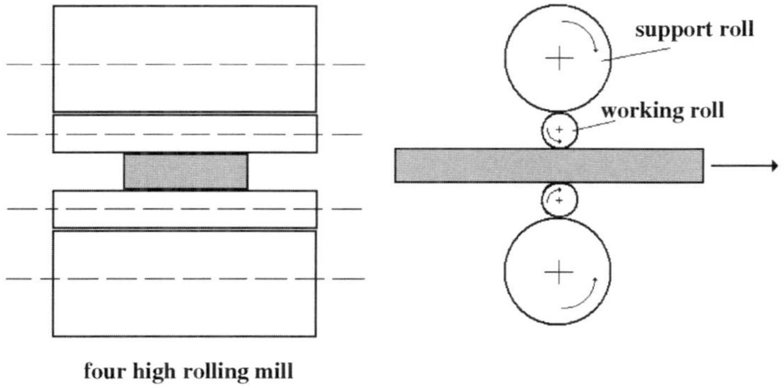

Figure 4.2 Scheme of a four-high rolling mill

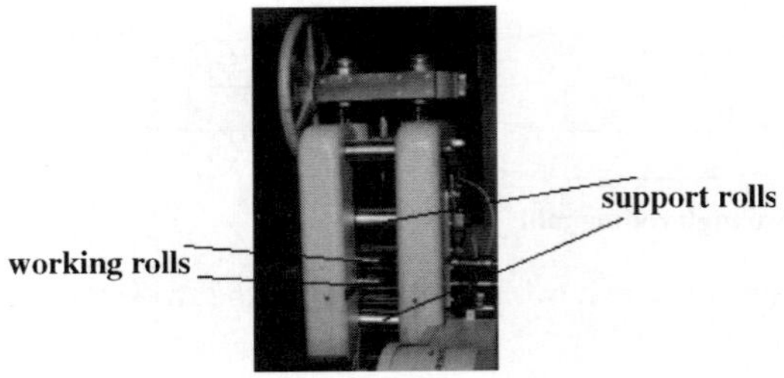

Figure 4.3 The four-high rolling mills at ENEA Frascati laoratories

surface. Accordingly, the annealing of the rolled sheets is performed at 800–1000 °C per 1–2 h under a flux of inert gas with a reducing agent (i.e., Ar with 5% of hydrogen) or under vacuum. The reducing atmosphere should be preferred because the silver evaporation under vacuum could significantly modify the composition of the Pd alloy.

4.2.2 Diffusion Welding

After cold rolling, the Pd-Ag foils are joined in order to form the permeator tubes. several welding procedure can be considered: brazing, autogenous welding, etc. Especially, brazing techniques may

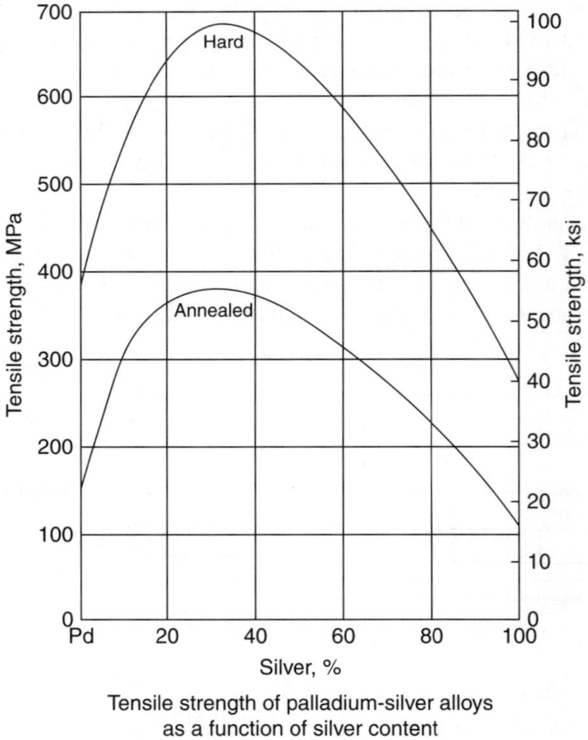

Figure 4.4 Tensile strength of Pd-Ag alloys versus their silver content [16]. Reprinted with permission of ASM International®. All rights reserved. www.asminternational.org

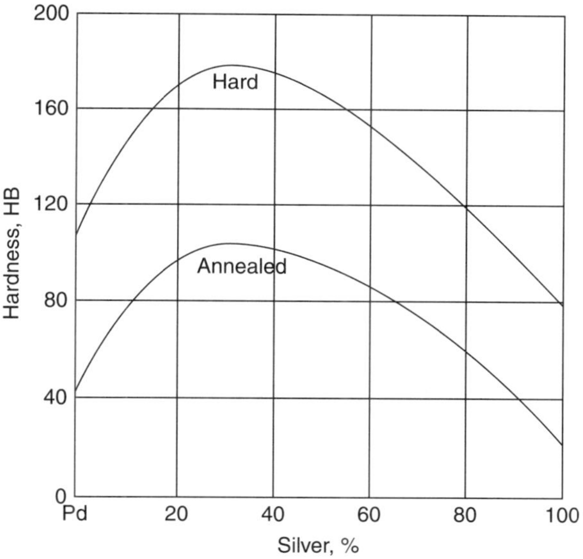

Figure 4.5 *Brinell hardness of Pd-Ag alloys versus their silver content [16]. Reprinted with permission of ASM International®. All rights reserved. www.asminternational.org*

cause both defects and penetration of metal impurities contaminating and altering the palladium–silver alloy thus both reducing the hydrogen permeability and the embrittlement resistance. Also the TIG (tungsten inert gas) welding has been applied [17]: such a technique produced thermally stressed parts of the permeator tubes where defects (cracks, microholes) took place under the operation (hydrogenation) of the material. Figures 4.6 and 4.7 show a Pd-Ag membrane tube TIG welded and a particular of a crack produced after hydrogenating the material, respectively.

In order to avoid this drawback, another joining procedure (diffusion welding) can be considered: it consists of pressing at high temperature the metal parts to be welded. In general, diffusion welding is a technique capable to join most metals and some nonmetals: usually, the load applied provokes no macroscopic deformation of the material while the bonding temperature is 50–75% of the metal melting point. Dunkerton describes the materials which can be joined

Figure 4.6 *TIG welded Pd-Ag permeator tube: the crack due to hydrogenation (circled) is shown in detail in Figure 4.7*

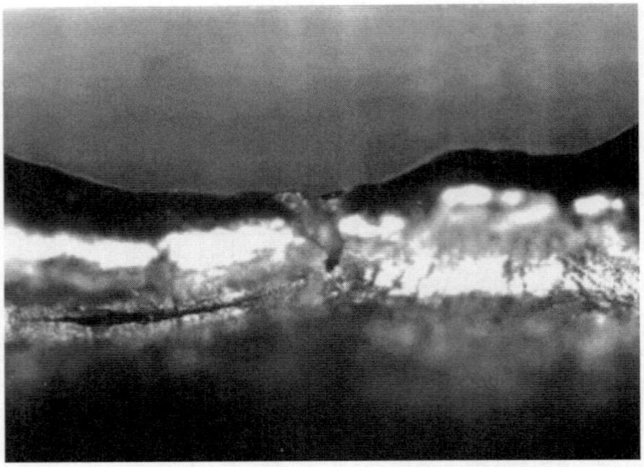

Figure 4.7 *Close view of the TIG welding of a Pd-Ag tube*

by diffusion welding and the main applications suggested are in the electronics, nuclear and space industry [18]. Deminet is the author of a patent which claims a diffusion welding method carried out under vacuum for bonding honeycomb structures to face sheets; such a composite material presents high performances and it is proposed for the aerospace industry [19].

Actually, the metal atoms diffuse into the lattice according to the Fick's law:

$$F = -D\frac{\partial c}{\partial x} \tag{4.2}$$

where F is the flux of the metal atoms (atoms m^{-2} s^{-1}), D is the diffusion coefficient (m^2 s^{-1}), c the metal concentration into the lattice (atoms m^{-3}) and x the spatial abscissa (m). Particularly, among the metals the silver presents very high values of the diffusion coefficient:

$$D = D_0\exp\left(\frac{-E_d}{RT}\right) \tag{4.3}$$

with $D_0 = 6.7$ 10–5 m^2 s^{-1} and $E_d = 45.2$ kcal mol^{-1} while R is the gas constant and T the absolute temperature (K) [20].

In the solid state diffusion through the lattice, starting from a concentration C_0, the metal atoms move along the metal thickness during the time as described in Figure 4.8.

Practically, a heat treatment of 2 h at 1000 °C under controlled atmosphere has been applied in order to perform the diffusion welding of Pd-Ag foils of thickness 50 µm accordingly to a procedure developed at ENEA Frascati laboratories [21,22].

A first procedure developed by ENEA used a pressure blade having a rounded alumina edge in order to apply a compression capable to weld the overlapped seams of a rolled Pd-Ag foil bent around an alumina bar, see Figure 4.9.

The resulting welding seam is shown in Figure 4.10 where an end of the thin wall Pd-Ag tube is reported.

According to a second diffusion welding procedure, the rolled Pd-Ag foils are wrapped around an alumina bar and then their limbs are pressed by using a special device (thermomechanical press). The resulting membrane is joined by brazing to two stainless steel tube ends, see Figure 4.11. The thermomechanical press shown in Figure 4.12 consists of two stainless steel plates and a threaded bar (screw) made of INVAR (a metal alloy having negligible thermal

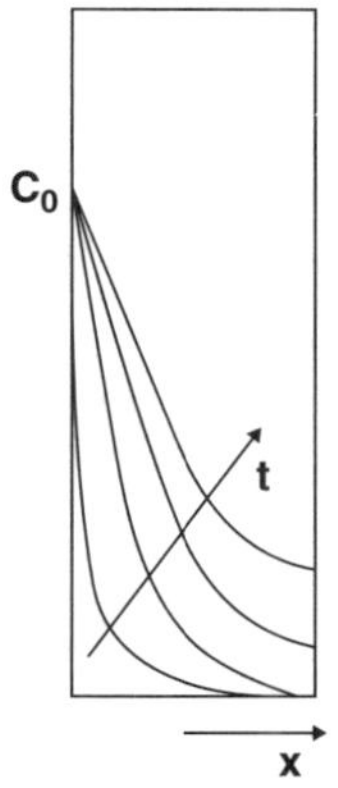

Figure 4.8 Concentration profile of atomic diffusion through a metal wall

expansion coefficient). At high temperature the thermal expansion of the steel plates and the other parts (Pd-Ag foils) is forced by the INVAR screw: the result is the compression of the Pd-Ag foils to be welded.

4.3 Applications

The main characteristic of the thin wall Pd-Ag tubes is their complete (infinite) hydrogen selectivity: therefore, these permeators permit the production of ultra-pure hydrogen. The hydrogen permeability and the chemical and physical stability have been verified in long term tests [23–25]. Under operating conditions of 300–350 °C and differential transmembrane

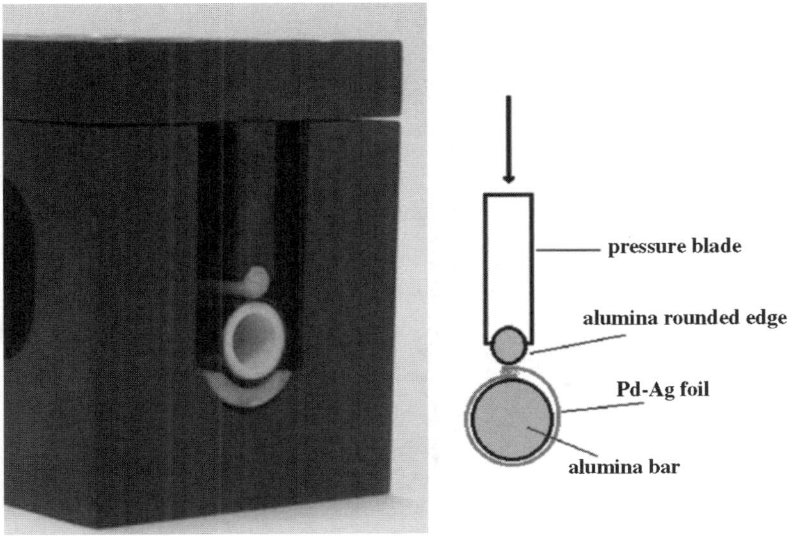

Figure 4.9 Picture and scheme of the diffusion welding device

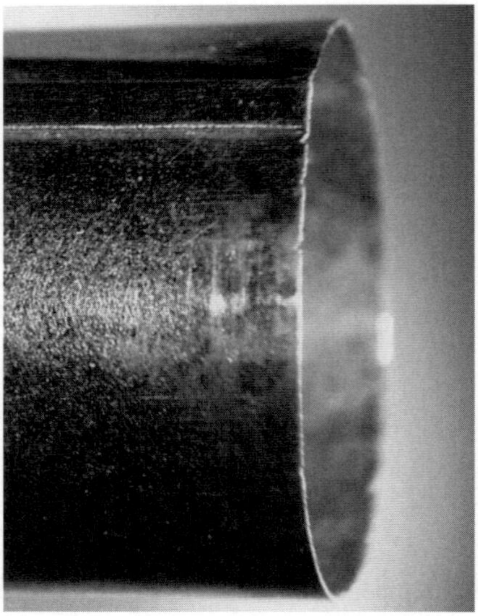

Figure 4.10 An end of a thin wall Pd-Ag tube: at the top, the welded zone is evident (overlapped junction lines)

pressure of 200 kPa, about 3 $Nm^3\,m^{-2}\,h^{-1}$ hydrogen permeation flow rates have been measured. By considering a Pd cost of about 10 € g^{-1}, the production of 1 $Nm^3\,h^{-1}$ of hydrogen involves a membrane cost of 1600 €.

Particularly, in order to ensure the long life of the permeator tubes an appropriate mechanical configuration of the membrane module has to be designed [23,26–28]. In fact, under hydrogenation the Pd alloy elongates significantly: the gas tight connection of the membrane tube to the module should permit its free elongation/contraction due to the thermal and hydrogenation

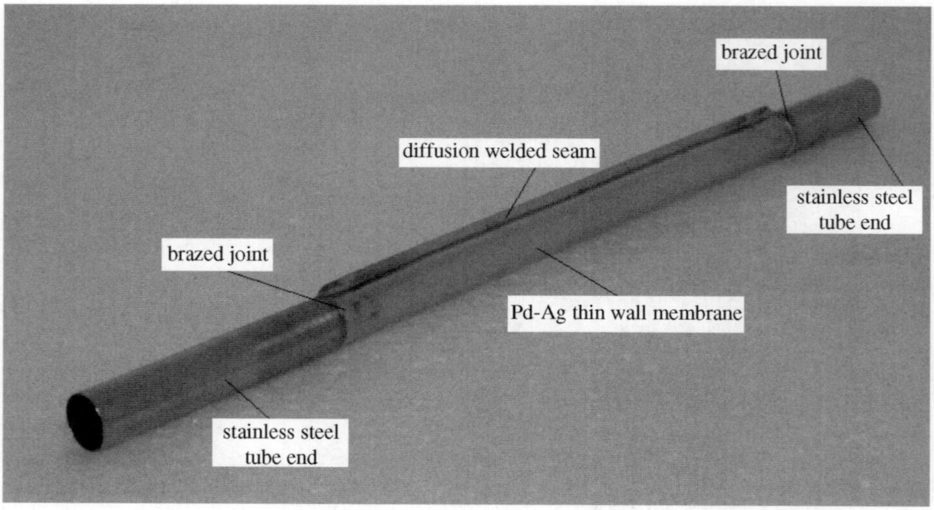

Figure 4.11 The thin wall permeator tube: the Pd-Ag membrane is joined to two stainless steel tube ends by brazing

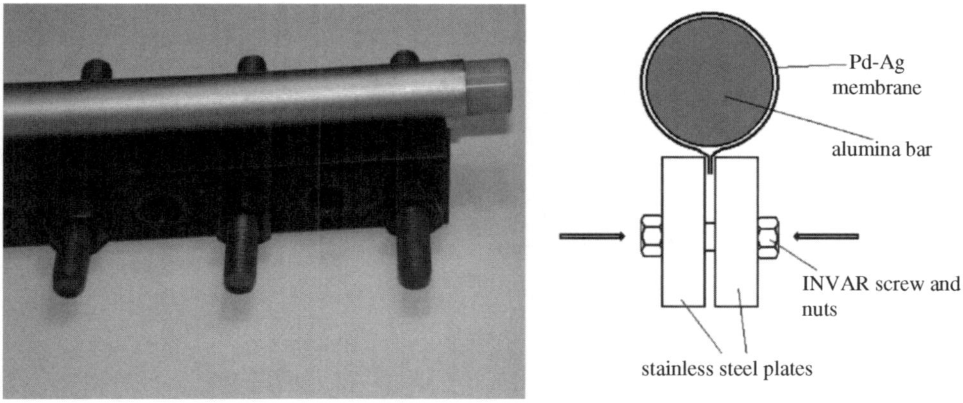

Figure 4.12 *Thermomechanical press used for the diffusion welding of thin wall Pd-Ag tubes*

cycling. Mainly, a finger-like (tube in tube) configuration has been adopted in order to avoid the rise of combined compressive and bending stresses of the Pd-Ag permeator tubes. In this mechanical design of the membrane modules, the Pd-Ag tube is tightly fixed to the module only at one its end. The other tube end is closed and the feed stream is sent through a stainless steel tube: in this way the permeator can elongate/contract under operation with hydrogen, see Figure 4.13.

Experimental campaigns have been carried out by using a Pyrex shell module containing the thin wall permeator: Figure 4.14 shows this testing device.

The finger-like configuration has been also adopted for manufacturing multi-tube membrane reactors: Figures 4.15 and 4.16 respectively show the scheme and a picture of a module with 19 thin wall permeators designed and tested for the water gas shift reaction [28].

Membrane reactors for dehydrogenation reactions are important applications of the thin wall Pd-Ag membrane tubes. Especially, the Pd-based membrane reactors combine a fixed bed catalytic reactor with a permselctive membrane: the hydrogen removal through the membrane promotes the reaction conversion beyond the thermodynamic equilibrium (shift effect) [1,6]. A

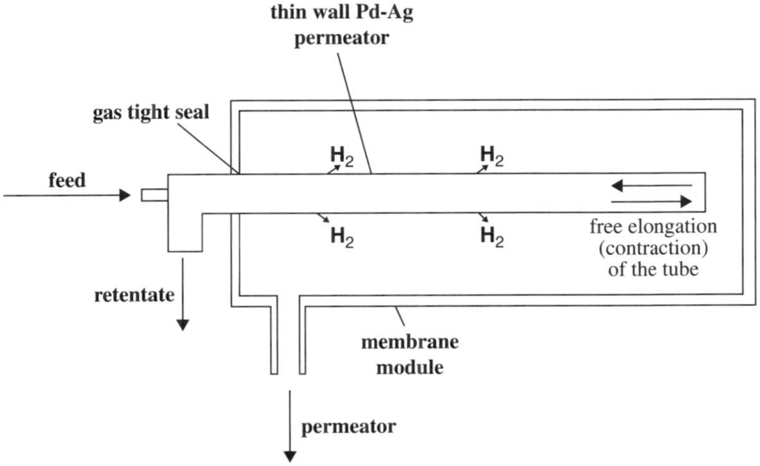

Figure 4.13 *Scheme of a finger-like assembly of a Pd-Ag thin wall permeator tube*

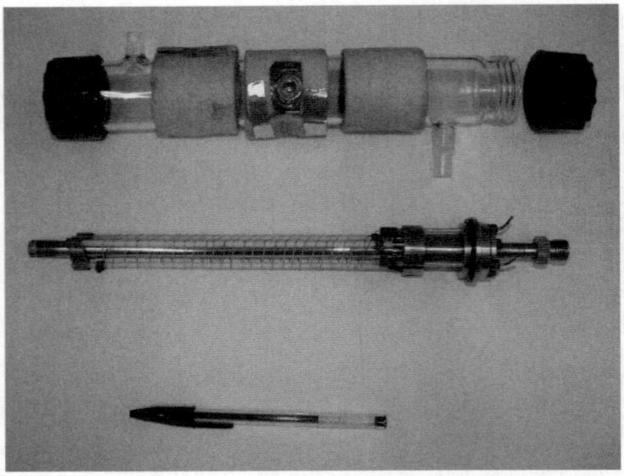

Figure 4.14 Experimental device: glass shell (above) and the thin wall Pd-Ag tube (below) with the heating system (an electrical heating wire)

wide experimental work has been carried out by membrane reactors which use the thin wall permeators: these tests have considered the production of ultra-pure hydrogen via steam reforming of ethanol [29–33] and acetic acid [34], methane dry reforming [35] and methanol steam reforming [36]. The capability of the thin wall membranes of producing ultra-pure hydrogen has been demonstrated: moreover, these membrane reactors permitted to attain reaction conversions and hydrogen yields higher than the traditional reformers do.

The thin Pd-Ag membranes has also been applied for building a membrane reactor (PERMCAT concept) used for purifying the hydrogen isotopes from the plasma exhausts of fusion reactors, see Figure 4.17. In this case a Pd-Ag tube of length 500 mm has been fixed to the steel shell by two metal bellows which compensate the elongation/contraction thus avoiding the mechanical stresses of the thin membrane [28].

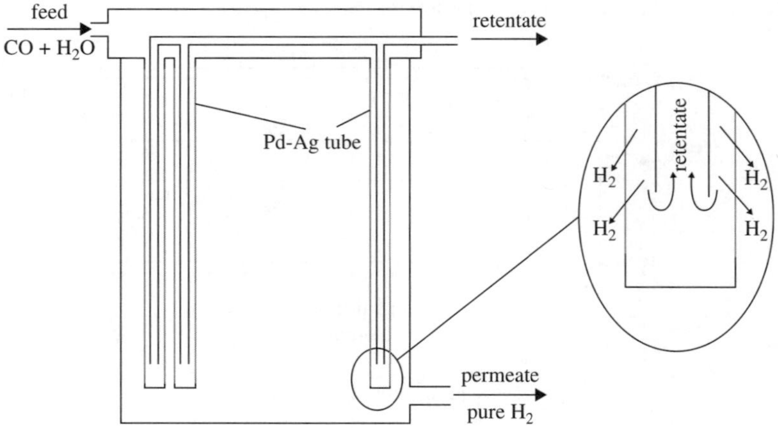

Figure 4.15 Scheme of the Pd-Ag multitube membrane module for the water gas shift reaction

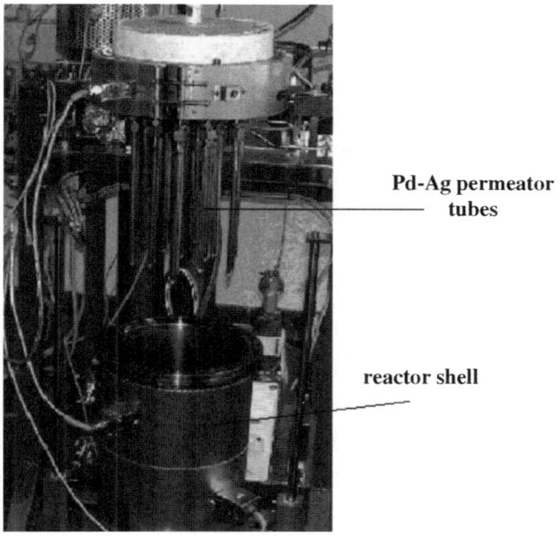

Figure 4.16 A multitube membrane module (opened)

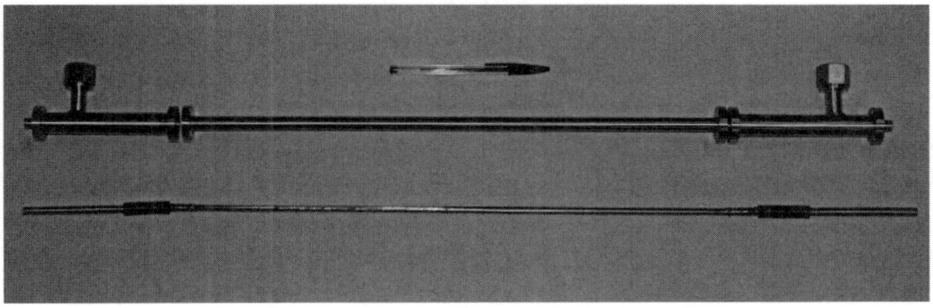

Figure 4.17 PERMCAT reactor: steel shell module (above) and the 500 mm long Pd-Ag tube (below)

4.4 Conclusions

A cold rolling and diffusion welding procedure has been developed for manufacturing Pd-Ag permeators of reduced wall thickness (50 μm). The commercial Pd alloy used for manufacturing hydrogen separator uses silver at 23–25 wt%; especially, the silver atoms have a high diffusivity into the metal lattice. Consequently, the diffusion welding of such a Pd alloy can be carried out at relatively low temperature (800–1000 °C) for 1–2 h by using a thermomechanical press into a controlled atmosphere oven.

These thin wall permeators are characterised by complete hydrogen selectivity and then are applied in the processes for producing ultra-pure hydrogen. A specific permeation capacity of about $3\,Nm^3\,m^{-2}\,h^{-1}$ has been measured in experimental campaigns which demonstrated also their hydrogen selectivity and chemical and physical stability.

Many applications have been tested at a laboratory scale: these experiments have mainly concerned Pd-based membrane reformers for producing hydrogen from methane, methanol, ethanol and acetic acid. The capability to move the reaction conversion beyond the thermodynamic equilibrium values (shift effect) has been demonstrated.

References

1. J. Shu, B.P.A. Grandjean, A. Van Neste, S. Kalaguine. Catalytic palladium-based membrane reactors: a review. *Can J Chem Eng*, **69**, 1036–1060 (1991)
2. J.N. Armor, Applications of catalytic inorganic membrane reactors to refinery products. *J Membr Sci*, **147**, 217–233 (1998).
3. E. Kikuchi. Membrane reactor application to hydrogen production. *Catal Today*, **56**, 97–101 (2000).
4. E. Drioli, E. Fontananova. Membrane technology and sustainable growth. *Chemical Engineering Research and Design*, **82** (A12) 1557–1562 (2004).
5. P.S. Apte, J.M. Schwartz, S.W. Callahan.Patent WO 2007/058913 A2 (2007).
6. Z. Dardas, Y. She, T.H. Vanderspurt, J. Yamanis, C. Walker.Patent Application PCT/US/2005/047047 (2005).
7. A. Friedbereger, G. Mueller, W. Jehle, C. Wolff,Patent DE102006051525 (2006).
8. F.A. Lewis. *The Palladium Hydrogen System*. Academic Press, London (1967).
9. A. Basile, F. Gallucci, S. Tosti. Synthesis, characterization, and applications of palladium membranes. *Membrane Science and Technology*, **13**, 255–323 (2008).
10. S.N. Paglieri, J.D. Way. Innovations in palladium membrane research. *Separation and Purification Methods*, **31** (1), 1–169 (2002).
11. S. Uemiya. State-of-the-art of supported metal membranes for gas separation. *Separation and Purification Methods*, **28** (1), 51–85 (1999).
12. O. Schramm, A. Seidel-Morgenstrern. Comparing porous and dense membranes for the application in membrane reactors. *Chem Eng Sci*, **54**, 1447–1453 (1999).
13. L.Q. Wu, N. Xu, J. Shu. Preparation of a palladium composite membrane by an improved electroless plating technique. *Ind Eng Chem Res*, **93**, 342–348 (2000).
14. S. Tosti, L. Bettinali, S. Castelli, F. Sarto, S. Scaglione, V. Violante. Sputtered, electroless, and rolled palladium-ceramic membranes. *J Membr Sci*, **196**, 241–249 (2002).
15. J. Shu, B.P.A. Grandjean E. Ghali, S. Kaliaguineet. Simultaneous depositing of Pd and Ag on porous stainless steel by electroless plating. *J Membrane Sci*, **77**, 181–195 (1993).
16. ASM, Handbook (formerly Metals Handbook, volume 2). Properties and selection: nonferrous alloys and special-purpose materials (2000).
17. S. Tosti, L. Bettinali, V. Violante. Rolled thin Pd and Pd-Ag membranes for hydrogen separation and production. *Int J Hydrogen Energy*, **25**, 319–325 (2000).
18. S.B. Dunkerton. Diffusion bonding – process and applications, *Welding and Metal Fabrication*, **59** (3), 132–136 (1991).
19. C. Deminet,Method of diffusion bonding, US Patent 4 013 210 (1977).
20. CRC. *Handbook of Chemistry and Physics*, 67th edition, CRC Press, New York (2000).
21. S. Tosti, L. Bettinali. Diffusion bonding of Pd-Ag membranes, *J Materials Science*, **39**, 3041–3046 (2004).
22. S. Tosti, L. Bettinali, D. Lecci, F. Marini, V. Violante.Method of bonding thin foils made of metal alloys selectively permeable to hydrogen, particularly providing membrane devices, and apparatus for carrying out the same. European Patent EP 1184125 (2001).
23. S. Tosti, A. Basile, L. Bettinali, F. Borgognoni, F. Gallucci, C. Rizzello. Design and process study of Pd membrane reactors. *International Journal of Hydrogen Energy*, **33**, 5098–5105 (2008).
24. S. Tosti, A. Adrover, A. Basile, V. Camilli, G. Chiappetta, V. Violante. Characterization of thin wall Pd-Ag rolled membranes. *Int J Hydrogen Energy*, **28**, 105–112 (2003).
25. S. Tosti, A. Basile, L. Bettinali, F. Borgognoni, F. Chiaravalloti, F. Gallucci. Long-term tests of Pd-Ag thin wall permeator tube. *J Membr Sci*, **284**, 393–397 (2006).

26. A. Basile, S. Tosti. Dispositivo portatile a membrana intercambiabile per valutazione di processi di permeazione e reazione in fase gassosa. Italian Patent RM2005U000107 (2005).
27. S. Tosti, A. Basile, L. Bettinali, D. Lecci, C. Rizzello.Dispositivo a membrana a fascio tubiero per la produzione di idrogeno ultrapuro. Italian Patent RM2005A000399 (2005).
28. S. Tosti, L. Bettinali, F. Borgognoni, D.K. Murdoch. Mechanical design of a PERMCAT reactor module. *Fusion Engineering and Design*, **82**, 153–161 (2007).
29. S. Tosti, A. Basile, F. Borgognoni, V. Capaldo, S. Cordiner, S. Di Cave, F. Gallucci, C. Rizzello, A. Santucci, E. Traversa. Low temperature ethanol steam reforming in a Pd-Ag membrane reactor – Part 1: Ru-based catalyst. *Journal of Membrane Science*, **308**, 250–257 (2008).
30. S. Tosti, A. Basile, F. Borgognoni, V. Capaldo, S. Cordiner, S. Di Cave, F. Gallucci, C. Rizzello, A. Santucci, E. Traversa. Low-temperature ethanol steam reforming in a Pd-Ag membrane reactor – Part 2. Pt-based and Ni-based catalysts and general comparison. *Journal of Membrane Science*, **308**, 258–263 (2008).
31. F. Gallucci, M. De Falco, S. Tosti, L. Marrelli, A. Basile. Ethanol steam reforming in a dense Pd-Ag membrane reactor: a modelling work. *Comparison with the traditional system. International Journal of Hydrogen Energy*, **33**, 644–651 (2008).
32. A. Basile, F. Gallucci, A. Iulianelli, S. Tosti. CO-free hydrogen production by ethanol steam reforming in a Pd-Ag membrane reactor. *Fuel Cells*, **08** (1), 62–68 (2008).
33. F. Gallucci, M. De Falco, S. Tosti, L. Marrelli, A. Basile. Co-current and counter-current configurations for ethanol steam reforming in a dense Pd-Ag membrane reactor. *International Journal of Hydrogen Energy*, **33**, 6165–6171 (2008).
34. A. Basile, F. Gallucci, A. Iulianelli, F. Borgognoni, S. Tosti. Acetic acid steam reforming in a Pd-Ag membrane reactor: the effect of the catalytic bed pattern. *Journal of Membrane Science*, **311**, 46–52 (2008).
35. F. Gallucci, S. Tosti, A. Basile, Pd-Ag tubular membrane reactors for methane dry reforming: a reactive method for CO_2 consumption and H_2 production. *Journal of Membrane Science*, **317**, 96–105 (2008).
36. A. Basile, A. Parmaliana, S. Tosti, A. Iulianelli, F. Gallucci, C. Espro, J. Spooren. Hydrogen production by methanol steam reforming carried out in membrane reactor on Cu/Zn/Mg-based catalyst. *Catalysis Today*, **137**, 17–22 (2008).

5

Preparation and Synthesis of Mixed Ionic and Electronic Conducting Ceramic Membranes for Oxygen Permeation

Jianhua Tong and Ryan O'Hayre

Metallurgical and Materials Engineering, Colorado School of Mines, Golden, Colorado, USA

5.1 Introduction

Mixed ionic and electronic conducting (MIEC) ceramic membranes are garnering significant scientific attention due to their remarkable combination of properties including extremely high oxygen selectivity, high operation temperature ($>600\,^{\circ}C$), continuous supply of lattice oxygen, long-term stability, and catalytic activity. MIEC ceramic membranes transport oxygen via a combined electrochemical surface reaction and solid-state ambipolar diffusion process involving both oxygen vacancies and electrons as schematically described in Figure 5.1. MIEC ceramic membranes are not only strong competitors to conventional oxygen producing techniques such as cryogenic distillation or pressure swing adsorption but are also good candidates for high-temperature oxygen-permeable membrane reactors [1–10]. Since the first report of MIEC ceramic membranes based on perovskite-type oxides [1], thousands of publications have appeared on this subject. Many of the important aspects of oxygen transport mechanism, materials development, microstructure characterisation, oxygen permeation properties, applications in oxygen-related reactions, and industrialisation status for these MIEC membranes have been reviewed in detail in recent book chapters and reviews [2–7]. However, a comprehensive discussion focusing on the preparation methods used to obtain MIEC ceramic membranes is lacking. This chapter therefore focuses almost exclusively on a review of this topic, because

Membranes for Membrane Reactors: Preparation, Optimization and Selection, Edited by Angelo Basile and Fausto Gallucci

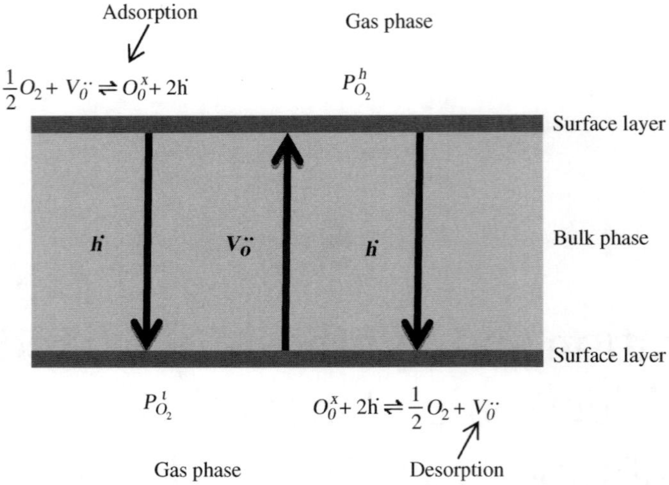

Figure 5.1 *Oxygen transport mechanism through MIEC ceramic membrane ($P_{O_2}^h > P_{O_2}^l$)*

successful materials development and membrane fabrication are crucial prerequisites for the investigation of the MIEC ceramic membranes.

5.2 Preparation of MIEC Ceramic Powders

It is strongly believed that the quality of the starting MIEC ceramic (multimetal perovskite-type oxide) powder is the most crucial factor for the successful preparation of a high-performance MIEC ceramic membrane. It is well known that perovskite-type oxide powders prepared using different methods can result in significantly different microstructures, thereby producing large discrepancies in oxygen permeation membrane performance [11–14]. The most common preparation methods for the multimetal perovskite-type oxide powders can be categorised into several main groups each with distinct pros and cons. These general methods, which will be discussed in further detail in the subsections that follow, include solid-state reaction, coprecipitation, conventional sol-gel, polymeric gelation, hydrothermal synthesis, spray pyrolysis, and combustion synthesis.

5.2.1 Conventional Solid-State Reaction

The conventional solid-state reaction method is the most commonly used route for preparing multimetal perovskite-type oxide powders. This method generally involves mixing raw materials such as carbonate, oxide, hydroxide or salt powders in appropriate stoichiometric quantities and firing the mixed powders at temperatures higher than two thirds of the melting points for a long period ($\geq$10 h) to facilitate a solid-state reaction of the precursors into the final desired phase-pure perovskite. It has been shown that the solid-state reaction process is both thermodynamically and kinetically controlled. Apart from the high temperatures and long reaction periods, the large contact area and short diffusion distance corresponding to the small particle size for the starting powders are also important factors to improve the quality of the multimetal perovskite-type oxide powders prepared by solid-state reaction. The particle sizes and diffusion distance can be markedly reduced by laborious mechanical mixing and grinding of the starting powders. However, it is technically difficult to mill the particle size below 1 μm. Therefore, the purity and homogeneity of perovskite powders synthesised by the solid-state reaction method are generally poor and show a broad particle size distribution. In addition, the high required reaction temperatures can introduce other serious problems, particularly if one component in the mixed powder shows a tendency to volatilise during a long-term reaction (e.g. barium). In spite of these challenges, the solid-state reaction method remains one the simplest and most cost-effective processes for the synthesis of multimetal oxides. A wide variety of different multimetal oxide powders for MIEC ceramic membranes have therefore been prepared by the solid-state reaction method [15–17].

Takeda *et al.* prepared $La_{0.8}Sr_{0.2}(Ga_{0.8}Mg_{0.2})_{1-x}Cr_xO_{3-\delta}$ ($0 \leq x \leq 1.0$) oxide powders by the solid-state reaction method [15]. Stoichiometric amounts of La_2O_3, $SrCO_3$, Cr_2O_3, Ga_2O_3, and MgO powders were ground and calcined at 1000 °C for 12 h. The resulting powders were reground and pressed into green disks at 150 MPa. Sintering temperatures of 1700, 1500, and 1450 °C for 12 h were required to obtain acceptable dense membrane disks for $La_{0.8}Sr_{0.2}CrO_{3-\delta}$, $La_{0.8}Sr_{0.2}(Ga_{0.8}Mg_{0.2})_{1-x}Cr_xO_{3-\delta}$ ($x = 0.9$–0.1), and $La_{0.8}Sr_{0.2}Ga_{0.8}Mg_{0.2}O_{3-\delta}$, respectively. This example demonstrates that the solid-state reaction method is suitable for the preparation of even highly complex multicomponent (five metal) powders. However, in order to obtain acceptable purity, the grinding and calcining cycles must generally be repeated several times and a long ball-milling period is generally necessary to improve powder quality and obtain good sintering ability.

5.2.2 Coprecipitation

Coprecipitation, which refers to the simultaneous precipitation of solids from solution, is one of the oldest methods for the preparation of multimetal perovskite-type oxide powders. The coprecipitation method begins with the preparation of a solution containing the desired cations in appropriate stoichiometric ratios, which is then mixed with a second solution containing a precipitation agent (or agents) which causes precipitation of a mixed solid powder. The solids are then filtered, dried, and thermally decomposed to form the desired ceramic powder. The experimental parameters of pH, mixing rate, temperature, and concentration must be carefully controlled in order to produce multimetal oxide powders with acceptable physical and chemical properties. The obtained composition, purity, particle size, and morphology are usually good if all involved cations have similar precipitation rates. However, large discrepancies in cation precipitation rates can lead to microscopic inhomogeneity. In addition, as with most solution techniques, aggregates are commonly formed. The coprecipitation technique is extensively used for synthesising MIEC perovskite-type oxide powders [18–20].

Lin *et al.* prepared prototypical MIEC ceramic powders of $La_{0.8}Sr_{0.2}Co_{0.6}Fe_{0.4}O_{3-\delta}$ by the coprecipitation method [18]. Stoichiometric amounts of La, Sr, Co, and Fe nitrate salts were first dissolved in water. The metal cations were then coprecipitated by slowly adding KOH solution while stirring. The coprecipitated solids were filtrated and washed several times using distilled water to remove potassium salt impurities. Then the solids were dried and calcined at 800 °C for 4 h to form $La_{0.8}Sr_{0.2}Co_{0.6}Fe_{0.4}O_{3-\delta}$ powder. $La_{0.8}Sr_{0.2}Co_{0.6}Fe_{0.4}O_{3-\delta}$ disks were sintered at 1275 °C for 2 h, yielding good relative density and morphology. However, actual composition from ICP measurement indicated a large Sr deficiency (85%), probably resulting from the filtration and washing procedures. The larger solubility of $Sr(OH)_2$ or the slower precipitation rate for Sr cation may have also contributed to the very low Sr amount in the final powders. As this example illustrates, although the coprecipitation method can obtain oxide powders at relatively low calcination temperatures with good sintering ability, deviations from desired stoichiometric composition can often occur, making it difficult to prepare complex mixed-metal oxides with precise and reproducible stoichiometry.

5.2.3 Conventional Sol-Gel Method

The conventional sol-gel method is a wet-chemical process widely used to prepare metal oxide powders starting from a chemical solution acting as the precursor for an integrated network (*gel*) of discrete particles, which in most cases involves in the metal alkoxides of the desired oxides. The exclusive property of the alkoxide route sol-gel process is that all the starting materials employed for the sol-gel process are in the form of metal alkoxides. Metal alkoxides are readily hydrolysed and condensed to hydroxides or oxides. The generalised hydrolysis and condensation reaction are as follows:

$$M(OR)_x + xH_2O \rightarrow M(OH)_x + xROH \tag{5.1}$$

$$2M(OH)_x \rightarrow M_2O_x + xH_2O \tag{5.2}$$

The negligibly small solubility of hydroxides or oxides in the resulting byproduct of aliphatic alcohol or the organic solvent for the alkoxide solution precursor ensures rapid nucleation of the hydroxide products, which permits the formation of nanosized solid particles homogeneously in the solvent. If proper drying methods are used to avoid aggregation by removing additional water during the gel drying process, nanosized powders can be easily prepared, yielding excellent powder properties for subsequent MIEC ceramic membrane fabrication. Several complex perovskite-type oxides have been synthesised using this technique [21–23]. Because the sol-gel route can provide nanoscale powders, dense perovskite membranes can typically be fabricated from this approach at far lower sintering temperatures than would be required using most other methods.

As one example of the power of the sol-gel route, Westin *et al.* demonstrated the fabrication of MIEC powder of $La_{0.5}Sr_{0.5}CoO_3$ using the alkoxide sol-gel method [21]. The metal alkoxide precursor solutions were prepared in house. A lanthanum methoxy-ethoxide solution was synthesised by dissolving a lanthanum chip with $HgCl_2$ catalyst in a toluene: moeH solvent at 50 °C for 30 h. A faint yellow lanthanum solution was subsequently obtained after removal of a fine green-black precipitate. The strontium solution was prepared by dissolving strontium metal in toluene: moeH solvent at room temperature for 24 h followed by centrifugating to separate the small residual precipitates. The cobalt solution was synthesised using metathesis by adding $CoCl_2$ to potassium dissolved in toluence:moeH solvent (Co/K = 0.5). After stirring for 24 h at room temperature, the KCl was separated to leave a cobalt solution. By mixing these alkoxides, the La-, Sr-, and Co-perovskite precursor solution was obtained that yielded nanosize powders upon precipitation. The reactive alkoxides allowed for full hydrolysis by air moisture into hydrated oxo-carbonate gels, which then could be converted to perovskites at 700–770 °C.

Despite the significant benefits of the alkoxide sol-gel process, there are several limitations which severely impact widespread adoption of this technique on a commercial scale. Firstly, the preparation of alkoxide precursors is technically and economically difficult. For example, metal alkoxides of groups I and II are solid and nonvolatile, making it difficult to purify by distillation. Therefore, metal salts or hydroxides are generally used in this case. This modified route requires the preparation of desired metal alkoxides for most of the metal cations other than the group I and II elements, which are then added as metal salts in an alcoholic or aqueous solution. In this process, hydrolysis must be specially controlled due to the significantly different reaction rates in the mixture of alkoxides and salts. Another limitation is that it is very difficult to find a proper solvent which can dissolve all metal alkoxides and/or salts for a desired multimetal oxide. Although the solubility is small, hydroxide or oxide formation rates are not exactly the same for all metals, which inevitably results in microscopic inhomogeneity, similar to the coprecipitation route. Moreover, proper drying of gel to form a non-aggregated mixed oxide powder is often difficult, requiring specialised or laborious drying techniques [24, 25].

5.2.4 Polymeric Gelation Method

The polymeric gelation method represents a variation on the gelation approach. Instead of forming colloidal particles of hydroxides or oxides in solution, in the polymeric gelation route, the desired cations are chelated by organic ligands, which allows for the homogenous formation of polymerised metal salts in the form of a sticky gel. After proper drying and calcining the gel can be successfully converted to the desired oxide powders with accurate composition and phase-pure structure at relatively low temperatures. Ethylenediaminetetraacetic acid (EDTA) and citric acid combined complexing (ECCC) is the most popular method for the preparation of phase-pure multimetal perovskite-type oxide powders, and this route has been utilised extensively to successfully prepare MIEC perovskite-type oxides for oxygen permeation membranes [26–33]. EDTA acid and citric acid are used as combined chelating agents because of their synergetic functions during the polymerisation process. The EDTA acid has a very strong chelating ability with metal ions. Because of its high denticity, this ligand has a high affinity for metal cations. Using M^{n+} ions as example, the chelating reaction can be described by Equation (5.3) to form the structure schematically shown in Figure 5.2a.

$$\left[M(H_2O)_6\right]^{n+} + H_4EDTA \leftrightarrow [M(EDTA)]^{-(4-n)} + 6H_2O + 4H^+ \qquad (5.3)$$

The addition of base to react with H^+ can shift this reaction towards the right; in other words, high pH increases the metal–EDTA complex stability. However, overly high pH should be avoided due to the possibility for the formation of hydroxide during the chelating process. Experimental results indicate that a pH range of 6–10 is good for chelating reactions between most metal cations and EDTA acid. Citric acid has significantly stronger polymerisation ability

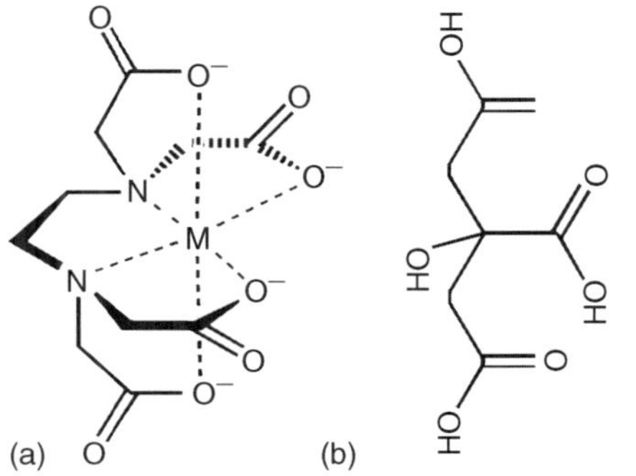

Figure 5.2 *Structure of (a) EDTA metal complex and (b) citric acid*

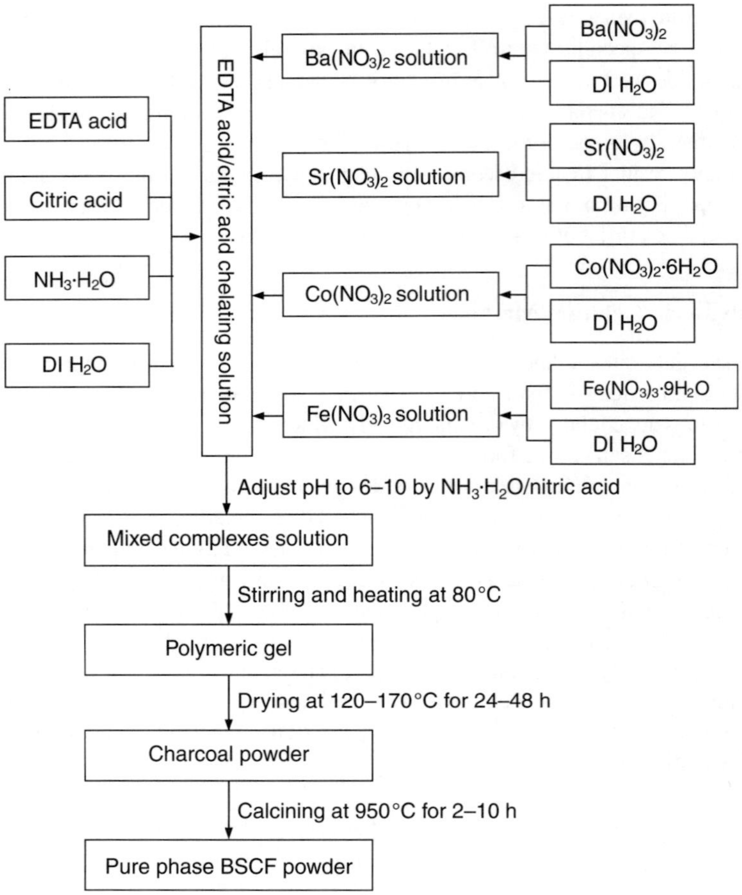

Figure 5.3 *Flow diagram of the preparation of $Ba_{0.5}Sr_{0.5}Co_{0.8}Fe_{0.2}O_{3-\delta}$ powder by the ECCC method*

because both an OH and a COOH group exist in its chemical structure (Figure 5.2b). The ECCC method involves the complexation of metal ions in the EDTA/citric acid solution, followed by evaporation of the water solvent and thermal decomposition of the complexes with subsequent formation of the perovskite phase.

The ECCC method is illustrated in Figure 5.3 using the preparation process for $Ba_{0.5}Sr_{0.5}Co_{0.8}Fe_{0.2}O_{3-\delta}$ powder, one of the most popular MIEC ceramic membrane materials, as an example. Stoichiometric amounts of $Ba(NO_3)_2$, $Sr(NO_3)_2$, $Co(NO_3)_2$, and $Fe(NO_3)_3$ are first dissolved into DI water to prepare clear aqueous solutions of the respective metal nitrates. Appropriate molar quantities of EDTA acid and citric acid are then prepared at 1.5- and 1.0-times the total moles of metal ions, respectively. The EDTA acid and citric acid are dissolved in ammonium hydroxide solution and the pH of the chelating solution is controlled to be ~6.0–10.0. While magnetically stirring, the metal ion solutions are slowly added into the chelating solution in the order $Ba(NO_3)_2$, $Sr(NO_3)_2$, $Co(NO_3)_2$, and $Fe(NO_3)_3$. Precipitation may occur after mixing, which can be eliminated by adjusting pH in the range 6.0–10.0 by the addition of either ammonium hydroxide or nitric acid. After evaporating the water by heating the mixed solution at 70–90 °C, a sticky gel is obtained. This gel is then dried at 120–170 °C for 24–48 h to make a charcoal powder. After calcination at 950 °C for 5 h, a phase-pure $Ba_{0.5}Sr_{0.5}Co_{0.8}Fe_{0.2}O_{3-\delta}$ powder is successfully synthesised. Most multimetal MIEC perovskite powders can be synthesised using this popular ECCC method provided they have good solubility in water and a large equilibrium quotient for the chelating reaction. This technique has therefore been one of the most widespread methods for the preparation of high-quality mixed oxide powders [26–33].

5.2.5 Hydrothermal Synthesis

The hydrothermal synthesis method has recently been applied to produce multimetal perovskite-type oxide powders. This process generally uses temperatures between the boiling point of water and its critical temperature of 374 °C, with pressures as high as 15 MPa. Consequently, the calcination step required by other synthesis methods can often be eliminated in this case. Hydrothermal synthesis can be used to assist the sol-gel technique by controlling the particle size and the crystallisation process. The materials used are generally inexpensive and easy to control in terms of size, shape and stoichiometry. Elimination of impurities associated with milling processes and the other advantages mentioned above result in very fine and highly reactive ceramic powders. Several perovskite oxides have been prepared using the hydrothermal synthesis technique [34–36], although it is still in the early stage of development.

For example, Lin *et al.* attempted the synthesis of $La_{0.8}Sr_{0.2}Co_{0.6}Fe_{0.4}O_{3-\delta}$ powder using the hydrothermal synthesis method [34]. A La-Co-Fe hydroxide gel was synthesised and reacted with strontium hydroxide under hydrothermal conditions. However, the perovskite structure was not obtained for the as-synthesised powders at temperatures lower than 300 °C. In some cases, calcination of the hydrothermal $La_{0.8}Sr_{0.2}Co_{0.6}Fe_{0.4}O_{3-\delta}$ powder produced perovskite-structured materials (with unknown stoichiometry). It was concluded that the relative stability of LaOOH and the solubility of $Sr(OH)_2$ prevent the desired reaction from taking place under commercially viable (i.e., nonsupercritical) conditions. Obviously, further work needs to be done in order to fully establish the viability of the hydrothermal synthesis process for mixed-metal perovskite oxides.

5.2.6 Spray Pyrolysis

Spray pyrolysis, a vaporisation method, has been successfully applied to the synthesis of multimetal ceramic powders [37, 38]. In this method, a misted stream of a precursor solution containing the desired components is continuously dried, precipitated, and decomposed in a tubular furnace reactor. Powders prepared by spray pyrolysis method are generally spherically shaped and non-aggregated with submicron size. For example, $La_{0.8}Sr_{0.2}Ga_{0.8}Mg_{0.2}O_{2.8}$ powders were prepared by the spray pyrolysis method and their powder morphology and sintering properties were investigated [37]. The spray pyrolysis system used in this study consisted of a droplet generator, a high-temperature tubular quartz reactor, and a powder collector. A 1.7 MHz ultrasonic spray generator equipped with six vibrators was used to generate a large number of droplets, which were subsequently introduced into the reactor by a carrier gas. Inside this reactor, the droplets and powders evaporated, decomposed, and/or crystallised. The length and diameter of the reactor were 1200 and 50 mm, respectively. The concentration of La, Sr, Ga, and Mg components was maintained at 0.5 M. The flow rate of the carrier gas (air) was 20 L min^{-1}. The as-prepared powders, obtained at 900 °C in the tubular quartz reactor, were subsequently post-treated in a box furnace at 1000 °C for 6 h. The powder was composed of spherically shaped and non-aggregated particles with submicron size and a fully densified disk was obtained from the spray pyrolysis powder.

5.2.7 Combustion Synthesis

Combustion synthesis is an energy-efficient method for the production of advanced materials. It is characterised by fast heating rates, high temperatures, and short reaction times, and typically occurs by one of two modes: self-propagating high temperature synthesis (SHS) or volume combustion synthesis (VCS). In the SHS mode, powder reactants are heated locally and the reaction follows a wave-like propagation through the medium, while in the VCS mode, reactants are heated uniformly and reaction occurs simultaneously throughout the mixture. A variant of the combustion method is aqueous combustion synthesis, which uses precursor aqueous solutions containing nitrates of the desired metals and a fuel, such as glycine, hydrazine, or urea. Powders prepared by aqueous combustion synthesis generally have high surface areas, as well as phase-purity and well controlled chemical composition. This process has been used to synthesise a variety of mixed oxides composed of fine particles [39–42].

Aqueous combustion synthesis of $La_{0.8}Sr_{0.2}CrO_3$ perovskite powder was conducted using the metal nitrate–glycine system [39]. The ratio between nitrates (oxidisers) was maintained constant corresponding to the desired stoichiometric composition, while the ratio between glycine (fuel) and the oxidisers was changed based on the extra oxygen amount. The fuel and oxidisers were mixed in distilled water, thoroughly stirred to reach complete dissolution of all solid reagents, and placed on a hot plate to initiate the reaction. As the solution temperature reached 100 °C, water began to boil and evaporate from the solution, increasing the solution viscosity substantially and culminating in a rapid reaction. The as-synthesised powders were calcined in air at 850 °C for 12 h to remove residual water and other easily volatilised components, and to improve the crystallinity of the powder. Pure perovskite phase and a high surface area of $24\,m^2\,g^{-1}$ resulted from this procedure. Although the aqueous combustion synthesis method is frequently used to synthesise mixed oxide powders, the water evaporation process generally results in some component segregation, thereby producing at least some local inhomogeneities.

5.3 Preparation of MIEC Membranes

5.3.1 Disk-Shaped Configuration

The disk-shaped MIEC ceramic membrane is the most common membrane configuration. It has been extensively used to study the fundamental properties of MIEC materials in laboratory settings due to the ease of preparing high-quality and fully densified membranes using this configuration [43–46]. The disadvantages of the disk-shaped membrane configuration include challenges associated with sealing at high temperature, small membrane area, and large membrane thickness.

Typically, the most simple and economic technique for the preparation of disk-shaped membranes is to dry press MIEC ceramic powders in a steel die with high pressure. Using this technique it is generally easy to obtain high green density (e.g. ~60%), which makes the subsequent sintering process straightforward and effective. For example, $La_{0.6}Sr_{0.4}Co_{0.2}Fe_{0.8}O_{3-\delta}$ powders prepared by the ECCC method as described in [43] were ball-milled and subsequently pressed into disks or bars by two steel dies (circular die diameter of 15 mm, and bar die dimensions of 5 × 8 × 25 mm) under a hydraulic pressure of 30 t. The resulting green disks or bars were sintered at temperatures of 1000–1300 °C for 5 h, respectively. The relative densities of the sintered products were measured to be >90% by Archimedes method. Figure 5.4 provides SEM micrographs of the fractured cross

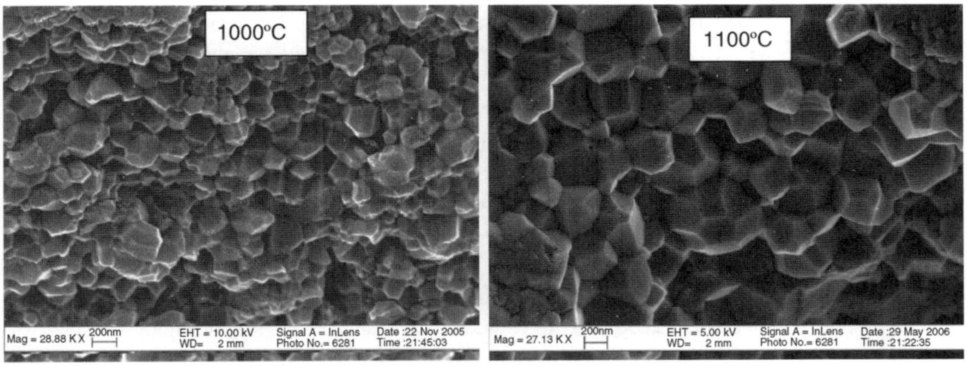

Figure 5.4 *SEM images for $La_{0.6}Sr_{0.4}Co_{0.2}Fe_{0.8}O_{3-\delta}$ disk sintered at 1000 and 1100 °C for 5 h. Reprinted from Journal of Membrane Science, Wang, H., et al., Investigation of phase structure, sintering, and permeability of perovskitetype $Ba_{0.5}Sr_{0.5}Co_{0.8}Fe_{0.2}O_{3-\delta}$ membranes. Vol. 262, 20–26. Copyright (2005) with permission from Elsevier*

sections of $La_{0.6}Sr_{0.4}Co_{0.2}Fe_{0.8}O_{3-\delta}$ disks sintered at 1000 and 1100 °C for 5 h, illustrating that full density was achieved using this approach at relatively low sintering temperatures.

5.3.2 Tubular-Shaped Configuration

The preparation of tubular membranes is generally more difficult than that of disk-shaped membranes. Nevertheless, the tubular configuration has several significant advantages. In particular, tubular configurations considerably simplify high-temperature sealing arrangements, can provide significantly larger effective membrane areas, and better reflect the typical commercial membrane configuration. There are several major techniques for tubular membrane production, including plastic extrusion, isostatic pressing, and slip casting.

5.3.2.1 Plastic Extrusion

Plastic extrusion is a popular technique for the preparation of tubular ceramic membranes. MIEC ceramic powders and certain organic additives are mixed together to form a paste, which generally has sufficient plasticity and strength to maintain the physical integrity of the molded tubes in their green states. The organic additives usually consist of binders, plasticisers, solvents, dispersants, and viscosity modifiers etc., which have to be experimentally optimised according to the specific properties of the MIEC ceramic powder (e.g. taking into account factors such as particle size, morphology, and density). The as-prepared plastic paste is then pressed through an extrusion die at high pressure to fabricate green tubes. The extruded tubes are then typically heated slowly at a relatively low temperature (e.g., 50–450 °C) range to controllably remove gases that result from the decomposition of the organic additives and avoid any damage to the tube structure. After holding for a sufficient period to remove most of the gases, the temperature is then escalated to the sintering range. Several hours are generally needed at the sintering temperature to densify the tubes before the temperature is finally decreased back to room temperature at a controllable rate to avoid cracking. By varying the organic additive loading and/

or the extrusion pressure, it is possible to make green tube with different dimensions from a single extrusion die. However, sensitive sintering programs are needed because of the significant shrinkage that can result from the decomposition of organic additives if they are present in high quantities. In most cases, the relative densities of tubes obtained by plastic extrusion are lower than for tubes obtained by isostatic pressing or for disks obtained by dry pressing [47–50].

Balachandran *et al.* first demonstrated the successful synthesis of $La_{0.2}Sr_{0.8}Fe_{0.2}Co_{0.8}O_{3-\delta}$ and $SrCo_{0.5}FeO_x$ tubular membranes by the plastic extrusion method [47]. In their experiments, powdered forms of $SrCo_{0.5}FeO_x$ and $La_{0.2}Sr_{0.8}Fe_{0.2}Co_{0.8}O_{3-\delta}$ were synthesised using solid-state reaction from the raw materials of $La(NO_3)_3$, $SrCO_3$, $Co(NO_3)_2 \cdot 6H_2O$, and Fe_2O_3 after calcination in air at 850 °C for 16 h. After grinding, the ceramic powders with average particle sizes ~7 μm were mixed with several organic additives (solvent, dispersant, binder, and plasticiser) to provide enough plasticity to form into various shapes while retaining satisfactory strength in the green state (i.e., before firing). After the slip was prepared (with sufficient solvent to ensure uniform mixing of all additives), some of the solvent was allowed to evaporate in order to produce a stiffer plastic paste that was forced through a die to produce hollow tubes. Tubes were extruded with an outside diameter of ~6.5 mm and lengths up to ~300 mm with wall thickness of 0.25–1.20 mm. The extruded green tubes were heated slowly (~5 °C h^{-1}) in the temperature range of 150–400 °C to facilitate the removal of gaseous species formed during decomposition of the organic additives. After the organics were removed, the heating rate was increased to ~60 °C h^{-1} and the tubes were sintered at 1200 °C for 5–10 h in air. The sintered tubes of $SrCo_{0.5}FeO_x$ and $La_{0.2}Sr_{0.8}Fe_{0.2}Co_{0.8}O_{3-\delta}$ were successfully used with good performance to supply pure oxygen in the partial oxidation of methane for syngas generation.

Yang *et al.* prepared $Ba_{0.5}Sr_{0.5}Co_{0.8}Fe_{0.2}O_{3-\delta}$ ceramic tubes by the plastic extrusion technique using a high-quality powder prepared by the ECCC method [49]. No detailed information about the organic additives was reported. The extruded green tubes were heated in the temperature range of 100–450 °C, heating slowly at a rate of 5 °C h^{-1} to controllably remove the gaseous species generated during the decomposition of the organic additives, then sintered at 1100–1200 °C in air for 3–5 h with the heating rate of 1 °C min^{-1}. Finally, the sintered tubes were cooled to room temperature at a rate of 2 °C min^{-1}. The sintered membrane tubes had outer diameters of ~8 mm, inner diameters of ~5 mm, and lengths up to 300 mm. The densities of the sintered tubular membranes were measured to be higher than 90%. Almost no pores could be found in the surface of these tubular $Ba_{0.5}Sr_{0.5}Co_{0.8}Fe_{0.2}O_{3-\delta}$ membranes (Figure 5.5).

5.3.2.2 *Isostatic Pressing*

The isostatic pressing method is used for the preparation of high-quality tubular MIEC membranes due to the complete uniformity of high pressure on all sides. If green tubes are already cast or extruded by other techniques, it is possible to obtain uniform densification by subsequently isostatically pressing them in a plastic or rubber case. Isostatic pressing of ceramic tubes is done by two methods: in a wet case for small, mainly experimental batches, and in the dry case for large industrial batches. In the first instance, the rubber or plastic sack is protected normally by a thin perforated steel jacket. With the second method, the jacket is absent and this accelerates the process of filling the powder, dropping the bag in the press chamber, and pressing. The liquid pressure for isostatic pressing usually varies between 350–1400 kg cm^{-2}. This method is also used for the direct preparation of tubular MIEC ceramic membranes [48, 51–53].

For example, Li *et al.* prepared $La_{0.6}Sr_{0.4}Co_{0.2}Fe_{0.8}O_{3-\delta}$ tubular membranes using the isostatic pressing technique [48, 51]. $La_{0.6}Sr_{0.4}Co_{0.2}Fe_{0.8}O_{3-\delta}$ powder was first prepared via solid-state reaction. Ball-milling, drying and calcining were repeated twice. For the isostatic pressing

Figure 5.5 SEM images for $Ba_{0.5}Sr_{0.5}Co_{0.8}Fe_{0.2}O_{3-\delta}$ plastic extrusion tube sintered at 1100–1200 °C for 3–5 h. Reprinted from Journal of Membrane Science, Wang, H., et al., Oxygen permeation study in a tubular $Ba_{0.5}Sr_{0.5}Co_{0.8}Fe_{0.2}O_{3-\delta}$ oxygen permeable membrane. Vol. 210, 259–271. Copyright (2002) with permission from Elsevier

process, poly(vinyl acetate) (PVA; 10 wt% in water) was added to increase the fluidity of the powders. Then the powders were sifted to obtain particle sizes between 40–60 meshes. Green tubes were prepared by powder loading, pressing at the pressure of 25 MPa, and ejecting. The green tubes prepared by this method were sintered in air at 1250 °C for 5 h with heating and cooling rates of 3 and 2 °C min^{-1}, respectively. The final $La_{0.6}Sr_{0.4}Co_{0.2}Fe_{0.8}O_{3-\delta}$ tubes had outer diameters of ~8 mm, inner diameters of ~5 mm, and lengths of ~150 mm. Relative densities as high as 95.2% were obtained.

5.3.2.3 Slip Casting

Slip casting is one of the most commonly used techniques in the preparation of ceramic membranes. However, the casting time is usually very long, making it difficult to control the wall thickness, which is therefore usually thick and uneven. When a well mixed ceramic slip is poured into a porous mould, solvent from the slip is extracted into the pores of the mould via capillary driving forces. The slip particles are, therefore, consolidated on the surface of the mould to form a layer of particles or a gel layer. It is important that the formation of the consolidated layer is fast so that particles will not penetrate into the pores of the mould. Processing parameters such as solid concentration, particle diameter, and slip viscosity must be well optimised for successful casting.

Tubular MIEC ceramic membranes have been prepared by slip casting in plaster molds using slips prepared from MIEC ceramic powders and a binder such as polyvinyl alcohol, which is widely used in ceramic processing to provide adequate green strength. The pH of the slip is usually controlled at 3–4. The slip is often degassed to remove any entrapped air before fabrication. The rheological behaviour of the slip depends on the type and dosage of the binders and sometimes viscosity modifiers are also added to control the rheology for ease of processing. MIEC ceramic powders (e.g., yttria-stabilised zirconia doped with titania/ceria) have been cast into tubular form from slips containing the constituent oxides. The cast tubes are then subject to sintering at 1200 to 1500 °C to render gas impervious [54].

5.3.3 Hollow Fibre Membrane

Although disk and tube-type MIEC membrane structures are by far the most common, other membrane structures have also been explored, including the hollow fibre type MIEC membrane structure. Hollow fibre MIEC ceramic membranes with asymmetric configuration have been successfully prepared using the phase inversion spinning/sintering technique [55–64]. Compared to disk-shaped and tubular-shaped configurations, the hollow fibre membrane possesses much larger membrane area per unit volume. Using long hollow fibres and keeping the two sealing ends away from the high-temperature zone, high-temperature sealing problems can also be alleviated in this configuration. Furthermore, due to the asymmetric configuration (a thin dense layer for separation integrated with porous layers on one or both sides), the oxygen permeation resistance is significantly smaller than that for a symmetric membrane. In addition, the integrated porous layers can provide much larger gas–membrane interfaces for oxygen exchange reactions, leading to an enhancement of surface oxygen exchange kinetics and thus to a higher oxygen permeation flux. All the advantages of the hollow fibre membranes could, in the foreseeable future, position these MIEC ceramic membrane structures into commercial applications, although their weak mechanical strength is still a significant issue.

Luyten *et al.* were the first to prepare $La_{0.6}Sr_{0.4}Co_{0.8}Fe_{0.2}O_{3-\delta}$ hollow fibre membranes [55]. They modified the pure polymer phase inversion technique in the sense that the desired amount of ceramic powder was included in the starting suspension and that the organic additives were later burned out. The suspension was made by dissolving polysulfone in a solvent of N-methyl-2-pyrrolidone (NMP) followed by dispersing $La_{0.6}Sr_{0.4}Co_{0.8}Fe_{0.2}O_{3-\delta}$ powder into the polymer solution. The suspension was pressed through a double cylinder nozzle mounted close to the coagulant bath. The phase inversion inside the hollow fibres was induced by a bore liquid flowing through the inner cylinder of the spinning head. Either pure water or a mixture of 50% NMP, 50% H_2O, and 1% polyvinyl pyrollidone (PVP360000) was used as the inside coagulant. To remove the polysulfone, the green hollow fibres were calcined at 600 °C for 1 h. Finally, the hollow fibres were densified by sintering at 1225 °C for 24 h. The sintered hollow fibres had OD of 1.5–2.0 mm and ID of 0.8–1.0 mm. As the first attempt to prepare MIEC hollow fibre membranes, the correct perovskite structure was successfully produced for the final hollow fibre membrane, although cross section morphology indicated that the hollow fibre membranes were not fully densified.

Li *et al.* described the experimental conditions in detail for preparing a MIEC hollow fibre membrane of $La_{0.6}Sr_{0.4}Co_{0.2}Fe_{0.8}O_{3-\delta}$ by the same phase inversion technique [56]. PESf and PVP were first dissolved in NMP. Then an appropriate amount of $La_{0.6}Sr_{0.4}Co_{0.2}Fe_{0.8}O_{3-\delta}$ powder was homogeneously dispersed in the formed polymer solution. Stirring was carried out continuously for at least 48 h before spinning. After degassing at room temperature for 30 min, the starting suspension was then transferred to a stainless steel reservoir and pressurised to 0.5 bar using nitrogen. A spinneret with an orifice diameter of 3.0 mm and an inner diameter of 1.2 mm was used to produce green hollow fibres. Deionised water and tap water were used as the internal and external coagulants, respectively. The green hollow fibres were immersed in a water bath for more than 24 h to complete the solidification process. Finally, a heat treatment at 600 °C for 2 h was used to remove the organic constituents. Sintering at 1100–1280 °C for 4 h resulted in the successful production of MIEC hollow fibre membranes. The detailed conditions employed in the preparation of these hollow fibre membranes are summarised in Table 5.1. The hollow fibre membranes sintered at 1280 °C for 4 h had OD ~1.3 mm and ID ~1.1 mm. The as-sintered hollow fibre membrane retains its asymmetric layer-like structure. The centre and top surface layers have been fully densified (Figure 5.6). At a temperature of 900 °C, the oxygen permeation

Table 5.1 *Preparation conditions for hollow fibre membranes. Reprinted from Industrial & Engineering Chemistry Research, Tan, X., et al., Preparation of LSCF ceramic hollow-fiber membranes for oxygen production by a phase-inversion/sintering technique. Vol. 44, 61–66. Copyright (2005) with permission from American Chemical Society.*

Parameter	Value
Dope composition (wt%)	
LSCF	66.33
PESf, Radel A-300	6.63
NMP	26.5
PVP K16-18	0.5
Dope temperature (°C)	20
Internal coagulant temperature (°C)	20
Injection rate of internal coagulant (ml min^{-1})	10
Nitrogen pressure (atm)	0.5
Air gap (cm)	1.5
Sintering temperature (°C)	1100–1280

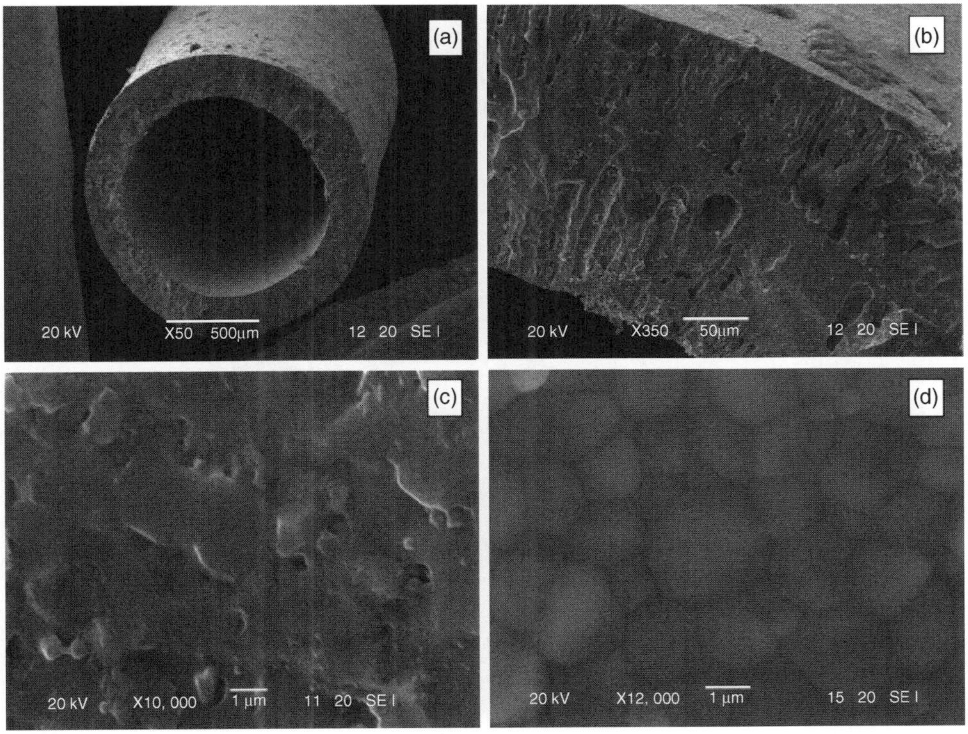

Figure 5.6 *SEM micrographs for* $La_{0.6}Sr_{0.4}Co_{0.2}Fe_{0.8}O_{3-\delta}$ *hollow fibre membranes sintered at 1280 °C for 4 h: (a) dimension, (b) the whole cross section, (c) centre of the cross section, (d) outer surface. Reprinted from Industrial & Engineering Chemistry Research, Tan, X., et al., Preparation of LSCF ceramic hollow-fiber membranes for oxygen production by a phase-inversion/sintering technique. Vol. 44, 61–66. Copyright (2005) with permission from American Chemical Society*

flux is 5.49×10^{-7} mol s^{-1} cm^{-2} for a downstream oxygen partial pressure of 0.017 atm, which is much higher than that obtained for a $La_{0.6}Sr_{0.4}Co_{0.2}Fe_{0.8}O_{3-\delta}$ tube membrane (1.57×10^{-7} mol s^{-1} cm^{-2}, at 900 °C with a downstream oxygen partial pressure of 1×10^{-3} atm).

5.3.4 Asymmetric Thin Film

Another promising approach to improve both the oxygen permeation flux and thermodynamic stability of the MIEC ceramic membrane focuses on the use of asymmetric thin film configurations. The asymmetric thin film approach is an area of active research in the field of MIEC ceramic membranes, and generally involves the fabrication of a dense and stable oxygen permeable layer on top of a thicker more oxygen permeable porous (or dense) layer for mechanical support. A variety of preparation techniques can be used to develop asymmetric thin film membranes, including dry pressing, tape casting, dip coating, spin coating, slip casting, slip brushing or painting, spray pyrolysis, chemical vapour deposition, and pulsed laser deposition (PLD) [65–86].

5.3.4.1 Dry Pressing

Dry pressing has been frequently used for the fabrication of asymmetric MIEC ceramic membranes for oxygen permeation because of its ease and low cost [65–69]. The preparation of a dual layer (dense thin layer for oxygen separation and thick layer for mechanical support) asymmetric configuration using this technique is schematically described in Figure 5.7 [65]. Firstly, a proper amount of substrate powder is dry pressed under a lower hydraulic pressure to shape the substrate. After holding for some time, the pressure is released and the top part of the die is removed to expose top surface of the substrate. MIEC oxide powder for the oxygen separation layer is then added onto the substrate surface. The top part of the die is subsequently re-positioned

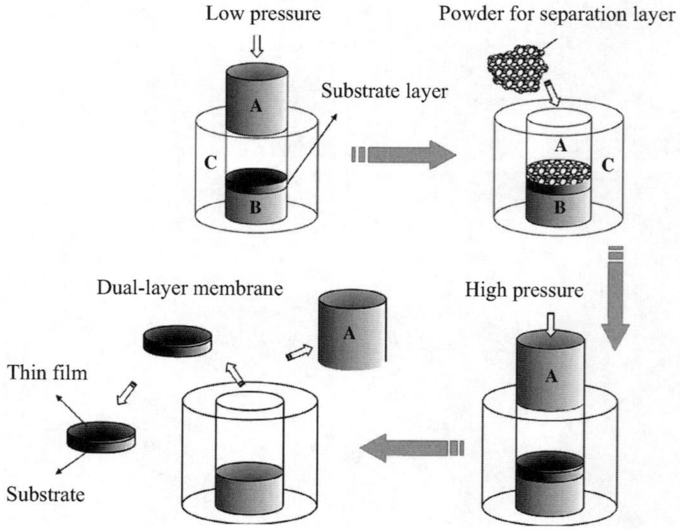

Figure 5.7 Schematic diagram for the preparation of a dual-layer oxygen separation membrane via the dry pressing technique. Reprinted from Journal of Membrane Science, Chen, Z., et al., A dense oxygen separation membrane with a layered morphologic structure. Vol. 300, 182–190. Copyright (2007) with permission from Elsevier

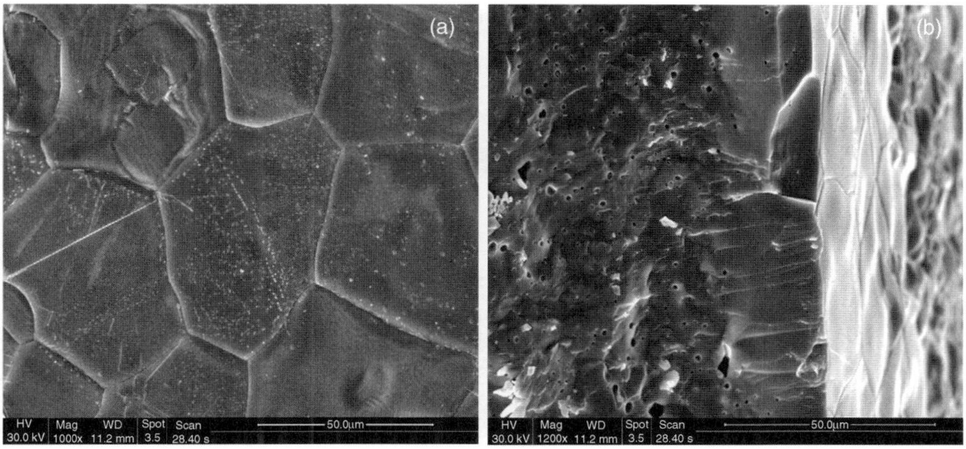

Figure 5.8 *SEM micrographs of a dual-layer membrane sintered at 1000–1100°C for 2–10 h. (a) Surface of top oxygen separation layer. (b) Cross section near the top layer. Reprinted from Journal of Membrane Science, Chen, Z., et al., A dense oxygen separation membrane with a layered morphologic structure. Vol. 300, 182–190. Copyright (2007) with permission from Elsevier*

and turned gently to evenly distribute the new powder. A higher pressure is then applied for a period of time in order to obtain the dual layer green disk. After releasing from the die, the green disk is generally sintered in a specially designed atmosphere with a programmed temperature schedule. Using this dry pressing method, it is relatively straightforward to obtain a dense top layer because of the high green density resulting from the high pressure applied during the second stage. However, it is difficult to disperse this second stage powder evenly across the substrate surface, which makes it difficult to prepare dense top layer films thinner than 5 μm. Additional challenges also often arise during the co-firing process due to the different sintering behaviours for the substrate layer and the top layer.

Using this dry pressing technique, Shao *et al.* synthesised an asymmetric MIEC ceramic membrane from two different MIEC powders [65]. The substrate was a thick dense layer with high oxygen permeability but relatively low chemical stability composed of $Ba_{0.5}Sr_{0.5}Co_{0.8}Fe_{0.2}O_{3-\delta}$. The top separation layer was a thin film composed of $Ba_{0.5}Sr_{0.5}Co_{0.2}Fe_{0.8}O_{3-\delta}$, which provided higher chemical and structural stability, although relatively lower oxygen permeability. SEM images of this asymmetric membrane structure (Figure 5.8) illustrate the successful creation of the two layer structure, with the fully densified $Ba_{0.5}Sr_{0.5}Co_{0.2}Fe_{0.8}O_{3-\delta}$ film (~25 μm thick) on the top of $Ba_{0.5}Sr_{0.5}Co_{0.8}Fe_{0.2}O_{3-\delta}$ substrate.

5.3.4.2 Tape Casting

The combination of tape casting, co-lamination, and co-firing, is the predominant method to produce multilayer ceramic membranes with controllable homogeneous microstructures. This technique is not only frequently used in the fabrication of asymmetric MIEC ceramic membranes for oxygen permeation, but is also commonly applied for solid oxide fuel cell stacks and other functional ceramic structures as well [70–75]. As schematically described in Figure 5.9, a green tape with designed composition can be prepared in large quantities by the tape casting method. The resulting functional tapes, with typical thickness from 50–1250 μm, can be cut into different

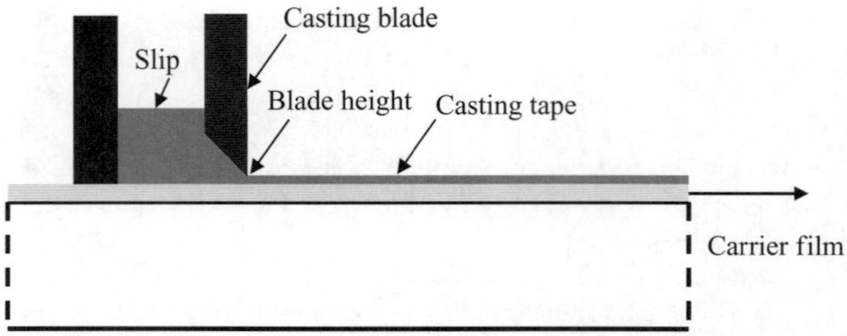

Figure 5.9 Schematic description of tape casting process

shapes and laminated into a multilayer composite configuration by high pressure. After firing under controlled atmospheres with programmed temperatures, the laminated multilayer composite configuration can be transferred into an asymmetric membrane. Delamination due to the different sintering behaviours of each layer generally represents the greatest challenge to the widespread use of this process for the preparation of asymmetric MIEC ceramic membranes.

Geffroy *et al.* prepared $La_{0.8}Sr_{0.2}Fe_{0.7}Ga_{0.3}O_{3-\delta}$/$La_{0.8}M_{0.2}FeO_{3-\delta}$ (M = Ca, Sr, Ba) asymmetric MIEC membranes by the tape-casting process [70]. $La_{0.8}Sr_{0.2}Fe_{0.7}Ga_{0.3}O_{3-\delta}$ powder synthesised by spray pyrolysis (PCF, France) was calcined and milled to obtain a specific surface area of $10\,m^2\,g^{-1}$ and dense particles with a mean grain size of 0.3 μm, suitable for the formation of thin dense layers via the tape-casting process. For the thicker, porous support layer, $La_{0.8}M_{0.2}FeO_{3-\delta}$ (M = Ca, Sr, Ba) powder was synthesised by the citrate method from nitrates. After calcining and milling, a specific surface area of 7–8 $m^2\,g^{-1}$ and mean particle size of 0.4 μm was obtained. Both powders were milled by planetary milling for 4 h in an azeotropic mixture of butanone-2 and ethanol (60/40, respectively) with the help of a dispersing agent (CP 213, Cerampilot, France). Then, an acrylic binder (methyl methacrylate, Degalan LP51/07 Degussa) and a phthalate plasticiser (dibutyl phthalate, Sigma–Aldrich) were added to the slurry, with a subsequent ball-milling of 12 h. Porosity was ensured for the support layer by adding corn starch (45 vol%) to the $La_{0.8}M_{0.2}FeO_{3-\delta}$ slurry after ball-milling, with an extra 4 h ball-milling. The as-synthesised slips were degassed and directly cast onto a silicone Mylar carrier film using a doctor blade. After solvent evaporation at room temperature under air, the 150 μm thick green tapes were easily handled. Disks of 30 mm in diameter were punched from the tapes, stacked and laminated under pressure 50 MPa at 85 °C. Asymmetric membranes were obtained by stacking seven disks of the $La_{0.8}M_{0.2}FeO_{3-\delta}$ tape containing starch and one disk of the $La_{0.8}Sr_{0.2}Fe_{0.7}Ga_{0.3}O_{3-\delta}$ tape without pore-forming agent. The binder was eliminated by heating to 650 °C under air; membranes were then sintered at 1250 °C for 2 h under air. Figure 5.10 provides the typical morphology for the resulting asymmetric membrane, with the dense $La_{0.8}Sr_{0.2}Fe_{0.7}Ga_{0.3}O_{3-\delta}$ layer (~80 μm thick) successfully supported on the porous $La_{0.8}Sr_{0.2}FeO_{3-\delta}$ substrate.

5.3.4.3 *Slurry/Slip Related Casting/Coating*

The *dip coating technique* is schematically illustrated in Figure 5.11. In this approach, an appropriate substrate is first immersed in a solution/slurry/sol of coating materials at a constant speed. After immersion for a certain time, the substrate is pulled out of the solution at a controlled speed, which is chosen to determine the thickness of the coated film. Dip coating is typically conducted in an atmosphere (or with proper solvents) that provides a relatively low solvent vapour pressure, thereby ensuring that the drying process starts simultaneously with the dip coating. The most critical parameters in dip coating are the viscosity of the particle suspension

Figure 5.10 *SEM micrographs of $La_{0.8}Sr_{0.2}Fe_{0.7}Ga_{0.3}O_{3-\delta}/La_{0.8}Sr_{0.2}FeO_{3-\delta}$ asymmetric MIEC membrane prepared by tape casting method after sintering at 1250 °C for 2 h. (a) cross section and (b) dense thin film surface. Reprinted from Journal of Membrane Science, Julian, A., et al., Elaboration of $La_{0.8}Sr_{0.2}Fe_{0.7}Ga_{0.3}O_{3-\delta}/La_{0.8}M_{0.2}FeO_{3-\delta}$ (M = Ca, Sr and Ba) asymmetric membranes by tape-casting and co-firing. Vol. 333, 132–140. Copyright (2009) with permission from Elsevier*

and the coating speed or time. Furthermore, to obtain a defect-free dense film the coating thickness per cycle must be very thin. So, a multistep process involving repeated cycles of dipping, drying and calcination is needed in order to prepare a defect-free dense MIEC ceramic film on a porous substrate using this technique [76–78].

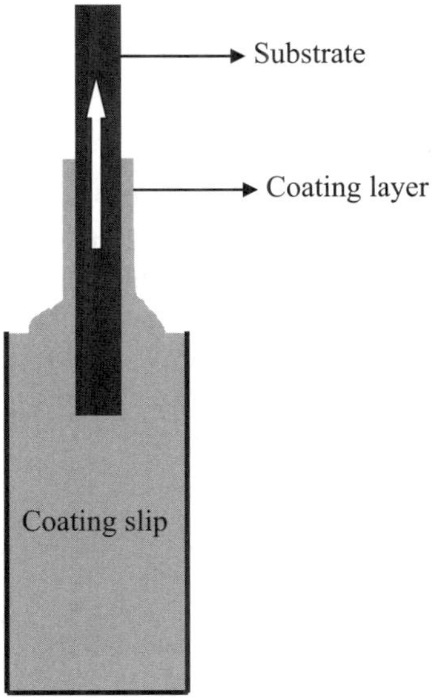

Figure 5.11 *Schematic diagram of dip coating*

Using the dip coating approach, for example, Yang *et al.* prepared a MIEC $La_2NiO_{4+\delta}$ asymmetric membrane on a porous α-Al_2O_3 substrate [76]. The coating slip was prepared from La and Ni nitrates by chelating with the ligands of EDTA acid. The overall starting concentration of metal ions was 1.0 M and the pH of the solution was adjusted to be ~6.0 by $NH_3 \cdot H_2O$. After aging at 60 °C for a period of time, a stable dark blue sol was obtained. Porous α-Al_2O_3 disks (with 40% porosity) were used as substrates. The preparation process of the asymmetric membrane involved the following cycle, repeated 10 times: dip coating, drying at <100 °C, and heat treatment at 400 °C. Final sintering was carried out at 850 °C. A final film thickness of ~40 μm was achieved and no cracks or leaks could be detected by gas chromatography.

Spin coating is a technique used to prepare uniform thin films on top of flat substrates. An appropriate quantity of slip is placed on the substrate, which is then rotated at high speed in order to spread the slip by centrifugal force. Specially designed machines used for this spin coating process are called a spin coaters or spinners. Rotation is continued while the excess slip spins off the edges of the substrate until the desired film thickness is achieved. Film thickness is controlled by the angular spin speed in combination with the slip solids concentration and viscosity. Spin coating is a good technique to prepare very thin films (thinner than 10 nm). A variety of asymmetric MIEC ceramic films have been prepared by the spin coating technique [79–81]. Unfortunately, the spin coating technique is generally restricted to flat, smooth substrates and it is difficult to coat over large areas.

As an example, dense films of $Ca_{0.8}Sr_{0.2}Ti_{0.7}Fe_{0.3}O_{3-\delta}$ (CSTF) perovskite-type oxide were successfully prepared on porous CSTF substrates by the spin coating technique [79]. The porous CSTF substrate was prepared using CSTF powder synthesised by the solid state reaction (SSR) method. This SSR powder was mixed with 0–30 wt% carbon black in isopropanol by ultrasonic agitation for 1 h. The resulting slurry was dried in air at 114 °C and pressed uniaxially into 13 mm diameter disks at ~0.2 MPa. The disks were partially sintered at 1150 °C in ambient air to remove carbon black using a temperature ramp rate of 0.087 °C/s. The disks were subsequently ground down to 0.5 mm thickness using waterproof sandpaper. For the dense top layer, high quality CSTF powder was prepared using the citrate method. A spin coating slurry was then prepared by ultrasonically dispersing the powder in isopropanol for 30 min. About 0.3 ml of the spin coating slurry was delivered dropwise on a porous substrate rotated at 800 rpm. After spin coating, the sample was dried at room temperature for 1 h, then pre-sintered in ambient air at 900 °C for 1 h. The coating–drying cycle was repeated several times to vary the film thickness. The film/substrate composite was subsequently sintered in ambient air with a slow ramp rate (0.060 $°C\,s^{-1}$) to 1450 °C, held for 15 h, and then cooled to ambient temperature at a still slower ramp rate (0.048 $°C\,s^{-1}$) to prevent thermal stress. Figure 5.12 provides the SEM micrograph for a representative asymmetric membrane prepared by this multicycle spin coating process. The film thickness was roughly controllable by adjustment of the number of spin coating cycles; the thickness of a film prepared using four spin coating cycles was ~35 μm. An oxygen permeation flux of 2.1 $ml\,min^{-1}\,cm^{-2}$, at 950 °C was obtained for this membrane, approximately 10 times better than the performance of a comparable 500 μm thick CSTF disk.

The *slip casting* method previously discussed in the context of homogeneous tubular membrane fabrication (Section 5.3.2.3) can also be used to prepare asymmetric tubular MIEC ceramic membranes. Rather than casting the slip or suspension directly into a porous cylindrical mold, it is instead to cast on a previously fabricated porous tubular substrate, which is used for mechanical support [82–84]. For example, asymmetric membranes were successfully prepared by slip casting a dense thin film of $La_{0.6}Sr_{0.4}Co_{0.2}Fe_{0.8}O_{3-\delta}$ on a porous tubular substrate of the same composition [82]. The $La_{0.6}Sr_{0.4}Co_{0.2}Fe_{0.8}O_{3-\delta}$ slip was cast directly on the surface of a green support tube, followed by sintering at 1300 °C. The quality and thickness of the top layer

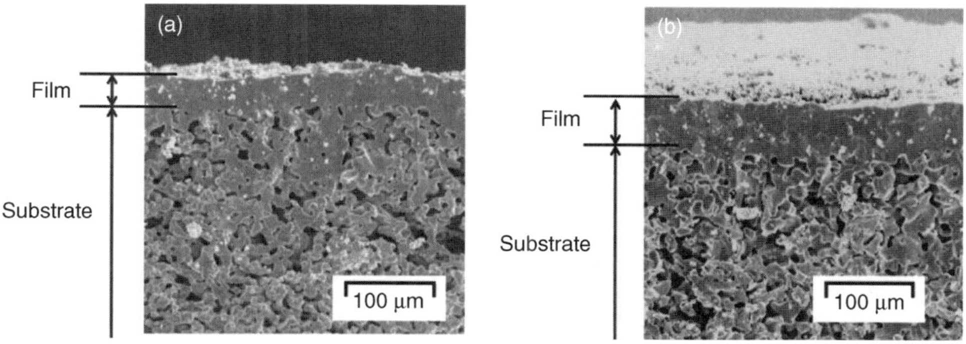

Figure 5.12 SEM micrographs of $Ca_{0.8}Sr_{0.2}Ti_{0.7}Fe_{0.3}O_{3-\delta}$ membranes prepared using: (a) four spin coating cycles and (b) eight spin coating cycles. Reprinted from Solid State Ionics, Araki, S., et al., Synthesis and characterization of mixed ionic–electronic conducting $Ca_{0.8}Sr_{0.2}Ti_{0.7}Fe_{0.3}O_{3-\delta}$ thin film. Vol. 178, 1740–1745. Copyright (2008) with permission from Elsevier

was controlled by the $La_{0.6}Sr_{0.4}Co_{0.2}Fe_{0.8}O_{3-\delta}$ concentration in the slip. Suitable slip concentration was in the range of 15–25 wt%. After sintering, the crystal phase of the top layer was the same as that of the bulk, indicating that there was no surface segregation in the asymmetric membrane. SEM results (Figure 5.13) indicated that the slip-cast outer layer had a thickness of $\sim$200 μm and was fully densified. Permeation testing demonstrated oxygen permeation fluxes about three or four times higher than the fluxes achieved from a comparable 2 mm thick dense symmetric disk.

Brushing/painting provides an alternative approach to coat a stable slip on a substrate surface. Due to the inhomogeneity of most brushing or painting processes, thicker coatings are typically required to achieve defect-free films compared to other coating techniques. Because of this limitation and the difficulty of achieving controllable or reproducible coatings, this technique has

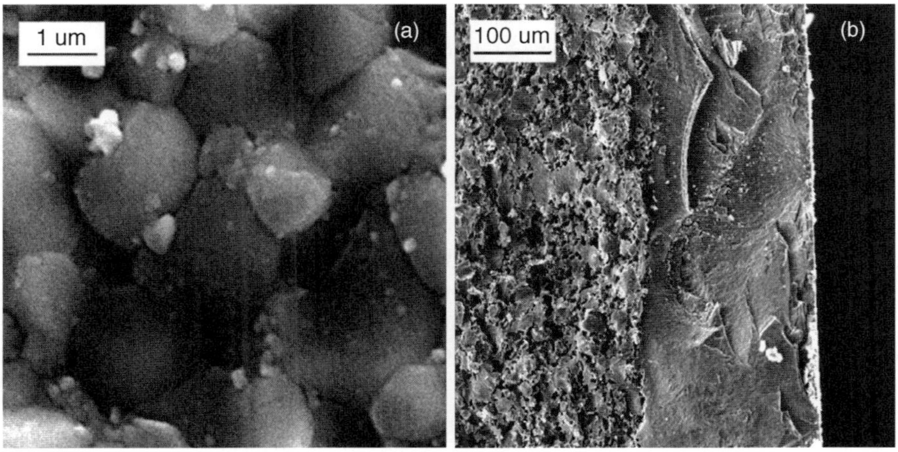

Figure 5.13 SEM micrographs of an asymmetric membrane (a) surface and (b) cross section. Reprinted from Journal of Membrane Science, Jin, W., et al., Preparation of an asymmetric perovskite-type membrane and its oxygen permeability. Vol 185, 237–243. Copyright (2001) with permission from Elsevier

only been used occasionally for the preparation of asymmetric MIEC ceramic membranes [85–87]. As an example, the slip brushing method was used to fabricate an asymmetric membrane tube composed of a porous CeO_2 support coated by a dense $La_{0.2}Co_{0.8}SrO_{3-\delta}/CeO_2$ composite film [85]. The slip was firstly prepared by adding 15 wt% $La_{0.2}Co_{0.8}SrO_{3-\delta}/CeO_2$ (1 : 1, v/v) mixed powder into an organic solution comprised of a variety of organic additives. An ink–like dispersion was formed by ball-milling the slip for 3 days. The slip was then brushed or sprayed on the external surface of porous CeO_2 tube. During the coating process, one end of the tube was blocked and the other end was vacuumed, resulting in the formation of a highly dense and uniformly packed $La_{0.2}Co_{0.8}SrO_{3-\delta}/CeO_2$ layer on the external tube surface. After drying overnight at room temperature, the coated CeO_2 tube was heated (at a heating rate of 0.5 °C min^{-1}) to 400 °C and held for 1 h, and followed by sintering at 1200 °C for 2 h. SEM characterisation confirmed the successful creation of a dense $La_{0.2}Co_{0.8}SrO_{3-\delta}/CeO_2$ film ~10 μm in thickness supported on the porous CeO_2 tube.

5.3.4.4 *Spray Pyrolysis*

Spray pyrolysis is extensively applied to fabricate thin films for semiconductor, solar cell, sensor, and photoactive devices [88, 89]. Solid oxide fuel cell electrolytes, interconnects, and cathodes have also been prepared using this technique [90, 91], as well as asymmetric MIEC ceramic membranes [92, 93]. As an example, dense MIEC ceramic films of cobaltites $La_{1-x}Sr_xCoO_3$, ferrites $La_{1-x}Sr_xFe_{1-y}(Co,Ni)_yO_3$, gallates $La_{1-x}Sr_xGa_{1-y}(Co,Ni,Fe)_yO_3$, manganites $La_{1-x}Sr_xMnO_3$ and perovskite-related oxides such as lanthanum nickelate La_2NiO_4 were successfully deposited on porous ceramic substrates by the atmospheric spray pyrolysis technique [92]. A deposition solution for each desired multimetal oxide composition is first synthesised by dissolving proper amounts of metal β-diketonates in monoglyme (1, 2-dimethoxyethane). Under well controlled conditions the solution is then aerosolised and transported to the deposition chamber by a N_2 or $N_2 + O_2$ carrier gas. Deposition was performed at 600–800 °C under atmospheric pressure, followed by *ex situ* annealing at 700–800 °C for 8–12 h in air. Relatively high growth rates (~4–6 μm h^{-1}) were obtained from this deposition process. Figure 5.14 provides several SEM micrographs for typical as-prepared films. Depending on the oxygen concentration in the aerosol flow, films of dense or porous morphology can be obtained. Dense films usually grew without oxygen, while the presence of oxygen in the gas flow led, in most cases (except for the gallates), to porous films with a disordered columnar structure.

5.3.4.5 *Chemical Vapour Deposition*

Chemical vapour deposition (CVD) is a commonly applied chemical process capable of producing a wide variety of high-purity and high-performance solid materials or thin films. In a typical CVD process, a substrate is exposed to one or more volatile precursors, which react and/or decompose on the substrate surface to produce a film. Frequently, volatile byproducts are also produced, which are removed by gas flow through the reaction chamber. Using the CVD method, dense thin films in monocrystalline, polycrystalline, amorphous, or epitaxial structure can be prepared. MIEC ceramic thin films are frequently prepared on porous substrates using the CVD technique [94–96]. However, several significant challenges limit the widespread application of CVD for MIEC films, including the costs and complexity associated with the required high-vacuum processing system and the difficulty in finding proper precursors for most MIEC ceramic membrane materials.

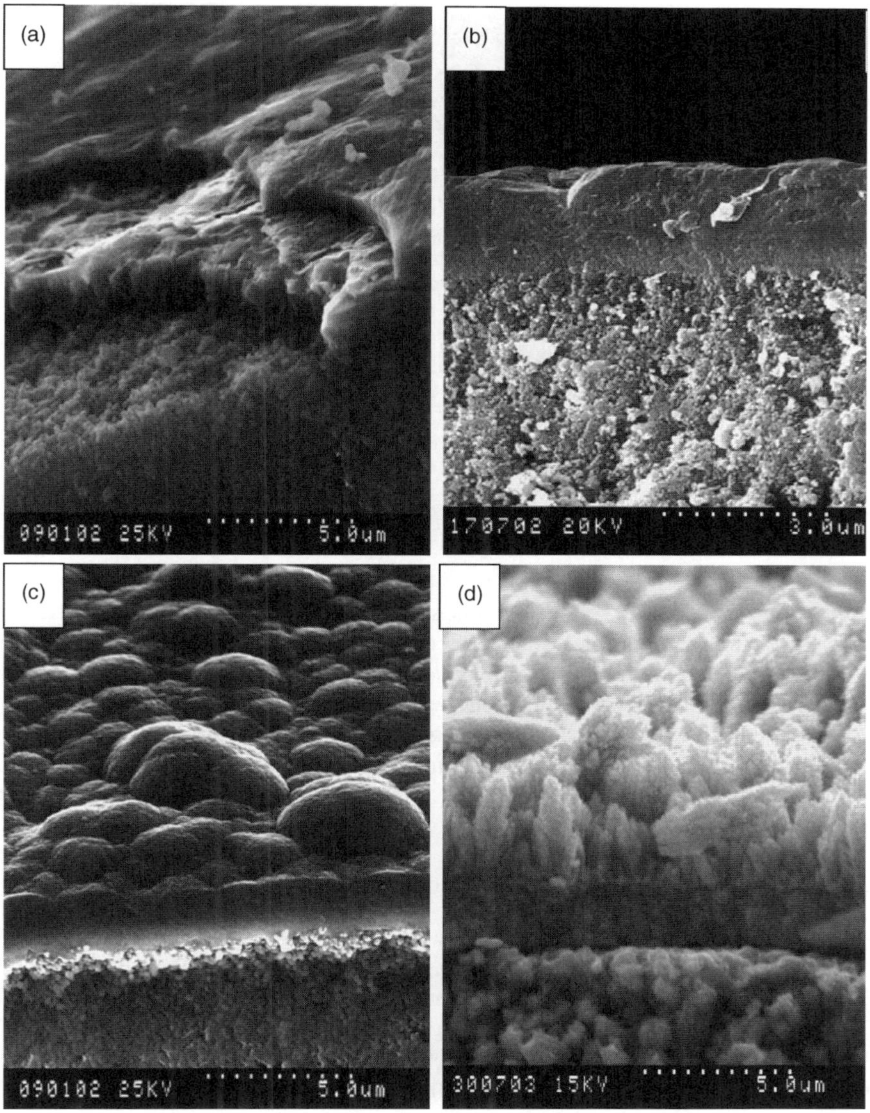

Figure 5.14 *SEM cross section micrographs of films grown without oxygen: (a)* $La_{1-x}Sr_xGa_{1-y}Co_yO_3$ *bylayer, (b)* $La_{1-x}Sr_xGa_{1-y}Ni_yO_3$*, (c)* $La_{1-x}Sr_xFe_{1-y}Co_yO_3$*; and grown with 5% oxygen, (d)* $La_{1-x}Sr_x$*-*MO_3*. Reprinted from Journal of Membrane Science, Preparation of dense, ultrathin MIEC ceramic membranes by atmospheric spray-pyrolysis technique. Vol. 240, 113–122. Copyright (2004) with permission from Elsevier*

Abrutis *et al.* examined the suitability of the pulsed injection metal–organic chemical vapour deposition (MOCVD) technique for the preparation of dense perovskite films for oxygen permeation membrane applications [94]. A variety of MIEC perovskite oxide thin films including manganites $La_{1-x}(Sr,Ca)_xMnO_3$, ferrates $La_{1-x}Sr_xFe_{1-y}Co_yO_3$, and gallates $La_{1-x}Sr_xGa_{1-y}(Co, Ni,Fe)_yO_3$ with a thickness of 0.8–2.0 μm, were deposited by pulsed injection MOCVD on a

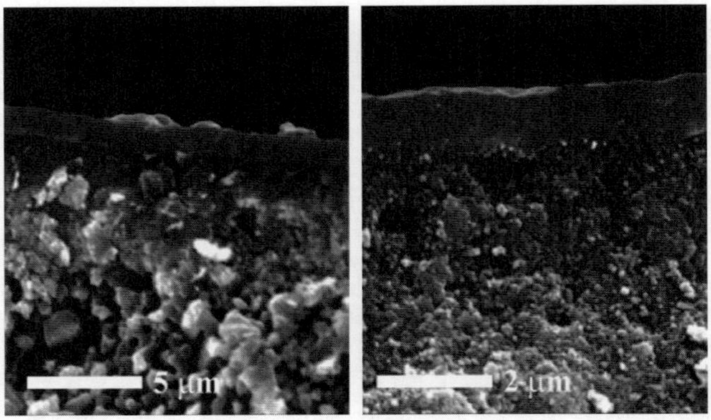

Figure 5.15 *SEM micrographs of a dense $La_{1-x}Sr_xGa_{1-y}Ni_yO_3$ membrane (~1.2 μm) on Tami 300 kDa substrate prepared by chemical vapour deposition followed by annealing at 700 °C in air for 10 h. Reprinted from Thin Solid Films, Abrutis, A., et al., Metal-organic chemical vapour deposition of mixedconducting perovskite oxide layers on monocrystalline and porous ceramic substrates. Vol. 240, 113–122. Copyright (2004) with permission from Elsevier*

variety of monocrystalline and porous ceramic substrates. The films were characterised in detail by X-ray diffraction, scanning electron microscopy, wavelength dispersion spectroscopy and by electrical measurements of total conductivity versus temperature. Initial experiments indicated that the pulsed injection MOCVD did not succeed in obtaining completely gas tight membranes when deposited on simple porous alumina substrates. Much better densities were obtained when the membranes were deposited onto a special multilayered TAMI 300 kDa substrate having an ultraporous ZrO_2(Y) top layer. Figure 5.15 provides a typical morphology of the resulting films on this special porous substrate.

5.3.4.6 Pulsed-Laser Deposition

Pulsed-laser deposition (PLD) is perhaps best known as the technique used to successfully deposit the well known superconducting $YBa_2Cu_3O_{7-\delta}$ thin film. Since then, PLD has been used to deposit high-quality thin films for many multicomponent oxides including MIEC ceramic oxides [97–102]. The main advantage of PLD is that the laser-induced expulsion produces a plume of materials with stoichiometry similar to the target. It is generally easier to obtain the desired film stoichiometry for multi-element materials using PLD than with other deposition technologies. Although PLD involves several complex physical phenomena including laser–material interaction, the formation of plasma plume with energetic species, and the transfer of the ablated material through the plasma plume onto the heated substrate surface, the actual system set up and operation is relatively simple, as shown schematically in Figure 5.16. The formation process of thin films using PLD can generally be divided into the following four stages: (i) laser radiation interaction with the target, (ii) ablation dynamics, (iii) deposition of the ablation materials on the substrate, and (iv) nucleation and growth of a thin film on the substrate surface.

Rojo *et al.* deposited $La_{0.6}Ca_{0.4}Fe_{0.8}Ni_{0.2}O_{3-\delta}$ perovskite oxide thin film on YSZ substrate by PLD technique and the structure and electrical conductivity of the as-deposited film were studied [97]. In their experiments, the $La_{0.6}Ca_{0.4}Fe_{0.8}Ni_{0.2}O_{3-\delta}$ perovskite oxide powder was prepared by the liquid mix process from respective nitrate salts of the componential metal cations. The resulting powder was pressed into pellets and sintered at 1200 °C for 10 h to obtain a

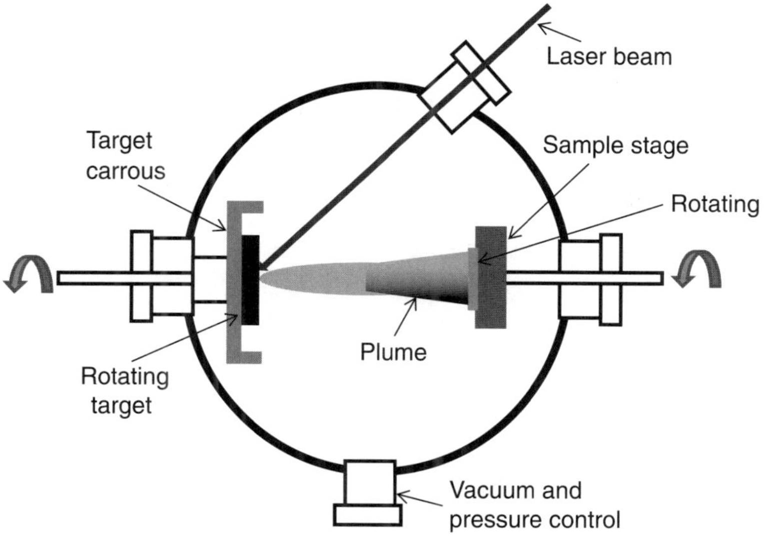

Figure 5.16 *Schematic diagram of pulsed laser deposition technique for thin film deposition*

target for PLD. $La_{0.6}Ca_{0.4}Fe_{0.8}Ni_{0.2}O_{3-\delta}$ films were deposited on single crystal (100) and polycrystalline YSZ substrates by PLD. The samples were deposited using a Lambda Physic Compex 102 KrF excimer laser (248 nm, 150 mJ pulse^{-1}) at a frequency of 15 Hz. The target was placed in a rotating target holder in a vacuum chamber with an initial pressure of 2×10^{-6} mbar. The single crystal or polycrystalline YSZ substrates were mounted on a heater. The films were deposited at a substrate temperature of 700 °C. The temperature of the substrate is one of the main parameters affecting atomic surface mobility during the deposition process. Oxygen gas was flowed into the chamber in a constant flux during deposition, keeping a pressure of 0.3 mbar. Deposition time was 120 min. Figure 5.17 shows the representative SEM micrographs of the surface and cross section of the resulting PLD deposited $La_{0.6}Ca_{0.4}Fe_{0.8}Ni_{0.2}O_{3-\delta}$ films on single crystal and polycrystalline YSZ substrates.

5.4 Example Applications of MIEC Membranes for the Partial Oxidation of Methane

By using the example of MIEC-based membrane reactors for the partial oxidation of methane (POM), in this section we discuss the general advantages and disadvantages of the various major MIEC membrane configurations. MIEC ceramic POM reactors can generally be found in three configurations: disk-shaped, tubular, and hollow fibre type. In general, disk-shaped membrane reactors give relatively low syngas production rates due to limited membrane area. Also, the requirement of rigorous high-temperature sealing makes scale up difficult. These issues are alleviated with the tubular membrane design, but the thickness required to achieve sufficient mechanical strength in the tubular configuration often leads to unsatisfactory oxygen permeation flux. Recently, hollow fibre membrane reactors have attracted much interest due to their advantages over the disc and tubular membranes including large membrane area/volume ratio and smaller membrane thickness. However, hollow fibre membranes have lower mechanical

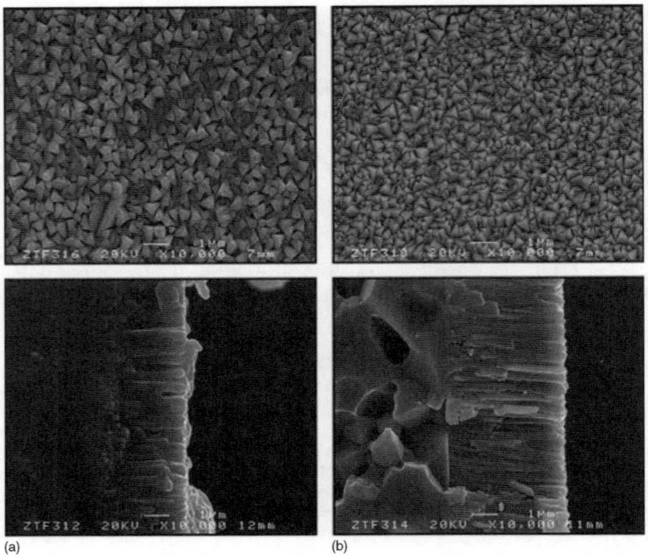

Figure 5.17 *SEM micrographs of the surface and cross section of $La_{0.6}Ca_{0.4}Fe_{0.8}Ni_{0.2}O_{3-\delta}$ films on (a) single crystal and (b) polycrystalline YSZ substrate. Reprinted from Journal of Power Sources, Ruiz de Larramendi, I., et al., Structure and impedance spectroscopy of $La_{0.6}Ca_{0.4}Fe_{0.8}Ni_{0.2}O_{3-\delta}$ thin films grown by pulsed laser deposition. Vol. 171, 747–753. Copyright (2007) with permission from Elsevier*

strength and thus are more difficult to assemble into membrane reactors. Finally, by combining the advantages of the tubular and the hollow fibre membranes, composite asymmetric tubular membranes composed of a thin dense perovskite film coated on a porous substrate tube are a good choice to obtain high POM selectivity yield so long as the thermal expansion mismatch between the coating and substrate ceramic can be minimised.

5.4.1 Disk-Shaped Membrane Reactor

In order to implement a disk-shaped POM membrane reactor, Dong *et al.* [103–105] prepared dense disks of the typical MIEC perovskite-type oxide $Ba_{0.5}Sr_{0.5}Co_{0.8}Fe_{0.2}O_{3-\delta}$ using the dry pressing method from a high-quality powder synthesised by the ECCC method [103]. The membrane reactor was constructed using these membrane disks with gold rings as sealants and LiLaNiOx/γ-Al_2O_3 as a catalyst. The partial oxidation of methane was carried out at 875 °C for 500 h in the as-constructed membrane reactor, demonstrating methane conversion >97% and CO selectivity >95% (Figure 5.18). An oxygen permeation flux as high as 11.5 ml min^{-1} cm^{-2} was achieved under an air/syngas gradient, which is almost 10 times greater than the flux achieved under an air/He gradient. Under membrane reaction conditions, the POM reaction mechanism was suggested to obey the CRR mechanism (complete combustion of CH_4 to CO_2 and H_2O with a subsequent reforming reaction of the residual CH_4 with CO_2 and H_2O to CO and H_2).

Tong *et al.* modified the disk-type membrane reactor design described above by improving the sealing effect of the gold ring using a large spring force [104]. Instead of $Ba_{0.5}Sr_{0.5}Co_{0.8}Fe_{0.2}O_{3-\delta}$, $BaCo_{0.4}Fe_{0.4}Zr_{0.2}O_{3-\delta}$ was used as it is believed to have much better

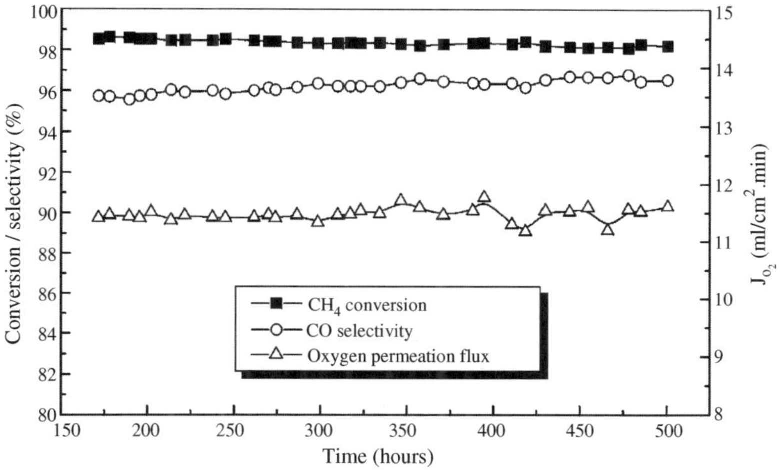

Figure 5.18 *Long-term performance for POM reaction at 875 °C in disk-type $Ba_{0.5}Sr_{0.5}Co_{0.8}Fe_{0.2}O_{3-\delta}$ membrane reactor. Air flow rate 300 ml min^{-1}, 50% diluted methane flow rate 42.8 ml min^{-1}. Reprinted from Catalysis Today, Dong, H., et al., Investigation on POM reaction in a new perovskite membrane reactor. Vol. 67, 3–13. Copyright (2001) with permission from Elsevier*

stability. Long term POM operation for over 2200 h was successfully carried out in this $BaCo_{0.4}Fe_{0.4}Zr_{0.2}O_{3-\delta}$ membrane reactor. Excellent performance of 96–98% methane conversion, 98–99% carbon monoxide selectivity and an oxygen permeation flux of 5.4–5.8 ml min^{-1} cm^{-2} at 850 °C was obtained. Moreover, a short induction period of 2 h was observed, compared to 21 h for the membrane reactor constructed from $Ba_{0.5}Sr_{0.5}Co_{0.8}Fe_{0.2}O_{3-\delta}$. These studies show that disk-type membrane can be successfully operated for long periods providing that high-temperature sealing problems are solved, enabling useful characterisation of MIEC ceramic membrane materials under real reaction conditions.

5.4.2 Tubular-Shaped Membrane Reactor

Balachandran *et al.* were first to report the partial oxidation of methane using a tubular MIEC ceramic membrane reactor [47, 106–108]. As schematically shown in Figure 5.19, they constructed their tubular membrane reactors using MIEC ceramic tubes of $La_{0.2}Sr_{0.8}Co_{0.8}Fe_{0.2}O_3$, $SrCo_{0.8}Fe_{0.2}O_{3-\delta}$, and $SrCo_{0.5}FeO_x$ synthesised by the plastic extrusion method. A Rh-based reforming catalyst (ca. 1 cm^3) was loaded adjacent to the tubular membrane. A gold wire mesh was then wrapped around the tube to prevent solid-state reaction between the catalyst and the membrane. Relatively short tubes with a membrane area of ~10 cm^2 or less were used for their initial investigations, although tubes with larger area could also be used. The $SrCo_{0.8}Fe_{0.2}O_{3-\delta}$ tubular membrane reactor survived only a few minutes at 850 °C during POM testing before breaking into several pieces. The most possible reason for failure appears to be lattice mismatch between the materials on the inner and outer walls of the tube. The difference in composition between the inner and outer zones leads to an expansion of 2%. In comparison, the $La_{0.2}Sr_{0.8}Co_{0.8}Fe_{0.2}O_3$ and $SrCo_{0.5}FeO_x$ tubes exhibited remarkable structural stability at high temperatures. Long-term operations (1000 and 500 h) under more severe conditions and in the presence of Rh-based catalyst were achieved for the $La_{0.2}Sr_{0.8}Co_{0.8}Fe_{0.2}O_3$ and $SrCo_{0.5}FeO_x$ membranes, respectively. Methane conversion >99% was obtained.

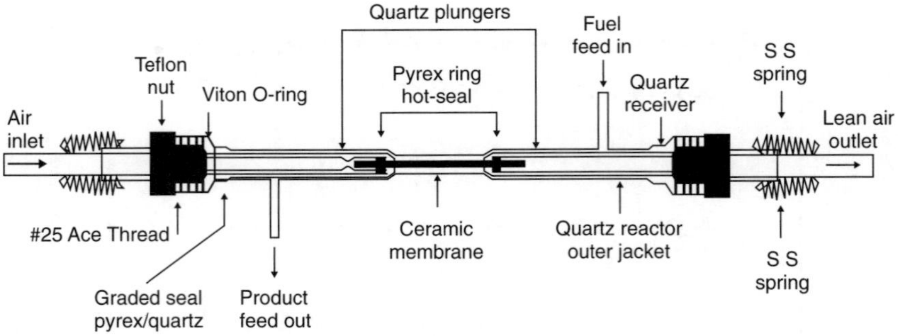

Figure 5.19 *Schematic diagram of tubular MIEC ceramic membrane reactor. Reprinted from Applied Catalysis A: General, Balachandran, U., et al., Dense ceramic membranes for partial oxidation of methane to syngas. Vol. 36, 265–272. Copyright (1995) with permission from Elsevier*

Wang *et al.* also prepared tube membranes of $Ba_{0.5}Sr_{0.5}Co_{0.8}Fe_{0.2}O_{3-\delta}$ with similar dimensions and tried POM [108]. It was shown that longer tube with larger diameters exhibited significantly greater failure rates, especially under the large oxygen gradients experienced in an air/syngas reaction configuration.

5.4.3 Hollow Fibre Membrane Reactor

Wang *et al.* [109–112] prepared a hollow fibre membrane reactor from $BaCo_{1-x-y}Fe_xZr_yO_{3-\delta}$ perovskite using phase inversion spinning followed by sintering, as schematically described in Figure 5.20 [109]. A long hollow fibre membrane was used and the ends of the hollow fibre were coated with a dense gold film, which was used for sealing in a lower temperature zone outside the furnace. Only a small length (membrane surface area of 3.3 cm^2) of the middle hollow fibre was put in the isothermal zone of the furnace for oxygen permeation. A Ni-based steam reforming catalyst was packed around and behind the hollow fibre membrane. Air was fed to the core side of the hollow fibre while a mixture of methane and helium was fed to the shell side. The POM reaction was carried out at 825–925 °C. The carbon monoxide selectivity and the ratio of H_2 to CO were ~100% and ~2, respectively. The highest methane conversion, (near 100%) was obtained at the highest temperature of 925 °C. It was also found that CO and H_2 were formed by reforming reactions of methane with CO_2 and H_2O in the hollow fibre membrane reactor on the catalyst bed. Additional catalyst was required behind the hollow fibre membrane to ensure complete reformation.

5.4.4 Asymmetric Membrane Reactor

In one demonstration of a tubular asymmetric membrane reactor design [113, 114], dense films (150 μm) of $La_{0.5}Sr_{0.5}Fe_{0.8}Ga_{0.2}O_{3-\delta}$, a very stable MIEC ceramic membrane material, were deposited on the exterior surfaces of porous α-alumina tubular substrates (20 mm OD, 15 mm ID, 64 mm long) by the spray deposition technique (with and without vacuum) [113]. The membrane tube was glued to two nonporous high-purity α-alumina tubes by Haldenwanger high-temperature alumina cement. A membrane reactor using this asymmetric MIEC membrane was constructed, which demonstrated a much larger active membrane area compared to the previously discussed membrane configurations. A POM catalyst containing 0.1% rhodium on

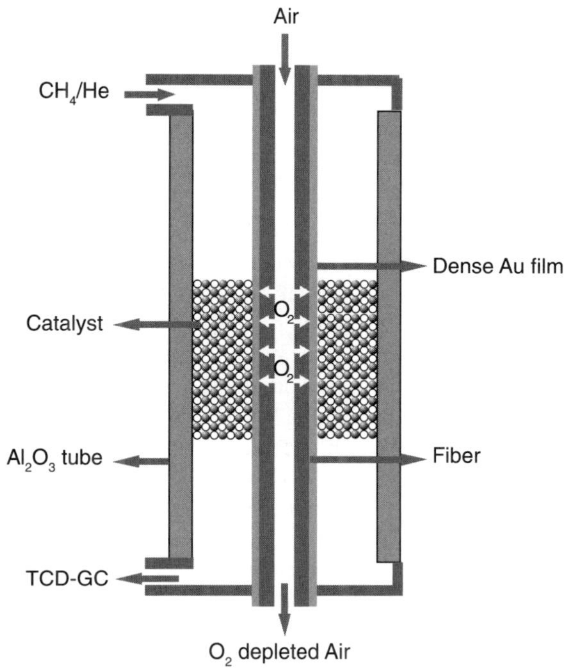

Figure 5.20 *Schematic diagram of* $BaCo_{1-x-y}Fe_xZr_yO_{3-\delta}$ *hollow fibre membrane reactor for POM reaction. Reprinted from Catalysis Communications, Wang, H., et al., Partial oxidation of methane to syngas in a perovskite hollow fiber membrane reactor. Vol. 7, 907–912. Copyright (2006) with permission from Elsevier*

3.2 mm cylindrical alumina supports was packed around the exterior of the tube and kept in place by quartz wool. A methane mixture was fed along the exterior of the tubular reactor while air was fed inside the tube. Good POM reaction performance including high methane conversion (98%), high H_2/CO ratio (1.76), and high CO selectivity (100%) was obtained using this membrane reactor. Furthermore, this membrane reactor exhibited good durability. The membrane film and seals did not fail following heating at a rate of $5\,°C\,min^{-1}$ or upon rapid cooling at a rate of $10\,°C\,min^{-1}$. The membrane reactor did not fail upon application of periodic tapping, which was used to simulate mild vibration in an industrial environment. Although, the top separation layer and the substrate were composed from different materials, short-term (29-day) testing did not lead to any degradation in the membrane performance. These preliminary experimental results suggest that it is possible to develop a commercial process for the production of syngas using this type of asymmetric tubular membrane reactor, although additional experiments will be needed to optimise performance. It is likely that this asymmetric tubular membrane reactor design can be applied other MIEC ceramic materials as well.

5.5 Final Remarks and Conclusions

This chapter reviews the typical synthesis and preparation methods used to obtain MIEC perovskite-type oxide powders and membrane structures. For MIEC perovskite powder production, the EDTA acid and citric acid complexing (ECCC) method is currently the most popular approach, although the conventional solid state reaction, coprecipitation, and alkoxide sol-gel method are also widely used in some cases. The new technique of hydrothermal synthesis has the potential to synthesise crystalline perovskite at low temperature and is therefore subject to vigorous current research, although more work is needed to be done before it can be extensively applied. Combustion synthesis and spray pyrolysis are also promising and increasingly popular synthesis techniques.

For MIEC perovskite membrane fabrication, the disk-shaped MIEC membrane is still the most popular and commonly used laboratory configuration due to ease of fabrication and its convenience for investigating the fundamental properties of membrane performance. Disk-shaped membranes are generally prepared by the dry pressing method. Tubular MIEC membranes enable larger permeation areas and greater compatibility with commercial membrane reactor designs. Tubular membranes are most commonly prepared by either the plastic extrusion or slip casting method. In some cases, high-quality tubes can also be prepared by the isostatic pressing method. While tubular designs offer larger membrane areas and easier sealing, it is usually difficult to operate a membrane reactor constructed from a long tube because of chemical instability of the membrane material and the lattice mismatch between the materials on the inner and outer walls of the tube. Even better performance can be achieved with hollow fibre MIEC ceramic membranes using an asymmetric configuration; this is an area of active current research. Most hollow fibre type membrane structures are prepared using the phase inversion spinning/sintering technique, although the fragile nature of these hollow fibre membrane structures can make device implementation difficult. Asymmetric membrane structures, both in planar and tubular configurations, achieve high permeation rates by using an ultra-thin dense MIEC layer integrated on top of a thicker support layer to provide mechanical strength and stability. Because they can combine strength, stability, and high performance, asymmetric membranes show great potential for commercial applications. Asymmetric membranes can be prepared by a variety of techniques, including dry pressing, tape casting/lamination, slip/slurry related casting/coating, CVD, PLD, and so on. The tape casting/lamination approach is of particular commercial interest for the preparation of planar type asymmetric MIEC membranes, while the slip casting approach is of interest for tubular type asymmetric membranes. PLD (pulsed laser deposition) represents a relatively new but powerful technique to prepare high-quality MIEC thin films with controllable microstructure. Although this technique has not been extensively used in MIEC membrane preparation, it enables both new scientific and commercial avenues of exploration for MIEC membrane design and preparation.

Finally, the application of MIEC membrane structures for the partial oxidation of methane (POM) was discussed in order to examine the relative merits of the various MIEC membrane configurations. The asymmetric-type membrane reactor design in particular exhibits significant commercial potential due to high oxygen permeance combined with excellent strength and durability.

Acknowledgements

The authors would like to acknowledge financial support from the Department of Energy, Office of Energy Efficiency and Renewable Energy, under Contract DEFG36-08GO88100 and the National Science Foundation MRSEC program under Grant DMR-0820518 at the Colorado School of Mines.

References

1. Y. Teraoka, H.M. Zhang, S. Furukawa *et al.*, *Chemistry Letters*, **11**, 1743–1746 (1985)
2. P.J. Gellings, H.J.M. Bouwmeester, *The CRC Handbook of Solid State Electrochemistry*, ISBN 0-8493-8956-9, CRC Press, Inc., New York, Chapter 14 (1997).
3. W. Yang, H. Wang, X. Zhu *et al.*, *Topic in Catalysis*, **35** (1/2) 155–167 (2005).
4. H.J.M. Bouwmeester, *Catalysis Today*, **82**, 141–150 (2003).
5. Y. Liu, X. Tan, K. Li, *Catalysis Reviews*, **48**, 145–198 (2006).
6. A.C. Bose, *Inorganic Membranes for Energy and Environmental Applications*, ISBN: 978-0-387-34524-6, Springer Science + Business Media, Heidelberg, Chapter 1, pp 3–25; Chapter 2, pp 26–51; Chapter 3, pp 52–65 (2009).
7. J. Sunarso, S. Baumann, J.M. Serra *et al.*, *Journal of Membrane Science*, **320**, 13–41 (2008).
8. J.E. ten Elshof, H.J.M. Bouwmeester, H. Verweij, *Applied Catalysis A: General*, **130**, 195–212 (1995).
9. C.Y. Tsai, A.G. Dixon, W.R. Moser *et al.*, *AIChE Journal*, **43** (11), 2741–2750 (1997)
10. Z. Shao, H. Dong, G. Xiong *et al.*, *Journal of Membrane Science*, **183**, 181–192 (2001).
11. R.A. Richardson, J.W. Cotton, R.M. Ormerod, *Dalton Transactions*, **2004**, 3110–3115 (2004).
12. L. Tan, X. Gu, L. Yang *et al.*, *Journal of Membrane Science*, **212**, 157–165 (2003).
13. J. Sfeir, S. Vaucher, P. Holtappels *et al.*, *Journal of the European Ceramic Society*, **25**, 1991–1995 (2005)
14. P. Cousin, R.A. Ross, *Materials Science and Engineering*, **A130** 119–125 (1990).
15. J. Ishida, K. Murata, T. Ichikawa *et al.*, *Solid State Ionics*, **180**, 1045–1049 (2009)
16. H. Lu, Y. Cong, W. Yang, *Materials Science and Engineering B*, **141**, 55–60 (2007).
17. X. Zhu, Y. Cong, W. Yang, *Journal of Membrane Science*, **283**, 158–163 (2006).
18. X. Qi, Y.S. Lin, S.L. Swartz, *Industrial Engineering Chemistry Research*, **29**, 646–653 (2000).
19. G. Xu, G. Han, *Materials Research Bulletin*, **39**, 1909–1915 (2004)
20. C.F. Kao, C.L. Jeng, *Ceramics International*, **26**, 237–243 (2000).
21. G. Westin, A. Pohl, M. Ottosson *et al.*, *Journal of Sol-Gel Science and Technology*, **48**, 194–202 (2008).
22. S. Bilger, G. Blaβ R. Förthmann, *Journal of the European Ceramic Society*, **17**, 1027–1031 (1997)
23. J. Xu, A.P. Wilkinson, S. Pattanaik, *Chemistry of Materials*, **12**, 3321–3330 (2000).
24. A. Pohl, G. Westin, *Journal of Sol-Gel Science and Technology*, **26**, 257–260 (2003).
25. W. Wang, D. Jia, Y. Zhou *et al.*, *Materials Research Bulletin*, **37**, 2517–2524 (2002).
26. Z. Shao, W. Yang, Y. Cong *et al.*, *Journal of Membrane Science*, **172**, 177–188 (2000).
27. H. Wang, C. Tablet, A. Feldhoff *et al.*, *Journal of Membrane Science*, **262**, 20–26 (2005).
28. J. Tong, W. Yang, B.C Zhu *et al.*, *Journal of Membrane Science*, **203**, 175–189 (2002).
29. A. Feldhoff, J. Martynczuk, H. Wang, *Progress in Solid State Chemistry*, **35**, 339–353 (2007).
30. W. Zhou, R. Ran, Z.P. Shao *et al.*, *Journal of Power Sources*, **174**, 237–245 (2007).
31. L. Ge, W. Zhou, R. Ran *et al.*, *Journal of Alloys and Compounds*, **450**, 338–347 (2008).
32. K. Efimov, M. Arnold, J. Martynczuk *et al.*, *Journal of American Ceramic Society*, **92** (4), 876–880 (2009).
33. L. Ge, R. Ran, Z. Shao *et al.*, *Ceramics International*, **35**, 2809, **2815** (2009).
34. Y. Zeng, Y.S. Lin, S.L. Swartz, *Journal of Membrane Science*, **150**, 87–98 (1998).
35. W.J. Dawson, S.L. Swartz, US Patent 5112433 (1992).
36. M.M. Amini, M. Mirzaee, *Ceramics International*, **35**, 2367–2372 (2009).
37. D.S. Jung, H.Y. Koo, H.C. Jang *et al.*, *Journal of Alloys and Compounds*, **487**, 693–697 (2009).
38. A. Gurava, T. Kodas, T. Pluym *et al.*, *Aerosol Science and Technology*, **19** (4), 411–452 (1993).

39. A.S. Mukasyan, C. Costello, K.P. Sherlock *et al.*, *Separation and Purification Technology*, **25**, 117–126 (2001).
40. R.G. Chandran, K.C. Patil, *Materials Letters*, **12**, 437–441 (1992).
41. K. Deshpande, A. Mukasyan, A. Varma, *Journal of American Ceramic Society*, **86** (7), 1149–1154 (2003).
42. A. Civera, M. Pavese, G. Saracco *et al.*, *Catalysis Today*, **83**, 199–211 (2003).
43. P. Zeng, R. Ran, Z. Chen *et al.*, *Journal of Membrane Science*, **302**, 171–179 (2007).
44. F.T. Akin, Y.S. Lin, Y. Zeng, *Industrial Engineering Chemistry Research*, **40**, 5908–5916 (2001).
45. A.A. Yaremchenko, A.A. Valente, V.V. Kharton *et al.*, *Catalysis Letters*, **91** (3/4) 169–174 (2003).
46. Z.Q. Deng, W.S. Yang, W. Liu *et al.*, *Journal of Solid State Chemistry*, **179**, 362–369 (2006).
47. U. Balachandran, J.T. Dusek, R.L. Mieville *et al.*, *Applied Catalysis A: General*, **133**, 19–29 (1995).
48. S.G. Li, H. Qi, N.P. Xu *et al.*, *Industrial and Engineering Chemistry Research*, **38**, 5028–5033 (1999).
49. H. Wang, Y. Cong, W. Yang, *Journal of Membrane Science*, **210**, 259–271 (2002).
50. M. Trunec, J. Cihlar, S. Diethelm *et al.*, *Journal of American Ceramic Society*, **89** (3), 949–954 (2006).
51. S. Li, W. Jin, P. Huang *et al.*, *Journal of Membrane Science*, **166**, 51–61 (2000).
52. V.V. Kharton, A.V. Kovalevsky, A.A. Yaremchenko *et al.*, *Journal of Solid State Electrochemistry*, **10**, 663–673 (2006).
53. F.T. Akin, Y.S. Lin, *AIChE Journal*, **48** (10), 2298–2306 (2002).
54. E.A. Hazbun,US Patent 4791079 (1988).
55. J. Luyten, A. Buekenhoudt, W. Adriansens *et al.*, *Solid State Ionics*, **135**, 637–642 (2000).
56. X. Tan, Y. Liu, K. Li, *Industrial Engineering Chemistry Research*, **44**, 61–66 (2005).
57. S. Liu, G.R. Gavalas, *Industrial Engineering Chemistry Research*, **44**, 7633–7637 (2005).
58. C. Tablet, G. Grubert, H. Wang *et al.*, *Catalysis Today*, **104**, 126–130 (2005).
59. K. Li, X. Tan, Y. Liu, *Journal of Membrane Science*, **272**, 1–5 (2006).
60. X. Tan, Y. Liu, K. Li, *AIChE Journal*, **51** (7), 1991–2000 (2005)
61. S. Liu, X. Tan, Z. Shao *et al.*, *AIChE Journal*, **52** (10), 3452–3461 (2006).
62. S. Liu, M. Liu, Z. Shao *et al.*, *Journal of American Ceramic Society*, **90** (1), 84–91 (2007).
63. S. Liu, G.R. Gavalas, *Journal of Membrane Science*, **246**, 103–108 (2005).
64. W. Li, J.J. Liu, C.S. Cheng, *Journal of Membrane Science*, **340**, 266–271 (2009).
65. Z. Chen, Z. Shao, R. Ran *et al.*, *Journal of Membrane Science*, **300**, 182–190 (2007).
66. A.V. Kovalevsky, V.V. Kharton, F.M.M. Snijkers *et al.*, *Solid State Ionics*, **179**, 61–65 (2008).
67. X. Chang, C. Zhang, X. Dong *et al.*, *Journal of Membrane Science*, **322**, 429–435 (2008).
68. S. Lee, J.H. Yu, D.W. Seo *et al.*, *Journal of Electroceramics*, **17**, 719–722 (2006).
69. A.V. Kovalevsky, V.V. Kharton, F.M.M. Snijkers *et al.*, *Journal of Membrane Science*, **301**, 238–244 (2007).
70. A. Julian, E. Juste, P.M. Geffroy *et al.*, *Journal of Membrane Science*, **333**, 132–140 (2009).
71. H. Raeder, C. Simon, *Journal of the European Ceramic Society*, **13**, 485–191 (1994).
72. S. Kaji, N. Sakamoto, *Solid State Ionics*, **108**, 235–240 (1998).
73. M. Shi, N. Liu, Y.D. Xu *et al.*, *Journal of Materials Processing Technology*, **169**, 179–183 (2005).
74. G. Etchegoyen, T. Chartier, P. Del-Gallo, *Journal of the European Ceramic Society*, **26**, 2807–2815 (2006)
75. H. Takamura, M. Ogawa, K. Suehiro *et al.*, *Solid State Ionics*, **179**, 1354–1359 (2008).
76. C. Li, T. Hu, H. Zhang *et al.*, *Journal of Membrane Science*, **226**, 1–7 (2003).
77. X. Yin, C. Choong, L. Hong *et al.*, *Journal of Solid State Electrochemistry*, **10**, 643–650 (2006).
78. S. Bhatia, C.Y. Thien, A.R. Mohamed, *Chemical Engineering Journal*, **148**, 525–532 (2009).

79. S. Araki, Y. Hoshi, S. Hamakawa *et al.*, *Solid State Ionics*, **178**, 1740–1745 (2008)
80. S. Hamakawa, L. Li, A. Li *et al.*, *Solid State Ionics*, **48**, 71–81 (2002).
81. I. Kagomiya, T. Iijima, H. Takamura, *Journal of Membrane Science*, **286**, 180–184 (2006).
82. W. Jin, S. Li, P. Huang *et al.*, *Journal of Membrane Science*, **185**, 237–243 (2001).
83. O. Buchler, J.M. Serra, W.A. Meulenberg *et al.*, *Solid State Ionics*, **178**, 91–99 (2007).
84. Y. Gu, G. Meng, *Journal of the European Ceramic Society*, **19**, 1961–1966 (1999)
85. X. Yin, L. Hong, Z.L. Liu, *Journal of Membrane Science*, **268**, 2–12 (2006).
86. X. Yin, L. Hong, Z.L. Liu, *Applied Catalysis A: General*, **300**, 75–84 (2006).
87. X. Yin, L. Hong, Z.L. Liu, *Journal of Membrane Science*, **311**, 89–97 (2008).
88. N. Kamoun, H. Bouzouita, B Rezig, *Thin Solid Films*, **515**, 5949–5952 (2007).
89. P.K. Siwach, H.K. Singh, O.N. Srivastava *et al.*, *Journal of Physics D: Applied Physics*, **39**, 3731–3737 (2006).
90. C.S. Hsu, B.H. Hwang, *Journal of the Electrochemical Society*, **153** (8), A1478–A1483 (2006).
91. D. Beckel, A. Dubach, A.R. Studart *et al.*, *Journal of Electroceramics*, **16**, 221–228 (2006).
92. A. Abrutis, A. Teiserskis, G. Garcia *et al.*, *Journal of Membrane Science*, **240**, 113–122 (2004).
93. A. Furusaki, H. Konno, R. Furuichi, *Journal of Materials Science*, **30**, 2829–2834 (1995)
94. A. Abrutis, A. Bartasyte, G. Garcia *et al.*, *Thin Solid Films*, **449**, 94–99 (2004).
95. V. Faucheux, M. Audier, L. Rapenne *et al.*, *Journal of Materials Processing Technology*, **204**, 248–254 (2008).
96. Y.S. Lin, K.J. de Vries, H.W. Brinkman *et al.*, *Journal of Membrane Science*, **66**, 211–226 (1992).
97. I. Ruiz de Larramendi, N. Ortiz, R. López-Antón *et al.*, *Journal of Power Sources*, **171**, 747–753 (2007).
98. D. Mori, H. Oka, Y. Suzuki *et al.*, *Solid State Ionics*, **177**, 535–540 (2006).
99. Y.W. Li, Z.G. Hu, F.Y. Yue *et al.*, *Applied Physics A*, **95**, 721–725 (2009).
100. N. Sata, K. Ikeda, F. Iguchi *et al.*, *Solid State Ionics*, **178**, 1563–1567 (2007).
101. M. Mosleh, N. Pryds, P.V. Hendriksen, *Materials Science and Engineering B*, **144**, 38–42 (2007).
102. F. S. Baumann, J. Fleig, G. Cristiani *et al.*, *Journal of The Electrochemical Society*, **154** (9), B931–B941 (2007).
103. H. Dong, Z. Shao, G. Xiong *et al.*, *Catalysis Today*, **67**, 3–13 (2001).
104. J. Tong, W. Yang, R. Cai *et al.*, *Catalysis Letters*, **78** (1/4) 129–137 (2002).
105. H. Lu, J. Tong, Y. Cong *et al.*, *Catalysis Today*, **104**, 154–159 (2005).
106. U. Balachandran, J.T. Dusek, P.S. Maiya *et al.*, *Catalysis Today*, **36**, 265–272 (1997).
107. H. Wang, Y. Cong, W. Yang, *Catalysis Today*, **82**, 157–166 (2003).
108. W. Jin, S. Li, P. Huang, N. Xu *et al.*, *Journal of Membrane Science*, **166**, 13–22 (2000).
109. H. Wang, C. Tablet, T. Schiestel *et al.*, *Catalysis Communications*, **7**, 907–912 (2006).
110. J. Caro, K.J. Caspary, C. Hamel *et al.*, *Industrial and Engineering Chemistry Research*, **46**, 2286–2294 (2007).
111. H. Wang, A. Feldhoff, J. Caro *et al.*, *AIChE Journal*, **55** (10), 2657–2664 (2009).
112. X. Tan, K. Li, *AIChE Journal*, **55** (10), 2675–2685 (2009).
113. J.T. Ritchie, J.T. Richardson, D. Luss, *AIChE Journal*, **47** (9), 2092–2101 (2001)
114. X. Yin, L. Hong, Z.L. Liu, *Journal of Physical Chemistry C*, **111**, 9194–9202 (2007).

6

Nanostructured Perovskites for the Fabrication of Thin Ceramic Membranes and Related Phenomena

V.V. Zyryanov[1], *A.P. Nemudry*[1] *and V.A. Sadykov*[2]

[1]*Institute of Solid State Chemistry and Mechanochemistry, SB RAS, Novosibirsk, Russia*
[2]*Boreskov Institute of Catalysis, SB RAS, Novosibirsk, Russia*

6.1 Introduction

High costs of infrastructure for transportation and storage of hydrogen fuel stimulated research and development efforts for membrane technologies to produce hydrogen on site from available sources (methane, methanol, dimethyl ether) [1]. For the creation of MR with a high oxygen flux ($>10\,cm^3\,cm^{-2}\,min^{-1}$) which could be competitive even with commercial technology of steam methane reforming, thin membranes comprised of materials with a high oxygen permeability possessing long-term chemical and mechanical stability under high oxygen chemical potential gradient are required. This can be provided by design of a supported membrane comprised of a macroporous substrate and compatible thin gas-tight layer providing selective oxygen permeability and high flux [2–5].

Advantages and drawbacks of supported membranes based on different substrates are summarised in Table 6.1. At different stages of MR fabrication, the nature of substrate should be properly taken into account. A mismatch of shrinkages of various materials during high temperature co-firing is the main problem which is to be solved at a stage of selection of a combination of compatible materials. Problem of mismatching in thermal expansion

Membranes for Membrane Reactors: Preparation, Optimization and Selection, Edited by Angelo Basile and Fausto Gallucci

Table 6.1 *Requirements to a multilayer ceramic membrane with selected geometry and material of a substrate*

Requirement	Material of support Alloy		Cermet		Ceramics		Glass–ceramic composite	
Geometry	Plate	Tube	Plate	Tube	Plate	Tube	Plate	Tube
Thermal resistance at 1600–1700 K	–	–	+^	+^	+	+	+	+
Accomodation of shrinkages	–	–	–	+/–	–	+/–	+/–	+
Accomodation of TECs	–	+/–	+/–	+	+/–	+	+/–	+
Optimisation of materials/operations, taking a seal into account	–	–	–	–	–	–	–	+°
Mechanical properties	+/–	+	+/–	+	–	+	–	+
Thermal cycling	+/–	+	+	+	+/–	+	–	+
Upscaling	+	+	+/–	+/–	–	+/–	–	+/–
Assembling	+	+/–	+	+/–	–	+/–	–	+/–
Convenience of manipulations	+	–	+/–	–	–	+/–	–	+
Long-term degradation	+/–	+/–	+/–	+/–	+/–	+	–	+/–

Comments: + = good, ^ = in reducing atmosphere only, ° = for dead-end tubes only.

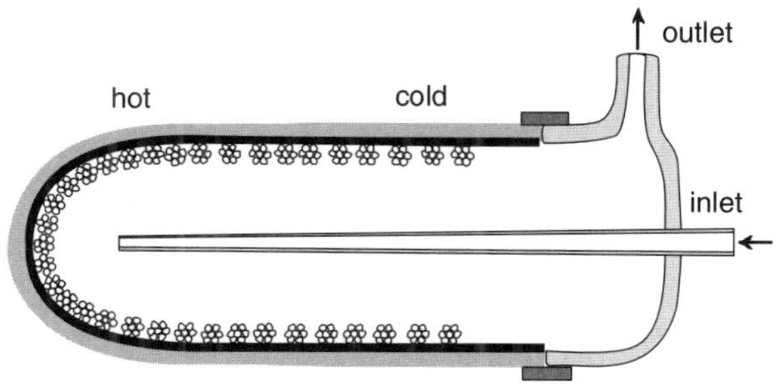

Figure 6.1 Dead-end tubular CMR with a cold end

coefficients (TECs) for potential membrane materials is well known, but less discussed problem of shrinkage misfit is much more significant. Without matching shrinkage curves for all membrane materials during sintering, delamination of a dense film and formation of cracks leading to membrane failure would occur.

For fast development of membranes, porous substrates with the shape of dead-end tubes have some advantage, since in this case their sealing can be easily made at the cold end (Figure 6.1). The recent progress in design of architecture for multilayer membranes is presented in Figure 6.2. The architecture of supported multilayer membranes provides an optimal distribution of functions for each structural element:

1. The substrate – integrity and mechanical strength of MR and facile oxygen transfer from the air stream through macropores to the microporous layer;

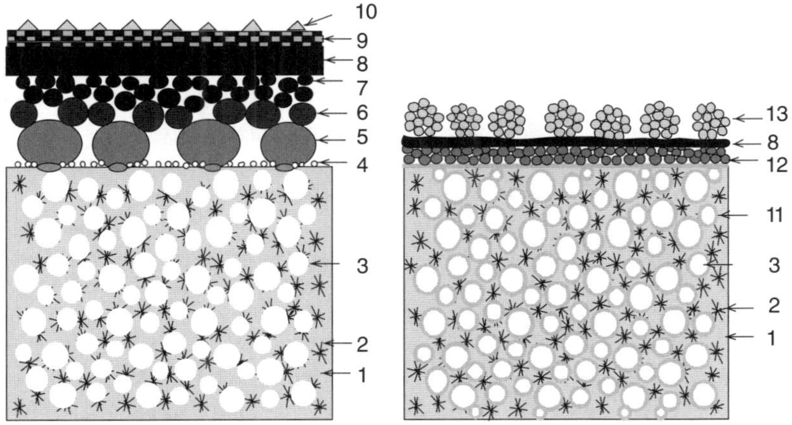

Figure 6.2 Progress made from 2005 (left) to 2009 (right) in the architecture of asymmetric multilayer ceramic membranes on glass–ceramic substrates. (1) Silica glass, (2) nanocrystallites of mullite $Al_6Si_2O_{13}$, (3) pores, (4) barrier layer from corundum nanoparticles, (5), (6), (7) macro-, meso-, microporous ceramic layers comprised of MIEC perovskites with a high incorporation rate k_s, (8) gas-tight perovskite layer with a high lattice oxygen mobility D_e, (9) protective layer from nanocomposite perovskite $(La,Sr)(Fe,Co)O_3$ + fluorite $Ce_{1-x}Gd_xO_y$, (10) porous catalytic layer, (11) ceramic coating by apatite $La_{10}Si_6O_{27}$, (12) microporous layer of layered perovskite $(La,Sr)_2(Fe,Co,Ni)O_4$, (13) microspherical catalyst of CH_4 steam and dry reforming

2. The microporous mixed ionic–electron conducting (MIEC) ceramic layer – efficient activation of oxygen molecules and incorporation of oxygen atoms into the structure;
3. A thin gastight perovskite layer with a high oxygen mobility – prevents nitrogen leakage and provides a high oxygen flux which preserves ceramics from degradation in the reducing atmosphere due to kinetic stabilisation.

Decrease of the MIEC ceramic layers thickness was achieved due to coating the surface of a silica glass substrate by thin La_2O_3 layer ~10 nm resulting in formation of well known oxygen conducting apatite $La_{10}Si_6O_{27}$ during high temperature firing, which is compatible with La-containing layered perovskite and silica glass [6]. Thin ceramic layers provide less strict requirements on matching of shrinkage curves for membrane materials due to anisotropy of shrinkage during sintering of supported thin films [5]. The critical value of thickness for green deposited films on a dense substrate with zero shrinkage potential is usually about 1 μm [7]. For a small shrinkage misfit (<2%) between ceramic layers and support, the critical thickness increases and can not be reached. Moreover, a tubular geometry of MR provides an additional densification of ceramic membrane due to effect of 2-dimensional pressing when the shrinkage of support is less than shrinkages of ceramic layers.

For making MR with optimal architecture (Figure 6.2, on the right), several problems are to be solved consecutively in iteration mode: (i) design of support, (ii) selection of MIEC ceramics with a high oxygen mobility D_{chem}, (iii) synthesis of ceramics with required sintering temperature T_s, (iv) selection of ceramics with a high rate of the oxygen surface exchange k_s, (v) collection of a combination of compatible materials and developing procedures of their successive supporting, (vi) design of a catalyst for selective transformation of methane into syngas.

6.2 Support

Development of a porous substrate with a high (≥15%) shrinkage potential, residual open porosity ≥30% after final firing, and the shrinkage curve close to the typical curves for MIEC ceramics was a demanding problem which took more than 5 years for its solution. Only glass–ceramic nanocomposites were found to be suitable for making a porous substrate retaining its porosity at high temperatures and possessing sufficient mechanical strength to endure subsequent manipulations. The detailed flow sheet of composite supports manufacturing is given elsewhere [8–10]. Such composites were made using optimised composition of selected kaolins and white firing clays with addition of 50–55% of microcrystalline cellulose (MCC) as burned-out pore former. The particles size distribution of MCC determines not only the size of macropores but also the total porosity. Selection of suitable samples from a vast collection of clay minerals from Russian and Ukrainian deposits was based on criteria of a high liquescence after addition of Na-electrolytes and a low thixotropy of water–clay slurries. Selected samples of clay minerals were washed and sieved to remove admixtures. After preparation of a slurry containing clay, MCC and water, it was ball milled for homogenisation. After degassing in vacuum, the slurry was casted into gypsum forms. After drying, green dead-end tubes were pre-calcined at 1220 K and then leached in the acid water solution of $HCl + HNO_3$ (pH ~ 1) to remove admixtures and increase further the microporosity. After washing and drying, substrates were leached in water solution of HF to create a gradient of SiO_2 content across their walls and further increase of microporosity (Figure 6.3). After washing and drying, substrates were modified by supporting a sol of lanthana with subsequent slow drying and firing at 1200 K to fix La_2O_3 on the

(a)

(b)

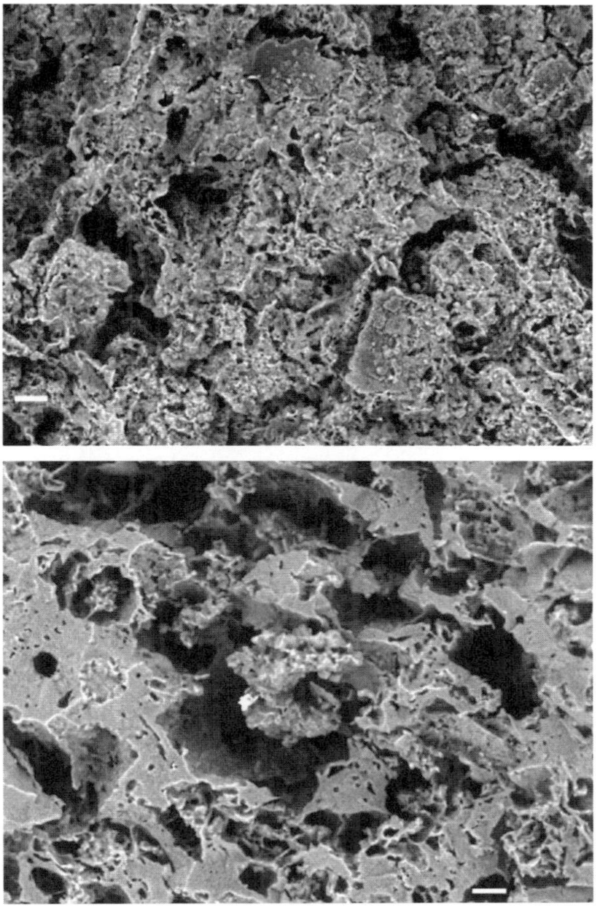

Figure 6.3 Green substrate (a) and after final sintering (b) with open porosity of 70 and 30%, respectively

silica surface. Modification of substrate by La_2O_3 provides improved chemical compatibility with ceramics and decreases shrinkage misfit with supported ceramic layers. Lanthana appears to be the best modifying agent among other refractory oxides for glass–ceramic substrate (Figure 6.4).

6.3 Selection of Ceramics with High Oxygen Mobility

MIEC perovskite-related oxides obtained by traditional ceramic techniques possess a high lattice oxygen mobility. Among studied compounds, ceramics based on $M(Co,Fe)O_{3-x}$ (M = Ca, Sr, Ba, Ln) perovskites provide the highest oxygen flux at intermediate temperatures 800–1000 K. Description of transport properties in oxides is usually based on the model of homogeneous diffusion of randomly distributed oxygen vacancies [11–13]. However, in several works, for

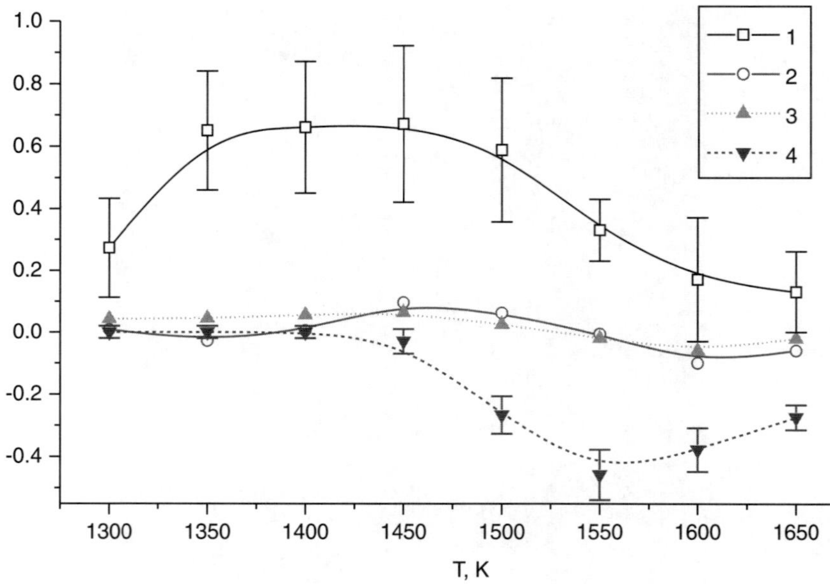

Figure 6.4 *Normalised effect of modifying agents (%/%) on the shrinkage curve of substrate: (1) La sol, (2) Zr salt, (3) ZrO_2 sol, (4) Al_2O_3 sol*

instance [14–16], some limitations of this approach are shown. The concept of domain microstructure for nonstoichiometric oxides was proposed by Ariya and Anderson [16, 17] and developed in works of Alario-Franco, Grenier, Hagenmuller *et al.* [18, 19]. Grossly nonstoichiometric oxides with a random distribution of defects possess a higher free energy as compared with those characterised by the presence of crystallographic shear planes and/or superstructures [16, 17]. However, in such a case description of the real structure of a solid in terms of defects is not applicable because the 'defect' becomes the element of a structure [17]. So, such compounds can be considered as homogeneous solid solutions with a high concentration of structural defects only at high temperatures exceeding that of the order–disorder transformation T_t. At $T < T_t$ the ordering of structural defects occur in such materials.

The ordering of defects (dopants, interstitials and oxygen vacancies) can be complete or incomplete (intermediate) depending on the initial state and cooling mode. This results in formation of defect clusters or discrete domains differing by composition and structure with the matrix. Reversible formation of domain/twin microstructure can also occur as a result of the phase transition between the paraelastic high-temperature high-symmetry phase and a ferroelastic low-temperature low-symmetry phase. Spontaneous mechanical stresses arising in the course of such phase transition can be minimised by the formation of a dense network of nanosized domains/twins [20]. Ferroelastic phase transitions with a formation of nano-twins were observed in the well known oxygen conductor – Mg, Sr-doped lanthanum gallate (LSGM) [21]. Studies of dielectric relaxation processes revealed that a high oxygen ionic conductivity along the domain walls provided by segregation of oxygen vacancies in their vicinity considerably reduces the total resistance of LSGM crystal. Ferroelastic phase transitions and nano-twinning are characteristics of $LaCoO_3$-based MIEC materials. Reversible formation

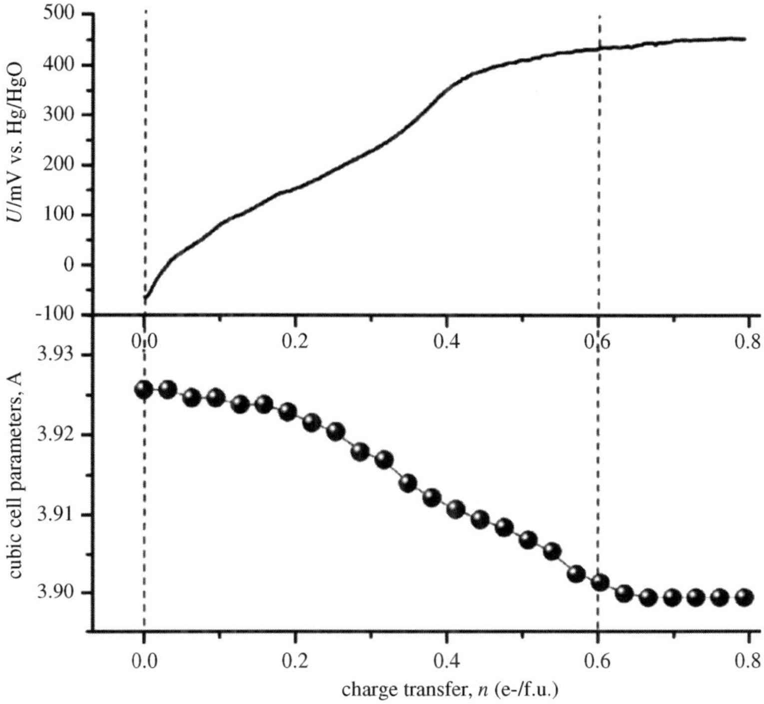

Figure 6.5 *In situ anodic oxidation of* $SrCo_{0.5}Fe_{0.2}Ta_{0.3}O_{2.7+x}$*: potential U versus charge transfer n (top) and change of the unit cell parameter with n, n/2 = x (bottom)*

of nano-twins under thermal cycling and their stability up to 1123 K in $La_{0.8}Ca_{0.2}CoO_3$ perovskite was shown by *in situ* TEM studies [22, 23].

In last decade the number of papers devoted to microheterogeneity and nanostructuring of nonstoichiometric complex oxides with MIEC properties is constantly growing [24–26]. Suggested in [21, 27–31] interrelation between nanostructuring of complex oxides and oxygen mobility could be applied for development of materials with high oxygen mobility. A simple phenomenological model of heterogeneous diffusion – a slow one in bulk of domains and a fast one along domain boundaries ($D_{db} \gg D_b$, with activation energy $E_{db} \sim {}^1/_2 E_b$), allowed to explain qualitatively a high reactivity and anomalous kinetics of oxygen uptake in perovskite-related compounds [27–31]. Electrochemical oxidation of such ceramics can be carried out even at ambient temperature (Figure 6.5) [30, 31]. Domain and domain boundary diffusion coefficients $D_d = 2 \times 10^{-13}$ and $D_{db} = 5 \times 10^{-10}\,cm^2\,s^{-1}$ respectively were estimated by fitting the experimental data on anodic oxidation. Fitting by the model of homogeneous diffusion gives much larger deviation from the experimental data (not shown for brevity) [31].

TEM studies of $SrCo_{1-x}Fe_xO_{2.5}$ perovskites show typical twinning of brownmillerite crystallites under cooling with formation of extended defects – domain boundaries (Figure 6.6). Domain boundaries as high diffusivity pathways with low activation energy can describe anomalous oxygen diffusion at low temperatures [28, 29].

Phase transformation perovskite–brownmillerite is accompanied by a large volume change resulting in cracking of ceramics. In addition, perovskites on the base of $SrCo_{1-x}Fe_xO_{2.5}$ system are not chemically stable in reducing atmosphere and cannot be used as membrane materials. An additional doping up to a certain degree stabilises the structure without deterioration of transport

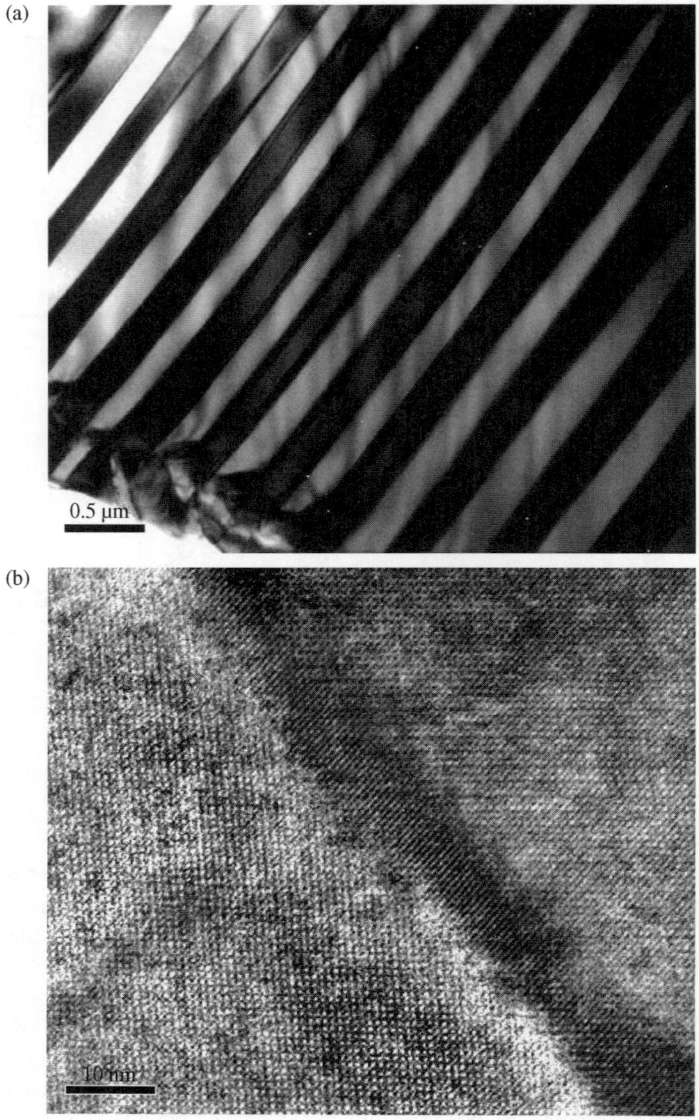

Figure 6.6 *SEM image (dark field) of $SrCo_{0.8}Fe_{0.2}O_{2.5}$ vacuum-quenched ceramic (a) and HRTEM of twin boundary (b)*

properties [31–33]. Domain microstructure is preserved in doped samples, but the sizes of domains decrease down to ~10 nm (Figure 6.7) [34]. Superposition of diffraction patterns from 90° orthorhombic domains of brownmillerite with the OTOT-type of octahedral and tetrahedral layers stacking (areas 1–2, size 10–20 nm and corresponding digital diffraction patterns) results in apparent cubic supercell with $a^* = 2a_p$ (area 3 , and corresponding digital diffraction pattern). The formation of 90° domain texture was observed as well by X-ray diffraction study on single-crystal samples (Figure 6.8) [35]. Weak spots in Figure 6.8 can be indexed in cubic supercell with $a^* = 2a_p$ and related to 90° domain formation. Similar results were obtained for

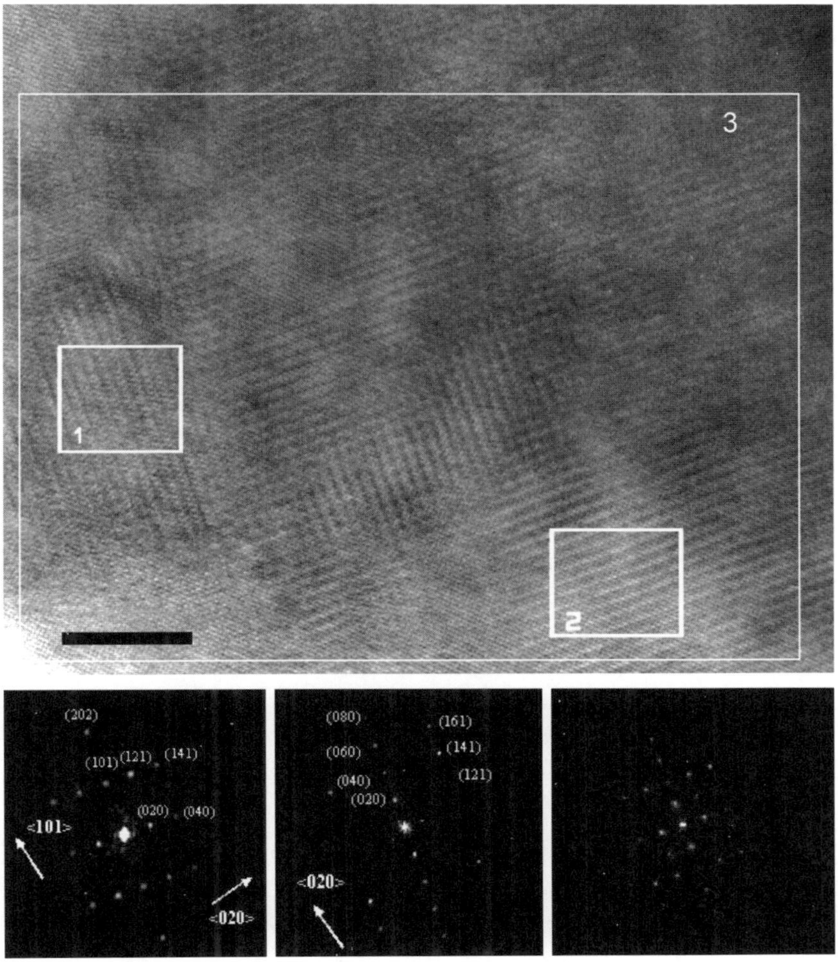

Figure 6.7 *HRTEM image of perovskite $SrFe_{0.95}Mo_{0.05}O_{3-x}$ after annealing in a reducing atmosphere of 5% H_2/Ar, and digital diffraction patterns of selected areas (1), (2) and total (3) respectively*

$La_{0.3}Sr_{0.7}Co_{0.5}Al_{0.3}Fe_{0.2}O_{2.5+z}$ and other perovskites as well (Figure 6.9, weak reflections marked by asterisks related to supercell with a*=2ap). In spite of brownmillerite composition of $SrCo_{0.8-y}Al_yFe_{0.2}O_{2.5}$ ($y \leq 0.3$) as a result of vacuum annealing and quenching, ceramics display cubic perovskite structure according to XRD patterns (Figure 6.10, Table 6.2).

In Figure 6.11 electron and X-ray (synchrotron radiation, $\lambda = 0.3675$ Å) diffraction patterns of $La_{0.3}Sr_{0.7}Co_{0.5}Al_{0.3}Fe_{0.2}O_{3-z}$ perovskite are shown. These techniques allowed to observe a fine changes in the structure which were not revealed by routine XRD method. In the diffraction pattern of quenched sample extra peaks appeared (shown by asterisks in Figure 6.11b). Such peaks can be assigned to ordering of oxygen vacancies and/or formation of domains. Indexing of extra peaks corresponds to superlattice with double perovskite unit cell ($a = 7.763$ Å). The electron diffraction pattern in Figure 6.11c is similar to those for samples with domain microstructure. Indeed, study of perovskite by HRTEM confirms this conclusion (Figure 6.12).

The use of synchrotron radiation allows to improve considerably the signal to noise ratio in diffraction patterns. As a result, in a number of SFC-based perovskites strong narrow peaks of

Figure 6.8 Single crystal X-ray diffraction pattern of $SrFe_{0.95}Mo_{0.05}O_{3-x}$ sample after annealing in reducing atmosphere

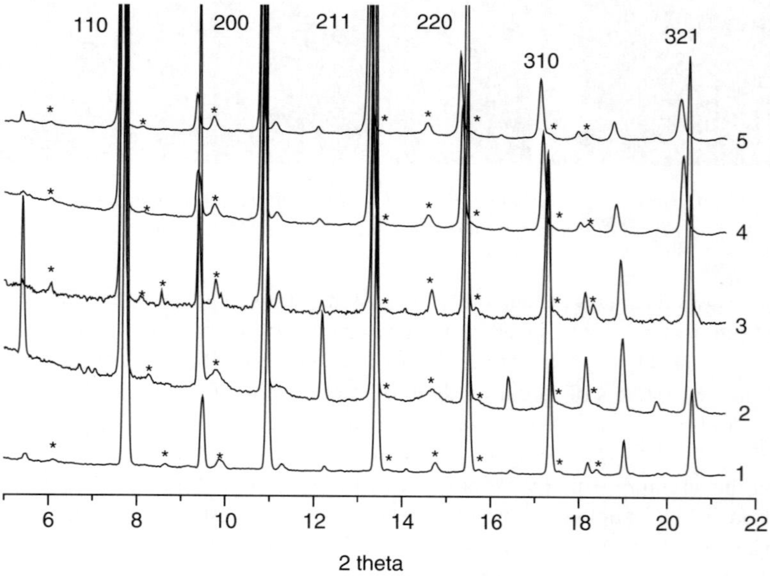

Figure 6.9 X-ray diffraction patterns (synchrotron radiation, $\lambda = 0.3675$ Å) of vacuum quenched perovskites: (1) $SrCo_{0.8}Fe_{0.2}O_{2.5+x}$, (2) $La_{0.3}Sr_{0.7}Co_{0.5}Fe_{0.2}Al_{0.3}O_{2.5+x}$, (3) $SrCo_{0.75}Fe_{0.2}Nb_{0.05}O_{2.5+x}$, (4) $SrCo_{0.7}Fe_{0.2}Nb_{0.1}O_{2.5+x}$, (5) $SrFe_{0.95}Mo_{0.05}O_{2.5+x}$

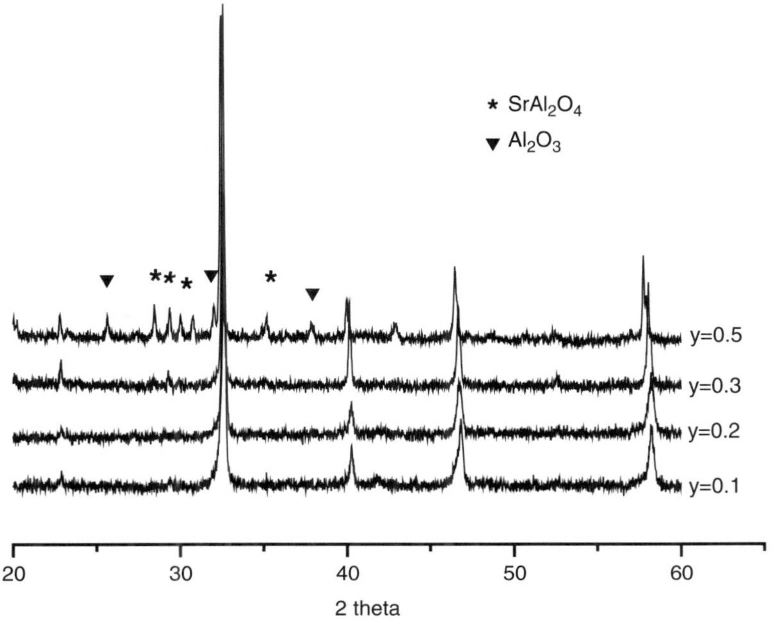

Figure 6.10 *XRD patterns (CuK_α) of $SrCo_{0.8-y}Al_yFe_{0.2}O_{2.5+\delta}$ solid solutions*

cubic perovskite (space group Pm3m, $a_p \sim 3.8$ Å) were observed simultaneously with a weak broadened extra peaks which can be attributed to double perovskite supercell with $a^* = 2a_p \sim 7.6$ Å related to 90° domain texture accommodating compositional variations and oxygen nonstoichiometry (Figure 6.11). The estimation of crystallites sizes from FWHM of extra peaks using Sherrer's formula gives a value ~20 nm which is consistent with HRTEM data, Figure 6.12.

It is believed, that nanostructuring of some perovskite ceramics is retained even at $T \sim 1200$ K. The direct study of structure by HRTEM is possible as a rule only at low temperatures, while operation temperature for membrane materials is relatively high. Preservation of domain microstructure at high temperatures is not obvious. In Figure 6.13 XRD patterns of perovskite $SrFe_{0.95}Mo_{0.05}O_{2.57}$ at various temperatures are presented. For this study, diffractometer D8 Bruker and high-temperature camera HTK-16 with control of atmosphere (vacuum/inert atmosphere) were used. Some extra peaks indexed in supercell a*=2ap and assigned to nanostructuring are observed up to $T \sim 1100$ K, but several peaks are visible even at $T \sim 1400$ K.

Table 6.2 *Unit cell parameter and oxygen content of $Sr_{1-x}M_xCo_{0.8-y}Al_yFe_{0.2}O_z$ perovskites*

M	x/y	Unit cell (Å)		z	
		Slow cooling	Quenching	Slow cooling	Quenching
Sr	x = 0; y = 0.1	3.873	3.914	2.66	2.52
	x = 0; y = 0.2	3.886	3.911	2.60	2.49
	x = 0; y = 0.3	3.888	3.910	2.58	2.45
La	y = 0.3; x = 0.3	3.821	3.848	2.68	2.54
	y = 0.3; x = 0.5	3.827	3.858	2.69	2.51

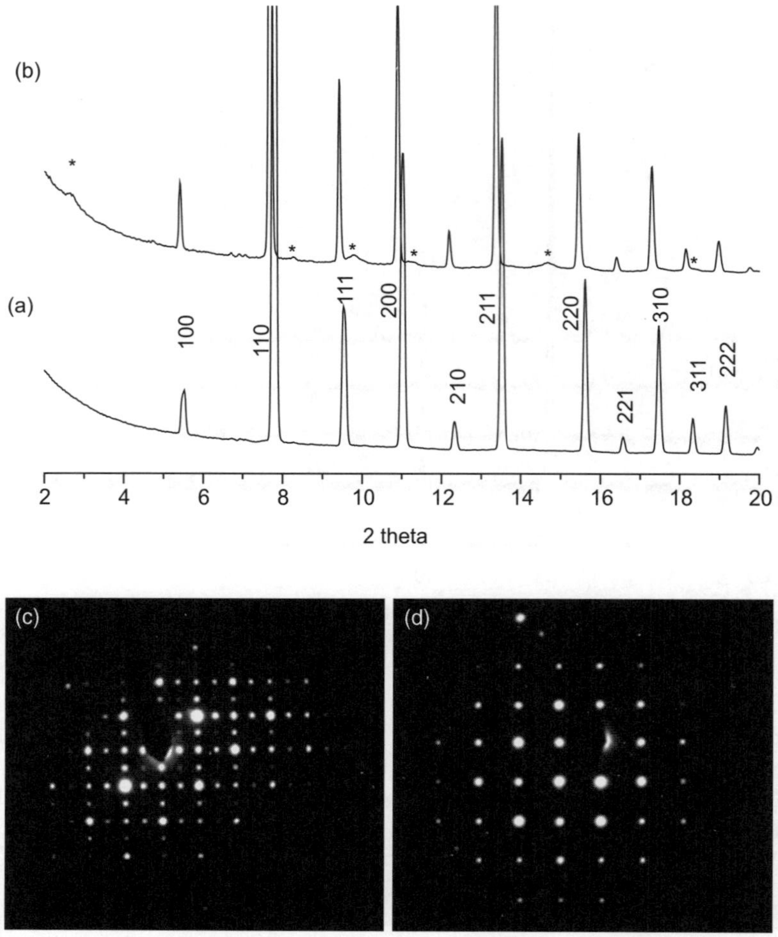

Figure 6.11 *X-ray (synchrotron radiation, $\lambda = 0.3675$ Å) and electron diffraction patterns for perovskite $Sr_{0.7}La_{0.3}Co_{0.5}Al_{0.3}Fe_{0.2}O_{3-z}$: (a) and (d) slow cooling in air, (b) and (c) vacuum quenching from 1173 K*

Thus, MIEC nonstoichiometric perovskites display nanostructuring phenomena resulting in formation of domains separated by interfaces. Domain walls/boundaries essentially differ from domain bulk by the local structure and composition [36]. The activation energy for the oxygen diffusion along domain boundaries is decreased [37], which provides high diffusivity paths at least at low temperatures. Hence, the synthesis of MIEC ceramics with a stable nanostructuring at $T \sim 1200$ K may be considered as necessary criteria to obtain membrane materials with a high oxygen flux.

6.4 Synthesis of Ceramics with Required T_s and a High Oxygen Permeability

Mechanochemical ceramic technique has been chosen for synthesis of ceramic nanopowders for membrane materials [38]. This technique is described in details in [39]. The technology of

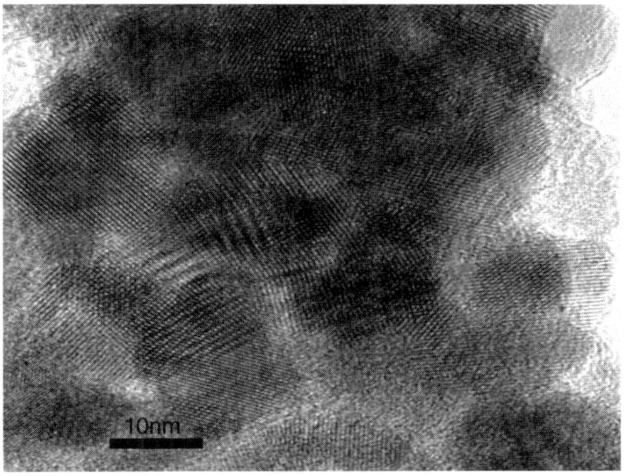

Figure 6.12 *HRTEM image of perovskite $Sr_{0.7}La_{0.3}Co_{0.5}Al_{0.3}Fe_{0.2}O_z$ after vacuum quenching from 1173 K*

ceramic nanopowders preparation, including synthesis of complex oxides with melting points up to 3000 K at ambient temperature, is based on high energy planetary mill and electromass classifier for separation to fractions in a closed volume. To diminish drastically a contamination and to enhance a uniformity of powders, a special procedure of mechanical treatment was developed including a coating of steel working surface of the balls and a vial by the same powder material and preserve this protective coating during milling by short time intervals ~30–150 s of

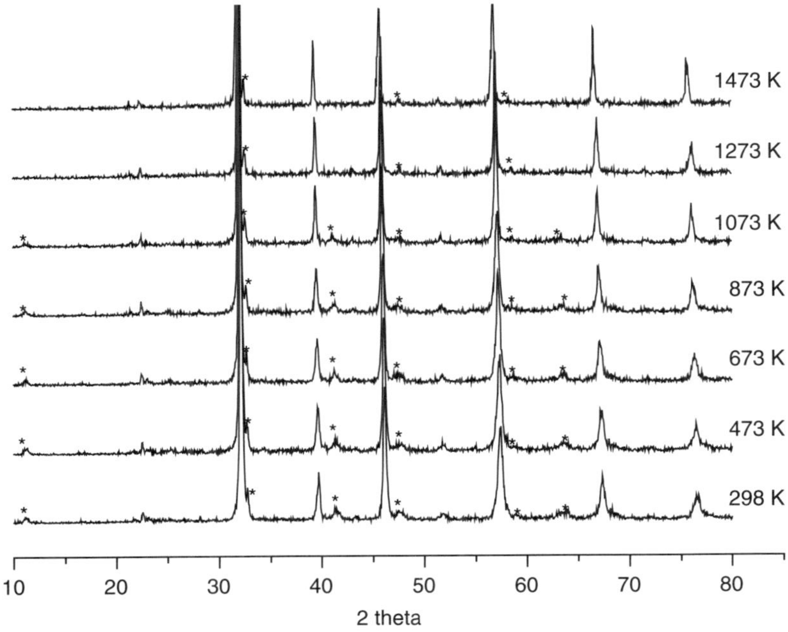

Figure 6.13 *XRD patterns (CuK_α) of perovskite $SrFe_{0.95}Mo_{0.05}O_{2.57}$ under heating in vacuum*

continuous treatment with a stop and remixing of powder/ball loading. Just unique features of mechanochemical nanopowders provide opportunity of development of asymmetric multilayer ceramic membranes. As received, mechanochemical powders represent an inhomogeneous mixture of strongly aggregated particles. Aggregates, consisting from nanocrystallites supersaturated by vacancy defects, have size $<2\,\mu m$ and relative density RD ~ 0.8. Agglomerates (RD ~ 0.7) with particles size distribution 2–100 μm represent secondary formations from aggregates. Submicron fraction of aggregates possessing the best molding ability and a high sintering activity is the most suitable for obtaining dense ceramics under relatively mild sintering conditions [40]. Fractions of agglomerates are the most appropriate for manufacturing supporting porous ceramic layers in multilayer membranes. It is very important that linear shrinkages of different fractions are very similar in spite of big differences in particles sizes.

For tailored synthesis of perovskites with required T_s, it is quite necessary to find a simple empirical rule for its estimation. However, temperature of sintering of single phase powders up to gas-tight ceramics strongly depends on a number of parameters: melting point of compound T_m, particle size, granulometric composition, atmosphere, green density, and so forth. In addition, no correlation was observed between T_m for perovskites ABO_3 and calculated mean melting point $T^*_m = {}^1\!/_2[T_m(AO_{1+x}) + T_m(BO_{1+y})]$. Nevertheless, in the case of submicron fraction separated from single phase complex perovskites obtained by mechanochemical synthesis we have found out a suitable linear correlation between $T^*_m = {}^1\!/_2[(1-x-y)T_m(A^1O_{1+a}) + xT_m(A^2O_{1+b}) + yT_m(A^3O_{1+c}) + (1-u-w)T_m(B^1O_{1+a}) + uT_m(B^2O_{1+b}) + wT_m(B^3O_{1+c})]$ and T_s (12%) – sintering temperature up to 12% shrinkage of pressed disks at 100 MPa [29]. After small correction of melting points for nickel and cobalt oxides (+600 and –500 K respectively), linear correlation with $k = 0.76$ was received in study of sintering of 32 complex perovskites. The choice of mentioned parameters allowed to decrease the influence of extraneous parameters to dependence $T_s \sim T^*_m$. Moreover, similar correlation was obtained for other structural types as well – fluorites and layered perovskites (Figure 6.14).

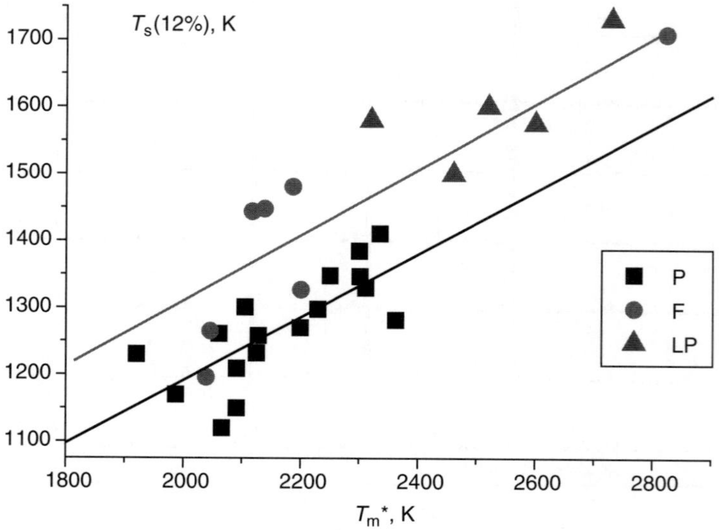

Figure 6.14 *Correlation between calculated T_m* for complex perovskites (P), fluorites (F) and layered perovskites (LP) and sintering temperature up to 12% shrinkage*

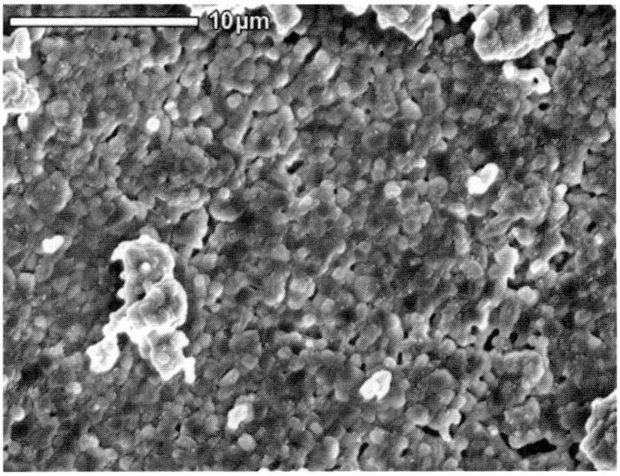

Figure 6.15 *SEM image of gastight ceramics $La_{0.4}Sr_{0.6}Fe_{0.9}Ni_{0.05}Co_{0.05}O_{2.65}$ derived from submicron fraction of aggregates separated from mechanochemical powder*

Formulated simple rule allows to calculate the composition of complex perovskites with required T_s obtained by mechanosynthesis. In such a way two perovskites with compositions $La_{0.6}Sr_{0.4}Fe_{0.9}Ni_{0.1}O_{2.8}$ and $La_{0.4}Sr_{0.6}Fe_{0.9}Ni_{0.05}Co_{0.05}O_{2.65}$ with $T_s = 1620$ and 1600 K – compatible with developed glass–ceramic substrate, were calculated and then obtained by mechanosynthesis to be used for manufacturing of microporous and dense ceramic layers, respectively (Figure 6.15). Dense ceramics sintered from submicron fraction of aggregates possesses a regular microstructure with mean grain size ~1 μm (Figure 6.15). However, ceramics from mechanochemical nanopowders retains nanostructuring even after high temperature sintering (Figure 6.16) [30]. Obtained by ultrafast mechanosynthesis at ambient temperature for 10^{-1} s of total mechanical loading, complex perovskites very often display a cubic structure in XRD patterns and nanostructuring in HRTEM images due to a composite structure with more ordered core and strongly disordered (up to amorphous state) shell [38, 43]. Even after

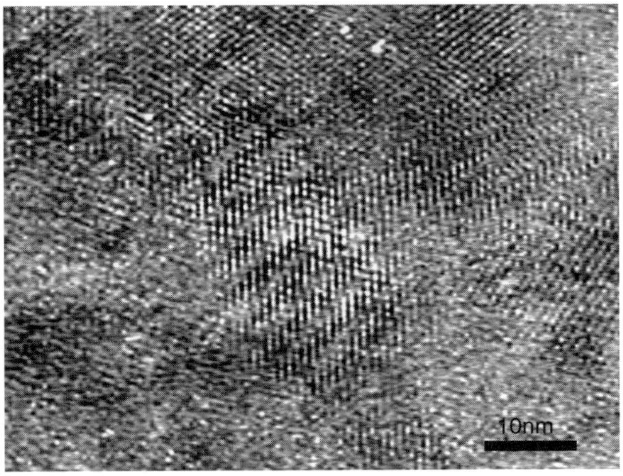

Figure 6.16 *Typical HRTEM image of perovskite $La_{0.7}Sr_{0.3}Co_{0.8}Fe_{0.2}O_{3-\delta}$ ceramics from mechanochemical powders sintered at 1400 K*

high-temperature thermal treatment perovskites keep a memory about nanostructuring in the form of nanoscale domains due to compositional inhomogeneity. However, we can not assert unequivocally that similar structure is beneficial for the oxygen conductivity at $T \sim 1000$ K.

The nanostructuring has different bases. Actually all phenomena aggregated by one term nanostructuring are a simple consequence of a complex technical problem in the characterisation of nanocrystalline solids. For example, as a result of intensive mechanical treatment of solid electrolyte $Zr_{0.88}Sc_{0.1}Ce_{0.01}O_{2-x}$ in conditions when mechanosynthesis occurs, it is formed as nanostructured powder. The best ceramics sintered from this powder has twice as low specific conductivity at $T \sim 1000$ K as compared with the reported data [44]. In the range of low temperatures $T \sim 500$ K, nanostructured Sc-stabilised zirconia ceramics obtained from mechanochemical powders show an ultrahigh conductivity exceeding by 3 orders of magnitude the levels of conductivity for ceramics obtained by traditional procedures [45]. We believe that the study of interrelation between reasons resulting in nanostructuring of a material, and oxygen-ion conducting properties is very important for the progress in fabrication of ceramic membranes with high permeability.

Under certain conditions, enhanced mobility of ions due to the presence of mobile electrons (holes) in MIEC materials (conjugate diffusion) could take place [46]. Optimal conditions for the maximal effect ($t_i \sim 0.1$–0.2) are realised in MIEC ceramics at low temperatures. In other words, anomalously high oxygen mobility in nonstoichiometric nanostructured perovskites can be caused by a combination of factors, and not just by formation of pathways for the easy diffusion along nanodomain boundaries and other extended defects.

Nanostructuring of oxygen conductors results in enhanced oxygen mobility at low temperatures due to a developed network of pathways with decreased activation energy. At high temperatures, irregularities (grain and domain boundaries, other extended defects) could act as barriers to ion conductivity. For instance, such conclusion was reached by molecular dynamics study of Y-stabilised zirconia [47].

For development of a MR with a high oxygen flux, materials with a high rate of oxygen incorporation into the lattice and high oxygen mobility are required for porous and dense layers respectively [48]. Hence, optimisation of multilayer supported membrane requires selection of perovskites possessing a high rate of surface oxygen exchange and/or oxygen mobility provided these materials are chemically compatible and have similar T_s. Some rules for determination of MIEC compounds with chemical compatibility were formulated in [49]. These rules are taken already into account in the new architecture of membranes (Figure 6.2, right). Many efforts were devoted to estimation of surface exchange rate k_s and oxygen diffusion coefficients D_{chem} in MIEC materials [50–55]. However, agreement in published data obtained by different techniques for materials with close compositions is not good enough. We studied the oxygen loss from dense (RD = 0.98–0.99) ceramic disks in He atmosphere by thermal gravimetric analysis technique. To prepare homogeneous samples, sintered ceramic disks were preliminary annealed in air for 8 h at temperatures of desorption experiments with subsequent slow cooling. Such experiments in relatively simple and available equipment Netzsch 4 allow to obtain basically kinetic curves of the weight loss for dense specimens with well defined geometry with reasonable accuracy provided uniformity of a sample. The fitting of kinetic curves gives evaluation of k_s and D_{chem}. However, in experiments where the initial step of the oxygen loss is described by a mixed mode of kinetics, k_s cannot be evaluated with a reasonable accuracy mainly due to transient processes at changing P and T [48]. The typical kinetic curve of the weight loss of ceramic disks has two characteristic areas intersected approximately at a point of a half-reaction (Figure 6.17). The initial stage corresponds to propagation of diffusion profiles from two sides of a disk up to their meeting. The kinetics at this stage in approximation of infinite plate is described by the formula:

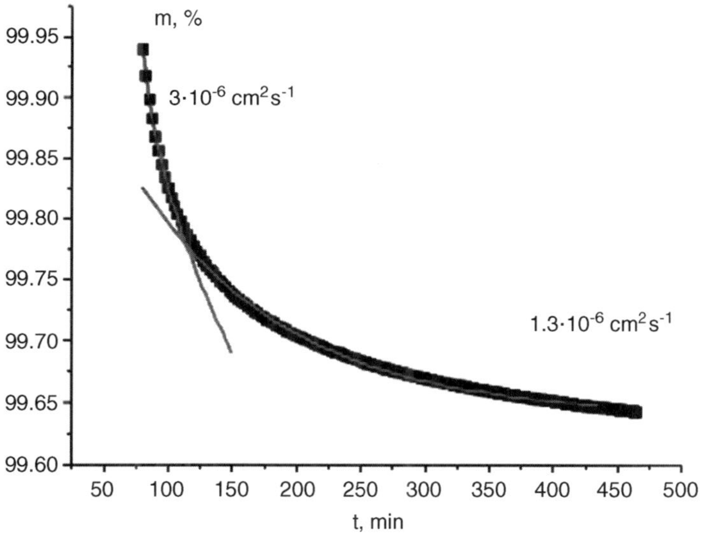

Figure 6.17 *Kinetic curve of reduction at 1073 K of complex perovskite with composition* $La_{0.25}Sr_{0.75}Fe_{0.35}Co_{0.15}Ga_{0.40}Al_{0.05}Mg_{0.05}O_{3-x}$, $d = 2.98$ *mm*

$$\Delta M = 4\Delta M_{\max}\sqrt{\frac{Dt}{\pi d^2}},$$

where ΔM is the change of a mass of a sample, $\Delta M_{\max}$ is the full change of mass, d is the thickness of a disk. At the final stage the kinetics is described by the formula:

$$\Delta M = \Delta M_{\max}\left\{1-\exp\left[-\frac{\pi^2 D(t-t_0)}{d^2}\right]\right\},$$

where t_0 is the efficient starting time of the oxygen removal process [48].

Values of diffusivity determined by kinetic data processing for 10 complex perovskites fall within a range of ±30% with lowering in the final stage of reduction [48]. For all studied perovskites – candidates for membrane materials with calculated T_s, a weak temperature dependence of D_{chem} was observed. Moreover, D_{chem} estimated by this approach strongly depends on the thickness of disk, Figure 6.18. This can be explained by a stronger deviation of the oxygen desorption process kinetics from a simple diffusion model for thinner pellets. Pronounced dependence of oxygen chemical diffusion coefficients on the perovskite stoichiometry controlled by the oxygen partial pressure could be important as well [56]. In addition, we observed a strong influence of small (10–100 ppm) contamination of starting reagents by the natural admixture of sulfur on the morphology, oxygen surface exchange, permeability (not shown for brevity) and mechanical strength of ceramics. On the surface of dense ceramic disks the formation of $SrSO_4$ film was observed directly by SEM with EDS analysis, Figure 6.19. Such specimens are often cracked after cooling. The usual image, that is, without formation of visible $SrSO_4$ film, for the same ceramics was observed for specimens with RD = 0.9–0.95 [48]. Hence, degree of segregation of sulfur on the surface of ceramics depends on RD and T_s (melting point of $SrSO_4$ = 1823 K). In a bi-layer membrane (Figure 6.20), the formation of microcracks with

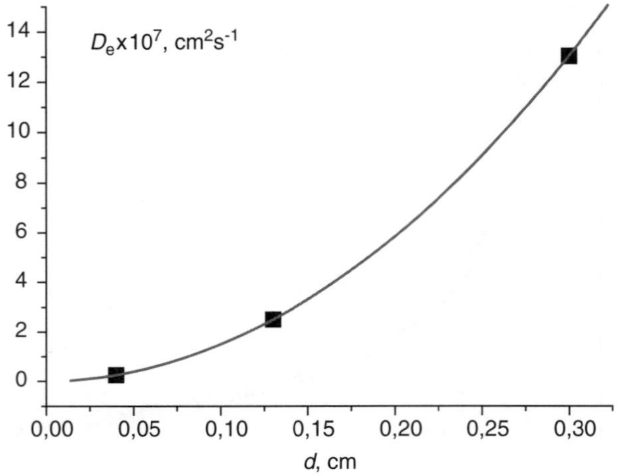

Figure 6.18 *Effect of thickness of ceramic disk with perovskite composition $La_{0.25}Sr_{0.75}Fe_{0.35}Co_{0.15}$-$Ga_{0.40}Al_{0.05}Mg_{0.05}O_{3-x}$ on oxygen D_{chem} at 1073 K, final stage. Curve describes a dependence $D_{chem} \sim d^2$*

subsequent loss of gas-tightness may be related not only to shrinkage misfit and inhomogeneity of ceramic coating, but to poor mechanical properties of ceramics due to sulfate segregation.

These facts help to understand mentioned above disagreements in kinetics of the oxygen loss and probably to explain the discrepancy between k_s and D values reported in the literature [50–55]. Starting from S-free reactants of another origin, for instance, $Sr(NO_3)_2$ and FeO_{1+x} prepared by oxidation of pure Fe metal powder, we received more encouraging results in MR production. The yield of MR with almost gas-tight membrane noticeably increased. Perovskite ceramics $La_{0.4}Sr_{0.6}Fe_{0.9}Ni_{0.05}Co_{0.05}O_{2.65}$ for dense ceramic layer with calculated

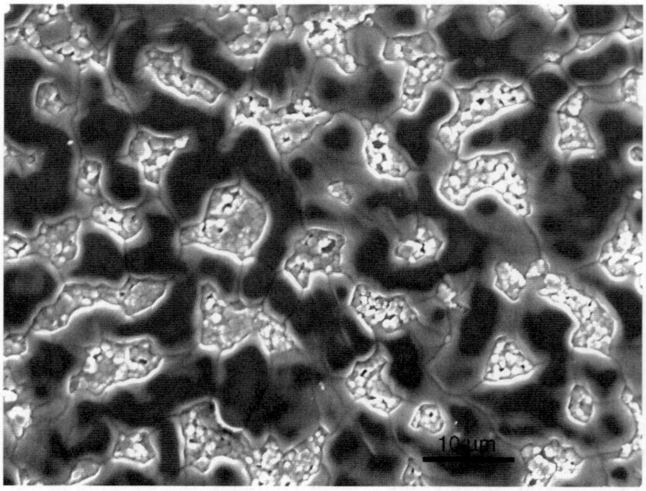

Figure 6.19 *SEM image of ceramics with composition: $La_{0.25}Sr_{0.75}Fe_{0.85}Co_{0.05}Ni_{0.1}O_{3-x}$, sintered at 1600 K, RD = 0.99, chemical composition of ceramic windows correspond to synthesised ceramics, and composition of surface matrix film – $SrSO_4$*

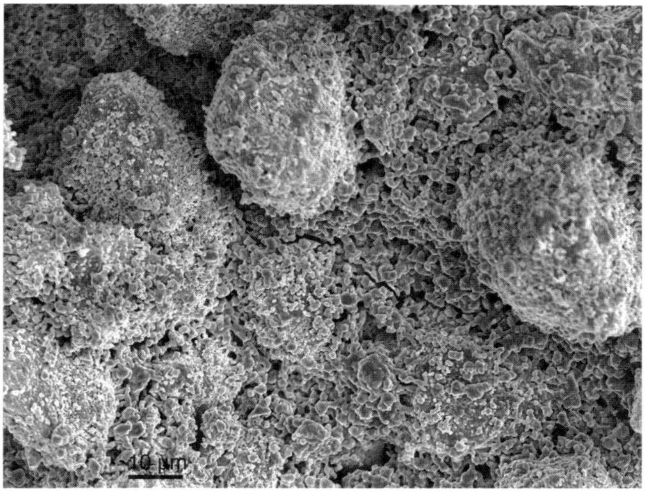

Figure 6.20 *SEM image of bilayer membrane with microcrack. S content on the surface 0.67(9) wt% according to EDS analysis*

$T_s = 1603$ K is shown in Figure 6.15. Indeed, in such a case SEM images are typical for oxide ceramics and EDS analysis display relatively small S contents on the surface of specimens. The remnants of S may be related to starting oxides Ni_2O_3 and CoO having admixtures of sulfur as well.

6.5 Combination of Compatible Materials and Operations

To collect a combination of fully compatible membrane materials and simultaneously to develop a flowsheet of MR fabrication is a challenge. A number of important parameters required for optimisation exceeds 200 [57]. In Figure 6.21 shrinkage curves for one possible combination are shown in different modes of thermal treatment. Shrinkage misfits for membrane materials measured separately and in a stack are presented in Figure 6.22 Measurement of shrinkage of the deposited thin ceramic layers remains unsolved problem. Shrinkage of green ceramic layers depends on the green density, cohesion and chemical interaction with other layers, anisotropy. Final optimisation of combinations is possible only in full stacks. The modes of preliminary firing, final sintering up to gas-tight top ceramic layer and annealing to achieve optimal nanostructuring of MIEC perovskites are the most sophisticated tasks.

Different fields in shrinkage misfit diagram are shown in Figure 6.22: (a) a field of wave-like surface of ceramics, (b) the most appropriate field with sintering conditions similar to 2-dimensional hot pressing, (c) a field of pulling stresses, (d) a dangerous field with a possibility of delamination and cracking of ceramic membrane. During a final firing of MR, a small co-agreement of shrinkages between substrate and thin ceramic layers can be observed for nanostructured perovskites with increasing of field (c). For instance, shrinkage value of free support at $T_s = 1600$ K was 14.0%, and shrinkage of MR with deposited ceramic layers after firing at the same conditions was 13.6%.

Cracking of membrane after cooling is shown in Figure 6.23. Mechanical stresses result from inevitable mismatching in TECs between substrate and ceramic layers. The solution of

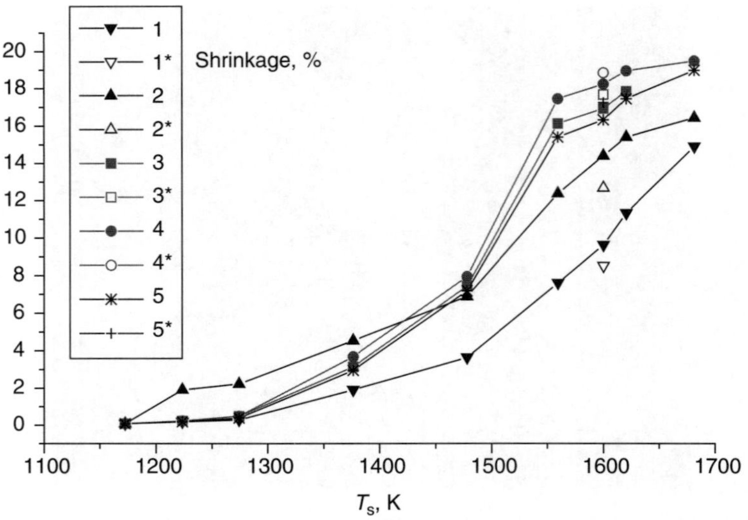

Figure 6.21 *Linear shrinkage curves for specimens after 1 min of sintering at maximal T: (1) pressed disk from submicron fraction of* $La_{0.6}Sr_{0.4}Fe_{0.9}Ni_{0.1}O_{3-x}$ *aggregates for porous ceramics layer after annealing at 950°C, (2) pressed disk from submicron fraction of* $La_{0.4}Sr_{0.6}Fe_{0.9}Ni_{0.05}Co_{0.05}O_{3-x}$ *aggregates for dense ceramics layer, (3) three-layer CMR on modified support, (4)* La_2O_3 *sol modified support, (5) unmodified support, * after one step firing*

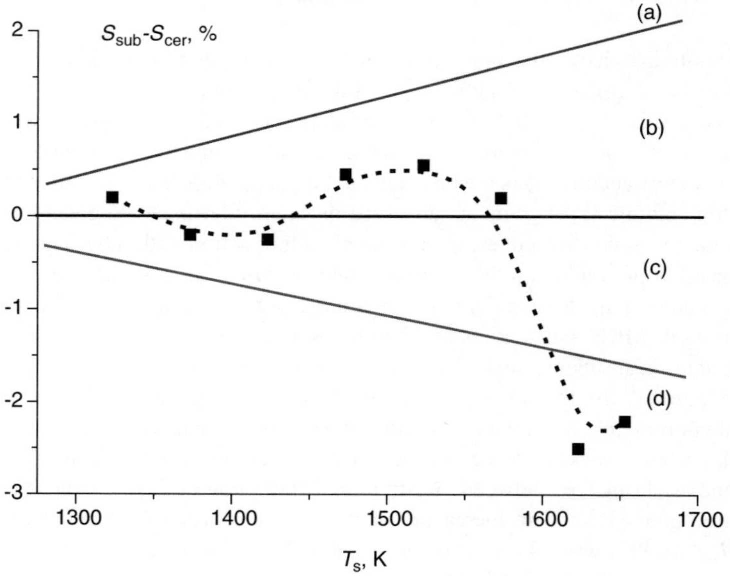

Figure 6.22 *Shrinkage misfits between substrate and pressed disks from ceramic nanopowders (for description, see text)*

Figure 6.23 *SEM image of three-layer membrane with top protective layer from composite*

membrane cracking problem can be found in a ways of increasing mechanical strength of ceramic membrane (decreasing S admixture in starting reactants, optimisation of thickness of ceramic layers, optimisation of porosity of microporous ceramic layer contacted with substrate for accommodating stresses), and optimisation of sintering and annealing modes. The glass–ceramic substrate possesses a weak viscosity at $T \sim 1000\,\mathrm{K}$ that can alleviate residual mechanical stresses.

The optimisation of chemical compositions of materials in their combination and technological operations for realisation of nanostructured ceramics with fast oxygen diffusion and required permeability at intermediate temperatures is in progress.

6.6 Design of Catalyst for Selective Reforming of Methane to Syngas

Preliminary results of tests for some reactors in the partial oxidation of methane (POM) process were obtained (Table 6.3) [5]. As a rule, tests show a relatively low selectivity to partial oxidation reaction, especially for reactors with the top layer from LSGM-based perovskites. Rather good results were obtained for reactors with protective layer comprised of ceria-based nanocomposites (9 in Figure 6.2) and catalysts developed in [6, 58]. A microspherical catalyst loaded into the MR strongly improved syngas selectivity of POM process. The optimisation of catalysts and procedures of its deposition on multilayer membranes are in progress.

The development of multilayer MR including low cost operations and small consumption of expensive reagents could be helped by the use of microspherical catalysts like magnetospheres [59]. The content of magnetospheres usually is $\sim$1–2% in fly ashes – combustion product of pulverised coals in power plants. However, their total resource is about 1 million t year^{-1} only in Russia [60]. Magnetospheres display selectivity in POM reaction [59]. Magnetospheres of a different origin present nanocomposites consisted mainly from nonstoichiometric spinel and hematite in silica glass matrix (Figure 6.24). The spinel composition strongly

Table 6.3 *The methane conversion tests for MRs with deposited catalyst*

MR Number	Catalyst	Input[a] content (%)		Output content at 1153 K (%)						CH_4 conversion (%)	
		CH_4	O_2	CO_2	CO	H_2	H_2O	O_2	CH_4	Total	To syngas
19	$Pt/Ce_{0.45}Zr_{0.45}La_{0.1}O_2$	3.4	1.7	0.72	0.42	0.92	1.5	0.43	2.33	31	16
75	$Pt/Ce_{0.45}Zr_{0.45}La_{0.1}O_2$	3.0	0.7	0.38	0.16	0	1.34	0.67	1.3	57	0
110	$Pt/LaNiO_3/Ce_{0.4}Zr_{0.4}La_{0.2}O_{2-x}$/+ microspherical catalyst	1.0	0.5	0.40	0.03	0.01	—	0.62	0.55	45	2
110	$Pt/LaNiO_3/Ce_{0.4}Zr_{0.4}La_{0.2}O_{2-x}/\gamma\text{-}Al_2O_3$	1.0	0.5	0.16	0.60	0.65	—	0.22	0.30	70	60

[a]Inlet feed 1.0–3.4% CH_4 in He; flow 2.78 $cm^3\,s^{-1}$. Oxygen content in the feed determined by the oxygen flux through membrane is estimated by passing pure He flow with the same flow rate at 1153 K.

Figure 6.24 *Magnetospheres produced in different power plants*

depends on the origin of magnetospheres and could vary between $Fe^{3+}_{0.97}Al_{0.03}[Fe^{2.5+}_{1.56}Mn^{2+}_{0.05}Mg_{0.02}Al_{0.17-0.2}]O_{3.775}$, $Fe^{3+}_{0.87}Mg_{0.07}Al_{0.06}[Fe^{3+}_{0.31}Fe^{2.5+}_{0.82}Mn^{2+}_{0.24}Mg_{0.19}Al_{0.19}Ca_{0.17}Ti_{0.08}]O_4$, $Fe^{3+}_{0.81}Mg_{0.15}Al_{0.04}[Fe^{3+}_{0.64}Fe^{2.5+}_{0.88}Mn^{2+}_{0.11}Mg_{0.30}Al_{0.07}]O_4$ and so forth [60]. Very cheap selective catalysts from separated and modified magnetospheres may help commercialisation of MR technology.

6.7 Conclusion

Research and development activity in the field of catalytic MR is based upon application of advanced nanomaterials and technologies of their manufacturing. Despite tremendous efforts worldwide in this area, a lot of critical scientific and technological problems remain unsolved. This includes: (i) design of nanostructured perovskites with a high oxygen mobility at intermediate temperatures; (ii) selection of materials with a high k_s for effective incorporation of oxygen into the lattice; (iii) co-sintering of thin green ceramic films deposited on porous substrates (including those based on a glass matrix); (iv) development of selective catalyst for the partial oxidation of methane by the lattice oxygen transferred through membrane. We shall note also importance of purity of reagents for design of MIEC ceramic membranes. However, all specified problems are conjugated, so they are to be solved simultaneously. This makes the solution of this problem – the design of advanced nanostructured materials and devices – a demanding task due to its interdisciplinary character.

Acknowledgement

This work is supported by RFBR, project 09-03-00364.

References

1. A. Basile, F. Gallucci, L. Paturzo. *Catal. Today*, **104**, 251–259 (2005).
2. V. Thangadurai, W. Weppner. *Ionics*, **8**, 360 (2002).
3. V.V. Zyryanov, N.F. Uvarov, V.A. Sadykov,*et al. Catal. Today*, **104**, 114–119 (2005).
4. V.V. Zyryanov.(in Russian) *Constructions from Composite Materials*, **1** 32–49 (2007).
5. V.V. Zyryanov, V.A. Sadykov, G.M. Alikina. *Sep. Sci. Technol.*, **42** (13), 2849–2861 (2007).
6. V.V. Zyryanov, V.A. Sadykov. In: *Abstr. 14th Intern. Congress on Catalysis, July 13-18,* COEX, Seoul, p. 390 (2008).
7. A.P. Safronov, E.G. Kalinina, Y.A. Kotov, A.M. Murzakaev, O.R. Timoshenkova. *Russian Nanotechnologies*, **1** (1/2), 162–169 (2006).
8. V.V. Zyryanov.Russian Federation Patent 2349373 (2009).
9. V.V. Zyryanov, M.S. Melgunov. *Inorg. Mater.*, **44** (10) 1130–1134 (2008).
10. V.V. Zyryanov, L.G. Karakchiev. *Inorg. Mater.*, **44** (4), 429–437 (2008).
11. F.A. Kroger. *The chemistry of imperfect crystals*. Wiley & Sons, Inc., New York, 1039 pp, (1964).
12. P. Kofstad. *Nonstoichiometry, Diffusion, and Electrical Conductivity in Binary Metal Oxides*. Wiley-Interscience, New York (1972).
13. H.J. Bouwmeester, A.J. Burggraaf. *Dense ceramic membranes for oxygen separation*, in: A.J. Burggraaf, L. Cot (eds), *Fundamentals of Inorganic Membrane Science and Technology*, Elsevier, Amsterdam, p. 435 (1996).
14. G.P. Kostikova, Y.P. Kostikov. *Chemical processes under doping of oxides*. Sankt-Peterburg, St Petersburg, 156 pp. (1997).
15. M.H.R. Lankhorst, H.J.M. Bouwmeester, H. Verweij. *J. Solid State Chem.*, **137**, 555–567 (1997).
16. S.M. Ariya, Y.G. Popov. *Russ. J. Gen. Chem.*, **32**, 2077 (1962).
17. J.S. Anderson. *The thermodynamics and theory of nonstoichiometric compounds*, in: A. Rabenau (ed.), *Problems of Nonstoichiometry*, North-Holland, Amsterdam, p. 1 (1970).
18. M.A. Alario-Franco, J.M. Gonzalez-Calbet, M. Vallet-Regi, J.-C. Grenier. *J. Solid State Chem.*, **49**, 219 (1983).
19. J.-C. Grenier, N. Ea, M. Pouchard, P. Hagenmuller. *J. Solid State Chem.*, **58**, 243 (1985).
20. E.K.H. Salje, S.A. Hayward, W.T. Lee. *Acta Cryst.*, **A61**, 3–18.

21. D.I. Savytskii, D.M. Trots, L.O. Vasylechko, N. Tamura, M. Berkowski. *J. Appl. Cryst.*, **36**, 1197–1203 (2003).
22. N. Orlovskaya, A. Nicholls, N. Browning. *Acta Materialia*, **51**, 5063 (2003).
23. N. Orlovskaya, Y. Gogotsi, M. Reece, B. Cheng, I. Gibson. *Acta Materialia*, **50**, 715–723 (2002).
24. J. Canales-Vazquez, F.M. Figueiredo, J.C. Waerenborgh, W. Zhou, J.R. Frade, J.T.S. Irvine. *J. Solid State Chem.*, **177**, 3105–3113 (2004).
25. F. Lindberg, G. Svensson, S.Ya. Istomin, S.V. Aleshinskaya, E.V. Antipov. *J. Solid State Chem.*, **177** 1592–1597 (2004).
26. J. Drennan, G. Auchterlonie. *Solid State Ionics*, **134**, 75–87 (2000).
27. E. Goldberg, A. Nemudry, V. Boldyrev, R. Schollhorn. *Solid State Ionics*, **110**, 223–233 (1998).
28. A. Nemudry, M. Weiss, I. Gainutdinov, V. Boldyrev, R. Schöllhorn. *Chem. Mat.*, **10**, 2403–2411 (1998).
29. A. Nemudry, P. Rudolf, R. Schoellhorn. *Chem. Mater.*, **81**, 2232–2238 (1996).
30. A. Nemudry, E. Goldberg, M. Aguirre, M.A. Alario-Franco. *Solid State Sci.*, **4**, 677 (2002).
31. I.L. Zhogin, A.P. Nemudry, P.V. Glyanenko, Y.M. Kamenetsky, H.J.M. Bouwmeester, Z.R. Ismagilov. *Catal. Today*, **118**, 151–157 (2006).
32. O.A. Savinskaya, A.P. Nemudry, N.Z. Lyakhov. *Inorg. Mater.*, **43** (12), 1350–1360 (2007).
33. E.B. Pyatiletova, A.P. Nemudry, N.Z. Lyakhov. *Inorg. Mater.*, **43** (10), 1134–1141 (2007).
34. A.A. Markov, O.A. Savinskaya, M.V. Patrakeev, A.P. Nemudry, I.A. Leonidov, Yu.T. Pavlyukhin, A.V. Ishchenko, V.L. Kozhevnikov. *J. Solid State Chem.*, **182**, 799–806 (2009).
35. O.A. Savinskaya, O. Hernandez, M. Ceretti, W. Paulus (in preparation).
36. J. Novak, J. Fousek, J. Maryska, M. Marvan. *Mater. Sci. Engin.*, **B 120**, 13–15 (2005).
37. W.T. Lee, E.K.H. Salje, U. Bismayer. *J. Phys.: Condens. Matter.*, **15**, 1353–1366 (2003).
38. V.V. Zyryanov. *Russ. Chem. Reviews*, **77** (2), 105–135 (2008).
39. V.V. Zyryanov. *Interceramica*, **52** (1), 22–26 (2003).
40. V.V. Zyryanov. *Sci. Sintering.*, **37**, 77–92 (2005).
41. V.V. Zyryanov, V.A. Sadykov. *Desalination*, **199** (1/3), 299–301 (2006).
42. V.V. Zyryanov, N.F. Uvarov, V.G. Kostrovskii, V.A. Sadykov, *et al. Mat. Res. Soc. Symp. Proc.*, **755**, DD.6.27.1 (2003).
43. V.V. Zyryanov. *Inorg. Mater.*, **41** (4), 438–464 (2005).
44. V.V. Zyryanov, N.F. Uvarov, A.S. Ulihin, V.A. Sadykov. *Mat. Res. Soc. Symp. Proc.*, **2009**, 1122-O03-10 (2009).
45. V.V. Zyryanov, N.F. Uvarov, V.A. Sadykov. *Glass Phys. Chem.*, **33** (2), 394–401 (2007).
46. M.A. Korzhuev. *Inorg. Mater.*, **34** (4), 504–508 (1998).
47. C.A.J. Fisher, H. Matsubara. *Solid State Ionics*, **113/115**, 311–318 (1998).
48. V.V. Zyryanov. *Asia-Pac. J. Chem. Eng.*, **4**, 285–290 (2009).
49. V.V. Zyryanov, V.A. Sadykov, N.F. Uvarov, G.M. Alikina, A.I. Lukashevich, S. Neophytides, J.M. Criado. *Solid State Ionics*, **176**, 2813–2818 (2005).
50. J. Maier. *Solid State Ionics*, **135**, 575 (2000).
51. J.A. Lane, J.A. Kilner. *Solid State Ionics*, **136/137**, 997 (2000).
52. S. Kim, Y.L. Yang, A.J. Jacobson, B. Abeles. *Solid State Ionics*, **106**, 189 (1998).
53. R. Bredesen, F. Mertins, T. Norby. *Catal. Today*, **56**, 315 (2000).
54. R.A. Souza. *Phys. Chem. Chem. Phys.*, **8**, 890 (2006).
55. D.S. Mebane, M. Liu. *J. Solid State Electrochem.*, **10**, 575 (2006).
56. M. Katsuki, S. Wang, M. Dokiya, T. Hashimoto. *Solid State Ionics*, **156**, 453–461 (2003).
57. V.V. Zyryanov, V.A. Sadykov, G.M. Alikina. *Sep. Sci. Technol.*, **42**, 2849–2861 (2007).
58. T.G. Kuznetsova, V.A. Sadykov, S.A. Veniaminov, *et al. Catal. Today*, **91/92**, 161 (2004).
59. A.G. Anshits, E.V. Kondratenko, E.V. Fomenko, *et al.*, *Catal. Today*, **64**, 59–67 (2001).
60. V.V. Zyryanov, D.V. Zyryanov. *Fly Ash – Technogenic Raw* (in Russian). Moscow (2009).

7

Compact Catalytic Membrane Reactors for Reforming Applications Based on an Integrated *Sandwiched* Catalyst Layer

Sreekumar Kurungot[1] *and Takeo Yamaguchi*[2]

[1]*Physical and Materials Chemistry Division, National Chemical Laboratory, Pune, India*
[2]*Chemical Resources Laboratory, Tokyo Institute of Technology, Yokohama, Japan*

7.1 Introduction

The subject of catalytic membrane reactors has attracted great attention especially with the tremendous boost in the development of novel and highly selective inorganic membranes. The process controllability to achieve particle distribution and pore openings down to nanosize has helped the researchers to come up with a large number of materials with excellent perm-selectivity capable of achieving separation of a verity of gases with different molecular weights [1–9]. Integration of these types of membranes with various catalysts by maintaining proper membrane–catalyst layer interface has been a subject of great technocommercial importance in the context shifting the chemical equilibrium in thermodynamically limited reactions. Such catalytic membrane reactors (CMR) combine the separation properties of membranes with the characteristics typical of catalytic reactions. A great number of reviews have come out on CMRs with an extensive description on their utility in a verity of applications, including dehydrogenation reactions (cyclohexane to benzene, ethane to ethylene, etc.), hydrogenation reactions (butadiene, acetylene, etc.), oxidative dehydrogenation reactions (1-butene to

Membranes for Membrane Reactors: Preparation, Optimization and Selection, Edited by Angelo Basile and Fausto Gallucci

butadiene, ethanol to acetaldehyde, etc.) and catalytic decomposition reactions (hydrogen sulfide, hydrogen iodide, etc.) [10–20].

In view of the existing energy crisis and the anticipated shift towards hydrogen economy, there has been a great interest in developing CMRs for producing hydrogen from the naturally abundant hydrogen feedstock such as CH_4. The commercially established processes meant for hydrogen generation such as steam reforming (STR), catalytic partial oxidation (CPO) and autothermal reforming (ATR), have severe limitations imposed by the thermodynamic equilibrium of the associated reactions [21–23]. A variety of CMRs has been developed to study the reforming reactions. These include steam reforming of CH_4 and other hydrocarbons, CO_2 reforming of CH_4 and also water–gas shift reaction [24–26]. In all these modes of reforming reactions, the selective removal of hydrogen with the help of appropriate hydrogen perm-selective membrane, with proper integration with the catalytic reactor unit, has been the key concept of enhancing reactant conversion by controlled alteration of the chemical equilibrium. However, insufficient perm-selectivity or permeance and thermochemical stability of membranes are often appeared to be the prevailing issues which question the credibility of such systems for catering various commercial requirements.

The development of CMRs for hydrogen generation has been a subject of one of our research areas in the past several years. As a practical way to circumvent the thermodynamic limitations, we have developed a compact CMR system based on the concept of direct integration of the catalyst layer with the hydrogen perm-selective silica membrane to function as a single unit with bifunctional characteristics. Microporous silica has been evolved as an efficient alternative to Pd systems due to its cost competitiveness and permeation characteristics [27,28]. Silica, in the form of a polymeric sol, is an excellent precursor to cast silica films on various substrates while retaining the required permeation characteristics. By taking a significant deviation from the conventional style of integrating a packed catalyst bed with a hydrogen perm-selective membrane, our strategy of integrating the reforming catalyst as a layer along with the hydrogen perm-selective membrane is expected to improve process efficiency and compactness at reduced cost. The work reported by Tsuru *et.al.* is worth mentioning here as the work is a direct deviation from the conventional approach of making CMRs [28]. The researchers dispersed NiO inside the pores of α-Al_2O_3 support tubes and subsequently a surface layer of silica membrane was formed by dip coating. However, NiO was prone to fast deactivation even at high steam/C ratios. In the present case, we have intentionally selected Rh catalyst, instead of NiO, due to its proven better coke resistance and high thermal stability due to low metal loading (usually $< 2\%$ metal) [29,30]. This significantly low metal loading is an added advantage in our system because metal incorporation can be done without creating large order deformation of the substrate morphology, leading to significant variations in pore size, porosity and surface roughness.

Since Rh supported on Al_2O_3 has been widely studied as a potential catalyst for converting CH_4 by STR and POX, we have developed an Rh incorporated γ-Al_2O_3 as a catalyst layer formed on a porous α-Al_2O_3 substrate tube in our integrated CMR unit. Further, the silica membrane is formed on the surface and in this way the Rh-γ-Al_2O_3 catalyst layer is virtually sandwiched between the membrane and α-Al_2O_3 substrate tube. As a further modification to this approach, we have also modified the catalyst layer with redox species such as CeO_2 to impart better oxygen binding capacity to the parent Rh-γ-Al_2O_3 system [31–34]. In addition to their proven ability to store and release oxygen, CeO_2 type materials stabilise alumina support and improve the dispersion state of precious metals [35–37]. Some of our previous publications highlight the advantages of such systems with special emphasis on system integration and performance evaluation [38,39]. Here, we provide a detailed illustration of the compact membrane reactors developed in our lab based on the aforementioned concept, by critically evaluating their merits in terms of activity and scope for wider applications. Silica as a material is weak at higher temperature and humid atmosphere. However, we have taken silica membrane as a model system for fabricating the compact CMR as it is a system which can be fabricated relatively easily. The durability of γ-Al_2O_3 under humid conditions and high temperatures is also a matter of concern in real operation conditions and durations. In general, the CMR developed from the aforementioned systems is only a model system which can provide useful information about the scope of the design under consideration, which will eventually lead to development of better systems by integrating promising materials and designs. Further to the results provided in the previous papers by these authors, more detailed information on the morphological control, structural integration and stability fine tuning will be provided to get a balanced outlook on these types of concept on compact CMRs for a wide spectrum of commercial requirements.

7.2 Experimental

7.2.1 Preparation of Silica-Rh-γ-Al_2O_3 Catalytic Membrane

A standard recipe reported by Burggraaf was used for making the boehmite (γ-AlOOH) sol [40]. 1 wt% each of $RhCl_3.2H_2O$ and poly(vinyl alcohol) (PVA) were added to the boehmite sol and appropriate sol concentration was maintained either by evaporation or by diluting with deionised water to get a Rh concentration of 1 wt% in 0.6 mol l^{-1} of boehmite in PVA. The α-Al_2O_3 substrate tube (pore size ~0.1 µm; length 3.5 cm) was procured from Noritake Corporation, Japan and fixed on a dense hollow ceramic tube by glass sealing. The top part of the α-Al_2O_3 tube, mounted on the ceramic tube, was closed and perfectly sealed by using a flat α-Al_2O_3 plate and glass sealing. The Rh incorporated boehmite sol was dip coated on the exposed outer surface of the α-Al_2O_3 substrate tube and subjected to initial drying at 50 °C, followed by calcination at 600 °C at a heating and cooling rate of 25 °C h^{-1}. The dip coating and calcination steps were repeated two times to maintain a crack-free layer of Rh-γ-Al_2O_3 (hereinafter RAL). Silica sol was prepared by acid hydrolysis (HNO_3, 1 mol l^{-1}) and condensation of tetrahydroorthosilicate (TEOS; Aldrich Chemical) in ethanol [41]. The HNO_3 solution was added slowly to a mixture of known dilution of TEOS in ethanol (Wako Pure Chemical Industries). Followed by the initial aging of 30 min; the temperature of the mixture was increased slowly to 70 °C and was refluxed for 2–5 h to get the silica sol of different quality levels. These sols were diluted to a concentration of 0.1 mol l^{-1} using ethanol and the RAL surface was immediately dipped (10 s) and dried at 40 °C in a humidity-controlled furnace with 60% RH. Subsequently, the samples were calcined at 600 °C to obtain thin layer of silica over the catalyst layer in each case. With respect to the duration of polymerisation of 2, 3, 4 and 5 h, as maintained during the experiments, the prepared samples are designated as S2-RAL, S3-RAL, S4-RAL and S5-R-AL, respectively. The course of fabrication of the catalyst layer integrated with the silica membrane can be visualised through the pictorial representation as shown in Figure 7.1. The details of the samples prepared are summarised in Table 7.1. Unless otherwise specified, the processing time of 4 h has been taken as the standard condition throughout the discussion in the following sections.

7.2.2 Preparation of Redox Modified S-RAL Systems

The redox modification of the catalyst layer (R-AL) was achieved by the incorporation of CeO_2 in the layer while keeping the Rh/Ce wt. ratios like 1/0.5, 1/1 and 1/2. This was achieved by

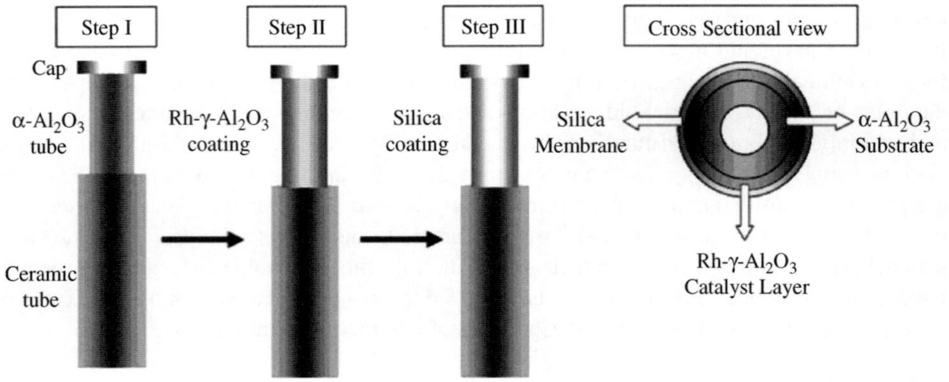

Figure 7.1 *Pictorial representation of the various steps involved in the process of fabrication of the unit consisting of the catalyst layer integrated with a hydrogen perm-selective silica membrane. A cross sectional view of the active region also can be seen*

incorporating stoichiometric amounts of $RhCl_3.2H_2O$ and $Ce(NO_3)_3.6H_2O$ in the boehmite sol prior to the coating on the α-Al_2O_3 tube. Heat treatment and silica membrane coating were done as per the procedure given in the previous section. Since our studies showed that the membrane characteristics were superior when the silica sol processing time was 4 h, this batch of silica sol was dip coated on the redox couple systems to get the respective catalytic membranes with different Rh/Ce ratios. The prepared sampled are designated as S-CRAL-0.5, S-CRAL-1.0 and S-CRAL-2.0, with respect to the Ce content in the systems. The sample designation and details are provided in Table 7.1.

7.2.3 Membrane Reactor

An indigenously developed membrane reactor set-up was used for carrying out permeation studies and reforming experiments. The unit has a furnace with provision for mounting catalytic membrane and has inlets for gas delivery using mass-flow controllers. The feed can be delivered through the tubular part of the CMR and the permeation occurs through the α-Al_2O_3-Rh(Ce)-γ-Al_2O_3-silica interface. The gas permeation rate can be measured by a Baratron pressure sensor

Table 7.1 *Compositions of the catalyst layer and membrane processing durations maintained for developing CMRs*

Sample designation	Catalyst layer composition (wt%)	Silica sol processing time (h)
Normal systems		
S2-RAL	Rh(1%)/γ-Al_2O_3	2
S3-RAL	Rh(1%)/γ-Al_2O_3	3
S4-RAL	Rh(1%)/γ-Al_2O_3	4
S5-RAL	Rh(1%)/γ-Al_2O_3	5
Redox systems		
S-CRAL-0.5	Ce(0.5%) Rh(1%)/γ-Al_2O_3	4
S-CRAL-1.0	Ce(1.0%) Rh(1%)/γ-Al_2O_3	4
S-CRAL-2.0	Ce(2.0%) Rh(1%)/γ-Al_2O_3	4

based on the rate of increase of pressure once permeate side vacuum condition is maintained. The permeation values were standardised with units of mol $m^{-2} s^{-1} Pa^{-1}$. An online gas chromatograph fitted with active carbon column and thermal conductivity detector was used to analyse the permeate and retentate fluxes separately during the CH_4 reforming experiments.

7.3 Results and Discussion

7.3.1 Physical Characteristics

A Hitachi S-900 FESEM was used to scan the α-Al_2O_3 substrate, catalyst layer and the silica membrane thickness and its continuity. Figure 7.2a shows the surface morphology of the α-Al_2O_3 surface. Grains with size distribution 40–60 μm are close-packed to give a relatively smooth surface characteristics. Figure 7.2b gives the cross sectional view of the Rh-γ-Al_2O_3-α-Al_2O_3 interface. The catalyst layer retains an average thickness of 9 μm with good continuity as evident from the figure. The analysis of the cross section of the silica-Rh-γ-Al_2O_3 interface displays excellent continuity of the silica membrane with a thickness of 1.5 μm. This indicates that γ-Al_2O_3 serves as an excellent substrate for forming the silica membrane and Rh incorporation has not imparted any visual changes to affect the continuity of the silica membrane. However, excess CeO_2 incorporation in the γ-Al_2O_3 matrix (2 wt%) shows poor adhesion of the catalyst layer with the α-Al_2O_3 surface, as shown by the detached nature of the catalyst layer in Figure 7.2d. N_2 sorption method was employed to monitor the pore size distribution characteristics, using a Micromeritics surface area analyser. A comparison of the pore size distribution is

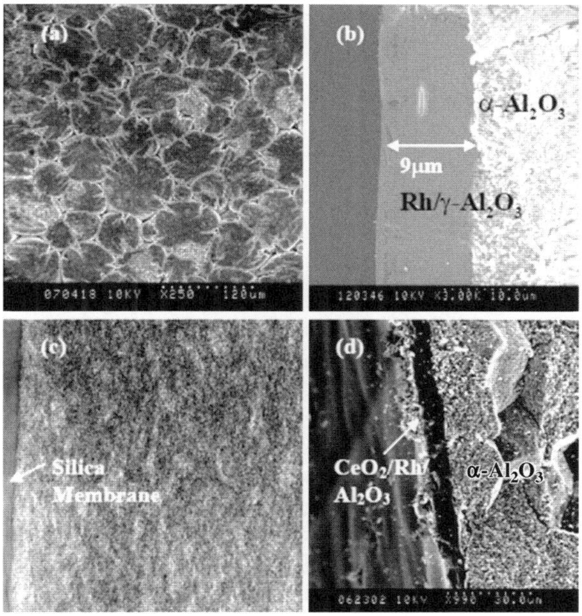

Figure 7.2 FESEM images of: (a) α-Al_2O_3 substrate surface, (b) Rh-γ-Al_2O_3-α-Al_2O_3 interface, (c) silica membrane-Rh-γ-Al_2O_3 interface, (d) detached portion of 2 wt% CeO_2 incorporated Rh-γ-Al_2O_3 layer

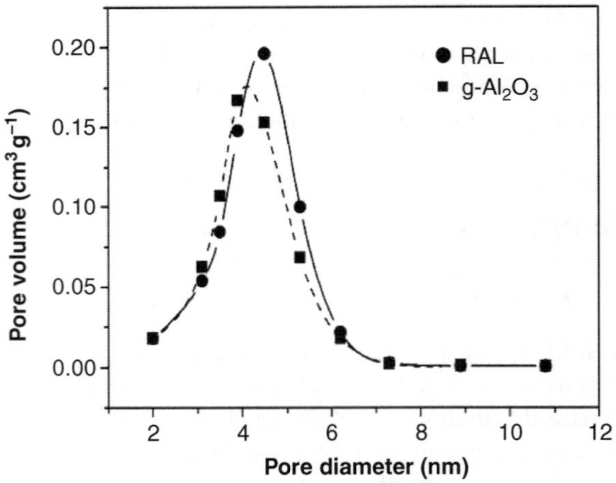

Figure 7.3 *Pore size distribution of γ-Al_2O_3 and Rh incorporated γ-Al_2O_3*

given in Figure 7.3. RAL and γ-Al_2O_3 display average pore diameters of 4.1 and 4.3 nm, respectively, indicating a good dispersion of Rh nano particles in the γ-Al_2O_3 matrix without creating significant microstructure variation. γ-Al_2O_3 sample displayed an average surface area of 272 $m^2 g^{-1}$, whereas, upon Rh incorporation this has shown a marginal improvement to 284 $m^2 g^{-1}$. The system could tolerate CeO_2 incorporation for lower level of loadings and unfavourable changes in pore size distribution occurred at a loading of 2 wt%. Table 7.2 summarises the pore diameter, pore volume and surface area values obtained on various compositions of the catalyst layers. A detailed discussion on the influence of CeO_2 addition on the Rh-γ-Al_2O_3 matrix is made in a previous paper by the same authors [39]. The basic conclusion from our works on CeO_2 doping is that CeO_2 incorporation can dramatically affect the quality and performance of the upper lying silica membrane and, therefore, its amount has to be retained well within the tolerance interval.

7.3.2 Gas Permeation Properties

Our previous studies give detailed information on the permeation characteristics of the CMRs consisting of silica membranes fabricated from the sols prepared by maintaining a polymerisation time of 3 h [38,39]. A maximum separation factor (H_2/CH_4) of 31 has been displayed by the integrated system at 525 °C. However, our continued efforts to improve the quality of the silica

Table 7.2 *Effect of catalyst layer composition on surface area, pore diameter and pore volume characteristics*

Layer composition	BET surface area ($m^2 g^{-1}$)	Pore diameter (nm)	Pore volume ($cm^3 g^{-1}$)
γ-Al_2O_3	270	4.1	0.33
Rh(1.0%)/γ-Al_2O_3	285	4.3	0.36
CeO_2(0.5%) Rh(1.0%)/γ-Al_2O_3	198	5.3	0.32
CeO_2(1.0%) Rh(1.0%)/γ-Al_2O_3	198	5.5	0.33
CeO_2(2.0%) Rh(1.0%)/γ-Al_2O_3	165	6.8	0.55

Table 7.3 *Gas transport data for H_2 and CH_4 at 525 °C over the different catalytic membranes prepared*

System designation	Permeation at 525 °C ($mol\ m^{-2}\ s^{-1}\ Pa$)		Separation factor
	H_2	CH_4	H_2/CH_4
Normal systems			
S2-RAL	4.4×10^{-7}	4.70×10^{-8}	9.4
S3-RAL	3.0×10^{-7}	0.96×10^{-8}	31.0
S4-RAL	2.5×10^{-7}	0.37×10^{-8}	67.6
S5-RAL	4.1×10^{-7}	0.85×10^{-8}	48.2
Redox systems			
S-CRAL-0.5	2.3×10^{-7}	0.35×10^{-8}	66.0
S-CRAL-1.0	2.6×10^{-7}	0.40×10^{-8}	65.0
S-CRAL-2.0	4.0×10^{-7}	0.84×10^{-8}	48.0

membrane lead to a superior membrane with twofold improvement in the perm-selectivity by increasing the polymerisation time to 4 h from the previously determined duration of 3 h. For example this sample which has given the designation S4-RAL, displayed a H_2/CH_4 separation factor of 68 at 525 °C, in comparison to the value of 31 when the polymerisation time was 3 h (S3-RAL). Further increase in the polymerisation temperature, however did not help to improve the permeation characteristics. S5-RAL, the sample with the polymerisation time of 5 h, displayed a H_2/CH_4 separation factor of 45. Therefore, for our detailed experiments on CH_4 reforming in CMRs, S4-RAL system has been employed. All the CeO_2 added systems possess silica membranes casted from the sols prepared under the polymerisation duration of 4 h. A brief illustration of the permeation behaviour of some of the selected systems is given below.

Table 7.3 comprises the permeation data of the catalytic membranes measured at a temperature of 525 °C using H_2 or CH_4 as the feed gas. In the case of the Rh-γ-Al_2O_3 systems, the results clearly reveal the significant role played by the polymerisation time. The H_2/CH_4 separation factor increased from 9.4 to 67.6 when the polymerisation time varied from 2 to 4 h. Further increase in time, however, reduced the separation factor. Generally, the H_2 permeation through silica membrane is considered to be activated transport. Since H_2 has a kinetic diameter of 2.8 Å, in comparison to 3.7 Å possessed by CH_4, the large dependence of polymerisation time on the H_2/CH_4 indicates that the microporous nature of the silica membrane faces a major modification to generate pores having dimensions < 3.7 Å. Considering the mechanism of growth of silica polymers, the hydrolysis and condensation of silica lead to structures with a variety of shapes and sizes depending upon the synthesis parameters. Acidic conditions (pH < 2) generally favour linear or weakly branched structures, whereas basic conditions (pH > 2) give significantly branched structures [42–44]. The polymerisation kinetics has been explained by different aggregation growth models by various researchers [45–47]. To get good microporous silica membranes, the ability of silica polymers to interpenetrate must be considerable. Generally, short-branched linear polymers are the best in this regard. However, at the same time excess inter-penetration must be controlled to prevent formation of dense structure. This in other words means that highly branched systems can lead to inefficient packing with resulting mesoporosity of the gels. There are a number of articles dealing with the branching and inter-penetration of silica polymers [48–51]. In the present case, a polymerisation time of 4 h seemed to impart the optimum microporous nature for the membrane, which helps the system to display

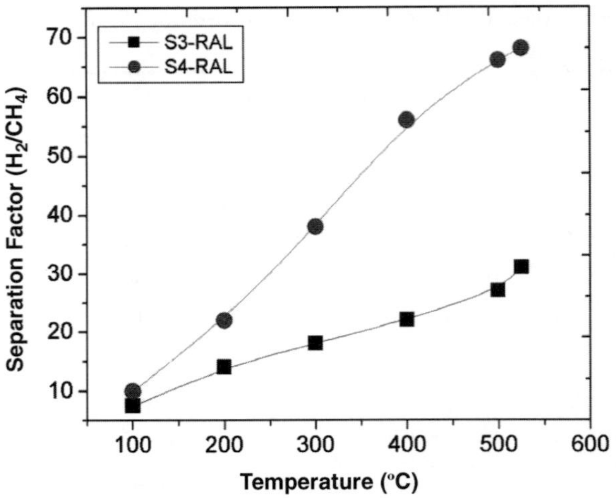

Figure 7.4 *H_2/CH_4 separation factor as a function of temperature on systems consisting of silica membranes fabricated from bases prepared under polymerisation durations of 3 and 4 h*

excellent selectivity towards H_2/CH_4 separation. Further increase in the process time seemed to generate mesoporosity in the system with a concomitant decrease in the separation factor.

The temperature dependence on the permeation behaviour on S3-RAL and S4-RAL is compared in Figure 7.4. In both the systems, the separation factor of H_2 to CH_4 increases with increase in temperature. However, S4-RAL outperforms S3-RAL throughout the temperature region. Since the H_2 transport in silica membrane is activated transport, the greater extent of enhancement in the separation factor in the case of S4-RAL reveals the better microporous characteristics achieved by this system under the process variation. The hydrogen permeation was activated with an apparent activation energy of 6.8 kJ mol^{-1} in the case of S3-RAL and 7.5 kJ mol^{-1} in the case of S4-RAL. The activation energy for permeance is considered as a useful extra parameter to explain the membrane quality; higher hydrogen activation energy in the case of S4-RAL reflects better extent of microporosity achieved by this system. The values of activation energy are roughly in agreement with those reported for high performance silica membranes [52,53].

All the CeO_2 incorporated systems are prepared by integrating silica membranes casted from the sols which gave the highest H_2/CH_4 perm-selectivity as explained in the previous section. It is interesting to note that, 0.5 (S-CRAL-0.5) and 1.0 wt% (S-CRAL-1.0) of CeO_2 incorporation in the γ-Al_2O_3 could be done successfully without abruptly changing the permeation characteristics of the upper casted silica membrane. For example, the original H_2/CH_4 perm-selectivity of 67.6 shows only a marginal decrease to 66.0 and 65.0 upon adding 0.5 and 1.0 wt% CeO_2, respectively, in the γ-Al_2O_3 matrix. However, an abrupt fall in selectivity is observed when the doping amount increased to 2.0 wt% (S-CRAL-2.0). The textural variations suffered by this system, which have already been identified from the difference in the pore size distribution characteristics, might have possibly influenced the quality of the silica membrane with a direct reflection in its permeation characteristics.

7.3.3 Hydrothermal Stability

The hydrothermal stability of the silica membrane is a matter of serious concern especially in the context of making use these systems for reforming applications where steam will be an inevitable

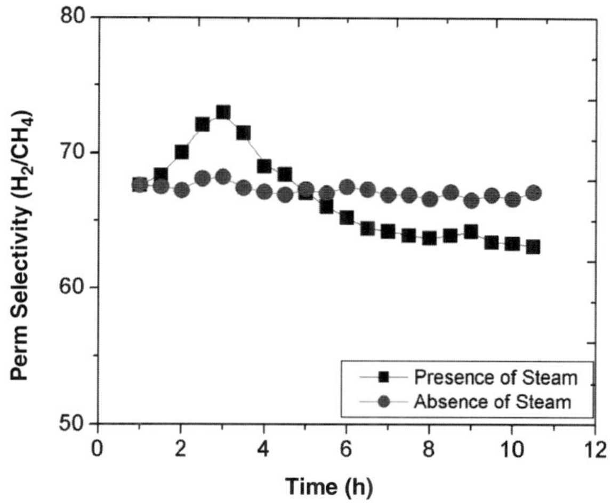

Figure 7.5 *Hydrothermal stability test using S4-RAL by monitoring H_2/CH_4 perm-selectivity as a function of time while maintaining a humid atmosphere in the system*

component. The hydrothermal stability of the silica membrane has been evaluated by using S4-RAL system by subjecting a humid atmosphere at 550 °C while maintaining the feed containing H_2 and CH_4. A H_2:CH_4:H_2O molar ratio of 1:1:3.5 was maintained and the H_2/CH_4 perm-selectivity was measured at intervals of 30 min. An identical experiment under dry condition also was carried out. The results are presented in Figure 7.5. Under the hydrothermal condition, the H_2/CH_4 selectivity shows an initial gradual increase up to 2.5 h. The maximum selectivity obtained at this stage is 73. Thereafter, the perm-selectivity steadily decreases until 6 hours and maintains a relatively slow performance loss. Probably, the swelling or microlevel restructuring of the matrix might have reduced the pore size under the initial hours of exposure of the humid atmosphere and helped the overall system to temporarily achieve textural characteristics suitable for maintaining high perm-selectivity. However, further morphological changes under the exposure conditions played a detrimental role and the perm-selectivity decreases afterwards. Irrespective of all such changes, by looking at the overall changes in the perm-selectivity under the period of investigation, it can be inferred that the system displays relatively good hydrothermal stability and retains fairly high value of perm-selectivity than the anticipated catastrophic failure. Under dry condition, the system displayed steady performance throughout the time period investigated.

7.3.4 Reforming of Methane

All the reforming experiments were carried out by delivering a diluted feed of CH_4 in Ar and air to a volume of 30% and O/C (where 'C' represents carbon in CH_4) molar ratio of 1.0 under atmospheric pressure. An N_2 sweep in the rate of 30 $cm^2\,min^{-1}$ was maintained in the permeate side to facilitate H_2 transfer. The experiments were performed in the temperature range of 400–575 °C by maintaining different H_2O/C molar ratios and contact times. In some cases, along with the tests by using the CMRs based on Rh-γ-Al_2O_3 and CeO_2 doped Rh-γ-Al_2O_3 systems, parallel investigations using only the catalyst layer, without the silica membrane casted on it, was also carried out to explore the role of the H_2 perm-selective membrane in the system.

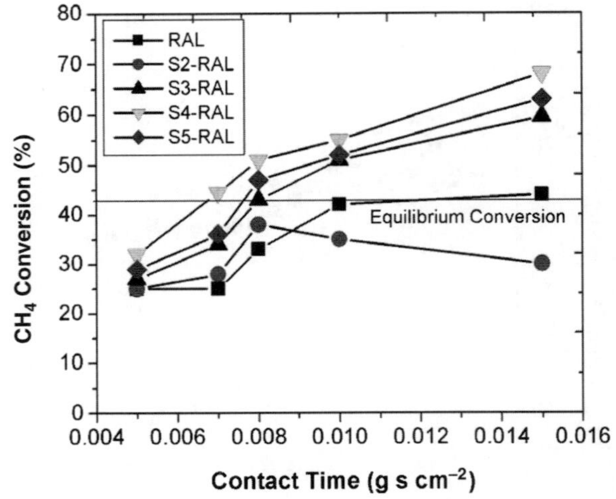

Figure 7.6 *Methane conversion as a function of contact time over silica based compact CMRs containing Rh-γ-Al_2O_3 catalyst layer. Reaction temperature: 525 °C; H_2O/C molar ratio: 3.5*

The basic optimisation of the reaction parameters is carried out by using the systems possessing the Rh-γ-Al_2O_3 catalyst layer. The influence of contact time (the time of contact of the reactant mixture per gram of catalyst) on CH_4 conversion at 525 °C using an H_2O/C molar ratio of 3.5 is depicted in Figure 7.6. From the performance profile, it can be inferred that the activity of the systems for CH_4 conversion is significantly influenced by the quality of the silica membrane. S4-RAL, which possesses the highest H_2/CH_4 perm-selectivity, outperforms all other systems. A general trend as reflected from the overall performance profile of the systems is that a high contact time always favours the reaction and helps the systems to deliver performances higher than the equilibrium restriction. In the case of S4-RAL, the CH_4 conversion, which is marginally higher than the equilibrium conversion level of 43% at a contact time of 0.007 g s^{-1} cm^{-3}, shows an overall improvement of 28% from the equilibrium value at a contact time of 0.01 g s^{-1} cm^{-3} leading to a net conversion of 55%. The CH_4 conversion further increases with increase in the contact time and the highest value of 68% has been obtained at a contact time of 0.015 g s^{-1} cm^{-3}, corresponding to 58% improvement from the equilibrium value. The quality of the silica membrane is a critical matter here. The S5-RAL, which displays a H_2/CH_4 separation factor of 48 gives a conversion value of 63% and S3-RAL with the separation factor of 31 leads to a conversion value of 59% at the contact time of 0.015 g s cm^{-3}. S2-RAL, on the other hand, failed to display useful performance improvement. Surprisingly, the performance even falls down to a level lower than RAL, where no silica membrane is integrated. A possible reason could be the pealing off of silica membrane parts in this system since the polymerisation time might be too short to maintain the required stable morphological characteristics. As the silica layer peels off, the catalyst particles adhered along with the membrane parts also can be delaminated, leading to a progressive decrease in the number of active sites. In all the cases, with further increase in contact time, we observed color change of the membrane from light yellow to gray, expected to be due to carbon deposition into the micropores as a result of increased rate of CH_4 deposition as a side reaction at very high contact time. On the other hand, no improvement from the equilibrium conversion is observed when the contact time is significantly low (0.005 g s cm^{-3}). In the absence of the silica layer, the conversion is either lower than or very close to the equilibrium value in the contact time region 0.005–0.015 g s cm^{-3},

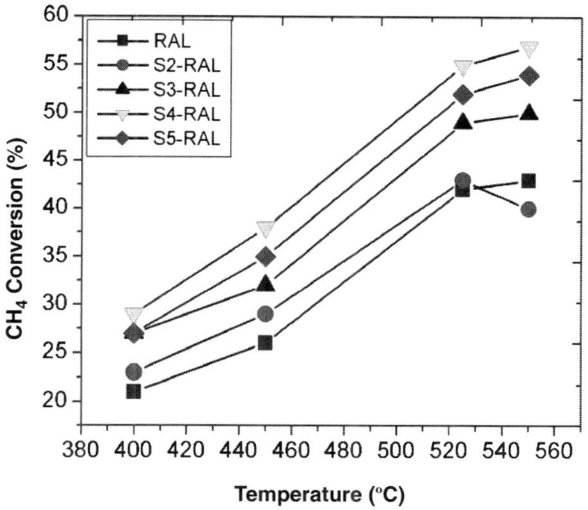

Figure 7.7 *Effect of reaction temperature on methane conversion during reforming of methane. H_2O/C molar ratio: 3.5; contact time: 0.01 g s cm^{-1}*

implying that the Knudsen selectivity possessed by the material in the absence of the silica membrane is not sufficient to drive the reaction above the thermodynamic equilibrium level. Thus, the aforementioned results successfully demonstrate the advantages of the integrated CMR unit for maintaining improved performance characteristics.

The CH_4 conversion activities as a function of reaction temperature are shown in Figure 7.7. The reaction temperature was varied from 400 to 550 °C, while maintaining the H_2O/C molar ratio of 3.5 and contact time of 0.01 g s cm^{-3}. It should be noted that we maintained a contact time lower than that of the value where we got maximum performance with an aim of minimising any possible deactivation effects, which otherwise will counteract the results since the experiment requires prolonged exposure time to cover the selected temperature window of 400–550 °C. Temperature shows a marked difference in the performance characteristics of the CMRs under study. The ones which possess integrated silica membranes with high H_2/CH_4 perm-selectivity display greater extent of activation with rise in the operating temperature. Especially, S4-RAL and S5-RAL display significantly higher performance at higher operating temperature, possibly facilitated by the activated transport of H_2 by the silica membrane as temperature increases. S3-RAL even though shows a performance always higher than the one without any silica membrane, the extent of improvement always limited to 5–6% in the temperature window of 400–525 °C. It is interesting to note that the margin of the performance improvement of all the aforementioned systems decrease progressively as the temperature goes beyond 525 °C. In addition to the activated transport facilitated by the silica membranes, the performance improvement with increase in temperature can be attributed to the possible increase in the resistance for Knudsen transport of the feed mixture, leading to an increase in the residence time of the reactants at the active sites with rise in temperature. Both the α-Al_2O_3 support and Rh/γ-Al_2O_3 intermediate layers play a dominant role in this context. A probable reason for the decrease in the extent of performance improvement at temperatures above 525 °C could be the drastic drop in water vapour permeation to a degree larger than the drop in CH_4 permeation in par with the Knudsen transport mechanism. Thus, the system could not maintain the required H_2O:C ratio to furnish higher conversion rates.

7.3.5 Stabilisation Effect by CeO_2 Incorporation

Even though the aforementioned discussion on the reforming of CH_4 in a compact sandwiched type CMR reveals the possibility of developing a potential system to redefine the performance levels higher that imposed due to thermodynamic restrictions, unfortunately, the extent of improvement from the equilibrium value did not display a consistent level. Generally, the performance steadily decreases after a few hours of stable performance. It has been already seen in a previous section that the change in the permeation properties of the silica membrane, under prolonged exposure to the reaction conditions, is relatively small. Therefore, the major cause of efficiency drop can be presumed to be due to the deactivation of the intermediate Rh/γ-Al_2O_3 catalyst layer. We have noticed higher deactivation rate at contact times higher than 0.015 g s cm^{-3}, implying the deposition of carbonaceous products as a possible root cause for the deactivation along with possible textural variations such as particle aggregation and associated reduction in the number of active sites for the reaction. The following sections illustrate the promising role played by CeO_2 as a promoter into the system to circumvent the deactivation related issues up to a large extent. In all the cases of the CMR compositions, a silica sol polymerisation time of 4 h has been maintained owing to the excellent permeation characteristics of the membrane derived from this sol. Generally, the discussion will be focused on S-CRAL-0.5 and S-CRAL-1.0, mainly due to their matching permeation characteristics to that of the unpromoted parent system (S4-RAL).

Figure 7.8 illustrates the CH_4 conversion as a function of time on stream over the catalytic membranes S4-RAL, S-CRAL-0.5 and S-CRAL-1.0 under the reaction conditions of contact time: 0.010 g s cm^{-3}, reaction temperature: 525 °C and H_2O/C ratio: 3.5. Here, we have intentionally avoided the contact time of 0.015 g s cm^{-3}, which leads to maximum CH_4 conversion, in order to maintain a relatively mild condition at the reactive interface of the sandwiched catalyst layers. S4-RAL, which displays an initial conversion of 51%, retains its performance level for 4 h and subsequently shows slow activity degradation. At the end of the 12 h run, the conversion reached to 43%, leading to a 16% fall from the initial value. On the other hand,

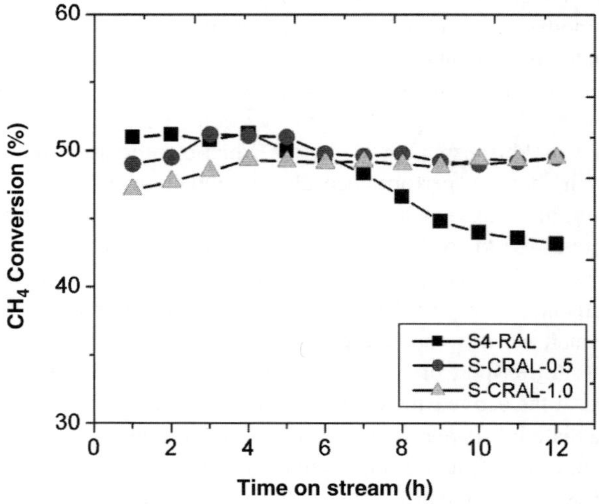

Figure 7.8 *Effect of time on stream on methane conversion during the reforming of methane. Reaction temperature: 525 °C; H_2O/C molar ratio: 3.5; contact time: 0.01 g s cm^{-1}*

S-CRAL-0.5, showed an initial performance of 49%, which is marginally lower than that of the performance displayed by S4-RAL. Interestingly, the performance shows a slight increase in the initial 2–3 h to 50% and maintained a relatively steady performance thereafter. S-CRAL-1.0, on the other hand, displayed an initial conversion level of 47%, followed by an increase of ~ 2% and retained a steady performance.

Coke deposition has been considered as the major culprit behind the deactivation of reforming catalysts which are exposed to hydrocarbon feeds. In par with many reported results on hydrocarbon reforming and recent results on reforming in CMRs, S4-RAL has been appeared to be vulnerable to deactivation caused by coke deposition under the experimental conditions. In the present case of a system involving an integrated CMR unit possessing sandwiched type formation of coke detrimentally affects the overall performance both by reducing the catalytic activity and permeation characteristics of the perm-selective membrane. We have noticed even faster deactivation of S4-RAL system at contact times higher than 0.010 g s cm^{-3} and H_2O/C molar ratios lower than 3.5. Other reasons like change of oxidation states of the catalytically active components and sintering and aggregation of particles can also result in catalyst deactivation. Development of the CeO_2 doped CMRs and enhanced stability obtained on these systems are significantly important in this context as this approach opens up great scope for fine tuning the stability characteristics of these types of materials. The role of CeO_2 in enhancing the stability, especially for noble metal systems, has been well studied and accounted on the basis of the effect of CeO_2 in modifying the thermal and structural stability of the catalyst carriers, dispersion of the supported metal and the oxygen storage and release capacities and so forth [54–60].

Looking at the performance profile of various systems as presented in Figure 7.8, one can realise that CeO_2 doping, even though has a profound influence in improving the stability pattern of the Rh/γ-Al_2O_3 system, has a detrimental effect in terms of maintaining the catalytic activity of the parent system based on its CH_4 conversion ability. This difference in terms of the conversion abilities of the systems with respect to the Rh/Ce ratios can be accounted on the basis of the differences in the potentialities of Rh and Ce towards the reaction. The incorporation of CeO_2 leads to progressive replacement of Rh with Ce or partial coverage of Rh_2O_3 particles by CeO_2 with a concomitant reduction in the activity of the system. The low surface areas of the promoted systems, as explained in a previous section, could also be another reason for the reduced activity of the promoted systems. At the same time, CeO_2 incorporation helps the system in another way by making the system capable for surviving better in the reaction environment. Thus a controlled interplay by the catalytically active component and additive helps to better define the overall performance of the system in terms of both CH_4 conversion activity and ability to survive in a drastic reaction environment. However, the existence of the additive on the active surface interface and the scope for further improvement to obtain promoted systems without making a penalty in terms of the number of active sites available to catalyse the reaction is a subject of extended study. The literature deals with fairly good explanations on the promotional influence of ceria and related systems on Rh catalysts meant for partial oxidation reactions [61,62]. In a previous paper, the authors have given a detailed explanation the promotional influence of CeO_2 in the catalyst layer based on a Mars–van Krevelen type redox cycle formed in the system by the mutual assistance of CeO_2 as a promoter and Rh as the active site of the catalyst [39]. The strong partial oxidation property of Rh together with its high CH_4 adsorption property facilitates decomposition of CH_4, leading to CO and H_2. This has been considered as the prominent reaction pathway under stoichiometric feed ratio. Even though consecutive oxidation of CO has been anticipated, the comparatively higher activation energy for the reaction for Rh (20 kcal mol^{-1}) compared to that of Pt (2.5 kcal mol^{-1}) does not highly favour consecutive

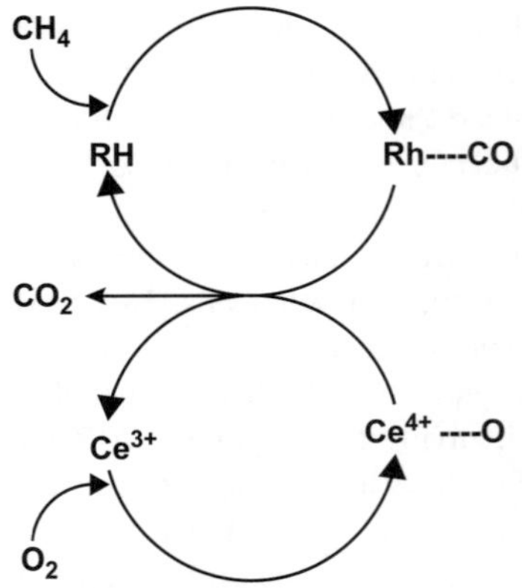

Figure 7.9 *Schematic representation of how ceria facilitates oxygen transport in a redox couple system*

reactions. The plausible routes of dissociative adsorption of CH_4 and the consequent reactions on Pt and Rh as specific examples have been well explained by Freni and coworkers [63]. The CH_4 adsorption and consecutive reduction and reoxidation of the metal oxide sites can be accounted on the basis of the Mars–van Krevelen type redox cycle, wherein the adsorbed CH_4 reduces the metal oxide, which gets reoxidised by the oxygen from the feed [64,65]. The role of ceria becomes influential at this context as it can serve as an oxygen reservoir and mobilise oxygen transport due to the fact that both Ce^{3+} and Ce^{4+} are stable, allowing the oxide to shift between CeO_2 and CeO_{2-x} [66]. A redox cycle established by the intervention of the CeO_2 moiety is shown in Figure 7.9. The released lattice oxygen reacts with CH_4 and CO and, thereby, the system retains better capability to survive under oxygen rich–lean transitions and fluctuations.

The results revealed through the above sections reflect the wide scope for the design aspects of the compact CMR unit considered for this study. Both silica and γ-Al_2O_3 are even though a part of this model system, may not be considered as the practical choice materials for developing CMRs for commercial applications, where durability of several years is often the matter of consideration. However, the development of this model system and the evaluation data unambiguously reveal the effectiveness of the integrated structure comprising of the perm-selective membrane and the 'sandwiched type' catalyst layer for simultaneously achieving system compactness and higher efficiencies. Durable membranes based on Pd and inorganic composites will better account the durability issues in real practical systems.

7.4 Conclusion

Efficient and compact catalytic membrane reactors (CMRs) can be developed by the judicious selection of the catalyst layer composition and appropriately integrating it with perm-selective membranes based on the nature of the reaction and system requirements. In the case of the compact CMR unit developed here for performing reforming of methane, the quality of the hydrogen perm-selective silica membrane and incorporation of a species which can improve the availability of oxygen in the reforming catalyst layer matrix appeared to be important requirements to simultaneously maintain activity and durability. Silica membranes with higher H_2/CH_4 perm-selectivity and good permeation characteristics help to push CH_4 conversion well beyond the equilibrium conversion limit. The Rh-γ-Al_2O_3 catalyst layer appeared to be sensitive to deactivation under the reforming environment. However, presence of CeO_2 as a dopant has significantly enhanced the durability of the system, probably due to the kinetic and oxidative stabilisation of the catalyst matrix with CeO_2. Oxygen mobilisation in the presence of CeO_2 as a promoter might have helped the catalyst matrix to maintain adequate oxygen concentration while concomitantly reducing the possibility of poisoning the active sites by adsorbed intermediates and carbonaceous residues.

Acknowledgement

The authors gratefully acknowledge NEDO, Japan, for financial support.

References

1. S. Fujisaki, K. Hataya, T. Saito, S. Arai, Y. Iwamoto, K. Kuroda, *J. Mater. Res.*, **24**, 372 (2009).
2. L. Luen, D.S. Lee, J. Tsai, *Am. Ceram. Soc.*, **82**, 2796 (1999).
3. S. Kim, G.R. Gavalas, *Ind. Eng. Chem. Res.*, **34** (1), 168.
4. C. Nistor, S. Shishatskiy, M. Popa, S.P. Nunes, *Sep. Sci. Tech.*, **44**, 3392 (2009).
5. H. Bai, W.S.W. Ho, *Ind. Eng. Chem. Res.*, **48** (5), 2344 (2009).
6. H. Ohya, K. Masaoka, M. Aihara, Y. Negishi, *J. Membr. Sci.*, **146** (1), 9 (1998).
7. S. Olsen, B. Tripathi, H. Mill, E. Gobina, *Membr. Tech.*, **2002** (6), 6 (2002).
8. J. S. Oklay, K. Hou, R. Hughes, *Appl. Catal. A*, **170**, 13 (1998).
9. K.A. Peterson, C.S. Nielsen, S.L. Laegsgaard, *Catal. Today*, **46**, 193 (1998).
10. R. Zhao, R. Govind, N. Itoh, *Sep. Sci. Tech.*, **25**, 1473 (1990).
11. W. H. Lin, H. F. Chang, *Catal. Today*, **97**, 181 (2004).
12. G. Sznejer, M. Sheintuch, *Chem. Eng. Sci.*, **59**, 2013 (2004).
13. A. Schmidt, R. Schomäcker, *J. Mol. Catal. A.*, **271**, 192 (2007).
14. T. Dwars, J. Haberland, I. Grassert, G. Oehme, U. Kragl, *J. Mol. Catal. A.*, **168**, 81 (2001).
15. J. Caro, K. J. Caspary, C. Hamel, B. Hoting, P. Kölsch, B. Langanke, K. Nassauer, T. Schiestel, A. Schmidt, R. Schomäcker, A. Seidel-Morgenstern, E. Tsotsas, I. Voigt, H. Wang, R. Warsitz, S. Werth, A. Wol, *Ind. Eng. Chem. Res.*, **46** (8), 2286 (2007).
16. X.L. Pan, B.J. Liu, G.X. Xiong, S.S. Sheng, J. Liu, W.S. Yang, *Catal. Lett.*, **66**, 125 (2000).
17. S.I. Hong, J.H. Jung, *Appl. Catal. A*, **156**, 239 (1997).
18. K. Sato, T. Hanaoka, S. Niwa, C. Stefan, T. Namba, F. Mizukami, *Catal. Today*, **104**, 260 (2005).
19. J.F. Trujillo, M.K. Hardiman, A.A. Adesina, *Chem. Eng. J.*, **143**, 273 (2008).
20. W. Jin, C. Zhang, X. Chang, Y. Fan, W. Xing, N. Xu, *Environ. Sci. Technol.*, **42**, 3064 (2008).
21. K. H. Hofstand, J. H. B. J. Hoebink, A. Holmen, G. B. Marin, *Catal. Today*, **40**, 157 (1998).
22. D.J. Moon, K. Sreekumar, S.D. Lee, B.G. Lee, H.S. Kim, *Appl. Catal. A*, **215**, 1 (2001).
23. J.N. Armor, *J. Membr. Sci.*, **147**, 217 (1998).
24. S.L. Jorgenson, P.E.H. Nielsen, P. Lehrmann, *Catal. Today*, **25**, 303 (1995).
25. E. Kikuchi, S. Umiya, Y. Chen, *Stud. Surf. Sci. Catal.*, **107**, 547 (1997).
26. C.H. Chang, R. Gopalan, Y.S. Lin, *J. Membr. Sci.*, **91**, 27 (1994).
27. R.M. de Vos, H. Verweij, *Science*, **279**, 1710 (1998).
28. T. Tsuru, T. Tsuge, S. Kubota, K. Yoshida, T. Yoshika, M. Asaeda, *Sep. Sci. Tech.*, **36**, 3721 (2001).

29. K. Seshan, H.W. Ten Barge, W. Halley, A.N.J. van Keulsen, J.R.H. Ross,in: *Natural Gas Coversion II*, H.E. Curry-Hide, R. F. Howe (eds), Elseiver, Amsterdam, pp 285–290 (1994).
30. J.R. Rostrup-Nielsen, J.-H.-B. Hasnen, *J. Catal.*, **144**, 38 (1993).
31. S. Imamura, T. Yamashita, R. Hamada, Y. Saito, Y. Nakao, N. Tsuda, C. Kaito, *J. Mol. Catal. A*, **129**, 249 (1998).
32. E.E. Lowenthal, L.F. Allard, M.T. Henry, H.C. Foley, *J. Mol. Catal. A*, **100**, 129 (1995).
33. N. Perkas, H. Rotter, L. Vradman, M.V. Landau, A. Gedanken, *Langmuir*, **22**, 7072 (2006).
34. J. Wang, J. Xi, Y. Bai, Y. Shen, J. Sun, L. Chen, W. Zhu, X. Qiu, *J. Power Sources*, **164**, 555 (2007).
35. B. Harrison, A.F. Diwell, C. Hallett, *Plat. Mat. Rev.*, **32**, 73 (1988).
36. M. Ozawa, M. Kimura, *J. Mater. Sci. Lett.*, **9**, 291 (1990).
37. A. Cook, A.G. Fitzgerald, J.A. Cairns, *Catalysis and Surface Characterization*, T.J. Denes, C.H. Rochester, J. Thomson (eds), Royal Society of Chemistry, Cambridge, p. 249 (1992).
38. S. Kurungot, T. Yamaguchi, S. Nakao, *Catal. Lett.*, **86**, 273 (2003).
39. S. Kurungot, T. Yamaguchi, *Catal. Lett.*, **92**, 181 (2004).
40. B.J.R. Uhhorn, K. Keizer, A.J. Burggrasf, *J. Membr. Sci.*, **66**, 259 (1992).
41. B.N. Nair, K. Keizer, W.J. Elferink, M.J. Gilde, H. Verweij, A.J. Burggraaf, *J. Membr. Sci.* **116**, 161 (1996).
42. R.K. Iler, *The Chemistry of Silica*, John Wiley & Sons, Inc., New York (1979).
43. R.E. Timms, *J. Chem. Soc. A*, **1971**, 1969 (1971).
44. J.C. Ro, I.J. Chung, *J. Non-Crystalline Solids*, **130**, 26 (1989).
45. G.H. Bogush, C.F. Zukoski, *J. Colloid Interface Sci.*, **142**, 19 (1991).
46. J.E. Martin, A. J. Hurd, *J. Appl. Cryst.*, **20**, 61 (1987).
47. K.D. Kafeer, D.W. Schaefer, *Phys. Rev. Lett.*, **56**, 2376 (1986).
48. B. Nair, W.J. Elferink, K. Keizer, H. Verweij, *J. Colloid Interface Sci.*, **178**, 565 (1996).
49. T.G. Movchan, N.B. Ur'ev, T.V. Khamova, E.V. Tarasyuk, A.N. Potapov, O.A. Shilova, N.S. Klimenko, V.V. Shevchenko, *Chem. Mater. Sci.*, **31**, 219 (2005).
50. S.E. Rankin, C.W. Macosko, A.V. McCormick, *AIChE J.*, **44**, 1141 (1998).
51. J. Sefcik, A.V. McCormick, *Catal. Today*, **35**, 205 (1997).
52. M. Kanezashi and M. Asaeda, *J. Membr. Sci.*, **271**, 86 (2006).
53. J.C. Daniz da Costa, G.Q. Lu, V. Rudolph, Y.S. Lin, *J. Membr.Sci.*, **198**, 9 (2002).
54. S. Imamura, T. Yamashita, R. Hamada, Y. Saito, Y. Nakao, N. Tsuda, C. Kaito, *J. Mol. Catal. A*, **129**, 249 (1998).
55. E.E. Lowenthal, L.F. Allard, M.T. Henry, H.C. Foley, *J. Mol. Catal. A*, **100**, 129–145 (1995).
56. T. Takeguchi, S. Manabe, R. Kikuchi, K. Eguchi, T. Kanazawa, S. Matsumoto, W. Ueda, *Appl. Catal. A*, **293**, 91 (2005).
57. J. Chen, C. Feng, B. Shu, J. Fan, *J. Rare Earths*, **24**, 54 (2006).
58. A. Cook, A.G. Fitzgerald, J.A. Cairns, *Catalysis and Surface Characterization*, T.J. Denes, C.H. Rochester, J. Thomson (eds), Royal Society of Chemistry, Cambridge, p. 249 (1992).
59. D. Srinivas, C.V.V. Satyanarayana, H.S. Potdar, P. Ratnasamy, *Appl. Catal. A.*, **246**, 323 (2003).
60. L.S.F. Feio, C.E. Hori, S. Damyanova, F.B. Noronha, W.H. Cassinelli, C.M.P. Marques, J.M.C. Bueno, *Appl. Catal. A.*, **316**, 107 (2007).
61. D. Wang, O. Dewaele, G.F. Froment, *J. Mol. Catal. A*, **136**, 301 (1998).
62. K.H. Hofstad, J.H.B.J. Hoebink, A. Holmen, G.B. Marin, *Catal. Today*, **40**, 157 (1998).
63. S. Freni, G. Calogero, S. Cavallaro, *J. Power Sources*, **87**, 28 (2000).
64. B. Grbic, N. Radic, A. Terlecki-Baricevic, *Appl. Catal. B*, **50**, 161 (2004).
65. S. D. M. Jacques, O. Leynaud, D. Strusevich, A. M. Beale, G. Sankar, C. M. Martin, P. Barnes, *Angew. Chem. Int. Ed.*, **45**, 445 (2006).
66. H. Cordatos, T. Bunluesin, J. Stubenrauch, J.M. Vohs, R.J. Gorte, *J. Phy. Chem.*, **100**, 785 (1996).

8

Zeolite Membrane Reactors

Carlos Téllez and Miguel Menéndez

Aragón Institute of Engineering Research, University of Zaragoza, Zaragoza, Spain

8.1 Introduction

Single crystal zeolite membranes were proposed by Barrer [1] but only during the last decades of the twentieth century were researchers able to synthesise continuous polycrystalline layers of zeolite that could be used as membranes. Zeolites are crystalline microporous alumino silicates with the following characteristics that make them very promising as membrane materials:

1. Their *well defined crystalline structure*, with pore sizes similar to the size of molecules [2]. This means they can be used as molecular sieves, that is, to allow the passage of some molecules while retaining others outside the porous structure. Figure 8.1 gives an idea of the relative size of some zeolites and molecules.
2. The *large number of zeolite structures*. The International Zeolite Association (IZA) has described 194 different structures [3] (until January 2010). This provides the possibility to select different pore sizes, pore shapes and pore network characteristics, adapted to the requirements of a given separation.
3. The many *options to change the chemical composition*. The chemical formula of a zeolite can be written as $Me^{m+}_{n/m}[Si_{1-n}Al_nO_2] \cdot x\,(H_2O)$, where the framework is between brackets, Me represents an extra-framework cation and water is the adsorbed phase. By decreasing the Si/Al ratio (i.e., by increasing the Al content) the zeolite becomes more hydrophilic. Therefore when affinity to nonpolar molecules is needed, high-silica zeolites such as the pure silica silicalite-1 are a suitable choice. In addition to the changes in Si/Al ratio, other compounds, named zeotypes, can be prepared with the same structure as zeolites, but substituting (totally or partially) some of the Si or Al ions in the framework by other ions (PO_4, Fe, V, Ti, etc.).

Membranes for Membrane Reactors: Preparation, Optimization and Selection, Edited by Angelo Basile and Fausto Gallucci

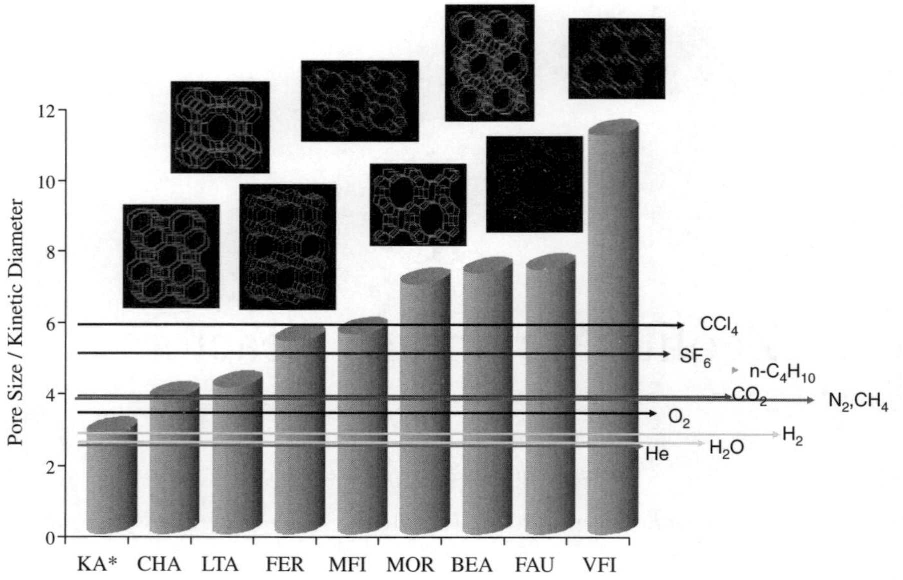

Figure 8.1 Pore size of different framework structures and kinetic diameter of selected molecules. Pore size and kinetic diameter are taken from Ref. [2] and framework structures from Ref. [3]. KA* = This is not a framework structure, but a zeolite A (LTA structure) with potassium instead of sodium as an extra-framework cation

This opens the way to an even larger number of compounds with a wider range of possible pore size and with new adsorption or catalytic activity properties.

4. The *ion exchange capability* given by the extra-framework cation. When a zeolite is in contact with a solution containing an ion different from that existing in the solid, equilibrium is achieved between the metal concentrations in the liquid and in the solid. This property is the basis for the well known use of zeolites as water softeners, by the exchange of Na from NaA zeolite and Ca and Mg from water. Some zeolite properties such as adsorption and pore size can be fine-tuned by ion exchange (see Figure 8.1 with zeolite A and potassium as the extra-framework cation).
5. The *catalytic activity* of zeolites. Several zeolites are widely used in the chemical and petroleum industry as catalysts. Under suitable conditions, a zeolite membrane can combine separation and catalytic activity for a given reaction.
6. The capability to withstand *harsh operating conditions*. In comparison with polymeric membranes, currently the most commonly used type of membrane, zeolite membranes can operate at higher temperatures and in the presence of some solvents (thanks to their inorganic character). Compared with some amorphous microporous materials (silica), their crystalline character provides additional thermal stability. It is worth remembering that, although sometimes zeolites are employed at very high temperatures (e.g., the zeolite catalyst employed in the FCC process supports temperatures of around 700 °C), several factors may affect the performance of zeolite membranes at temperatures lower than those needed to destroy the crystal structure: appearance of defects between crystals, different thermal dilatation coefficients between support and zeolite layer or lack of adsorption of the molecule to be separated.

7. Compared with other inorganic membranes (metallic or dense ceramic) they offer the *chance to permeate selectively other molecules*, not only H_2 or O_2. As an example, the most successful application of zeolite membranes up to now, and the only one that has achieved use on an industrial scale, is the separation of water [4], where zeolite A membranes are currently employed, although other zeolites (T, sodalite) can also provide high selectivities.

Given the numerous opportunities that zeolite membranes may provide, including some difficult separations, it is not surprising that many research teams have been involved in the search of new and better zeolite membranes since the first patent [5] and articles [6,7] were published. This large body of research has been reviewed in several papers [8–20], most of which include descriptions of the state of the art of zeolite membrane synthesis and its uses. The above characteristics are advantageous for many applications of membrane reactors, and thus it is not surprising that the ingenuity of researchers has enabled them to devise many possible uses for zeolite membranes in combination with chemical reactions (i.e., in zeolite membrane reactors) in an effort to achieve a synergic effect. The first attempts at using a ZMR were made by Dalmon and coworkers [21,22], since when several reviews focused mainly on ZMRs have been published [17,23]. Information on other applications of zeolite layers, such as in sensors or corrosion resistant coatings, can be found elsewhere [24,25].

8.2 Zeolite Membrane Preparation Outlines

A key factor in the development of a ZMR is the quality of the zeolite material. Given the large number of types of zeolite membranes that have been described in the literature, it is beyond the scope of this chapter to describe the preparation of each of them in detail. We will concentrate in this section on a brief overview of the preparation of zeolite membranes in general, and in the next section provide a detailed description of the synthesis method for two types. Figure 8.2 shows schematically the most commonly used routes for the synthesis of zeolite membranes.

8.2.1 Support

Zeolite membranes are usually synthesised on a support made of a mechanically resistant material, although self-supported membranes have also been prepared [14]. Supports with different geometries, chemical compositions and physical structures have been used to synthesise zeolite membranes. The most common support suppliers have recently been listed [16]. Flat or tubular supports are the most frequently used geometries for zeolite layers. The synthesis of the zeolite layer in a tubular support is preferably carried out in the inner part because the layer is thus protected from physical damage and is also desirable in high area applications such as capillary or multichannel supports [26]. The chemical composition of the supports is variable: ceramic (e.g., alumina, titanium oxide, zirconia, mullite), carbon, glass or stainless steel membranes have been used [13,16].

The zeolite synthesis includes two main stages, the nucleation and the crystal growth. The nucleation stage is sensitive to synthesis conditions (composition and temperature among others), support properties (physical structure and chemical nature) and impurities; the last two factors can generate heterogeneous nucleation. Therefore, the nucleation process is difficult to control and sometimes the crystallisation of undesirable zeolite phases occurs and poor reproducibility

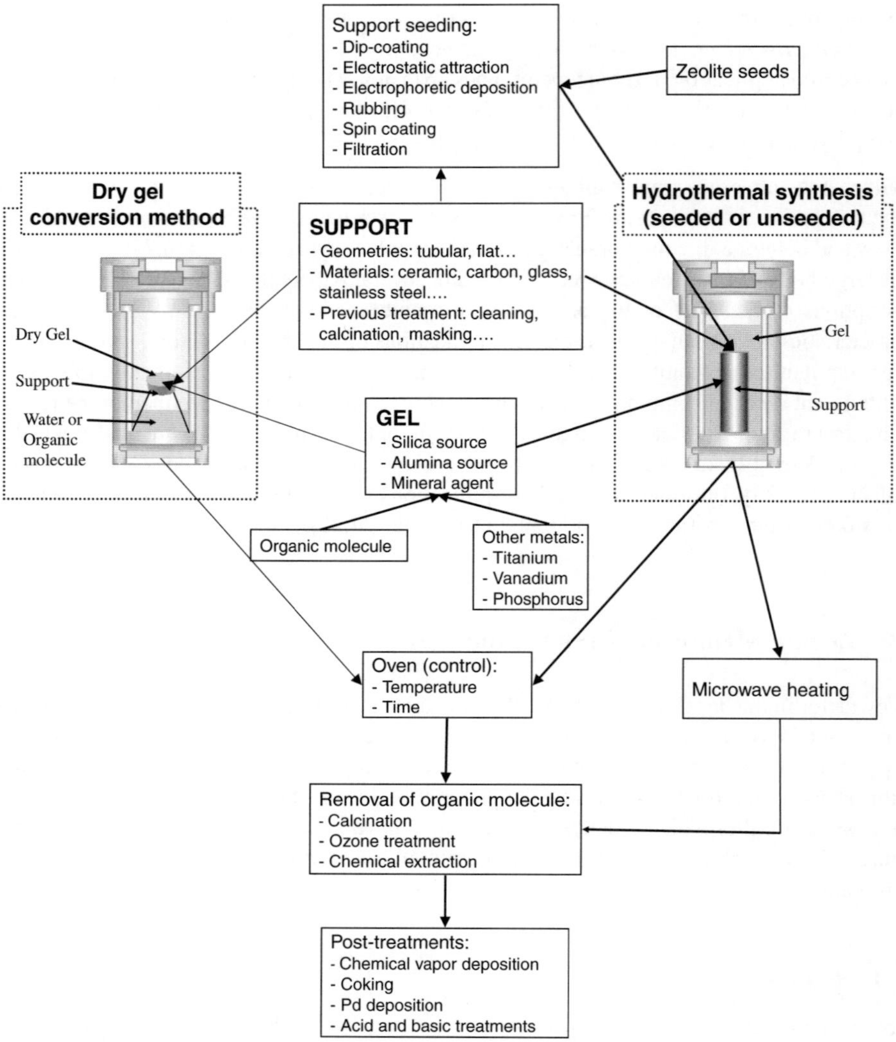

Figure 8.2 *Scheme of synthesis routes for zeolite membranes*

of the microstructure of the zeolite layer may also result [27]. Thus a pre-treatment of the support, for example, by acid treatment, deposition of metals and metal oxides, mechanical polishing, adsorption of surfactant molecules or seeding, can sometimes improve the control of the zeolite layer growth [27]. Another interesting pre-treatment of the support for the preparation of high quality zeolite membranes with high flux is the masking technique [28]. This procedure consists of filling all the support pores with wax, leaving the top surface free for deposition of the zeolite. The wax is stable during membrane synthesis, avoiding deposits inside the support pores and reducing leaching of Al that could change the gel composition in the vicinity of the growing zeolite, and consequently the Si/Al ratio of the formed zeolite, the growth habits and even the zeolite phase formed. The wax can be removed after synthesis by calcination.

8.2.2 Zeolite Synthesis by Hydrothermal Synthesis

Zeolite membranes are commonly synthesised by hydrothermal synthesis onto a given type of support, with or without previous seeding. The latter process (unseeded) is strictly speaking called *in situ* or direct synthesis and was the first synthesis process used when Suzuki [5] patented a method to prepare zeolite membranes, opening the way for the preparation of polycrystalline rather than the single crystal zeolite membranes that Barrer proposed [1]. The seeded process is called secondary growth and was first employed by Kita *et al.* to prepare zeolite A membranes [29] and later by Tsapatsis and coworkers who synthesised different kinds of membranes, including silicalite-1 [30], zeolite L [31] and zeolite A [32].

The *in situ* synthesis process is simple and requires only minor modifications of the usual synthesis recipe for zeolite powder [33], sometimes being based on trial and error [15]. The support is introduced directly into the synthesis mixture in an autoclave and under controlled conditions, usually in an oven, the zeolite is allowed to nucleate and grow to form a layer on the support surface. Temperatures from 90 °C (zeolite A) to 180 °C (silicalite-1) and times ranging from hours to days have been used in the preparation of zeolite membranes. To prepare the gel requires a silica source [tetraethyl orthosilicate (TEOS), sodium silicate, silica sol, fumed silica, etc.], together with an aluminium source if the zeolite contains aluminium (sodium aluminate, aluminium foil, aluminium sulphate, aluminium chloride, alumina, etc.). A mineralising agent (NaOH, NaF, KOH) is also needed to control the pH and provide the cations needed to balance the structure. Also, as occurs in the synthesis of zeolite powder, organic molecules are often added to serve as structure-directing or as space–filling agents [34]. For example, molecules such as tetrapropylammonium hydroxide (TPAOH) and tetrapropylammonium bromide (TPABr) are often used to direct the MFI crystal structure and facilitate crystallisation in the zeolite membrane [33]. Another interesting case is the synthesis of Al-free LTA-type zeolite membranes (ITQ–29) that needs a supramolecular organic structure-directing agent [35].

In the direct synthesis, the synthesis mixture is commonly a milk-like gel where the concentration of silica and alumina is high enough to sustain nucleation and growth processes. The use of a clear solution, where the reactants are strongly diluted with water to inhibit nucleation processes, is more frequent in secondary growth.

The use of organic molecules in the synthesis provides better control of zeolite crystallisation, but the organic template molecules are trapped within the zeolite pores and must be removed after synthesis. The most usual method for removing the organic compound requires higher temperatures, which in the case of a zeolite membrane could promote the formation of cracks and defects. The mechanism responsible for this process has not yet been clarified, but for MFI membranes shrinkage and expansions in the zeolite framework probably occur during template removal and posterior cooling, while the support also suffers changes but in contrary directions [36,37] or with different dilatation/shrinkage paths. Thus, the control of heating and cooling rates and annealing steps has been used during the calcination process. Some attempts have also been made to avoid the formation of cracks, for example by the use of low temperature ozone treatment for organic template removal from MFI zeolite membranes [38], which provides excellent gas permeance and perm-selectivity together with more reproducible membrane properties.

Another method is the so-called dry gel conversion method (DGM). The process consists of depositing a layer of dry aluminosilicate gel (parent gel) on the support and then to transform this gel to the zeolite in the presence of vapours. Two different routes can be distinguished: the so-called vapour phase transport (VPT) method, when the organic structure-directing agents are not included in the parent dry gel, and steam assisted crystallisation (SAC), where only steam is supplied from the vapour phase. This process has the advantages of avoiding homogeneous nucleation and reducing the consumption of reactives. Xu *et al.* [39] reported the dry gel conversion method for the first time to obtain MFI type zeolite membranes by putting the parent gel in contact with water and amine vapours. Latter Matsukata *et al.* [8,40], using the same procedure, prepared ZSM-5, ferrierite, mordenite and analcime membranes, always on flat supports. Subsequent attempts using tubular supports have been described by Alfaro *et al.* [41], preparing MFI membranes using the SAC method.

8.2.3 Seeding

In the secondary growth method, the support is seeded with zeolite crystals using different procedures. The seeding of the support and generally any pre-treatment of the support leads to a higher reproducibility in the synthesis of zeolite membranes or layers [27]. For example, a mordenite membrane prepared with seeding is shown in Figure 8.3a, b. Details of the preparation of this membrane can be found in the work of Navajas *et al.* [42] and is explained in greater detail in the next section. To attain statistically significant results, they prepared a total of 71 membranes under the same synthesis conditions using secondary growth, and a 73.2% yield of high-quality membranes was reported.

Several seeding techniques have been used including dip coating, filtering, spin coating (only for flat support), rubbing and pulsed laser deposition. The last two methods are mostly restricted to the outer surface when a tubular support is used.

Dip coating is a simple and frequently used seeding procedure. It basically consists of preparing a suspension of zeolite seeds with a determined concentration and then dipping the substrate into the suspension and drying. It is usually necessary to repeat the procedure several

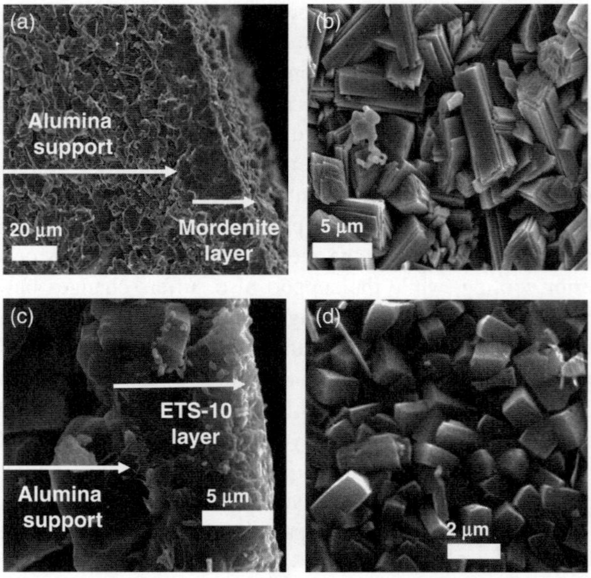

Figure 8.3 SEM photographs: (a) cross section of mordenite membrane, (b) top view of mordenite membrane, (c) cross section of ETS-10 membrane, (d) top view of ETS-10 membrane

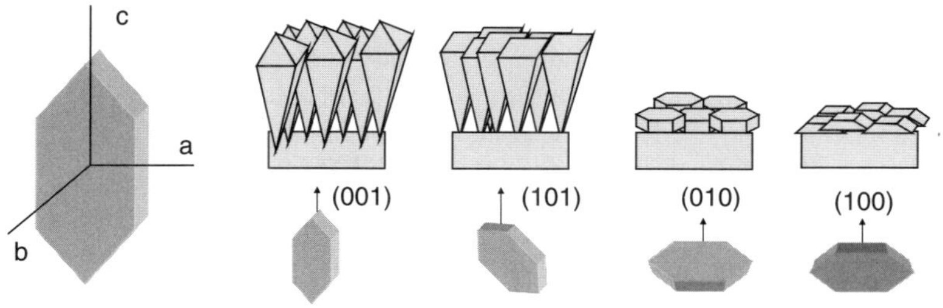

Figure 8.4 *Geometry and directions of MFI crystals and possible orientations of MFI crystals on a support*

times. The dip coating process can be improved via electrostatic attraction of support and seeds, by modifying the surface charges of the support with a cationic electrolyte or by electrophoretic deposition (EPD) [43] which enables coating with the zeolite seeds. For example, silicalite-1 seeds are negatively charged and if the surface is charged positively with PDDA [poly (diallyldimethylammonium) chloride] then the seeds are attached to the surface [44]. In the filtration procedure, the zeolite seeds move towards the support under a pressure gradient; both dead-end [45] and crossflow configurations [46,47] have been used. Rubbing [29] is an easy seeding procedure, where zeolite seeds are attached to the support, usually with a small brush; several steps are needed to reach optimal results and reproducibility of the process is low.

Seeding can control the zeolite phase obtained. Kumakiri *et al.* [48] prepared zeolite A and faujasite membranes under identical conditions from solutions of the same composition, the only difference being the seed crystal used. The zeolite A membrane was formed on a substrate seeded with zeolite A crystals, while the FAU membrane was formed with zeolite Y seeds. Similar behaviour was observed by Li *et al.* [49] with ZSM-5 and mordenite membranes.

A crucial feature of zeolite membranes is the crystal orientation because obviously to favour the molecule movement the zeolite channels must be in the right direction in relation to the support. The orientation of silicalite-1 (MFI structure) has been studied in depth. The MFI structure consists of straight pores (0.53×0.56 nm) along [010] direction interconnected with sinusoidal and elliptic pores (0.51×0.54 nm) along [100] direction. A tortuous path exists along [001] direction. Figure 8.4 shows the typical morphology of a MFI crystal and possible orientation in a MFI membrane. Usually, using a seeding procedure, a preferred orientation can be achieved, and if an oriented seed monolayer can be deposited in the support then the final membrane can retain this orientation [50]. In any case, it should be taken in account that other factors can control the crystal orientation, for example in mordenite membranes it can be manipulated by simply changing the water content in a secondary growth mixture [49].

Finally, it should be noted that a zeolite layer obtained with simply a seeding procedure can act as a membrane ready to be used in some applications. In this context, the stacking of zeolite nanoblocks has recently been used to prepare interesting zeolite membranes [13,51].

8.2.4 Improvements and Achievements in Synthesis of Zeolite Membranes

Zeolite layer synthesis is usually carried out using static hydrothermal processes; however, rotation or stirring of the autoclave has sometimes been used to achieve homogeneity of the mixture [52,53]. The rotation enables nutrients to be supplied to the growth interface at a sufficiently high rate, since the zeolite growth depletes the nutrients in the boundary layer.

Microwave heating has also been studied for synthesising several zeolite membranes (MFI, FAU, LTA, among others) [14] to attain a better uniformity and a higher rate of synthesis reactions [54]. An interesting system for the synthesis of zeolite membranes has been reported by Tiscareño–Lechuga *et al.* [55]. These authors prepared zeolite A membranes under a centrifugal force field, following the ideas of Xu *et al.* [56]. In this system, the centrifugal forces achieved under a sufficiently high rotational speed drive the crystals and crystal nuclei formed in the homogeneous phase towards the support surface, promoting the formation of a more continuous and dense layer.

Yan *et al.* [57] have shown that the in situ synthesis method can be improved by pre-ageing the support material in the synthesis gel before zeolite crystallisation. The pre-ageing is done with a slow zeolite crystallisation rate (usually at a low temperature). In the ageing period, the nucleation can act without competition from crystal growth, and the length of the ageing time controls the number of zeolite nuclei formed. After ageing, the crystallisation rate is increased (e.g., by increasing temperature) to allow zeolite crystal growth. Another synthesis procedure described in some studies involves reactants being continuously supplied to the synthesis vessel while this is maintained at a constant temperature. This type of procedure provides a better control of the synthesis and crystallisation conditions, avoids depletion of nutrients in the liquid phase and is easier to implement on an industrial scale. With this idea, zeolite NaA [58–60] and MFI [61,62] membranes have been synthesised.

A new strategy for the synthesis of zeolites [63,64] is preparing two solutions with different nutrients and/or concentrations (that upon mixing give rise to the zeolite precursor gel). The two solutions are initially separated by means of the support. Silicalite-1 membranes have been synthesised when silicon source and template precursors that diffused along the support channels from opposite directions reacted when they encountered each other and formed a dense membrane [64]. Mateo *et al.* [63] used two solutions of nutrients in different concentrations at each side of the support to be able to govern reactant diffusion, controlling with it the growth of the zeolitic material.

Another method is the pore–plugging method [65]. In this strategy, zeolite crystals grow within the pores of a ceramic alumina substrate until the pores are completely blocked by the zeolitic material. Nanocomposites are formed by the support together with discrete, small zeolite crystals with a size in the same order as the pore diameter of the support. The advantages of this method are the protection of the separative layer, reduction of crack formations due to long–range stresses and greater reproducibility.

8.2.5 Types of Zeolites

The literature describes several kinds of zeolite membranes synthesised with previously described methods [8–16]: LTA-type (zeolite A, ITQ-29), faujasite (zeolite X, zeolite Y), mordenite, MFI (ZSM-5, silicalite-1), BEA (beta zeolite), DDR (DD3R zeolite), sodalite, type MEL, ferrierite, Chabazite, and zeolite T. Other microporous membrane groups synthesised include pseudo–zeolitic materials containing tetrahedrally coordinated phosphorus, such as $AlPO_4$ [66] and $SAPO_4$ [67], and zeolite membranes containing Ti and V in their structure. These membranes, such as TS-1 [68] and VS-1 [69], are potentially interesting from the point of view of catalytic applications. Another microporous silicate system attractive for membrane applications is based on octahedral–pentahedral–tetrahedral microporous (OPT) siliceous frameworks [70]. This group includes: (i) ETS-10 membranes [71,72] (Figure 8.3c, d) where titanium is coordinated octahedricaly and silicon tetrahedrically, (ii) ETS-4 membranes [73,74] possessing a mixed tetrahedral(Si)–pentahedral(Ti)–octahedral(Ti) framework and (iii) the umbite membrane [75] which possesses framework structures built of TiO_6. OPT materials present novel possibilities of isomorphous framework substitution. For instance, in synthetic umbite membrane, titanium has been substituted by tin [75]. These framework substitutions allow fine tuning of the catalytic and adsorption properties of a given membrane, while preserving its microporous structure. In the case of pseudo–zeolitic materials, TS-1, VS-1 and OPT, the preparation of inorganic membranes is similar to the synthesis of zeolite membranes previously described (see Figure 8.2), including in the gel the corresponding source of the metal to be incorporated and obviously, when secondary growth is employed, seeding with the corresponding material.

Multilayer zeolite membranes have also been synthesised to achieve novel properties by combining layers of different types. The synthesis method of these membranes does not differ from single-layer zeolite synthesis; after the synthesis of the first kind of zeolite the second kind of zeolite is synthesised over the previous one. The main problem is usually related to the dissolution of the initially formed layer. For example, a silicalite-1/mordenite bi-layered self-supporting membrane has been synthesised by two dry gel conversion processes consecutively [76]. Also, two-layered mordenite-H-ZSM-5 [77] and NaA-silicalite-1 [78] zeolite membranes have been prepared by two consecutive hydrothermal syntheses.

The synthesis of membranes made of mesoporous silicates and aluminosilicates, such as MCM-22 [79], MCM-41 [80] and MCM-48 [81,82], has also attracted interest because bulky molecules can be separated. With this material, calcinations are most frequently used to remove the organic molecules used to form the mesoporous structure but a solvent or chemical extraction has also been used to avoid calcination, for example, in MCM-48 membranes, a solution containing ethanol and hydrochloric acid was used at 55 °C for 8 h under reflux [83].

8.2.6 Post-Treatment of Zeolite Membranes

All the above contributions attempted to improve the properties of membranes by acting upon the synthesis conditions or the support. Unfortunately, it is often found that zeolite membranes present inter-crystalline defects which are sufficiently large (a few nanometres are enough) to prevent selective separation in the size-exclusion regime. In spite of this, high selectivity separations can still be achieved when the separation mechanisms involve selective sorption on zeolitic and nonzeolitic pores, and/or pore blocking by capillary condensation. However, other investigations aim to increase the quality of zeolite membranes by means of post-synthetic treatments. Some authors have tried to eliminate intercrystalline defects by means of chemical vapour deposition [84], coking [85] or Pd deposition [86] as defect-plugging processes. Finally, another possibility is the use of post-synthetic alkaline [87] or acid [88–90] treatments to improve the properties of mordenite or ZSM-5 membranes. In theses cases the improvement is related to several factors such as zeolite recrystallisations or changes in the Si/Al zeolite composition. Sometimes mesopores are created that increase the flux in pervaporation processes.

8.3 Detailed Preparation Method of a Zeolite Membrane

In this section, the preparation of a zeolite membrane by seeded hydrothermal synthesis on the outer surface of an alumina tubular support (e.g., microfiltration membranes from Inocermic with 1.9 μm pores, 7 mm i.d. and 10 mm o.d.) is explained in detail. In some steps, a specific example such a mordenite membrane is cited because its synthesis has been reported with a high degree of reproducibility [42], or a more hydrophobic membrane (ZSM-5 membrane) because such membranes have been successfully prepared in several works [91,92]. The procedure for the synthesis of a zeolite membrane has the following steps:

1. *Cleaning the support.* Previous to any handling of the support, it is important to bear in mind some previous steps such as support cleaning. The cleaning procedure can include several stages such as washing with organic solvents and/or water using ultrasonic baths or high temperatures, sometimes including calcinations. These processes remove impurities or any organic contaminants adsorbed within the pores and in case of calcination serve to thermally stabilise the support. A typical procedure can involve the following steps:
   (a) Immerse the whole support in boiling distilled water during a period of 1–2 h.
   (b) Replace water by acetone and place in an ultrasonic bath for at least 1 h.
   (c) Dry in oven at 100 °C for 24 h.
2. *Enameling the support.* This procedure aims to control the permeable area and assure the sealing between the permeate and retentate sides. Enameling may be carried out with different enamel products. One commonly used is commercial enamel (e.g., IN 1001 Clear, Envision Glaze, Duncan) able to support high temperatures. A typical procedure is as follows:
   (a) Enamel with a brush the ends of the support on the inside and outside part to delimit the desired length of the porous support.
   (b) Leave the support horizontally for 2 h for the enamel to dry.
   (c) Calcine in a furnace with a heating ramp of less than 3 °C min^{-1}, to avoid the appearance of cracks, up to 800 °C, then allow to cool to room temperature.
   (d) Repeat the enameling and calcination process to ensure impermeability of the support ends. To check this, flow a stream of nitrogen along the membrane submerged in water and with one end covered; if bubbles are observed in the enameled area then another layer of enamel is required.
   (e) Repeat the cleaning procedure.
3. *Seeding procedure.* Several seeding procedures have been used in zeolite synthesis, as explained in the previous section. Here a simple rubbing procedure is described, which is limited to the outer surface of the support. This procedure has been used to synthesise mordenite membranes using commercial mordenite seeds supplied by the company Tosoh Co. (H/mordenite, Si/Al = 5.1), with an average size of approximately 1 micrometre. Seeding is done by the following steps:
   (a) Put a small amount of seeds on a smooth plastic surface (e.g. vinyl acetate) and press the whole support surface on them.
   (b) Press seeds on the support with the finger sheathed with a latex glove.
   (c) Repeat the process 2–5 times to obtain a homogeneous seed layer.
4. *Gel preparation.* In the literature, there are several gel preparations for obtaining different zeolite membranes, as described in the previous section. Here a mordenite membrane and a ZSM-5 membrane are taken as examples.
   (a) *Mordenite membrane:*
      - The quantities of reactants are calculated to obtain the following molar composition: $H_2O:SiO_2:Na_2O:Al_2O_3 = 80:1:0.38:0.025$.
      - Sodium aluminate (e.g. $Na_2Al_2O_4$, Carlo Erba Reagenti, 99% purity), NaOH (e.g. Panreac, 97% purity) and deionised water are consecutively added in a polypropylene vessel. The mixture is stirred for 5 h.
      - This solution is filtered in a vacuum to prevent small lumps forming on the surface which could act as nucleation points in the subsequent synthesis.

   - The silica source (e.g., Ludox® AS–40) is poured over the filtered solution and the mixture is stirred for at least 20 h until the complete dissolution and homogenisation of the reagents within the gel.
   (b) *ZSM-5 membrane:*
   - The quantities of reactants are calculated to obtain the following molar composition: 21 SiO_2 : 987 H_2O : 3 NaOH : 1 TPAOH : 0.105 $Na_2Al_2O_4$
   - Make a first solution with NaOH, 1 M TPAOH (e.g., TPAOH, Aldrich 1 M), sodium aluminate (e.g., $Na_2Al_2O_4$, Carlo Erba Reagenti, 99% purity) and approximately one-fourth of the distilled water. Divide this solution into five aliquots.
   - Prepare a second solution with fumed silica (Aerosil 300) and the remaining distilled water.
   - Pour over the second solution the aliquots of the first solution over a period of 3 h.
   - Age the mixture stirring for three days at ambient temperature, after which the gel becomes cloudy and thin.

5. *Zeolite membrane synthesis.*
   (a) Cover the enameled part of the seeded support with Teflon to prevent deterioration during synthesis, given that the gel is very basic and high temperatures can dissolve it.
   (b) Place a Teflon ring at one end of the seeded support so that it is vertical within the vessel of the Teflon autoclave. Teflon caps could be added to both ends of the membrane to avoid any synthesis in the inner part of the support.
   (c) Pour the previously prepared gel over the seeded support. The seeded support must remain submerged in the gel and the Teflon autoclave must not be completely full to avoid a drastic increase in pressure during synthesis.
   (d) *As an alternative to steps (b) and (c)*, plug with a Teflon tape one end of the support and fill the inside of the support and then plug the other end of the support. In this case add approximately 1 ml of water to the autoclave. This alternative method is useful for synthesising the membrane only in the inner part and requires a support with sufficiently low permeability to avoid the loss of the synthesis gel.
   (e) Close the autoclave and place it in the synthesis oven during the required temperature and time. A mordenite membrane requires a temperature of 180 °C for between 8 and 24 h. A ZSM-5 membrane requires 170 °C for 15–24 h.
   (f) After the hydrothermal synthesis is finished, quench the autoclave in running water.
   (g) In the case of the ZSM-5 membrane, the zeolite membrane synthesis is repeated until the membrane is impermeable to N_2 (or permeation is below detectable limits). Since the template blocks the zeolite pores, this condition assures the lack of intercrystalline defects.
   (h) Wash the membrane with water at room temperature and dry overnight at 100 °C. The mordenite membrane is ready to be characterised. Figure 8.3a, b shows a mordenite membrane prepared with this procedure. In the case of the ZSM-5 membrane, the template must be removed from the zeolitic micropores by calcination. A suitable procedure is to heat in air at 480 °C for 8 h, using a heating rate of 0.5 °C min^{-1}.

8.4 Types of Zeolite Membrane Reactors

Figures 8.5 and 8.6 show the different ways in which zeolite membranes have been employed as part of a chemical reactor. This section will describe each of these cases, showing examples of applications and discussing the advantages and limitations of using zeolite membranes.

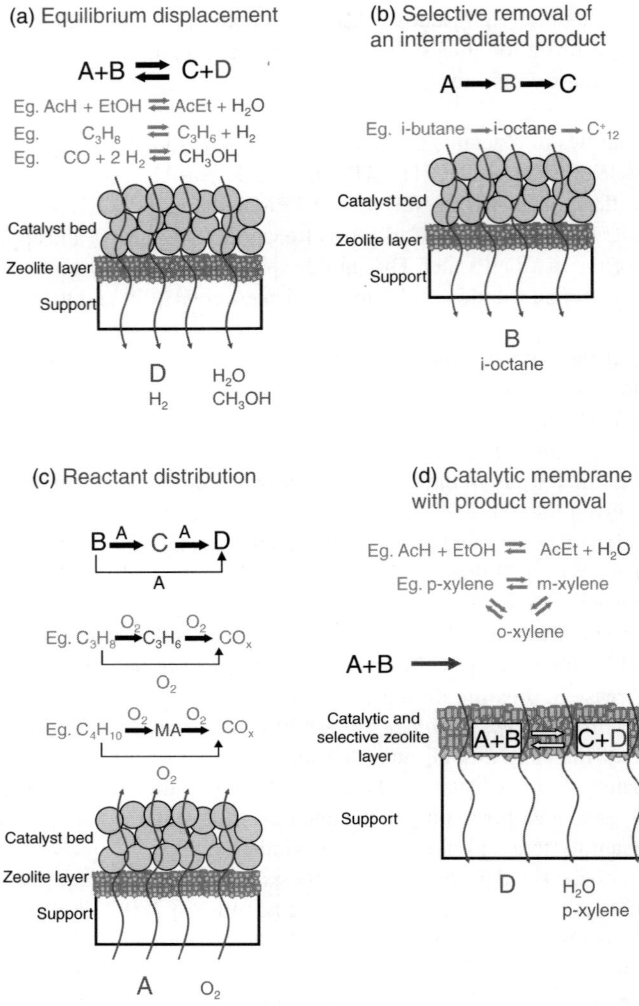

Figure 8.5 *Scheme I of the types of zeolite membrane reactor*

8.4.1 Equilibrium Displacement

In this case the removal of a product in the reaction whose conversion in a conventional reactor is limited by the thermodynamic equilibrium allows a larger conversion to be achieved (according to the Le Chatelier's principle). Three principal groups will be considered:

1. Equilibrium displacement by water removal;
2. Removal of hydrogen;
3. Other processes.

8.4.1.1 Equilibrium Displacement by Water Removal

The integration of a membrane and a reactor in the so-called packed bed enclosed membrane reactor is based on the separation of a product while it is formed and is the most commonly

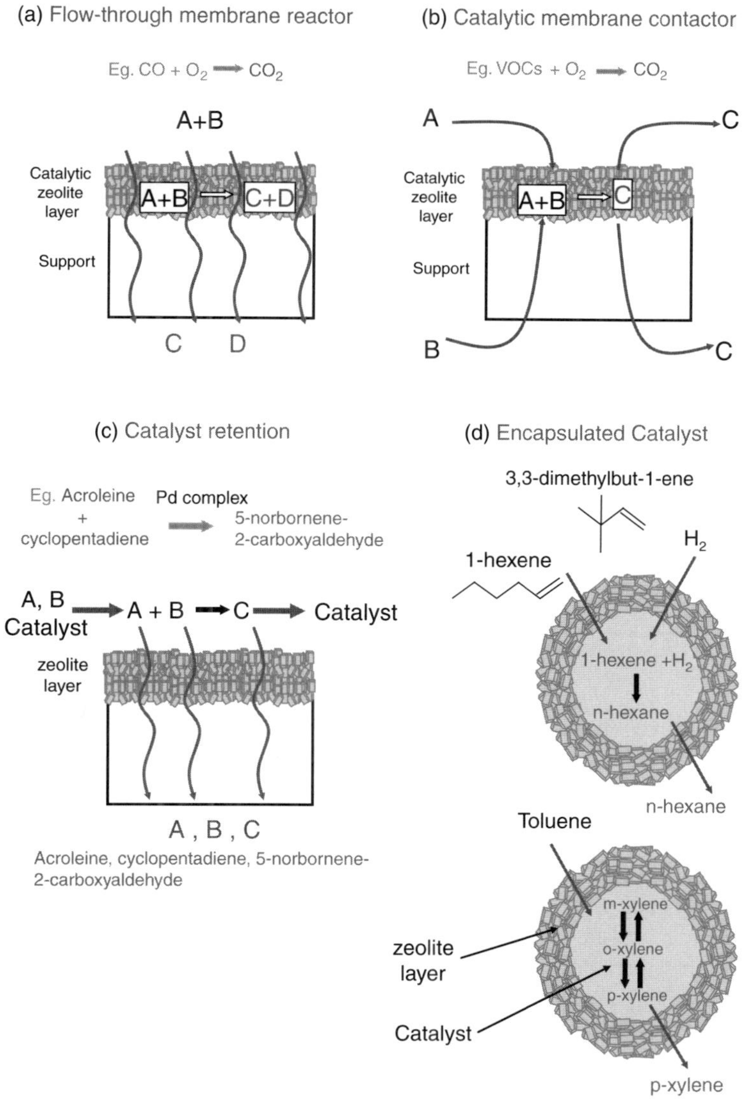

Figure 8.6 *Scheme II of the types of zeolite membrane reactor*

studied case of a membrane reactor (a scheme is shown in Figure 8.5a). Since hydrophilic zeolite membranes are well known for their capability to selectively separate water and in fact their only industrial application up to now is the purification of solvents by removing water by pervaporation [93], it is not surprising that ZMRs have been widely studied in reactions with equilibrium–limited conversion where water is formed. Table 8.1 lists such studies [94–103]. In most cases, esterifications are employed as a test reaction with a variety of organic acids and alcohols. In many cases zeolite A membranes are employed, since they provide high water flux and selectivity. However, this zeolite has a low stability in acidic or basic media and thus some other zeolites have been proposed as alternative membrane materials [96–98].

Table 8.1 ZMR with equilibrium displacement by water removal

Reference	Zeolite membrane	Reaction	Special features and reactor type
[94]	Zeolite A	Esterification (oleic acid with ethanol)	Vapour permeation
[95]	Zeolite A	Esterification of lactic acid with ethanol	Vapour permeation
[96,97]	Zeolite A and zeolite T	Esterification (acetic or oleic acid with ethanol)	Zeolite T stable in contact with liquid
[98]	Zeolite A and mordenite	Esterification of acetic acid with ethanol	Instability of zeolite A in acidic conditions
[99]	Hydrophilic ZSM-5 (MFI)	Knoevenagel condensation	Microreactor
[100]	NaA (LTA)	Knoevenagel condensation	Microreactor
[101]	Cs-X (FAU)	Knoevenagel condensation	Microreactor
[102]	Mordenite and zeolite A	MTBE from tert-butanol and methanol	Amberlyst catalyst
[103]	Zeolite A	di-*n*-pentyl ether by *n*-pentanol dehydration	Experiments, model and economical analysis

Several reactor configurations have been employed [104], as shown in Figure 8.7. In the simplest configuration the zeolite membrane is submerged in the reacting liquid [96] (Figure 8.7a) in which, if needed, some catalyst is suspended. Water permeates through the membrane with the help of a sweep gas stream or by maintaining a vacuum in the permeate side. A permeate is easily obtained with a cold trap in which water condensates and with the help of a vacuum pump to remove noncondensable products. The second system is similar, also with a batch reactor, but in this case the membrane is kept in vapour phase (Figure 8.7b) [95]. Since the chemical potential of water is the same in the liquid and vapour phase (at least in theory), the driving force for water permeation should be the same irrespective of the membrane being in the liquid or in the vapour phase. The advantage is that the risk of damage to the zeolite in the acid reaction environment is avoided. This is especially interesting for the highly selective and permeable zeolite A membranes, which are unstable in acidic media. It is worth commenting that other membranes, although not yet used industrially, can provide good performances in water pervaporation and are more resistant to acidic media, such as ZSM-5 [105], mordenite [106] or zeolite T [97]. In first and second mode the liquid could be in batch or continuous mode. The third mode (Figure 8.7c), continuous operation, with a packed bed membrane enclosed reactor is a bit more complicated since the required adjustment between permeation and reaction, that is, it has one less degree of freedom (the amount of catalyst or reaction volume is limited by the volume inside the membrane tube). Design issues for this type of ZMR have been studied by Lim *et al.* [103]. The last type of reactor combines three functions (Figure 8.7d; reaction, separation by distillation, separation by membrane) [107]. Given the success achieved in process integration by reactive distillation [108] and by the combination of zeolite membranes and distillation, it may be expected that this tri-functional reactor will provide a successful solution to problematical reacting systems. Although reactive distillation coupled with pervaporation membranes is still a nascent topic, the modelling of this complex system is a challenging task. It is worth

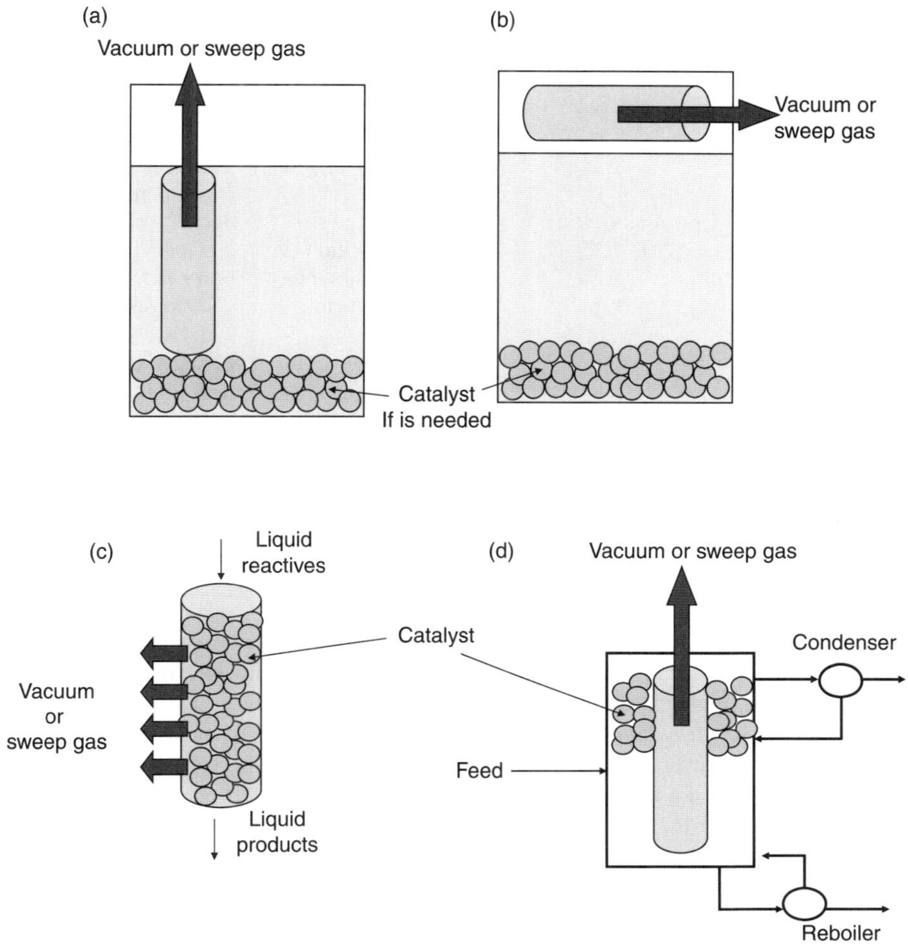

Figure 8.7 *Reaction configurations for water removal*

remembering that for applications in which water is removed at not too high a temperature, polymeric membranes compete with the zeolite counterparts (e.g., Figueiredo *et al.* [109]).

Finally, a step forward could be achieved by using the increased reactant conversion in a reaction (removing water with a zeolite membrane) to increase the conversion of a second reaction for which the product is the reactant for the first reaction. This has been applied to the following reaction system [110]:

$$\text{Methyl ester} + \text{acetone} \leftrightarrow \text{solketal} + H_2O$$

$$\text{glycerol ester(oil)} + \text{MetOH} \leftrightarrow \text{Methyl ester} + \text{glycerol}$$

8.4.1.2 Removal of Hydrogen

The equilibrium displacement of hydrogen–forming reactions has been widely studied using metallic membranes. Zeolite membranes have also been used sometimes (see Table 8.2), since

Table 8.2 *ZMR with H_2 removal*

References	Zeolite membrane	Reaction	Special issues or results
[21,113–117]	Silicalite membrane	Isobutane dehydrogenation	Catalyst deactivation by increased coking; increased yield compared with PBR
[118]	FAU-type zeolite	Cyclohexane dehydrogenation	Benzene also removed selectively
[119]	Silicalite-1	Dehydrogenation of ethylbenzene to styrene	Iron catalyst; conversion increased from 67.5 to 74.8%
[120]	Silicalite-1	C3 aromatisation	Changes in product distribution
[121]	NaA and La_2NiO_4/NaA composite membranes	CO_2 reforming of methane	Improved CO and H_2 yield
[122]	Silicalite-1 membrane	Partial oxidation of methane	Fluidised bed reactor

the hydrogen permeation is faster than that of hydrocarbons. However, at the temperatures required for alkane dehydrogenation (typically 300–600 °C) the perm-selectivity of hydrogen is usually only slightly higher than in the Knudsen flow, i.e. that given by the square root of the ratio of molecular weights. It is known [111] that with such a low selectivity as that given by the Knudsen factor, the conversion improvements would be small and in many cases the reported conversion increase is not larger than that achievable by a pure dilution effect, that is, if the reactant was mixed with the inert gas used as sweep gas. (Note that from the equilibrium point of view, dilution with an inert gas is equivalent to lower operation pressure and since the number of moles increases in dehydrogenation, lower pressure favours higher equilibrium conversion.) Moreover, due to the low selectivity, loss of reactant occurs through the membrane. In exceptional cases [112], an H_2/iso-C_4H_{10} selectivity higher than Knudsen has been reported.

The removal of hydrogen in zeolite membranes has been commonly used for equilibrium displacement of dehydrogenation reactions [21,113–119], but other reactions have also been tested [120–122] (see Table 8.2).

8.4.1.3 Other Processes

The production of methanol from syngas is a typical example of an equilibrium limited reaction, where an increase in pressure helps to improve the achievable conversion. In spite of the high pressure employed in industrial processes (30–80 bar), the achieved conversion is not very high (e.g., 15% in the ICI process, operating at 30 bar) and thus a large recirculation of unconverted reactants is needed. Previous works used Nafion membranes [123], showing that it was possible to increase the conversion by removing methanol, but the maximum temperature achievable with those membranes (200 °C) resulted in a low reaction rate. The use of a ZMR was proposed as a way to avoid the temperature limit [124]. Experimental trials show that some methanol separation was possible [125,126] and improvements in conversion of syngas were achieved but the performance was lower than that expected from the simulation [127]. The main difficulty of this system lies in the loss of perm-selectivity with the tested membranes at temperatures higher than the critical temperature of methanol. This problem may be due to the low thermal stability of the zeolite usually used and may be overcome with other zeolites, such as mordenite [128] or ZSM-5 [129].

Another kind of reaction is comprised by those where several isomers are formed, with different kinetic diameters, such as linear and ramified hydrocarbons. The capability of membranes to separate both types of hydrocarbon has often been explained by the pore size value, being between the kinetic diameters of the two hydrocarbons (although in fact the reason is preferential adsorption). A typical example is xylene isomerisation using a MFI-type membrane, whose structure has been described in Section 8.2.3, where p-xylene (0.58 nm) can diffuse much faster than its bulkier isomers (0.68 nm) [130]. It should be noted that with this zeolite it is very important to synthesise an oriented membrane where the pores useful for separating the isomers are oriented in the correct direction [50]. Improvements in p–xylene productivity and selectivity were achieved in the membrane reactor compared with a conventional fixed bed reactor [131,132]. A similar system is toluene methylation, where Tanaka *et al.* [133] obtained an increased selectivity to p-xylene in a ZMR.

Another related example is the beneficial use of a silicalite-1 membrane in two reactions involving butene [134,135], the geometrical isomerisation of cis-2-butene, where the separation of trans-2- and cis-2-butene is governed by differences in shape, and the metathesis of propene, where the separation is based on differences in adsorption between these components. Quite satisfactory permeselectivity has been found experimentally in both cases. An economical evaluation of a process for heptane hydroisomerisation in which the reactor was combined with a zeolite membrane separation unit showed that further developments were needed and that the expected high cost of the zeolite membranes was the main drawback [136].

8.4.2 Product Removal (In Non–Equilibrium Limited Reactors)

In other cases, even if the achievable conversion is not limited by the thermodynamic equilibrium, the removal of a reaction product may provide advantages in terms of increased conversion and selectivity. Two cases will be considered.

1. Removal of a product that inhibits the reaction;
2. Removal of intermediate products.

8.4.2.1 Removal of a Product that Inhibits the Reaction

Water removal in Fischer–Tropsch. The Fischer–Tropsch process is widely considered as one of the most promising to obtain liquid fuels from other sources (biomass, coal or natural gas). In this reaction synthesis gas reacts to give hydrocarbons and water according to the following reaction:

$$n\mathrm{CO} + 3n\mathrm{H_2} \rightarrow \mathrm{C}_n\mathrm{H}_{2n} + n\mathrm{H_2O}$$

Large amounts of water are formed in this reaction and therefore the partial pressure of water becomes very high (the pressure may be between 1800 and 5000 kPa, typically about 2000–4500 kPa). The presence of water has several negative effects:

1. It deactivates the catalyst (Fe based ones) and therefore its removal could improve the stability of the reactor.
2. The high partial pressure of water decreases the reaction rate (having an inhibitory effect).

Experiments simulating the composition, temperatures and pressures of both fluid bed and slurry reactors showed that water can be removed with high flux (up to 12 kg h^{-1} m^{-2}) and good selectivity [$S(H_2O/CH_4)$ up to 72] [137,138]. The simulation of the performance of a membrane reactor was satisfactory [139]. Further studies on the use of membrane reactors for Fisher–Tropsch have been done with a silica membrane [140] and a zeolite A has been also proposed [141], although the most promising material is sodalite [142].

Ethanol removal in Fermentation broth. One of the disadvantages of ethanol fermentation is that the reaction rate is low because fermentation is inhibited by ethanol. This self-poisoning phenomenon is well known in alcoholic fermentation, and the decrease in the reaction rate results in lower ethanol productivity. A silicalite-1 membrane was employed to extract ethanol from the fermentation broth in order to obtain a quicker fermentation and at the same time a product with a high concentration of ethanol [143,144], instead of the diluted mixture that the production system normally provides.

8.4.2.2 Removal of Intermediate Products

The problem of improving the selectivity to the intermediate product in a series reaction ($A \rightarrow B \rightarrow C$) is a common one in chemical engineering. If a membrane could selectively remove the intermediate product (B), its conversion to the undesired product (C) would be avoided and the process selectivity could be greatly improved (Figure 8.5b). A process where this has been done is the oligomerisation of butene [145]. The dimer is a suitable octane enhancer, but the trimer or tetramer are not so useful. It was found that an MFI membrane can selectively remove C_8 hydrocarbons and the yield for a given conversion was larger in the ZMR than in the conventional reactor. This perm-selectivity to C_8 was explained by the stronger adsorption of C_8 compared to C_4, thus hindering the adsorption of the latter, and by the large size of C_{12} and C_{16}, which were probably unable to enter the zeolite channels.

8.4.3 Reactant Distribution

The distributed feed of one reactant to a fixed bed can be traced back to classical *chemical reaction engineering* textbooks [146] (Figure 8.5c). The use of a porous membrane was first proposed for the oxidative coupling of methane [147] and later for a large variety of catalytic oxidations [148]. A lower partial pressure of a reactant, as is obtained in this way, is expected to improve the selectivity to the desired product if the reaction order is lower for the desired than for the undesired product. Zeolite membranes have been used in this way for butane oxidation to maleic anhydride [149,150] and oxidative dehydrogenation of propane [22,151] using silicalite membranes to distribute the oxygen [152]. The oxidation to CO_x is thus decreased and selectivities to desired products increased. In addition to the benefits of higher selectivity at a given conversion due to oxygen distribution, heat generation is distributed more evenly along the bed, thereby decreasing the formation of hot spots and the probability of runaway [153]. Finally, the membrane reactor makes it possible to work outside the flammability region. This is important in butane oxidation to maleic anhydride, which requires working with a feed rich in n-butane.

In addition to oxygen, other gases can be distributed. For example, Chommeloux *et al.* [154] used a MFI zeolite membrane to distribute hydrogen in a packed bed membrane reactor in the selective hydrogenation of butadiene. A ZSM-5 membrane has been used for ozone distribution for the removal of total organic carbon (TOC) in water [155]. It was demonstrated that ozonisation with traditional sparger in a batch reactor yielded a 1.1% decrease in TOC, and replacing the sparger with the ZSM-5 membrane distributor gave a 20% TOC removal.

Although a zeolite membrane can distribute homogenously a reactant to a fixed bed of catalyst, it seems doubtful that so sophisticated a material was really needed to achieve such a simple task.

8.4.4 Catalytic Membrane with Product Removal

In the previous cases the zeolite membrane was catalytically inert (or at least its catalytic activity was not deliberately employed). There are several ways in which the catalytic and separation properties of zeolite membranes have been combined. Here two examples are given:

1. Bernal *et al.* [156] used a H-ZSM-5 membrane to carry out simultaneously an esterification reaction and water removal, taking advantage of the catalytic properties of the zeolite (Figure 8.5d). It was found that with the same amount of catalytic material the conversion was larger than using a packed bed membrane enclosed reactor, with the H-ZSM-5 zeolite as powder and an inert Na-ZSM-5 membrane carrying the separation. This effect was attributed to the integration of reaction and separation at microscopic level, that is, to the avoidance of a diffusion step through the bed of catalyst to the membrane surface. A similar approach was employed by de la Iglesia *et al.* [77], using a bi-functional H-ZMS-5/mordenite membrane.
2. Haag *et al.* [157] also used a H-ZSM-5 membrane that was catalytically active for xylene isomerisation and more perm-selective to p-xylene than to the other xylene isomers (Figure 8.5d). The results show an increase in m-xylene conversion and higher p-xylene selectivity, although the differences in conversion could be due to the different activity of the zeolitic powder and of the zeolite in the membrane. Although the selectivity improvement was only around 10 perceptual points (e.g., from 52.3 to 63.6%), this improvement shows the good quality of the membrane, since the difference in kinetic diameter of m- and p-xylene is very small. In addition, during the separation of xylene isomers, selectivity is only achieved at low partial pressures, but disappears at normal partial pressures (near atmospheric). This is explained by the phenomenon known as 'single file diffusion': in the narrow channels of zeolite the molecule with larger diffusivity finds its flow hindered by the presence of another slower molecule. This effect disappears at low partial pressures (and thus low surface coverage) where the mutual influence of adsorbed molecules becomes negligible. Tarditi *et al.* [158] used Pt/H-ZSM-5 and Ba-ZSM-5 catalytic membranes for the same reaction. Finally, Fong *et al.* [159] have reviewed the developments of functionalised zeolite membrane for p-xylene production from xylene isomers.

8.4.5 Flow-Through Membrane Reactor

In this kind of reactor the membrane does not have a separation function, and in some cases does not even have to be perm-selective. In this operation mode all the reactants are fed through the membrane (Figure 8.6a). This avoids the problems derived from internal or external mass transfer that may appear in a conventional fixed bed. A review of flow-through catalytic membrane reactors has been published recently [160].

The most interesting results using this configuration with zeolite membranes have been observed in the selective oxidation of CO in the presence of H_2. This reaction is employed to remove CO from hydrogen to be used in a polymer electrode membrane (PEM) fuel cell. A key factor for these fuel cells is that the CO content in the feed must be lower than 10 ppm to avoid catalyst deactivation. Several authors [161–164] have found that the CO conversion achieved in a flow-through configuration with a Pt-zeolite Y membrane is larger than in a conventional reactor.

Another process where the flow-through ZMR provided a clear improvement in the reactor performance was methanol to olefins (MTO). This reaction is a key step of a very interesting alternative process for the production of liquid products from natural gas and is a series reaction (methanol $\rightarrow$ dimethylether $\rightarrow$ olefins $\rightarrow$ parafins + aromatics). Tago *et al.* [165] and Masuda *et al.* [166], using a ZSM-5 membrane, found a substantial increase in selectivity to olefins (the most valuable product) compared with a conventional fixed bed reactor. This increase was explained by the catalytic membrane unifying the contact time of the molecules with the active sites, leading to the high selectivity of intermediate species being achieved in series reactions. A further treatment of the membrane (catalytic cracking of silane, CCS) provided even more selectivity by deactivating preferentially the outer surface but not the acid sites inside the pore network.

Another interesting case of a flow-through membrane reactor was reported by Torres *et al.* [167], who studied i-butene oligomerisation using a beta-zeolite membrane. They found that the membrane reactor was stable for 168 h, whereas the fixed bed reactor was completely deactivated after only 4 h. A further interesting reaction studied is the gas–phase photocatalytic oxidation of trichloroethylene (TCE) using a catalytic titanium silicalite-1 (TS-1) membrane [168]. In this work the conversion of TCE and selectivity to CO_2 was increased by passing the mixture reaction through the membrane, instead of a contact in parallel mode between reactives and membrane. This fact is explained as the molecular sieve effect of the membrane where the smaller product molecules may pass through the membrane while the larger molecules remain for further conversion.

Finally, other reactions such as oxidative dehydrogenation on MFI and vanadium–MFI membranes [169] have been tested in flow-through contactors.

8.4.6 Catalytic Membrane Contactor

In this configuration (Figure 8.6b) the reactants are located at different sides of the membrane and must diffuse through the zeolite layer to react. The procedure can be traced back to the early work of Sloot *et al.* [170] on the Claus process, a reaction requiring strict stoichiometric feed rates of premixed reactants. The reactants are fed in the membrane reactor to the different sides of the catalytic membrane. If the reaction rate is fast compared to the diffusion rates of the reactants, the reaction occurs in the catalytic layer, thus preventing the mixing of the reactants. Zeolite membranes (Pt-ZSM-5) have been employed [92] for the removal of volatile organic compounds (VOCs). A surprising result from this work was that an intermediate quality of the membrane (as measured in terms of perm-selectivity) was better for this operation than a perfect membrane. Probably the existence of some defects facilitates the diffusion of reactants.

Composite (mixed matrix) membranes containing a polymer and a zeolite have been used at relatively low temperatures. Wu *et al.* [171] reported the application of a composite membrane TS-1/PDMS as catalytic contactor in the oxyfuntionalisation of n–hexane by hydrogen peroxide (a biphasic reaction).

8.4.7 Catalyst Retention

The use of transition metal catalysts (TMCs) for homogeneous catalysis in organic reactions is problematic because commonly used TMCs are quite expensive and generate residues, making it necessary to reuse the catalysts after synthesis. The separation of the catalyst from the products and the unconverted reactives is often difficult. A membrane can be used to recover a TMC efficiently from the reaction mixture (Figure 8.6c). This concept has been successfully applied to the Diels–Alder catalysed by a cationic dinuclear Pd(II) complex using a silicalite-1 membrane [172].

In enantioselective hydrolytic reactions, the homogeneous catalysts can be immobilised on one side of the membrane in contact with water, while on the other side of the zeolite membrane is the organic reactive [173]. The catalyst is retained because it is not able to diffuse through the pores of the ZSM-5 layer and is insoluble in water. Furthermore, in this system the products are separated because the obtained epoxide product remains in the organic phase and the diols formed (which are water soluble) diffuse to the aqueous phase.

8.4.8 Encapsulated Catalyst

This combination consists of coating a zeolite onto catalyst particles, usually spherical, and takes advantage of the molecular sieving or the catalytic properties of the zeolite layer (Figure 8.6d). The molecular sieve effect has been proved in two different ways. First, in so-called reactant selectivity, the diffusivity of one reactive is much greater than that of the other reactive (top of Figure 8.6d). In the second, the so-called product selectivity, the reaction is limited by the equilibrium according to the thermodynamics and the diffusivity one of the products is greater than the others and a displacement of the equilibrium occurs (bottom of Figure 8.6d). The hydrogenation of a mixture of linear and ramified alkenes using a Pt/TiO_2 catalyst coated with a 1-silicalite [174,175] layer is an example of reactant selectivity. Toluene disproportionation to give xylene isomers using a silica–alumina catalyst [176] or a H–ZSM-5 catalyst [177] covered with a silicalite layer are examples of product selectivity.

Finally, the catalytic properties of a catalyst capsule have been used in the Fischer–Tropsch synthesis using a Co/SiO_2 catalyst coated with a HZSM-5 membrane [178]. The feed gas ($CO + H_2$) can diffuse through the HZSM-5 membrane and react in the capsule core. When the hydrocarbons formed diffuse through the zeolite membrane, they are cracked and isomerised by acid sites inside the zeolite channels. The low diffusion of long-chain hydrocarbons causes them to stay in the zeolite pores longer and their formation is suppressed. Furthermore, the capsule catalyst has a larger membrane area per unit reactor volume compared to conventional membrane reactors.

8.5 Concluding Remarks

ZMRs appear to be interesting candidates for process integration in the chemical industry to achieve more efficient processes. Consequently, they have been used at laboratory scale for several applications. However, industrial application still seems to be far off. Certain aspects still require further attention, such as reproducibility, long-term stability, trade-off permeability–selectivity, up-scaling, increased membrane area/pore volume ratio by using capillary or multichannel modules, sealing at high temperature and pressure and, of course, cost. For a zeolite module, the cost of which has been published as US\$ 3000 m^{-2} of installed membrane, 10–20% of the cost is attributed to the zeolite membrane itself [14]. Tennison has indicated that costs of more than € 1000 m^{-2} will not be tolerable for zeolite materials in industrial processes [179]. Only the industrial development of pervaporation-assisted esterification, where hydrophilic zeolite membranes are used for removing water, seems a feasible reality in the short term, whenever the stability of the prepared zeolite membrane in an acid medium is increased. The use of a ZMR for displacement of the equilibrium in esterification reactions does not only increase the selectivity but also helps to save the high-energy costs ascribed to reactive distillation (membranes might help save up to 85% of the energy demands ascribed to distillation involving a reduction of 30–50% of the energy costs [103]). Pervaporation is also advantageous compared to distillation when dealing with temperature-sensitive chemicals. If the application of ZMR in water removal applications opens the way, and therefore zeolite membranes become more established in the chemical industry, it may be expected that in the mid–term other types of ZMR will find application niches where they will compete with conventional reactors. A special case is the encapsulated catalyst particle (i.e., like a membrane covering each particle). Since its use will not require large changes in installed reactors, its application would be much easier.

Researchers have made tremendous efforts in the preparation of supported zeolite membranes, studying different synthesis conditions, pre-treatment of the support, post-treatment of the synthesised membrane, strategies, approaches and new materials. As a result there is extensive knowledge of synthesis procedures and formation mechanisms (nucleation and growth) of zeolite layers. Even if zeolite membranes do not find industrial applications in the field of membrane reactors, they can provide benefits in other fields (e.g., corrosion, sensors, medical devices, microreactors, gas separation, liquid separation).

Future trends, in addition to overcoming the limitations of ZMRs by means of a more profound knowledge of reactor engineering and improved material preparation methods, will probably concentrate on the study of new materials. A possible alternative to zeolite membranes at low temperatures are mixed matrix membranes (MMMs) of zeolites and polymers. MMMs have the advantage of combining the benefits of both phases: the superior transport, the catalytic properties and the thermal resistance of zeolites with the desirable mechanical properties, low price, and good processability of polymers. Furthermore, recently developed new porous materials such as metal–organic frameworks (MOFs), with a variety of porous structures and excellent catalytic properties, appear to be excellent candidates for new membrane reactors.

References

1. R.M. Barrer, Porous crystal membranes, *J. Chem. Soc. Faraday Trans.*, **86**, 1123–1130 (1990).
2. S.M. Auerbach, K.A. Carrado and P.K. Dutta (eds), *Handbook of Zeolites Science and Technology*, Marcel Dekker, New York (2003).
3. C. Baerlocher and L.B. McCusker, *Database of zeolite structures,* http://www.iza–structure.org/databases/ (visited on 21 October 2010).
4. Y. Morigami, M. Kondo, J. Abe, H Kita and K. Okamoto, The first large–scale pervaporation plant using tubular-type module with zeolite NaA membrane, *Sep. Pur. Technol.*, **25**, 251–260 (2001).
5. H. Suzuki, Composite membrane having a surface layer of an ultra thin film of cage–shaped zeolite and processes for production thereof, US Patent 4,699,892 (1987).
6. A. Ishikawa, T. Chiang and F. Toda, Separation of water alcohol mixtures by permeation through a zeolite membrane on porous–glass, *J. Chem. Soc. – Chem. Commun.* 764–765 (1989).
7. T. Sano, Y. Kiyozumi, M. Kawamura, F. Mizukami, H. Takaya, T. Mouri, W. Inaoka, Y. Toida, M. Watanabe and K. Toyoda, Preparation and characterization of ZSM-5 zeolite film, *Zeolites*, **11**, 842–845 (1991).
8. M. Tsapatsis, G. Xomeritakis, H.W. Hillhouse, S. Nair, V. Nikolakis, G. Bonilla and Z. Lai, Zeolite membranes – a review, *Cattech*, **3**, 148–163 (2000).

9. T. Bein, Synthesis and applications of molecular sieve layers and membranes, *Chem. Mater.*, **8**, 1636–1653 (1996).
10. M. Matsukata and E. Kikuchi, Zeolitic membranes: synthesis, properties, and prospects, *Bull. Chem. Soc. Jpn*, **70**, 2341–2356 (1997).
11. J. Coronas and J. Santamaría, Separations using zeolite membranes, *Sep. Pur. Methods*, **28**, 127–177 (1999).
12. A. Tavolaro and E. Drioli, Zeolite Membranes, *Adv. Mater.*, **11**, 975–996 (1999).
13. J. Caro and M. Noack, Zeolite membranes – Recent developments and progress, *Micropor. Mesopor. Mater.*, **115**, 215–233 (2008).
14. J. Caro, M. Noack, P. Kölsch and R. Schäfer, Zeolite membranes–state of their development and perspective, *Micropor. Mesopor. Mater.*, **38**, 3–24 (2000).
15. S.T. Chiang and K.-J. Chao, Membranes and films of zeolite and zeolite-like materials, *J. Phys. Chem. Solids*, **62**, 1899–1910 (2001).
16. A. Julbe, Zeolite membranes – synthesis, characterization and application, *Stud. Surf. Sci. Catal.*, **168**, 181–219 (2007).
17. E.E. McLeary, J.C. Jansen, F. Kapteijn, Zeolite based films, membranes and membrane reactors: progress and prospects, *Micropor. Mesopor. Mater.*, **90**, 198–220 (2006).
18. P. Gorgojo, O. de la Iglesia, J. Coronas, Preparation and characterization of zeolite membranes, in *Inorganic Membranes Synthesis, Characterization and Applications (Membrane Science and Technology Series 13)*, R. Mallada and M. Menéndez (eds), Elsevier, Rotterdam (2008).
19. J.D.F. Ramsay and S. Kalls, Zeolite membranes, in *Membrane Science and Technology, Vol 6: Recent Advances in Gas Separation by Microporous Ceramic Membranes*, p. 373–395, N.K. Kanellopoulos (ed), Elsevier, Rotterdam (2000).
20. K. Jansen, T. Maschmeyer, Progress in zeolitic membranes, *Top. Catal.*, **9**, 113–122 (1999).
21. D. Casanave, A. Giroir–Fendler, J. Sanchez, R. Loutaty and J.A. Dalmon, Control of transport–properties with a microporous membrane reactor to enhance yields in dehydrogenation reactions, *Catal. Today*, **25**, 309–314 (1995).
22. A. Pantazidis, J.A. Dalmon and C. Mirodatos, Oxidative dehydrogenation of propane on catalytic membrane reactors, *Catal. Today*, **25**, 403–408 (1995).
23. J. Coronas and J. Santamaría, State–of–the–art in zeolite membrane reactors, *Top. Catal.*, **29**, 29–44 (2004).
24. K. Sahner, G. Hagen, D. Schoenauer, S. Reiss and R. Moos, Zeolites – Versatile materials for gas sensors, *Solid State Ionics*, **179**, 2416–2423 (2008).
25. R. Cai and Y. Yan, Corrosion–resistant zeolite coatings, *Corrosion*, **64**, 271–278 (2008).
26. S. Aguado, J. Gascón, J. C. Jansen and F. Kapteijn, Continuous synthesis of NaA zeolite membranes, *Micropor. Mesopor. Mater.*, **120**, 170–176 (2009).
27. J.L.H. Chau, C. Téllez, K.L. Yeung and K. Ho, The role of surface chemistry in zeolite membrane formation, *J. Membr. Sci.*, **164**, 257–275 (2000).
28. J. Hedlund, F. Jareman, A.–J. Bons and M. Anthonis, A masking technique for high quality MFI membranes, *J. Membr. Sci.*, **222**, 163–179 (2003).
29. H. Kita, K Horii, Y. Ohtoshi, K. Tanaka and K.I. Okamoto, Synthesis of a zeolite NaA membrane for pervaporation of water–organic liquid–mixtures, *J. Mater. Sci. Lett.*, **14**, 206–208 (1995).
30. M.C. Lovallo and M. Tsapatsis, Preferentially oriented submicron silicalite membranes, *AIChE J.*, **42**, 3020–3029 (1996).
31. M.C. Lovallo and M. Tsapatsis and T. Okubo, Preparation of an asymmetric zeolite L film, *Chem. Mater.*, **8**, 1579–1583 (1996).
32. L.C. Boudreau and M. Tsapatsis, A highly oriented thin film of zeolite A, *Chem. Mater.*, **9**, 1705–1709 (1997).

33. L.T.Y. Au, W.Y. Mui, P.S. Lau, C. Tellez and K.L. Yeung, Engineering the shape of zeolite crystal grain in MFI membranes and their effects on the gas permeation properties, *Micropor. Mesopor. Mater.*, **47**, 203–216 (2001).
34. S.L. Burkett and M.E. Davis, Mechanisms of structure direction in the synthesis of pure–silica zeolites.1. Synthesis of TPA/Si–ZSM-5 *Chem. Mater.*, **7**, 920–928 (1995).
35. I. Tiscornia, S. Valencia, A. Corma, C. Téllez, J. Coronas and J. Santamaría, Preparation of ITQ–29 (Al-free zeolite A) membranes, *Micropor. Mesopor. Mater.*, **110**, 303–309 (2008).
36. J. Dong, Y.S. Lin, M.Z.–C. Hu, R.A. Peascoe, E.A. Payzant, Template–removal–associated microstructural development of porous–ceramic–supported MFI zeolite membranes, *Micropor. Mesopor. Mater.*, **34**, 241–253 (2000).
37. M.J. den Exter, H. van Bekkum, C.J.M. Rijn, F. Kapteijn, J.A. Moulijn, H. Schellevis and C.I.N. Beenakker. Stability of oriented silicalite-1 films in view of zeolite membrane preparation, *Zeolites* **19**, 13–20 (1997).
38. S. Heng, P.P.S. Lau, K.L. Yeung, M. Djafer and J.–C. Schrotter, Low–temperature ozone treatment for organic template removal from zeolite membrane, *J. Membr. Sci.*, **243**, 69–78 (2004).
39. W. Xu, J. Dong, J. Li, W. Li and F. Wu, A novel method for the preparation of zeolite ZSM-5, *J. Chem. Soc. Chem. Commun.*, **1990**, 755–756 (1990).
40. M. Matsukata, N. Nishiyama and K. Ueyama, Zeolitic membrane synthesized on a porous alumina support, *J. Chem. Soc. Chem. Commun.*, **1994**, 339–340 (1994).
41. S. Alfaro, M. Arruebo, J. Coronas, M. Menéndez and J. Santamaría, Preparation of MFI type tubular membranes by steam–assisted crystallization, *Micropor. Mesopor. Mater.*, **50**, 195–200 (2001).
42. A. Navajas, R. Mallada, C. Téllez, J. Coronas, M. Menéndez and J. Santamaría, Study on the reproducibility of mordenite tubular membranes used in the dehydration of ethanol, *J. Membr. Sci.*, **299**, 166–173 (2007).
43. T. Seike, M. Matsuda and M. Miyake, Preparation of FAU type zeolite membranes by electrophoretic deposition and their separation properties, *J. Mater. Chem.*, **12**, 366–368 (2002).
44. N. Navascués, C. Téllez and J. Coronas, Synthesis and adsorption properties of hollow silicalite-1 spheres, *Micropor. Mesopor. Mater.*, **112**, 561–572 (2008).
45. A. Huang, Y.S. Lin and W. Yang, Synthesis and properties of A-type zeolite membranes by secondary growth method with vacuum seeding, *J. Membr. Sci.*, **245**, 41–51 (2004).
46. C. Algieri, P. Bernardo, G. Barbieri and E. Drioli, A novel seeding procedure for preparing tubular NaY zeolite membranes, *Micropor. Mesopor. Mater.*, **119**, 129–136 (2009).
47. M. Pera–Titus, J. Llorens, F. Cunill, R. Mallada and J. Santamaría, Preparation of zeolite NaA membranes on the inner side of tubular supports by means of a controlled seeding technique, *Catal. Today*, **104**, 281–287 (2005).
48. I. Kumakiri, T. Yamaguchi and S. Nakao, Preparation of zeolite A and faujasite membranes from a clear solution, *Ind. Eng. Chem. Res.*, **38**, 4682–4688 (1999).
49. G. Li, E. Kikuchi and M. Matsukata, The control of phase and orientation in zeolite membranes by the secondary growth method, *Micropor. Mesopor. Mater.*, **62**, 211–220 (2003).
50. Z. Lai, G. Bonilla, I. Diaz, J.G. Nery, K. Sujaoti, M.A. Amat, E. Kokkoli, O. Terasaki, R.W. Thompson, M. Tsapatsis and D.G. Vlachos, Microstructural optimization of a zeolite membrane for organic vapor separation, *Science*, **300**, 456–460 (2003).
51. N. Nishiyama, M. Yamaguchi, T. Katayama, Y. Hirota, M. Miyamoto, Y. Egashira, K. Ueyama, K. Nakanishi, T. Ohta, A. Mizusawa and T. Satoh, Hydrogen–permeable membranes composed of zeolite nano–blocks, *J. Membr. Sci.*, **306**, 349–354 (2007).
52. K. Aoki, K. Kusakabe and S. Morooka, Separation of gases with an A-type zeolite membrane, *Ind. Eng. Chem. Res.*, **39**, 2245–2251 (2000).

53. K. Aoki, K. Kusakabe and S. Morooka, Preparation of oriented A-type zeolite membranes, *AIChE J.*, **46**, 221–224 (2000).
54. X. Xu, W. Yang, J. Liu and L. Lin, Synthesis of NaA zeolite membrane by microwave heating, *Sep. Pur. Technol.*, **25**, 241–249 (2001).
55. F. Tiscareño–Lechuga, C. Téllez, M. Menéndez and J. Santamaría, A novel device for preparing zeolite–A membranes under a centrifugal force field, *J. Membr. Sci.*, **212**, 135–146 (2003).
56. X.C. Xu, W.S. Yang, J. Liu and L.W. Lin, Synthesis of NaA zeolite membranes from clear solution, *Micropor. Mesopor. Mater.*, **43**, 299–311 (2001).
57. Y. Yan, S.R. Chaudhuri and A. Sarkar, Synthesis of oriented zeolite molecular sieve films with controlled morphologies, *Chem. Mater.*, **8**, 473–479 (1996).
58. M.P. Pina, M. Arruebo, A. Felipe, F. Fleta, M.P. Bernal, J. Coronas, M. Menéndez and J. Santamaría, A semi–continuous method for the synthesis of NaA zeolite membranes on tubular supports, *J. Membr. Sci.*, **244**, 141–150 (2004).
59. S. Aguado, J. Gascón, J.C. Jansen and F. Kapteijn, Continuous synthesis of NaA zeolite membranes, *Micropor. Mesopor. Mater.*, **120**, 170–176 (2009).
60. S. Yamazaki and K. Tsutsumi, Synthesis of A-type zeolite membrane using a plate heater and its formation mechanism, *Micropor. Mesopor. Mater.*, **37**, 67–80 (2000).
61. H. Richter, I. Voight, P. Fischer and P. Puhlfurb, Preparation of zeolite membranes on the inner surface of ceramic tubes and capillaries, *Sep. Pur. Technol.*, **32**, 133–138 (2003).
62. P.Z. Culfaz, A. Culfaz and H. Kalipcilar, Preparation of MFI type zeolite membranes in a flow system with circulation of the synthesis solution, *Micropor. Mesopor. Mater.*, **92**, 134–144 (2006).
63. E. Mateo, R. Lahoz, G.F. de la Fuente, A. Paniagua, J. Coronas and J. Santamaría, Growth of silicalite-1 by a method involving separation of reactants, *Chem. Mater.*, **19**, 594–599 (2007).
64. W. Xiao, J. Yang, J. Lu and J. Wang, A novel method to synthesize high performance silicalite-1 membrane, *Sep. Pur. Tech.*, **67**, 58–63 (2009).
65. S. Miachon, P. Ciavarella, L. van Dyk, I. Kumakiri, K. Fiaty, Y. Schuurman and J.A. Dalmon, Nanocomposite MFI–alumina membranes via pore–plugging synthesis: specific transport and separation properties, *J. Membr. Sci.*, **298**, 71–79 (2007).
66. G. Guan, T. Tanaka, K. Kusakabe, K. Sotowa and S. Morooka, Characterization of AlPO(4)-type molecular sieving membranes formed on a porous alpha–alumina tube, *J. Membrane Sci.*, **214**, 191–198 (2003).
67. J.C. Poshusta, V.A. Tuan, J.L. Falconer and R.D. Noble, Synthesis and permeation properties of SAPO–34 tubular membranes, *Ind. Eng. Chem. Res.*, **37**, 3924–3929 (1998).
68. L.T.Y. Au, J.L. Chau, C. Tellez and K.L. Yeung, Preparation of supported Sil–1, TS–1 and VS–1 membranes – Effects of Ti and V metal ions on the membrane synthesis and permeation properties *J. Membr. Sci.*, **183**, 269–291 (2001).
69. A. Tavolaro, VS–1 composite membrane: preparation and characterization, *Desalination*, **147**, 333–338 (2002).
70. J. Rocha and Z. Lin, Microporous mixed octahedral–pentahedral–tetrahedral framework silicates, *Rev. Mineral. Geochem.*, **57**, 173–201 (2005).
71. Z. Lin, J. Rocha, A. Navajas, C. Téllez, J. Coronas and J. Santamaría, Synthesis and characterisation of titanosilicate ETS–10 membranes, *Micropor. Mesopor. Mater.*, **67**, 79–86 (2004).
72. I. Tiscornia, S. Irusta, C. Tellez, J. Coronas and J Santamaría, Separation of propylene/propane mixtures by titanosilicate ETS–10 membranes prepared in one–step seeded hydrothermal synthesis, *J. Membr. Sci.*, **311**, 326–335 (2008).
73. C.M. Braunbarth, L.C. Boudreau and M. Tsapatsis, Synthesis of ETS–4/TiO_2 composite membranes and their pervaporation performance, *J. Membr. Sci.*, **174**, 31–42 (2000).
74. B. Yilmaz, K.G. Shattuck, J. Warzywoda and A. Sacco, Oriented growth of ETS–4 films using the method of secondary growth, *Chem. Mater.*, **18**, 1107–1112 (2006).

75. V. Sebastián, Z. Lin, J. Rocha, C. Téllez, J. Santamaría and J. Coronas, Synthesis, characterization, and separation properties of Sn– and Ti–silicate umbite membranes, *Chem. Mater.*, **18**, 2472–2479 (2006).
76. Y. Kiyozumi, T. Nagase, Y. Hasegawa and F. Mizukami, A process for synthesizing bilayer zeolite membranes, *Mater. Lett.*, **62**, 436–439 (2008).
77. O. de la Iglesia, S. Irusta, R. Mallada, M. Menéndez, J. Coronas and J. Santamaría, Preparation and characterization of two–layered mordenite–ZSM-5 bi–functional membranes, *Micropor. Mesopor. Mater.*, **93**, 318–324 (2006).
78. X. Zhang, H. Liu and K.L. Yeung, Novel two–layered zeolite NaA–silicalite-1 membranes, *J. Phys. Chem. Solids*, **66**, 1034–1038 (2005).
79. M. Cheng, G. Hu, D. Tan and X. Bao, On the formation of vertically oriented MCM–22 zeolite crystal films, *Micropor. Mesopor. Mater.*, **50**, 69–76 (2001).
80. C. Liu, J. Wang and B. Li, Preparation and characterization of ordered mesoporous silica membrane, *J. Non–Crystalline Solids*, **351**, 409–412 (2005).
81. N. Nishiyama, D.H. Park, A. Koide, Y. Egashira and K. Ueyama, A mesoporous silica (MCM–48) membrane: preparation and characterization, *J. Membr. Sci.*, **182**, 235–244 (2001).
82. N. Nishiyama, A. Koide, Y. Egashira and K. Ueyama, Mesoporous MCM–48 membrane synthesized on a porous stainless steel support, *Chem. Commun.* 2147–2148 (1998).
83. O. de la Iglesia, M. Pedernera, R. Mallada, Z. Lin, J. Rocha, J. Coronas and J. Santamaría, Synthesis and characterization of MCM–48 tubular membranes, *J. Membr. Sci.*, **280**, 867–875 (2006).
84. M. Nomura, T. Yamaguchi and S. Nakao, Silicalite membranes modified by counterdiffusion CVD technique, *Ind. Eng. Chem. Res.*, **36**, 4217–4223 (1997).
85. Y. Yan, M.E. Davis and G.R. Gavalas, Preparation of highly selective zeolite ZSM-5 membranes by a post-synthetic coking treatment, *J. Membr. Sci.*, **123**, 95–103 (1997).
86. F. Morón, M.P. Pina, E. Urriolabeitia, M. Menéndez and J. Santamaría, Preparation and characterization of Pd–zeolite composite membranes for hydrogen separation, *Desalination*, **147**, 425–431 (2002).
87. A. Navajas, R. Mallada, C. Téllez, J. Coronas, M. Menéndez and J. Santamaría, The use of post-synthetic treatments to improve the pervaporation performance of mordenite membranes, *J. Membr. Sci.*, **270**, 32–41 (2006).
88. G. Li, X.H. Su and R.S. Lin, Highly water–selective mordenite membranes prepared via hydrothermal post-treatment in a dilute hydrochloric acid solution, *Sep. Pur. Tech.*, **61**, 225–228 (2008).
89. G. Li, X.–H. Su and R.–S. Lin, Preparation of highly water–selective mordenite membranes via post-synthetic treatment with oxalic acid, *Mater. Lett.*, **61**, 4576–4578 (2007).
90. G. Li, X.H. Su and R.S. Lin, Effect of post-treatment in the hydrochloric acid solution on the pervaporation performance of mordenite membranes, *Sep. Pur. Technol.*, **56**, 378–382 (2007).
91. J. Coronas, J.L. Falconer, R.D. Noble, Characterization and permeation properties of ZSM-5 tubular membranes, *AIChE J.*, **43**, 1797–1812 (1997).
92. S. Aguado, J. Coronas and J. Santamaría, Use of zeolite membrane reactors for the combustion of VOCs present in air at low concentrations, *Chem. Eng. Res. Des.*, **83**, 295–301 (2005).
93. S.L. Wee, C.T. Tye and S. Bhatia, Membrane separation process – Pervaporation through zeolite membrane, *Sep. Pur. Technol.*, **63**, 500–516 (2008).
94. M. Yamamoto, N. Munehisa, M. Kaibara, K. Horii, K. Tanaka, H. Kita and K.–I. Okamoto, Vapor–permeation–aided esterification of oleic acid: application of pressurized vapor circulation system and zeolite membrane. *Maku (Membrane)*, **20**, 143–148 (1995).
95. J.J. Jafar, P.M. Budd and R. Hughes, Enhancement of esterification reaction yield using zeolite – a vapour permeation membrane, *J. Membr. Sci.*, **199**, 117–123 (2002).

96. K. Tanaka, R. Yoshikawa, C. Ying, H. Kita and K. Okamoto, Application of zeolite membranes to esterification reactions, *Catal. Today*, **67**, 121–125 (2001).
97. K. Tanaka, R. Yoshikawa, C. Ying, H. Kita and K. Okamoto, Application of zeolite T membrane to vapor–permeation–aided esterification of lactic acid with ethanol, *Chem. Eng. Sci.*, **57**, 1577–1584 (2002).
98. O. de la Iglesia, R. Mallada, M. Menéndez and J. Coronas, Continuous zeolite membrane reactor for esterification of ethanol and acetic acid, *Chem. Eng. Sci.*, **131**, 35–39 (2007).
99. S.M. Lai, C.P. Ng, R. Martin–Aranda and K.L. Yeung, Knoevenagel condensation reaction in zeolite membrane microreactor, *Micropor. Mesopor. Mater.*, **66**, 239–252 (2003).
100. X.F. Zhang, E.S.M. Lai, R. Martin–Aranda and K.L. Yeung, An investigation of Knoevenagel condensation reaction in microreactors using a new zeolite catalyst, *Appl. Catal. A – Gen.*, **261**, 109–118 (2004).
101. W.N. Lau, K.L. Yeung and R. Martin–Aranda, Knoevenagel condensation reaction between benzaldehyde and ethyl acetoacetate in microreactor and membrane microreactor, *Micropor. Mesopor. Mater.*, **115**, 156–163 (2008).
102. M.A. Salomon, J. Coronas, M. Menéndez and J. Santamaría, Synthesis of MTBE in zeolite membrane reactors, *Appl. Catal. A – Gen.*, **200**, 201–210 (2000).
103. M. Pera–Titus, J. Llorens and F. Cunill, Technical and economical feasibility of zeolite NaA membrane–based reactors in liquid–phase etherification reactions, *Chem. Eng. Process.*, **48**, 1072–1079 (2009).
104. S.Y. Lim, B. Park, F. Hung, M. Sahimi, T.T. Tsotsis, Design issues of pervaporation membrane reactors for esterification, *Chem. Eng. Sci.*, **57**, 4933–4946 (2002).
105. G. Li, E. Kikuchi and M. Matsukata, A study on the pervaporation of water–acetic acid mixtures through ZSM-5 zeolite membranes, *J. Membr. Sci.*, **218**, 185–194 (2003).
106. L. Casado, R. Mallada, C. Tellez, J. Coronas, M. Menéndez and J. Santamaría, Preparation, characterization and pervaporation performance of mordenite membranes *J. Membr. Sci.*, **216**, 135–147 (2003).
107. F. Aiouache and S. Goto, Reactive distillation–pervaporation hybrid column for tert–amyl alcohol etherificatin with ethanol, *Chem. Eng. Sci.*, **58**, 2465–2477 (2003).
108. K. Sundmacher and A. Kienle (eds), *Reactive Distillation, Status and Future Directions*, Wiley–VCH, Weinheim (2003).
109. K.C.S. Figueiredo, V.M. M. Salim, C.P. Borges, Synthesis and characterization of a catalytic membrane for pervaporation–assisted esterification reactors, *Catal. Today*, **133/135**, 809–814 (2008).
110. J.M. Fraile, R. Mallada, J.A. Mayoral, M. Menéndez, L. Roldán, Shift of multiple incompatible equilibriums by a combination of heterogeneous catalysis and membranes, *Chem Eur. J.*, **16**, 3296–3299 (2010).
111. F. Tiscareño–Lechuga, M.A. Anderson and G.C. Hill Jr, Effect of dilution in the experimental dehydrogenation of cyclohexane in hybrid membrane reactors, *J. Membr. Sci.*, **118**, 85–92 (1996).
112. P. Ciavarella, H. Moueddeb, S. Miachon, K. Fiaty and J.A. Dalmon, Experimental study and numerical simulation of hydrogen/isobutane permeation and separation using MFI–zeolite membrane reactor, *Catal. Today*, **56**, 253–264 (2000).
113. D. Casanave, P. Ciavarella, K. Fiaty and J.A. Dalmon, Zeolite membrane reactor for isobutane dehydrogenation: experimental results and theoretical modelling, *Chem. Eng. Sci.*, **54**, 2807–2815 (1999).
114. A. Giroir-Fendler, J. Peureux, H. Mozzanega, J.A. Dalmon, Characterization of a zeolite membrane for catalytic membrane reactor application, *Stud. Surf. Sci. Catal.*, **101**, 127–136 (1996).

115. P. Ciavarella, D. Casanave, H. Moueddeb, S. Miachon, K. Fiaty and J.A. Dalmon. Isobutane dehydrogenation in a membrane reactor: Influence of the operating conditions on the performance, *Catal. Today*, **67**, 177–184 (2001).
116. U. Illgen, R. Schäfer, M. Noack, P. Kölsch, A. Kühnle and J. Caro. Membrane supported catalytic dehydrogenation of iso-butane using an MFI zeolite membrane reactor, *Catal. Comm.*, **2**, 339–345 (2001).
117. L. van Dyk, S. Miachon, L. Lorenzen, M. Torres, K. Fiaty and J.A. Dalmon, Comparison of microporous MFI and dense Pd membrane performances in an extractor-type CMR, *Catal. Today*, **82**, 167–177 (2003).
118. B.H. Jeong, K.I. Sotowa and K. Kusakabe, Catalytic dehydrogenation of cyclohexane in an FAU-type zeolite membrane reactor, *J. Membr. Sci.*, **224**, 151–158 (2003).
119. C.L. Kong, J.M. Lu, H.H. Yang and J.Q. Wang, Catalytic dehydrogenation of ethylbenzene to styrene in a zeolite silicalite-1 membrane reactor, *J. Membr. Sci.*, **306**, 29–35 (2007).
120. W. Yang, P. Yang, X. Xu and L. Lin, Propane aromatization in a silicalite-1 membrane reactor, *Stud. Surf. Sci. Catal.*, **130**, 2699–2704 (2000).
121. B.S. Liu, L.Z. Gao and C.T. Au, Preparation, characterization and application of a catalytic NaA membrane for CH_4/CO_2 reforming to syngas, *Appl. Catal. A – Gen.*, **235**, 193–206 (2002).
122. T. Ostrowski, A. Giroir–Fendler, C. Mirodatos and L. Mleczko, Comparative study of the partial oxidation of methane to synthesis gas in fixed–bed and fluidized–bed membrane reactors. Part II – Development of membranes and catalytic measurements *Catal. Today*, **40**, 191–200 (1998).
123. R. P. W. J. Struis, S. Stucki and M. Wiedorn, A membrane reactor for methanol synthesis, *J. Membr. Sci.*, **113**, 93–100 (1996).
124. M. Menéndez, E. Piera, J. Coronas and J. Santamaría, A zeolite membrane reactor to obtain methanol and other alcohols from synthesis gas, Spanish Patent application 9901722 (1999).
125. F. Gallucci, L. Paturzo and A. Basile, An experimental study of CO_2 hydrogenation into methanol involving a zeolite membrane reactor, *Chem. Eng. Proc.*, **43**, 1029–1036 (2004).
126. A. Tavolaro and P. Tavolaro, LTA zeolite composite membrane preparation, characterization and application in a zeolitic membrane reactor, *Catal. Commun.*, **8**, 789–794 (2007).
127. G. Barbieri, G. Marigliano, G. Golemme and E. Drioli, Simulation of CO_2 hydrogenation with CH_3OH removal in a zeolite membrane reactor, *Chem. Eng. J.*, **85**, 53–59 (2002).
128. K.–I. Sawamura, T. Shirai, M. Takada, Y. Sekine, E. Kikuchi and M. Matsukata, Selective permeation and separation of steam from water–methanol–hydrogen gas mixtures through mordenite membrane, *Catal. Today*, **132**, 182–187 (2008).
129. E. Sjöberg, L. Sandström and J. Hedlund, Membrane processes for effective methanol synthesis in the forest based biorefinery, *Int. Conf. Catal. Membr. Reactors*, Lyon (France), June/July (2009).
130. A.M. Tarditi, S. Irusta and E.A. Lombardo, Xylene isomerization in a membrane reactor – Part I: The synthesis of MFI membranes for the p–xylene separation, *Chem. Eng. J.*, **122**, 167–174 (2006).
131. L. van Dyk, L. Lorenzen, S. Miachon, J.A. Dalmon, Xylene isomerization in an extractor type catalytic membrane reactor. *Catal. Today*, **104**, 274–280 (2005).
132. C. Zhang, Z. Hong, X. Gu, Z. Zhong, W. Jin and N. Xu, Silicalite-1 zeolite membrane reactor packed with HZSM-5 catalyst for meta–xylene isomerization, *Ind. Eng. Chem. Res.*, **48**, 4293–4299 (2009).
133. Y. Takata, T. Tsuru, T. Yoshioka and M. Asaeda, Gas permeation properties of MFI zeolite membranes prepared by the secondary growth of colloidal silicalite and application to the methylation of toluene, *Micropor. Mesopor. Mater.*, **54**, 257–268 (2002).
134. J.M. van de Graaf, M. Zwiep, F. Kapteijn and J.A. Moulijn, Application of a zeolite membrane reactor in the metathesis of propene, *Chem. Eng. Sci.*, **54**, 1441–1445 (1999).

135. J.M. van de Graaf, M. Zwiep, F. Kapteijn and J.A. Moulijn, Application of a silicalite-1 membrane reactor in metathesis reactions, *Appl. Catal. A – Gen.*, **178**, 225–241 (1999).
136. M.L. Maloncy, T. Maschmeyer and J.C. Jansen, Technical and economical evaluation of a zeolite membrane based heptane hydroisomerization process, *Chem. Eng. J.*, **106**, 187–195 (2005).
137. R.L. Espinoza, E. du Toit, J. Santamaría, M. Menéndez, J. Coronas and S. Irusta, Use of membranes in Fischer–Tropsch reactors, *Stud. Surf. Sci. Catal.*, **130**, 389–394 (2000).
138. R. Espinoza, J. Santamaría, M. Menéndez, J. Coronas and S. Irusta, Production of hydrocarbons, US Patent 6403660–B (1998).
139. R. Espinoza, J. Santamaría, M. Menéndez, J. Coronas and S. Irusta, Removal of water from a Fischer–Tropsch system using zeolite–based membranes, Sud–african Patent application 985093 (1998).
140. M. P. Rohde, D. Unruh and G. Schaub, Membrane application in Fischer–Tropsch synthesis to enhance CO_2 hydrogenation, *Ind. Eng. Chem. Res.*, **44**, 9653–9658 (2005).
141. W. Zhu, L. Gora, A.W.C. van den Berg, F. Kapteijn, J.C. Jansen and J.A. Moulijn, Water vapour separation from permanent gases by a zeolite–4A membrane, *J. Membr. Sci.*, **253**, 57–66 (2005).
142. M.P. Rhode, G. Schaub, S. Khajavi, J.C. Jansen and F. Kapteijn, Fischer–Tropsch synthesis with in–situ removal. Directions of membrane development, *Micropor. Mesopor. Mater.*, **115**, 123–126 (2008).
143. M. Nomura, T. Bin, S.–I. Nakao, Selective ethanol extraction from fermentation broth using a silicalite membrane, *Sep. Pur. Tech.*, **27**, 59–66 (2002).
144. T. Ikegami, H. Yanagishita, D. Kitamoto, K. Haraya, T. Nakane, H. Matsuda, N. Koura and T. Sano, Highly concentrated aqueous ethanol solutions by pervaporation using silicalite membrane – Improvement of ethanol selectivity by addition of sugars to ethanol solution, *Biotech. Lett.*, **21**, 1037–1041 (1999).
145. E. Piera, C. Téllez, J. Coronas, M. Menéndez and J. Santamaría, Use of zeolite membrane reactors for selectivity enhancement: application to the liquid–phase oligomerization of i–butene, *Catal. Today*, **67**, 127–138 (2001).
146. O. Levenspiel, *Chemical Reaction Engineering* (2nd edn), John Wiley & Sons, Inc., New York (1972).
147. D. Lafarga, J. Santamaría, M. Menéndez, Methane oxidative coupling using porous ceramic membrane reactors. 1. reactor development *Chem. Eng. Sci.*, **49**, 2005–2013 (1994).
148. J. Coronas and J. Santamaría, Catalytic reactors based on porous ceramic membranes, *Catal. Today*, **51**, 377–389 (1999).
149. A. Cruz–Lopez, N. Guilhaume, S. Miachon and J.A. Dalmon, Selective oxidation of butane to maleic anhydride in a catalytic membrane reactor adapted to rich butane feed, *Catal. Today*, **107/108**, 949–956 (2005).
150. S. Mota, S. Miachon, J.C. Volta and J.A. Dalmon, Membrane reactor for selective oxidation of butane to maleic anhydride, *Catal. Today*, **67**, 169–176 (2001).
151. D. Farrusseng, A. Julbe and C. Guizard, Evaluation of porous ceramic membranes as O_2 distributors for the partial oxidation of alkanes in inert membrane reactors, *Sep. Pur. Technol.*, **25**, 137–149 (2001).
152. J.A. Dalmon, A. Cruz–López, D. Farrusseng, N. Guilhaume, E. Iojoiu, J. Jalibert, S. Miachon, C. Mirodatos, A. Pantazidis, M. Rebeilleau–Dassonneville, Y. Schuurman, and A.C. van Veen, Oxidation in catalytic membrane reactors, *Appl. Catal. A: Gen.*, **325**, 198–204 (2007).
153. J. Coronas, M. Menéndez and J. Santamaría, The porous–wall ceramic membrane reactor: an inherently safer contacting device for gas–phase oxidation of hydrocarbons, *J. Loss Prev. Process Ind.*, **8**, 97–101 (1995).
154. B. Chommeleoux, S. Cimaomo, E. Jolimaitre, D. Uzio, P. Magnoux and J. Sanchez, New membrane for use as hydrogen distributor for hydrocarbon selective hydrogenation, *Micropor. Mesopor. Mater.*, **109**, 28–37 (2008).

155. S. Heng, K.L. Yeung, A. Julbe, A. Ayral and J.C. Schrotter, Preparation of composite zeolite membrane separator/contactor for ozone water treatment, *Micropor. Mesopor. Mater.*, **115**, 137–146 (2008).
156. M.P. Bernal, J. Coronas, M. Menéndez and J. Santamaría, Coupling of reaction and separation at the microscopic level: esterification processes in a H–ZSM-5 membrane reactor, *Chem. Eng. Sci.*, **57**, 1557–1562 (2002).
157. S. Haag, M. Hanebuth, G.T.P. Mabande, A. Avhale, W. Schwieger and R. Dittmeyer, On the use of a catalytic H–ZSM-5 membrane for xylene isomerization, *Micropor. Mesopor. Mater.*, **96**, 168–176 (2006).
158. A.M. Tarditi, G. I. Horowitz and E.A. Lombardo, Xylene isomerization in a ZSM-5/SS membrane reactor, *Catal. Lett.*, **123**, 7–15 (2008).
159. Y.Y. Fong, A.Z. Abdullah, A.L. Ahmad and S. Bhatia, Development of functionalized zeolite membrane and its potential role as reactor combined separator for para–xylene production from xylene isomers, *Chem. Eng. J.*, **139**, 172–193 (2008).
160. T. Westermann and T. Melin, Flow-through catalytic membrane reactors – Principles and applications, *Chem. Eng. Proc.*, **48**, 17–28 (2009).
161. Y. Hasegawa, K. Kusakabe and S. Morooka, Selective oxidation of carbon monoxide in hydrogen–rich mixtures by permeation through a platinum–loaded Y-type zeolite membrane, *J. Membr. Sci.*, **190**, 1–8 (2001).
162. Y. Hasegawa, K. Sotowa, K. Kusakabe and S. Morooka, The influence of feed composition on CO oxidation using zeolite membranes loaded with metal catalysts, *Micropor. Mesopor. Mater.*, **53**, 37–43 (2002).
163. K. Sotowa, Y. Hasegawa, K. Kusakabe and S. Morooka, Enhancement of CO oxidation by use of H_2–selective membranes impregnated with noble–metal catalysts, *Int. J. Hydrogen Energy*, **27**, 339–346 (2002).
164. P. Bernardo, C. Algieri, G. Barbieri and E. Drioli, Hydrogen purification from carbon monoxide by means of selective oxidation using zeolite catalytic membranes, *Sep. Pur. Technol.*, **62**, 629–635 (2008).
165. T. Tago, K. Iwakai, K. Morita, K. Tanaka and T. Masuda, Control of acid–site location of ZSM-5 zeolite membrane and its application to the MTO reaction, *Catal. Today*, **105**, 662–666 (2005).
166. T. Masuda, T. Asanuma, M. Shouji, S.R. Mukai, M. Kawase and K. Hashimoto, Methanol to olefins using ZSM-5 zeolite catalyst membrane reactor, *Chem. Eng. Sci.*, **58**, 649–656 (2003).
167. M. Torres, L. Lopez, J.M. Dominguez, A. Mantilla, G. Ferrat, M. Gutierrez and M. Maubert, Olefins catalytic oligomerization on new composites of beta–zeolite films supported on [alpha]–Al_2O_3 membranes, *Chem. Eng. J.*, **92**, 1–6 (2003).
168. A.J. Maira, W.N. Lau, C.Y. Lee, P.L. Yue, C.K. Chan and K.L. Yeung, Performance of a membrane–catalyst for photocatalytic oxidation of volatile organic compounds, *Chem. Eng. Sci.*, **58**, 959–962 (2003).
169. A. Julbe, D. Farrusseng, J.C. Jalibert, C. Mirodatos and C. Guizard, Characteristics and performance in the oxidative dehydrogenation of propane of MFI and V–MFI zeolite membranes, *Catal. Today*, **56**, 199–209 (2000).
170. H.J. Sloot, G.F. Versteeg and W.P.M. Van Swaaij, A non–perm-selective membrane reactor for chemical processes normally requiring strict stoichiometric feed rates of reactants, *Chem. Eng. Sci.*, **45**, 2415–2421 (1990).
171. S. Wu, J.E. Gallot, M. Bousmina, C. Bouchard and S. Kaliaguine, Zeolite containing catalytic membranes as interphase contactors, *Catal. Today*, **56**, 113–129 (2000).
172. I. Gallego, R. Mallada, E.P. Urriolabeitia, R. Navarro, M. Menéndez and J. Santamaría, Selective separation of homogeneous catalysts using silicalite membranes, *Inorganica Chimica Acta*, **357**, 4577–4581 (2004).

173. S.D. Choi, G.J. Kim, Enantioselective hydrolytic kinetic resolution of epoxides catalyzed by chiral Co(III) salen complexes immobilized in the membrane reactor, *Catal. Lett.*, **92**, 35–40 (2004).
174. N. Nishiyama, K. Ichioka, D.H. Park, Y. Egashira, K. Ueyama, L. Gora, W.D. Zhu, F. Kapteijn and J.A. Moulijn, Reactant–selective hydrogenation over composite silicalite-1–coated Pt/TiO_2 particles, *Ind. Eng. Chem. Res.*, **43**, 1211–1215 (2004).
175. N. van der Puil, E.J. Creyghton, E.C. Rodenburg, T.S. Sie, H. van Bekkum and J.C. Jansen, Catalytic testing of TiO_2/platinum/silicalite-1 composites, *J. Chem Soc – Far. Trans.*, **92**, 4609–4615 (1996).
176. N. Nishiyama, M. Miyamoto, Y. Egashira and K. Ueyama, Zeolite membrane on catalyst particles for selective formation of p–xylene in the disproportionation of toluene, *Chem. Commun.*, 1746–1747 (2001).
177. D.V. Vu, M. Miyamoto, N. Nishiyama, S. Ichikawa, Y. Egashira and K. Ueyama. Catalytic activities and structures of silicalite-1/H–ZSM-5 zeolite composites, *Micropor. Mesopor. Mater.*, **115**, 106–112 (2008).
178. J. He, Y. Yoneyama, B. Xu, N. Nishiyama and N. Tsubaki, Designing a capsule catalyst and its application for direct synthesis of middle isoparaffins, *Langmuir*, **21**, 1699–1702 (2005).
179. S. Tennison, Current hurdles in the commercial development of inorganic membrane reactors, *Membr. Technol.*, **128**, 4–9 (2000).

9

Metal Supported and Laminated Pd-Based Membranes

Silvano Tosti[1], Angelo Basile[2] and Fausto Gallucci[3]

[1]ENEA, Unità Tecnica Fusione, CR ENEA Frascati, Frascati (RM), Italy
[2]Institute of Membrane Technology, CNR, c/o University of Calabria, Rende, CS, Italy
[3]Chemical Process Intensification, Faculty of Chemical Engineering and Chemistry, Eindhoven University of Technology, Eindhoven, The Netherlands

9.1 Introduction

As introduced in Chapter 4, the Pd-H system has been largely studied [1, 2]: particularly, the applications of Pd alloys to the production of membrane devices for producing hydrogen have been experimentally investigated and reported by several authors [3–7].

The applications of Pd-based membranes to the hydrogen production are braked by the cost of the expensive metal. Furthermore, due to its limited production over the world the Pd price oscillates very much: in the last years, the cost of Pd was in the range 5–10 € g^{-1}. As a matter of fact, the reduction of the Pd material used in the membranes manufacturing has been the aim of several research studies. Thin layers of Pd alloys have been coated over porous supports by means of several deposition techniques [6–12]: these composite membranes exhibit high permeation fluxes but, generally, not complete hydrogen selectivity. Particularly, higher selectivity is attained by increasing the thickness of the Pd layers: however, in this case the composite membrane present poor chemical and physical stability. In fact, the shear stresses at the Pd-ceramic interface due to the thermal and hydrogenation cycling dramatically increase with the Pd thickness thus involving the detachment of the active layer from the support and the loss of membrane selectivity [11].

Membranes for Membrane Reactors: Preparation, Optimization and Selection, Edited by Angelo Basile and Fausto Gallucci

The Pd-Ag thin wall tubes described in the chapter IV represent a compromise situation: a Pd alloy thickness as thin as possible is applied in order to both guarantee the absence of defects on the Pd-Ag layer and to withstand the trasmembrane differential pressure needed for driving the hydrogen permeation. In this way, both the hydrogen selectivity and the long-life of the permeators is obtained beside good performances in terms of high hydrogen production ($3\,Nm^3\,m^{-2}\,h^{-1}$) and acceptable cost (the production of $1\,Nm^3\,h^{-1}$ of hydrogen involves a membrane cost of about 1600 €).

Compared to the thin wall permeator tubes, a further reduction of the Pd alloy thickness is aimed at lowering the membranes costs and further increasing the hydrogen permeation fluxes. However, as a practical aspect the Pd-Ag layer thickness of about 50 μm is considered to be the minimum value capable to assure the mechanical resistance of the membrane under the typical operating conditions of the separation processes considered (temperature of 300–400 °C and pressure difference of 200–300 kPa). The study of metal supported and laminated Pd-based membranes is aimed at separating the two characteristics required for a thin wall membrane: high hydrogen permeation/selectivity and good transmembrane pressure resistance. These membranes have been designed in order to give to the Pd-Ag layer the function of selectively separating hydrogen and to the metal support the function of withstand the mechanical stresses [13].

9.2 Preparation Method

The diffusion welding technique has been applied for preparing two kinds of Pd-based composite membranes in which thin Pd-Ag layers are joined to metal structures such as grids (metal supported membranes) or are covering thick non-noble metals bulk (laminated membranes).

9.2.1 Metal Supported Membranes

The use of metal support for Pd-based membranes is mainly aimed at overcoming the drawbacks given by the ceramic supports. In fact, because of the different metal–ceramic thermal expansion coefficients, the thermal cycling of the membranes can produce at the metal/ceramic interface significant mechanical stresses with formation of defects and cracks and loss of membrane selectivity.

Among the techniques of joining the Pd alloy layers to the metal supports, the diffusion welding has been considered by different groups. Iniotakis *et al.* developed hydrogen permeation membranes consisting of a thin Pd alloy foil enclosed between two fine-mesh metal fabrics [14]: the presence of a bilateral support permits to significantly reduce the Pd thickness (10–30 μm) thus giving high hydrogen permeability. For applications at high temperature (>600 °C), in order to avoid the interdiffusion of metal between the surfaces in contact (Pd alloy/support metal) then increasing the membrane stability, a diffusion-blocking coating (for an example Ni) can be deposited over the fabrics. As a practical apsect, such a supported membrane can easily be worked and manufactured.

Juda *et al.* described a method for hermetically bonding Pd-Cu membranes to a metallic frame in order to have a pressure tight seal [15]: this method can be applied for connecting the Pd-Cu membranes to the membrane module or to manufacture sandwiched membranes. Particularly, the diffusion welding process is carried out at about 300 °C for several hours under hydrogen atmosphere.

Tosti *et al.* studied the diffusion welding for producing Pd-Ag thin wall permeator tubes [16, 17]. As described in Chapter 4, the silver atoms exhibit a very high diffusivity into the metal lattice: this means that the welding of metal parts containing silver can be obtained at relatively low temperature. Especially, at 800–1000 °C the diffusion of silver atoms through a metal thickness of some tens of microns takes place within 1–2 h. The diffusion bonding process has been carried out in an oven operating under controlled atmosphere (vacuum, inert gas, inert gas with hydrogen) by packing together the metal parts to be welded through a thermomechanical press (see Figure 9.1). This device consists of stainless steel plates which compress the supported membrane via threatened bars made of INVAR, a metal alloy of negligible thermal expansion coefficient. When increasing the temperature, all the materials (steel plates, Pd-Ag foil, metal support) expand except for the INVAR screws which compress the parts to be joined.

As metal supports both stainless steel grids and nickel perforated metals have been used as a consequence of their low cost, good mechanical properties and good chemical inertia. In particular, in order to improve the adhesion of the Pd-Ag foil, the metal supports are covered by a very thin layer (flash) of silver via electrochemical deposition. In Figure 9.2, the steel grid with the flash of silver before the diffusion welding is shown, while Figure 9.3 reports in detail the Pd-Ag foil joined to the steel support. In particular, this picture shows the cross section of the composite membrane: the welded zones where the metal interdiffusion took place are put in evidence.

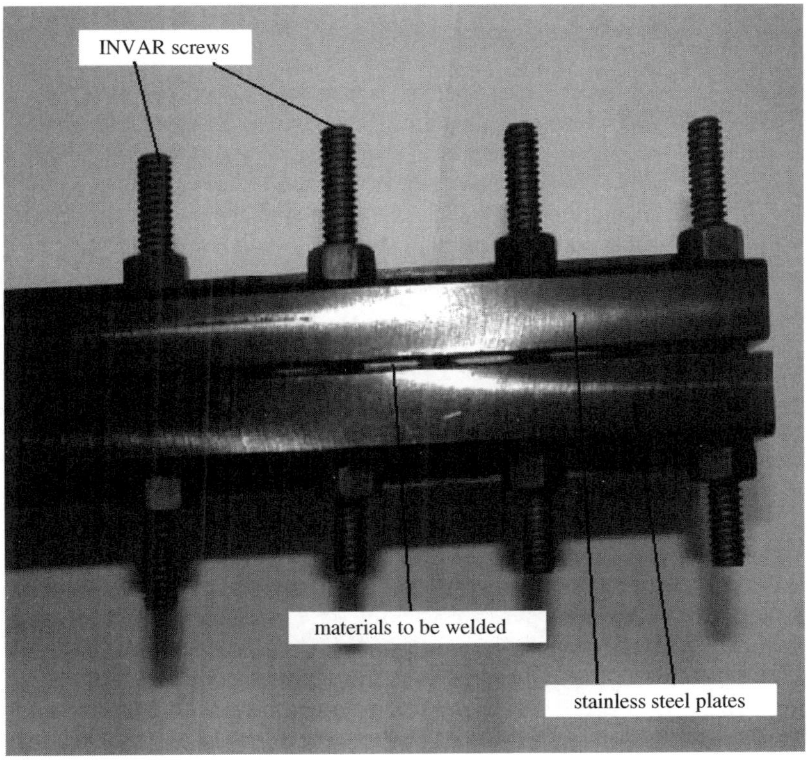

Figure 9.1 *The thermomechanical press used for preparing metal-supported membranes via diffusion welding*

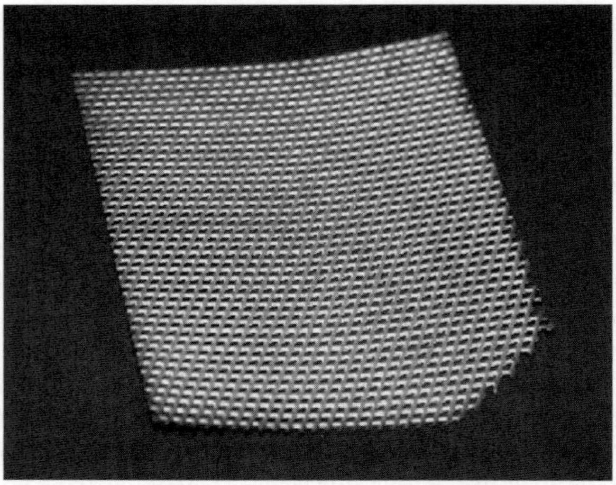

Figure 9.2 The stainless steel grid (silver coated) before welding the Pd-Ag foil

The adhesion of Pd alloy layer over the metal support can be evaluated by a peel test in which a strength is applied in order to pull apart the welded layer. The above cited sample is shown after the peel test in Figure 9.4: after pulling, the delaminated parts of the Pd foil are attached to the support grid thus demonstrating a sound adhesion (see the details of Figure 9.4).

A composite membrane consisting of a porous stainless steel support in form of a disc has been produced, too: Figure 9.5 shows the sample before and after the membrane preparation via diffusion welding. Such a kind of support is cheap and very stiff: the membrane disc can withstand relatively high operating transmembrane pressure.

A nickel perforated metal support (thickness 210 μm, diameter of holes 2.5 mm) has been also used for preparing a supported membrane tube which is reported in Figure 9.6 (general view) and Figure 9.7 (detail of the tube external surface). A flat composite membrane has been produced via diffusion welding of a Pd-Ag foil of thickness 42 μm by using the thermomechanical press. Afterwards, the composite membrane has been bent and the seams have been joined via diffusion welding by using the device described in Chapter 4 for the

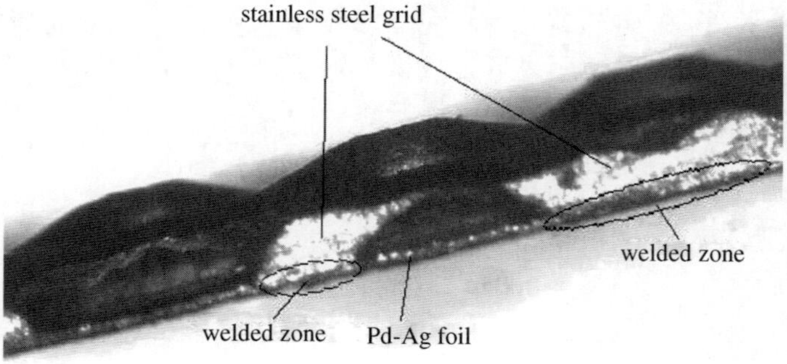

Figure 9.3 The supported membrane: Pd-Ag foil has been joined to the steel grid by diffusion welding

Figure 9.4 *The Pd-Ag foil supported by the steel grid after the peel test*

Figure 9.5 *A stainless steel porous support: virgin (left) and after welding a Pd-Ag foil (right)*

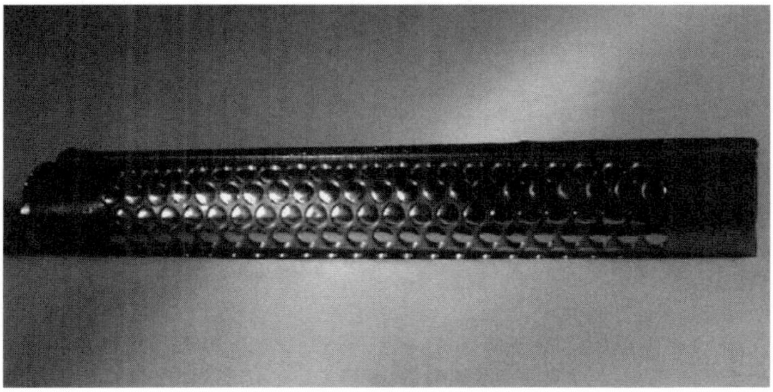

Figure 9.6 *Permeator tube consisting of a Pd-Ag membrane supported by a perforated nickel sheet*

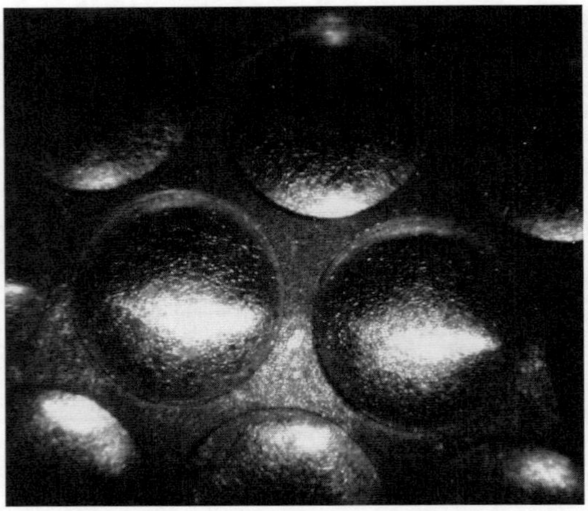

Figure 9.7 Particular of the external surface of the Ni-supported membrane

production of the thin wall tubes. Especially, in this case a Pd-Ag strip has been interposed between the metal limbs to be welded in order to ensure the diffusion of the silver (see Figure 9.8). During testing, hydrogen permeability values (1.6×10^{-9} mol m^{-1} s^{-1} Pa$^{-0.5}$ at 350 °C) in agreement with the literature have been measured [18] while a good durability has not been observed. In fact, the formation of defects (cracks, holes) of the Pd alloy surface along the contact area with the support took place: especially, the sharp edges of the perforated support produced defects of the Pd layer during the pressing phase of the diffusion welding process. As a main result, due to the negligible mass transfer resistance of the metal support, the reduced Pd layer (40 μm) of the supported membranes compared to the thin wall tubes (50–60 μm) involves a higher hydrogen permeance: conversely, the drawbacks are the higher manufacturing cost and the shorter lifetime.

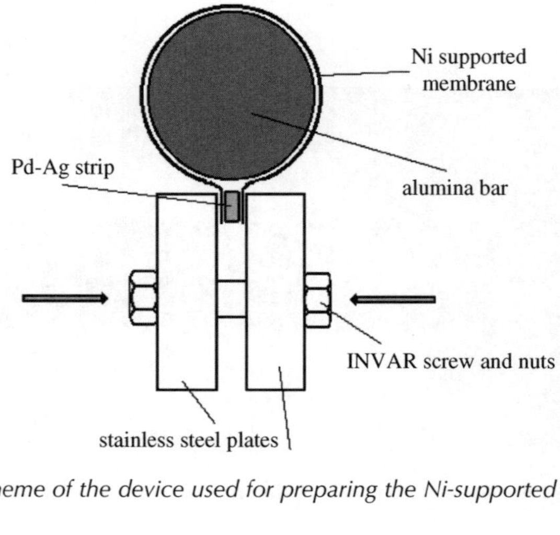

Figure 9.8 Scheme of the device used for preparing the Ni-supported membrane tube

9.2.2 Laminated Membranes

The use of low cost metals instead of Pd and its alloy represents a possible way to reduce the costs of metal permeators for producing pure hydrogen. The hydrogen permeation is a mass transfer mechanism characteristic of all the metals: in particular, the refractory metals such as Nb, Ta and V exhibits hydrogen permeability higher than Pd [6]. However, under the usual operating conditions of the processes for separating hydrogen the non-noble metals strongly react with gases (oxygen, nitrogen, etc.) by forming on their surfaces layers which reduce or block the hydrogen permeation. A wide variety of work has been carried out on the synthesis of V, Ta and Nb alloys membranes. For an example, a vanadium alloy (V-15%Ni) covered by Pd thin layers has been studied by Nishimura *et al.* [19] and Zhang *et al.* [20] while T. Ozaki *et al.* [21] considered in their work a V-Ni-Al alloy which exhibited at 623 K hydrogen permeability values of 6.29×10^{-8} mol m^{-1} s^{-1} Pa$^{-0.5}$ (higher than the V-15%Ni alloy by a factor of two). Other studies have taken into consideration the addition of Nb to the Ni-Ti alloy: in such a manner, the hydrogen permeability can significantly be increased [22]. Ta foils of thickness 13 μm have been coated with Pd films of thickness 1 μm by vapour deposition: these composite metal membranes showed hydrogen permeability values lower than Pd [23].

Tosti *et al.* developed a diffusion welding procedure for covering the surfaces of non-noble metals with (thin) Pd layers. In this case, two commercial Pd-Ag foils have been welded over thick sheets of Ni and Nb, respectively. In the resulting bilayered Ni or Nb membrane, the oxidation of the bulk material is avoided by the presence of the Pd alloy on its surface: the increase of the hydrogen permeance has been obtained by the thickness reduction. Accordingly, the composite membrane has been cold rolled so that the overall thickness of the composite membrane has been reduced (Figure 9.9). As a matter of fact, both the Ni or Nb bulk and the Pd-Ag thickness reduced proportionally: in order to reduce the cost, a well sized composite membrane should have a final Pd-Ag thickness of a few microns while the bulk material should be as thin as possible for assuring the mechanical stiffness (practically, 100–200 μm depending on the operating pressures). As a main advantage, in case of damage of the Pd alloy thin layer this

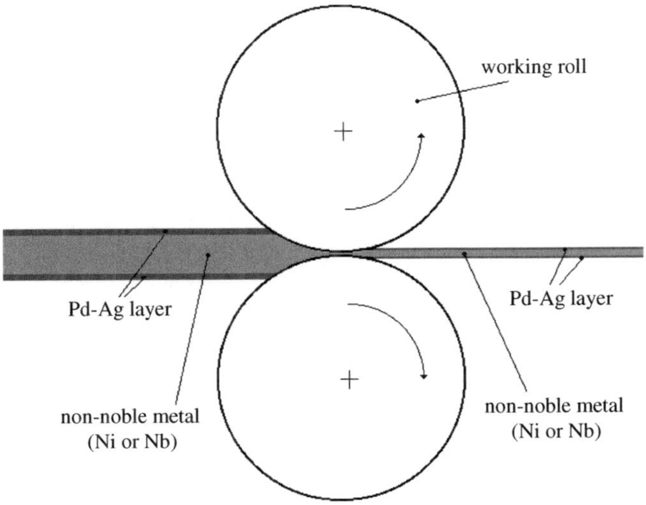

Figure 9.9 *Cold rolling of composite Pd-based membranes*

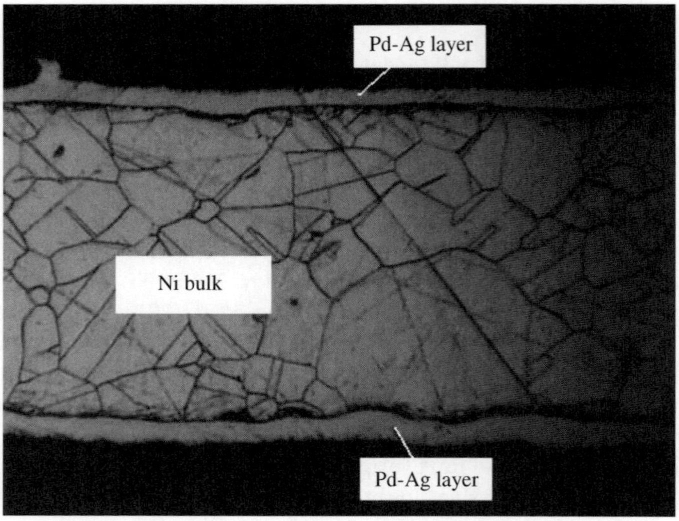

Figure 9.10 *Cross section of the laminated Ni membrane*

kind of composite membranes reduces its permeance without losing its perm-selective: in fact, the presence of the dense bulk material can anyway guarantee the hydrogen selectivity.

9.2.2.1 Laminated Ni Membrane

The cross section of a composite consisting of a Ni bulk covered by Pd-Ag layers is reported in Figure 9.10. By diffusion welding two Pd-Ag foils of thickness 28 μm have been joined to a 0.500 mm thick Ni sheet: the following cold rolling step reduced the overall thickness of the composite membrane down to 141 μm (about 127 μm of Ni bulk with two Pd-Ag layers of 7 μm). Then, such flat membrane has been used for preparing a permeator tube by the above described technique (device in Figure 9.8): the membrane tube has been joined by brazing to a stainless steel end in order to perform permeation tests, see the permeator details in Figure 9.11. In the temperature range 250–400 °C, hydrogen permeance values from 1×10^{-8} to 1×10^{-9} mol m^{-2} s^{-1} Pa$^{-0.5}$ have been measured. These values of permeance are at 2–3 orders of magnitude

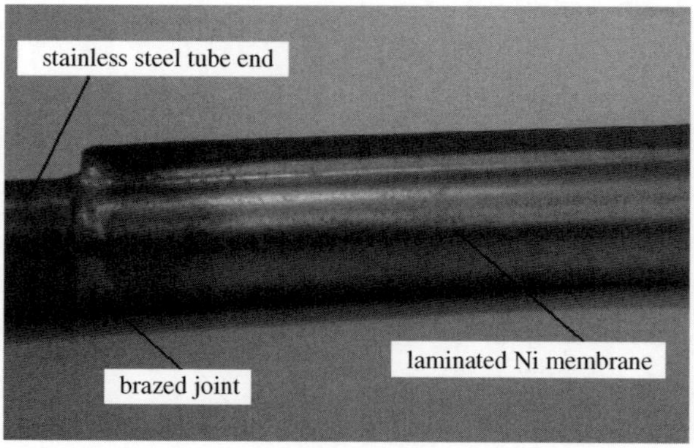

Figure 9.11 *The permeator tube produced by the laminated Ni membrane*

lower than the Pd alloy ones: however, the results demonstrated the applicability of such technique to produce laminated membranes with low-cost metals.

As an alternative, a porous Ni support could be used instead of a dense bulk. Due to the reduced mass transfer resistance of the porous support compared to the dense one, such a composite membrane has much higher hydrogen permeance. Conversely, due to the absence of the dense bulk material the hydrogen selectivity cannot be maintained in the case of damage of the Pd-Ag layers.

9.2.2.2 Laminated Nb Membrane

In this case, a 1 mm thick Nb plate has been covered by two Pd-Ag foils of thickness 25 μm; after rolling a composite membrane of overall thickness 128 μm has been obtained (about 122 μm of Nb bulk with two very thin Pd-Ag layers of 3 μm). In order to characterise this composite membrane a permeator tube has been prepared: however, the hydrogenation test (180 °C and 200 kPa) produced a quick damage of the membrane as shown in Figure 9.12. As a matter of fact, the Nb as well as most of the refractory metals exhibit very intense hydrogen uploading: particularly, at relatively low temperature large amounts of hydrogen are solubilised into the lattice of these metals. Consequently, the significant expansion of the lattice produce embrittlement and rupture of the materials.

9.2.3 Non Pd-Based or Low Pd Content-Based Membranes

Cold rolling of metal alloys different than Pd ones has been used to produce also new types of dense metal membranes. In fact, it is strongly desired to develop nonpalladium-based or at least low palladium content-based and low-cost hydrogen permeation alloys. Among the different possible alloys, it has been demonstrated that Nb-Ti-Ni alloy consisting of only the primary phase and the eutectic phase shows a high hydrogen permeability and resistance to the hydrogen embrittlement as described by Luo *et al.* [24]. On the other hand, equiatomic Ti-Ni alloy presents high thermal and chemical stability and has infinite perm-selectivity towards hydrogen while the permeation is quite lower than Pd-based membranes as demonstrated by Orekhova *et al* [25]. Based on these observations, Tereschenko and coworkers [26, 27] produced new stable and

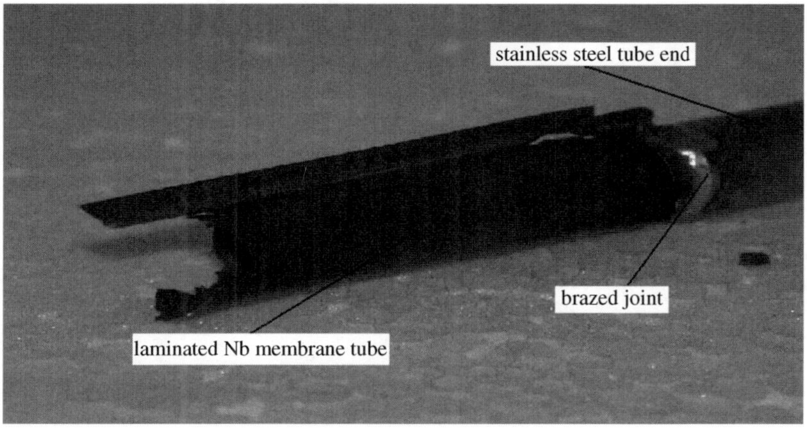

Figure 9.12 *The laminated Nb membrane after hydrogenation*

highly hydrogen permeable membranes on Ti–Ni basis by adding a low content of Pd in concentration of 2, 5 and 9 wt% by rolling technique.

Three alloys of Ti, Ni and Pd containing 2, 5 and 9 wt% Pd (over Ni) and having the Ti/(Ni + Pd) atomic ratio 1:1 were made from 99.9% pure starting materials by arc melting. Each alloy button was re-melted at least five times to ensure the compositional homogeneity. The specimens were then subjected to a thermomechanical treatment, consisting of cold rolling followed by annealing at 1000–1300 K. The final thickness of foils was about 40 μm. The membranes were used in permeation tests with different H_2 rich mixtures [27] as well as in flat membrane reactor for dehydrogenation reactions showing good performances in terms of permeability and resistance to poisoning and good results in terms of shift effect on the reaction system.

9.3 Applications

The Pd-based metal supported and laminated membranes have been studied in order to reduce the amount of precious metal compared to the Pd-Ag thin wall membrane but anyway ensuring the complete hydrogen selectivity. Therefore, these membranes are considered for the production of ultra-pure hydrogen and are expected to exhibit higher permeance than the thin wall Pd-Ag tubes do. Presently, only preliminary characterisation of these membranes have been carried out: Pd-Ag membranes supported on stainless steel and nickel supports exhibited high hydrogen permeation fluxes but long term tests are expected to verify their durability.

The potential applications of the metal supported and laminated concern the production of ultra-pure hydrogen: especially, compared to the thin wall permeators the reduced cost of these composite membranes could enlarge the field of applications.

For the cases of tubular permeators, the membrane module design could take into account the same configuration studied for the Pd-Ag thin wall tubes (see Chapter 4). As a matter of fact, the

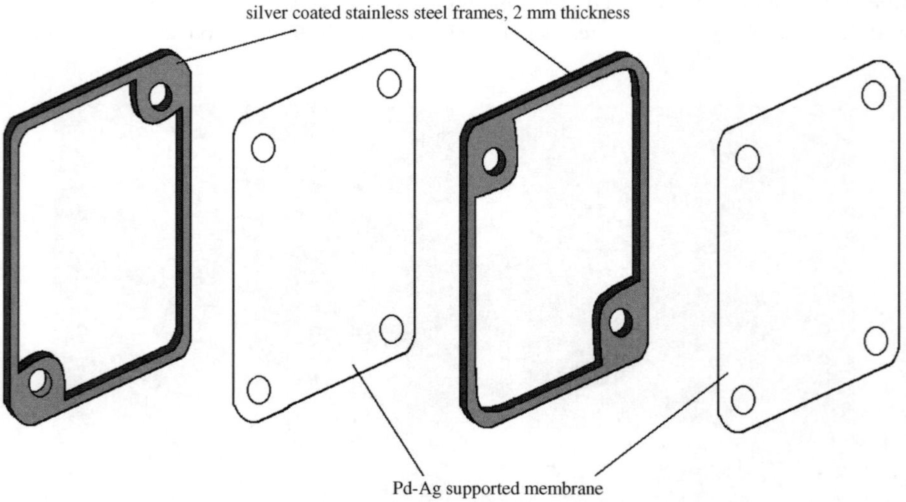

Figure 9.13 Stainless steel frames (silver-coated) and Pd-Ag supported membranes to be used in the flat and frame configuration design module

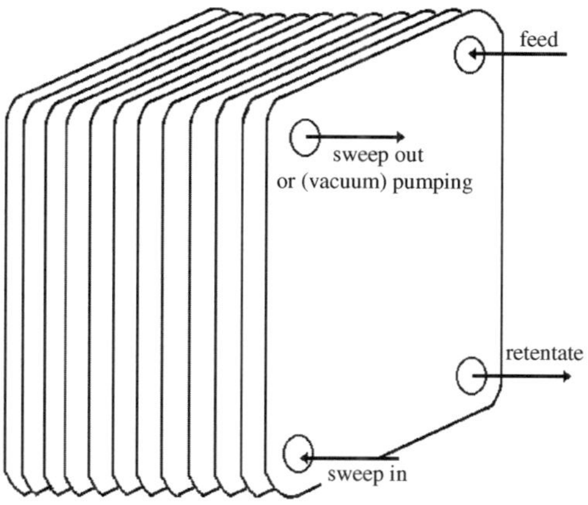

Figure 9.14 *Flat and frame membrane module*

finger-like design of membrane modules should be applied in order to avoid mechanical stresses due to thermal expansion differences. Particularly, due their low costs, the metal supported membranes could be used for manufacturing membrane separators when a large permeation area is required beside a small size. In these cases, a special configuration similar to the design of flat and frame heat exchanger can be proposed [28]. The Pd-Ag composite membranes are produced in form of thin sheets supported over stainless steel grids and welded to stainless steel frames previously covered by silver, see the scheme of Figure 9.13. The diffusion welding procedure can further be operated in order to tightly join the Pd-Ag membranes to the stainless steel frames. The resulting membrane module is compact, see Figure 9.14: a permeator of permeation surface area of $10\,m^2$ could be about $1000 \times 120 \times 180$ mm.

The preparation of laminated membranes has been aimed at using low costs metals as bulk material for Pd-based membranes. However, presently the possible applications of these membranes seem to be unpractical: in fact, the laminated Ni membrane exhibited very low permeance values while Nb one failed after very short testing as a consequence of the hydrogen embrittlement. Future studied should be addressed to increase the hydrogen diffusivity of metals such as Ni as well as to reduce the hydrogen solubility of the Nb series metals (i.e., the refractory metals).

9.4 Conclusions

In the applications for producing ultra-pure hydrogen, the Pd-Ag tubes (wall thickness 50 μm) described in Chapter 4 represent a good compromise in terms of performances (hydrogen permeance) and cost. A further cost reduction has been attained by preparing Pd-based composite membranes where the thickness of the Pd alloy layer is reduced below 50 μm and the mechanical stiffness is guaranteed by a well designed metal support.

Generally, the diffusion welding procedure has been applied in order to join the Pd alloy layers to the metal supports as well as to tightly connect Pd alloy foils to metal frames. Several configurations of composite metal membranes have been studied (i.e., single or bilateral support, discs or tubes). Especially, metal supported and laminated have been considered.

The metal supported membranes which use stainless steel or nickel structures have exhibited high permeation fluxes: in particular, a permeability of 1.6×10^{-9} mol m^{-1} s^{-1} Pa$^{-0.5}$ has been measured at 350 °C with a Pd-Ag foil (thickness 42 μm) joined to a Ni support. Their application for separating pure hydrogen could take into consideration the design of special compact membrane modules (flat and frame) which present high permeation surfaces into a small size.

Conversely, at the state of art the laminated membranes seem be of impractical applications. In fact, when metals of low hydrogen permeability have been considered (i.e., Ni), the effectiveness of the proposed technique of depositing Pd-Ag thin films and avoiding the oxidation of the poor metal surfaces has been demonstrated: however, very low hydrogen permeance have been measured (about 2 or 3 orders of magnitude lower than Pd alloys). The laminated Nb membranes have shown a very short durability as a consequence of the hydrogen embrittlement: in the case of the refractory series metals (Nb, Ta, V) future activities should be addressed to study alloys with lower hydrogen solubility.

Aimed at the cost reduction, other interesting researches have been focused on nonPd-based or low Pd content-based membranes made of Ni, Ti, Nb and their alloys. These studies characterised the new membranes in terms of hydrogen permeance and resistance to the embrittlement.

References

1. F.A. Lewis. *The Palladium Hydrogen System*. Academic Press, London (1967).
2. J. Shu, B.P.A. Grandjean, A. Van Neste, S. Kalaguine. Catalytic palladium-based membrane reactors: a review. *Can J Chem Eng*, **69**, 1036–1060 (1991).
3. J.N. Armor, Applications of catalytic inorganic membrane reactors to refinery products. *J Membr Sci*, **147**, 217–233 (1998).
4. E. Kikuchi. Membrane reactor application to hydrogen production. *Catal Today*, **56**, 97–101 (2000).
5. E. Drioli, E. Fontananova. Membrane technology and sustainable growth. *Chem Eng Res Des*, **82** (A12) 1557–1562 (2004).
6. A. Basile, F. Gallucci, S. Tosti. Synthesis, characterization, and applications of palladium membranes. *Membrane Sci Technol*, **13**, 255–323 (2008).
7. S.N. Paglieri, J.D. Way. Innovations in palladium membrane research. *Sep Purif Methods*, **31** (1), 1–169 (2002).
8. S. Uemiya. State-of-the-art of supported metal membranes for gas separation. *Sep Purif Methods*, **28** (1), 51–85 (1999).
9. O. Schramm, A. Seidel-Morgenstrern. Comparing porous and dense membranes for the application in membrane reactors. *Chem Eng Sci*, **54**, 1447–1453 (1999).
10. L.Q. Wu, N. Xu, J. Shu. Preparation od a palladium composite membrane by an improved electroless plating technique. *Ind Eng Chem Res*, **93**, 342–348 (2000).
11. S. Tosti, L. Bettinali, S. Castelli, F. Sarto, S. Scaglione, V. Violante. Sputtered, electroless, and rolled palladium-ceramic membranes. *J Membrane Sci*, **196**, 241–249 (2002).
12. J. Shu, B.P.A. Grandjean, E. Ghali, S. Kaliaguineet. Simultaneous depositing of Pd and Ag on porous stainless steel by electroless plating. *J Membrane Sci*, **77**, 181–195 (1993).
13. S. Tosti. Supported and laminated Pd-based metallic membranes. *Int J Hydrogen Energy*, **28**, 1445–1454 (2003).
14. N. Iniotakis.*et al. Hydrogen permeation membrane.* US Patent 4 699 637 (1987).
15. W. Juda, C.W. Krueger, R.T. Bombard,*Diffusion-bonded palladium-copper membrane for pure hydrogen and the like and the method of preparing the same.* US Patent 5 904 754 (1999).

16. S. Tosti, L. Bettinali, Diffusion bonding of Pd-Ag membranes. *J Mat Sci*, **39**, 3041–3046 (2004).
17. S. Tosti, L. Bettinali, D. Lecci, F. Marini, V. Violante.*Method of bonding thin foils made of metal alloys selectively permeable to hydrogen, particularly providing membrane devices, and apparatus for carrying out the same.* European Patent EP 1184125 (2001).
18. H. Yoshida, S. Konishi, Y. Naruse. Preliminary design of a fusion reactor fuel cleanup system by the palladium-alloy membrane method. *Nucl Technol/Fusion*, **3**, 471–484 (1983).
19. C. Nishimura, M. Komaki, S. Hwang, M. Amano, V-Ni alloy membranes for hydrogen purification. *J Alloy Compounds*, **330/332** 902–906 (2002).
20. Y. Zhang, T. Ozaki, M. Komaki, C. Nishimura. Hydrogen permeation of Pd-Ag alloy coated V-15Ni composite membrane: effects of overlayer composition. *J Membrane Sci*, **224**, 81–91 (2003).
21. T. Ozaki, Y. Zhang, M. Komaki, C. Nishimura. Hydrogen permeation characteristics of V-Ni-Al alloys. *Int J Hydrogen Energy*, **28**, 1229–1235 (2003).
22. K. Hashi, K. Ishikawa, T. Matsuda, K. Aoki. Hydrogen permeation characteristics of multi-phase Ni-Ti-Nb alloys. *J Alloy Compounds*, **368**, 215–220 (2004).
23. N.M. Peachey, R.C. Snow, R.C. Dye. Composite Pd/Ta metal membranes for hydrogen separation. *J Membrane Sci*, **111** (1), 123–133 (1996).
24. W. Luo, K. Ishikawa, K. Aoki. High hydrogen permeability in the Nb-rich Nb–Ti–Ni alloy. *J Alloys Comp*, **407**, 115–117 (2006).
25. N.V. Orekhova, M.M. Ermilova, V.M. Gryaznov,*et al.* Abstracts of Russian conference Membrane-95 Moscow, p. 186 (1995).
26. G.F. Tereschenko, M.M. Ermilova, V.P. Mordovin, N.V. Orekhova, V.M. Gryaznov, A. Iulianelli, F. Gallucci, A. Basile. New Ti-Ni dense membranes with low palladium content. *Int J Hydrogen Energy*, **32** (16), 4016–4022 (2007).
27. A. Basile, F. Gallucci, A. Iulianelli, G.F. Tereschenko, M.M. Ermilova, N.V. Orekhova. Ti-Ni-Pd dense membranes – The effect of the gas mixtures on the hydrogen permeation. *J Membrane Sci*, **310**, 44–50 (2008).
28. CRC. *Handbook of Chemistry and Physics*, 67th edn. CRC Press, London (2005).

10

PVD Techniques for Metallic Membrane Reactors

R. Checchetto[1], R.S. Brusa[1], A. Miotello[1] and A. Basile[2]

[1]Dipartimento di Fisica, Università di Trento, TN, Italy
[2]Institute of Membrane Technology, CNR, c/o University of Calabria, Rende, CS, Italy

10.1 Introduction

The upward trend of the petroleum price and the increase in the emission of greenhouses gases in the atmosphere, which is one of the causes of the global warming, are pushing forward the use of hydrogen as relevant energy carrier [1].

High purity hydrogen can be generated from natural gas or from other hydrocarbons with high efficiency using membrane reactors [2]. In a membrane reactor a catalyst filled reaction chamber is combined with a membrane that selectively removes reaction products. In the steam reforming process, for example, steam (H_2O) reacts at high temperature (700–1100 °C) with methane to yield syngas: $CH_4 + H_2O \rightarrow 3H_2 + CO + 191.7\,kJ\,mol^{-1}$.

The heat required for the reaction is obtained by burning part of the methane. Additional hydrogen can be obtained adding more water at low temperature, 130 °C, through the water gas shift reaction: $CO + H_2O \rightarrow H_2 + CO_2 - 40.4\,kJ\,mol^{-1}$ [3].

Hydrogen generated from these reactions is extracted by the use H_2 perm-selective membranes. A further advantage of the use of the membrane reactor is the fact that even if the hydrocarbon conversion reactions release carbon dioxide, it is separated from H_2 and remains in the reactor at high pressure, at disposal for capture and storage [4].

Hydrogen selective membranes have to be defect-free, exhibit a definitely high selectivity along with high hydrogen fluxes, resistance to poisoning, long operational lifetimes, and low cost. Polymers, metals, zeolite, silica, and carbon are currently investigated as membrane

Membranes for Membrane Reactors: Preparation, Optimization and Selection, Edited by Angelo Basile and Fausto Gallucci

materials: the hydrogen separation performed by each membrane type is based upon different chemicophysical processes which could imply advantages and drawbacks [5].

Gas permeation in polymeric membranes involves the solution of the gas molecules at the upstream side of the membrane and their diffusion through the polymeric layers. The separation of H_2 from other gases is achieved because of the extreme higher diffusion coefficient of H_2 relative to all the other molecules. Unfortunately, in hydrogen selective polymeric membranes the hydrogen permeation fluxes are limited by the low H_2 solubility in the membrane material. Polymeric membranes also present temperature stability problems that impede some application as, for example, in membrane reactors for hydrogen production [6].

Silica, silicates, and zeolites are nanoporous materials presenting a connected network of pores with size in the 0.5 nm range: H_2 selectivity is obtained by a proper combination of sorption and mobility in the membrane materials. Usually porous membranes consist of a functional nanoporous layer with a thickness of about 100 nm, grown on a thick porous support with high permeance. Their operative performance depends on the morphology of the functional layer (that is, pore size and interconnectivity): less dense structures have higher H_2 permeance but, inevitably, lower selectivity. The advantages of inorganic membranes rely in their operational stability and inertness under different conditions of temperature and chemical environment [7].

In dense metallic membranes, the hydrogen permeation occurs by a solution–diffusion process. Here the diffusing species is atomic H: the permeation process thus requires H_2 dissociative adsorption on the metal surface exposed to the H_2-rich gas and hydrogen recombination before desorption in the downstream side [8]. These membranes don't exhibit the trade-off between selectivity and permeability which is common for polymeric and nanoporous silica membranes [9]. The membranes can achieve very high selectivity for H_2 because the transport of species different from hydrogen can only occur in presence of microdefects such as pinholes or cracks. The currently studied membranes are made of very expensive materials such as Pd or Pd-based alloys deposited in form of thick films on inorganic porous supports such as alumina, silica or stainlees steel [10]. Good selectivity for hydrogen is obtained when metallic film is continuous and defect-free. Depending on the roughness of the support and on the deposition technique, this occurs when its thickness ranges from a few μm up to 20 μm. This high thickness implies high working temperature to obtain large H_2 permeation fluxes. Thus it is necessary to make progress in the development of fabrication routes of selective membranes with Pd-based functional layers [11].

Methods for preparing thin films can be divided essentially in three main categories: physical vapour deposition (PVD), chemical vapour deposition (CVD) [12] and chemical methods [13]. PVD techniques include thermal evaporation, physical sputtering, cathodic arc deposition and pulsed laser deposition (PLD): these techniques are atomic in nature in the sense that films grow from the condensation of single atoms (cluster deposition may be obtained with PLD for some specific choice of the laser pulse parameters). In CVD techniques molecular species in the gas phase chemically react at the surface of the substrate resulting in the formation of the film layers and in the production (and emission) of volatile species. Chemical methods include electrolytic deposition and electroless deposition and make use of a solution (electrolyte) where the atoms that form the film are present in form of positively charged metal cations.

The discussion in this chapter is dedicated on the preparation of Pd-based membranes by PVD techniques: their structure and operative performances are also discussed. Compared to other techniques, physical methods, in fact, allow the preparation of high purity thin films with well controlled nanostructure. In the first part of the presentation we give general remarks on PVD techniques for thin film deposition, namely thermal evaporation, pulsed laser deposition and physical sputtering. We also indicate the respective advantages and limits when used for the deposition of Pd-based thin films. In the second part we present a review on the Pd-based membranes prepared by physical techniques and their performances are compared with those of Pd-based membranes prepared by different methods. Finally in the conclusion we indicate open points and routes for the development of better performing PVD-prepared membranes.

10.2 Physical Vapour Deposition Techniques

A thin film is a thin material layer, grown on a substrate, having thickness from few atomic monolayers to several micrometers. PVD is the commonly used method for the deposition of metallic thin films: their most important application is in the microelectronic industry as metal contacts, barriers and interconnects for the fabrication of advanced silicon integrated circuits. This chapter will give emphasis on the sputter deposition techniques as it is the most common method for the preparation of membranes with Pd-based thin films as functional layer. The aim of this presentation is to provide only an introduction to the PVD techniques and is thus not exhaustive: detailed information on thin film deposition and characterisation as well as the description of their applications in modern technologies can be found in reference texts and handbooks [12, 14, 15].

PVD techniques used in the modern thin film technology require the creation of vacuum conditions in the apparatus where the deposition process occurs. There are many reasons for depositing films in vacuum. First, vacuum conditions increase the mean free path of the evaporated particles permitting them to reach the substrate from the evaporation source without collisions. At 25 °C in air for pressures ranging from 10^{-4} to 10^{-6} mbar the mean free path of a particle is in the range 50 to 5000 cm: for a standard apparatus with dimension in the 100 cm range it is thus necessary to use residual vacuum at pressures lower than 10^{-5} mbar to avoid that the dispersion of the evaporated particles, due to particle–particle collisions, impedes the film formation. Vacuum conditions reduce the presence of gases that could react with the deposited material. At 10^{-5} mbar vacuum level (assuming that the residual vacuum is composed by air at 25 °C) when all impinging molecules condense on the growing film, a monolayer of impurity is formed in times ~0.1 s. This means, for example, that a metallic film could result completely oxidised: from this simple evaluation it follows that high purity levels of the deposited film require Ultra High Vacuum (UHV) conditions ($P < 10^{-8}$ mbar). The interested reader can find useful information on the design and use of Ultra High Vacuum systems as well as the underlying physical and chemical processes in the reference text of Redhead *et al.* [16].

10.2.1 Evaporation

Thermal evaporation is the simplest PVD technique for thin film preparation [12, 15]. Nevertheless it has the advantage of high deposition rates and produces thin film with a very high purity level. This deposition technique is commonly used for the deposition of conducting materials in electronic circuits and for optical coatings. The deposition process consists of several stages: (i) the transformation of the material to be deposited (source material) in gaseous state by heating until it evaporates or sublimates, (ii) the transport of atoms from the evaporation source to the substrate and (iii) condensation of the evaporated atoms on the substrate or on the atomic layers of the growing film. The evaporation rates is proportional to the difference between the equilibrium pressure $P_e(T)$ of the evaporant at a given temperature T and the hydrostatic pressure P_h surrounding the evaporant.

The mass evaporation rate Γ_e of the evaporant material per unit area of the source surface at a given temperature T can be expressed by the following equation:

$$\frac{\Gamma_e}{N_A} = \alpha_e \frac{P_e(T) - P_h}{\sqrt{2\pi MRT}}$$

where N_A is the Avogadro number, R is the gas constant, α_e is the sticking coefficient (which assumes values between 0 and 1) and M and $P_e(T)$ are the molecular mass and the vapour pressure of the evaporant, respectively [17]. The maximum evaporation rate is obtained in vacuum conditions ($P_h = 0$) and for sticking coefficient $\alpha_e = 1$. A useful expansion of the formula for the maximum value of Γ_e is given by:

$$\Gamma_e = 5.834 \times 10^{-2} \sqrt{\frac{M}{T}} P_e(T)$$

(Γ_e units: $g\,cm^{-2}\,s^{-1}$) where P_e is expresses in torr. At pressures of 10^{-2} torr, Γ_e is in the order of $10^{-4}\,g\,cm^{-2}\,s^{-1}$. The film deposition rates also depend on the source–substrate distance r and on the substrate to source geometry as specified by the emission angle φ and the incidence angle θ, see Figure 10.1. A simple evaporation source is the Knudsen cell, consisting of an isothermal enclosure of the evaporant material with a very small aperture (of area A_e) through which the evaporated atoms effuse. If the aperture is small compared with the source to substrate distance the source can be considered as point-like and the mass deposition rate per unit area of the substrate R ($g\,cm^{-2}\,s^{-1}$) is given by [12]:

$$R = \frac{\Gamma_e A_e}{\pi r^2} \cos\theta \cos\varphi = 1.856 \times 10^{-2} A_e \sqrt{\frac{M}{T}} P_e(T) \frac{\cos\theta \cos\varphi}{r^2}$$

This equation reveals that the deposition rate varies across the substrate. Assuming that the substrate has diameter W and that it is placed directly over the source ($\theta = \varphi$), see

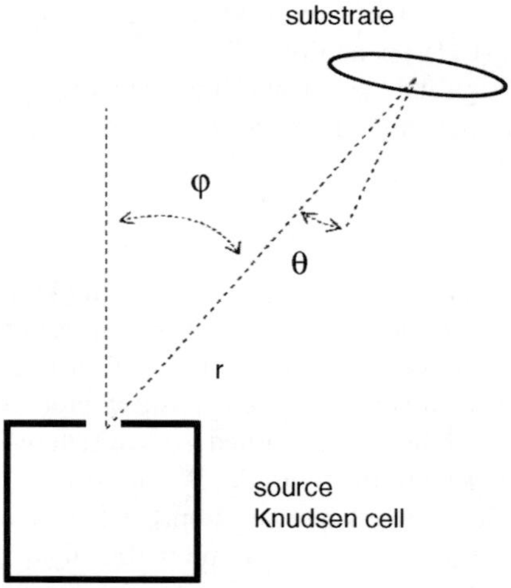

Figure 10.1 *Schematic view of an evaporation apparatus (Joule effect)*

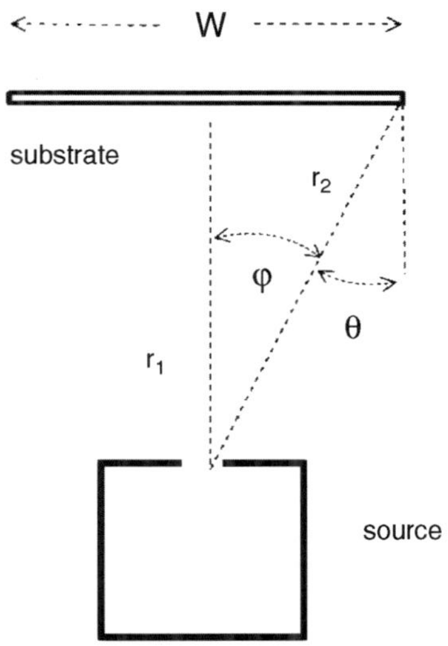

Figure 10.2 *Substrate to source geometry when the substrate is in front of the source (incidence angle θ equal to the emission angle φ)*

Figure 10.2, than the mass deposition rate on the centre of the substrate R_1 and on the edge R_2 are proportional to:

$$R_1 \propto \frac{1}{r_1^2}$$

$$R_2 \propto \frac{1}{r_2^2}\cos^2\theta = \frac{r_1^2}{r_2^4}$$

Defining the uniformity as:

$$\sigma(\%) = \frac{R_1 - R_2}{R_1}(\%)$$

then we have:

$$\sigma(\%) = 1 - \left[1 + \left(\frac{W}{2r_1}\right)^2\right]^{-2} \approx \frac{W^2}{2r_1^2}.$$

Thickness uniformity better than 1% on a 10 cm size diameter thus requires r_1 of at least 70 cm: note that for practical reasons this distance has to be at least doubled. The large r values needed for thickness uniformity of the deposited film thus require large chambers and pumps with high pumping speed: consequently deposition rates are low and large amounts of evaporant material are wasted.

The grow rate of the film dh/dt (cm s^{-1}) can be expressed as $\frac{dh}{dt} = R/\rho$ where ρ is the density of the evaporant material (g cm^{-3}). Let us consider aluminum having $M = 27$ and $\rho = 2.7$ g cm^{-3}. For a substrate to source distance of 50 cm and a source aperture area $A_e = 10^{-2}$ cm^2 the deposition

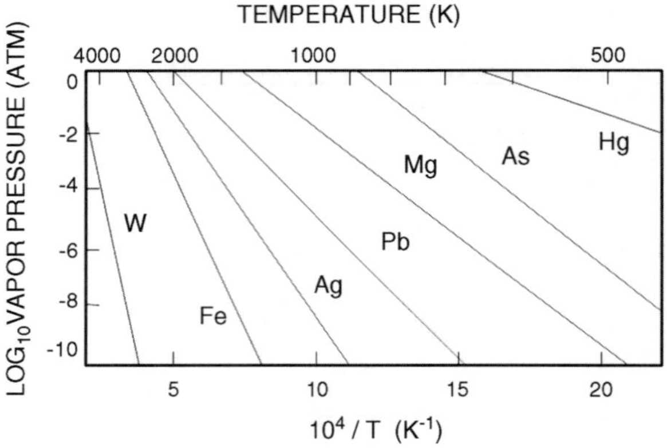

Figure 10.3 *Temperature dependence of vapour pressure for various metals (data are taken from Ref. [18])*

rate, in $\mathrm{\AA\,s^{-1}}$, is given by $\frac{dh}{dt} = 50\,P_e(T)$. We thus observe that deposition rates in the order of 1 to $10\,\mathrm{\AA\,s^{-1}}$ order requires a vapour pressure for the evaporant in the order of 100 mtorr.

To obtain significantly large deposition rates, heating the source material is required because the vapour pressure P of a substance increases nonlinearly with the temperature T, see Figure 10.3, as described by the Clausius–Claperyon equation:

$$\frac{dP}{dT} = \frac{L_V}{T\Delta v}$$

where L_V is the latent heat and Δv is the volume change involved in the phase transition [18].

A typical deposition system consists of: 1) a vacuum chamber generally made of stainless steel because its high strength, ease of welding and machinability, ii) a pumping system consisting of a rough pump with pumping velocity larger than $100\,\mathrm{l\,s^{-1}}$ and high vacuum pump to reach UHV conditions, iii) evaporation sources, monitoring equipment (pressure gauges and quartz crystal to monitor deposition rate and film thickness) and iv) substrate holder provided with a substrate heater to improve film adhesion, to control the grain structure and to reduce the surface roughness.

There are two primary types of evaporation sources: thermal evaporation and electron beam evaporation.

In thermal evaporation, a schematic diagram of the apparatus is presented in Figure 10.4, a crucible, a boat or wire coil is used to hold the source material. Heat is produced by passing an electrical current through the boat or the external wire coil. The temperature of the evaporating material can be controlled by changing the intensity of the current through the heating element. This system is of simple implementation and low cost but has the disadvantage that the entire source material as well as its containment structure (that is the crucible or the boat) have to be heated to the temperature required to evaporate the source material. The maximum temperatures are thus limited, the deposition rates cannot be rapidly changed and the material of the crucible has to be properly selected in order to avoid reactions or alloying with the source material. The crucible material has also to be impurity free to avoid contaminations of the deposited film. Crucibles are generally made of refractory metals such as tungsten, tantalum and molybdenum having a high melting point, T_m (3380 °C for W; 3000 °C for Ta; 2620 °C for Mo) and low vapour pressure($P_e \sim 10^{-2}$ Pa at 3230, 3060 and 2530 °C for W, Ta and Mo) or chemically inert

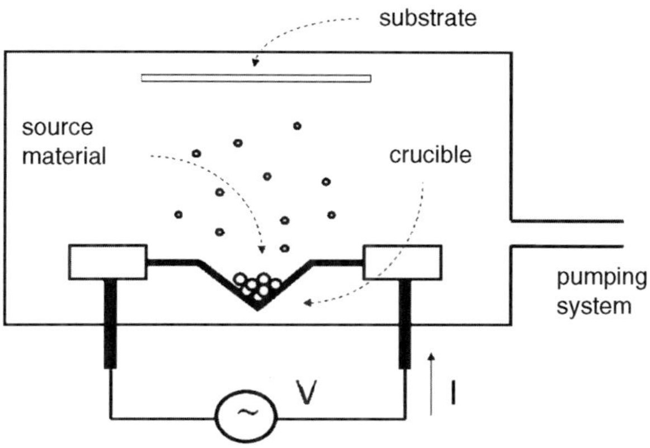

Figure 10.4 *Schematic view of an evaporation apparatus (Joule effect)*

refractory ceramics such as graphitic carbon, alumina and boron nitride ($T_m = 3800$, 2030 and 2500 °C for g-C, Al_2O_3 and BN; $P_e \sim 10^{-2}$ Pa at 2600, 1900 and 1600 °C for g-C, Al_2O_3 and BN).

Many of the problems related to the thermal evaporation can be reduced using electron beam sources. An electron beam is thermoionically emitted from a heated cathode, accelerated through 5–10 kV potential, magnetically deflected using fixed electromagnets and focused on a small area of the source material, see Figure 10.5. The crucible can be thus water-cooled: consequently there is no alloying of the source material with the crucible and the degassing of impurities contained in the crucible is strongly reduced. This ensures high purity levels of the deposited film. Because the energy of the electron beam is deposited in a small area of the source material, large temperatures can be locally obtained allowing the deposition of high temperature melting materials and high deposition rates.

There are several reasons that impede the practical use of thermal evaporation for the deposition of Pd-based thin films as H_2 selective membrane.

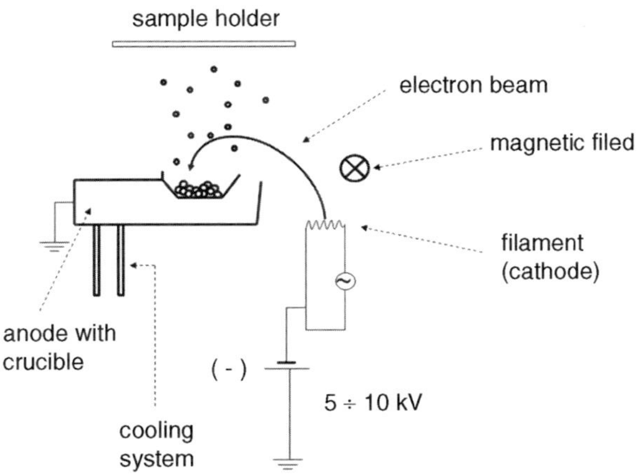

Figure 10.5 *Schematic view of an evaporation apparatus (electron gun)*

Vacuum evaporation is a low energy process because the evaporating particles condensate on the substrate with kinetic energy lower than 0.5 eV: the deposited film thus grows with large grains (in the micrometre range), open structures and low adhesion level of the deposited layer to the substrate.

The film deposition is strictly line of sight and the vapour condensates only onto the exposed surfaces of the substrate: this effectively impedes the deposition of an homogeneous and pin-holes free thin metallic layer on the porous substrates which are the support for the H_2 perm-selective membrane.

Finally a practical problem of evaporation comes when a single multicomponent source is used for the deposition of alloys or compounds. Components with different saturated vapour pressures will evaporate, in fact, at different rates. The composition of the film will be thus different from that of the source material: compared to the source material the deposited film will be, in fact, richer in the component with higher vapour pressure. If the substance to be evaporated consists of component A and B with molecular mass M_A and M_B and equilibrium pressure P_A and P_B it follows that the atomic ratio n_A/n_B of deposited particles of component A and B is:

$$\frac{n_A}{n_B} = \frac{C_A}{C_B}\frac{\alpha_A P_A}{\alpha_B P_B}\sqrt{\frac{M_B}{M_A}}$$

where C_A and C_B are the fraction of components in the material source. For a two-component system the evaporation process can be conducted using separate evaporation sources each kept at a different temperature while maintaining the substrate at elevated temperature. This method, if properly conducted, can produce films with a defined composition. It anyway requires accurate calibrations, an optimal control of the temperature and evaporation rates of the two sources and does not assure uniform composition of the deposited film when large substrates are used.

10.2.2 Pulsed Laser Deposition

Pulsed laser deposition (PLD) uses a high-power laser pulse to ablate a small area of the target material [19]. A high-power laser is used as an external energy source: the cloud of the target material generated by the laser pulse (plume) is then deposited onto a substrate (see Figure 10.6). The process chamber is provided by a window for the entry of the laser pulse generated by an external laser source while a set of optical components is used to focus and raster the beam over the target surface in order to consume uniformly the target material. Laser ablation systems generally use ultraviolet excimer laser that provide pulses of 0.1 to 1.0 J cm^{-2} with duration of 15 to 45 ns at repetition rates of 1 to 100 Hz. The laser-target interaction is a very complex process and the mechanism that leads to the material ablation depends on the laser pulse characteristics, as well as the optical, topological and thermodynamic properties of the target material [20]. When the laser radiation is absorbed by the surface of the material, the electromagnetic energy of the beam is converted into electronic excitations and then into thermal, chemical and mechanical energy to cause vaporisation, phase explosion, hydrodynamic, and photomechanical effects like exfoliation and spallation. The ablated species form a plume consisting of atoms, molecules, electrons, ions, clusters and micro-sized particulates. Because the mean free path inside the plume is very short, after irradiation the plume rapidly expands in vacuum from the target to form a noodle jet with hydrodynamic flow characteristic. This process has many advantages as well as disadvantage for thin film processes. Advantage are the simple experimental setup and flexibility. The primary advantage of the laser ablation process is anyway the ability to deposit thin

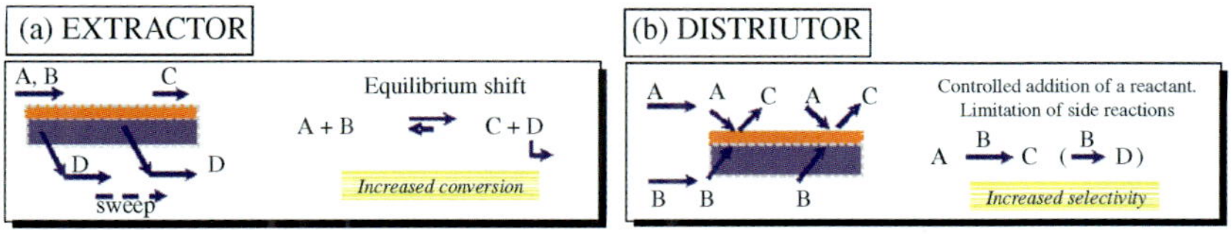

Figure 11 *Two approaches in membrane reactors [257]*

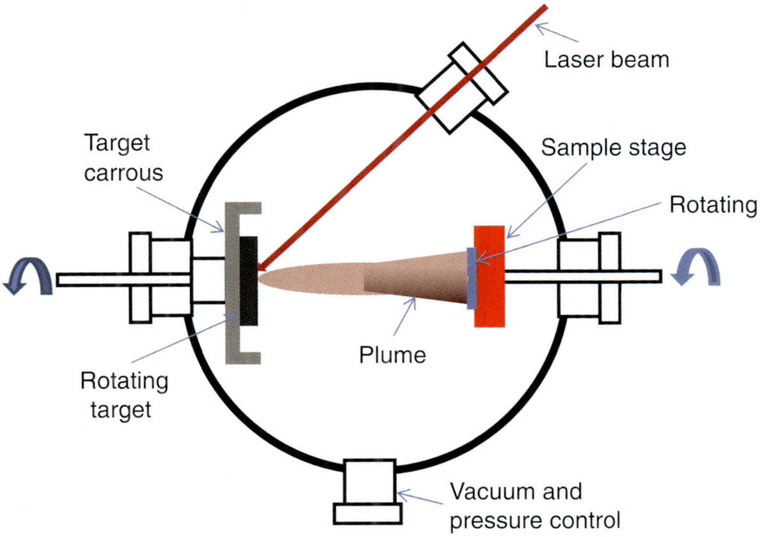

Figure 5.16 *Schematic diagram of pulsed laser deposition technique for thin film deposition*

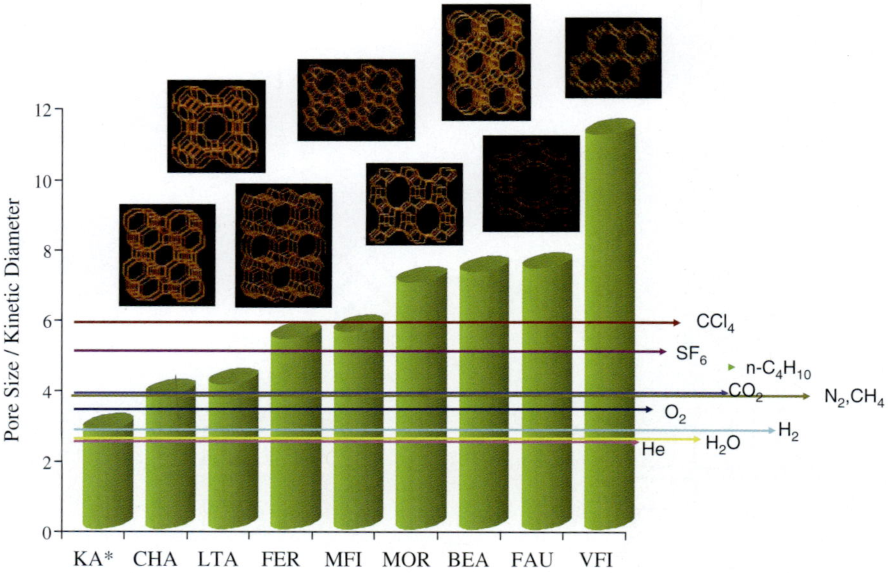

Figure 8.1 *Pore size of different framework structures and kinetic diameter of selected molecules. Pore size and kinetic diameter are taken from Ref. [2] and framework structures from Ref. [3]. KA* = This is not a framework structure, but a zeolite A (LTA structure) with potassium instead of sodium as an extra-framework cation*

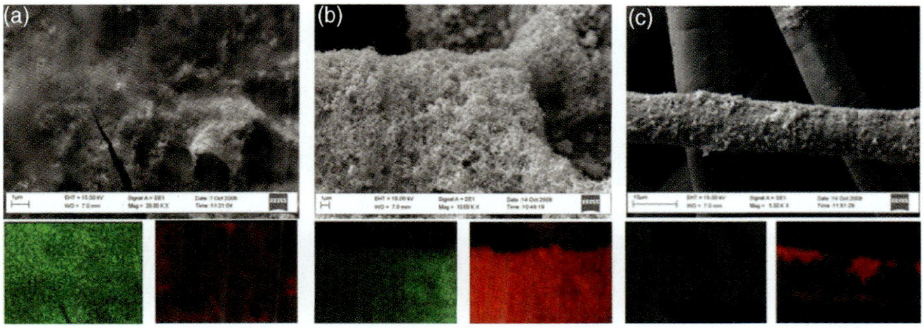

Figure 17.4 *SEM-EDAX images of a metallic fibre plate coated with two layers formed by: (a) sequential* $Al(OH)_3$ *(−1.2 V for 600 s) and Pd° (at −1.3 V for 600 s) synthesis, (b) sequential* $Al(OH)_3$ *(−1.2 V for 600 s) calcinations and Pd° synthesis (−1.3 V for 600 s), (c) simultaneous coprecipitation of* $Al(OH)_3$ *and Pd° (−1.3 V and 600 s)*

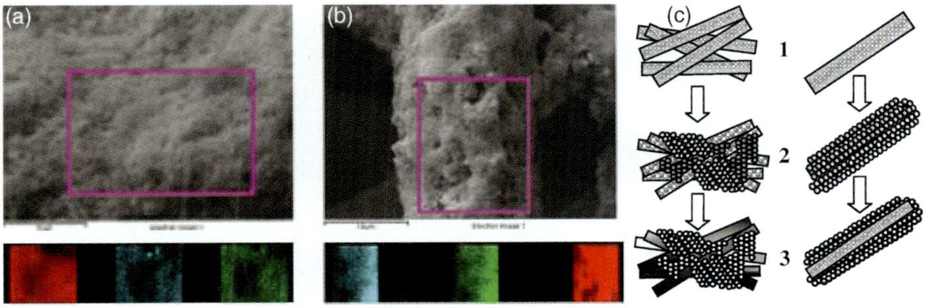

Figure 17.5 *Multilayer sequential synthesis of Mg/Al HT (−1.2 V for 1000 s) and Pd° (−1.3 V for 600 s) on a metallic fibre plate, with SEM EDAX of the multilayer (a) and within interpore (b) on the fibre. (c) Schematic representation of the preparation steps for the multilayer synthesis on the fibres and within the interpores (among the fibres) showing: (1) inconel support, (2) inconel covered by Mg/Al and (3) Inconel-Mg/Al-Pd seeds*

Figure 19.1 *Metal membrane produced by planar flow casting at CSIRO (courtesy M. Fergus and D. Liang, CSIRO).*

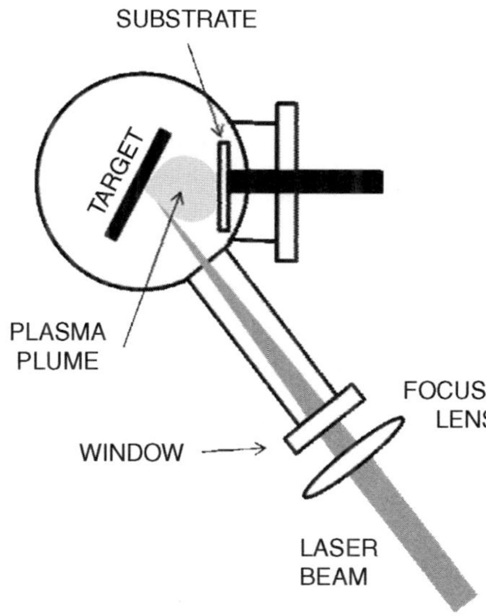

Figure 10.6 *Schematic view of a pulsed laser deposition system*

films with the same composition as that of the target material. Disadvantages are the presence of micro-sized particulates in the deposited films and the narrow forward angular distribution that makes difficult to obtain homogeneous large-area thin films.

10.2.3 Sputter Deposition

The sputter deposition is a PVD process involving the removal of atoms from a solid target bombarding the target with positive ions. Because the sputtering mechanism has mechanical nature, refractory materials can be easily deposited at temperatures well below their melting point. In addition it is very useful for alloys and compounds because the resulting film composition, generally, matches that of the source material. Finally the film structure can be tailored using the proper deposition parameters. The interested reader can find a detailed presentation of the technique in Ref. [12] and Ref. [21].

10.2.3.1 The Sputtering Process

In the sputtering process an energetic particle bombards a target material with sufficient energy to produce the ejection of one or more atoms from the atomic surface layers. The ejected species, mostly neutral atoms, are then transported in form of vapour to the substrate where they condense forming the deposited film.

The parameter describing the sputtering process is the sputtering yield Y: it is defined as the ratio of the number of ejected particles to the number of bombarding ones. It's worthy to mention that just before the impact with the target material ions are neutralised by capture of an electron of the surface atoms: the sputtering effect given by ions or neutral of the same species and energy is thus identical. Ions are anyway preferred in thin film deposition processes because: (i) large ion fluxes having defined energies can be easily produced and controlled and (ii) the electrical current to the target due to ion bombardment can be easily measured.

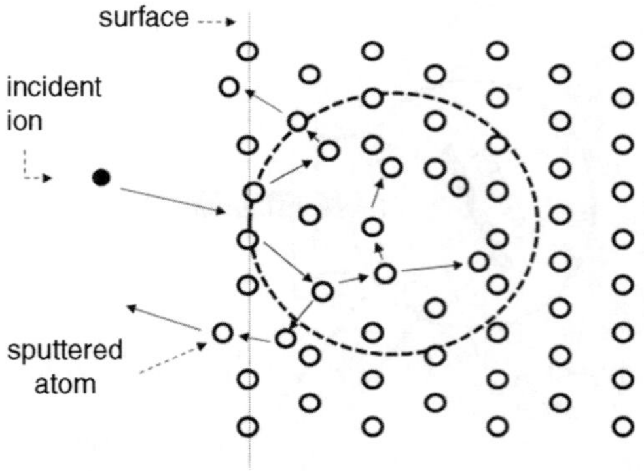

Figure 10.7 *Schematic of the physical sputtering process*

Microscopically, the sputtering process is consequence of momentum transfer from the impinging particles to the near surface atoms of the crystal lattice. The kinetic collisions occur first between the energetic incident particles and surface atoms of the target: after this collision the incident and the impacted particles move into the material thus causing more and more collisions. After many subsurface collisions one or more surface atoms can acquire kinetic energy exceeding the surface binding energy and will be thus ejected from the target material. In Figure 10.7 we present a schematic of the physical sputtering process.

Depending on the kinetic energy E of the incident ions four different regimes occur [22]:

1. Low energy regime ($0 < E < 20$–50 eV). The incident ion has not enough energy to eject a target atom.
2. Moderate energy regime ($50\,\text{eV} < E < 1000\,\text{eV}$). This is the energy range used in PVD techniques. The sputtering yield depends on the incident ion energy, on the substrate mass (see Figure 10.8 and Ref. [23]). For a given material *Y*, increases with the mass of the incident particles (see Figure 10.9 and Ref. [24]) and with the angle of incidence from normal incidence up to about 50°.
3. High energy regime ($1\,\text{keV} < E < 50\,\text{keV}$). The incident particle has enough energy to break the bonds between atoms in a region close to the impact site thus causing the formation of a dense cascade of secondary particles. *Y* is only a little larger than in the previous regime but the higher voltages required make the sputtering process of not practical use for thin film deposition.
4. Very high energy regime ($E > 50\,\text{keV}$). The incident ion penetrates into the target material many layers before producing a significant number of collisions: the affected volume is well below the target surface and thus very few atoms are emitted.

10.2.3.2 Plasma and Sputtering Systems

The sputter deposition technique makes use of a plasma formed very close to the target surface as the source of energetic, bombarding ions. A plasma is a collection of charged and neutral particles resulting from the partial ionisation of gas atoms or molecules, generally by an external electric field. In thin film deposition the plasma is obtained mostly using argon gas (or mixtures of Ar and oxygen or nitrogen): the degree of ionisation is low, 10^{-4}, so that the plasma is mostly neutral. The electron density and ion density are equal on average and their number which is much less than

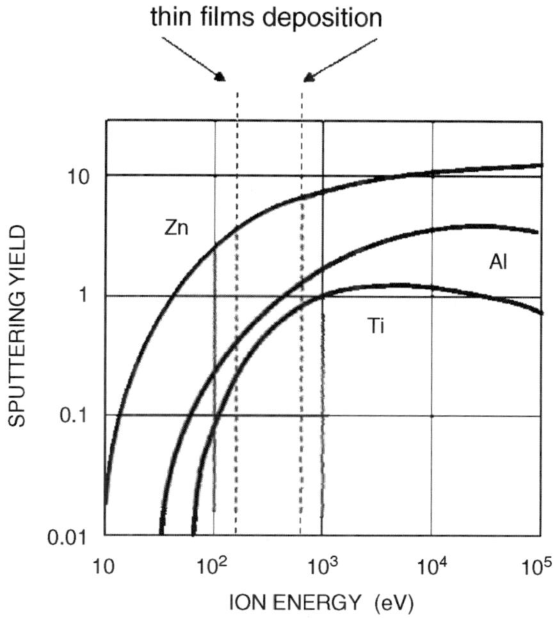

Figure 10.8 *Sputtering yield as function of the ion energy for Ar^+ bombardment of common metals (after Ref. [23])*

the number of neutral, is called plasma density. In thin film technology the plasma density is in the $10^8\,cm^{-3}$ level (at 5 mbar Ar pressure, the Ar density is $10^{13}\,cm^{-3}$). The electron energies are in the range 2–10 eV, which corresponds to electron temperatures of 10^4–10^5 K (cold plasmas). Due to collisions with neutral atoms which are the dominant species of the plasma, ions have temperatures which are in the 500 K order so that there is a non-equilibrium nature of the plasma [25].

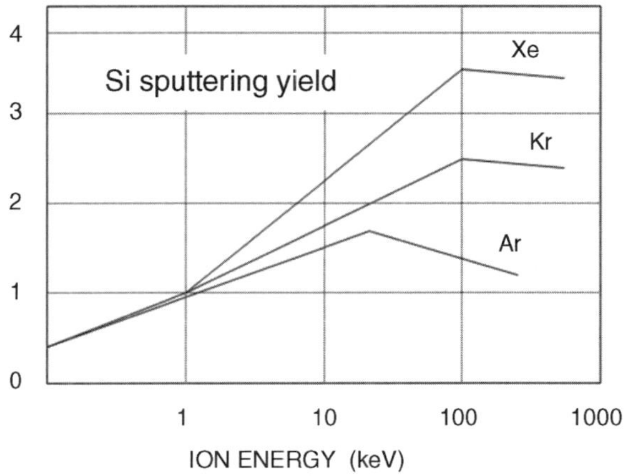

Figure 10.9 *Sputtering yield for Si as function of the ion energy for several inert gas ions (after Ref. [24])*

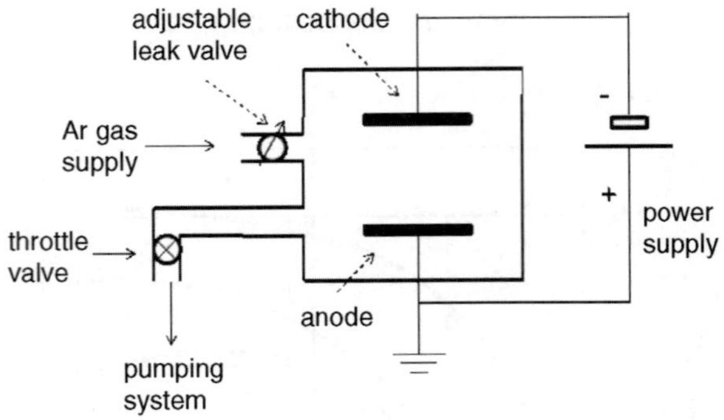

Figure 10.10 *Typical diode plasma system with grounded substrate electrode (anode)*

The most common method to produce a plasma for sputtering applications is to place a moderate DC voltage (from $\sim 10^2$ to $\sim 10^3$ V) between two metal electrodes in a evacuated chamber containing Ar or other working gas at low pressure, see Figure 10.10. In the gas there is always a small number of ions and electrons due to cosmic radiations: the application of the DC potential causes these charges to move and giving rise to a burst of constant current. The cathode emits electrons because the electric field near its surface extracts electrons (this process is called field emission). Increasing the voltage the charged particles acquire energy and undergo collisions with electrodes and neutral atoms thus producing more ions. If the voltage between the electrodes is high enough ions striking the cathode can release electrons (secondary electrons). These electrons make the current to increase at a constant voltage called breakdown voltage V_{m}. The secondary electron yield depends on both the bombarding ion species and the substrate material but, for a given material, it is rather insensitive to the incoming ion energy up to about 1 keV (it is of the order of 0.1) so that many ions have to bombard the cathode before a secondary electron is emitted. When the number of secondary electrons is enough to produce ions that regenerate the same number of electrons, a self-sustaining discharge is obtained: here the gas begins to glow, the voltage drops and the current rises abruptly. The magnitude of the breakdown voltage V_m depends on the electrode distance and the gas pressure which determines in fact the electron free mean path. Because of too high pressures or too large distances, ions generated in the plasma are slowed down by inelastic collisions and strike the cathode with too low energy to produce secondary electrons. At too low pressures or too small electrode distances the secondary electrons cannot undergo a sufficient number of collisions before they are collected at the anode. Initially the ion bombardment is concentrated near irregularities of the cathode surface: increasing the power the bombardment covers all the surface and a constant current is obtained. A further increase in power gives rise to both an increase in current and voltage, see Figure 10.11 [14]: this condition is usually used in all glow discharge processes for thin film deposition.

Plasma has roughly two specific regions: the bulk of the plasma and the sheet (dark space), located between the bulk of the plasma and the cathode. Because the plasma is a conductor, its potential is fairly constant across its width and all of the voltage drop between cathode and anode occurs in the thin dark space at the cathode, see Figure 10.12. In the dark space near the cathode there is thus a high electric field: here ions are accelerated rapidly and strike the cathode. The electrons are light compared to the ions and so it is a worthy approximation to consider the ions as immobile: in the bulk of the plasma the loss rate of the electrons from the plasma edge is larger than the loss rate of ions. This produces a slightly positive charging of the

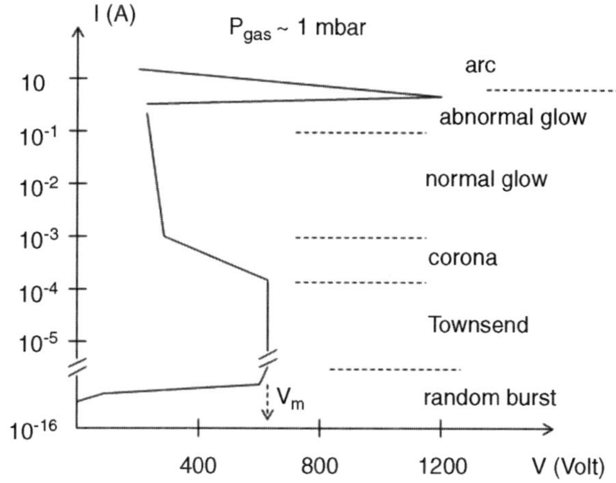

Figure 10.11 *Voltage–current relationship in a DC glow discharge [14]*

plasma giving rise to a small positive potential that reduces the rate of electron loss to be the same as that of the ion loss. The plasma potential is virtually always positive and typically a few volts more positive than the potential of the anode.

For most sputtering applications the cathode serves as source of the sputtered atoms while the substrate is placed over the anode: ions from the plasma in fact will bombard the cathode (which is generally called target) with sufficient energy to cause sputter emission of cathode atoms. Typical value of the working gas pressure in DC plasmas are between 10 and 100 Pa: the operating pressure of the working gas is controlled by the rate of gas passing into the chamber and by a throttle valve placed between the vacuum system and the deposition chamber. The gas flow into the chamber is adjusted by this valve and the effective pumping speed by the opening of the throttle valve. Electrodes have dimensions of 1–10 cm and the separation of cathode and anode is between 1 and 10 cm. The anode is in electrical contact with the chamber walls: since the chamber walls are grounded (for safety reasons) the cathode potential is then hundreds of

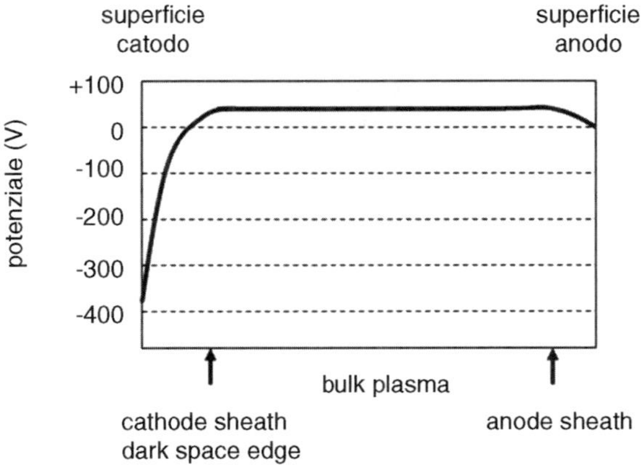

Figure 10.12 *Spatial distribution of potential in a DC glow discharge tube*

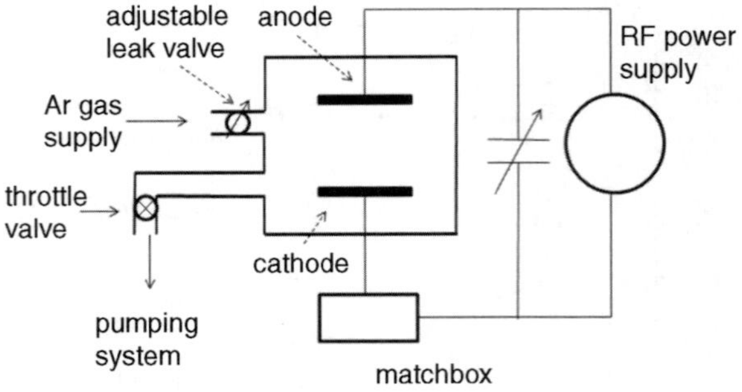

Figure 10.13 *RF system with nongrounded sample electrode*

volts negative. This two electrode system in called diode plasma because is characterised by a flux of ions that strikes the cathode (negative electrode) and a flux of electrons that move toward the anode.

The practical use of DC diode plasmas for sputter deposition applications is limited by the low ion currents to the cathode (which causes low deposition rates): the low ion currents are given by the low cross section for electron-collision ionisation of most gas species: most secondary electrons pass right through the plasma and hit the anode without producing new ions.

Another problem occurs when a reactive gas such as oxygen is added to the gas mixture: the cathode becomes rapidly oxidised and as a result the net DC current to the insulating oxydised layer drops to negligible values because of charging of the surface. The deposition rates will be thus too small to be useful: the same problem obviously occurs when insulating targets are used.

Insulating targets can be sputtered by using a radio frequency (RF) potential, typically 13.5 MHz, in place of a DC potential. The RF sputtering apparatus is similar to the DC diode system except for the addition of an impedance matching network between power supply and electrodes, see Figure 10.13. In the RF apparatus the second electrode (the anode) is also powered by the RF power supply and not grounded as in the DC case.

The RF powered electrode oscillates between positive and negative: because the higher mobility of the electrons the RF powered electrode (the cathode) pick up a much greater electron than ion current in one cycle: consequently the electrode will accumulate a negative charge. A capacitor placed in the impedance network blocks this DC potential from the power supply. On each successive RF cycle the electrode is charging more negatively: the ion collection time in each cycle will consequently increase while the electron collection time decreases. The net result is that, at regime, after several cycles a negative DC potential set up on the electrode at a value which is approximately close to one half of the applied RF peak to peak voltage, see Figure 10.14. The consequence of this negative DC component is that only for a very short time of each RF cycle the electrode is positive and collects electrons while for the rest of the cycle is negative and collects ions: since there is no net current flow (the electron and ion currents cancel out in a cycle) there is no charging at the insulating surface of the target.

A further consequence of the RF power is the increase of the ion current and thus of the deposition rates as compared to the DC case. The rapid change of the voltage polarity traps, in fact, the electrons (plasma electrons and secondary electrons) within the plasma: this trapping increases the ionisation rates and thus the plasma density results larger than that obtained with DC plasmas. Consequently the system can operate at pressures as low as 1–10 mPa.

The use of RF plasmas requires an additional component of the deposition apparatus, the matching network (matchbox). RF plasmas have in fact complex impedance with resistive,

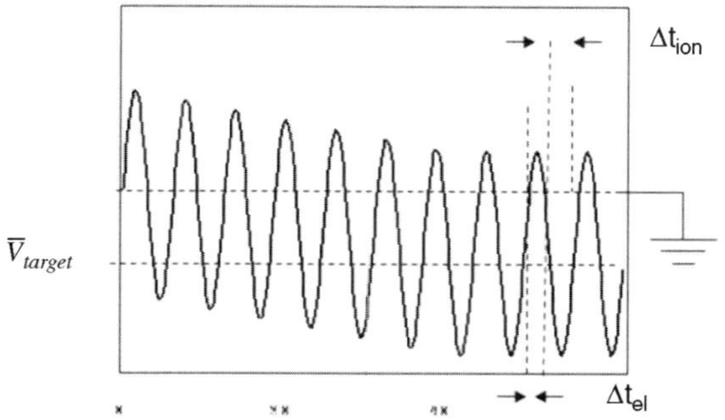

Figure 10.14 *Applied RF voltage as function of time to the target electrode.* Δt_{ion} *= fraction of the RF cycle where the cathode is bombarded by ions (the positive charge on the target surface increases),* Δt_{el} *= fraction of the RF cycle where the cathode is bombarded by electrons (the positive charge on the target surface decreases)*

inductive and capacitive component. Power supply are designed to efficiently transfer power to a 50 ohm load. The matching network optimises the impedance of the plasma and network system to maximise the amount of power that can be delivered to the plasma. In the most common network, the 'L' system, the incoming power is split upon entering the network: one side goes through a variable capacitor to ground (load capacitor) the second side goes through an inductor and a variable capacitor (tuning capacitor) to the cathode, see Figure 10.15. The circuit is then closed through the chamber walls (ground). The matchbox is to matches the net impedance of the plasma side of the circuit to the load capacitor within the matchbox: half of the applied power goes into the and thus the reflected power will be minimized.

It is important to remark that since an AC voltage is involved both electrode should be sputtered by using RF plasmas. When the sputter target is an insulator capacitively coupled with the RF generator, the equivalent circuit of the sputtering system is given by two capacitors in series with the applied voltage divided between them, one at the target sheath region the other at the substrate. Since the capacitive reactance is inversely proportional to the capacitance and thus the capacitor area, more voltage will be dropped across the capacitor of a smaller surface area: for efficient sputtering the target area has to be small compared with the area of the other electrode. In practice this second electrode consists on the substrate holder and system ground including the chamber walls. It can be shown that the ratio of the voltage across the sheath at the small

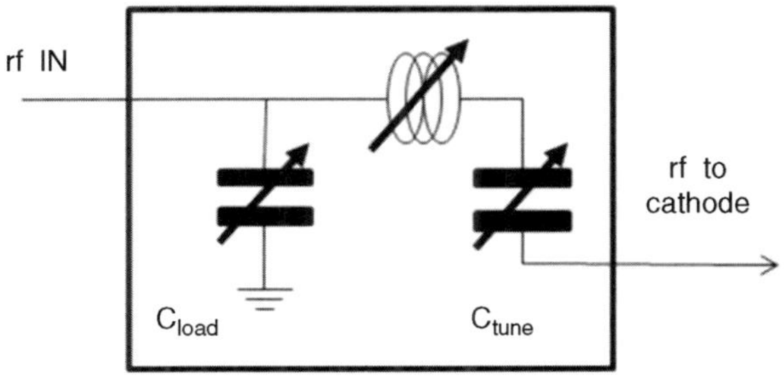

Figure 10.15 *L-type matching network*

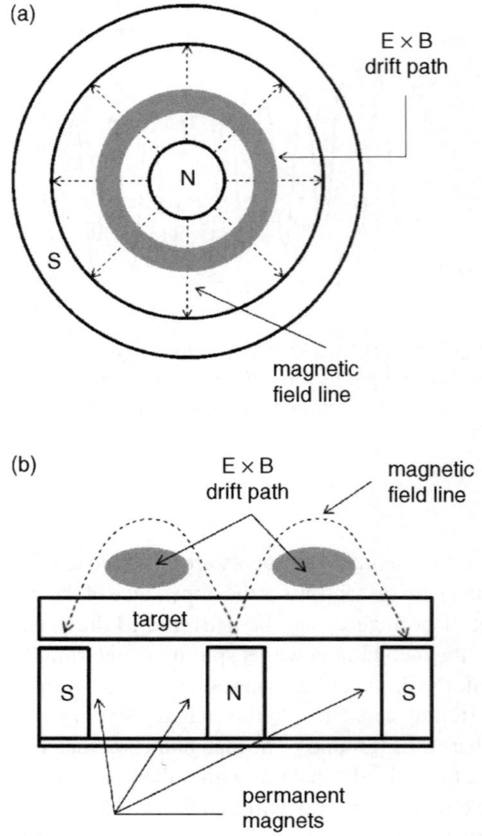

Figure 10.16 *Top (a) and side (b) views of the target in a circular planar magnetron cathode*

capacitively coupled electrode (V_C) to the voltage across the large directly coupled electrode (V_D) is given by $(V_C/V_D) = (A_D/A_C)^4$ where A_C and A_D are the respective electrode areas [26]. The fourth power dependence means that the large value of A_D effectively increases the target sheath potential while minimises the ion bombardment of the grounded substrate holder.

An important way to alter the plasma and increase the deposition rates is to place a magnetic field into the plasma region close to the cathode (magnetron sputtering). The most common configuration is that presented in Figure 10.16a, top view, and Figure 10.16b, front view: the target material is backed by a permanent magnet that provides a **B** field with the field lines forming a closed tunnel on the target surface (this particular configuration is called circular planar magnetron). The electric field **E** is normal to the surface: the application of the magnetic field **B,** through the **E** × **B** Lorentz force, gives rise to trapping of the secondary electrons in the region close to the cathode, indicated by the grey colour. This results in a large number of collisions of the electrons with the gas atoms close to the target material and, for a given applied power, to a local increase of the plasma density. Consequence is that the system can operate at working gas pressures in the 10^{-2} Pa order while maintaining high deposition rates. The ions experience the same force as the electrons but due to the much higher mass their motion is not confined and the sputter bombardment proceeds as in a normal diode.

Magnetron sources introduce anyway some problems. The ion bombardment is localised on the cathode directly under the $\mathbf{E} \times \mathbf{B}$ closed path: this results in a poor utilisation of the cathode as deep groves, called etch tracks, are formed. This effect is unwanted particularly when precious metals such as Pd are used for the deposition. A consequence of this localised sputter emission is the fact that the film deposition is not uniform on large substrates. It will be thus necessary to move the substrate or to increase the working pressure to smear out the distribution of sputtered atoms through collisions with Ar gas atoms.

In bias sputtering the electric field near the substrate is altered in order to vary the flux and energy of the incident charged particles. This is done by applying either a negative DC or RF bias to the substrate holder. With a target voltage of –1000 to –3000 V, bias voltages of –50 to –300 V are generally used. The bias sputtering is effective in changing many properties of the growing film as, for example, hardness, residual stresses, morphology (generally columnar microstructure are replaced by compacted fine-grained structures) and adhesion [27].

10.2.3.3 Film Structure

Sputtering differs from evaporation because atoms ejected from the target surface have significantly more kinetic energy. Cu sputtered atoms using 500 eV Ar ions have an average energy close to 8 eV much larger than that of Cu atoms evaporated at 1500 K which is lower than 0.5 eV. The peak in the kinetic energy distribution of sputtered atoms differs from each ion target system and slightly depends on the kinetic energy of the bombarding ions.

Sputtered atoms travel a distance which is in the cm order before their impact with the substrate surface where the deposited film grows. The typical pressure for most sputtering applications ranges from 10 mPa to 10 Pa corresponding to a mean free path of $\sim 10^2$ cm to ~ 1 mm. This influences the modality of the atoms transport to the substrate.

At low pressure, 10 Pa or less, the sputtered atoms reach the substrate without collisions prior to the deposition and arrive there with their original kinetic energy. This ballistic transport gives rise to an energetic deposition process with the atoms itself implanting in the near surface layers of the film. Film deposited in this regime are generally small-grained, dense, show good adhesion to the substrate, have compressive stress and are reach of defects. At pressures larger than 1000 Pa, the ejected atoms lose their kinetic energy in collisions with the gas atoms and thermalise before reaching the substrate (diffusive transport): the resulting films are thus similar to those grown in thermal evaporation processes and, compared to films grown in ballistic transport conditions, show larger grain size, lower defects density, worst adhesion to the substrate and stress having tensile character. The film morphology can be also controlled by the deposition temperature. The microstructure of Al thin films, for example, changes from highly porous and columnar, to densely columnar and finally to dense grained by increasing the substrate temperature. A general description of the microstructure of metallic thin films deposited by sputtering as function of the working gas pressure and the deposition temperature is provided by the Thornton diagram [28].

A further significant advantage over evaporation is the fact that by sputtering it is possible to deposit films which have the same composition as the target material even if the sputtering yield is different for the different components. When, in fact, a component of the target material is preferentially sputtered then a surface depletion of this component occurs. This depletion continues until the concentration of the component with the lower sputtering rates compensated for the lower rate of removal owing to the surface enrichment of the target in this component. A dynamic state of equilibrium set up and the rate of sputtering of both component at any instant is proportional to the target composition. We note anyway that large difference between target and film composition can occurs when the sticking coefficients of the two components are different.

10.3 Pd-Based Metallic Membranes

This section will be dedicated to a literature review on the preparation and operative performances of composite membranes where the Pd-based functional layer was prepared by PVD techniques. In the first part, to understand the underlying physical processes, we will present some note on the hydrogen permeation through dense metallic membranes, the so called solution–diffusion mechanisms [8], and then we will indicate the required properties on the structure of Pd-based membranes for its proper working.

10.3.1 Hydrogen Permeation Through Metallic Membranes

The hydrogen transport through membranes is controlled by microscopic processes occurring in the bulk of the material, at the surface–bulk and at the surface–gas interfaces. Let us consider a thin membrane having thickness x much lower than the lateral dimension. One side of the membrane is exposed to H_2 gas at pressure P (high pressure side, HPS), while the other side (low pressure side, LPS) is exposed to vacuum.

The permeation process is satisfactorily described by the solution–diffusion mechanism [8] and a relationship between the hydrogen permeation flux φ_P through a metallic membrane can be obtained considering the balance of all the involved microscopic processes. The hydrogen molecule is adsorbed on the metallic surface where its dissociation into atomic hydrogen occurs: the fractional surface area containing two adjacent vacant surface sites will be $(1-\theta)^2$, where θ is the fractional surface coverage. Calling α_m the adsorption probability for incident molecules, the flux of adsorbing atoms is given by:

$$\varphi_{ads} = 2\alpha_m(1-\theta)^2\,\mu P \tag{10.1}$$

where the quantity μ is defined by the kinetic theory of gases $\mu = (2\pi M k_B T_m)^{-1/2}$ and thus μP is the incident molecular flux at the surface: in this equation M is the H_2 molecular mass, k_B the Boltzmann's constant and T_m the temperature of the H_2 gas.

The desorbing flux is given by:

$$\varphi_{des} = 2\delta\theta^2 \tag{10.2}$$

where δ is the rate constant for desorption. The factor 2 in the previous equations means that two H atoms are involved for each adsorbed (desorbed) molecule.

Adsorbed hydrogen atoms then enter into the metallic lattice: the flux of absorbing atoms is given by:

$$\varphi_{ab} = \gamma\theta \tag{10.3}$$

where γ is the rate constant. The reverse process will be proportional to the bulk concentration c of H atoms just below the surface and to the fraction of unoccupied surface sites $(1-\theta)$ and the flux is given by:

$$\varphi_{dsb} = \beta(1-\theta)c \tag{10.4}$$

where β is the rate constant. The bulk H concentration between the HPS and the LPS surfaces of the membrane is given by the Fick's law for diffusion:

$$\varphi_{\text{diff}} = -D\,dc(x')/\text{d}x' \tag{10.5}$$

where x' is the permeation coordinate and D the H diffusion coefficient. In the LPS side of the membrane hydrogen atoms are first transferred from subsurface bulk sites to the membrane surface where desorption occurs after their surface recombination. The respective fluxes are described by equations similar to the previous ones but the reverse of Equation (10.1) is absent because the pressure in the LPS is assumed negligible. In steady state conditions the permeation flux φ_{P} is given by the net flux across each interface:

$$\varphi_{\text{P}} = 2\alpha_{\text{m}}(1-\theta_1)^2\mu P - 2\delta\theta_1^2$$

$$\varphi_{\text{P}} = \gamma\theta_1 - \beta(1-\theta_1)c_1$$

$$\varphi_{\text{P}} = -Ddc(x')/dx'$$

$$\varphi_{\text{P}} = \beta(1-\theta_2)c_2 - \gamma\theta_2$$

$$\varphi_{\text{P}} = 2\delta\theta_2^2$$

where the subscript 1 and 2 refer to the HPS and LPS of the membrane, respectively. Assuming that each rate constant is equal for the HPS and LPS (symmetric membrane) the previous equations can be used to obtain a relation between the permeation flux φ_{P} and the H_2 pressure P.

Two special limiting cases are of interest [8]:

1. When the desorption process is much slower than the hydrogen diffusion, the hydrogen permeation flux φ_{P} can be approximated by the relation:

$$\varphi_{\text{P}} = \alpha_{\text{m}}\mu P.$$

   This equation indicates that half of the absorbed atoms are desorbed on either surface of the membrane. This regime is revealed by the linear dependence of the permeation flux with the H_2 pressure. Moreover φ_{P} does not depend on the membrane thickness.
2. When hydrogen diffusion is the rate-limiting process, the permeation flux is given by:

$$\varphi_{\text{P}} = (D/x)(\gamma/\beta)(\alpha_{\text{m}}\mu/\delta)^{1/2}P^{1/2} = SDP^{1/2}/x$$

   where S is the hydrogen solubility, also called Sieverts' constant [29]. The quantity $\Phi = S\,D$ is the hydrogen permeability of the material [30]. In the diffusion-limited regime the hydrogen permeation flux is inversely proportional to the membrane thickness: this regime is experimentally evidenced by the $P^{1/2}$ dependence of the permeation flux.

The hydrogen permeation through thick Pd based membranes (bulk material) occurs in this regime as observed, for example, by Tosti *et al.* with thick membranes made of rolled Pd or Pd-Ag foils with thickness of 50–70 μm at 673 K [31] and by Wang *et al.* with 140 μm thick Pd foils at 473 K [32].

The interest to the use of PVD techniques for Pd-based membrane preparation is related to the possibility of preparing thin, nanostructured materials having advantages both in terms of cost of the system and membrane performances. In these polycrystalline solids, when the crystal size is in the 10 nm range, 20–50% atoms are located at grain boundaries where the atomic packing density is lower than in the regular lattice. Compared to lattice diffusion, the H diffusion through grain boundaries is thus a much faster process: the presence of this fast diffusion channel can increase the H diffusivity by a factor of 10 compared to that of the polycrystalline materials without compromising the membrane selectivity [33, 34]. In this case the rate-limiting step can turn from bulk diffusion to surface reactions thus ensuring the highest permeation rates [35].

10.3.2 Requirements for a Pd-Based Membrane

Pure palladium membranes cannot be used because Pd shows a transition from the α-phase (hydrogen poor) to the β-phase (hydrogen rich) at temperatures below 573 K and pressures below 2 MPa: since the lattice constant of the β-phase is about 3% larger than that of the α-phase this transition leads to lattice strain and consequently after a few cycles to the embrittlement of the membrane [36]. Alloying Pd with Ag or other group 1d metals reduces the critical temperature for this embrittlement and increases the hydrogen permeability [37]. The optimal Ag concentration is obtained for additive content of approximately 23 at% [38].

Investigations have evidenced poisioning effects of Pd membranes by gaseous components of reaction mixtures. Poisoning can occurs in two ways. The first is a chemical reaction of gaseous components with surface atoms, as occurs with gases containing sulfur or chlorine. At elevated temperatures sulfur attack, for example, is evidenced by a discoloration of the surface and a rapid disintegration of the metal structure [39]. A study by Antoniazzi *et al.* has demonstrated that sulfur compounds, such as H_2S are irreversible poisons causing a reduction in the hydrogen permeation through pure Pd foils by 1% for each ppm of H_2S present in the hydrogen feed [40]. The second way poisoning can happen from the chemisorption of gases such as water vapour, oxygen, carbon dioxide and carbon monoxide that block the available adsorption sites for hydrogen [41]. The addition of CO to hydrogen gas at concentration between 10 and 50%, for example, was observed to cause the reduction of the H_2 permeability of Pd membranes at temperatures below 250 °C whereas CO_2 at similar concentrations has little effect [42]. Anyway as soon as the impurity gas is turned off, the transfer rate immediately returns to normal. Saturated hydrocarbons, methanol, benzene, and cyclohexane vapours mixed with hydrogen also appear to have no permanent effect on the hydrogen permeation.

Thus, the development of a robust, poison resistant and economical hydrogen membrane material is necessary for successful implementation of advanced fossil energy plants. Recent studies have reported that palladium–copper alloy materials possess the high permeability and resistance to sulfur poisoning needed to be considered as viable candidates for hydrogen separation in aggressive environments [43, 44].

10.3.3 Pd-Based Membranes Prepared by PVD Techniques

Pd-Ag metal composite membranes for hydrogen purification were prepared by Athayde *et al.* coating a conventional polymeric gas separation membrane with a Pd-Ag film deposited by sputter deposition. The deposition process was carried out using Ar as working gas and as target material an alloy with this composition: 76 at% Pd−24 at% Ag. The authors prepared membranes with thicknesses of the metal layer between 25 and 100 nm under different growth rates and observed that the best membranes were obtained at high metal deposition rates. These membranes have a hydrogen flux of $\sim 10^{-9}$ mol m^{-2} s^{-1} Pa^{-1} at room temperature and a $H_2:CO_2$ selectivity greater than 100 [45].

Jayaraman *et al.* deposited ultrathin Pd films on porous ceramic support by RF sputter deposition and studied the influence of the support temperature and surface roughness on the structure of the deposited film. It was observed that the grain size of the Pd film increases by increasing the substrate temperature and that the N_2 gas tightness increases by decreasing the surface roughness of the ceramic support. Then optimum coating temperature was 673 K when Pd grains with 30 nm were produced: at this temperature a good adhesion level was observed between the γ-alumina support and the Pd layer up to thicknesses of 300 nm. At larger thickness, the peeling off the support was observed probably due to lattice stresses involved in the deposition and cooling process [46]. In a successive work the authors deposited Pd-Ag films with thickness ranging from 250–500 nm on the surface of 3 nm pore sol-gel derived γ-alumina support by RF magnetron sputtering and studied the separation performances of the composite membrane. The $H_2:N_2$ separation factor of the ultrathin membrane was 5.7 at 523 K and increased by increasing the temperature thus indicating the presence of pinholes defects [47]. At larger membrane thicknesses, 0.1–1.5 μm, the Pd and Pd-Ag membranes showed H_2 permeance in the range 1.0 to 2.0×10^{-7} mol m^{-2} s^{-1} Pa^{-1} and H_2:He selectivity between 30 and 200 at 573 K. The H_2 permeance of the membrane was quite independent on the membrane thickness, indicating that the rate-limiting step was connected to surface reactions [48]. The composition dependence of the hydrogen permeation flux was studied preparing submicron thick samples sputter deposited on mesoporous γ-alumina substrates. The hydrogen permeance was found in the range of 3.0×10^{-8} to 1.0×10^{-7} mol m^{-2} s^{-1} Pa^{-1} and the $H_2:He$ selectivity was in the range of 4 to 4000 depending on the Ag concentration and microstructure of the Pd-Ag films: the best selectivity and H_2 permeability were obtained with film samples deposited at low RF power (20 W) having Ag content close to 18 at% [49].

Pd-Ag membranes on porous alumina substrates were deposired by O'Brien *et al.* using unbalanced magnetron sputtering: Pd-Ag films were prepared with thickness from 0.7 to 1.1 μm at 0.05 μm min^{-1} deposition rate by changing the level of substrate bias and the substrate temperature [50]. At room temperature and 1 bar pressure the samples were N_2 permeable and the gas flow ranged from 8×10^{-9} to 1×10^{-5} mol m^{-2} s^{-1} Pa^{-1} depending on the deposition parameters. The $H_2:N_2$ separation factor ranged from 4 to 80. The sample analysis revealed that coatings contain pinholes and it was observed that their number and size mostly depended on the substrate conditions before deposition.

Zhao *et al.* deposited Pd and Pd-Ag (24 wt%) membranes with thickness <1 μm by magnetron sputtering on commercial microfiltration membranes coated with a sol-gel derived γ-Al_2O_3 layer using a Pd-24 wt% Ag alloy as target material [51]. H_2 permeation tests showed that the rate-limiting step in the hydrogen permeation was due to surface reactions and that the membrane H_2/N_2 selectivity was larger than 60 at temperature of 587 K. Increasing the temperature the selectivity decreased due to the formation of macroscopic defects (cracks).

In a recent study Zhang *et al.* deposited Pd-Ag films with thickness of about 100 nm on V-15Ni support by DC sputtering in Ar atmosphere using separate Pd and Ag targets independently controlled. It was observed that in the 423 to 673 K temperature range and 10^3 to 10^5 Pa pressure the hydrogen permeation was controlled by the hydrogen diffusion through the Pd-Ag overlayer and the permeability increased with the Ag content up to 30 at%: the highest H_2 permeability was of 1.53×10^{-8} mol m^{-1} s^{-1} Pa$^{-1/2}$ at 673 K [52]. In a successive study the authors examined the structural changes of 100 nm thick Pd-25Ag and Pd membranes upon H_2 permeation. It was observed that after long term operation at 573 K there was no metallic interdiffusion between the Pd-25Ag membrane and the V-15Ni support and that the oxidation of the Pd-25Ag film was limited. On the contrary after long-term operation at 673 K severe interdiffusion occurred between the film and the support. The interdiffusion was accompanied by the selective oxidation of vanadium thus causing the deterioration of the membrane performances [53].

To overcome the problems related to the formation of pinholes and to the resistance to mass transfer of the porous support Gielensa *et al.* coated the smooth surface of a silicon wafer with Pd, Pd-Ag and Pd-Ta-Pd films (thickness between 0.5 and 1.2 μm) and then etched the support to obtain the composite membrane [54]. The highest H_2 flux, 3.6 mol H_2 m^{-2} s^{-1}, was found with a defect-free 1 μm thick Pd-Ag film at 723 K and 0.83 bar H_2 pressure: this flux is approximately one order of magnitude higher than the fluxes reported in the literature for Pd or Pd alloy membranes deposited on porous supports.

Checchetto *et al.* deposited Pd films ~5 μm thick by RF magnetron sputtering on porous stainless steel discs. In order to fill the pores and prepare a flat surface for the Pd coating a polymeric layer of commercial polycarbonate was deposited on the steel surface by spin coating. The Pd deposition produced rough and pinholes free coating well adherent to the substrate. The hydrogen permeance at 373 K was of 5×10^{-7} mol m^{-2} s^{-1} Pa^{-1} with high H_2: N_2 selectivity. Long term operation produced macroscopic defects such as cracks thus reducing the membrane performances [55]. In a successive study Checchetto *et al.* [56] prepared a composite membrane consisting of 100 nm nanoporous silica [57] coated with a 150 nm Pd-Ag layer. An alumina disc having periodic micro sieves structure was used as support for the bilayer. The hydrogen transport through this nanocomposite membrane was controlled by the dissociation of molecular hydrogen at the surface of the Pd-Ag functional layer. When operating at 573 K, the membrane exhibits high $H_2 : N_2$ selectivity (a factor as high as 600–900), high H_2 permeance (~10^{-6} mol m^{-2} s^{-1} Pa^{-1}), and operative stability upon long-term operations [56].

Pd (77 wt%) – Cu (23 wt%) alloy films with 0.75 μm thickness were deposited by DC sputtering with Ar gas using pure Pd and Cu targets independently controlled. Analysis showed that the deposited film has polycrystalline structure with grain size of 14 nm when the substrate temperature was 300 K and 240 nm at substrate temperature of 600 K. With membrane thickness between 200 and 750 nm the H_2 permeance was 1.4×10^{-5} mol m^{-2} s^{-1} Pa^{-1} and the $H_2 : He$ selectivity was close to 500 at 698 K [58]. Pd-Cu-Ni ternary alloy membrane were prepared by Hoang *et al.* by magnetron sputtering and Cu reflow on porous nickel support. The H_2 permeance of 4 μm thick films at 773 K temperature was 2.3×10^{-7} mol m^{-2} s^{-1} Pa^{-1} with very large $H_2 : N_2$ selectivity due to defect-free surface. The hydrogen permeation process was controlled by surface reactions [59].

A recent work by Bryden *et al.* has shown that nanostructured Pd films deposited by DC sputtering on porous Vycor glass delaminated from the substrate when exposed to hydrogen at room temperature owing to the $\alpha \rightarrow \beta$ phase transition and exhibited grain growth at temperatures larger than 473 K [60].

10.3.4 Pd-Based Membranes Prepared by NonPVD Techniques

Alternative techniques for the deposition of Pd-based membranes are the electroless plating (ELP) and the Chemical Vapour Deposition (CVD).

ELP is the most common method for preparing Pd-based composite membranes. This chemical process, also known as auto-catalytic plating, is a nongalvanic type of plating method that involves several simultaneous reactions. The process occurs at moderate temperature immerging the support in an aqueous solution containing Pd salts, EDTA (EDTA is a polyamino carboxylic acid with the formula $\{CH_2N[CH_2CO_2H]_2\}_2$), ammonia and hydrazine without the use of external electrical power [13]. Palladium particles are produced by reduction of the palladium precursors. To increase the adhesion with the support, prior to the Pd deposition, the support surface has to be activated by seeding with Pd nuclei generally using a chloroform solution of Pd complexes as precursor of the Pd seed combined with alkaline hydrazine solution as reducing agent [61]. These particles grow on the palladium seeds which also act as a catalyst for the reduction of the palladium compounds. The preparation of Pd-based alloy membrane is done in two steps: (i) layers of Pd and (for example) Ag are sequentially deposited by electroless deposition on the activated porous substrate, (ii) atomic mixing by high-temperature annealing ($T > 600\,K$) in H_2 atmosphere to form an homogeneous alloy [62].

The advantage of this technique is due to the simple equipment, cost performance and applicability also to nonconductive materials of any complicated shape [13]. By this technique the Pd-based films are deposited on tubular supports made of macroporous stainless steel or alumina. Because the pore size of the support is in the $0.1\,\mu m$ range, to avoid macroscopic defects (pinholes) it is necessary to deposit Pd-based layers with thickness larger than several micrometers [63, 64]. The hydrogen permeation process is thus controlled by the H diffusion process as indicated by the square root dependence of the permeation flux from the H_2 pressure. These membranes show very high selectivity but permeation fluxes comparable to those obtained with PVD prepared membranes can be obtained only at process temperatures larger than 600 K [65–69].

In the chemical vapour deposition (CVD) process a volatile compound of the material to be deposited chemically reacts with other gases to produce a nonvolatile solid that deposit atomically on the substrate. It is a high temperature process which has found wide applications in the epitaxial growth of semiconductors films [12]. With Pd-based coating the major problem lies in the unavailability of proper Pd precursors with high volatility and good thermal stability which are necessary requirements for high metal deposition rates [70]. Dense Pd films have been deposited by Morooka *et al.* using CVD by thermal decomposition at reduced pressure of a metallorganic Pd precursor (palladium acetate) inside the pores of α-Al_2O_3 tubes [71]. A reactive CVD process was used by Xomeritakis *et al.* employing $PdCl_2$ as precursor and H_2 gas to grow thin Pd films on a composite support consisting of a γ-Al_2O_3 layer with average pore size of 4 nm on macroporous α-Al_2O_3 disk [72]: the $0.5\,\mu m$ thick Pd film at 673 K showed permeance of $1.0 \times 10^{-6}\,mol\,m^{-2}\,s^{-1}\,Pa^{-1}$, high $H_2:He$ selectivity and the permeation process was controlled by surface reactions. A modification of the CVD techniques was the forced-flow CVD: due to a pressure difference between the outside and inside of the support tube the vapours of a metal organic salt can enter the porous layer of the support and the membrane formation proceeds with mending pores and pinholes preferentially thus giving rise to a defect-free membrane [73]: the composite membranes exhibited H_2 permeance of $6.7 \times 10^{-6}\,mol\,m^{-2}\,s^{-1}\,Pa^{-1}$, high $H_2:He$ selectivity and the permeation process was controlled by H diffusion as consequence of the larger membrane thickness close to $4\,\mu m$.

10.4 Conclusions

PVD is mature a technique with commercial character for thin film deposition. The underlying physical processes are well known and the experimental technique well developed. The easy control of the deposition parameters allows the preparation of ultrathin films with low impurity content, controlled nanostructure and good adhesion to the support. Pd-based membranes prepared by PVD techniques are considered a potential alternative to the commercial membranes

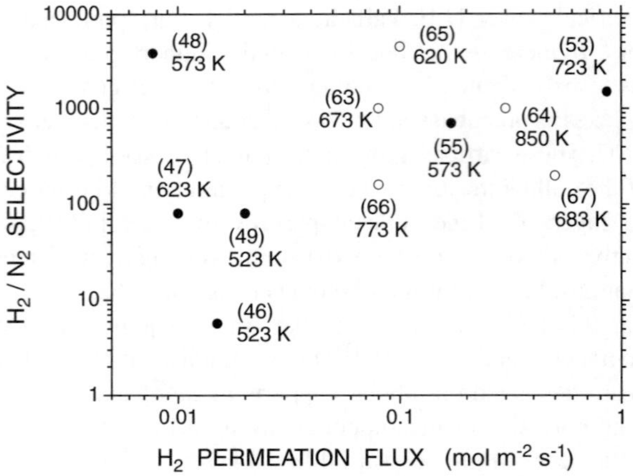

Figure 10.17 *Comparison of H_2 selectivity and permeation flux of some Pd-based membranes. Solid symbols: membranes prepared by PVD techniques. Open symbols: membranes prepared by electroless plating. The H_2 fluxes of the Pd-based membranes have been evaluated at 10^5 Pa pressure by using the experimental data presented by the authors, see the reference reported close to each symbol. In the figure we also indicate the process temperature. The selectivity values of Refs. [48, 49] are related to H_2 : He separation*

consisting of thick Pd or Pd-Ag foils and to the composite membranes prepared by electroless plating [74].

In Figure 10.17 we present a figure of merit describing the operative performances of PVD prepared membranes (solid symbols): the reported values of the permeation flux have been evaluated assuming 10^5 Pa trans-membrane pressure and using the H_2 permeability values indicated by the authors. Symbols report, at the indicated temperature, the H_2 selectivity (vertical scale) and the H_2 permeation flux (horizontal scale). Close to each symbol we report the reference. For comparison in the figure the performances of membranes prepared by electroless plating are also indicated (open symbols).

The figure clearly indicates that membranes prepared by PVD techniques will be competitive with those prepared by electroless plating: thin membranes offer, in fact, comparable permeation fluxes at lower temperatures and reduced material cost. Progresses are required to routinely avoid the formation of macroscopic defects such as pinholes during the fabrication process and cracks upon exposure to hydrogen gas that reduce the H_2 selectivity [75].

References

1. L. Schlapbach and A. Zuettel, *Nature*, **414**, 353 (2001).
2. H. S. Fogler, *Elements of Chemical Reaction Engineering*, Prentice-Hall, Upper Saddle River, NJ (2005).
3. K. Aasberg-Petersen, C. Stub Nielsen and S. Laegsgaard Jorgensen, *Catal. Today*, **46**, 193 (1998).
4. B. Metz, O. Davidson, H. de Coninck, M. Loos and L. Meyer (eds), *Carbon Dioxide Capture and Storage*, Cambridge University Press, Cambridge (2005).
5. N. W. Ockwig and T. M. Nenoff, *Chem Rev.*, **107**, 4078 (2007).
6. T. N. Nenoff, R. J. Spontak and C. M. Aberg, *MRS Bull.*, **31**, 735 (2006).

7. H. Verweij, Y. S. Lin and J. Dong, *MRS Bull.*, **31**, 756 (2006).
8. P. L. Andrew and A. A. Hassz, *J. Appl. Phys.*, **72**, 2749 (1992).
9. D. S. Sholl and Y. H. Ma, *MRS Bull.*, **31**, 770 (2006).
10. S. N. Paglieri and J. D. Way, *Sep. Purif. Methods*, **31**, 1 (2002).
11. G.Q. Lu, J.C. Diniz da Costa, M. Duke, S. Giessler, R. Socolow, R.H. Williams and T. Kreutz, *J. Coll. Int. Sci.*, **314**, 589 (2007).
12. M. Ohring, *The Materials Science of Thin Films*, Academic Press Inc., San Diego (1992).
13. G. O. Mallory and J. B. Hajdu, *Electroless Plating – Fundamentals and Applications*, William Andrew Publishing/Noyes, New York (1990).
14. K. J. Klabunde (ed.), *Thin Films from Free Atoms and Particles*, Academic Press Inc., Orlando (1985).
15. Aicha Elshabini-Riad and F. D. Barlow, III (eds), *Thin Film Technology Handbook*, McGraw-Hill, New York (1997).
16. P. A. Redhead, J. P. Hobson and E. V. Kornelsen, *The Physical Basis of Ultrahigh Vacuum*, American Institute of Physics, New York (1993).
17. L. I. Maissel and R. Glang (eds), *Handbook of Thin Film Technology*, McGraw-Hill, Ney York (1970).
18. H. H. P. Lupis, *Chemical Thermodynamic of Materials*, North Holland, Amsterdam (1983).
19. D. B. Chrisey and G. K. Hubler (eds), *Pulsed Laser Deposition of Thin Films*, John Wiley & Sons, Inc., New York (1994).
20. A. Miotello, R. Kelly, *Appl. Phys. A: Mater. Sci. Process*, **69**, S67 (1999).
21. R. A. Powell and S. M. Rossnagel (eds) *Thin Films, vol. 26, PVD for Microelectronic: Sputter Deposition Applied to Semiconductor Manufacturing*, Academic Press, San Diego (1999).
22. D. M. Mattox, *J. Vac. Sci. Technol. A*, **7**, 1105 (1989).
23. D. W. Hoffman, *J. Vac. Sci. Technol. A*, **12**, 953 (1984).
24. H. F. Winters, H. Coufal, C. T. Renner and D. S. Bethune, *Phys. Rev. B*, **41**, 6240 (1990).
25. B. N. Chapman, *Glow Discharge Processes*, John Wiley & Sons, Inc., New York, (1980).
26. H. R. Koenig and L. I. Maissel, *IBM J. Res. Dev.*, **14**, 1968 (1970).
27. D. M. Mattox, *J. Vac. Sci. Technol. A*, **7**, 1105 (1989).
28. J. A. Thornton, *J. Vac. Sci. Technol. A*, **11**, 666 (1974).
29. A. Sieverts, *Z. Electrochem.*, **16**, 707 (1910).
30. O. V. Richardson, J. Nicol and T. Parnell, *Philos. Mag.*, **8**, 1 (1904).
31. S. Tosti, L. Bettinali, S. Castelli, F. Sarto, S. Scaglione and V. Violante, *J. Membrane Sci.*, **196**, 241 (2002).
32. E. Wang, T. B. Flanagan and K. L. Shanahan, *J. Alloys Compd.*, **372**, 158 (2004).
33. H. Gleiter, *Prog. Mater. Sci.*, **33**, 233 (1988).
34. D. R. Arantes, X. Y. Huang, C. Marte and R. Kirchheim, *Acta Metall. Mater.*, **41**, 3215 (1993).
35. T. S. Moss, N. M. Peachey, R. C. Snow and R. C. Dye, *Int. J. Hydrogen Energy*, **23**, 99 (1998).
36. J. B. Hunter, *Platinum Metal Rev.*, **4**, 130 (1960).
37. G. Bohmholdt, *Z. Phys. Chem. Neue Folge*, **56**, 133 (1967).
38. S. Uemiya, T. Matsuda and E. Kikuchi, *J. Membrane Sci.*, **56**, 315 (1996).
39. J. Philipott and D. R. Couplan, in: Z. Paal and P. G. Menon (eds), *H_2 Effects in Catalysis*, Marcel Dekker, Basel, p. 697 (1988).
40. A. B. Antoniazzi and A. A. Haasz, *J. Nucl. Mater.*, **162/164**, 1065 (1989).
41. A. Li, W. Liand and R. Hughes, *J. Membrane. Sci.*, **165**, 135 (2000).
42. M. Amano, C. Nishimura and M. Komaki, *Mater. Trans. J. J. M.*, **31**, 404 (1990).
43. B.D. Morreale, M.V. Ciocco, R.M. Enick, B.I. Morsi, B.H. Howard, A.V. Cugini and K.S. Rothenberger, The permeability of hydrogen in bulk palladium at elevated temperatures and pressures. *J. Membr. Sci.*, **212**, 87 (2003).
44. S. E. Nam and K.H. Lee, *J. Membr. Sci.*, **192**, 177 (2001).

45. A. L. Athayde, R. K. Baker and P. Nguyen, *J. Membrane Sci.*, **94**, 299 (1994).
46. V. Jayaraman, Y. S. Lin, M. Pakala and R. Y. Lin, *J. Membrane Sci.*, **99**, 89 (1995).
47. V. Jayaraman and Y. S. Lin, *J. Membrane Sci.*, **104**, 251 (1995).
48. G. Xomeritakis and Y. S. Lin, *J. Membrane Sci.*, **133**, 217 (1997).
49. B. McCool, G. Xomeritakis and Y. S. Lin, *J. Membrane Sci.*, **161**, 67 (1999).
50. J. O'Brien, R. Hughes and J. Hisek, *Surf. Coat. Technol.*, **142/144**, 253 (2001).
51. H. B. Zhao, G. X. Xiong and G. V. Baron, *Catalysis Today*, **56**, 89 (2000).
52. Y. Zhang, T. Ozaki, M. Komaki and C. Nishimura, *J. Membrane Sci.*, **356/357**, 553 (2003).
53. Y. Zhang, M. Komaki and C. Nishimura, *J. Membrane Sci.*, **246**, 173 (2005).
54. F. C. Gielens, H.D. Tong, C.J.M. van Rijnc, M.A.G. Vorstman, J.T.F. Keurentjes, *Desalination*, **147**, 525 (2002).
55. R. Checchetto, N. Bazzanella, B. Patton and A. Miotello, *Surf. Coat. Technol.*, **177/178**, 73 (2004).
56. R. Checchetto, N. Patel, A. Miotello and R. S. Brusa, *J. Appl. Phys.*, **105**, 034502 (2009).
57. N. Patel, S Mariazzi, L. Toniutti, R. Checchetto, A. Miotello, S. Dirè and R. S. Brusa, *J. Phys. D: Appl. Phys.*, **40**, 5266 (2007).
58. S. K. Ryi, J. S. Park, S. H. Kim, S. H. Cho, D. W. Kim and K. Y. Um, *Sep. Purif. Technol.*, **50**, 82 (2006).
59. H. T. Hoang, H. D. Tong, F. C. Gielens, H. V. Jansen and M. C. Elwenspoek, *Mater. Lett.*, **58**, 525 (2004).
60. K. J. Bryden and J. Y. Ying, *Mat. Sci. Eng. A*, **204**, 140 (1995).
61. D. A. Pacheco-Tanaka, M. A. Llosa Tanco, A. Niwa, Y. Wakui, F. Mizukami, T. Namba and T. M. Suzuki, *J. Membrane Sci.*, **247**, 21 (2005).
62. F. Roa, J. D. Way, R. L. McCormick and S. N. Paglieri, *Chem. Eng. J.*, **93**, 11 (2003).
63. G.Q. Lu, J.C. Diniz da Costa, M. Duke, S. Giessler, R. Socolow, R.H. Williams and T. Kreutz, *J. Coll. Int. Sci.*, **314**, 589 (2007).
64. D. S. Sholl and Y. H. Ma, *MRS Bull.*, **31**, 770 (2006).
65. R. Dittmeier, V. Hollein and K. Daub, *J. Mol. Catal. A: Chem.*, **173**, 135 (2001).
66. A. Li, W. Liang, and R. Hughes, *Thin Solid Film*, **350**, 106 (1999).
67. K. Huo and R. Hughes, *J. Membr. Sci.*, **214**, 43 (2003).
68. D. Wang, J. Tong, H. Xu and Y. Matsumura, *Catal. Today*, **93/95**, 689 (2004).
69. J. N. Keuler and L. Lorenzen, *J. Membrane Sci.*, **195**, 203 (2002).
70. J. E. Gozum, D. M. Pollina, J. A. Jensen and G. S. Girolami, *J. Am. Chem. Soc.*, **110**, 2688 (1988).
71. S. Morooka, S. Yan, S. Yokoyama and K. Kusakabe, *Sep. Sci. Technol.*, **30**, 2877 (1995).
72. G. Xomeritakis and Y. S. Lin, *J. Membr. Sci.*, **120**, 261 (1996).
73. N. Itoh, T. Akiha and T. Sato, *Catal. Today*, **104**, 231 (2005).
74. T. A. Peters, M. Stange, H. Klette and R. Bredsen, *J. Membr. Sci.*, **316**, 119 (2008).
75. T. A. Peters, W. M. Tucho, A. Ramachandran, M. Stange, J. C. Walmsley, R. Holmestad, A. Borg and R. Bredesen, *J. Membr. Sci.*, **326**, 572 (2009).

11

Membranes Prepared via Electroless Plating

M. Broglia[1], *P. Pinacci*[1] *and A. Basile*[2]

[1]*ERSE S.p.A., Milano, Italy*

[2]*Institute of Membrane Technology, CNR, c/o University of Calabria, Rende, CS, Italy*

11.1 Introduction

Palladium alloy membranes are widely studied for hydrogen separation in membrane reactors [1–4]. Due to the considerable cost of palladium, research efforts have been addressed in obtaining low thickness deposits on cheaper support materials such as ceramics and metals by various techniques including electroplating [5], sputtering [6,7], chemical vapor deposition [8,9], and electroless plating [10–12].

Electroless plating, in particular, is an autocatalytic process which allows obtaining thin films with good adhesion characteristics: deposition occurs by the reaction in solution of a metal salt and a chemical reducing agent, without the use of external electric power. As a consequence, the deposition process occurs on both conductive and nonconductive materials, such as ceramics and plastics; further it does not require expensive set up and is relatively easy to scale up from laboratory to industrial scale [13].

Due to the above characteristics, electroless plating is nowadays a wide spread industrial practice: it is commonly used in engineering coating applications where wear resistance, hardness and corrosion protection is required. The most common plated metals include nickel, copper, silver, gold, zinc, chrome, palladium and tin. For example gold plating is used in electronic industry to provide a corrosion-resistant conductive layer on copper, in printed circuits and electrical connectors. Similarly thin chromium deposits (approximately 10 μm) are used to provide a mirror like finish to metal surfaces, while thinner chromium deposits (up to 1000 μm)

Membranes for Membrane Reactors: Preparation, Optimization and Selection, Edited by Angelo Basile and Fausto Gallucci

are used for antiwear applications in automotive industry [13]. Concerning palladium deposits, Rhoda was the first to develop hydrazine-based plating baths by using Pd in the form of tetra-amine dichloride, $Pd(NH_3)_4Cl_2$, although plating baths were rather unstable [14]. Over the years hydrazine plating bath composition and plating conditions where well established and, starting from the 1990s, they have been utilised to prepare dense Pd and Pd-alloy membranes on porous supports such as glasses [10], ceramic [15,16] and metal sintered alloys [17].

Porous ceramics substrates such as α-alumina have aroused a great attention since early times because of the relatively smooth surface and narrow pore distribution. These characteristics allow to obtain dense composite Pd-alloy membrane with a thickness as low as 5 μm [3]. Ceramic substrates, however, present the problem of obtaining leak tight seals at the ceramic–metal interface, as well as an intrinsic fragility which, in perspective, makes it difficult to assemble them into a module, affecting their potential industrial application as well. As a consequence many research efforts have been addressed to porous metals, which have higher robustness and mechanical stability, and can be easily welded and assembled in a module [18]. Commercially available porous metals, however, present high roughness and large pore dimensions (microns range); this determines high thickness of the Pd deposit (up to 20–30 μm) in order to obtain dense membranes [19]. Moreover, high operating temperatures (higher than 400 °C) and long exposure times can determine the diffusion of substrate metals (such as iron) into the palladium layer and, subsequently, a significant decrease of hydrogen permeance and selectivity. To overcome the above problems, the support should be modified, introducing a suitable barrier layer. Several types of barrier layers have been considered, including mixed oxides grown by oxidation of the support [20], a bi-metallic deposited multilayer [21], ceramic materials such as zirconia oxide [22], zirconia–yttria [23], cerium oxide [24], and alumina [25–27]. Ceramic materials, in particular, could allow obtaining smooth surfaces with narrow pore size, resulting in the preparation of dense membrane with lower thickness (5–10 μm). The challenge, for these membranes, is related to stability under thermal cycling, due to a mismatch in the thermal expansion coefficient between the ceramic and the palladium layer, especially when hydrogenated [21].

Based on the above considerations, the first part of this chapter will be focused on the preparation procedure of Pd and Pd-alloy membranes via electroless plating by evidencing critical issues linked to the specific support characteristics; morphological characteristics of the deposits will also be described. In a second part membrane performances will be presented and significant developments up to the pre-industrial scale will be indicated; performances in membrane reactors will be also discussed.

11.2 Description of the Electroless Plating Process

11.2.1 Introduction

The electroless plating bath generally consists of a metallic salt, a complexing agent, a reducing agent and a stabilising agent. Electroless plating, indeed, involves the reduction of the metallic salt on a catalytic surface. The reaction is initiated by the oxidation of the reducing agent in the solution with the simultaneous release of electrons, which reduce metal ions on the target surface. However, common substrates such as metals, ceramics and glasses need very long induction times to start the nucleation process; therefore an activation of the support is necessary to initiate the autocatalytic deposition process. As a consequence the preparation procedure of palladium alloy membranes consists in the following main subsequent steps (Figure 11.1): cleaning of the support, sensitization and activation of the support and metal plating.

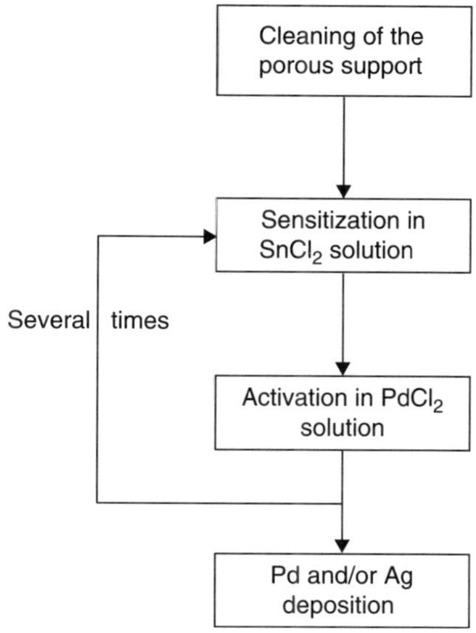

Figure 11.1 *Steps for preparation of Pd/Pd-alloy membranes via electroless plating*

The first two steps can be optimised according to the support characteristics (pore size, surface morphology and type of material), while, in the plating step, geometric and morphological characteristics of the support should be also considered. In case a bi-metallic alloy, such as silver palladium, is prepared, either sequential deposition of metals or co-deposition is carried out; with both procedures, however, heat treatment of deposits should be performed, in order to obtain alloying by inter-metallic diffusion. Deposition procedure, as well, can be optimised depending on alloy composition.

A detailed description of single steps of the preparation procedure is reported below.

11.2.2 Cleaning of the Support

Cleaning of the porous support is carried out in order to remove contaminants, such as grease oil, dirt, corrosion products, from the surface of the support. Support cleaning should be carried out very carefully since impurities can inhibit, at various extent and way, autocatalytic sites formation in the subsequent activation phase. This, in turn, can adversely affect the formation of a pinhole-free, adherent film on the support. Membrane cleaning of ceramic supports is usually carried out in the following phases: cleaning in an organic media in an ultrasound bath, basic cleaning, acid cleaning and final rinsing in deionised water [28].

Concerning metallic supports, cleaning can be performed in an ultrasonic bath with an alkaline solution at 60 °C, followed by rinsing in tap water and in deionised water to remove all the alkaline solution and, finally, by cleansing in an organic solvent, such as isopropanol [29].

At the end of the cleaning procedure supports are dried overnight in an oven at a temperature of about 120 °C.

11.2.3 Activation of the Support

Activation is usually performed by dipping the support first into a stannous chloride solution (*sensitization*) and, thereafter, in a palladium chloride solution (*activation*), for several times.

Table 11.1 *Typical composition of solutions used for sensitisation/activation of the support [21]*

	Chemical	Concentration
Sensitization	$SnCl_2 \cdot 2H_2O$	0.1 g l^{-1}
	HCl (37%)	1 ml l^{-1}
Activation	$PdCl_2$ 99.9% (metal basis)	0.1 g l^{-1}
	HCl (37%)	1 ml l^{-1}
Room temperature		

A typical composition of the sensitization and activation solutions is shown in Table 11.1: it can be noted that activation is performed at acidic pH in order to maintain a good salt solubility. The overall reaction is:

$$SnCl_2 + PdCl_2 \rightarrow SnCl_4 + Pd$$

The sensitising/activation cycle is performed at room temperature and the number of dipping cycles depends on the support material, its pore size distribution and surface morphology. Stannous solution, in fact, enters the pores of the support and reduces palladium ions, thus creating autocatalytic sites where palladium plating can occur. Autocatalytic sites must be uniformly distributed on the surface of the support and should be in sufficient number to allow the subsequent palladium deposition. Typically sensitisation/activation cycle is repeated from two up to ten times [28,30].

Rinsing with deionised water between sensitization and activation is carried out in order to avoid deposition/adsorption of the products of stannous ions hydrolysis such as $SnOH_{1.5}Cl_{0.5}$ and other hydroxyl chlorides on the surface of the support. An excess of stannous ions, in fact, can determine the formation of a loose palladium layer, while the deficiency of stannous ions can lead to a non uniform seeding of palladium nuclei. In some works rinsing with a 0.01 N HCl solution after the activation step is also performed to prevent the hydrolysis of palladium ions [30].

Paglieri *et al.* proposed an alternative activation method base on the use of palladium acetate [31]. Ceramic substrates were dip-coated in a chloroform solution of Pd-acetate, dried, calcined, and then reduced in flowing H_2. Pd membranes prepared with such a procedure were more stable over the time than those activated with the Sn-Pd procedure. The lack of stability in this latter case was attributed to the presence of Sn^{4+} at the interface between the Pd layer and the support. In a recent work, however, EDX and ICP-AES analysis indicate that, after six cycles of the standard Sn-Pd activation procedure, no Sn impurity has been detected over the surface of a porous stainless steel support [32].

An important modification of the standard sensitization–activation cycle based $SnCl_2/PdCl_2$ consists in applying an activated layer on the support. This approach has been first suggested by Zhao *et al.* [33]: a Pd (II)-modified bohemite sol gel was applied on a ceramic support; after calcination and reduction under H_2 at 500 °C, Pd nuclei were obtained on the substrates.

This approach has grown in importance over the years, particularly while dealing with the development of composite membranes where the support consists in a porous metal coated with a ceramic material such as zirconia oxide [22], cerium oxide [24], and alumina [25–27]. Broglia *et al.* [34] showed that conventional activation procedure, when applied to a porous stainless steel coated with α-alumina, determines the partial dissolution of the alumina layer; moreover Pd seeds are not distributed evenly on the surface. On the contrary pre-activation of the α-alumina external layer, consisting in the addition of palladium chloride to the alumina gel and palladium reduction in sodium boron hydride solution after calcination, was effective in producing a smooth activated surface where Pd seeds are evenly distributed on the surface.

11.2.4 Palladium Deposition

The plating bath consists in the following main components: a palladium ion source ($PdCl_2$, $Pd(NH_3)_4Cl_2$, $Pd(NH_3)_4(NO_3)_2$, a complexant (ethylene di-amine tetra acetic acid, EDTA; ethylenediamine, EDA), a reducing agent (hydrazine, sodium hypophosphite), a pH controller (ammonia) [18].

Hydrazine is the most common and suitable reducing agent for electroless plating of palladium. Hypophosphite-based baths with the use of EDA as complexant have been also tried in the past, since they have a better efficiency than hydrazine-based baths; however hydrogen evolution during the plating process is absorbed in the palladium layer and determines the formation of cracks and delamination [12].

A typical composition of a hydrazine-based bath is shown in Table 11.2. Deposition is carried out at basic pH (10–12) and at a controlled temperature ranging between 40 and 60 °C. More in detail the electroless technique is based on the following redox reactions, which occur simultaneously in the solution [28]:

$$\text{Anodic: } N_2H_4 + 4OH^- \rightarrow N_2 + 4H_2O + 4e^- \qquad E^0 = 1.12\,V \tag{11.1}$$

$$\text{Cathodic: } 2Pd[NH_3]_4^{2+} + 4e^- \rightarrow 2Pd^0 + 8NH_3 \qquad E^0 = 0.95\,V \tag{11.2}$$

The overall reaction is:

$$2Pd[NH_3]_4^{2+} + N_2H_4 + 4OH^- \rightarrow 2Pd^0 + 8NH_3 + N_2 + 4H_2O \qquad E^0 = 2.07\,V \tag{11.3}$$

The reaction occurs at the surface of the support, preferentially at the Pd seeds. Hydrazine reacts with hydroxide ions, forming nitrogen and water and with the release of electrons which are used to reduce the Pd^{2+} complex into Pd metal. Nitrogen gas is released as bubbles during the process.

The rate of palladium deposition and hydrazine consumption are given, respectively, by the following kinetic equations [12]:

$$-r_{Pd} = -\frac{d[Pd]}{dt} = k_1[Pd]^{\alpha}[N_2H_4]^{\beta} \tag{11.4}$$

$$-r_{N_2H_4} = -\frac{d[N_2H_4]}{dt} = \frac{1}{2}\frac{d[Pd]}{dt} + k_2[N_2H_4]^4 \tag{11.5}$$

Table 11.2 *Typical compositions of palladium plating baths reported in literature*

	Component	Concentration
Pd ion source	$Pd(NH_3)_4Cl_2$ (g l^{-1})	4–5
Complexant	Na_2EDTA (g l^{-1})	40–80
Reducing agent	N_2H_4 1 M (ml l^{-1})	6–10
Buffer, stabiliser	NH_4 (OH) 28% (ml l^{-1})	200–650
pH		10–12
T (K)		313–333

where: r_{Pd} and $r_{N_2H_4}$ are the rates of palladium and hydrazine consumption; [Pd] and [N_2H_4] are the concentration of palladium and hydrazine in the plating bath, k_1 and k_2 are the rate constants, α and β are constant values, usually less than 1. Constant values can vary accordingly to the plating conditions such as temperature concentration of reagents, plating surface/volume of plating solution ratio.

According to Equation (11.5), hydrazine is consumed both by plating (first term) and by decomposition (second term). It has been observed experimentally that hydrazine decomposes rapidly in the presence of palladium metal [35]; this reaction is responsible of the low efficiency, typically less than 20% [28], of the hydrazine based plating baths. Plating efficiency, defined as the ratio between deposited palladium and the initial amount of palladium in the plating bath, can be increased by increasing palladium concentration in the bath. However an excess of hydrazine can result in bulk precipitation and determine a non-uniform coating. From a practical point of view, hydrazine can be added several times during the plating period in order to maintain a high concentration in the bath and increase plating rate.

Electroless plating should be performed in controlled conditions in order to obtain the formation of pinhole-free, adherent palladium film; a typical laboratory apparatus is shown in Figure 11.2. It can be noted that the membrane support is inserted in a reactor and maintained in rotation at constant velocity by a variable speed motor to allow the removal from the reaction zone of nitrogen produced by the electroless reaction [see Equation 11.4). The reactor is put inside a thermostatic bath to keep the temperature at a constant value. Hydrazine can be periodically added from the tube inserted in the middle of the reactor, while nitrogen can be evacuated from the reactor through the tube on the right side.

A picture of a Pd membrane obtained in the above described apparatus is shown in Figure 11.3; for comparison the bare support of macroporous AISI 316L is shown, too. It can be noted that two nonporous tubes have been welded at the ends in order to allow a leak tight sealing of the membrane when used in permeation tests at high temperature.

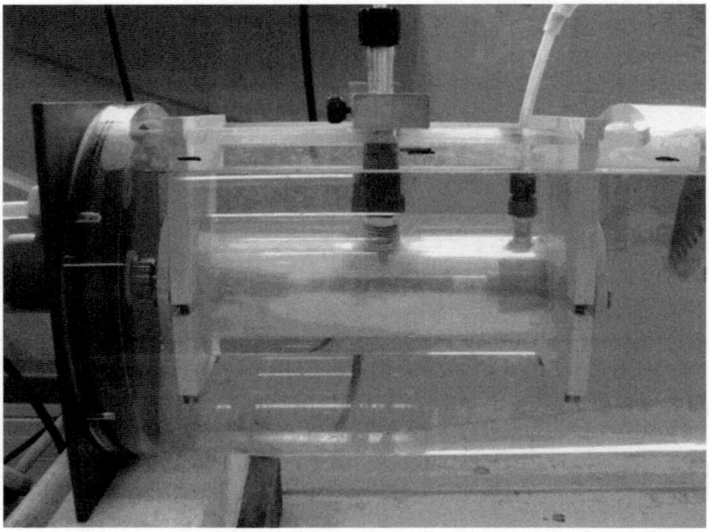

Figure 11.2 *Apparatus used for Pd membrane preparation by electroless plating [38]*

Figure 11.3 *Macroporous metallic support with welded ends of nonporous metal: before Pd deposition (left) and after Pd deposition (right) [38]*

11.3 Morphology of Palladium Deposits

The morphology of the palladium deposits on porous substrates depends on several factors such as the activity of palladium nuclei, concentration of hydrazine and metal ion in the bath, bath temperature and agitation. The initial growth of a film is strongly influenced by the support material, surface roughness, surface imperfection and contaminants, while subsequent stages of growth are more dependent on bath physicochemical parameters [17]. For example Shu *et al.* [36] compared Pd layers obtained by deposition on a porous stainless steel in concentrated hydrazine bath and in a 100× diluted bath. Diluted baths led to the formation of a porous Pd layer with a mean grain size of 420 nm, while concentrated bath resulted in the formation of a dense layer with a grain size of 11 nm.

Typically the grain size to obtain a dense membrane on a metallic support can vary between 50 and 100 nm [37]. During deposition, however, single crystal can agglomerate to form Pd clusters up to 1–2 μm in size and give to the surface a *cauliflower-type* structure.

Figure 11.4 shows the SEM images of a Pd membrane obtained by electroless plating on the external surface of an α-alumina support with a 200 nm pore size [28]. For comparison a Pd membrane obtained on a stainless steel support, with a mean pore size around 2 μm, is shown in Figure 11.5 [38]. It can be noted that Pd deeply penetrates in the pores of the support, providing a good anchoring of the deposited film.

Electroless plating has also been used in conjunction with osmotic pressure in order to better control the morphology of the Pd deposit on glass (Vycor) [39], alumina [40], and porous stainless steel [41,42]. All these authors showed that the combined use of electroless plating and osmosis produces a smaller grain size dense Pd film.

11.4 Pd-Alloy Preparation

Pure palladium membranes present the problem of hydrogen embrittlement caused by the transition between the α and β phases of palladium hydride which occurs below the critical

Figure 11.4 *SEM images of Pd deposited on a ceramic support: cross section (left) and surface (right) [after 28]*

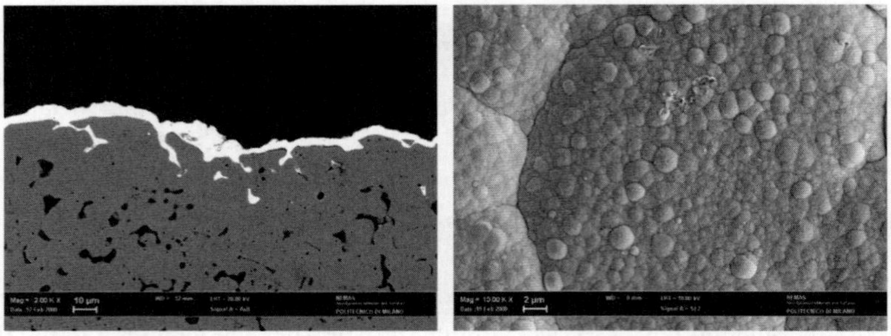

Figure 11.5 SEM images of Pd deposited on a metallic support: cross section (left) and surface (right) [38]

temperature of about 300 °C [18]. Phase transition and hydrogen embrittlement can be avoided by the addition to palladium of alloying elements such as Ag, Au, Cu, Y, Ce. Moreover the addition of such elements in a proper amount can increase hydrogen permeability as shown by Knapton [43] and Gryazov [44], while the addition of Cu and Au is known to increase the resistance to sulphur poisoning [45]. From a practical point of view, however, only Pd-Ag alloys, with Ag up the 30%, and Pd-Cu alloys, with Cu in the 10–40% range, have been extensively studied.

The formation of Pd–alloy films can be obtained either by co-deposition or by sequential deposition of metals, followed by high temperature annealing in an inert gas (i.e. Ar, He, N_2), in order to obtain the alloy formation by intermetallic diffusion.

Co-deposition of silver and palladium has been studied by several researchers on various materials [17,46,47]; a typical composition of the bath is shown in Table 11.3, for a Pd_{90}-Ag_{10} and $Pd_{77}Ag_{23}$ alloy, respectively, on an α-alumina support. Silver, however, tends to have a higher deposition rate due to its higher oxidation–reduction potential ($E^\circ = 1.73$ V) respect to Pd ($E^\circ = 1.12$ V); this determines a nonhomogeneous distribution of the elements along the deposited layer and, ultimately, a percentage composition differing from the target alloy.

For this reason sequential deposition has been preferred for the alloy preparation [48–50]. With such a method, silver and palladium can be deposited in the target amount. However, while Pd penetrates into valleys and crevices of the surface, Ag plating tends to occur on easily accessible sites resulting in formations which grow perpendicularly to the surface. This characteristic of the Ag deposits can lead to a nonhomogenous coverage of the surface and to concentration gradients along the entire film which can adversely affect the formation of an alloy with homogeneous composition during the subsequent annealing treatment [20]. Annealing

Table 11.3 Typical composition of electroless plating baths for palladium and silver co-deposition on porous alumina [after 47]

Component	$Pd_{90}Ag_{10}$	$Pd_{77}Ag_{23}$
$PdCl_2$ (g l^{-1})	0.297	0.207
$AgNO_3$ (g l^{-1})	0.03	0.085
Na_2EDTA (g l^{-1})	40	40
NH_4 (OH) 28% (ml l^{-1})	200	200
N_2H_4 1 M (ml l^{-1})	5.6	5.6
T (K)	333	333

conditions (temperature and time) required to obtain a complete mixing and homogeneity of the Pd-Ag film depends both on thickness and on composition of the deposited metal layers. However there is not a full consensus in literature on the conditions of alloy treatment: temperature can vary from 500 to 900 °C and times from a few hours up to dozens of hours [15,20,46]. Generally speaking ceramic supports can be considered more suitable than glass and metallic supports in order to obtain homogenous alloying, since they can stand high annealing temperatures. Annealing temperatures, in fact, are limited by the collapse of pore structure in the case of the glass supports [10] and by the inter-metal diffusion of elements from the porous metallic support to the membrane layer [20].

In more detail, Uemiya *et al.* [15] firstly studied systematically annealing conditions of thin film Pd-Ag alloys. He determined the distribution of palladium and silver along the depth of a 5.8 μm thick Pd_{77}-Ag_{23} membrane on an alumina support, by using an electron probe microanalyser (EPMA). Annealing at 500 °C for 12 h in an argon stream led to a composition gradient across the membrane and to the presence of double layer consisting of palladium and silver–palladium. Higher temperatures, above 800 °C, allowed obtaining a complete alloying.

More recently a study on annealing condition of a 10 μm thick Pd_{70}-Ag_{30} membrane on an α-alumina support was performed in the 550–700 °C temperature range in nitrogen atmosphere [51]. X–ray diffraction measurements indicated that a complete formation of the alloy has been achieved at 600 °C after 14 hours, and at 700 °C after 4 hours, respectively. SEM/EDS analysis, however, evidenced the non uniform composition of the alloy along membrane thickness and length. Such a problem has been evidenced even on a very thin Pd_{75}-Ag_{25} layer, 1.9 μm thick, deposited on an α-alumina support, after annealing at 500 °C for 32 hours in helium [52].

Besides silver, copper deposition occurs in rather uniform way; composite Pd-Cu membranes, with Cu concentration in the 10–30% range, have been successfully prepared by sequential deposition of the two metals on ceramic supports such as symmetric α-alumina, asymmetric zirconia on α-alumina, and asymmetric γ-alumina on α-alumina [53], according to the bath composition shown in Table 11.4. The resulting composite membranes were annealed in H_2 atmosphere at temperatures between 350 and 700 °C for times ranging from 6 to 25 days in order to obtain a homogenous alloy film.

Table 11.4 *Typical composition of electroless plating baths of palladium and copper on porous ceramics [52]*

Component	Pd bath	Cu bath
$PdCl_2$ (g l^{-1})	5.45	
$CuSO_4$ $5H_2O$ (g l^{-1})		6.225
Na_2EDTA (g l^{-1})	70	20.1
$NH_4(OH)$ 28% (ml l^{-1})	389.6	
N_2H_4 1 M (ml l^{-1})	10	
HCl (ml l^{-1})	10.9	
Formaldeyde 37% (ml l^{-1})		14.04
NaOH (g l^{-1})		20
Triton X-100 (mg l^{-1})		25
2-2 bipyridyl (mg l^{-1})		5
T (K)	338	338

Table 11.5 *Typical composition of electroless plating baths of palladium and copper on porous metals [20]*

Component	Pd bath	Cu bath
$PdCl_2$ ($g\,l^{-1}$)	4	
$CuSO_4\ 5H_2O$ ($g\,l^{-1}$)		20
Na_2EDTA ($g\,l^{-1}$)	40.1	30
$NH_4(OH)$ 28% ($ml\,l^{-1}$)	198	
N_2H_4 1 M ($ml\,l^{-1}$)	5.6	
Formaldeyde 37% ($ml\,l^{-1}$)		14
EDA ($mg\,l^{-1}$)		100
$K_4Fe(CN)_6\ 3H_2O$ ($mg\,l^{-1}$)		35
$(C_2H_5)_2NCS_2Na\ 3H_2O$ ($mg\,l^{-1}$)		5
T (K)	338	293–298

Similarly Pd-Cu membranes have been obtained by sequential deposition on porous metals such as stainless steel [20] and Hastelloy with an alumina anti-diffusion layer [37]; the corresponding bath composition is shown in Table 11.5.

11.5 Membrane Performances and Integration in Membrane Reactors

Evaluation of the performances of Pd/Pd-alloy membranes obtained by electroless plating and, in general, by any other physical or chemical deposition method, requires an important premise. No membrane is perfectly dense, for example, some defects are present in the deposited layer and/or they can develop as a consequence of operation at high temperature and pressure.

In presence of defects, permselectivity of hydrogen versus other non-absorbable gases exhibit a decreasing trend as a function of the total pressure difference between feed and permeate, due to the predominant contribution of the viscous flow through defects. This behaviour has been very well shown by Rothenberger *et al.* while evaluating the performances of a Pd membrane, 22 μm thick, deposited on a porous stainless steel support by electroless plating [54]. As shown in Figure 11.6, at pressure differences of 1–2 bar, selectivity of H_2 versus He can be considered as infinite (e.g., helium could not be detected in the permeate); however, while increasing pressure difference at 5 bar, selectivity drops to 100–150 and continues to decrease while increasing pressure values.

It is important, therefore, while comparing membrane performances, to consider experimental conditions and measurement procedures, including operating pressure and duration of tests.

The first works in the 1990s were mainly focused on feasibility of the membrane preparation method and related permeation mechanisms, and performances where measured by permeation tests in single gas or gas mixtures at rather low transmembrane pressures. Nevertheless this data can be important to address some trends concerning membrane behaviour such as correlation between hydrogen flux and membrane thickness [48]. A comprehensive review of permeance data of palladium membrane up to 2003 as a function of thickness and temperature is shown in Figure 11.7.

In the next years research progressively focused on more specific applications, where membranes have been tested in gas mixtures closely simulating operating conditions; moreover

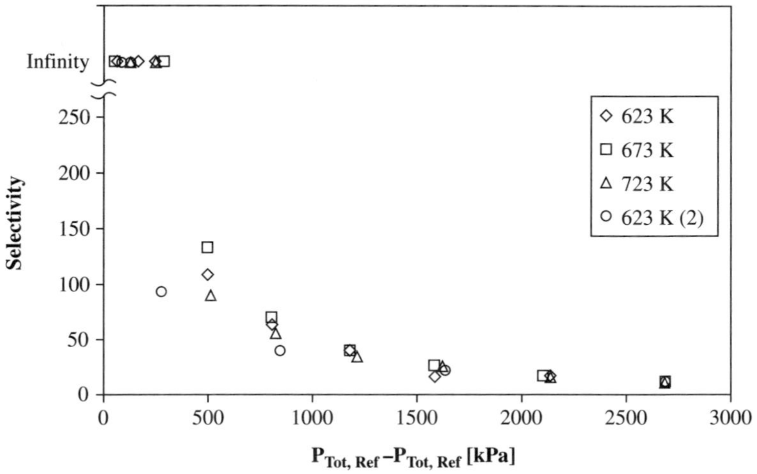

Figure 11.6 *Selectivity of a 22 μm Pd membrane on a porous stainless steel support as a function of total pressure drop across the membrane. Reprinted from Journal of Membrane Science, Rothenberger, K. S., et al., High pressure hydrogen permeance of porous stainless steel coated with a thin palladium film via electroless plating. Vol. 244, 55-68. Copyright (2004) with permission from Elsevier*

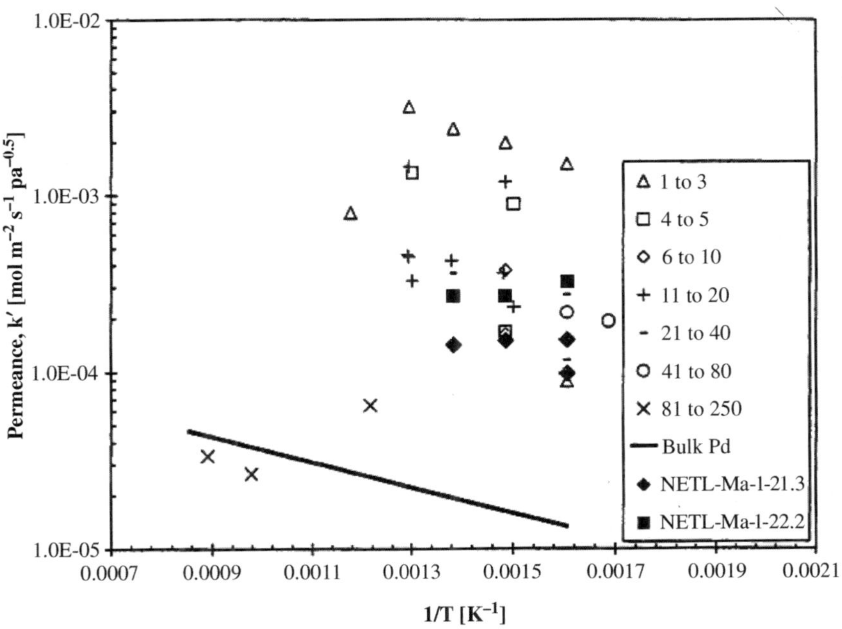

Figure 11.7 *Permeance of palladium membranes as function of thickness (microns) and reciprocal of temperature: summary of literature data up to 2003. Reprinted from Journal of Membrane Science, Rothenberger, K. S., et al., High pressure hydrogen permeance of porous stainless steel coated with a thin palladium film via electroless plating. Vol. 244, 55–68. Copyright (2004) with permission from Elsevier*

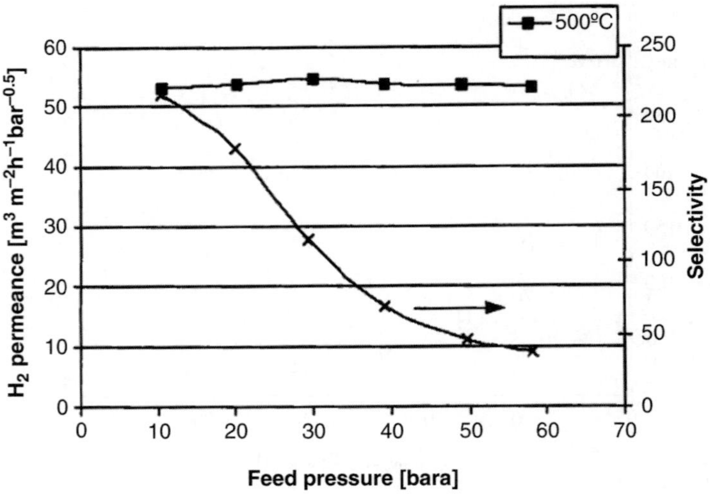

Figure 11.8 *Permeance and selectivity of a $Pd_{70}Ag_{30}$ membrane on a ceramic support as a function of feed pressure [after 3]*

membrane performances have been evaluated in comparison of target performances such as purity of separated hydrogen. Also emphasis has been put on long-term behaviour.

Concerning Pd and Pd-alloy membranes on ceramic supports, two important developments carried out at ECN and at the Dalian Institute of Chemical Physics, respectively, can be mentioned. The first concerns $Pd_{70}Ag_{30}$ composite membranes, 3–5 μm thick, obtained by electroless plating (sequential deposition followed by annealing treatment) on an asymmetric α-alumina support [3]. These membranes have been tested up to 500 °C and 60 bar in gas mixtures simulating reformate composition (see Figure 11.8), obtaining high hydrogen permeances of $50\,m^3\,m^{-2}\,h^{-1}\,bar^{-0.5}$; selectivity, however, was rather low, due to leakages through the sealing at high pressures. Recently, however, ECN claimed to have overcome such a problem by the development of leak tight sealing designed to operate up to 700 °C and 39 bar [55]. The scale up of the preparation method resulted in the manufacturing of membrane tubes with a length up to 80 cm and a diameter of 14 mm [56] and in a spin off from the manufacturing technology (trademark HYSEP). Hydrogen separation modules, of different sizes, consisting from 1 to 13 tubular membranes with a total area ranging from 0.04 to 0.5 m^2 are currently available for operation with reformate gases up to 480 °C and 25 bar.

Similarly Pd membranes 50 cm long were produced at Dalian Institute on a ceramic support made at ECN. Prior to deposition of the Pd layer by electroless plating, the pores of the support tubes were pre-filled with an inorganic gel to obtain a smooth surface and prevent Pd from entering into the pore structures [57]. These pores were re-opened after completion of the Pd deposition by decomposing the gel at 500 °C in H_2. Membranes 2–10 μm thick have been tested in gas mixtures simulating both water gas shift and reformate (see Figure 11.9) [58].

Concerning Pd and Pd-alloy membranes on metallic supports, important developments are carried out at CRI/Criterion and at Pall Corporation, respectively.

Membranes under development at CRI are based on the studies performed at the Worchester Polytechnic Institute, already quoted in this chapter [18–21,25]. More in detail Pd and Pd-Ag membranes, 5–12 μm thick, have been obtained on a porous Inconel support with a bimetallic

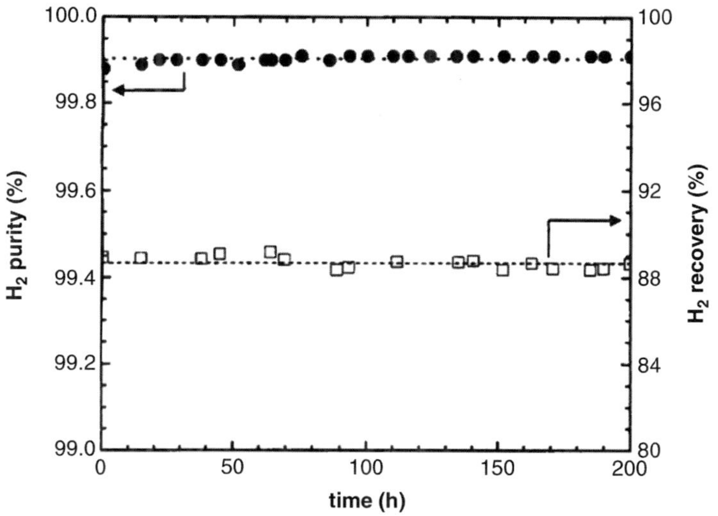

Figure 11.9 *Performance of a 3 μm Pd membrane on a ceramic support, during a 200 h test at 400 °C and 1.1 MPa in an unshifted reformate [after 58]*

anti-diffusion barrier layer [59,60]. CRI has produced tubular membranes of this type, as large as 2 inches O.D. and up to 48 inches (122 cm) long [61]. These membranes have been tested in separations tests at temperatures of 300–500 °C and differential pressures of 26–42 bar, for periods exceeding 2000 hours, obtaining stable hydrogen flux and selectivity; hydrogen permeance in the range of 50–70 $Nm^3\, m^{-2}\, h^{-1}\, bar^{-0.5}$ and hydrogen purity exceeding 99% have been reported [62].

Composite membranes under development at Pall Corporation consists in a Pd-alloy film deposited on a porous stainless support (trademark Accusep) coated with yttria-stabilised zirconia [63]. The deposited layer can be either Pd-Cu [63] or Pd-Au [64], a few microns thick. Pd-Au membranes, with Au in the 1–30% range, have been tested in mixtures simulating WGS composition, at 400 °C and with a transmembrane pressure up to 12 bar [65]. The composite membranes preparation process is currently being scaled up to 30-inch (76-cm) tubular elements [66].

As regards manufacturing process, projections usually assume that raw material, namely palladium, is the main factor in determining membrane cost [1,2]. Furthermore electroless plating equipments can be considered relatively inexpensive, if compared to other methods such as magnetron sputtering. As much as the deposited layer thickness is reduced, however, the support material tend to become an important component in membrane cost; this is particularly true, when porous metallic materials for high temperature applications, such as Inconel or Hastelloy, are used [62].

As far as concerns membrane reactor technology, in early times Kikuchi *et al.* [67] showed that, by using composite Pd and Pd-Ag membranes 5.0–22.5 μm thick, obtained by electroless plating on α-alumina asymmetric supports, methane conversion in steam reforming was well above the equilibrium thermodynamic values (see Figure 11.10). Similar results were found for the water–gas shift (WGS) reaction [68], dehydrogenation of isobutane [69] and dehydrocyclisation of propane to aromatics [70].

Since then a few experimental works regarding WGS in a palladium membrane reactor have been performed, mainly at low temperature (e.g., 250–300 °C), by using membranes prepared by

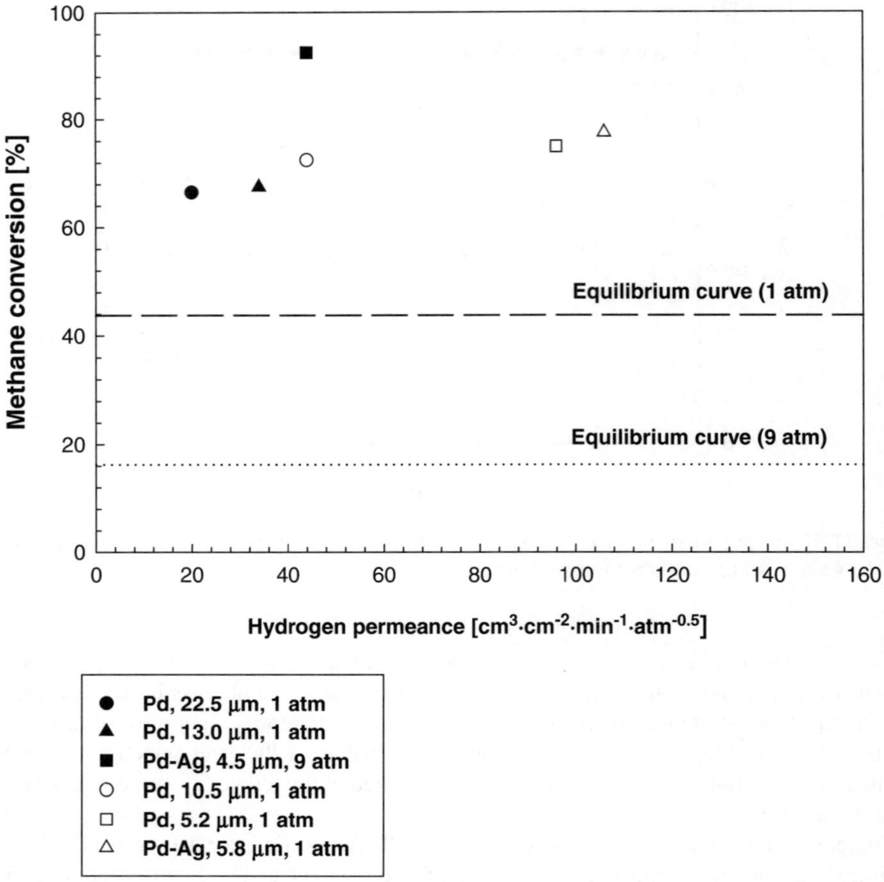

Figure 11.10 *Effect of hydrogen permeation on the conversion of methane in a membrane reactor; test performed at $T = 773$ K [after 67]. Reprinted from Catal. Today, E. Kikuchi, Palladium/ceramic membranes for selective hydrogen permeation and their application to membrane reactor, 25, 333-337. Copyright (1995) with permission from Elsevier*

different methods [71,72]; only recently WGS tests have been carried out at higher temperature (350–450 °C), due to the fact that the stability of composite Pd membranes, both on metallic supports and on ceramic supports, in this temperature range has been sufficiently proved [73–75].

As an example Figure 11.11 shows the results of WGS tests carried out at 410 °C in a Pd membrane reactor with gas mixtures simulating a syngas produced in IGCC plants [73]. While increasing the feed pressure, CO conversion increases up to 78% due to the hydrogen permeation through the membrane. This value is well above the maximum conversion achievable with a traditional reactor at the thermodynamic equilibrium (about 18%).

Similar results have been reported by Augustine *et al.* concerning WGS tests, conducted with a Pd membrane 12.5 μm thick on a Inconel support, over a temperature range of 300–450 °C, pressures of 4.4–14.6 bar, steam/CO ratios of 1.2–2.2, GHSVs of 800–3300 h^{-1} [74].

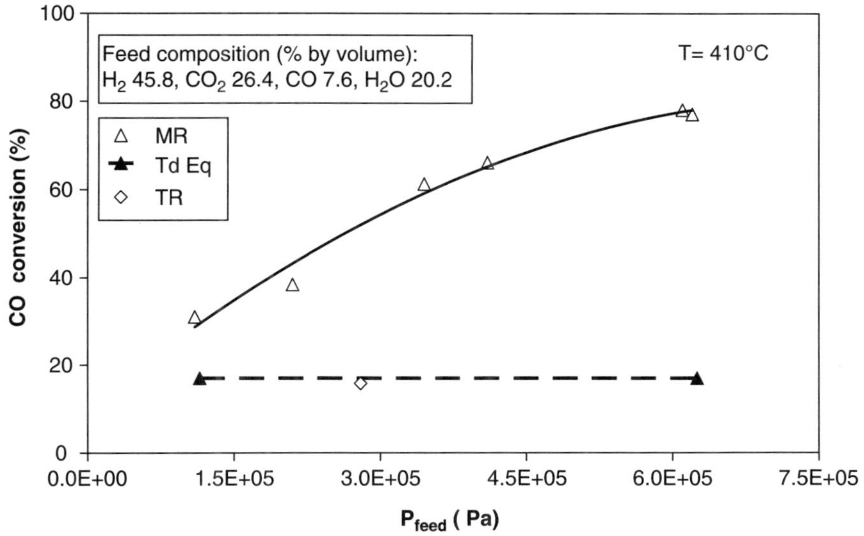

Figure 11.11 *CO conversion as a function of feed pressure in WGS tests with a membrane reactor (MR): comparison with a traditional reactor (TR) and thermodynamic equilibrium values (Td Eq) [73]*

WGS tests have been also performed with gas mixtures simulating a syngas produced by autothermal reforming of natural gas [75]. The performances of two different catalysts (Fe-Cr versus $Pt/Ce_{0.6}Zr_{0.4}O_2$) have been evaluated in a wide range of operating conditions (see Figures 11.12, 11.13), in a membrane reactor equipped with a 1.4 mm thick Pd membrane on a ceramic support. It is interesting to note that, in all the above experiments, attention has

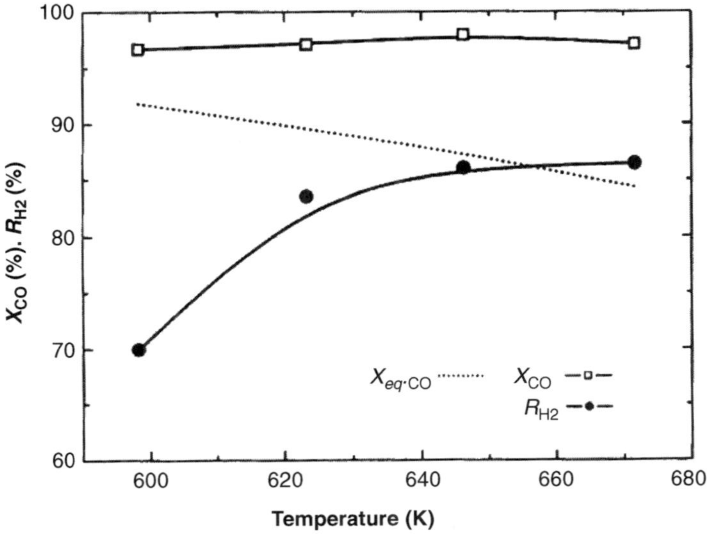

Figure 11.12 *WGS tests in a Pd membrane reactor: influence of reaction temperature on CO conversion (X_{co}) and H_2 recovery (R_{H2}) at $P_{tot} = 1.2$ Mpa, GHSV = 4050 l kg^{-1} h^{-1} and S/C = 3 [after 75]*

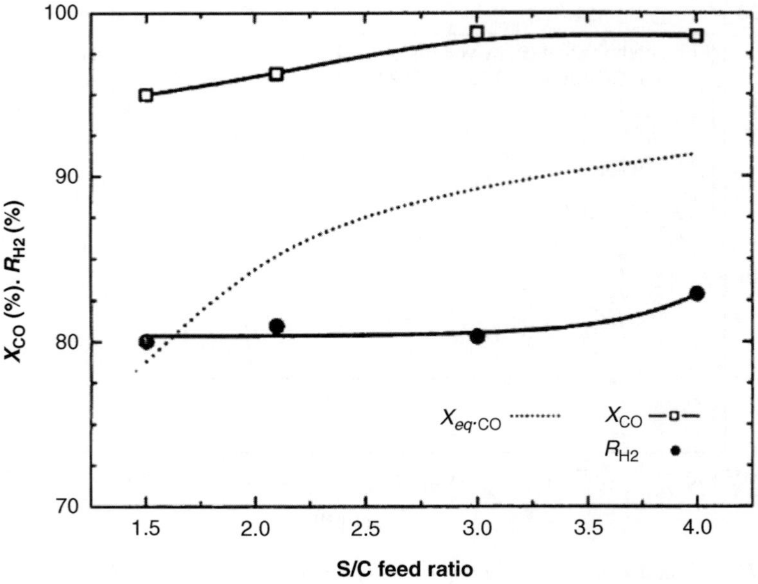

Figure 11.13 *WGS tests in a Pd membrane reactor: influence of steam: CO ratio on conversion (X_{co}) and H_2 recovery (R_{H2}) at P_{tot} = 1.2 Mpa, GHSV = 4050 l kg^{-1} h^{-1} and T = 623 K [after 75]*

been focused on specific phenomena which can decrease H_2 flux through the membrane such as concentration polarization in the boundary layer and competitive adsorption of other gases on the membrane surface; these aspects are currently object of specific studies [76,77].

Finally, results of steam reforming experiments with a 1-inch (2.54-cm) OD, 6-inch (15.2-cm) long Pd membrane on a metallic support, were obtained by Engwall *et al.* [61]. Tests have been run at 500 °C and 30 bar in an electrically heated shell and tube reactor, with a commercial catalyst packed in the annular chamber between membrane and reactor. Membrane performances were quite stable for a period of 51 days, and resulted in a conversion of about 94% and hydrogen purity exceeding 98%.

11.6 Conclusions

In this chapter the preparation methodology of Pd and Pd-alloy membranes via electroless plating has been reviewed. Morphological characteristics of the deposits have been described and critical issues linked to the specific support characteristics have been discussed. Results of laboratory tests both in separation experiments and in membrane reactor applications have been presented.

It can be concluded that, after almost twenty years from the first laboratory feasibility tests, prototype membranes at pre-industrial scale are now available. Consequently, development is now focused on use of membranes in specific applications, such as water gas shift and steam reforming, in order to evaluate long-term performances in operating conditions and assess economic viability of the technology.

References

1. A. Middleton, P. Hurst, P. Walker and P. Grace,in D. C. Thomas, S. M. Benson, *Electroless Plating*, Vol. 1, Chap. 27, Elsevier, New York (2005).
2. P. Chiesa, G. Manzolini, F. Vigano, P. Savoldelli, P. Pinacci, L. Mazzocchi, CO_2 capture in IGCC power plants by means of Pd-Ag membranes *Proc. GHGT8*, Trondheim, 18–23 June (2006).
3. J.V. Dijkstra, Y.C. van Delft, D. Jansen, P. Pex, proc. of GHGT 8, Development of hydrogen membrane reactor for power production with pre-combustion decarbonisation, *Proc. GHGT8*, Trondheim, 18–23 June (2006).
4. R. Bredesen, K. Jordal, O. Bolland, *Chem. Eng. Process.*, **43**, 1129–1158 (2004).
5. S. E. Nam, K. H. Lee, *J. Membr. Sci.*, **192**, 1977 (2001).
6. B. McCool, G. Xomeritakis, Y. S. Lin, *J. Membr. Sci.*, **161**, 67–76 (1999).
7. V. Jayaraman, Y. S. Lin, *J. Membr. Sci.*, **104**, 251–262 (1995).
8. N. Itoh, T. Akiha, T. Sato, *Catal. Today*, **104**, 231 (2005).
9. S. Morooka, S. Yan, K. Aoki, S. Yokoyama, K. Kusakabe, *Sep. Sci. Technol.*, **30**, 2877–2889 (1995).
10. S. Uemiya, N. Sato, H. Ando, Y. Kude, T. Matsuda, N. Kikuchi, *J. Membr. Sci.*, **56**, 303–313 (1991).
11. A. Li, W. Liang, R. Hughes, *Sep. Purif. Technol.*, **15**, 113–119 (1999).
12. Y. S. Cheng and K. L. Yeung, *J. Membr. Sci.*, **182**, 195–203 (2001).
13. G. O. Mallory and J. B. Hajdu, *Electroless Plating – Fundamentals and Applications*, William Andrew Publishing/Noyes, New York (1990).
14. R. N. Rhoda, N. J. Westfield, A. M. Madison,US Patent 2,915,406 (1959).
15. S. Uemiya, T. Matsuda and N. Kikuchi, *J. Membr. Sci.*, **56**, 315–325 (1991).
16. J. P. Collins and D. Way, *Ind. Eng. Chem. Res.*, **32**, 3006–3013 (1993).
17. J. Shu, B. P. A. Grandjean, E. Ghali, S. Kaliaguine, *J. Membr. Sci.*, **77**, 181–195 (1993).
18. Y.H. Ma, *Proc. ICIM 9*, Lillehammer, 25–29 June (2006).
19. I. P. Mardilovich, E. Engwall, Y. H. Ma, *Desalination*, **144**, 85 (2002).
20. Y. H. Ma, B. C. Akis, M. E. Ayturk, F. Guazzone, E. E. Engwall I. P. Mardilovich, *Ind. Eng. Chem. Res.*, **43**, 2936–2945 (2004).
21. M. E. Ayturk, I. P. Mardilovich, E. E. Engwall, Y. H. Ma, *J. Membr. Sci.*, **285**, 385–394 (2006).
22. H. Gao, Y. Li, J. Y. S. Lin, B. Zhang, *J. Porous Mater.*, **13**, 419 (2006).
23. R. P. Singh, T. Powers, K. Rekciz, S. Opkins, C. Love, Supports for high temperature gas separation membranes, *Proc. ICIM 9*, Lillehammer, 25–29 June (2006).
24. J. Tong, C. Su, K. Kuraoka, H. Suda, Y. Matsumura, *J. Membr. Sci.*, **269**, 101 (2006).
25. Y.H. Ma, F. Guazzone,*Composite gas separation modules having a layer of particles with a uniform binder distribution,* US patent application, PA US2006/0016332 (2006).
26. D. A. Pacheco Tanaka, M. A. Llosa Tanco, S. Niwa, Y. Wakui, F. Mizukami, T. Namba, T.M. Suzuki, *J. Membr. Sci.*, **247**, 21 (2005).
27. J. Tong, H. Suda, K. Haraya, Y. Matsumura, *J. Membr. Sci.*, **260**, 10 (2005).
28. A. Bottino, G. Capannelli, A. Comite, R. Di Felice M. Broglia, P. Pinacci, Pd and Pd/Ag composite membranes prepared by the electroless plating technique: preparation, characterization and interpretation of the H_2 transport mechanism, *Proc. ICIM 9*, Lillehammer, 25–29 June (2006).
29. P. P. Mardilovich, Y. She, Y. H. Ma, M. H. Rei, *AIChE J.*, **44**, 310–322 (1998).
30. Y. H. Ma, P.P. Mardilovich, Y. She,US Patent 6,152,987 (2000).
31. S. N. Paglieri, K. Y. Foo, J. D. Way, J. P. Collins, D.L. Harper-Nixon, *Ind. Eng. Chem. Res.*, **38**, 1925–1936 (1999).
32. J. A. Calles, R. Sanz, D. Alique, Study of preparation fo Pd based membranes by electroless palting over PSS supports, *Proc. Euromembrane 2009*, Montpellier, 6–10 September (2009).

33. H. B. Zhao, K. Planz, J.H. Gu, A. V. Li, N. Stroh, H. Brunner, G. X. Xiong, *J. Membr. Sci.*, **142**, 147 (1998).
34. M. Broglia, P. Pinacci, M. Radaelli, A. Bottino, G. Capannelli, A. Comite, G. Vanacore, M. Zani, *Desalination*, **245**, 508–515 (2009).
35. K. L: Yeng, S. C. Christiansen, A. Varma, *J. Membr. Sci.*, **159**, 107 (1999).
36. J. Shu, B. P. A. Grandjean, S. Kaliaguine, *Thin Solid Films*, **252**, 26–31 (1994).
37. F. Guazzone, PhDThesis, Worchester Polytechnic Institute, Worchester, USA (2005).
38. M. Broglia, S. Concari, M. Radaelli, *CESI RICERCA report n. 08000983* (2008).
39. K. L. Yeung, J. M. Sebastian, A. Varma, *Cat. Today*, **25**, 231–236 (1995).
40. K. L. Yeung, A. Varma, *AIChE J.*, **41**, 2131–2139 (1995).
41. A. Li, W. Liang, R. Hughes, *J. Membr. Sci.*, **149**, 259–268 (1998).
42. R. S. Soulemanova, A. S. Mukasyan, A. Varma, *J. Membr. Sci.*, **166**, 249 (2000).
43. A. G. Knapton, *Plat. Met. Rev.*, **21**, 44–50 (1977).
44. V. Gryaznov, *Sep. Pur. Methods*, **29**, 171 (2000).
45. D. L. Mc Kinley,US Patent 3,350,845 (1967).
46. Y. S. Cheng, K. L. Yeung, *J. Membr. Sci.*, **158**, 127–141 (1999).
47. T. C. Huang, M. C. Wei, H. Y. Chen, *Sep. Pur. Tech.*, **32**, 239–245 (2003).
48. E. Kikuchi, S. Uemiya, *Gas Sep. Purif.*, **5**, 261–266 (1991).
49. J. N. Keuler, L. Lorenzen, R.D. Sanderson, V. Prozesky, W. J. Pryzbylowicz, *Thin Solid Films*, **347**, 91–98 (1999).
50. J. N. Keuler, L. Lorenzen, *J. Membr. Sci.*, **195**, 203–213 (2002).
51. P. Pinacci, M. Broglia, M. Radaelli, A. Bottino, G. Capannelli, A. Comite, Optimization of palladium alloy membranes for H_2 separation, *Permea 2007*, Siofok, 2–6 September (2007).
52. L.C. Witjens, J. H. Bitter, A. J. Van Dillen, W. M. Arnoldbik, F. H. P. M. Habraken, K. P. de Jong, *J. Membr. Sci.*, **254**, 241–248 (2005).
53. F. Roa, J. D. Way, R. L. Mc Cormick, S. Paglieri, *Chem. Eng. J.*, **93**, 11–22 (2003).
54. K. S. Rothenberger, A. V. Cugini, B. H. Howard, R. P. Killemeyer, M. V. Ciocco, B. D. Morreale, R. M. Enick, F. Bustamante, I. P. Mardilovich, Y. H. Ma, *J. Membr. Sci.*, **244**, 55–68 (2004).
55. D. Jensen, J. W. Dijkstra, R. Van den Brink, T. A. Peters, M. Stange, R. Bredesen, A. Goldbach, H. Y. Xu, A. gottschalk, D. Doukelis, *Energy Procedia*, **1**, 253–260 (2009).
56. Y. C. van Delft, L. A. Correia, J. P. Overbeek, D. F. Meyer, A de Groot, J. W. Dijkstra, D. Jansen, palladium membrane reactors for large scale production of hydrogen, *Proc. 8th Int. Conf. on Catalysis in Membrane Reactors*, Kolkata, 18–21 December (2007).
57. S. F. Hou, K. Jiang, W.Z. Li, H.Y. Xu, L.X. Yuan,WO patent 2005/065806 (2005).
58. K. Jiang, A. Goldbach, C. Bao, C.Y. Yu and H.Y. Xu, *Proc. 10th Int. Conf. on Inorganic Membranes ICIM 10*, Tokyo, paper 1a07 (2008).
59. M. E. Ayturk, Y. H. Ma, *J. Membr. Sci.*, **330**, 233–245 (2009).
60. Y. H. Ma, I. P. Mardilovich, E. E. Engwall,US patent 7,255,726 (2007).
61. E. Engwall, J. Saukaitis, M. Joshi, A. Matzakos, A. Del Paggio, Hydrogen production with a membrane steam reformer, *Proc. AIChE Nat. Meet.*, Nashville, November (2007).
62. E. Engwall, J. Saukaitis, G. Komplin, M. Joshi, A. Del Paggio, A. Nijmeijer, Reducing CO_2 footprint with hydrogen separation membranes, *Proc. Euromembrane 2009*, Montpellier, 6–10 September (2009).
63. S. K. Gade, M. K. Keeling, D. K. Steele, J. D. Way, P. M. Thoen, High flux Pd membranes deposited on stainless steel supports by electroless plating, *Proc. ICIM 9*, Lillehammer, 25–29 June (2006).
64. S. K. Gade, M. K. Keeling, D. K. Steele, P. M. Thoen, J. D. Way, S. DeVoss, G. Alptekin, Sulphur resistant Pd-Au composite membranes for H_2 separation, *Proc. ICIM 9*, Lillehammer, 25–29 June (2006).

65. J. Acquaviva, High performance, durable, palladium alloy membrane for hydrogen separation and purification, *DOE Annu. Merit Rev. Meet.*, 19 May (2009).
66. A.S. Damle, S. Opkins, K. Rekczis, H. Zhao, D. Henkel, C. Love, K. Stark, M. Keeling, Development of Pd–alloy composite membranes for hydrogen separation, *Proc. AIChE Annu. Meet.*, Nashville, 8–13 November (2009).
67. E. Kikuchi, *Catal. Today*, **25**, 333–337 (1995).
68. S. Uemiya, N. Sato, H. Ando, E. Kikuchi, *Ind. Eng. Chem. Res.*, **30**, 589–591 (1991).
69. T. Matsuda, I. Koike, N. Kubo, E. Kikuchi, *Appl. Catal.*, **96**, 3 (1993).
70. S. Uemiya, I. Koike, E. Kikuchi, *Appl. Catal.*, **76**, 171 (1991).
71. B. Arstad, H. Venvik, H. Klette, J. C. Walmsley, W. M. Tucho, R. Holmestad, H. Holmen, R. Bredesen, *Catal. Today*, **118**, 63–72 (2006).
72. P. Pinacci, M. Broglia, M. Radaelli, A. Bottino, G. Capannelli, A. Comite, Development of palladium alloy membranes for H_2 separation from synthesis gas in membrane reactors, *3rd Int. Conf. on Clean Coal Technologies for our Future*, Cagliari, 15–17 May (2007).
73. M. Broglia. P. Pinacci, C. Valli, Evaluation of the water gas shift reaction in a palladium membrane reactor, *Proc. ICCMR9*, Lyon, 28 June–2 July (2009).
74. A. S. Augustine, M. E. Ayturk, N. Kazantzis, Y. H. Ma, Composite palladium membrane reactor for the water gas shift reaction, *Proc. AIChE Annu. Meet.*, Nashville, 8–13 November (2009).
75. Y. Bi, H. Xu, W. Li, A. Goldbach, *Int. J. Hydrogen Energy*, **34**, 2965–2971 (2009).
76. T. A. Peters, M. Stange, H. Klette, R. Bredesen, *J. Membr. Sci.*, **316**, 119–127 (2008).
77. H. Li, A. Golbach, W. Li, H. Xu, *J. Membr. Sci.*, **299**, 130–137 (2007).

12

Silica Membranes – Preparation by Chemical Vapour Deposition and Characteristics

J. Galuszka and T. Giddings

Natural Resources Canada, CanmetENERGY, Ontario, Canada

12.1 Introduction

Advanced inorganic membranes for hydrogen separation have evolved from emerging to enabling technology essential to intensification of the strategic processes related to sustainable clean energy supply, biotechnology and climate change. In this category, Pd membranes have been researched for the longest time, as the first recorded use of palladium for hydrogen separation dates back to 1866 [1] and the first patent for a Pd membrane reactor application for hydrogen was issued back in 1916 [2]. The first small-scale commercial application could be ascribed to Johnson Matthey in 1964 [3], when a 23 wt% Pd-Ag membrane was used for hydrogen purification. Although the Pd membrane has evolved significantly [4], it still suffers from problems related mostly to high palladium costs, long-term stability and reactivity with hydrocarbons or sulfur.

Microporous silica hydrogen perm-selective membranes (H-membranes) have also been extensively studied as a potentially more practical alternative to Pd membranes. Several techniques for the fabrication of silica membranes have been developed but chemical vapour deposition (CVD) and the sol-gel method are the most commonly used, producing amorphous silica films with similar characteristics. The sol-gel routes seem to offer more flexibility for tailoring the porosity and composition of the separating layer. However, the final porosity

Membranes for Membrane Reactors: Preparation, Optimization and Selection, Edited by Angelo Basile and Fausto Gallucci

depends rather drastically on the various synthesis parameters, making reproducibility a more challenging issue [5].

CVD was introduced in the 1880s [6] for the production of incandescent lamps and further developed mostly for use by the electronics industry [7]. Application of CVD to membrane synthesis dates back to the late 1980s, when the group of Gavalas synthesised a silica membrane having H_2/N_2 selectivity between 500 and 1000 on a Vycor support by thermal decomposition of TEOS [8]. However, the hydrogen permeance of these membranes was hindered by a high resistance of the Vycor support pore structure, with a mean pore diameter of about 4 nm in an approximately 1-mm thick wall. Consequently, the CVD reduced the hydrogen permeance of Vycor by only 30% [9]. In 1994 Mooroka [10] overcame this limitation by using an asymmetric ceramic support having an α-alumina porous tube coated with γ-alumina. Although the pore size of γ-alumina was similar to that of Vycor, the γ-alumina layer was only about 1.5 to 2.5 μm thick, offering almost no resistance to hydrogen flow through the membrane and providing appropriate bridging between the silica membrane-forming layer and the open structure of α-alumina.

A number of excellent reviews have been published since then, tackling various aspects of membrane synthesis, application and economics [5,9–15]. It is evident from this mass of literature that state of the art silica membranes have good hydrogen flux and separation as well as respectable thermal stability extending approximately up to 600 °C. However, the hydrothermal stability of a silica hydrogen perm-selective membrane (H-membrane) is a key factor in determining its suitability for a commercial application for membrane-assisted processes. Although there seems to be general agreement that water has a detrimental effect on silica membrane performance, understanding the microporosity of the amorphous silica that forms the membrane and its hydrothermal densification mechanism is not complete [16–24].

In this chapter, the general principles of CVD applied to silica H-membrane preparation will be outlined. Also, the membrane-forming amorphous silica structure will be discussed and the mechanism of gas transport through the membrane and the possible mechanism of hydrothermal densification will be offered. Finally, a few examples of the catalytic silica membrane reactor applications for potentially important commercial processes will be presented.

12.2 Fundamentals of Chemical Vapour Deposition

In general terms, chemical vapour deposition (CVD) can be defined as a condensation of precursor compound, or compounds from the gas phase, onto a substrate to produce a solid deposit. Precursor compounds most often diluted in carrier gases are delivered into the reaction chamber at approximately ambient temperatures. As they are heated in the reaction chamber, they react or decompose forming a solid phase that is deposited onto the substrate. The reaction chamber and substrate temperature are critical and can influence which reactions will take place. Besides heat, the energy required to get the precursor to react or decompose can also be delivered by laser (laser chemical vapour deposition; LCVD), plasma formation (plasma-assisted chemical vapour deposition; PACVD; or plasma-enhanced chemical vapour deposition; PECVD) or photo energy (photochemical vapour deposition; PCVD). All of these can be combined with atmospheric pressure chemical vapour deposition (APCVD), low pressure chemical vapour deposition (LPCVD), metal–organic chemical vapour deposition (MOCVD) or chemical vapour infiltration (CVI), producing a sometimes almost overwhelming maze of acronyms that were thoroughly reviewed, together with the general principles and applications of CVD technology, in a number of older and new reviews [5–7,25–27]. Here, we concentrate on CVD aspects pertinent to silica H-membrane formation on porous substrates.

The CVD processing sequence to obtain a silica thin film on a porous substrate depends mostly on the choice of a gaseous SiO_2 precursor. The desirable aspects of any SiO_2 precursor are safety (nontoxicity), ease of handling and chemical stability. Consequently, the list of the practically used precursors is not very long. The prime choice for silica membrane preparation by CVD is tetraethylorthosilicate (TEOS). Thermally stimulated deposition of TEOS occurs at an optimum temperature of around 750 °C but could be lowered to below 600 °C under LPCVD conditions [28]. Other possible choices include tetramethylorthosilicate (TMOS) [28–31], triisopropylsilane (TPS) [32], silicon tetrachloride ($SiCl_4$) [24] or silane (SiH_4) [16]. In order to either lower the decomposition temperature of the precursors or provide a co-reagent, O_2, O_3, N_2O or water vapour are co-fed to the CVD chamber.

12.3 CVD Apparatus

A CVD apparatus consists of several basic components, as shown in Figures 12.1, and 12.2. The CVD reactor chamber within which the deposition takes place forms a central piece of the

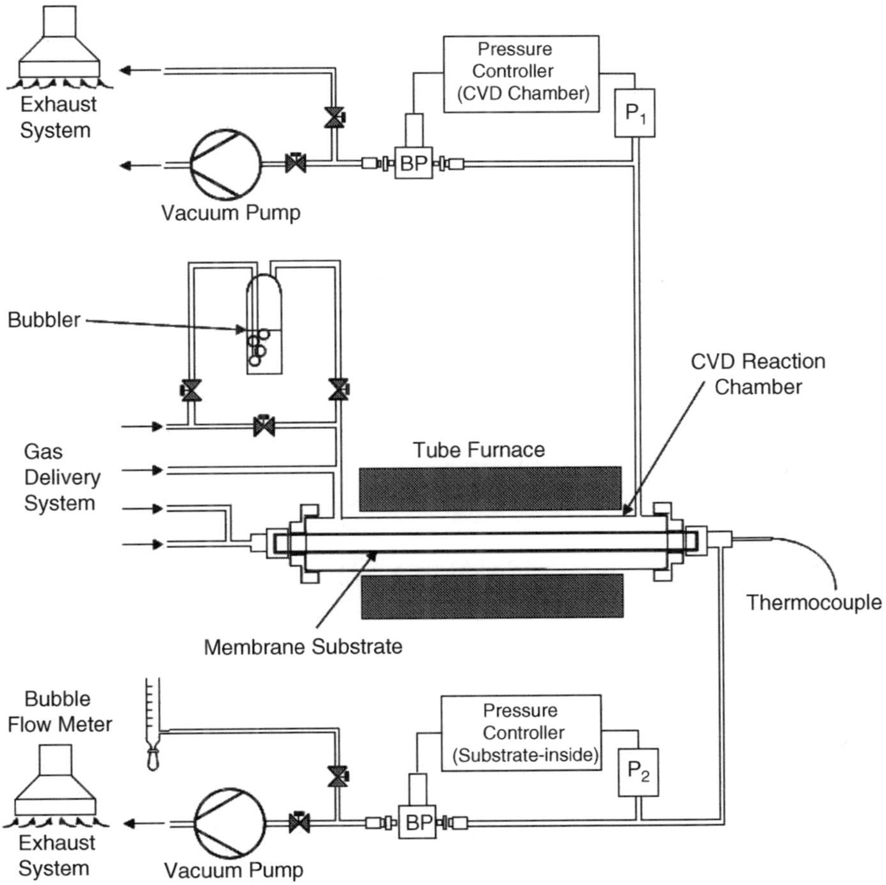

Figure 12.1 *Schematic of CVD semi-automated system at CanmetENERGY for tubular silica H membrane preparation*

Figure 12.2 Semi-automated CVD system at CanmetENERGY

apparatus. The reactor is connected to the gas delivery system, to supply the CVD precursors and co-reagents, and to a vacuum system to remove the gaseous species other than those required for the reaction/deposition. An efficient exhaust system is required for the removal of volatile byproducts from the reaction chamber. In some instances, exhaust gases may not be suitable for release into the atmosphere and may require treatment or conversion into safe/harmless compounds. The CVD reactor chamber must allow for an uncomplicated way of introducing and removing the membrane substrates. The most common energy source for getting the precursors to react/decompose is heat delivered by resistive heating in tube furnaces. The CVD system needs to be equipped with process control equipment like gauges and controls, to monitor process parameters, such as pressure, gas flows, temperature and time. Alarms and safety devices would also be included in this category.

12.4 Silica H-Membranes Produced by CVD

There are numerous descriptions of the CVD experimental procedures used for silica membrane formation by thermal- [33–37], oxygen- [18,30,31,38–41] or ozone- [42–44] aided decomposition of the most popular silica precursors: TEOS [18,33,34,41,43,45] or the much more thermally stable TMOS [30,31,39,43,46]. Practically, however, two main approaches could be distinguished that basically differ by the reagents' delivery method to the reaction chamber in relation to the porous substrate that divides the CVD reactor into two separate volumes. In so-called one- or single-sided CVD, the precursor or the precursor and co-reagents are delivered on the same side of the substrate surface [18,40,43,47,48]. Silica membrane formation by thermal decomposition of TEOS would, of course, fall into this category.

In the other approach, TEOS or TMOS is delivered into the CVD chamber from one side of the substrate and the oxidising or reactive agent enters from the other side [30,31,40,41,47].

Therefore, the two reagents are forced to diffuse from opposite sides of the substrate and react inside the pores, forming a thin silica deposit and eventually plugging all the pores. This approach is known as two-sided CVD or chemical vapour infiltration (CVI).

Both these approaches seem to be equally effective yielding membranes with similar characteristics. Consequently, making any attempt of ranking would be rather difficult, especially, that a correlation between the final property of a membrane and the key experimental parameters during the CVD process is generally rather obscure.

CanmetENERGY membranes synthesised on 60-cm long tubular substrates with 10-mm OD could serve as a good example of a one-sided CVD [49]. The α-alumina macropores of this asymmetric substrate were reduced to about 50 nm by a coat of γ-alumina, which formed a top bridging layer approximately 0.5 μm thick, as shown in Figure 12.3. TEOS diluted in an inert carrier gas entered the CVD reactor on a shell side, and the products of decomposition were vented on the tube side of the substrate until the pores were closed by silica and the membrane was formed. The TEOS concentration was regulated by the amount of the inert carrier gas and the temperature of the bubbler. If an oxidant was used, its amount was regulated by the mass flow controllers. The progress of CVD coating was monitored *in situ* by periodically measuring single-gas He and N_2 permeances. He is used instead of H_2 for a safety reason [50]. The typical range of He permeances for CanmetENERGY membranes at 500 °C was between 0.3 and 0.7×10^{-6} mol m^{-2} s^{-1} Pa^{-1}, and He/N_2 separations were between 250 and 750. The He/H_2 permeance ratios (separation) reported in the literature are between 0.6 and 3 [18,34,51–53] and our experimentally verified ratio is

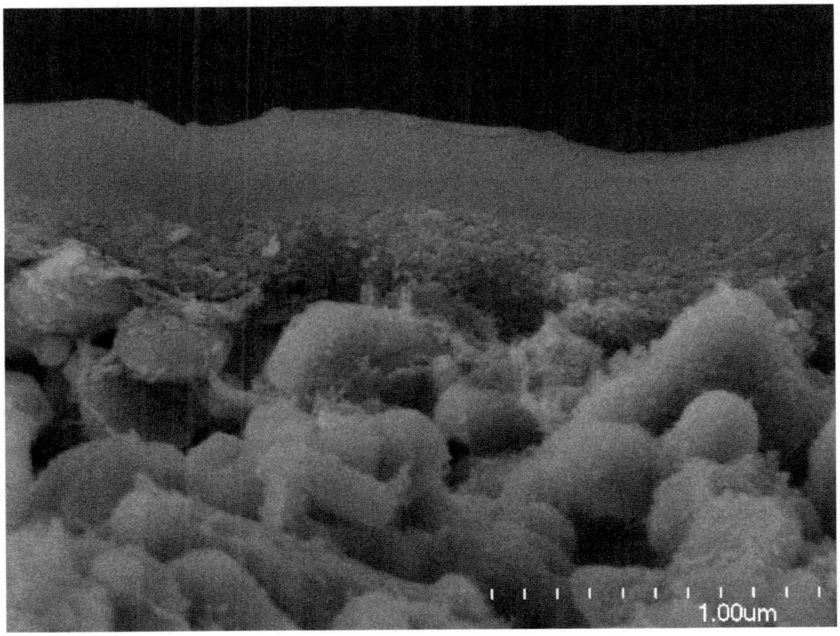

Figure 12.3 SEM micrograph showing cross section of the outer edge of a CanmetENERGY TEOS-derived silica H membrane. Significant infiltration of silica into the layer of γ-alumina resting on top of α-alumina substrate is clearly visible

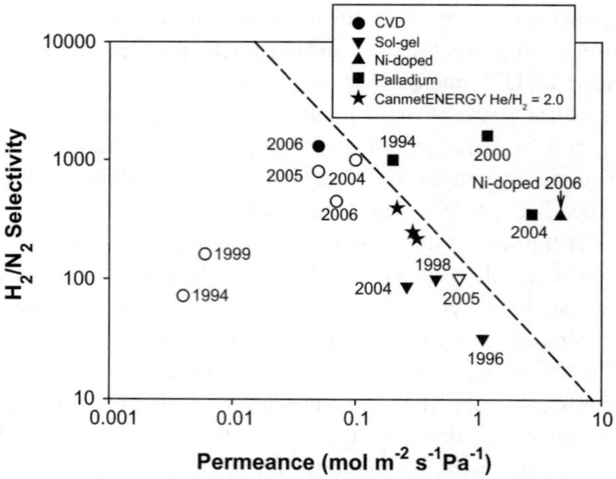

Figure 12.4 *Comparison of H_2 permeances and H_2/N_2 selectivities measured at 500 °C (solid points) or 600 °C (open points) for sol-gel- and CVD-derived state of the art silica H membranes reported in literature with palladium-based membranes (reprinted from [5] with permission from Elsevier). Top performing CanmetENERGY, CVD-derived silica H membranes were added*

about 2. Clearly, these membranes are at the leading edge of the best known silica membranes reported recently by Ayral, *et al.* [5], as can be seen in Figure 12.4.

The He and N_2 permeances of silica membranes measured between 100 °C and 500 °C are linearly temperature-dependent, as shown in Figures 12.5, and 12.6 for the two CanmetENERGY membranes. However, the He permeance increased with temperature, indicating activated transport, whereas N_2 permeance decreased following the Knudsen diffusion pattern [33]. A substantial difference in He/N_2 separations between these two membranes having similar He

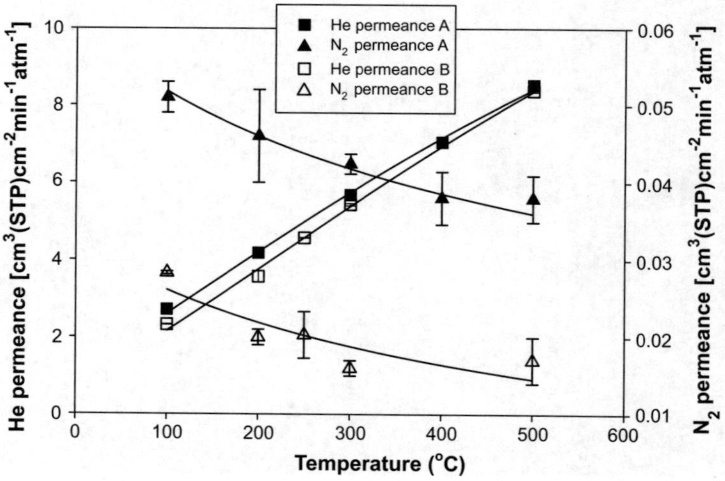

Figure 12.5 *Temperature dependence of He and N_2 permeances for two CanmetENERGY silica H membranes. He permeances for both of these membranes are almost identical, whereas N_2 permeance of membrane A is about two times greater as compared to membrane B*

Table 12.1 Performance of CanmetENERGY membranes and energy activation for helium transport

Membrane	E_a (kJ mol^{-1})	He permeance at 500 °C cm^3 (STP) cm^{-2} min^{-1} atm^{-1}	Separation
A	7.12	8.5	225
B	8.06	8.4	504

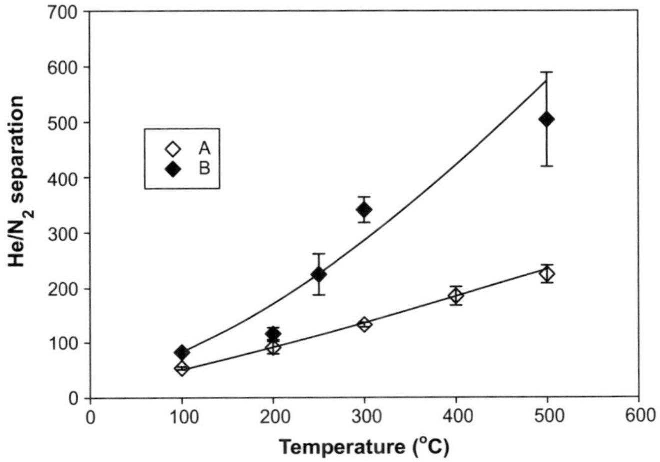

Figure 12.6 Temperature dependence of the He/N_2 separations for the two CanmetENERGY silica H-membranes (A and B) shown in Figure 12.5

permeances did not affect the energy activation of He transport estimated by fitting the data to the Arrhenius equation, as shown in Table 12.1.

12.5 Silica Membrane Structure and Transport Mechanism

It is assumed that the amorphous silica structure could be approximated by a disordered structure of β-cristobalite with multimember siloxane rings that contain between two and eight silicons per ring [33,35,54,55]. These rings are truncated at the surface and stabilised by surface hydroxyls that come in three configurations, namely vicinal, geminal and isolated [56,57]. It is believed that these surface hydroxyls are essential to maintain the microporosity suitable for hydrogen transport through a silica membrane.

Thermal treatment definitely decreases the number of surface hydroxyls [54,57,58]. Vicinal hydroxyls are the easiest to remove through a condensation reaction that forms Si–O–Si bridges and densifies the structure. For silica membranes, this results in a decreased pore size and consequently, decreased H_2 permeance. Above 300 °C the isolated hydroxyls are present [19,57] and dehydroxylation becomes increasingly difficult as the distance between the neighbouring hydroxyls increases. It was determined that the isolated hydroxyls account for about 30–50% of the total hydroxyls existing on the amorphous silica surface [54]. The surface curvature in the smaller pores promotes hydrogen bonding by reducing the distance between neighbour hydroxyls. Consequently, these hydroxyls are more strongly bonded, requiring a higher dehydroxylation temperature [54,58,59]. Since it was shown that surface hydroxyls are the principal sites for water adsorption, dehydroxylation makes the silica surface more hydrophobic [54].

The thermal stability of silica is very high as crystallisation does not occur below 900 °C [58]. However, the presence of water at elevated temperatures may result in the breaking of Si–O–Si bonds and the formation of surface hydroxyls [60]. Water vapour is presumed to attack strained Si–O–Si bonds as in the three- or fourfold rings inducing a cleavage by the hydrolysis reaction [20,61]. Cleavage of the bonds promotes the rearrangement of the network [20,61]. Thus, during hydrothermal treatment the silica surface is repopulated with OH and rearranged, and the number of neighbouring hydroxyls increases. Consequently, a greater amount of hydroxyls can be removed by a condensation reaction promoting further densification of the amorphous silica structure and causing further deterioration of silica membrane performance [19,20,61].

The microporosity of amorphous silica is difficult to study and, for a long time, there has been no clear concept of the origin of the microporosity responsible for the transport of H_2, He and larger molecules like N_2 or the associated transport mechanism. The numerous BET measurements [19] have not been particularly helpful as the probe molecules are too large to reveal the relevant microporosity [54,62]. Also, a popular belief that N_2 permeates through microcracks or defects in the amorphous silica layer does not seem to be convincing [19,22].

Recently it was suggested that in an amorphous silica film, the Si-O sixfold ring, having a 0.3 nm diameter [35,55,56], provides the main selective permeation path for small-sized molecules (He = 0.26 nm, H_2 = 0.289 nm), while excluding transport of the larger ones (CO_2 = 0.33 nm, CO = 0. 376 nm, N_2 = 0.364 nm). The assumed mechanism of transport was a molecular diffusion in which gas molecules adsorbed and jumped between the solubility sites under the driving force of a concentration gradient. However, it was observed that although the jump analysis gave physically realistic values, the experimental permeances for H_2 and He were significantly higher than those simulated for vitreous silica. The discrepancy was explained by assuming that the silica structure obtained by CVD was more open than that in vitreous glass [35,55].

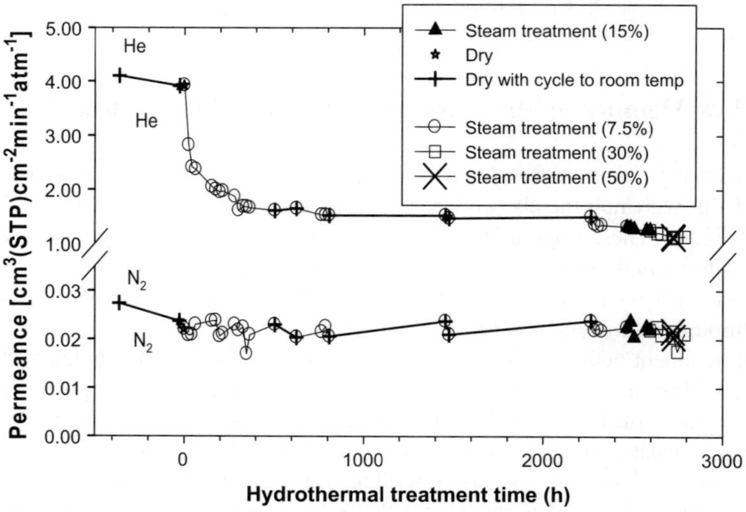

Figure 12.7 *Time course of He and N_2 permeances of a CanmetENERGY silica H-membrane during hydrothermal treatment at 500 °C*

Our study of the hydrothermal stability of silica membranes showed convincingly that the observed decrease in He(H_2)/N_2 separation was caused mainly by the decrease of He(H_2) permeance. The permeance of N_2 was practically unchanged, as shown in Figure 12.7. Therefore, it was concluded that the shrinkage of silica porosity that provided passage for hydrogen or helium molecules does not affect the porosity responsible for nitrogen transport. In 2006, a bimodal structure of amorphous silica, shown schematically in Figure 12.8, was proposed [50] to explain the mechanism of gas transport through the CVD-formed silica membrane. In 2008, an

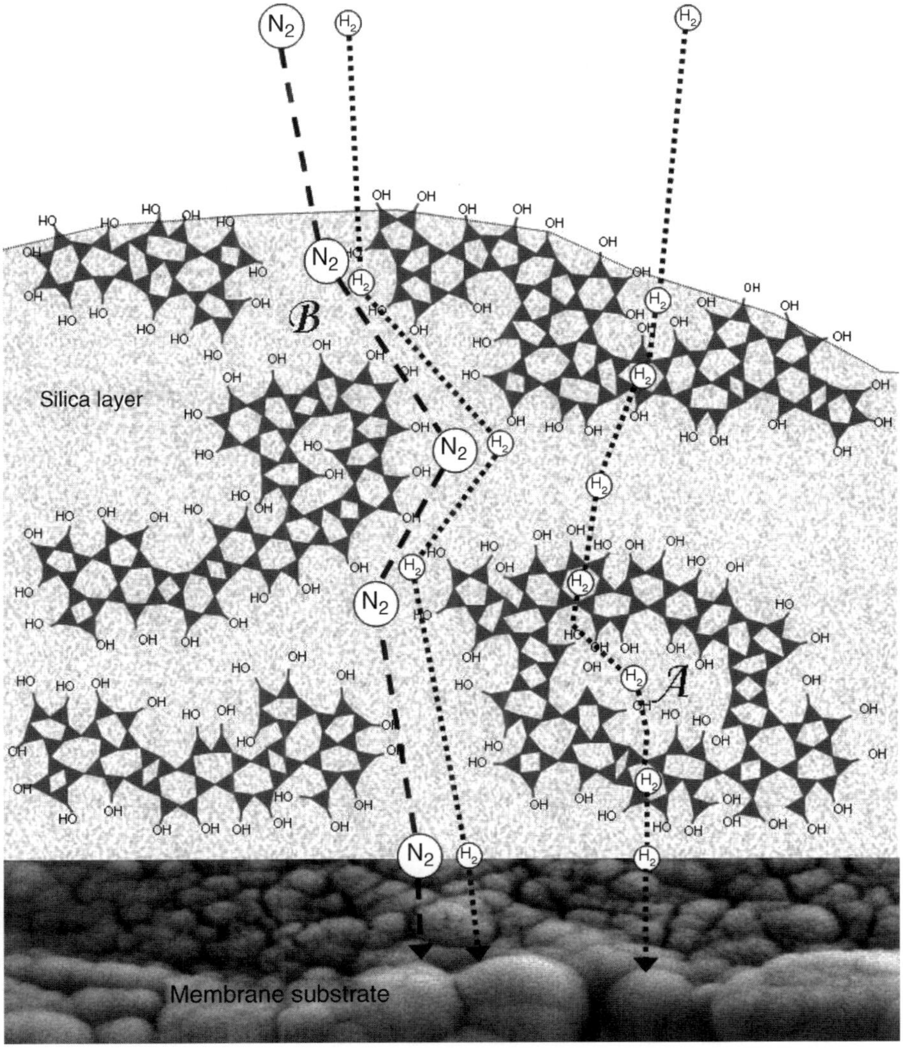

Figure 12.8 *Schematic representation of silica membrane bimodal pore structure showing H_2 and N_2 molecular transport mechanism through the membrane. Only small molecules like H_2 or He can penetrate siloxane rings that form the wall of the three-dimensional closed cavity (A) in the silica structure. The pores between silica clusters (B) are accessible to larger molecules like N_2, CO, CO_2 and CH_4. Transport of small molecules through these pores in a fully developed membrane is insignificant*

attempt was made to echo this concept [5], however, the main features were not accurately conveyed.

The most important feature of the proposed silica structure [38] is the existence of the numerous, three-dimensional closed cavities, the outside walls of which, are formed from the Si-O n-fold rings. Rings having $n = 6$ with a diameter of about 0.3 nm provide the main selective permeation path for small-sized molecules, therefore, excluding the transport of larger ones. The molecules – after entering the mesoporous-sized cavity through a Si-O sixfold ring – move fast to the opposite wall under the driving force of a concentration gradient and exit through another Si-O six-fold ring, as shown in Figure 12.8. This structure significantly decreases the effective thickness of the silica layer or makes it looser, as assumed in [35,55], which explains the higher H_2 and He permeances than those simulated for vitreous silica.

The main transport path for larger molecules like N_2, CO, CO_2, and so forth, would occur through the between-grain porosity. The 2×6 nm grains were identified by TEM in the membrane-forming amorphous silica [62], and this finding could be discussed in terms of the between-grain porosity and its role in the membrane transport mechanism [51]. Of course, smaller molecules could pass through the between-grain channels as well but the contribution of this nonselective transport to the overall H_2(He) permeance would be insignificant in a fully developed membrane. These two porosities (bimodal structure) are fully independent and that is why the hydrothermal densification does not affect the permeance of N_2.

The decrease of H_2(He) permeance during the hydrothermal treatment can now be explained by the wall densification of the cavities in the silica structure caused by the surface OH removal in a condensation reaction. This reaction forms new Si-O-Si bridges, as shown in Figure 12.9. Consequently, an average diameter of the siloxane rings forming the wall of cavities decreases, restraining access to the inside of the cavities and significantly decreasing the small-molecule

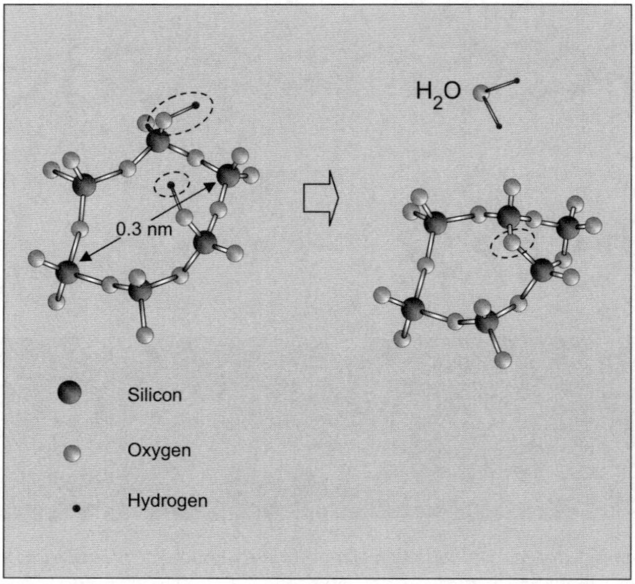

Figure 12.9 *Example of hydrothermal densification mechanism of a silica membrane resulting from the condensation reaction of neighbouring hydroxyls causing the decrease of siloxane ring aperture in the wall of closed cavities in silica structure shown schematically in Figure 12.8*

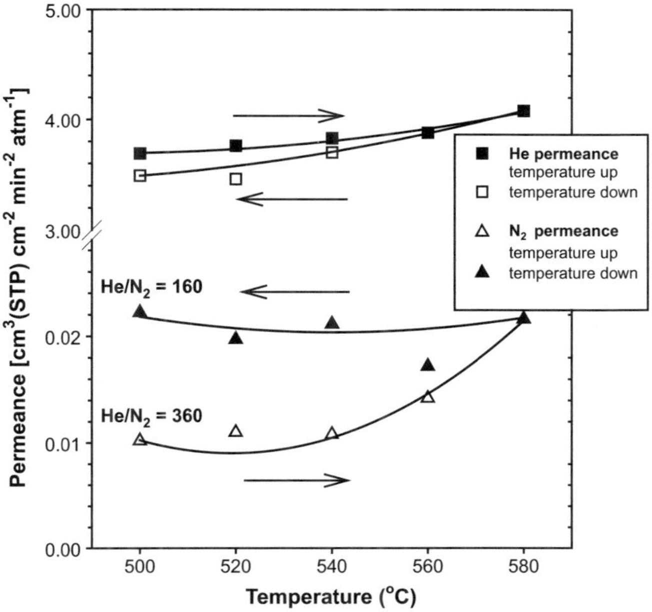

Figure 12.10 *Temperature dependence of He and N_2 permeances of H-membrane heated between 500 and 580 °C and cooled back to 500 °C indicating morphological changes in amorphous silica. Only the pores responsible for N_2 transport were significantly and irreversibly affected by the temperature increase, enhancing N_2 permeance by about 100%*

transport rate without affecting N_2 permeance. It was observed that although siloxane rings with fewer than six members do not have the Si-O-Si equilibrium bond angle of 150°, the large ring assemblies on the silica surface allow puckering and the angle could be adjusted to form rings that are well preserved in the stiff silicate matrix [54].

The proposed bimodal pore structure of silica agrees very well with the experimental results presented in Figure 12.10. A thermal treatment of a silica membrane from 500 to 580 °C and back to 500 °C resulted in a decrease of He/N_2 separation, but this time mostly owing to the increase in N_2 permeance measured at 500 °C. This indicated a rearrangement of the amorphous silica morphology, causing an increase in either the number or the size of the between-grain pores. The observed decrease in He permeance resulting from this thermal treatment was rather insignificant.

The origin of the proposed bimodal structure seems to be a logical consequence of the well researched TEOS thermal decomposition mechanism leading to silica film formation. Adachi *et al.* [63,64], Desu [65] and Smolik *et al.* [66] studied the conditions of silica film formation and particle generation during the thermal [63] and oxygen-aided decomposition [64,66] of TEOS. The size distribution and number of particles were measured by differential mobility particle sizing (DMPS). It was concluded that the intermediate species are deposited on the substrate and form an SiO_2 layer by β-elimination, releasing H_2O and C_2H_4. Simultaneously, ultrafine particles with a diameter of about 50 nm and a log-normal size distribution [64] are generated in the gas phase by homogeneous nucleation. A similar conclusion was also reached by Shareef *et al.* [67]. These particles are drawn to the substrate and attach themselves to the surface and to each other most likely by a hydroxyl condensation reaction, forming a bimodal silica structure, as shown in Figure 12.8 and described above.

12.6 Hydrothermal Stability of Silica Membranes

The hydrothermal stability of a number of CanmetENERGY silica H-membranes was assessed at 500 °C, and Figure 12.7 shows an example of an approximately 3000 h on stream test in which one of these membranes was exposed to nitrogen containing between 7.5 and 30.0% steam. The hydrothermal treatment had a very negligible effect on N_2 permeance, whereas He permeance decreased to about 30% of its initial 3.9 cm^3(STP) $cm^{-2}\,min^{-1}\,atm^{-1}$ ($0.29 \times 10^{-6}\,mol\,m^{-2}\,s^{-1}\,Pa^{-1}$) value, producing the final He/N_2 separation of about 60. About 75% of the total He-permeance decrease took place in the first 350 h on stream. In the next 2600 h, the silica membrane gave a very stable performance even at 30% steam concentration, and several heating and cooling cycles did not have an effect on this membrane's performance and stability. The presented example of the CanmetENERGY membrane seems to be one of the longest hydrothermal stability tests ever reported in the literature for a silica H-membrane in which the membrane, after about 3000 h on stream, still maintained a respectable He/N_2 selectivity of 60.

It has been confirmed for CanmetENERGY membranes that the initial rapid He(H_2) permeance decrease during a hydrothermal treatment at 500 °C seemed to be always in the range of 60% to 70% regardless of the wide array of the initial He(H_2) permeances and He(H_2)/N_2 separations of the silica membranes used in the hydrothermal treatment experiments. Similar behaviour of silica membranes was also reported in the literature [29,52,53,68–71], although the range of the decrease varied according to hydrothermal treatment conditions and various initial pre-treatment procedures applied to the membranes prior to steam exposure. It is postulated here that this 60% to 70% loss of the initial silica membrane performance is not accidental and stems from the chemistry of the amorphous silica surface itself. It is very likely that there is a stoichiometric correlation between all forms of the *n*-member siloxane rings on the dehydroxylated silica surface. Thus, after all of the strained Si-O bonds are broken and new bonds are formed in the process facilitated by a high temperature and the presence of water, the new rearranged surface reaches equilibrium and stability at a level of 60–70% denser than the original one, as measured by He or H_2 permeance.

This observation seems to be supported by the experimentally verified behaviour of another hydrothermally treated CanmetENERGY membrane. After being on stream, containing 7.5% water at 500 °C for about 600 h and becoming about 50% denser, the membrane was used as a membrane substrate and was recoated with silica by applying a standard CVD procedure. After recoating the He/N_2 separation measured at 500 °C was increased from about 20 to 90. This membrane was subjected again to a hydrothermal treatment at 500 °C, similar to the one previously applied. However, this time, after about 500 h on stream, the He permeance decreased by only 28%. Apparently, the silica surface of the previously hydrothermally treated membrane provided a different *rearranged* template for a new coat of silica. Effectively, the dehydroxylated silica surface of a reused membrane had a relaxed morphology that inhibited the densification process under the similar hydrothermal conditions.

There have been numerous attempts to improve the hydrothermal stability of silica membranes, as recently summarised by Ayrai *et al.* [5] or Castricum *et al.* [72]. The most common strategies relied on doping amorphous silica with various oxides, such as TiO_2 [69,73,74], ZrO_2 [75–77], Al_2O_3 [34,74,78] and more recently NiO [53], Nb_2O_5 [52] and Co_3O_4 [79,80].

Also, the incorporation of hydrophobic methyl [21,81] or carbon moieties [72,82,83] into a silica structure was reported. These approaches were tried for membranes synthesised by sol-gel as this methodology offers more practical flexibility with doping than CVD. In this group, the Ni-SiO_2 membrane showed superior performance and stability [53]. However, the transport mechanism and resistance to poisoning by sulfur, hydrocarbons or carbon monoxide still need to be studied and demonstrated for this membrane.

The effective strategies for improving the hydrothermal stability of silica membranes synthesised by CVD were less numerous. A composite membrane synthesised by dual-element silica–alumina CVD was reported by Gu *et al.* [34]. The H_2 permeance of this membrane showed about 50% stronger resistance to the 520 h of hydrothermal treatment at 600 °C but the H_2/CO_2 selectivity of the composite membrane with a high content of alumina was almost 40 times lower after only 130 h on stream as compared to the pure silica membrane. It was concluded that a compromise between permeance and stability was needed for these membranes [34].

More recently, Gu and Oyama [69] reported another dual-element membrane prepared by the CVD of titanium isopropoxide (TIP) and TEOS at temperatures between 500 and 600 °C. This composite silica–titania membrane was treated hydrothermally and, after 130 h on stream at 650 °C, presented a somewhat better stability compared to an alumina–silica membrane and significantly better than a pure silica membrane. However, it needs to be observed that the starting H_2 permeance for the silica–titania membrane was the lowest as compared to the two other membranes and, consequently, the selectivities to H_2/CH_4 and H_2/CO, making the comparison between these three membranes somewhat complicated.

Silica membranes prepared by counter-diffusion CVD of TMOS and oxygen at 600 °C were reported by Nomura *et al.* [44,70] and more recently by Akamatsu *et al.* [29]. A high hydrothermal stability of the resultant silica membranes was claimed even at 800 °C [29]. However, a practical meaning of these claims is not obvious as these membranes were kept on the hydrothermal stream for only 12 h [29] or 80 h [70]. Also, the methodology of on stream H_2 and N_2 permeance assessment leaves some doubts as to whether the *wet* membrane performance is representative and connects to the single-gas measurements for dry membranes normally reported in the literature.

12.7 Examples of Silica Membrane Application

A common feature of all catalytic membrane reactors incorporating a hydrogen perm-selective silica membrane could be defined as an ability of such a reactor to circumvent thermodynamic limitation of an equilibrium-controlled process by separating H_2 as it is produced. It is expected that membranes with high selectivity and permeability towards hydrogen – if successfully integrated with advanced catalysts – would offer a breakthrough in process efficiency and economics by enabling production at a much lower temperature with greater intensity. The application of various membrane reactors was comprehensively reviewed by Sanchez and Tsotsis [84]. Here, only a few of the most promising applications of a membrane reactor incorporating the H-silica membrane are discussed.

12.7.1 Dehydrogenation of Light Paraffins

The current technology for ethylene production involves thermal cracking of ethane at very high temperatures (up to 900 °C) and is very energy intensive. Therefore, membrane reactor application has been contemplated since the early nineties to circumvent thermodynamic limitation of the equilibrium-controlled ethane dehydrogenation and enable ethylene and high-purity hydrogen production at much lower temperatures. Early reports claimed that the membrane reactor increases the thermodynamically allowed ethane conversion between six to eight times [85,86]. However, as the quality of membranes and membrane reactor assessment increased, rather moderate gains in membrane-assisted conversion of paraffins were reported [87].

Ethane dehydrogenation in a membrane reactor was studied by Galuszka *et al.* [88] and although about 50% of the H_2 produced passed through the membrane, only moderate enhancement of ethane conversion to ethylene was observed. The slow processes on the catalyst surface were thought to counterbalance the positive effect of membrane-assisted hydrogen removal. It was concluded that a commercial realisation of the membrane reactor concept for ethylene production would require a new dehydrogenation catalyst designed to function with the membrane. This idea was proven in a more recent study when chromium-based catalysts with different formulations were operated in a membrane reactor with a silica membrane having He/N_2 separation of about 500 and He permeance of about $6.4\,cm^3(STP)\ cm^{-2}\,min^{-1}\,atm^{-1}$ ($0.48 \times 10^{-6}\,mol\,m^{-2}\,s^{-1}\,Pa^{-1}$) measured at 500 °C [89,90]. The achieved ethane conversion at 555 °C was 10% higher than the thermodynamically allowed conversion, as shown in Figure 12.11. However, ethylene selectivity decreased with time on stream producing the highest ethylene yield in the membrane reactor: about 8% greater than for the conventional one at 555 and 600 °C.

Silica membrane-assisted dehydrogenation of propane was studied by Schafer *et al.* [87] using Cr_2O_3/Al_2O_3 at 535 °C and Pt-Sn/Al_2O_3 at 500 °C as dehydrogenation catalysts. The membrane reactor produced an approximately 5% propylene yield increase as compared to a conventional packed-bed reactor. The importance of a synergy between the membrane and catalyst to counter the loss of selectivity was emphasised again in the conclusion of this paper. This relatively low propylene yield increase was comparable to the results of Dittmeyer *et al.* [91] reported for a Pd/Ag membrane reactor.

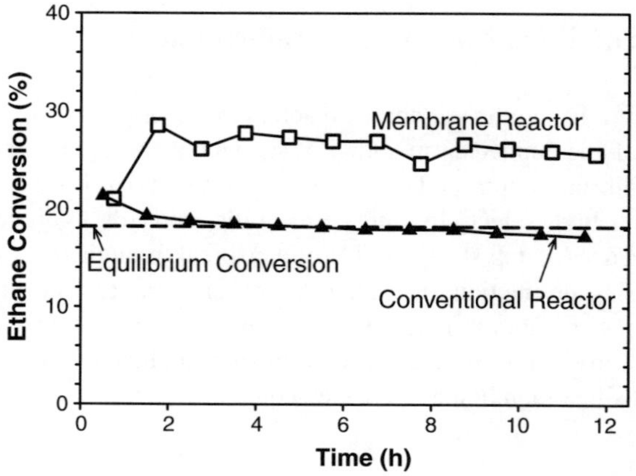

Figure 12.11 *Comparison of catalytic ethane conversion in conventional and membrane reactor at 555 °C. About 50% of H_2 produced during dehydrogenation of ethane was removed through the membrane*

12.7.2 Water Gas Shift Reaction

It is considered that combining the integrated gasification combined cycle (IGCC) and hydrogen membrane reactor technologies could radically improve the commercial outlook for the IGCC scheme application to power production with zero CO_2 emissions, leading to elimination of the coal-fired power plants [92–94]. The expected significant process simplification and intensification capitalises on a paradigm offered by membrane reactors that allows for the reaction and separation to be combined in one step. Effectively, the products of the gasification of low-cost opportunity feedstocks could be converted in a membrane-assisted one-step water gas shift reaction (WGSR) process to a clean-hydrogen and a clean-CO_2 stream sequestration-ready. The proposed approach seems to be far superior [94] to the contemplated development of CO_2 perm-selective membranes for achieving a similar target.

Membrane-assisted WGSR was studied by Galuszka *et al.* [95] using a commercial catalyst at 450 °C in a silica membrane and a conventional fixed-bed reactor. The H_2O/CO ratio was kept at 1.75. The catalyst gave a steady performance and up to 80% of the H_2 produced during WGSR in the membrane reactor was removed through the membrane that had an He/N_2 separation of ~300 and an He permeance of ~5 cm^3 (STP) $cm^{-2}\,min^{-1}\,atm^{-1}$ ($0.37 \times 10^{-6}\,mol\,m^{-2}\,s^{-1}\,Pa^{-1}$). A 6–10% increase in CO conversion was achieved in a membrane reactor as compared to a conventional reactor for a simulated feed stream from a gasifier. After about 10 days on stream, the H_2/N_2 separation of the membrane usually decreased to about 50% of the initial value. After that, the further decrease in membrane performance was barely noticeable. The overall excellent performance of the CanmetENERGY membrane evoked great confidence in the demonstrated feasibility of a novel WGS reactor concept. However, hydrothermal stability of the ceramic silica membrane needs further improvement before commercialisation could be considered.

Application of a molecular sieve silica membrane to the low-temperature WGS at 280 °C was reported by Giessler *et al.* [68]. Although an almost complete conversion of CO was claimed, the low H_2/N_2 separation of the membrane opened up a possibility for CO to cross the membrane to the sweep stream and for the sweep gas to enter the product stream, artificially boosting the conversion. This may also explain the observed CO conversion decrease for the high sweep rates. But more importantly, application of a membrane reactor in a kinetically controlled range of the WGS reaction is perhaps not the best choice of conditions to demonstrate a paradigm offered by membrane reactors, as the equilibrium conversion is already high but the reaction rate is low.

A similar conclusion was reached by Battersby *et al.* [96] in their WGSR study carried out in the membrane reactor with cobalt–silica membrane below 300 °C. It looks like the hydrothermal stability of a membrane was an issue at higher temperatures and dictated the choice of these less-representative conditions.

An economic feasibility study of the membrane-assisted WGSR conducted to quantify the advantage over conventional technology supported the use of ceramic membranes [94] but disproved the concept for Pd membrane reactors [97].

12.7.3 H_2S Decomposition

Hydrogen sulfide is a byproduct of sour natural gas (NG) sweetening, the hydrodesulfurisation of light hydrocarbons and the upgrading of heavy oils, bitumens and coals. Since this H_2S has a limited industrial application, it is viewed as a pollutant requiring treatment and removal. Currently, this H_2S is processed by Claus oxidation technology yielding sulfur and water. However, H_2S potentially has a much higher economic value if not only sulfur, but also hydrogen could be recovered. It is highly desirable to convert H_2S to sulfur and hydrogen within the refinery, and return the recovered hydrogen to the heavy oil hydrogenation step. This would significantly improve hydrogen inventory and reduce associated carbon dioxide emissions. However, there is currently no process for a direct conversion of H_2S to hydrogen and sulfur owing to a need for a very high temperature ($< 1000\,^\circ C$) to achieve a suitable level of conversion. Consequently, the cost of such a process would be prohibitive because of energy and exotic metallurgy requirements.

The application of a membrane reactor with a hydrogen perm-selective membrane to intensify H_2S decomposition to hydrogen and sulfur at a lower temperature was first considered in the eighties [8,98,99]. However, the expected positive effects on H_2S conversion observed with the early Vycor-based membranes were rather insignificant. Also, a zirconia–silica membrane developed more recently and applied to decomposition of H_2S by Ohashi *et al.* [100] had only Knudsen selectivity and produced similar results.

A catalytic membrane reactor having a tubular ceramic membrane for H_2S decomposition was patented by Vizoso [101]. It was claimed that operation of a membrane reactor between 400 and 700 °C containing a molybdenum sulfide catalyst deposited directly on the surface of the ceramic membrane gave an approximately 20% increase in the conversion of a 4% H_2S stream, though, few experimental details were provided.

A multilayer metallic membrane reactor for H_2S decomposition was patented by Edlund and Friesen [102] and, later, further described by Edlund and Pledger [103,104]. It was reported that at 700 °C, the membrane-assisted thermolysis of 1.5% of H_2S in nitrogen under a total pressure of about 8 atm was practically driven to completion, whereas the equilibrium conversion under similar conditions without hydrogen removal was only 13%. However, the required contact time to reach this conversion was about 12 min. which is far from being practical.

The most recent application of a silica membrane – prepared by counter-diffusion CVD of TMOS and oxygen – to H_2S decomposition was reported by Akamatsu *et al.* [105]. It was claimed that using this membrane and a commercial desulfurisation catalyst at 600 °C, about 70% H_2S diluted in 99% nitrogen was converted in a relatively short residence time of 7 s. Nonetheless, this claim is not supported by a material balance, and the amount of hydrogen passing through the membrane in a very diluted feed stream is not reported.

Clearly, further development is needed before a successful and practical application of a membrane reactor to hydrogen production through catalytic H_2S decomposition could be claimed.

12.8 Conclusions

The application of CVD to the synthesis of silica membranes has been practised for about 20 years now, and silica membranes' ability to transport hydrogen perm-selectively has attracted considerable attention. There is no question that in spite of significant progress in refining the CVD methodology, a correlation between the final property of a membrane and the key experimental parameters during the CVD process is still somewhat obscure. However, substantial progress has recently been made in understanding the mechanism of hydrogen transport and the origin of the perm-selectivity of silica membranes, and a plausible relationship was identified between the membrane-forming amorphous silica structure and the morphological changes during hydrothermal treatment. The bimodal porosity of amorphous silica containing numerous three-dimensional closed cavities having the outside wall formed from the Si-O n-fold rings and the between-grain porosity is thought to be an adequate representation of the silica membrane morphology. Nevertheless, the stability of silica membranes in the presence of steam at high temperatures needs to be further improved before a commercial application can be considered. However, the leading silica membranes – with H_2 permeances in the range of 10^{-7} mol m^{-2} s^{-1} Pa^{-1} and H_2/N_2 selectivity ≥ 200 presenting an adequate hydrothermal performance – have evoked great confidence in the demonstrated feasibility of novel membrane reactor concepts to important processes such as the dehydrogenation of light paraffins or WGSR, thereby forming a solid base for economic feasibility studies and concept optimisation in order to quantify the advantages of membrane-assisted applications over conventional technologies.

References

1. T. Graham, On the absorption and dialytic separation of gases by colloid septa, *Philosophical Transactions of the Royal Society of London*, **156**, 399–439 (1866).
2. W.O. Snelling, *Apparatus for Separating Gases,* US Patent 1174631 (1916).
3. J.S.C. Booth, M.L. Doyle, S.M. Miller, L.A. Scholtz and P.A. Walker, Advanced hydrogen separation via thin supported Pd membranes, in *Hydrogen Energy Progress* XI, **1** (1996).
4. A. Basile, F. Gallucci and S. Tosti, Synthesis, characterization, and application of palladium membranes, in *Membrane Science and Technology*, **13**, *Inorganic Membranes: Synthesis, Characterization and Applications*, R. Mallada and M. Menendez (eds), Elsevier B.V., Rotterdam (2008).
5. A. Ayral, A. Julbe, V. Rouessac, S. Rousaldes and J. Dureand, Microporous silica membrane: basic principles and recent advances, in *Membrane Science and Technology Series*, **13**, *Inorganic Membranes: Synthesis, Characterization and Applications*, R. Mallada and M. Menendez (eds), Elsevier B.V., Rotterdam (2008).
6. C.F. Powell, J.H. Oxley and J.M. Blocher, Jr., (eds), *Chemical Vapour Deposition*, John Wiley & Sons, Inc., New York (1996).
7. H.O. Pierson, *Handbook of Chemical Vapour Deposition (CVD)*, 2nd edn, Noyes Publications/ William Andrew Publishing, Norwich, New York (1999).
8. G.R. Gavalas and C. Megiris, *Synthesis of SiO_2 Membrane on Porous Support and Method of Use of Same*, US Patent 4,902,307 (1990).
9. G.R. Gavalas, R.A. Johnson and V.K. Venkataraman, New developments in hydrogen permselective membranes (US DOE Contract No. DE-AC21-90MC26365) *Proceedings of the Coal-Fired Power Systems 94*, Advances in IGCC and PFBC Review Meeting, **2** (1994).
10. S. Yan, H. Maeda, K. Kusakabe, S. Morooka and Y. Akiyama, Hydrogen-permselective SiO_2 membrane formed in pores of alumina support tube by chemical vapor deposition with tetraethylorthosilicate, *Industrial and Engineering Chemistry Research*, **33**, 2096–2101 (1994).
11. J.N. Armor, Challenges in membrane catalysis, *Chemtech*, **9**, 557–563 (1992).
12. G. Sarraco and V. Specchia, *Catalytic inorganic membrane reactors: Present experience and future opportunities*, Catalysis Reviews, Science and Engineering, **36**(2), 305–384 (1994).
13. G. Saracco, G.F. Versteeg and W.P.M. van Swaaij, Current hurdles to the success of high temperature membrane reactors, *Journal of Membrane Science*, **95**, 105–123 (1994).
14. K. Keizer and H. Verveij, Progress in inorganic membranes, *Chemtech*, 37–41 (1996).
15. A.J. Burggraaf and L. Cot, General overview, trends and prospects, in *Fundamentals of Inorganic Membrane Science and Technology*, A.J. Burgraffand L. Cot (eds), Elsevier, Amsterdam (1996).
16. G.R. Gavalas, C. Megiris and S.W. Nam, Deposition of hydrogen perm-selective silica films, *Chemical Engineering Science*, **44**(9), 1829–1835 (1989).

17. R.J.R. Ulhorn, K. Keizer and A.J. Burggraaf, Gas transport and separation with ceramic membranes, Part II, *Journal of Membrane Science*, **66**, 271–287 (1992).
18. J.C.S. Wu, H. Sabol, G.W. Smith, D.L. Flowers and P.K.T. Liu, Characterization of hydrogen-permselective microporous ceramic membranes, *Journal of Membrane Science*, **96**, 275–287 (1995).
19. R.S.A. de Lange, K. Keizer, and A.J. Burggraph, Aging and stability of microporous sol-gel-modified ceramic membranes, *Industrial and Engineering Chemistry Research*, **34**, 3838–3847 (1995).
20. H. Imai, H. Morimoto, A. Tominaga and H. Hirashima, Structural changes in sol-gel derived SiO_2 and TiO_2 films by exposure to water vapor, *Journal of Sol-Gel Science and Technology*, **10**, 45–54 (1997).
21. R.M. de Vos, *High-Selectivity, High-Flux Silica Membranes for Gas Separation: Synthesis, Transport and Stability*, Thesis Enschede, ISBN 90 365 11410 (1999).
22. M. Asaeda and M. Kashimoto, Sol-gel derived silica membranes for separation of hydrogen at high temperature – separation performance and stability against steam, *Proc. Int. Conf. Inorg. Membr.*, **5**, 172 (1998).
23. C.J. Brinker and G. Cao, Inorganic Polymer-Derived Ceramic Membranes, Final Technical Report to Electric Power Research Institute, EPRI Agreement No. RP8007-15, (1996).
24. S. Kim and G.R. Gavalas, Preparation of H_2 permselective silica membranes by alternating reactant vapor deposition, *Industrial and Engineering Chemistry Research*, **34**, 168–176 (1995).
25. W.A. Bryant, Review, The fundamentals of chemical vapour deposition, *Journal of Material Science*, **12**, 1285–1306 (1977).
26. *Chemical Vapour Deposition*, Surface Engineering Series, **2**, Jong-Hee Parkand T.S. Sudarshan (eds), ASM International, Singapore (2001).
27. *Chemical Vapour Deposition, Precursors, Processes and Applications*, A.C. Jonesand M.L. Hitchman (eds), RSC Publishing, London (2009).
28. E.J. Kim and W.N. Gill, Low pressure chemical vapor deposition of silicon dioxide films by thermal decomposition of tetra-alkoxysilanes, *Journal of the Electrochemical Society*, **142**(2), 976–682 (1995).
29. K. Akamatsu, M. Nakane, T. Sugawara and S. Nakao, Performance under thermal and hydrothermal condition of amorphous silica membrane prepared by chemical vapor deposition, *AIChE J.*, **55** (8), 2197–2200 (2009).
30. S. Araki, N. Mohri, Y. Yoshimitsu and Y. Miyake, Synthesis, characterization and gas permeation properties of a silica membrane prepared by high-pressure chemical vapor deposition, *Journal of Membrane Science*, **290**, 138–145 (2007).
31. S. Gopalakrishman, M. Nomura, T. Sugawara and S. Nakao, Preparation of a multi-membrane module for high-temperature hydrogen separation, *Desalination*, **193**, 230–235 (2006).
32. C.E. Megiris and J.H.E. Glezer, Preparation of silicon dioxide films by low-pressure chemical vapor deposition on dense and porous alumina substrates, *Chemical Engineering Science*, **47** (15/16) 3925–3934 (1992).
33. D. Lee, L. Zhang, S.T. Oyama, S. Niu and R.F. Saraf, Synthesis, characterization, and gas permeation properties of a hydrogen permeable silica membrane supported on porous alumina, *Journal of Membrane Science*, **231**, 117–126 (2004).
34. Y. Gu, P. Hacarlioglu and S.T. Oyama, Hydrothermally stable silica-alumina composite membranes for hydrogen separation, *Journal of Membrane Science*, **310**, 28–37 (2008).
35. S.T. Oyama, D. Lee, P. Hacarlioglu and R.F. Saraf, Theory of hydrogen permeability in nonporous silica membranes, *Journal of Membrane Science*, **244**, 45–53 (2004).

36. S.-S. Kim and B.-K. Sea, Gas permeation characteristics of silica/alumina composite membrane prepared by chemical vapour deposition, *Korean Journal of Chemical Engineering*, **18**(3), 322–329 (2001).
37. G.-J. Hwang, J.-W. Kim, H.-S. Choi and K. Onuki, Stability of a silica membrane prepared by CVD using gamma- and alpha-alumina tube as the support tube in the HI-H_2O gaseous mixture, *Journal of Membrane Science*, **215**, 293–302 (2003).
38. T. Okubo and H. Inoue, Introduction of specific gas selectivity to porous glass membranes by treatment with tetraethoxysilane, *Journal of Membrane Science*, **42**, 109–117 (1989).
39. J.C.S. Chu, J. Breslin, N.S. Wang and M.C. Lin, Relative stabilities of tetramethylorthosilicate and tetraethylorthosilicate in the gas phase, *Materials Letters* **12**(3), 179–184 (1991).
40. Y. Ohta, K. Akamatsu, T. Sugawara, A. Nakao, A. Miyoshi, and S.I. Nakao, Development of pore size-controlled silica membranes for gas separation by chemical vapour deposition, *Journal of Membrane Science*, **315**, 93–99 (2008).
41. T. Nagano, S. Fjjisaki, K. Sato, K. Hatay and Y. Iwamoto, Relationship between the mesoporous intermediate layer structure and the gas permeation property of an amorphous silica membrane synthesized by counter diffusion chemical vapor deposition, *Journal of the American Ceramic Society*, **91**(1), 71–76 (2008).
42. S.-S. Nakao, T. Suzuki, T. Sugawara, T. Tsuru and S. Kimura, Preparation of microporous membranes by TEOS/O3 CVD in the opposing reactants geometry, *Microporous and Mesoporous Materials*, **37**, 145–452 (2000).
43. A.I. Labropoulos, G.E. Romanos, G.N. Karanikolos, F.K. Katsaros, N.K. Kakizis and N.K. Kanellopoulos, Comparative study of the rate and locality of silica deposition during the CVD treatment of porous membranes with TEOS and TMOS, *Microporous and Mesoporous Materials*, **120**, 177–185 (2009).
44. M. Nomura, K. Ono, S. Gopalakrishnan, T. Sugawara and S.-I. Nakao, Preparation of a stable silica membrane by a counter diffusion chemical vapour deposition method, *Journal of Membrane Science*, **251**, 151–158 (2005).
45. S. Gopalakrishnan and J. Diniz da Costa, Hydrogen gas mixture separation by CVD silica membrane, *Journal of Membrane Science*, **323**, 144–147 (2008).
46. Y. Yoshino, T. Suzuki, T. Hisatomi, M. Nomura, S.-I. Nakao and N. Itoh, Development of an all-ceramic module with silica membrane tubes for high temperature hydrogen separation, *Separation Science and Technology*, **43**, 3432–3447 (2008).
47. H.Y. Ha, S.W. Nam, S.-A. Hong and W.K. Lee, Chemical vapour deposition of hydrogen-permselective silica films on porous glass supports from tetraethylorthosilicate, *Journal of Membrane Science*, **85**, 279–290 (1993).
48. K. Kuraoka, T. Kakitani, T. Suetsugu and T. Yazawa, Methanol vapour separation through the silica membrane prepared by the CVD method with the aid of evacuation, *Separation and Purification Technology*, **25**, 161–166 (2001).
49. J. Galuszka and D. Liu, Methane to syngas: development of non-coking catalyst and hydrogen-permselective membrane, in *Studies in Surface Science and Catalysis*, 136, *Natural Gas Conversion VI*, E. Iglesia, J. J. Spiveyand T. H. Fleish (eds), *Proceedings of the 6th Natural Gas Conversion Symposium, Alaska, 2001*, 363–368, Elsevier (2001).
50. J. Galuszka, T. Giddings and I. Clelland, Silica H-membrane thermal and hydrothermal stability, *in Proceedings of the 9th International Conference on Inorganic Membranes, ICIM9* R. Bredesen and H. Raeder (eds), Lillehammer (2006).
51. M. Kanezashi, K. Yada, T. Yoshioka and T. Tsuru, Design of silica networks for development of highly permeable hydrogen separation membranes with hydrothermal Stability, *Journal of the American Chemical Society* **131**, 414–415 (2009).

52. V. Boffa, D.H.A. Blank and J.E. ten Elshof, Hydrothermal stability of microporous silica and niobia-silica membranes, *Journal of Membrane Science*, **319**, 256–263 (2008).
53. M. Kanezashi and M. Asaeda, Hydrogen permeation characteristics and stability of Ni-Doped silica membranes in steam at high temperature, *Journal of Membrane Science*, **271**, 86–93 (2006).
54. C.J. Brinker and G.W. Scherer, Surface chemistry and chemical modification, in *Sol-Gel Science, The Physics and Chemistry of Sol-Gel Processing*, Academic Press, New York, (1990).
55. D. Lee and S.T. Oyama, Gas permeation characteristics of a hydrogen selective hydrogen permeable silica membrane, *Journal of Membrane Science*, **210**, 291–306 (2002).
56. R.M. Barrer and D.E.W. Vaughan, Solution and diffusion of helium and neon in tridymite and cristobalite, *Transactions of the Faraday Society*, **63**, 2275–2290 (1967).
57. E.F. Vansant, P. Voort and K.C. Vrancken, *Characterization and Chemical Modification of the Silica Surface, Studies in Surface Science and Catalysis*, 93, Elsevier Science B.V., Rotterdam (1995).
58. R.K. Iler, *The Chemistry of Silica; Solubility, Polymerization, Colloid and Surface Properties and Biochemistry*, John Wiley & Sons, Inc., New York (1979).
59. A. Anedda, C.M. Carbonaro, F. Clemente, R. Corpino and P.C. Ricci, Raman investigation of surface OH-species in porous silica, *The Journal of Physical Chemistry B*, **107**(13), 661–664 (2003).
60. H. Scholze, Chemical durability of glasses, *Journal of Non-Crystalline Solids*, **52**, 91–103 (1982).
61. H. Imai, M. Yasumori, H. Hirashima, K. Awazu and H. Onuki, Significant densification of sol-gel derived amorphous silica films by vacuum ultraviolet irradiation, *Journal of Applied Physics*, **79** (11), 8304–8309 (1996).
62. A. Julbe, C. Guizard, A. Larbot, L. Cot and A. Giroir-Fendler, The sol-gel approach to prepare candidate microporous inorganic membranes for membrane reactor, *Journal of Membrane Science*, **77**, 137–153 (1993).
63. M. Adachi, K. Okuyama, N. Tohge, M. Shimada, J. Satoh and M. Muroyama, Gas-phase nucleation in an atmospheric pressure CVD process for SiO_2 films using tetraethylorthosilicate (TEOS), *Japanese Journal of Applied Physics, Part I: Regular Papers and Short Notes and Review Papers*, **31**(10A), 1439–1442 (1992).
64. M. Adachi, K. Okuyama, N. Tohge, M. Shimada, J. Satoh and M. Muroyama, Particle generation and film formation in an atmospheric-pressure CVD reactor using the tetraethylorthosilicate (TEOS)/He, TEOS/O_2/He, and TEOS/O_2/He systems, *Japanese Journal of Applied Physics, Part I: Regular Papers and Short Notes and Review Papers*, **32**(5B), 748–751 (1993).
65. S.B. Desu, Decomposition chemistry of tetraethoxysilane, *Journal of the American Ceramic Society*, **72**(9), 1615–1621 (1989).
66. J. Smolik and P. Moravec, Gas phase synthesis of fine silica particles by oxidation of tetraethylorthosilicate vapour, *Journal of Materials Science Letters*, **14**, 387–389 (1995).
67. I.A. Shareef, G.W. Rubloff and W.N. Gill, Role of gas phase reactions in subatmospheric chemical-vapor deposition ozone/TEOS processes for oxide deposition, *Journal of Vacuum Science and Technology B*, **14**(2), 772–774 (1996).
68. S. Giessler, L. Jordan, J.C. Diniz da Costa and G.Q. (Max) Lu, Performance of hydrophobic and hydrophilic silica membrane reactors for the water gas shift reaction, *Separation and Purification Technology*, **32**, 255–264 (2003).
69. Y. Gu and S.T. Oyama, Permeation properties and hydrothermal stability of silica-titania membranes supported on porous alumina substrates, *Journal of Membrane Science*, **345**, 267–275 (2009).

70. M. Nomura, H. Aida, S. Gopalakrishan, T. Sugawara, S. Nakao, S. Yamazaki, T. Inada and Y. Iwamoto, Steam stability of silica membrane prepared by counter-diffusion chemical vapor deposition, *Desalination*, **193**, 1–7 (2006).
71. P.P.A.C. Pex and Y.C. van Delft, Silica membranes for hydrogen fuel production by membrane water gas shift reaction and development of a mathematical model for a membrane reactor, in *Carbon Dioxide Capture for Storage. in Deep Geologic Formations*, David C. Thomas (ed.), Elsevier, B.V., Rotterdam (2005).
72. H.L. Castricum, A. Sah, J.A.J. Geenevasen, R. Kreiter, D.H.A. Blank, J.F. Vente and J.E. ten Elshof, Structure of hybrid organic-inorganic sols for the preparation of hydrothermally stable membranes, *Journal of sol-gel science and technology*, **48**, 11–17 (2008).
73. D.-W. Lee, S.-K. Ihm and K.-H. Lee, Synthesis of a mesoporous titania/silica and its application to a composite membrane with high hydrothermal stability, *Desalination*, **199**, 363–364 (2006).
74. R.S.A. de Lange, J.H.A. Hekkink, K. Keizer and A.J. Burggraaf, Formation and characterization of supported microporous ceramic membranes prepared by sol-gel modification techniques, *Journal of Membrane Science*, **99**, 57–75 (1995).
75. K. Yoshida, Y. Hirano, H. Fuji, T. Tsuru and M. Asaeda, Hydrothermal stability and performance of silica-zirconia membranes for hydrogen separation in hydrothermal Conditions, *Journal of Chemical Engineering of Japan*, **34**(4), 523–530 (2001).
76. W. Liu, B. Zhang, X. Liu and L. Xu, Thermal stability of silica-zirconia membranes, *Chinese Journal of Chemical Engineering*, **14**(1), 31–36 (2006).
77. M. Asaeda, Y. Sakou, J. Yang and K. Shimasaki, Stability and performance of porous silica-zirconia composite membranes for pervaporation of aqueous organic solutions, *Journal of Membrane Science*, **209**, 163–175 (2002).
78. G.P. Fotou, Y.S. Lin and S.E. Pratsinis, Hydrothermal stability of pure and modified microporous silica membranes, *Journal of Materials Science*, **30**, 2803–2808 (1995).
79. R. Igi, T. Yoshioka, Y.H. Ikuhara, Y. Iwamoto and Y. Tsuru, Characterization of Co doped silica for improved hydrothermal stability and application to hydrogen separation membranes at high temperatures, *Journal of the American Ceramic Society*, **91**(9), 2975–2981 (2008).
80. D. Ulhmann, S. Liu, B.P. Ladewig and J.C. Diniz da Costa, Cobalt-doped silica membranes for gas separation, *Journal of Membrane Science*, **326**(2), 316–321 (2009).
81. R.M. de Vos, W.F. Maier and H. Verweij, Hydrophobic silica membranes for gas separation, *Journal of Membrane Science*, **158**, 277–288 (1999).
82. M. Kanezashi, K. Yada, T. Yoshioka and T. Tsuru, Design of silica networks for development of highly permeable hydrogen separation membranes with hydrothermal Stability, *Journal of the American Chemical Society* **131**, 414–415 (2008).
83. H.B. Park and Y.M. Lee, Pyrolitic carbon-silica membrane: a promising membrane material for improved gas separation, *Journal of Membrane Science*, **213**, 263–272 (2003).
84. J.G. Sanchez and T.T. Tsotsis, *Catalytic Membranes and Membrane Reactors*, Wiley-VCH Verlag GmbH, Weinheim (2002).
85. A.M. Champagnie, T.T. Tsotsis, R.G. Minet and E. Wagner, The study of ethane dehydrogenation in a catalytic membrane reactor, *Journal of Catalysis*, **134**, 713–730 (1992).
86. E. Gobina, and R. Hughes, Ethane dehydrogenation using a high-temperature catalystic membrane reactor, *Journal of Membrane Science*, **90**, 11–19 (1994).
87. R. Schafer, M. Noack, P. Kölsch, M. Stöhr and J. Caro, Comparison of different catalysts in the membrane-supported dehydrogenation of propane, *Catalysis Today*, **82**, 15–23 (2003).
88. J. Galuszka, T. Giddings and I. Clelland, Catalytic dehydrogenation of ethane in hydrogen membrane reactor, in *Inorganic Membranes for Energy and Environmental Applications*, A.C. Bose (ed.), Springer, New York (2009).

89. J. Galuszka and T. Giddings, *Book of Abstracts, 9th International Conference on Catalysis in Membrane Reactors*, Lyon, 47–48 (2009).
90. T. Giddings, L. Nossova and J. Galuszka, *Abstracts, Euromebrane 2009 The EMS Conference, Montpellier*, 83 (2009).
91. R. Dittmeyer, V. Hollein and K. Daub, Membrane reactors for hydrogenation and dehydrogenation processes based on supported palladium, *Journal of Molecular Catalysis A*, **173**, 135–184 (2001).
92. S.J. Doong, F. Lau, M. Roberts and E. Ong, GTI's solid fuel gasification to hydrogen program, *The 3rd Natural Gas Technology Conference*, Orlando (2005).
93. R. Bredesen, K. Jordalb and O. Bollandb, High-temperature membranes in power generation with CO_2 capture, *Chemical Engineering and Processing*, **43**, 1129–1158 (2004).
94. M. Bracht, P.T. Alderliesten, R. Kloster, R. Pruschek, G. Haupt, E. Xue, J.R.H. Ross, M.K. Koukou and N. Papayannakos, Water gas shift membrane reactor for CO_2 control in IGCC systems: techno-economic feasibility study, *Energy Conversion and Management*, **38** (Supplement I) S159–S164 (1997).
95. J. Galuszka, T. Giddings, I. Clelland, D. Lu and R. Hughes, *Book of Abstracts, 8th International Conference on Catalysis in Membrane Reactors*, Kolkata, 47–48 (2007).
96. S. Battersby, S. Smart, B. Ladewig, S. Liu, M.C. Duke, V. Rudolpf, J.C. Diniz da Costa, Hydrothermal stability of cobalt silica membranes in a water gas shift membrane reactor, *Separation and Purification Technology*, **66**, 299–305 (2009).
97. A. Criscuoli, A. Basile, E. Drioli and O. Loiacono, An economic feasibility study for water gas shift membrane reactor, *Journal of Membrane Science*, **181**, 21–27 (2001).
98. T. Kameyama, M. Dokiya, M. Fujishige, H. Yokokawa and K. Fukuda, Possibility for effective production of hydrogen from hydrogen sulphide by means of a porous vycor glass membrane, *Industrial and Engineering Chemistry Fundamentals*, **20**, 97–99 (1981).
99. T. Kameyama, M. Dokiya, M. Fujishige, H. Yokokawa and K. Fukuda, Production of hydrogen from hydrogen sulphide by means of selective diffusion membranes, *International Journal of Hydrogen Energy*, **8**, 5–13 (1983).
100. H. Ohashi, H. Ohya, M. Aihara, Y. Negeshi and S.I. Semenova, Hydrogen production from hydrogen sulphide using membrane reactor integrated with porous membrane having thermal and corrosion resistance, *Journal of Membrane Science*, **146**, 39–52 (1998).
101. R.B. Vizoso, *Catalytic membrane reactor for breaking down hydrogen sulphide into hydrogen and sulfur and separating the products of this breakdown*, US Patent Application US2004/0141910 A1 (2004).
102. D.J. Edlund and D.T. Friesen, *Hydrogen-permeable composite metal membrane and uses thereof,* US Patent 52175006 (1993).
103. D.J. Edlund and W.A. Pledger, Thermolysis of hydrogen sulfide in a metal-membrane reactor, *Journal of Membrane Science*, **77**, 255–264 (1993).
104. D.J. Edlund and W.A. Pledger, Catalytic platinum-platinum based membrane reactor for removal of H_2S from natural gas stream, *Journal of Membrane Science*, **94**, 111–119 (1994).
105. K. Akamatsu, M. Nakane, T. Sugawara, T. Hattori and S. Nakao, Development of a membrane reactor for decomposing hydrogen sulphide into hydrogen using a high- performance amorphous silica membrane, *Journal of Membrane Science*, **325**, 16–19 (2008).

13

Membranes Prepared via Molecular Layering Method

A.A. Malygin[1], A.A. Malkov[1], S.V. Mikhaylovskiy[1], S.D. Dubrovensky[1], N.L. Basov[2], M.M. Ermilova[2], N.V. Orekhova[2] and G.F. Tereschenko[2]

[1]St. Petersburg State Institute of Technology, St. Petersburg, Russia
[2]Topchiev Institute of Petrochemical Synthesis, Moscow, Russia

13.1 Introduction

The application of selectively permeable inorganic membranes in catalytic reactors has been under the intensive study for the past two decades. Such membranes can be used for controlling the introduction and distribution of the reagents in a reactor (membrane contactors) and for selectively extracting one of the reaction products (membrane extractors). The insertion of membranes in a catalytic reactor leads to an increase in the productivity, and in many cases also gives a rise of selectivity against the target product. Owing to the high thermal stability and chemical resistance inorganic membranes are more suitable for catalytic processes. As it is known, the membranes can either be catalytically inert or serve as a component of composite membrane catalyst.

The continuous (compact) palladium-containing membranes were the first to be used in the membrane catalysis due to the palladium ability of the selective hydrogen transfer and its catalytic activity in reactions with hydrogen [7].

Micro- and nanoporous inorganic membranes can be produced on the basis of a wide range of materials (ceramics, metals, and carbon) which are far less expensive and have a longer lifetime as compared to palladium. The main drawback of the membranes is their failure to reach a 100% selectivity. However, the membranes with the pore diameter less than 2 nm can secure the separation factors high enough to isolate the target product. The advantage of using the

Membranes for Membrane Reactors: Preparation, Optimization and Selection, Edited by Angelo Basile and Fausto Gallucci

membranes in a catalytic reactor may be actual also in the case of the absence of high selectivity of permeability of reagents [6].

The key problem in the synthesis of micro- and nanoscale inorganic membranes for gas separation is the development of methods for modification of macro- and mesoporous inorganic materials (e.g., porous oxides of aluminium, silicon, titanium, zirconium) in order to give them the desired selectivity without a decrease in productivity. What is more, the thermal stability and chemical resistance as well as the phase stability in wet and corrosive medium present the significant problems.

The most widely used methods for forming selective membranes on the porous substrates are the sol-gel method [11], the method of chemical vapour deposition (CVD) [12,13], the pyrolysis of silicon-containing polymers [14,15], along with the synthesis of mesoporous films [16–19].

In the past few decades the molecular layering method (ML) became more and more popular for optimisation of structural parameters and catalytic activity of MRs. It is free of most of the disadvantages of other methods for the formation of coatings with a desired composition and structure.

In the suggested review special attention is given to perspectives of application of the molecular layering nanotechnology to establish the control over the structural characteristics of MRs and their catalytic activity. Examples are given which demonstrate the application of inorganic membranes modified by the ML method in catalytic MRs.

13.2 Molecular Layering: Principles, Synthesis Possibilities and Fields of Application

As far as the ML is concerned, it is rather extensively covered in scientific literature. In the papers of western authors this method has been called either atomic layer epitaxy (ALE) or atomic layer deposition (ALD) [20–23].

The principles of the ML are based on the *framework hypothesis* of V.B. Aleskovskii. They were originally put forward by V.B. Aleskovskii and S.I. Koltsov in the beginning of the 1960s [22,24–31].

The main idea of the ML method consists in the sequential growth of monolayers of structural units of the preassigned chemical composition and texture on the surface of a solid matrix, by the realisation of chemical reactions between the functional groups of a solid and the reagents brought to it under conditions being far from equilibrium.

The scheme of process of the chemical assembly of nanostructures on the surface of solid matrix by the ML method, given in Figure 13.1, and analysis of the available experimental data testify, that the ML method makes it possible to synthesise on the surface of a solid phase matrix the nanostructures of different chemical composition (monolayers, including multicomponent monolayers, see Figure 13.1a, b) and carry out the chemical assembly of the surface nano-, micro- and macrostructures. It is achieved through the repetitive alternation of chemical reactions according to predefined program (see Figure 13.1a, b). It is necessary to underline that the main requirement to be met for carrying out reproductive synthesis by the ML method consists in accomplishing the different stages of the reactants interaction (AC_4, AB_4, NB_4, NC_4, MC_4, Figure 13.1) with functional groups (FG) of solid body (B, C, Figure 13.1) under conditions maximal far from equilibrium.

It should be noted that the ML method provides the formation of nanolayers on a solid substrate with the accuracy of one monomolecular layer (Figure 13.1). The thickness of nanolayer depends

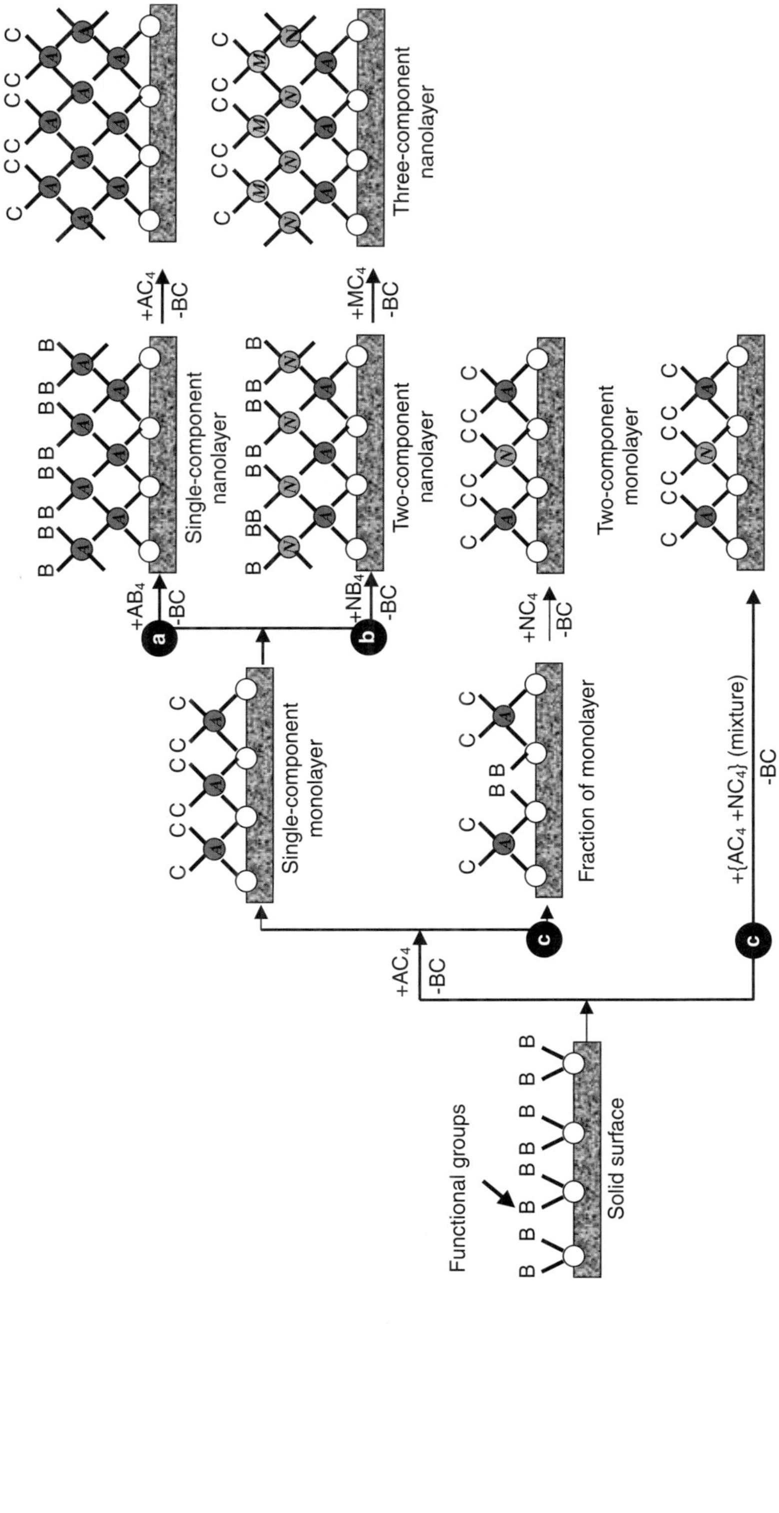

Figure 13.1 *The chemical assembly of solid surface by the molecular layering method*

not on the time of reactants flow over the solid surface, but on the number of the ML cycles. Each cycle includes a set of a certain chemical reactions. Thus, the process of the ML comprises the elements of self organisation occurring during the formation of the first and subsequent layers.

Reactions of highly volatile halogenides with the OH groups on the surface of solid matrices have been studied rather extensively. As an example, let us consider the surface chemical reactions which take place during formation of a monolayer of certain element oxychloride groups on the surface of silica. In each case the reaction is carried out until the entire replacement of all the accessible reactive hydroxyl groups under experimental conditions:

$$2(\equiv\text{Si-OH}) \xrightarrow[-2\text{HCl}]{+\text{TiCl}_4} (\equiv\text{Si-O-})_2\text{TiCl}_2 \tag{13.1}$$

$$3(\equiv\text{Si-OH}) \xrightarrow[-3\text{HCl}]{+\text{PCl}_3} (\equiv\text{Si-O-})_3\text{P} \tag{13.2}$$

$$2(\equiv\text{Si-OH}) \xrightarrow[-2\text{HCl}]{+\text{CrO}_2\text{Cl}_2} (\equiv\text{Si-O-})_2\text{CrO}_2 \tag{13.3}$$

In this way the new functional groups were synthesised, which can be generally represent as:

$$(>\text{R-O-})_n\text{ECl}_{m-n}\text{O}_k$$

where ($>$R-O) is a fragment of the solid matrix surface (organic or inorganic, crystalline or amorphous); E is an element in the composition of a new functional group (Ti, V, Cr, Zn, Fe, Al, W, Ta, Zr, B and others); O is oxygen; n, m, k are stoichiometric coefficients ($k = 0$, provided that chloride is used as low molecular reagent and not oxychloride). The synthesis of such structures was considered in detail in the works of V.B. Aleskovskii, S.I. Koltsov and others [24–27].

Depending on the composition of grafted groups and their chemical properties, there are the several ways for the secondary transformations on the surface. For example, to obtain the oxide nanolayer on the surface from ($\equiv$Si-O-)$_2$TiCl$_2$ groups, it is necessary to chose water vapour, the reaction proceeding on the surface can be presented schematically, for example as follows:

$$(\equiv\text{Si-O-})_2\text{TiCl}_2 + 2\text{H}_2\text{O} \rightarrow (\equiv\text{Si-O-})_2\text{Ti(OH)}_2 + 2\text{HCl} \tag{13.4}$$

At that stage of the ML process we again obtain the hydroxylated surface.

Then, the hydroxylated surface is again treated with vapours of the appropriate chloride, water vapour etc. Using the repeated and subsequent treatment of silica with titanium tetrachloride and water in accordance with the principles of the ML, it is possible to form on the silica surface the titanium oxide nanolayer, the thickness of which being determined not by the time of the reagent supply, but by the number of the ML cycles.

Depending on the chemical composition and structure of monolayer of the new functional groups, different chemical reactions may be chosen after the completion of the first cycle, to carry out the second and subsequent stages of the synthesis.

For example, to obtain hydroxyl groups on the surface which are able to react with chromium oxychloride (Equation 13.3) the sample may be processed by a reducing agent – hydrogen. The reaction taking place on the surface can be represented as follows:

$$(\equiv\text{Si-O-})_2\text{CrO}_2 + 3/2\text{H}_2 \rightarrow (\equiv\text{Si-O-})_2\text{CrOH} + \text{H}_2\text{O} \tag{13.5}$$

To create the multicomponent monolayers as well as nanostructures containing the alternative mono(nano) layers with the set chemical nature, it is necessary to apply different reagents on the different stages of the ML (see Figure 13.1b, c). The choice and sequence of the reagents supply is defined by the properties of new functional groups on the surface of the solid matrix synthesised in the previous cycle of the ML [30,31]. Synthesis of the multicomponent element oxide monolayers can be realised in accordance with the schemes shown in Figure 13.1: (i) by the consequent treatment of the matrix with the vapour of elements chlorides, (ii) by the replacement of the surface element oxychloride groups in reactions with chlorides of elements of different chemical nature, (iii) by treatment of the substrate with the mixture of different elements [30].

So far, the ML method has been used for the synthesis of oxide layers of titanium, aluminium, chromium, silicon, phosphorous and a number of other elements on the surface of the porous oxides of silicon and aluminium. The volatile chlorides or oxychlorides of specified elements and water vapour are used as low molecular reagents [24–35].

For the growth of the oxide layer, different types of surface chemical reactions are used, and not only the replacement of proton in the hydroxyl groups, but also, for example, redox reactions [30,35].

As a result of cyclic alternate treatment of porous substrate with the vapour of element chloride and the water (or other agents), the formation of the conform oxide layer is observed on the surface of the pores. And the regular changing of porous structure of the substrate takes place with the increase of its thickness, that is, the specific surface and the volume of pores are decreased. Depending on the chemical behaviour of the reagents and the number of the ML cycles, it is possible to control the average radius of pores with accuracy to several nanometres [27,31–33].

It was shown by example of experimental data and theoretical considerations for cyclic alternative treatment of the silica surface that the fitting out of silicon oxygen framework by the element oxide nanostructures occurred. Due to that fact, it is possible to directionally control the porous structure of the used substrate (Figure 13.2).

Using redox reactions it was also possible to carry out the synthesis of two-component oxide layers on the surface of silica.

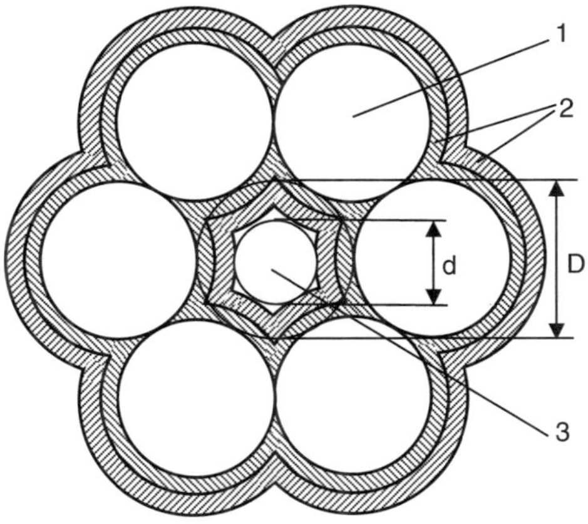

Figure 13.2 *Scheme of the consequent decrease of the specific surface volume and the average radius in the process of titanium oxide nanostructures layering on the surface of the silica gel globule [24]. 1 – globule, 2 – layered titanium oxide nanostructures, 3 – pore, D – diameter of the original pore, d – diameter of the pore after layering*

Table 13.1 *The chemical composition of the products of repeated alternate treatment of silica gel with CrO_2Cl_2 and hydrogen*

ML cycle number	Cr^{3+} content (mol g^{-1})	Volume of pores (cm^3 g^{-1})	Specific surface area (m^2 g^{-1})
0		0.96	246
I	1.04	0.79	229
II	2.04	0.77	209
III	3.19	0.64	193
IV	4.84	—	166
V	6.32	0.54	155
VI	8.23	—	133

As an example, the multiple cyclic alternate treatment of silica by vapours of CrO_2Cl_2, PCl_3 and H_2O results in the regular increasing of chromium and phosphorous content in the samples.

Thus, analysis of the literature data and our own results of the experimental work testifies that the ML method can be applied for fine control of the structural characteristics of porous systems (Tables 13.1–13.4). This is of certain interest for optimisation of transport pores in ceramic MRs.

At the same time it is necessary to take into consideration a number of points related to the preparation and identification of composition as well as to the structure of the formed in the process of the ML surface functional groups and nanostructures. Nonobservance of the main principles of the ML in the process of synthesis can cause alternative reactions to proceed depending on the technological modes, properties of the substrate and the reagents used (Figure 13.3).

The other important point is the reliability of identification of the surface groups. In a number of cases even the up to date physicochemical testing methods do not allow one to interpret

Table 13.2 *Properties of commercial silica gels KSK-2 (Russia), initial and modified by titanium oxide nanolayers*

Number of ML cycles	Content of Ti^{4+} (mmol g^{-1})	Density, ρ (g cm^{-3})	Globule diameter, d (nm)	Specific surface area, S_{BET} (m^2 g^{-1})	Pore volume, V (cm^3 g^{-1})	Pore radius, r (nm)
0	—	2.20	7.2	377	0.98	0.52
1	0.65	2.25	7.6	349	0.88	0.50
4	1.90	2.35	9.1	281	0.69	0.49
8	3.05	2.45	10.7	228	0.54	0.48

Table 13.3 *Chemical composition of the products of repeated alternate treatment of silica gel with vapours of CrO_2Cl_2, PCl_3 and H_2O*

Sample	Content (mmol g^{-1})				ΔCr_o	ΔP
	Cr(VI)	Cr(III)	Cr_o	P(V)		
Cr	0.98	0.05	1.03	0	1.03	0
CrP	0	0.82	0.82	1.0	—	1.0
CrPCr	1.11	1.03	2.14	1.0	1.32	—
2(CrP)	0.21	1.83	2.04	1.73	—	0.73
2(CrP) Cr	0.92	2.22	3.14	1.72	1.10	—

Table 13.4 Changes in chemical composition and specific surface caused by repeated alternate treatment of Al_2O_3 by vapours of CrO_2Cl_2, PCl_3, H_2O

Sample	Content (mmol g^{-1})				Specific surface area, S_{BET} (m^2 g^{-1})
	P(V)	P(III)	P_{sum}	Cr_{sum}	
0	0	0	0	0	280
1 (CrP)	0.95	0.02	0.94	1.30	172
2 (CrP)	1.7	0.06	1.76	2.1	128
3 (CrP)	2.27	0.06	2.33	2.65	58

unambiguously their structure and composition. And it should be kept in mind, that they can have an essential influence on the materials functional properties (catalytic, sorptional, and others), besides that they affect the route of the transformations on the subsequent ML stages. Therefore, special attention is given to application of the quantum-chemical approach for identification and prediction of the composition and structure of local centres on the surface in the process of the ML.

For example, in case of the layered vanadium oxide systems, the catalytic activity of isolated trifunctional structures (≡VO) is one order higher than the activity of mono[-$VO(OH)_2$] and di[-VO(OH)] functional groups [36,37]. The latter two also can enter the reactions of polycondensation with formation of oligomers [38,39]. The stability of synthesised groups in the regard to the molecular reagents (in particularly water) also depends on the functionality [40,41].

In order to choose the synthesis conditions that can provide the synthesis of functional groups of desired structure and composition, it is necessary to have the means for identification of the local structure of the surface centres. To solve pointed task for vanadium oxide systems the

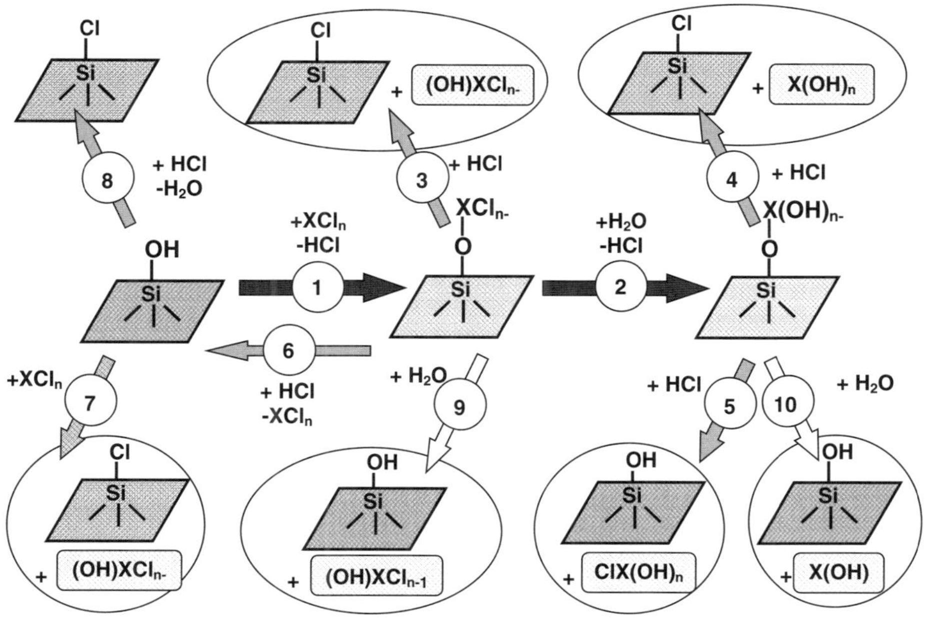

Figure 13.3 Possible routes of chemical transformations of ML process on the surface of silica

chemical analysis [39,41], vibrational spectroscopy [38–40] and NMR [36,42] were used. However, due to the complicated character of the objects even the combination of different physicochemical methods of investigation could not solve the problem of identification unambiguously.

In connection with this, it seems to be promising to use quantum chemistry as one of the key means of the identification [40,43,44]. Using the clusters models, it was shown that the presence and the number of the covalent bonds of vanadium atoms with the silica surface should be reflected in the vibration spectra in the form of the peaks of stretch vibrations in the range of 900–1100 cm^{-1}. Taking into consideration the anharmonic effects, it is possible to achieve the quantitative accordance between the calculated and the experimentally obtained data [41] related to the position of the single band νSi-O-V of 960 cm^{-1} for monofunctional centres. As to the functional groups containing the di- and tridentate vanadium oxide groups, they reveal the splitting into symmetrical (980 and 1000 cm^{-1}) and asymmetrical (935 and 925 cm^{-1}) stretch modes [44], wave numbers of which are rather close to the numerous published data [36–40,43], related to the presence of the bands in the range of 940–910 cm^{-1} in the vibration spectra of vanadium containing products. The analysis of the values of matrix elements of fundamental transitions shows that the IR spectra are characterised only by asymmetric modes and Raman spectra – by symmetric ones. Besides vibrations of bonds between modificator and substrate, according to the calculations it is also possible to determine in the spectrum the stretch and bend vibrations of centres V-O-H: the O-H stretch in the region of 3620–3660 cm^{-1}; the V-OH stretches (760–780 cm^{-1}) and the V-O-H bend in the range of 600–650 cm^{-1}.

Thus, the quantum chemical predictions can be served as a basis for the identification of the local structure of the synthesised element oxide systems. However, the application of the models to prognosis the routes of the local chemical transformations, including the destructive ones, is also of the certain interest [45].

That line of the research can lead to the creation of a kind of *virtual chemical stand*, allowing calculations to determine the optimal modes of the process of the ML and reveal the ways to control the composition and properties of the obtained products.

The study of the products obtained by the ML method gave the possibility to discover a number of the fundamental regularities which are distinctive from the ones of the conventional synthetic techniques used for preparation of the similar structures [26,27,34,35,46].

13.3 Optimisation of MR Structure and Catalytic Properties by the ML Method

One of the first examples for the application of the ML to modify in a well directed way the membrane porosity was the layering of aluminium oxide on the membrane from macroporous anode aluminium oxide (AAO) – Anodisc [47–50]. The other type of nanostructured membranes obtained by the ML is described in [51]. An interesting modification of the ML method for regulating the pores size of the structured mesoporous membranes from the silicon oxide was suggested in [52].

The authors of the present review studied the effect of the number of cycles (from 50 to 450) in the process of titanium oxide layering by the ML on the gas permeability of the individual gases through the oxide aluminium asymmetrical tubes (PaL-Schumacher, Germany) with the average transport pores diameter of 60 nm. In the contrast to the above mentioned examples, the ML method was applied for modification of the complicated system of multilayer asymmetrical membrane.

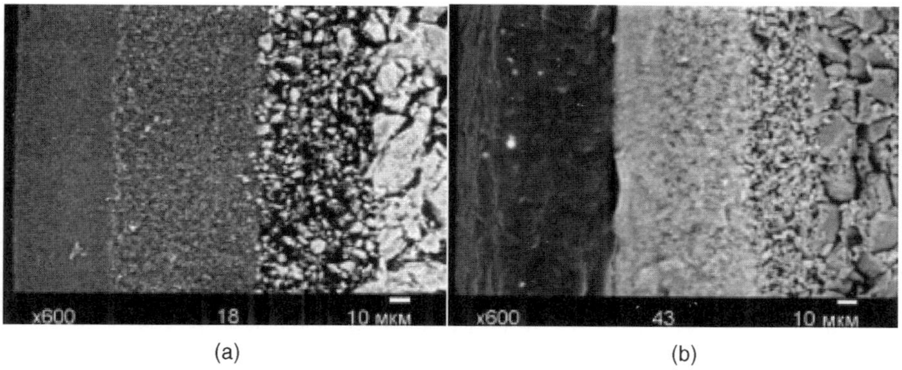

Figure 13.4 *SEM image of the cross section of the fragments of the initial asymmetric membrane of aluminium oxide (a) and a membrane modified by titanium oxide for 450 cycles (b)*

As is seen from Figure 13.4, which gives electron micrographs of the cross section of the fragments of the initial membrane and the membrane modified by the 450 cycles of titanium oxide molecular layering, the main mass of the titanium oxide is layered on the two the most narrow porous layers of the membrane. This is evident from the data of layer X-ray microanalysis (see Figure 13.5).

Applying similar approaches, it is possible to use the ML method in order to create catalytic systems, including multi component systems, *in situ* on the surface and in the pores of the membranes. Following the principles of the molecular layering, it is possible to carry out

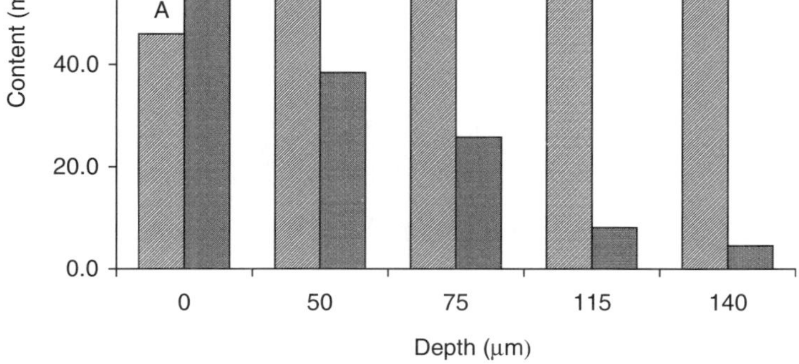

Figure 13.5 *The change of the content of the aluminium (A) and titanium (B) oxides (according to X-ray spectroscopic microanalysis) along the cross section of the asymmetrical OAO membrane modified by 450 cycles of titanium oxide layering, using the ML method*

the chemical assembly of multilayered structures in which monolayers of different nature are altered.

Authors of the present article were the first to achieve success in applying the ML method for obtaining nanostructured composite membrane catalysts by layering the metal structures on the surface of the porous substrate [53]. Low dimensional chromium – phosphorous oxide structures were used as catalytically active compounds. The modification of the porous membrane surface from stainless steel of the trademark SS316L was carried by repeated three times subsequent treatment with the vapours of CrO_2Cl_2, PCl_3 and water, and subsequent treatment at 773 K in the nitrogen flow.

Methane partial oxidation, carried out at the temperatures of 500–700 K in the membrane reactor divided into two chambers by Pd-Ru alloy foil membrane pressed closely to the composite CrP oxide membrane, gave [53] formaldehyde and hydrogen as main products with the traces of carbon dioxide at the temperatures above 650 K.

Recently we have applied successfully the ML method for obtaining the nanostructured composite membrane catalysts by layering on the surface of the porous substrate the low-scale vanadium–phosphorous oxide structures [54–56].

Vanadium is the key component of many catalysts of selective oxidation [57]. Vanadium oxide both in its pure form and in a supported form provides the high selectivity of the formaldehyde synthesis under the oxidative dehydrogenation of methanol [58]. The vanadium surrounding plays an important role in its activity and selectivity in this reaction. It was shown [59], that the highly dispersed vanadium oxide layered on the mesoporous silicon oxide was very selective in the reaction of methanol oxidative transformation into formaldehyde. Vanadium–phosphorous structures are also of the great interest in the reactions of selective oxidation [60].

To prepare a membrane catalyst of oxidative dehydrogenation of methanol into formaldehyde, vanadium and vanadium–phosphorous oxides were synthesised in the pores of the asymmetrical tube membrane from α-Al_2O_3 with the thin layer of narrow porous γ-Al_2O_3. The length of the working part of the tube was 13 cm, the inner diameter was 7 mm. The average diameter of the narrow pores was 50 nm. The synthesis was carried out by the ML method *in situ* in the membrane reactor at 453 K by the vapours of $VOCl_3$ and water, or $VOCl_3$, PCl_3 and water, by pumping their vapours through the membrane pores. The hydrolysis in the vapour phase of chlorine-containing substances leads to the replacement of the chlorine atoms by hydroxyl groups with their subsequent condensation.

Two kinds of oxide layers were synthesised on the surface of the aluminium oxide (samples V-O = Cat 1 and V-O-P = Cat 2). Chemical reactions which take place may be represented by the following schemes:

Sample V-O

$$n(=\text{Al-OH}) + \text{VOCl}_3 \rightarrow (\text{Al-O-})_n\text{VOCl}_{3-n} + n\text{HCl} \quad (13.6)$$

$$(\text{Al-O-})_n\text{VOCl}_{3-n} + (3-n)\text{H}_2\text{O} \rightarrow (\text{Al-O-})_n\text{VO(OH)}_{3-n} + (3-n)\text{HCl (Cat-1)} \quad (13.7)$$

Sample V-O-P

$$3(=\text{Al-OH}) + \text{VOCl}_3 \rightarrow (=\text{Al-O-})_3\text{VO} + 3\text{HCl} \quad (13.8)$$

$$(=\text{Al-O-})_3\text{VO} + \text{PCl}_3 \rightarrow (=\text{Al-O-})_3\text{V} \leftarrow \text{OPCl}_3 \quad (13.9)$$

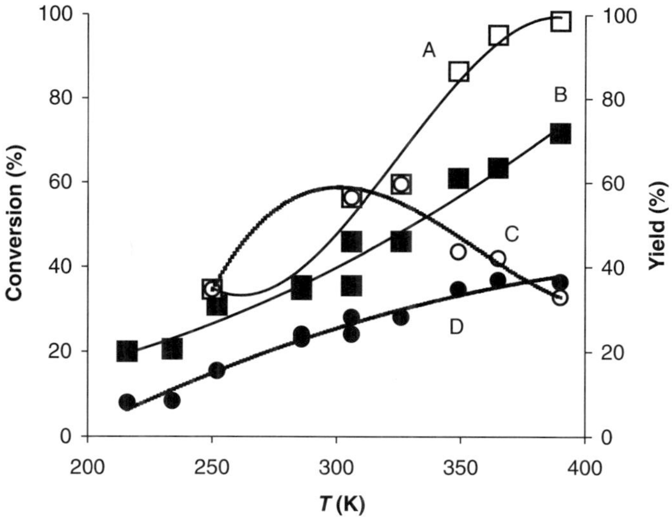

__Figure 13.6__ The temperature dependences of the conversion of methanol (A, B, solid lines, squares) and the yield of formaldehyde (C, D, dotted line, circles) on the phosphorous composite catalysts obtained by the ML method at diffusion though the membrane of methanol (A, C, open symbols) or air (B, D, solid symbols)

$$(=\text{Al-O-})_3\text{V} \leftarrow \text{OPCl}_3 + \text{H}_2\text{O} \rightarrow (=\text{Al-O-})_3\text{V} \leftarrow \text{OP(OH)}_3 + 3\text{HCl (Cat-2)} \qquad (13.10)$$

The tests of the composite membrane catalysts of the both types in the reaction of the oxidative dehydrogenation of methanol was carried out using two ways of introducing the reagents into the catalyst layer, that is, by diffusion of methanol through the membrane (MR-M) and by the diffusion of air (MR-O) and in the direct flow of the streams containing methanol and oxygen.

Figure 13.6 gives as an example of the temperature dependence of the methanol conversion and the yield of formaldehyde with using vanadium–phosphorous composite catalyst (V-O-P) for the two ways of the supply of the reagents.

As one can see from the dependences shown in the Figure 13.6, the mode of MR-M provided the greater reaction yield than the MR-O. This so-called *asymmetry effect* [61] developed to greater extend on the membrane V-O-P than on the V-O.

The results described in the works [7,54–56] for the first time showed, that the ML method is simple and flexible way of modification of the porous inert membrane substrate and provides the layering of the active component not only on the surface but also in micro- and nanopores.

Hence, the submitted results allow us to draw the conclusion that using the ML nanotechnology, it is possible in the united chemical–technological cycle to carry out the assembly of nanostructures of various functional purpose on the surface of porous support: for the regulating of the pores size of substrates and for providing the system with the desired catalytic properties.

References

1. Zaman J., Chakma A., *J. Membr. Sci.*, **92**, 1 (1994).

2. Saracco G., Specchia V., *Catal. Rev. Sci. Eng.*, **36** (2), 305 (1994).
3. Coronas J., Santamaria J., *Catal. Today*, **51**, 377 (1999).
4. Julbe A., *J. Membr. Sci.*, **181**, 3 (2001).
5. McLeary E.E., Jansen J.C., *Top. Catal.*, **29** (1/2), 85 (2004).
6. Specchia S., Fino D., Saracco G., Speccia V., Inorganic membrane reactors. In: Structured Catalysts and Reactors, 2nd edn, A. Cybulski and J. A. Moulijn (eds), Taylor and Francis, London, p. 615 (2005).
7. Gryaznov V.M., Ermilova M.M., Orekhova N.V., Tereschenko G., Reactors with metal and metal-containing membranes. In: Structured Catalysts and Reactors, 2nd edn, A. Cybulski and J. A. Moulijn (eds), Taylor and Francis, London, p. 579 (2005).
8. Larbot A., Fabre J.P., Guizard C., Cot L.J., *Membr. Sci.*, **39**, 203 (1988).
9. Leenaars A.F.M., Keizer K., *J. Mat. Sci.*, **19**, 1077 (1984).
10. Keizer K., Ulhorn R.J.R., Van Vuren R.J., Burggraaf A.J., *J. Membr. Sci.*, **39**, 285 (1988).
11. Meinema H.A., Dirrix R.W.J., Brinkman H.W., Terpstra R.A., Kösters P.H., *Interceramica*, **54** (2), 86 (2005).
12. Gavalas G.R., Megeris C.E., Nam S.W., *Chem.Eng.Sci.*, **44**, 68 (1989).
13. Tsaspatsis M., Kim S., Nam S.W., Gavalas G., *Ind. Eng. Chem. Res.*, **30**, 2152 (1991).
14. Koresh J.E., Sofer A., *Sep. Sci. Technol.*, **18**, 723 (1983).
15. Lee K.H., Khang S.J., *Ceram. Eng. Sci. Proc.*, **8**, 85 (1987).
16. Beck J.S., Vartuli J.C., Roth W.J., Leonowicz M.E., Kresge C.T., Schmitt K.D., Chu C.T.-W., Olson D.H., Sheppard E.W., McCullen S.B., Higgins J.B., Schlenker J.L., *J. Am. Chem. Soc.*, **114**, 10834 (1992).
17. C.-M. Yang, B. Zibrowius, F. Schüth, *Chem. Commun.*, **2003**, 1772 (2003).
18. Y. Lu, R. Ganguli, C.A. Drewien, M.T. Anderson, C.J. Brinker, W. Gong, Y. Guo, H. Soyez, B. Dunn, M.H. Huang, J.I. Zink, *Nature*, **389**, 364 (1997).
19. van Water L.G.A., Maschmeyer T., *Top. Catal.*, **29**, 67 (2004).
20. Suntola T., Antson M.J.,Finland Patent 51742 (1977).
21. Nishizava J., *J. Electrochem. Soc.*, **132**, 1197 (1985).
22. Malygin A.A., Smirnov V.M., *Solid State Technol.*, **2002**, 14 (2002).
23. Puurunen R., *J. Appl. Phys.*, **97**, 121301 (2005).
24. Koltzov S.I., Aleskovskii V.B., *J. Phys. Chem.* (Russian edn), **42**, 1210 (1968).
25. Aleskovskii V.B., *J. Appl.Chem.* (Russian edn), **47** (10), 2145 (1974).
26. Aleskovskii V.B. *Khimia nadmolecularnykh soedinenii.* Leningrad State University Publishing, Leningrad, 256 pp (1996).
27. Dubrovenskii S.D., Malkov A.A., Malygin A.A., The chemical basis of surface modification technology of silica and alumina by molecular layering method. In: *Adsorption on New and Modified Inorganic Sorbents.* A. Dabrowski and V. A. Tertykh (eds), Studies in Surface Science and Catalysis, vol **99**. Elsevier, Amsterdam, **99**, 213 (1996).
28. Koltzov S.I., Aleskovskii V.B. *J. Appl. Chem* (Russian ed.), 1967. V. **40**, No 4, P. 907.
29. Koltzov S.I., Aleskovskii V.B., Silicagel, ego stroenie I khimicheskie svoystva (in Russian). *Goskhimizdat*, **1963**, 10 (1963).
30. Malygin A.A., *J. Common Chem.* (Russian edn), **72** (4), 617 (2002).
31. Aleskovskii V.B., *Vestnik USSR Acad. Sci.*, **1975** (6), 48 (1975).
32. Leboda R., Turov V.V., Marcyniak M., Malygin A.A., Malkov A.A., *Langmuir*, **15**, 8441 (1999).
33. Koltzov S.I., Malygin A.A., Aleskovskii V.B., *J. Phys.Chem.* (Russian edn), **54** (9), 2331 (1980).
34. Malygin A.A., *J. Ind. Eng. Chem.*, **12** (1), 1 (2006).
35. Malygin A.A., Rossiyskie Nanotekhnologii (Russian edn), **2** (3/4), 87 (2007).
36. Wachs I.E., *Catal. Today*, **100**, 79 (2005).
37. Weckhuysen B.M., Keller D.E., *Catal. Today*, **78**, 25 (2003).
38. Lee E.L., Wachs I.E., *J. Phys. Chem. C*, **111** (39), 14410 (2007).

39. Osipenkova O.V., Malkov A.A., Malygin A.A., *J. Common Chem.* (Russian edn), **66** (1), 7 (1996).
40. Keller D.E., Visser T., Soulimani F., Koningsberger D.C., Weckhuysen B.M., *Vibrat. Spectrosc.*, **43**, 140 (2007).
41. Rice G.L., Scott S.L., *Langmuir*, **13**, 1545 (1997).
42. Das N., Eckert H., Hangchun Hu, Wachs I.E., Walzer J.F., Feher F.J., *J. Phys. Chem.*, **97** (31), 8240 (1993).
43. Magg N., Immaraporn B., Giorgi J.B., Schroeder T., Baumer M., Dobler J., Wu Z., Kondratenko E., Cherian M., Baerns M., Stair P.C., Sauer J., Freund H.J., *J. Catal.*, **226**, 88 (2004).
44. Dubrovenskii S.D., Kvantovo-khimicheskoe modelirovanie element-oksidnykh structur na poverkhnosti kremnezema (in Russian). In: Khimia Poverkhnosti I Nanotekhnologiya Vysokoorganizovannykh Veshchestv, St Petersburg State Institute of Technology, St Petersburg, p. 253 (2007).
45. Dubrovenskii S.D., Kulakov N.V., Malygin A.A., *J. Common Chem.*, **79** (2), 177 (2006).
46. Gusarov V.V., Ishutina Zh.N., Malkov A.A., Malygin A.A., *Doclady RAS*, **357**, 203 (1997).
47. Ott A.W., Klaus J.W., Johnson J.M., George S.M., *Chem. Mater.*, **9**, 707 (1997).
48. Pellin M.J., Stair P.C., Xiong G., Elam J.W., Birell J., Curtiss L., George S.M., Han C.Y., Iton L., Kung H., Kung M., Wang H.-H., *Catal. Lett.*, **102**, 127 (2005).
49. Stair P.C., Marshall C., Xiong G., Feng H., Pellin M.J., Elam J.W., Curtis L., Iton L., Kung H., Kung M., Wang H.-H., *Top. Catal.*, **39**, (3/4), 181 (2006).
50. Velleman L., Triani G., Evans P.J., Shapter J.G., Losic D., *Microporous Mesoporous Mater.*, **126**, 87 (2009).
51. Triani G., Evans P.J., Attard D.J., Prince K.E., Barlett J., Tan S., Burford R.P., *J. Mater. Sci.*, **16**, 1355 (2006).
52. McCool B.A., DeSisto W.J., *Ind. Eng. Chem. Res.*, **43**, 2478 (2004).
53. Malygin A.A., Ermilova M.M., Gryaznov V.M., Orekhova N.V., Malkov A.A., *Desalination*, **144**, 433 (2002).
54. Orekhova N.V., Ermilova M.M., Malygin A.A., Orlova A.I., Tereshchenko G.F., *Catal. Today*, **118** (1/2), 85 (2006).
55. Ermilova M.M. Orekhova N.V., Tereshchenko G.F., Malygin A.A., Malkov A.A., Basile A., Gallucci F., de Luca G., *Desalination*, **200**, 692 (2006).
56. Ermilova M.M., Orekhova N.V., Tereshchenko G.F., Malygin A.A., Malkov A.A., Basile A., Gallucci F.J. *Membr. Sci.*, **317**, 88 (2008).
57. Satterfield C.N., *Heterogeneous Catalysis in Industrial Practice*, 2nd edn, McGraw Hill, New York (1991).
58. Forzatti P., Tronconi E., Elmi A. S., Busca G., *Appl. Catal. A Gen.*, **157**, 387 (1997).
59. Hess C., Hooefelmyer J.D., Tilley T.D., *J. Phys. Chem. B*, **108**, 9703 (2004).
60. Vedrine J.C., *Top. Catal.*, **12**, 147 (2000).
61. Tereshchenko G.F., Malygin A.A., Ermilova M.M., Orekhova N.V., Volkov V.V., Lebedeva V.I., Petrova I.V., Tsodikov M.V., Teplyakov V.V., Trusov L.I., Moiseev I.I., Catalys v promyshlennosti (in Russian), *Nanotechnol. Catal.*, **2008**, 36–43 (2008).

14

Solvated Metal Atoms in the Preparation of Catalytic Membranes

Emanuela Pitzalis[1], *Claudio Evangelisti*[2], *Nicoletta Panziera*[2], *Angelo Basile*[3], *Gustavo Capannelli*[4] *and Giovanni Vitulli*[5]

[1]*CNR, Institute of Chemistry of Organometallic Compounds, Pisa UOS, Pisa, Italy*
[2]*Department of Chemistry and Industrial Chemistry, University of Pisa, Pisa, Italy*
[3]*Institute of Membrane Technology, CNR, c/o University of Calabria, Rende, CS, Italy*
[4]*Department of Chemistry and Industrial Chemistry, University of Genoa, Genoa, Italy*
[5]*Advanced Catalysts Srl, Pisa, Italy*

14.1 Introduction

Among various preparative routes, the metal vaporisation chemistry provides a valuable synthetic route to weakly stabilised nanostructured homo- and heterometallic particles [1,2]. The co-condensation at low temperature of metal vapour (commonly produced on lab scale by resistance heating and on larger scale by electron beam vaporisation) with vapour of weakly stabilising organic ligands (such as *n*-pentane, toluene, tetrahydrofuran, acetone, or acetonitrile), using commercially available reactors [3], affords solvent-stabilised metal microclusters (solvated metal atoms, SMA) soluble in the excess of ligand [4]. SMA can be handled at low temperature (223–243 K) under inert atmosphere and they are suitable precursors for the deposition of nanostructured metal particles, allowing a good control on their size [5,6]. This technique is employed for the preparation of heterogeneous mono- and bimetallic catalysts containing metal nanoparticles supported on a wide range of organic and inorganic materials [7–10]. In recent years the method has been applied with many advantages also to the

Membranes for Membrane Reactors: Preparation, Optimization and Selection, Edited by Angelo Basile and Fausto Gallucci

deposition of metals onto the surface of porous ceramic membranes, as an alternative to the well known traditional methods, such as impregnation, ion exchange, chemical vapour deposition (CVD), sol-gel techniques or physical methods like sputtering and physical vapour deposition techniques [11–13].

The main advantage of this deposition process is that the metal is deposited directly in its active form, so that, calcination and activation processes of the conventional wet deposition method are not required. This reduces the probability to introduce defects on the thin layer of the membrane. As a consequence, the effectiveness of the deposition technique does not depend on the material of the support, either organic or inorganic, or on the metal that can be employed, provided the choice of the suitable operating conditions, like vaporisation temperature or stabilising solvent. Moreover, the preparation of membranes modified by bimetallic catalysts is affordable either by depositing two metals sequentially from two different metal atom solutions, or by vaporising two metals at the same time in a multi-electrode reactor [14], employing a suitable stabilising solvent or a mixture of them. Figure 14.1a shows a multi-electrode MVS reactor, while Figures 14.1b,c respectively show the frozen matrix inside the reactor before and during the vaporisation of the metal (note that the liquid nitrogen bath has been removed to take the picture).

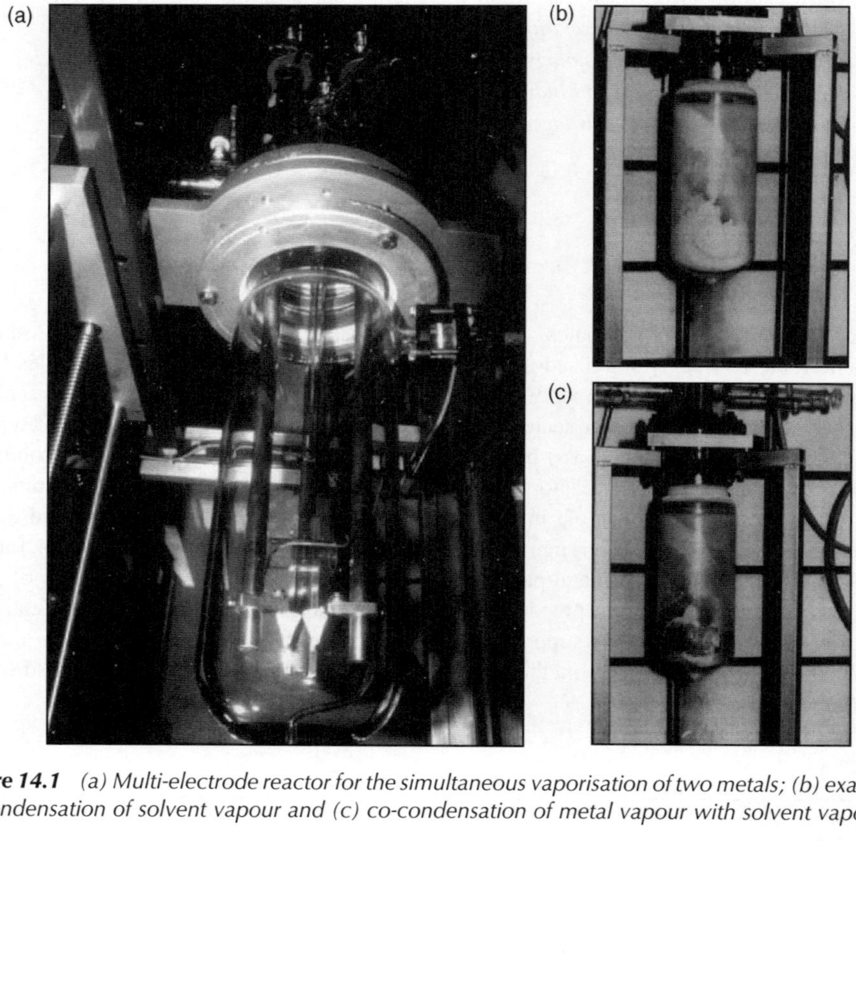

Figure 14.1 *(a) Multi-electrode reactor for the simultaneous vaporisation of two metals; (b) example of condensation of solvent vapour and (c) co-condensation of metal vapour with solvent vapour*

14.2 Preparation of Catalytic Membranes

14.2.1 Platinum on γ-Alumina Membranes

Porous Pt/γ-Al_2O_3 catalytic membranes (Societé des Céramiques Techniques, SCT, Bazet, France) were prepared using mesitylene solvated Pt atoms [15,16]. The support for the catalytic phase was a multilayered ceramic tubular membrane of inner diameter 6.7 mm and outer diameter 10.2 mm. The membrane was 150 mm long but the permeable porous part, located in the central region of the tube, was only 50 mm long since the membrane presented the extreme parts sealed by a vitrification process for an extent of 50 mm at both ends. The walls of the membrane were composed by four layers: three α-Al_2O_3 layers, on the external part of the membrane, with different thickness and decreasing porosity going from the external to the internal part, while the fourth layer, of γ-Al_2O_3, had a thickness of about 4 μm and a nominal average pore diameter of 5 nm. Therefore, all the resistance to flow occurred in this fourth layer, and the three γ-Al_2O_3 layers simply acted as a macroporous support for the thin γ-Al_2O_3 layer.

The platinum phase was deposited on the permeable part of the γ-Al_2O_3 layer by using mesitylene solvated Pt atoms as source of active particles. Platinum vapour and mesitylene were co-condensed yielding a red brown solid matrix. The flask was heated up to 233 K and the resulting yellow brown mesitylene solvated Pt atoms solution was isolated under argon and handled at low temperature. Platinum particles were rapidly deposited by shaking the membrane tube (fitted with teflon stoppers at both ends) filled with the above solution and heating up to room temperature. The deposition process is reported in Figure 14.2.

Regarding the distribution of the Pt on the membrane, SEM/X-ray analysis showed that the metal was deposited on the γ-Al_2O_3 layer and no deposition occurred on the α-Al_2O_3 support layer, where the catalytic action of the metal is unnecessary owing to the lack of selectivity of this layer. The total Pt loading was 1.65 wt% of the γ-Al_2O_3 layer. The Pt loading was measured via a colorimetric technique after reducing the membrane to powder, followed by chemical attack with a mixture of hydrofluoric acid and aqua regia. The size distribution of the Pt particles was evaluated by transmission electron microscope (TEM, Jeol JEM 2000 EX/T). TEM micrographs showed that most particles were 1.5–3.0 nm in diameter, and volume/area particle size, DVA [17], was 2.4 nm, which is comparable with the average particle size obtained with traditional methods [18].

The permeability of the catalytic membrane so prepared was only slightly modified with respect to the one in the absence of the metal: the pore size distribution, evaluated before and after the deposition of the metal by liquid–liquid displacement porometry (LLDP), showed that the partial occlusion of some pores led to the formation of smaller ones.

Porous Pt/γ-Al_2O_3 catalytic membranes were prepared by MVS technique also starting from 1-hexene solvated Pt atoms solutions, in order to verify if the solvent could play a role in affecting the membrane features [19]. TEM analysis showed that the particles were still monodispersed, with size around 1.25 nm; however, the Pt content, as a percentage of the γ-Al_2O_3 layer, was higher than in case of Pt/mesitylene, with a different distribution profile (Figure 14.3) though the metal did not penetrate the α-Al_2O_3 layer.

14.2.2 Platinum on Silica Membranes

Porous Pt/SiO_2 catalytic membranes [20] were prepared using mesitylene solvated Pt atoms following the same procedure described above. The SiO_2 surface inner layer of the membrane was prepared by the sol-gel process, starting from a commercial colloidal solution (LUDOX^R

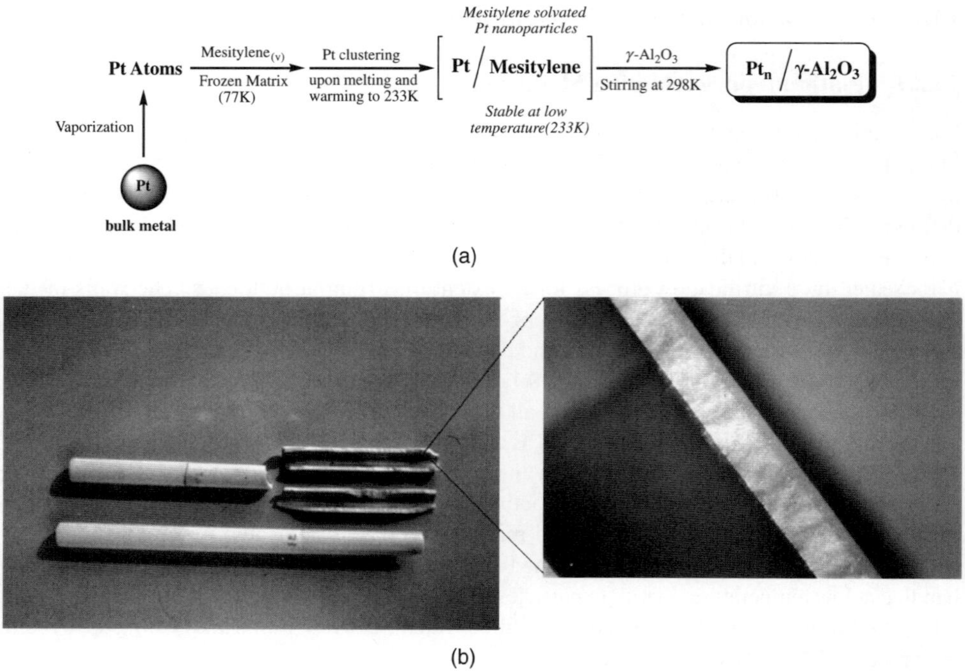

Figure 14.2 (a) Deposition method scheme from solvated Pt atoms. (b) Metal film deposited on the inner layer of a 150 mm long ceramic membrane. The enlargement in the inset shows the homogeneity of the distribution. Reprinted from G. Vitulli, E. Pitzalis, P. Salvadori, et al., Porous Pt/SiO_2 catalytic membranes prepared using mesitylene solvated Pt atoms as source of Pt particles, Catal. Today, 25, 249–253. Copyright (1995) with permission from Elsevier

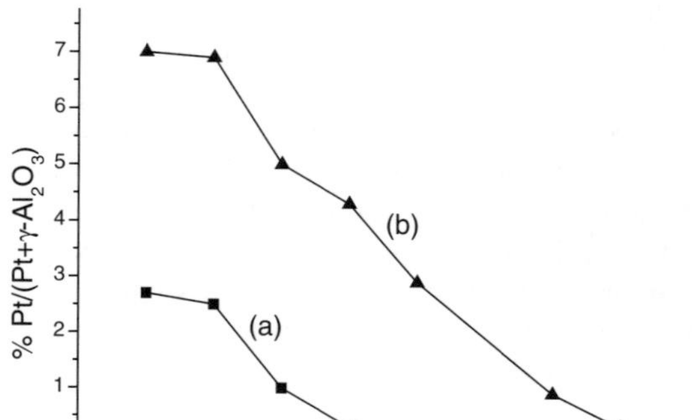

Figure 14.3 Comparison of Pt distribution profiles on γ-Al_2O_3 membrane selective layer starting from (a) Pt/mesitylene and (b) Pt/1-hexene SMA

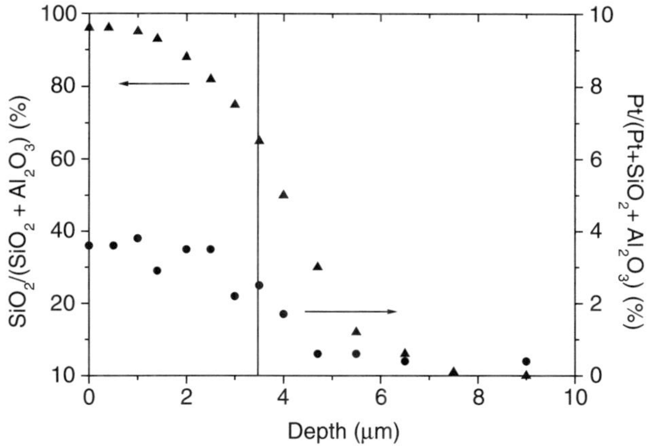

Figure 14.4 *Silica and platinum distribution within the membrane*

ASGI-Dupont), by depositing a thin sol layer by slip casting on the asimmetric α-alumina tubular support, and then dried. Also in this case the membrane was sealed by a vitrification process at both ends, for an extent of 25 mm. The platinum distribution on the SiO_2 inner layer was evaluated by SEM/EDS, in order to control also the silica interdispersion within the α-Al_2O_3 support. The results indicated that, besides the metal particles on the silica inner layer, a small amounts of Pt was found even in the α-Al_2O_3 support layer, due to the lack of a sharp interface between the silica thin layer and the support: in fact, SiO_2 small grains (diameter 20–30 nm, measured by SEM) could enter α-Al_2O_3 pores, which had a nominal size of 0.2 μm (Figure 14.4), and diffuse within them.

Regarding the catalyst dispersion, TEM analysis showed a DVA of 2.4 nm and a percentage dispersion of 47%: these results were comparable with the ones obtained in the case of Pt/ γ-Al_2O_3 membranes, indicating that this method ensures high Pt loading without affecting the catalyst dispersion.

14.2.3 Palladium on Alumina Membranes

Pd/γ-Al_2O_3 catalytic membranes were prepared following a similar procedure [21]: palladium vapour, obtained from Pd shots heated by Joule effect on an alumina crucible, were co-condensed with a 1-hexene/mesitylene 1/1 mixture; after melting, the resulting yellow-brown solution was kept at 233 K in inert atmosphere. Pd particles were deposited onto the inner γ-Al_2O_3 surface as reported above with the Pd/1-hexene/mesitylene solution. The γ-Al_2O_3 tubular membrane was similar to the one described above, except for the length, which in this case was 25 cm. In the same work, a comparison of characteristics was made with other two deposition techniques, magnetron sputtering (MS) and physical vapour deposition (PVD) technique. These two techniques allowed to deposit Pd only onto the outer surface of the membrane, which had an average pore size greater than 10 nm, and high rugosity, so that the films covered only partially the grain surface of the ceramic membrane without creating an uniform thin film, as revealed by SEM analysis. On the contrary, the membrane prepared using solvated metal atoms showed a very homogeneous and compact thin film, with thickness about 0.1 μm, difficult to be obtained with other techniques; moreover, the metal was present only on the top layer of the inner surface, indicating that the pores of the γ-Al_2O_3 layer were filled with the metal, which did not penetrate the α-Al_2O_3 layers.

14.2.4 Palladium–Silver on Titania–Alumina Membranes

Bimetallic Pd-Ag/TiO_2-Al_2O_3 catalytic membranes from asymmetric porous commercial membranes (Inoceramic, Germany) were produced by MVS method in two different ways [22]: concerning the first method, from a Pd/mesitylene/1-hexene solution Pd particles were deposited at room temperature onto the inner surface of an asymmetric porous commercial TiO_2-Al_2O_3 membrane, then Ag particles from a Ag/acetone solution were deposited on the Pd film by evaporating the solvent. Both Pd and Ag solutions were prepared by MVS. In a different way, Pd and Ag were deposited at the same time on the membrane support from a Pd-Ag/acetone solution, prepared by MVS in a multi-electrode reactor, by co-condensing Pd and Ag vapour together with acetone vapour on its walls, and then siphoning the solution after melting, at low temperature. From permeation experiments with pure gases, H_2/other gases selectivity α was calculated ($\alpha_{H_2/\text{other gases}} = P_{H_2}/P_{\text{other gases}}$); CO_2, CH_4, CO, N_2 gases were chosen for a comparison with H_2 behaviour. In both membranes, $\alpha_{H_2/\text{other gases}}$ resulted between 2.0 and 4.5.

14.2.5 Palladium and Platinum on Polymeric Membranes

Metal particles deposition onto polymeric membranes is usually carried out by chemical reduction from metal salts water solutions [23]. This method, though simple and inexpensive, sometimes fails to achieve homogeneous particles distribution and complete metal reduction. Pd and Pt solvated metal atoms were successfully employed also to deposit metal particles onto polymeric membranes [24]. A versatile technique to prepare porous flat polymeric membranes is the phase inversion method [25]. The polymer is dissolved in a suitable solvent, and then the flat membrane is obtained by casting on a plate to form a thin film which is finally immersed in a nonsolvent (e.g., water), washed and dried.

The deposition and dispersion of metal particles in the pore structure of the polymeric membrane was influenced by the preparation condition adopted during the metal particles deposition (e.g., the type of solvent) and by the membrane properties (e.g., hydrofobicity). In particular with polyvinylidene fluoride (PVDF) porous membranes, it was observed that the aggregation of the metal particles was more favoured by using acetone as a solvent rather than mesitylene [24].

Concerning flat polymeric composite membranes prepared by the phase inversion process from a DMF mixture of PVDF and γ-Al_2O_3, metal particles were deposited at room temperature in inert atmosphere by dipping the membranes into Pt/mesitylene or Pd/mesitylene-1-hexene solvated metal atoms until the solutions discoloured [25].

As a comparison, Pt- or Pd-modified PVDF membranes were prepared by reduction starting from H_2PtCl_6 or $PdCl_2$ water solutions (impregnation method). HRTEM analysis of Pt/PVDF membranes showed a homogeneous dispersion of metal particles with 1–2 nm diameter for MVS systems, a less homogeneous dispersion with 2–6 nm diameter for the traditional ones. Pd particles deposited onto PVDF membranes from solvated metal atoms were regularly distributed, with size <4 nm, while those from $PdCl_2$ had a size range of 1.5–7.0 nm. All the membranes so prepared were tested in some three phase model reactions (e.g., hydrogenation of methylencyclohexane or hydrogenation of cinnamic aldehyde) where the membrane was used as a catalytic interface between the gas phase (e.g., hydrogen) and the liquid phase (e.g., substrates in a solvent).

14.3 Catalytic Exploitation

The catalytic activity of Pt/γ-Al_2O_3 and Pt/SiO_2 membranes prepared from mesitylene solvated platinum atoms was studied in the benzene hydrogenation reaction in gas phase [15], and in *p*-chloronitrobenzene hydrogenation in liquid phase [16] as model reactions.

Different feeding configurations of reagents were explored in the hydrogenation of benzene to cyclohexane in a membrane reactor, and the results showed a membrane effect due to different degrees of accessibility of pores of different size; the nonuniform pore size distribution resulted in a 60% decrease of the activation energy with respect to the one evaluated with Pt/γ-Al_2O_3 powder employed in a conventional packed bed reactor.

A similar reaction, toluene hydrogenation, was examined to test Pt/SiO_2 catalytic membranes activity [20]. The apparent activation energy resulted slightly smaller than the ones generally reported ($E_a = 9.05$ kcal mol^{-1} vs 11–12 kcal mol^{-1}) [26] probably because of the wide pore size distribution. These membranes however appeared to have a much higher specific activity than the Pt/γ-Al_2O_3 ones (947 vs 87.6, expressed as molecule of consumed toluene per atom of Pt per hour): this was addressed to the different pore size in the inner support, as the silica membranes have pore diameter of 10 nm, while the γ-Al_2O_3 membranes have pores with nominal diameter of 5 nm, and this fact could account for a different accessibility of the catalyst for the substrate.

As an example of liquid phase reaction, *p*-chloronitrobenzene hydrogenation in ethanol was performed under mild reaction conditions ($P_{H_2} = 1$ bar) using Pt/γ-Al_2O_3 catalytic membranes [16], and compared with some Pt/γ-Al_2O_3 powder catalytic systems, under the same reaction conditions (Scheme 14.1).

In the reaction performed with Pt/γ-Al_2O_3 catalytic membrane, *p*-chloronitrobenzene was rapidly reduced to *p*-chloroaniline which was no further hydrogenated even at long reaction time. Completely different results were obtained with Pt/γ-Al_2O_3 powder catalyst, as a complex mixture of products was observed, ranging from the dehydrodechlorination product, aniline, to the completely hydrogenated ciclohexylamine, in mixture with the main product, diciclohexylamine, and with other partially hydrogenated alkyl-arylamines. The dependence of the observed selectivity on the metal particles size was excluded in this case, as TEM analysis showed a very similar Pt particle size distribution in the compared systems. Instead, this fact was addressed to the idea that a catalytic membrane can supply more H_2 with respect the powder system, improving the contact between gas, liquid and solid because the volatile reactant does not have to diffuse through a liquid film covering the solid catalysts, as in conventional reactors [13].

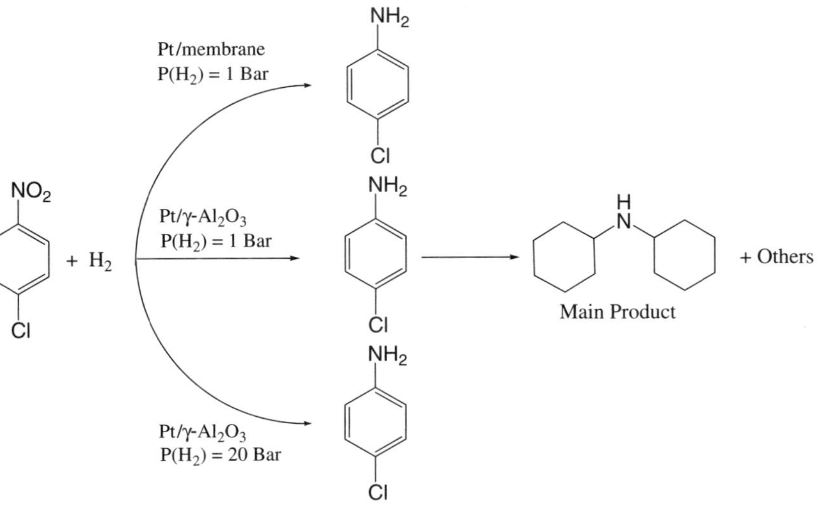

Scheme 14.1 *p-Chloronitrobenzene hydrogenation in ethanol*

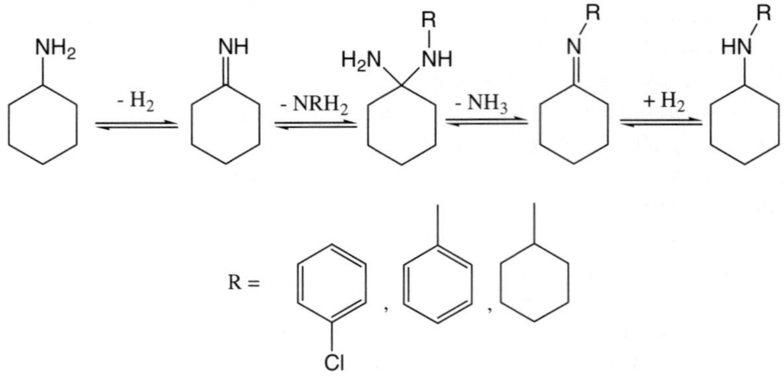

Scheme 14.2 *Dehydrogenation of amines to imines*

The same behaviour was observed in the reaction performed at a higher H_2 pressure (20 bar) in the presence of a Pt/γ-Al_2O_3 powder catalyst: in this case, a greater amount of *p*-chloroaniline was produced, and a lower concentration of all those products formed at H_2 atmospheric pressure, whose formation would involve a dehydrogenation step of the amines to imines (Scheme 14.2), unlikely to occur at higher H_2 pressure.

A Pd/γ-Al_2O_3 catalytic membrane prepared by MVS was employed in the water gas shift reaction and its performance was compared with the other two membranes prepared with different methods, MS and PVD, described in the previous section, which allowed to deposit Pd onto the outer surface of the membrane [21]. The conversion of CO was reported versus the feed flow rate of CO at different H_2/CO ratios: owing to the macropores in the composite membranes, the CO conversion data were under the thermodynamic equilibrium values of a TR working at the same conditions for both membranes prepared via MS and PVD. For the membrane prepared via MVS, experimental results showed a higher degree of CO conversion, compared to the equilibrium, due to the hydrogen permeation through the composite Pd membrane: in fact, the hydrogen flux through the Pd layer was very high, owing to the low thickness of the metal film. The membrane showed a certain instability for long reaction times, because after 42 h some decline in selectivity was observed: anyway, every catalytic membrane prepared by MVS can be easily reconditioned by performing a new deposition from solvated metal atoms without any further process. The results obtained with these membranes are compared in this work with the ones obtained by Uemiya [27] with membranes modified by electroless plating technique.

Regarding the two bimetallic Pd-Ag/TiO_2-Al_2O_3 membranes prepared via MVS with two different procedures, described above, they were employed in MRs in methanol steam reforming reaction [22]; good results were obtained in particular for the Pd-Ag supported porous membrane prepared through sequential deposition of the two metals from Pd/mesitylene-1-hexene and Ag/acetone solutions: this membrane was able to increase the methanol conversion and the hydrogen production, as well as to decrease the carbon monoxide selectivity with respect to the TR.

Catalytic polymeric membranes, applied in three phase reactions, showed an higher catalytic activity than similar catalysts in traditional three-phase reactors (e.g., slurry reactor). The catalytic particles prepared by the MVS technique resulted well dispersed and easily available for the meeting of reactants, due to the open pore structure nature of the polymeric membrane.

14.4 Conclusions

In recent years the metal vapour synthesis has been found to be a very useful technique for the preparation of modified catalytic membranes. Solvated metal atoms have been proven to be very efficient precursors for the deposition of highly dispersed metal particles, catalytically active even without activation processes generally required for other conventional wet depositions; SMA can be conveniently used in the preparation of inorganic and polymeric catalytic membranes. The MVS technique has been applied successfully also to the preparation of bimetallic systems with peculiar structure and catalytic activity. In fact, depending on the desired structure, the technique allows to obtain bimetallic nanostructures derived from different vaporisation of two metals as well as contemporaneous vaporisation of the two metals which leads to a closer interaction between them during the growth of their aggregates.

Recently, in the scaling up of MVS technique the amount of metal which can be vaporised from 0.1–1.0 g (laboratory scale) has increased to 5–10 g of metal allowing to obtain larger quantity of solvated metal atoms: this improvement can make MVS a powerful technique for the preparation of catalytic membranes on a great scale.

References

1. K.J. Klabunde, *Free Atoms, Clusters and Nanoscale Particles*, Academic Press, New York (1994).
2. G. Vitulli, C. Evangelisti, A.M. Caporusso, P. Pertici, N. Panziera, S. Bertozzi, P. Salvadori in *Metal Nanoclusters in Catalysis and Materials Science: the Issue of Size-Control*, B. Corain, G. Schmid, N. Toshima (eds), Elsevier, Amsterdam, Chapter 32 (2008).
3. K.J. Klabunde, Y.X. Li, B.J. Tan, Solvated metal atom dispersed catalysts, *Chem. Mater.*, **3**, 30–39 (1991).
4. J. S. Bradleyin *Cluster and Colloids: From Theory to Applications*, E. Schmid (ed.), VCH, Weinheim (1994).
5. G. Vitulli, E. Pitzalis, P. Pertici, P. Salvadori, S. Coluccia, G. Martra, L. Lampugnani, M. Mascherpa, Preparation via solvated metal atoms and surface reactivity of homo- and hetero-nanostructured metal particles: supported and unsupported Rh nanoaggregates, *Mater. Sci. Eng. C*, **15**, 207–209 (2001).
6. G. Uccello-Barretta, C. Evangelisti, P. Raffa, F. Balzano, S. Nazzi, G. Martra, G. Vitulli, P. Salvadori, The control of the growth of Pt clusters in solution: a way to prepare Pt particles of tailored size, *J. Organomet. Chem.*, **694**, 1813–1817 (2009).
7. G. Vitulli, M. Bernini, S. Bertozzi, E. Pitzalis, P. Salvadori, S. Coluccia, G. Martra, Nanoscale copper particles derived from solvated Cu atoms in the activation of molecular oxygen, *Chem. Mater.*, **14**, 1183–1186 (2002).
8. N. Panziera, P. Pertici, L., A.M. Caporusso, G. Vitulli, P. Salvadori, S. Borsacchi, M. Geppi, C.A. Veracini, G. Martra, L. Bertinetti, MVS-derived palladium nanoparticles deposited on polydimethylphosphazene as recyclable catalysts for Heck-type reactions: preparation, structural study and catalytic activity, *J. Catal.*, **246**, 351–361 (2007).
9. C. Evangelisti, N. Panziera, P. Pertici, G. Vitulli, P. Salvadori, C. Battocchio, G. Polzonetti, Palladium nanoparticles supported on polyvinylpyridine: catalytic activity in Heck-type reactions and XPS structural studies, *J. Catal.*, **262**, 287–293 (2009).
10. C. Evangelisti, N. Panziera, M. Vitulli, P. Pertici, F. Balzano, G. Uccello-Barretta, P. Salvadori, Supported rhodium nanoparticles obtained by metal vapour synthesis as catalysts in the preparation of valuable organic compounds, *Appl. Catal. A: Gen.*, **339**, 84–92 (2008).

11. V. Gryaznov, Membrane catalysis, *Catal. Today*, **51**, 391–395 (1999).
12. J. Coronas, J. Santamaria, Catalytic reactors based on porous ceramic membranes, *Catal. Today*, **51**, 377–389 (1999).
13. A. Julbe, D. Farrusseng, C. Guizard, Porous ceramic membranes for catalytic reactors – overview and new ideas, *J. Membr. Sci.*, **181**, 3–20 (2001).
14. G. Vitulli, E. Pitzalis, R. Lazzaroni, P. Pertici, P. Salvadori, O. Salvetti, S. Coluccia, G. Martra, Nanoscale platinum particles by clustering of arene solvated Pt atoms, *Mater. Sci. Forum*, **195**, 93–102 (1995).
15. G. Capannelli, A. Bottino, G. Gao, A. Servida, G. Vitulli, A. Mastantuono, R. Lazzaroni, P. Salvadori, Porous Pt/γ-$A1_2 0_3$ catalytic membrane reactors prepared using mesitylene solvated Pt atoms, *Catal. Lett.*, **20**, 287–297 (1993).
16. G. Vitulli, A. Verrazzani, E. Pitzalis, P. Salvadori, Pt/γ-Al_2O_3 catalytic membranes vs. Pt on γ-Al_2O_3 powders in the selective hydrogenation of p-chloronitrobenzene, *Catal. Lett.*, **44**, 205–210 (1997).
17. O. Monticelli, S. Russo, R. Campagna, B. Voit, Preparation and characterisation of blends based on polyamide 6 and hyperbranched aramids as palladium nanoparticle supports, *Polymer*, **46**, 3597–3606 (2005).
18. D. Uzio, J. Peureux, A. Giroir-Fendler, A Torres, M. Ramsy and J.A. Dalmon, Platinum/γ-Al_2O_3 catalytic membrane: preparation, morphological and catalytic characterizations, *Appl. Catal.*, **96**, 83–97 (1993).
19. E. Pitzalis,Homo and heteronuclear nanometric aggregates of transition metals: preparation, characterization and examination of their physico-chemical properties, *PhD Thesis,* University of Pisa (1995).
20. G. Vitulli, E. Pitzalis, P. Salvadori, G. Capannelli, O. Monticelli, A. Servida, A. Julbe, Porous Pt/SiO_2 catalytic membranes prepared using mesitylene solvated Pt atoms as source of Pt particles, *Catal. Today*, **25**, 249–253 (1995).
21. A. Basile, E. Drioli, F. Santella, V. Violante, G. Capannelli, G. Vitulli, A study on catalytic membrane reactors for water gas shift reaction, *Gas Sep. Purif.*, **10**, 53–61 (1996).
22. A. Basile, S. Tosti, G. Capannelli, G. Vitulli, A. Iulianelli, F. Gallucci, E. Drioli, Co-current and counter-current modes for methanol steam reforming membrane reactor: Experimental study, *Catal. Today*, **118**, 237–245 (2006).
23. A. Bottino, G. Capannelli, A. Comite, R. Di Felice, Polymeric and ceramic membranes in three-phase catalytic membrane reactors for the hydrogenation of methylenecyclohexane, *Desalination*, **144**, 411–416 (2002).
24. E. Sarno, Atomi di metalli solvatati nella preparazione di nuovi catalizzatori, *Degree Thesis,* University of Pisa (2001).
25. A. Bottino, G. Camera-Roda, G. Capannelli, S. Munari, The formation of microporous polyvinylidene difluoride membranes by phase-separation, *J. Membr. Sci.*, **57**, 1–20 (1991).
26. S.D. Lin, M.A. Vannice, Hydrogenation of aromatic hydrocarbons over supported catalysts II, *J. Catal.*, **143**, 554–562 (1993).
27. S. Uemiya, N. Sato, H. Ando, E. Kikuchi, The water gas shift reaction assisted by a palladium membrane reactor, *Ind. Eng. Chem. Res.*, **30**, 585–589 (1991).

15

Electrophoretic Deposition for the Synthesis of Inorganic Membranes

F.J. Varela-Gandía[1], A. Berenguer-Murcia[1], A. Linares-Solano[1], E. Morallón[2] and D. Cazorla-Amorós[1]

[1]Universidad de Alicante, Departamento de Química Inorgánica, Alicante, Spain
[2]Universidad de Alicante, Departamento de Química Física and Instituto Universitario de Materiales, Alicante, Spain

15.1 Introduction

15.1.1 Electrophoretic Deposition: Basic Principles

Of all the electrokinetic phenomena, perhaps that which is best known is Electrophoresis, as first reported by Reuss in 1809 [1]. Conceptually, electrophoresis is described as the motion of charged dispersed particles in a liquid under the influence of a uniform electric field. Electrophoresis takes place due to the surface electric charge that most particles generate, and that is thus affected by the Coulomb forces of the electric field. The charged particle is surrounded by ions of opposite charge forming the so-called double layer. According to the double layer theory, the charge of this double layer is equal to that of the surface charge of the particle. In the presence of an electric field the force that such electric field would exert on this double layer (in principle equal in magnitude and of opposite direction to the surface charged particle) would make electrophoresis unfeasible due to this force compensation. However, the ions are also attracted by the particle, and a fraction of them will not move in the opposite direction but move along with the particle. The potential of the surface of shear is defined as zeta-potential or electrokinetic potential. In equilibrium, the speed of the particle is determined by four forces [2]: (i) the interaction between the surface charge and the electric field which accelerates the particle, (ii) the viscous drag from the liquid according to the Stoke's law, (iii) the force of the electric field of

Membranes for Membrane Reactors: Preparation, Optimization and Selection, Edited by Angelo Basile and Fausto Gallucci

the double layer surrounding the particle (retardation) and (iv) the distortion of the double layer caused by a displacement between center of the negative and positive charge (relaxation). These three last forces slow the particle.

Electrophoresis has found a significant number of applications in different fields such as in biochemistry and biotechnology. Here, affinity electrophoresis is used both as a qualitative and quantitative method for the separation of proteins or nucleic acids, or capillary electrophoresis which can be used to separate molecules attending to their size to charge ratio in the interior of a small capillary filled with an electrolyte. In addition to these biochemical applications, a novel application for electrophoresis has arisen during the past decade, known as electrophoretic deposition (EPD) [3–5]. This methodology consists on the application of a known electric field so that species with the appropriate surface charge in a liquid solution/suspension are attracted to an electrode (electrophoresis) and then, the particles collected at this electrode form a coherent deposition. The present chapter will explain how EPD can be used in conjunction with other procedures for the preparation of zeolite-based inorganic membranes. We present some examples which, whenever applicable, focus on our experience in the field in order to illustrate the applicability of the methodologies described in this chapter.

15.1.2 Electrophoretic Deposition as a Seeding Technique: Seeding Methods

In zeolite membrane synthesis (see below), direct synthesis on a given support was the first methodology. Two main steps are involved in this protocol: (i) a zeolite synthesis solution is prepared and the support is immersed in it, (ii) the mixture is submitted to hydrothermal treatment under the zeolite synthesis conditions [6,7]. Several reports have dealt with this topic, despite its disadvantages such as the necessity of long synthesis times and several synthesis steps which in turn lead to thick zeolite layers [8–10]. Recently, a new procedure called seeded (or secondary) growth [11–16], has been developed for the preparation of zeolite membranes. The main advantages are the control of thin oriented layers [8], improved zeolite growth on different supports [11,14] and the need for only one hydrothermal synthesis step.

On the whole, zeolite synthesis can be separated in two main steps: nucleation and crystal growth. The former is the rate-limiting step in the synthesis [10] and it determines the subsequent crystal growth. Bearing in mind the two aforementioned methods (direct and seeded growth) the former needs long synthesis times and there is a limited intergrowth between the crystals which form the membrane. The latter 'skips' the nucleation step and then a thin zeolite film can be obtained with comparatively higher quality crystals and better intergrowth with reduced synthesis times.

Concerning other seeding procedures, different techniques have been developed in the last years. Some of the methods are summarised in Table 15.1.

Table 15.1 *Seeding methods reported in the literature*

Seeding method	References
Metal surface modification with organic polymer	[15,50]
Dip coating	[12,13,51–53]
Rubbing	[54,55]
Suction	[11,56]
Ultrasound	[57]
Spin coating	[58]
Electrophoretic deposition (EPD)	[14]

15.1.3 Zeolites as Electrophoretic Species: 'Role' of the Templating Agent

A zeolite is a crystalline aluminosilicate with a 3D regular open structure [17]. This characteristic structure is based on silicon and aluminium tetrahedra, with the cations occupying their centre and the apical oxygen ions being always shared by two adjacent tetrahedra. A zeolite is usually synthesised in the presence of a 'template' which acts as a structure-directing agent towards the formation of the final zeolite structure. This template is removed by thermal decomposition (i.e., calcination) or leaching after synthesis in order to activate the zeolite.

In this sense, organic templates such as tetramethylammonium (TMA^+) ions, [12] trapped in the β cages of the LTA structure of zeolite A or tetrapropylammonium ions (TPA^+), in silicalite [14], confer a positive charge to the as-synthesised zeolite. As an example, Domínguez-Domínguez *et al.* [18] have prepared LTA zeolite membranes over carbon supports by the secondary growth method using as-synthesised LTA zeolite nanocrystals as growth seeds. To sum up, the template agent has an interesting role in providing the backbone for zeolite formation, and it also results in a certain network charge, which has significant implications in their final applications. Therefore, this phenomenon may be of relevance in the preparation of seeded supports using electrophoretic deposition.

15.2 State of the Art

15.2.1 Methodologies Employed

As it has been explained previously, EPD is a fairly novel method that permits the coating of a selected support with selected zeolite seed crystals by the application of an electric potential. The coating method presented here, based on the electrophoretic deposition of seed crystals, can be easily performed by assembling a very simple electrochemical cell and yields homogeneous films of seed crystals, as shown below. The amount of seed crystals can be easily controlled by adjusting either the seeding time or the current density applied to the circuit, thus making possible a better control of the nanocrystals deposition rate. This novel method is (to a certain extent) comparable to the seed film method, where electrostatic interactions between the support surface and the zeolite crystals are responsible for the seeding, although in the case of EPD, a more precise control of the seed deposition can be obtained thanks to the control over the current density and the deposition time.

Regarding the methodology, the required equipment for the preparation of seeded supports by EPD is an electrochemical cell [14,18,19]. In principle, the procedure is available when the support is electron conducting, although this is not mandatory, since the support maybe located near an electrode. First, a colloidal suspension of the selected material is prepared. Then, the support is immersed in the solution as a working electrode and a second electrode (i.e., Pt or graphite) acts as counter-electrode. Then, a direct current (DC) is applied for a given time for the preparation of the seeded support. Concerning the liquid media, W. Shan *et al.* [19], have studied the seeding of different nanosized zeolite on stainless steel grid substrates in a non-aqueous media. This study was motivated by the use of organic solvent in order to avoid water electrolysis, which results in an efficiency loss. Nevertheless, our results on the EPD on silicalite [14] and LTA [18] performed in an aqueous media were been carried out successfully despite water electrolysis.

Apart from the conventional method, other ways have been developed to apply EPD to non-conductive materials. Abdollahi and coworkers, have set up a new system in order to seed alumina discs [20]. The disc is placed between the surface of a stainless steel electrode and a Pt rod which is used as a counter electrode. The alumina is situated near the stainless steel electrode and therefore, it is seeded with the crystal zeolites which move towards the working electrode. Kanamura and Hamagami [21] have used EPD for preparing continuous deposits on several materials. Their purpose was the preparation of ceramic membrane filters for water purification as described above. Secondly, they prepared a membrane electrode assembly for its use as a polymer electrolyte fuel cell. The employed EPD cell consisted of two compartments separated with a high ionic conducting polymer. The last of the methods performed in their study is the use of a micro-EPD method that permitted to prepare a micro-deposit on the support with an excellent ordered microstructure. It consisted of the application of a local electric field on a very small area of the support towards deposition of the particles.

Having elucidated the methodology of EPD, we now focus on its application in the seeding of different supports with a variety of zeolites.

15.2.2 Application of EPD to Aluminium-Free Zeolites

We have already explained the importance of the species known as 'templating agent' (see Section 15.1.3) in the preparation of synthetic zeolites, and the direct impact this feature has on their electrochemical nature. Then again, taking into account that zeolites are aluminosilicates, their isoelectric point is by no means neutral due to the presence of trivalent aluminium cations, which cause the crystalline network to adsorb cations on its surface in order to compensate for an excess in negative framework charge, which is in turn responsible for their outstanding properties as ion-exchangers. This behaviour, however, does not apply to all-silica counterparts (which are not zeolites, but 'zeotypes'), in which each tetravalent silicon ion compensates the charge provided by its four surrounding oxygen ions. As a novel development within our study of MFI zeotype membranes (i.e., silicalite-1) [7], we considered that the preparation of nanocrystalline deposits by EPD should be made possible by the fact that the as-synthesised zeolite crystals contain four molecules of the templating agent (for instance, tetrapropylammonium or TPA^+) per zeolite unit cell. As a consequence, the use of uncalcined colloidal zeolite crystals in EPD became a possibility that we successfully tested for the development of zeolitic coatings on macroporous carbon supports [14], which has resulted in other groups following (and expanding) similar research pathways.

15.2.3 Continuous Zeolite Deposits on Different Materials

Following up on the materials, zeolites and methods which have been described above (See Sections 15.2.1 and 15.2.2), several research groups have successfully prepared continuous deposits of zeolites of different chemical nature on a wide range of supports. Table 15.2 illustrates the variety of deposits that have been reported in the literature.

15.3 Experimental

15.3.1 Instrumentation and Reactants

The first step consists on the preparation of a stable zeolite suspension to use as seeding solution. There are some excellent reports on the preparation of nanosized zeolite suspensions. Among

Table 15.2 *Some examples of EPD deposits on different materials and some tentative applications*

Sample	Support	Application	Reference
ZSM-5	Porous alumina	Pervaporation in i-propanol/ water mixtures	[29]
Mesoporous silica	Stainless steel	Water adsorption	[59]
LTA	Porous graphite	N_2/O_2 separation	[18]
Hydrotalcite	Macroporous alumina	N_2/CO_2 separation	[60]
Clay films montmorillonite	Tin-doped In_2O_3-coated glass substrates	—	[61]
3A, 4A, 5A, 13X, FAU	Nafion electrode	Heavy metals separation	[62]
13X	Glassy carbon	—	[63]
Y	Stainless steel	—	[64]
Silicalite	Porous graphite	—	[14]
Silicalite, ZSM-5, BEA, SOD, LTL, Na-LTA	Stainless steel	—	[19]
3A/5A	Stainless steel	—	[65]

these we based our research on the works of Hedlund *et al.* [22] and Persson *et al.* [23] for LTA and MFI zeolites, respectively. Table 15.3 illustrates the conditions under which the zeolite seeds were synthesised.

The zeolite nanocrystals were recovered by high-speed centrifugation (20 k rpm, 30 min) and re-suspended in distilled water until further use. The pH of the resulting suspensions was 9.0–9.5.

The electrochemical cell consists in a bath of the zeolite suspension in which the two electrodes are introduced (the carbon support as working electrode and a platinum counter-electrode). The current between the two electrodes was applied by DC power.

For the hydrothermal treatment following EPD required to grow a continuous zeolite film, Teflon-lined stainless steel autoclaves are used in order to work under autogenous pressure conditions at the reaction temperature.

15.3.2 Procedure

In spite of the seeding suspensions being stable for several months, it is highly advisable to sonicate the suspensions prior to their use due to the fact that the zeolite nanocrystals tend to agglomerate into larger entities while in suspension for a long time.

A carbon disc was placed in a Teflon holder and a copper wire is used to deliver the electric current to the carbon disc. The Teflon holder was immersed in the zeolite suspension with a tilt angle of approximately 30° (see below). A coiled platinum wire was immersed in the suspension

Table 15.3 *Synthesis conditions of the employed zeolite seeds. Note: synthesis requires the use of a Teflon-lined steel autoclave*

Zeolite	Synthesis solution	Temperature (K)	Time (h)
LTA	$0.22Na_2O{\cdot}5.0SiO_2{\cdot}1.0Al_2O_3{\cdot}8.0(TMA)_2O^a{\cdot}400H_2O$	336	63
MFI	$1SiO_2{\cdot}4EtOH{\cdot}1TPA\text{-}OH^b{\cdot}49H_2O$	333	336

[a]Tetramethylammonium oxide
[b]Tetrapropylammonium hydroxide

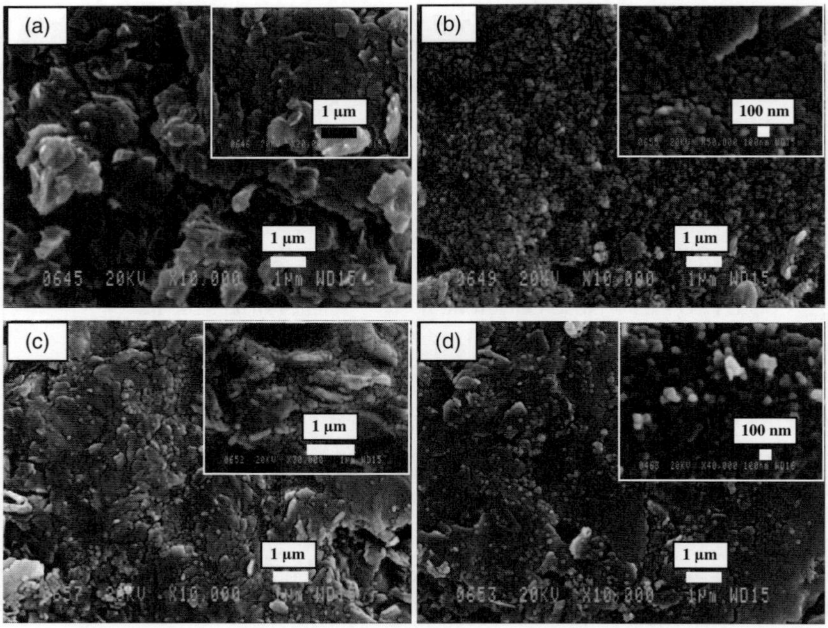

Figure 15.1 An example of the tunability of electrophoresis for the deposition of zeolites (in this case, silicalite-1). (a) Deposition time 30 min, current density 5 mA cm^{-2}. (b) Deposition time: 30 min, current density 9 mA cm^{-2}. (c) Deposition time 20 min, current density 9 mA cm^{-2}. (d) Deposition time 10 min, current density 9 mA cm^{-2}. Note how increasing both current density and deposition times results in more extensive zeolite seed deposits, reaching full coverage in the case of Figure 15.1b

to act as counter-electrode. The potential between the two electrodes was measured with a multimeter. A continuous current was then passed through the circuit with the desired value for a given time in order to perform the EPD. The current density refers to the geometric area of the carbon support exposed to the zeolite solution (0.57 cm^2) that in all cases was the same. During the experiments, the current applied causes the reduction of water on the cathode, and as a consequence hydrogen bubbles are evolved. If the hydrogen bubbles grow too much in size they prevent EPD from happening by blocking the circuit, which is the reason why the disc holder is inserted at an angle of approximately 30° to favour the removal of hydrogen bubbles. Figure 15.1 shows some SEM images of the carbon supports after EPD under a variety of conditions.

After EPD, the samples were washed with distilled water and dried in a furnace at 373 K for 1 h. Zeolite layers were prepared at various temperatures for different times for comparison purposes. Silicalite-1 and zeolite A layers were grown on the carbon supports by standard hydrothermal synthesis, as described in Refs. [24–26]. The silicalite-1 precursor solutions were obtained from fumed silica (Aldrich 112945-52-5) or tetraethyl orthosilicate (TEOS; Aldrich 13,190-3), high-purity NaOH (Aldrich 30,657-6) and using ultrapure (Millipore MilliQ) water as solvent. Tetrapropylammonium hydroxide (TPA-OH; Aldrich 25,453-3) and tetrapropylammonium bromide (TPA-Br; Aldrich 2,556-8), were used as the templating agents. The solution was prepared by adding the appropriate amounts of NaOH and templating agents to the ultrapure water until a clear solution was obtained. The silica precursors were then added carefully and the solution was left to age for 6 h.

The composition of this zeolite A precursor solution, based on the work of Zhu *et al.* [27], was prepared from the mixture of two previously prepared solutions. One of the solutions was obtained dissolving, at room temperature, sodium hydroxide (Sigma-Aldrich, 306576, 99.99%) in part of the high-purity water. Sodium aluminate (Riedel de Häen, 13404, anhydrous, technical) was subsequently added to this mixture. The other solution was prepared dissolving the rest of the sodium hydroxide in the other part of the high-purity water, and adding sodium silicate (Panreac, 211714, neutral solution, chemically pure). Both solutions were separately stirred for 60 min. The solutions were then mixed and stirred for a further 60 min.

The typical synthesis solution compositions for the aforementioned membranes are:

Silicalite-1 with TPA-OH: $1.0SiO_2{\cdot}0.05TPA\text{-}OH{\cdot}0.22NaOH{\cdot}41.5H_2O$
Silicalite-1 with TPA-Br: $1.0SiO_2{\cdot}0.047TPA\text{-}Br{\cdot}0.278NaOH{\cdot}41.48H_2O$
Zeolite A: $50Na_2O{\cdot}5.0Si_2O{\cdot}1.0Al_2O_3{\cdot}1000H_2O$

The carbon supports were held vertical inside the autoclaves by means of a teflon piece inserted inside the autoclave to reduce sedimentation of zeolite crystals on the support. The synthesis solution was poured inside the autoclave with the disc in it and the hydrothermal synthesis was carried out, after the ageing time, under static conditions. After the synthesis, the composites were washed and tested for mechanical stability in an ultrasonic bath for 15 min. The resulting materials were dried at 398 K overnight.

15.3.3 Sample Treatment

After cleaning, the prepared silicalite-1 composites were thermally treated by one of these two protocols:

1. Heating up to 973 K in a horizontal furnace with a heating rate of 1 K min^{-1} under a flow of N_2 (100 ml min^{-1}). The temperature was kept for 12 h and finally the samples were cooled down to room temperature with a heating rate of 1K min^{-1}.
2. Heating up of the sample up to 673 K in a N_2 atmosphere (flow 100 ml min^{-1}) at a rate of 0.5 K min^{-1} (hold time 4 h). The gas flow was then switched from N_2 to synthetic air (flow 100 ml min^{-1}; hold time 16 h). Cooling down of the sample to room temperature was done at a rate of 0.5 K min^{-1}.

The crystallinity and zeolitic phase identification of the synthesised samples were ascertained by powder X-ray diffraction (XRD) using a SEIFERT 2002 diffractometer with Cu-Kα radiation (1.54 Å). Each diffractogram was taken in the 2–50° 2θ angle range, and the scanning velocity was 2° min^{-1}.

The morphology, thickness and a qualitative evaluation of the continuity of the LTA/carbon membranes was obtained by scanning electron microscopy (SEM), using a Hitachi S-3000N instrument. Imaging was achieved under high vacuum at 20 kV acceleration voltage. The samples were analysed both for their top view and for their cross-sections. An energy dispersion spectroscopy (EDS) apparatus coupled to the SEM microscope, Link QX-200, was used to corroborate the chemical composition of the samples.

Considering the final application of the prepared composites, their performance as membranes stands out as a test of the utmost importance. In order to analyse the permeation characteristics of the membranes, the LTA/carbon composites were sealed in a stainless steel module by means of silicone O-rings, leaving an exposed area of the membrane of 0.28 cm^2. This permeation cell was attached to a permeation set up known as the Wicke–Kallenbach (WK) cell [28]. In this kind of equipment a gas or mixture of gases is fed to one of the sides of the membrane on a continuous basis. This feed gas is divided into two streams upon reaching the membrane: the permeate, which corresponds to the fraction that flows though the membrane, and the retentate, which corresponds to the fraction that does not and is removed from the system on a continuous basis. The other side of the membrane is flushed continuously by an inert gas (which ideally does not back-permeate to the membrane feed side) named sweep gas. Thus, quasi-steady state is reached in a WK cell once the permeate flux has stabilised.

Prior to the measurements the membrane unit was heated at 403 K flushing the system with helium at a flow rate of 100 ml min^{-1} at both sides of the membrane to remove adsorbed impurities. The flow rates of all streams were controlled by automated mass flow controllers (Digital Bronkhorst Hi-Tec) and kept at 100 ml min^{-1} on both sides of the membrane. The total pressure of both sides of the membrane was regulated at 1–2 bar. Permeation experiments were performed between 293 and 373 K. The concentration of the permeating gases was analysed with a mass spectrometer (Balzers, Thermostar GSD 301T).

15.4 Discussion and Applications

15.4.1 MFI Zeolite Membranes

One of the main applications of zeolitic materials is the separation of mixtures based on different mechanisms, such as steric, equilibrium, and kinetic ones [29]. Considering quasi-equilibrium conditions (as it is the case when using a WK cell), zeolites have micropores with different abilities to accommodate different molecules. As a rule of thumb, the stronger adsorbing species are preferentially removed by the adsorbent. [30]. Out of the many existing zeolite structures, MFI is perhaps the most studied, best characterised and the most widely used in practice [31]. In particular, silicalite-1, the all-silica MFI zeotype, is hydrophobic and stable up to high temperatures. It provides a nonpolar structure for adsorption of relatively small molecules. These characteristics can be used for environmental applications, such as the adsorption of nonpolar hydrocarbon molecules in water [32]. The pore structure of such zeolite consists of a system of intersecting channels composed of zigzag channels along the x axis, cross-linked by straight channels along the y axis. Both channels are defined by 10-membered rings (a member of such rings would be the unit comprising an oxygen atom and either a silicon or an aluminium atom). The straight channels are approximately elliptical in shape having a 0.53×0.56 nm cross section while the zigzag channels have a 0.51×0.55 nm cross section.

The ability of this zeolite to interact with relatively small, nonpolar molecules opened new avenues for the separation of linear from branched hydrocarbons. The importance of boosting the octane number in light hydrocarbon mixtures in modern petrochemical industry acts as a key factor due to the environmental restrictions that are being imposed on fossil fuels. Thus, one very important application of membranes in general is their use in industry as a means to separate linear from mono- and dibranched hydrocarbons in C4–C6 mixtures, as other authors have already noted (e.g., see Refs. [33,34]). In the case of MFI zeolites (more specifically, silicalite-1), separation of such mixtures occurs mostly due to the preferential adsorption of the linear alkanes on the zeolite channels, which ultimately results in higher linear alkane fluxes through the membrane. In this respect, numerous studies have been carried out for the separation of n-butane from i-butane [35] or the separation of n-hexane from 2-methylpentane [36]. Furthermore, these processes have also been employed as standard tests for the assessment of the quality of synthesised membranes. More recently, the separation of xylene isomers has stood out as an

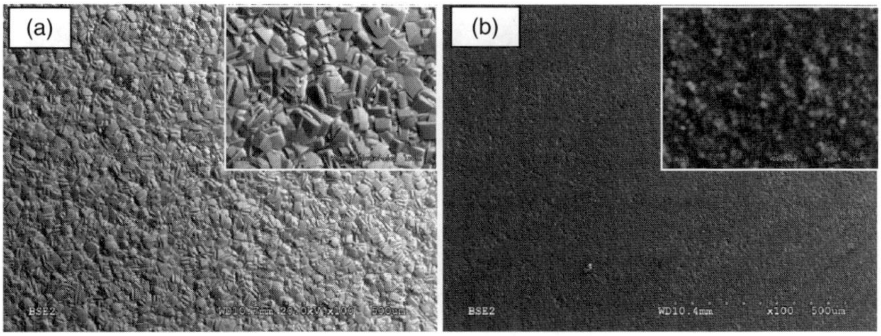

Figure 15.2 *Two silicalite-1 membranes synthesised by (a) direct synthesis (without EPD) and (b) secondary synthesis (EPD followed by hydrothermal treatment). Note the marked differences in crystal size when the two methods are used. Larger crystal sizes mean longer diffusion paths and thus hindered membrane properties*

alternative for the testing of membrane quality. In this respect, Tsapatsis and coworkers reported a very elegant strategy for the synthesis of high-performance silicalite-1 membranes by using multimers of the templating agent, obtaining *o*-xylene/*p*-xylene separation factors in the range of several hundreds [37].

Our first results in the separation of light alkanes [7], which were dedicated to the separation of *n*-C4/*i*-C4 mixtures, resulted in separation factors over 10 at room temperature (which is taken as an indicative of good membrane quality) using silicalite-1 membranes prepared on carbon discs by direct synthesis. These results were in good agreement with previous results obtained using silicalite-1 membranes supported on alumina [38] or porous metals [39], although lower fluxes were observed in our case, probably due to template residues present from incomplete burn-off during the calcination stage or to the thickness of the separating layer. In this sense, the use of EPD permits the preparation of thinner zeolitic layers, which constitutes an advantage for this kind of applications. As Figure 15.2 clearly shows, the use of EPD for the preparation of silicalite-1 membranes supported on porous carbon discs [14] results in zeolite crystals with a substantially decreased size.

15.4.2 LTA Zeolite Membranes

According to their molecular sieving properties, zeolites act as selective regenerable adsorbents for purification of gaseous streams and separation of vapours and gases. In this respect, type LTA zeolites gained significant industrial relevance for drying of technical gases and liquids and for the n/i-alkane separation in discontinuous sorption processes thanks to its high hydrophilic character. The sodium exchanged form, Na-LTA, besides being largely employed as an ion exchanger in the field of detergency [40], is used for air, methane, natural gas and nitrogen purification. The separation of the two main components of air has been attracting increasing attention in recent years due to the industrial importance of both N_2 and O_2 [41]. The two most popular approaches for their separation include cryogenic distillation and pressure swing adsorption. The former is an energy-consuming process which is only economically viable in the large scale, whereas the latter requires packed beds of a substantially large volume. Thus, the development of membranes capable of performing continuous air enrichment still remains a very interesting prospect that has been discussed and tested by previous authors [42–44]. Nevertheless, it must be pointed out that most of the works reported so far refer to mixed matrix

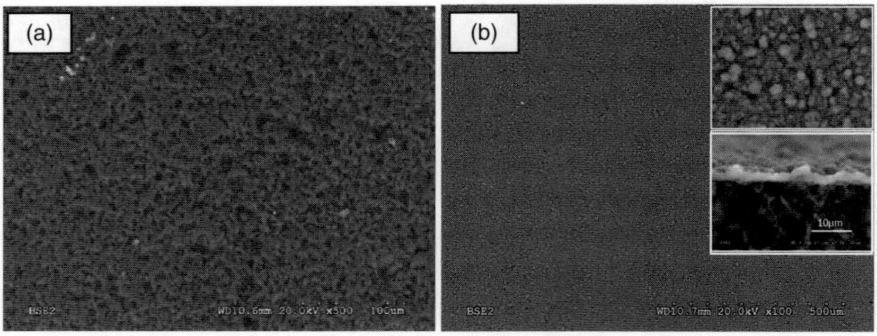

Figure 15.3 *SEM images of (a) a macroporous carbon disc after direct synthesis using a zeolite A precursor solution and (b) the same macroporous carbon disc after secondary growth (EPD followed by hydrothermal treatment) using an identical precursor solution. The insets in (b) show a detailed view of the grown crystals (top) and a cross section of the prepared composite (bottom)*

composites, in which one of the components is a polymer. Few reports on zeolite membranes for such application exist (e.g., see Ref. [45]). Our study focused on the use of carbon supports for the deposition of a zeolite layer on its surface in order to prepare a new material with novel properties. As well as in the case of silicalite-1, where seeding by EPD resulted in a substantial improvement from the methodologies established in our laboratories, the use of the secondary growth method in the case of strongly hydrophilic zeolites (i.e., with high aluminium contents) proved to be a turning point in the growth of continuous thin films on carbon supports. In this respect, while using an unseeded support yielded a very small amount of zeolite A crystals covering the surface of the carbon disc after hydrothermal treatment, the use of a seeded disc resulted in a continuous layer of LTA crystals with good intergrowth (see Figure 15.3).

Our results indicated that it was possible to separate oxygen from nitrogen using synthetic air as the feed gas mixture (i.e., 80% N_2, 20% O_2) using a zeolite A membrane synthesised at 373 K following the secondary growth methodology, employing EPD as the seeding technique. The prepared membranes showed permeance values in good agreement with those found in the literature together with O_2/N_2 separation factors as high as 2.5 at room temperature (see Figure 15.4). As expected, this value decreased with increasing temperature due to an increase in both the zeolite framework flexibility and the diffusion coefficients. The permeance flux and the separation factor values obtained were very similar to other materials in the literature [46], but lower than other composites, in which polymers are used in the preparation of the zeolite/carbon composites [47], and which have achieved O_2/N_2 separation factors between 10 and 16. It should be noted, however, that in the latter example the zeolite did not form a film on the support, but was embedded within a polymer matrix, which then rendered a carbon network by pyrolysis.

15.4.3 Outlook on Zeolite Membranes

At this point, the potential of zeolite membranes in a wide range of applications becomes clear. A couple of aspects should be nevertheless addressed for clarity and perspective purposes:

1. The questions arises whether there is an 'ideal' membrane which is capable of efficiently separating gases upon demand. The answer is, unfortunately, there is no such material. Bearing in mind the span of molecular sizes, polarities, diffusivities, and so forth of some of the molecules at play (from gases such as hydrogen or nitrogen to vapours such as xylene or

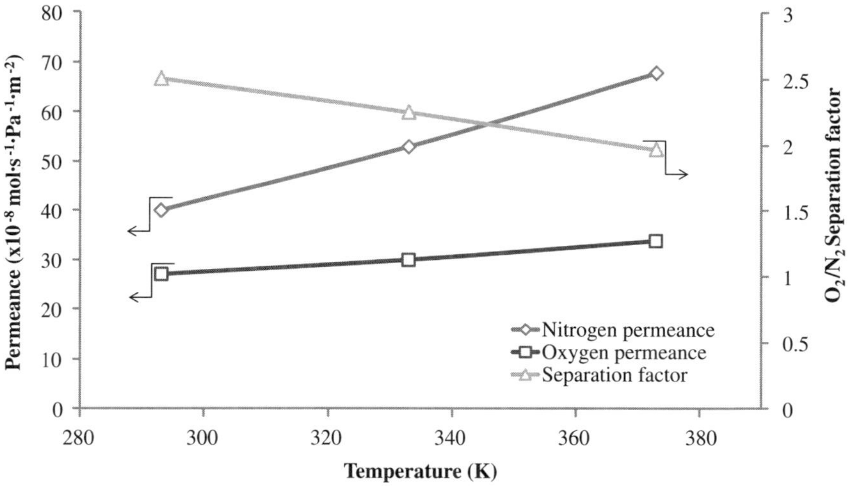

Figure 15.4 *Dependence of nitrogen and oxygen permeances and the O_2/N_2 separation with temperature using a zeolite A/carbon membrane (adapted from Ref. [18])*

hexane), the variety of zeolites at our disposal becomes an indispensable tool to achieve the desired separation. In this respect, it is important to note that the zeolite layer thickness is a parameter of great importance. Gases with high diffusitivies, at least in principle, require thicker zeolite layers to make up for the presence of defects, intercrystalline gaps, and/or defects in order to retain an acceptable separation factor. When vapours are considered, however, condensation of the species in such defects may improve the performance of the membrane, thus requiring thinner zeolitic layers. All things considered, in most cases, selecting the appropriate zeolite membranes for a given mixture will rely of a compromise solution.

2. Considering the temperature range which has been normally been studied for supported zeolite membranes [48], the question of their applicability has been studied by some authors with promising results (e.g., see Ref. [49]). The thermal stability of zeolitic materials makes them promising candidates for high-temperature separation processes, but their anisotropic thermal expansion coefficient [34] must be taken into account in order to prevent crack formation and the subsequent loss in membrane performance.

15.5 Conclusions

Electrophoretic deposition has provided the possibility of coating materials with a wide variety of particulate deposits. Amongst the possible deposits, zeolites have sprung as a very interesting candidate due not only to their remarkable properties, but also due to the implications in the preparation of continuous zeolite films for the development of membranes with novel properties. In this sense, our group has worked extensively for the preparation of different types of zeolite coatings on carbon materials in which a seeding methodology based on EPD of nanoparticles has resulted in high quality coatings. In this chapter two examples of the synthesis of zeolite coatings (i.e., silicalite, LTA zeolite) have been described. The results presented have shown the great potential of these systems by effective combination of EPD and hydrothermal treatment towards the development of integrated zeolite membranes.

Acknowledgements

The authors would like to thank the Spanish Ministerio de Ciencia e Innovación and FEDER (Projects CTQ2006-08958/PPQ, CTQ2009-10813/PPQ, RyC-2009-03813 and MAT2007-60621) and Generalitat Valenciana (ACOMP/2009/174 and PROMETEO/2009/047) for financial support. FJVG thanks the University of Alicante for a PhD studentship.

References

1. F.F. Reuss, *Mem. Soc. Imp. Nat. Moscow*, **2**, 1809 (1809).
2. D.A. Salville, *Annu. Rev. Fluid Mech.*, **9**, 321 (1977).
3. O.O. Van der Biest, L.J. Vandeperre, *Annu. Rev. Mater. Sci.*, **29**, 327 (1999).
4. L. Besra, M. Liu, *Prog. Mater. Sci.*, **52**, 1 (2007).
5. A.R. Boccaccini, J.A. Roether, B.J.C. Thomas, M.S.P. Shaffer, E. Chavez, E. Stoll, E.J. Minay, *J. Ceram. Soc. Jpn*, **114** (1), 1 (2006).
6. Y. Yan, M.E. Davis, G.R. Gavalas, *Ind. Eng. Chem. Res.* **34**, 1652 (1995).
7. A. Berenguer-Murcia, J. Garcia-Martinez, D. Cazorla-Amoros, A. Linares-Solano, A.B. fuertes, *Micropor. Mesopor. Mater.*, **59**, 147 (2003).
8. B.J. Schoeman, A. Erdem-Senatalar, J. Hedlund, J. Sterte, *Zeolites*, **19**, 21 (1997).
9. Z.A.E.P. Vroon, K. Keizer, A.J. Burggraaf, H. Verweij, *J. Membr. Sci.*, **144**, 65 (1998).
10. J. H. Koegler, H. Van Bekkum, J.C. Jansen, *Zeolites* **19**, 262 (1997).
11. Y. Takata, T. Tsuru, T. Yoshioka, M. Asaeda. *Micropor. Mesopor. Mater.*, **54**, 257 (2002).
12. H. van Bekkum, E.M. Flanigen, P.A. Jacobs, J.C. Jansen, *Introduction to Zeolite Science and Practice*, Elsevier, Amsterdam (2001).
13. S. Mintova, J. Hedlund, V. Valtchev, B. Schoeman, J. Sterte. *J. Mater. Chem.*, **8** (10), 2217 (1998).
14. Á. Berenguer-Murcia, E. Morallón, D. Cazorla-Amorós, A. Linares-Solano. *Micropor. Mesopor. Mater.*, **66**, 331 (2003).
15. J. Hedlund, J. Sterte. *Nanostruct. Mater.*, **12**, 413 (1999).
16. M.P. Bernal, G. Xomeritakis, M. Tsapatsis. *Catal. Today*, **67**, 101 (2001).
17. Á. Berenguer Murcia, Ph. D Thesis. University of Alicante (2005).
18. S. Domínguez-Domínguez, Á. Berenguer-Murcia, E. Morallón, Á. Linares-Solano, D. Cazorla-Amorós. *Micropor. Mesopor. Mater.*, **115**, 51 (2008).
19. W. Shan, Y. Zhang, W. Yang, C. Ke, Z. Gao, Y. Ye, Y. Tang, *Micropor. Mesopor. Mater.*, **69** (1/2), 35 (2004).
20. M. Abdollahi, S.N. Ashrafizadeh, A. Malekpour, *Micropor. Mesopor. Mater.*, **106** (1/3), 192 (2007).
21. K. Kanamura, J. Hamagami, *Solid State Ionics*, **172**, 303 (2004).
22. J. Hedlund, B. Schoeman, J. Sterte, *Chem. Comm.* 1997, 1193 (1997).
23. A.E. Persson, B.J. Schoeman, J. Sterte, J.E. Otterstedt, *Zeolites* **14** (7), 557 (1994).
24. V. Valtchev, B. Schoeman, J. Hedlund, S. Mintova, J. Sterte, *Zeolites* **17**, 408 (1996).
25. M.C. Lovallo, M. Tsapatsis, *AIChE J.*, **42**, 3020 (1996).
26. J. Caro, M. Noack, P. Kölsch, R. Schäfer, *Microporous Mesoporous Mater.*, **38**, 3 (2000).

27. W. Zhu, L. Gora, A.W.C. van den Berg, F. Kapteijn, J.C. Jansen, J.A. Moulijn, *J. Membr. Sci.*, **253**, 57 (2005).
28. K. Hou, M. Fowles, R. Hughes, *Chem. Eng. Res. Des.*, **77**, 55 (1999).
29. D.D. Do, *Adsorption Analysis: Equilibria and Kinetics*, Imperial College Press, London (1998).
30. W. Zhu, PhD Thesis, TU Delft (2001).
31. R. Krishna, D. Paschek, *Phys. Chem. Chem. Phys.*, **3**, 453 (2001).
32. M.S. Sun, O. Talu, D.B. Shah, *J. Phys. Chem.*, **100**, 17276 (1996).
33. E.E. McLeary, J.C. Jansen, F. Kapteijn, *Microporous Mesoporous Mater.*, **90**, 198 (2006).
34. Á. Berenguer-Murcia, J. García-Martínez, D. Cazorla-Amorós, Á. Linares-Solano, A.B. Fuertes, *Microporous Mesoporous Mater.*, **59**, 147 (2003).
35. Z.A.E.P. Vroon, K. Keizer, A.J. Burggraaf, H. Verweij, *J. Membr. Sci.* **144**, 65 (1998).
36. J. Coronas, R.D. Noble, J.L. Falconer, *Ind. Eng. Chem. Res.*, **37**, 166 (1998).
37. Z. Lai, G. Bonilla, I. Diaz, J.G. Nery, K. Sujaoti, M.A. Amat, E. Kokkoli, O. Terasaki, R.W. Thompson, M. Tsapatsis, *Science* **300** (5618), 456 (2003).
38. C. Bai, M.D. Jia, J.L. Falconer, R.D. Noble, *J. Membr. Sci.* **43**, 105 (1995).
39. W.J.W. Bakker, F. Kapteijn, J. Poppe, J.A. Moulijn, *J. Membr. Sci.* **117**, 57 (1996).
40. W. Bückner, R. Schliebs, G. Winter, K.H. Büchel, *Industrial Inorganic Chemistry*, VCH, Weinheim (1989).
41. G. Q. Guan, K. Kusakabe, S. Morooka, *J. Chem. Eng. Jpn*, **34**, 990 (2001).
42. G. R. Rigby, H. C. Watson, *J. Membr. Sci.*, **87**, 159 (1994).
43. Z. Huang, Y. Li, R. Wen, M. M. Teoh, S. Kulprathipanja, *J. Appl. Pol. Sci.*, **101**, 3800 (2006).
44. X. J. Yin, G. S. Zhu, W. S. Yang, Y. S. Li, G. Q. Zhu, R. Xu, J. Y. Sun, S. L. Qiu, R. R. Xu, *Adv. Mater.*, **17**, 2006 (2005).
45. G. Q. Guan, K. Kusakabe, S. Morooka, *Sep. Sci. Tech.*, **37**, 1031 (2002).
46. X.C. Xu, W.S. Yang, J. Liu, L.W. Lin, *Microporous Mesoporous Mater.*, **43**, 299 (2001).
47. Q.L. Liu, T.H. Wang, C.H. Liang, B. Zhang, S.L. Liu, Y.M. Cao, J.S. Qiu, *Chem. Mater.*, **18**, 6283 (2006).
48. Á. Berenguer-Murcia, L. Gora, W. Zhu, J.C. Jansen, F. Kapteijn, D. Cazorla-Amoros, Á. Linares-Solano, *Ind. Eng. Chem. Res.*, **46**, 3997 (2007).
49. M. Kanezashi, Y.S. Lin, *J. Phys. Chem. C*, **113** (9), 3767 (2009).
50. S. Mintova, J. Hedlund, B. Schoeman, V. Valtchev, J. Sterte, *Chem. Commun.*, **13**, 1193 (1997).
51. J. Hedlund, M. Noack, P. Kölsch, D. Creaser, J. Caro, J. Sterte, *J. Membr. Sci.*, **159**, 263 (1999).
52. M. C. Lovallo, M. Tsapatsis, *AIChE J.*, **42**, 3020 (1996).
53. K. Weh, M. Noack, I. Sieber, J. Caro, *Micropor. Mesopor. Mater.*, **54**, 27 (2002).
54. K. Kusakabe, T. Kuroda, S. Morooka, *J. Membr. Sci.*, **148**, 13 (1998).
55. I. Lee, J. Buday, H. Jeong, *Micropor. Mesopor. Mater.*, **122**, 288 (2009).
56. C. Algieri, P. Bernardo, G. Barbieri, E. Drioli, *Micropor. Mesopor. Mater.*, **119**, 129 (2009).
57. S. M. Holmes, C. Markert, R. J. Plaisted, J. O. Forrest, J. R. Agger, M. W. Anderson, C. S. Cundy, J. Dwyer, *Chem. Mater.*, **11**, 3329 (1999).
58. S. Mintova, T. Bein, *Adv. Mater.*, **13**, 1880 (2001).
59. H. Negishi, A. Endo, T. Ohmori, K. Sakaki, *Ind. Eng. Chem. Res.*, **47**, 7326 (2008).
60. T.W. Kim, M. Sahimi, T.T. Tsotsis, *Ind. Eng. Chem. Res.*, **47**, 9127 (2008).
61. C. Song, G. Villemure, *J. Electroanal. Chem.*, **462**, 143 (1999).
62. C.B. Ahlers, J.B. Talbot, *Electrochimica Acta*, **45**, 3379 (2000).
63. B. Yu, S.B. Khoo, *Electrochem. Comm.*, **4**, 737 (2002).
64. T. Seike, M. Matsuda, M. Miyake, *Solid State Ionics*, **151**, 123 (2002).
65. X. Wang, E. A. Olevsky, E. Bruce, M. B. Stern, D. T. Hayhurst, *Surf. Eng.*, **23** (6), 443 (2007).

16

Electrochemical Preparation of Nanoparticle Deposits: Application to Membranes and Catalysis

J. Arias-Pardilla[1], *A. Berenguer-Murcia*[2], *D. Cazorla-Amorós*[2] *and E. Morallón*[3]

[1]*Centro de Electroquímica y Materiales Inteligentes. Universidad Politécnica de Cartagena, Cartagena, Spain*

[2]*Universidad de Alicante, Departamento de Química Inorgánica, Alicante, Spain*

[3]*Universidad de Alicante, Departamento de Química Física and Instituto Universitario de Materiales, Alicante, Spain*

16.1 Introduction

16.1.1 Principles of Electrochemical Deposition

Electrocrystallisation (or electrochemical deposition) constitutes a process by which deposition of metals or other substances at electrodes takes place as a consequence of an electrodic process exhibiting a number of specific features. Electrochemical deposition of metals is an area of considerable interest both from a fundamental and applied point of view and is used in different areas such as coatings, analysis of solutions or preparation of catalysts. The process takes place under the influence of an electric field and includes phenomena of formation of a new phase and interphase, which complicates its study. Therefore, if electrochemical properties are to be studied, transient methods should be employed that allow measurements to be made before major changes in surface morphology (structure) occur.

The two steps that take place during electrocrystallisation are, essentially, discharge of ions at the electrode and the incorporation of the discharged ions into the crystal lattice. These steps are

Membranes for Membrane Reactors: Preparation, Optimization and Selection, Edited by Angelo Basile and Fausto Gallucci

separated both in time and space, and intermediate species called adatoms exist at the surface. These species are not strongly bound to the surface and can thus move freely on it.

Problems of the crystal-building step in electrocrystallisation are in many aspects identical to those of crystal growth from the gas phase. Crystal building can occur in any of next three possible ways:

1. Growth from kinks, which constitute a suitable half-lattice position at which an atom is surrounded by one-half of the number of atoms that would coordinate it in the bulk of the metal. There, adatoms can be successively trapped and thus the crystal lattice is extended along a crystal edge and further on across the surface.
2. Mechanisms associated with screw dislocations, or twinning edges, can provide for a continuous growth of crystals (i.e. the screw dislocation mechanism).
3. If growth sites are rare, or if the support at which deposition should take place is very different from the depositing metal, the charge transfer results in supersaturation (i.e., accumulation of adatoms to a concentration considerably larger than that which can exist at equilibrium conditions with the crystal lattice). In such a situation, agglomeration of adatoms to form crystal nuclei is favoured.

There is still controversy about the basic principles involved in modelling these systems although the first experimental studies in this field were done in the early nineteenth century. In any case, there is a general agreement that electrodeposition occurs through a process of nucleation and growth, that is, the nuclei appear in active sites of the substrate by any law of nucleation rate, and then grow through the incorporation of other ions from dissolution. The basic thermodynamic concepts of nucleation and crystal growth were described in 1878 by Gibbs. In the beginning of the twentieth century, these concepts were developed by different authors as Volmer, Kossel, Stranski, Kaischew, Becker and Döring introducing statistical approaches and molecular kinetics [1].

16.1.2 Choice of Methods and Deposited Metals

There are numerous well documented methods for the deposition of particles (catalytic or not) on different supports, which include wet impregnation [2], microwave irradiation [3], microemulsion [4], polyol process [5], or two-step spray pyrolysis [6]. On the other hand, electrochemical deposition is an efficient method for the controlled preparation of nanostructured materials as metallic particles, alloys or composites. Compared to vacuum-based techniques, the advantages of electrodeposition include high deposition rates, a simple experimental setup, and low cost. Additionally, these techniques remove the restriction of high thermal resistance of the substrate and the deposition can be carried out even at room temperature, minimising interdiffusional limitations or chemical reactions, and the deposited quantity can be accurately controlled by monitoring the consumed electric charge. Furthermore, the tunability and versatility of the technique is further enhanced by the fact that it can be easily adapted by selecting the metal that may be deposited and the material on which deposition would take place (see Section 16.2.2 – Supports and deposited metals).

16.2 State of the Art

16.2.1 Methodologies for Electrochemical Deposition and Theoretical Models

The methodologies that are nowadays most widespread are:

1. *Current step* [7,8]: A constant current is applied to the support electrode; this provides the possibility to get a qualitative insight into the participation of parallel reduction processes. Using this methodology, the growth of the metal particles occurs under varying overpotential and may result in broader size distributions and various growth shapes of the metal crystals.
2. *Potential sweep* [9,10]: In this case the potential of the support electrode is changed following a time program, usually linear, during the experiment. As in the case of current step, the growth of the metal particles occurs under varying overpotential, obtaining broader size distributions.
3. *Potential step*: Used as single potential step [11–14] or double-pulse deposition [15–17]. In this case, the same overpotential is used offering the possibility for initiating instantaneous nucleation. This would result in narrower size distributions of the metal particles and in low overpotential experiments in equilibrium-shaped crystals. Especially in the case of deposition through two or more pulses or potential step deposition (PSD), we have a powerful tool to fine tune the amount of metal deposited, the number of metallic sites and their size by adjusting the initial and final step potentials and its duration. This method offers the possibility of applying a given pulse program in such a fashion that allows for the nucleation and growth to be performed (and controlled) separately. Another advantage of the potentiostatic technique is that it provides additional information on the kinetics of the metal nucleation and growth, through the theoretical modelling of current transients obtained in the course of metal electrocrystallisation.

As the experimental techniques for the electrochemical deposition of different metals have been refined, several theoretical models have appeared in order to predict and assess the nucleation and growth pathways that the species will follow during the electrodeposition process. Some of those models are: Bewick–Fleischmann–Thirsk model (BFT) [18], Armstrong Fleischmann–Thirsk model (AFT) [19], Scharifker–Hills model (SH) [20], and Mirkin–Nilov–Heerman–Tarallo model (MNHT) [21,22].

16.2.2 Supports and Deposited Metals: Membrane Reactors

In the literature we can find a large number of materials used as supports for electrochemical deposition of metals, alloys, composites, etc. The only requirement is its electrical conductivity. The electrochemical deposition can be done on any kind of metallic support, carbon materials such as graphite, vitreous carbon, and so forth. Porous nonconducting materials can also be used as templates for electrochemical deposition, if previously one face of the template is coated with a conductive film, usually via either ion sputtering or thermal vapour deposition. In this way, anodised aluminium oxide (AAO) and track etched polycarbonate membranes for one-dimensional nanomaterials or ordered porous materials fabrication have been used. The main advantage of AAO membranes is that the pores are practically perpendicular to the surface; however, this does not occur in polycarbonate membranes because, due to the manufacturing method, the channels have an angle of 34° with respect to the surface normal. In this way, Au [23], Pd [24,25] and Ni [26] nanoparticles have been obtained, although it is difficult to generate macroscopic quantities of nanoparticles using this method due to the fact that growth is limited to the template pores arrayed in a two-dimensional plane.

Many different metals and alloys can be electrodeposited. Due to its catalytic properties, platinum [27,28] is the most studied metal, but other important noble metals have been deposited like palladium [29,30], gold [31,32], silver [33,34], copper [35,36], and other metals [37–44], as well as alloys of platinum with another metal as Pt-Ru [45,46], Pt-Sn [47,48], Pt-Pb [49], Pt-Os [50] and Pt-Mo [50].

A research field that has gained increasing importance during the past decade is that of catalytic membrane reactors (CMRs) [51] that can operate under continuous conditions. In contrast to *classic* packed-bed reactors (PBRs), CMRs constitute an outstanding example of a means to reach more efficient systems by providing a reactor with either a product or poison/inhibitor removal unit (thus acting as a conversion enhancer) or with a reactant distributor system (thus acting as a selectivity enhancer) avoiding the large pressure drops that may occur in PBRs. Modern design of CMRs include the control of residence time or reactant traffic, and also constitute good examples of selectivity control. More specifically, the developments in the area of zeolite membrane reactors have been impressive during the last few years (a tentative example is given below). Thus, the procedures available nowadays for the preparation of zeolite membranes and layers have reached enough quality and reliability to allow the development of the first industrial plant based on zeolite membranes, which are employed for ethanol dehydration [52].

16.3 Experimental

16.3.1 Instrumentation and Reactants

In this chapter, three different carbon materials have been used as working electrode (WE): graphite (Ellor +35) and glassy carbon (CV25) both supplied by Carbone Lorraine and a macroporous carbon prepared by the agglomeration of natural expanded graphite (DFP-1 supplied by Poco Graphite). In this last case, discs were obtained from a carbon sheet (thickness 0.3 mm, mean pore size 0.7 μm, exposed geometric area 2.9 cm^2).

All electrodeposition experiments were performed in a conventional electrochemical cell consisting of three electrodes kept at room temperature [53,54]. For all experiments, the counter electrode (CE) was a platinum wire and the reference electrode (RE) was a reversible hydrogen electrode (RHE) immersed in a 0.5 M H_2SO_4 solution, which was connected to the working electrode compartment by means of a Luggin capillary.

All solutions were prepared using high-purity water (resistivity $\geq$ 18 MΩ cm) obtained from an Elga Labwater Purelab Ultra system. The solutions used were 0.5 M H_2SO_4 (Merck, Suprapur), 5 mM H_2PtCl_6 (from the solid purchased from Sigma–Aldrich) in 0.5 M H_2SO_4 and 0.1 M CH_3OH (Merck) in 0.5 M H_2SO_4. All solutions must be saturated in nitrogen before their use and inert atmosphere was maintained during the experiments.

16.3.2 Procedure

The metallic particles were electrochemically deposited under potentiostatic conditions. Thus, in order to study the influence of different parameters on the electrodeposition of platinum, four different experimental procedures were used:

1. Single step from an initial potential (where no deposition of platinum occurs, in our case 0.80 V) to different final potentials (0.20, 0.15, 0.10, and 0.05 V in which the platinum is deposited) for different times (from 0.1 to 60.0 s).
2. Several consecutive steps from an initial potential (0.80 V) to a final potential (0.15 V) for different times, being the total time the same as in procedure (1).

3. Single step from an initial potential (0.80 V) to an intermediate potential (–0.35, –0.25, –0.15, 0.0 V) for 1 s, and immediately to several final potentials (0.20, 0.15, 0.10, 0.05 V) for 6 s.
4. Multiple, consecutive steps applied for a short time (5 ms) from an initial potential (0.80 V) to a final potential (0.00 or –0.15 V) for 5 ms and being the total time 5 s. All runs were repeated three times in order to ensure reproducibility of the results.

The electrochemical properties of the prepared Pt-supported electrodes were tested in the methanol oxidation reaction, which was measured in a 0.1 M $CH_3OH + 0.5$ M H_2SO_4 solution by cyclic voltammetry, over a potential range of 0.06–1.0 V with a scan rate of 10 mV s^{-1}.

16.3.3 Sample Treatment

In order to analyse the physicochemical properties of the prepared materials, they were analysed by X-ray photoelectron spectroscopy (XPS) in a VG-Microtech Multilab electron spectrometer to assess the oxidation state of the deposited particles. No reductive treatment was applied to the Pt/C samples prior to XPS analyses. The source was Mg Kα ($h\nu = 1253.6$ eV) radiation (generated from twin anodes in constant analyser energy mode) and the pressure of the analysis chamber was maintained below 5×10^{-10} mbar.

The binding energy was adjusted by setting the C 1s transition at 284.6 eV with ± 0.2 eV accuracy. The intensities of the peaks were estimated by determining the integral of each peak, after subtraction of an S-shaped background and fitting of the experimental peak to Lorentzian/ Gaussian lines (L : G = 70 : 30).

The active surface area of electrodeposited Pt was measured by comparing the charge corresponding to the adsorption/desorption processes at Pt sites between 0.05 and 0.45 V of the cyclic voltammograms (CVs) of the synthesised electrodes in 0.5 M H_2SO_4 aqueous solutions, before and after platinum deposition.

Scanning electron microscopy (JEOL JSM-840) was used in order to analyse the surface morphology of the samples, whereas transmission electron microscopy (JEOL JEM 2010 operating at 200 kV with a structural spatial resolution of 0.5 nm), was used to observe the size of the smallest platinum particles that could not be detected with SEM images. Small quantities of sample in powder form were collected by scratching and subsequently dispersed in ethanol. The suspension was then placed onto a copper grid covered with a Lacie carbon film.

16.4 Discussion and Applications

16.4.1 Electrodeposition of Platinum on Carbon Materials

Nowadays the need for a new clean energy vector has driven the research focused on hydrogen and its related technologies to a remarkable extent. In this sense, fuel cells have been attracting increasing attention during the past decade (a fact which is reflected by the thousands of publications that have appeared in the literature since the publication in 1967 on the first report of a hydrazine–air fuel cell [55]). Pursuing such an interest, we aimed at the development of platinum catalysts for the electrooxidation of methanol. In this sense, as is often the case with supported catalysts, it is of crucial importance to be able to control the particle size (and consequently the dispersion) of the active phase. Thus electrodeposition offers a key tool due to the large number of parameters that may be fine tuned in order to fit specific needs. Additionally,

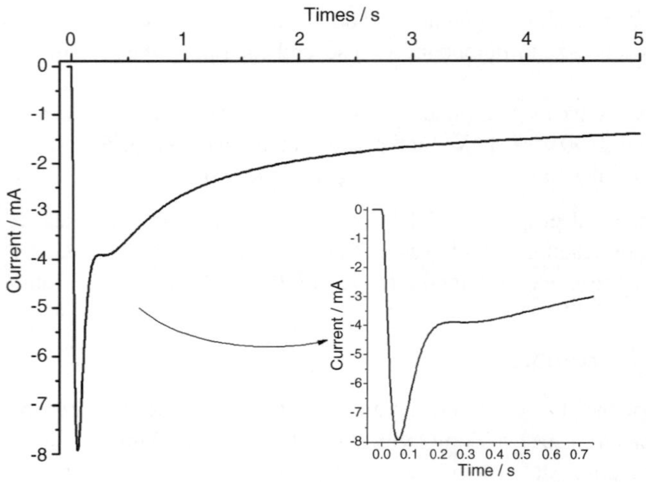

Figure 16.1 *Current versus time plot of a macroporous carbon disc sample with platinum deposited following methodology (1) – single step deposition. Inset: magnification of the lowest time scale*

electrodeposited metal nanoparticles can be useful for zeolite films growth and membrane reactor applications.

A typical chronoamperogram is shown in Figure 16.1, in which the three zones observed are similar to those reported in the deposition of platinum on synthetic boron-doped diamond and carbon electrodes [56]:

1. A sharp decrease in current from the charging of the double layer and a rise in current due to the initiation of the nucleation process (see inset in Figure 16.1).
2. Increase in the current due to both the growth of independent nuclei and the creation of new nuclei. At this point, the observed current is simply the deposition current, without any overlapping effects.
3. In the third zone, two opposite effects coexist: growth of independent nuclei and overlap, reaching a maximum of the current. Afterwards a decrease in current is observed. In this region the nuclei complete their growth, and may even coalesce during this step.

Assuming optimal current efficiency and that any side effects were negligible (partial reductions of Pt(IV), hydrogen evolution at the electrodeposited Pt and double layer charging), it is possible to determine the amount of noble metal deposited from the chronoamperograms. Furthermore, assuming that the only contributing reaction was the Faradaic process:

$$PtCl_6^{2-} + 4e^- \rightarrow Pt + 6Cl^- \tag{16.1}$$

the amount of Pt deposited is:

$$m_{Pt} = \frac{Q_{Pt}M}{4F} \tag{16.2}$$

where Q_{Pt} is the electrical charge employed during the process (C), M is the molecular weight of platinum (195.1 g mol^{-1}), and F is Faraday's constant (96 485.3 C mol^{-1}).

From our observations, there was a direct correlation between the time during which the electrochemical treatment is applied and the amount of metal deposited on the support surface. Thus, upon using one single pulse there is a tendency to obtain fewer, larger nuclei, whereas the replacement of a single, large step (method 1) for several short steps with the same final duration using the same initial and final potentials (method 2), yields a larger number of nuclei with smaller average size. Thus, the quantity of platinum deposited is observed to increase with step time and the number of steps.

In order to ascertain the mechanism through which the process of nucleation and growth take place, the theory of Scharifker and Hills [20] was applied. According to their model, there is a theoretical relationship for the determination of the nucleation mechanism, distinguishing between: (i) instantaneous nucleation, in which all nuclei are rapidly created and their number remains constant throughout the growth process, and (ii) progressive nucleation, in which the nucleation rate is low and new nuclei are continuously being formed during the whole deposition process. Furthermore, the model is also able to predict the formation of 2D islands or 3D clusters following either one of the aforementioned mechanisms. In agreement with previous results [57], the model that best fitted our chronoamperograms was that of 3D progressive nucleation.

XPS studies performed on the fresh samples revealed the presence of platinum in its metallic state from the presence of two overlapping peaks at binding energy values of 71.1 and 74.3 eV, discarding any traces of Pt^{2+} or Pt^{4+} in the prepared samples. Table 16.1 shows some of the characteristics derived from the study of the cyclic voltammograms obtained in 0.5 M H_2SO_4 aqueous solution from a start and end potential of 0.1 and 1.1 V, respectively, and a scan rate of 50 mV s^{-1}.

Electron microscopy studies (both scanning and transmission) were performed on the samples in order to verify the particle size and morphology of the deposited particles. Figures 16.2, 16.3 show some SEM and TEM illustrative images, respectively.

From the images, it can be observed that, in general, the particles exhibit uniform size and spherical shape and appear homogeneously distributed over the entire support surface. From

Table 16.1 *Amount of platinum deposited, platinum surface area, and particle size obtained for Pt/macroporous carbon electrodes prepared in different PSD conditions. In all cases the initial potential was 0.8 V*

PSD method	Conditions	m_{Pt} ($\mu g \cdot cm^{-2}$)[a]	S_{Pt} ($m^2 \cdot g^{-1}$)[b]	Particle size (nm)[c]
1	1 step (5 s) 0.15 V	1.8	12.7	22
1	1 step (15 s) 0.15 V	2.9	12.2	23
1	1 step (10 s) 0 V	11.4	10.5	27
1	1 step (10 s) 0.20 V	4.7	9.5	29
2	5 steps (1 s) 0.15 V	3.3	10.5	27
2	3 steps (5 s) 0.15 V	8.2	12.8	21
3	1 s –0.35 V/6 s 0.10 V	11.5	17.9	16
3	1 s –0.25 V/6 s 0.10 V	9.4	19.4	14
3	1 s –0.15 V/6 s 0.10 V	5.7	21.4	13
3	1 s 0 V/6 s 0.10 V	5.4	19.3	14
4	–0.15 V/0.80 V 5 ms (5 s)	3.7	29.7	9
4	0 V/0.80 V 5 ms (5 s)	2.7	20.4	14

[a]Obtained using Equation (16.2).
[b]The electrochemically active surface area is estimated from the voltammograms following the assumption that 1 cm^2 of smooth Pt requires 210 µC for the adsorption of one electron per Pt site.
[c]Calculated from the specific surface area, assuming that this value is the ratio between the area and the weight of one spherical particle.

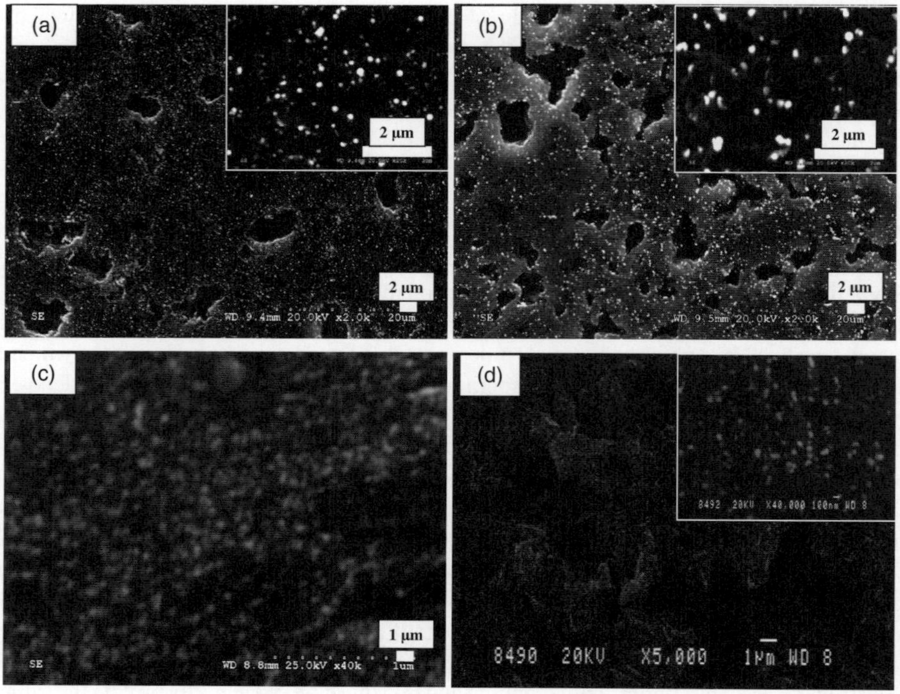

Figure 16.2 SEM images of platinum nanoparticle deposits prepared on macroporous carbon discs following: (a) procedure (1), (b) procedure (2), (c) procedure (3), (d) procedure (4). Experimental conditions: (a) 15 s/0.8–0.1 V, (b) 3 × 15 s/0.8–0.1 V, (c) 1 s/0.8–0.25 V//6 s/–0.25 to 0.1 V, (d) 5 s (in 5 ms steps)/0.8 to –0.15 V

the SEM images the observed Pt particle size appears bigger than that calculated from cyclic voltammetry studies (see Table 16.1). This is attributed to the lack of resolution of the SEM used in our study or to the possibility of the particles being aggregates of smaller crystals. TEM studies shows that the particle size is in close agreement with the one predicted from

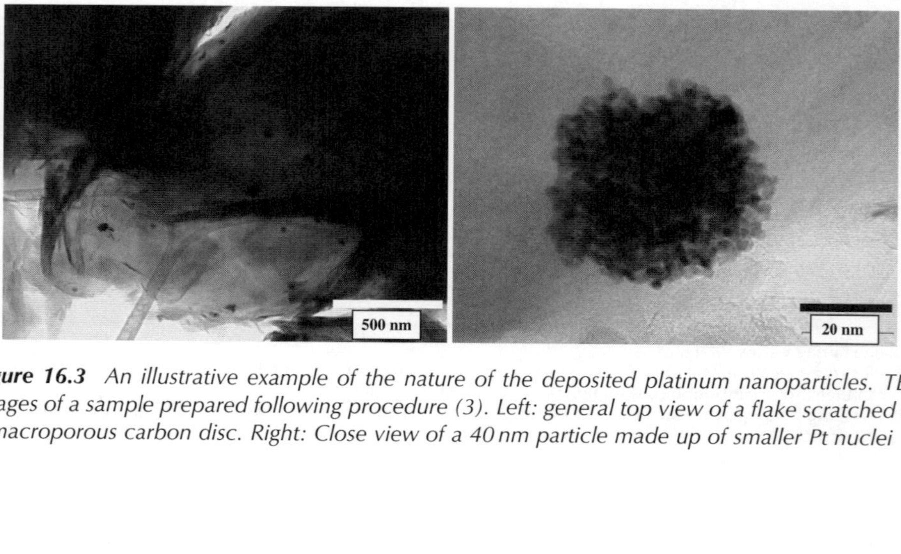

Figure 16.3 An illustrative example of the nature of the deposited platinum nanoparticles. TEM images of a sample prepared following procedure (3). Left: general top view of a flake scratched off a macroporous carbon disc. Right: Close view of a 40 nm particle made up of smaller Pt nuclei

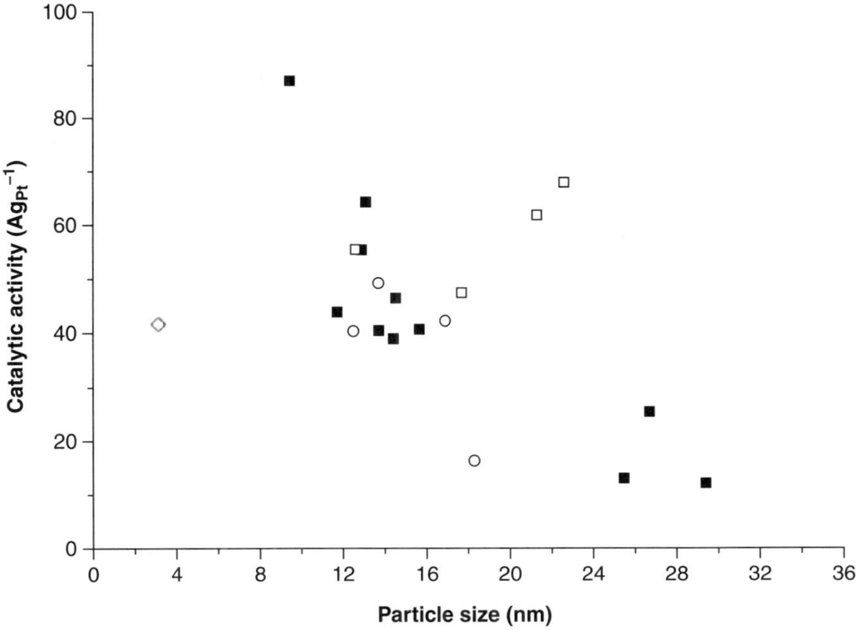

Figure 16.4 *Catalytic activity in methanol oxidation versus Pt particle size for different Pt supported on different carbon catalysts: (◇) Pt/vulcan, (■) Pt/macroporous carbon, (○) Pt/graphite, (□) Pt/glassy carbon electrodes*

voltammetry studies and that, indeed, the particles of about 40 nm diameter were agglomerates of smaller particles of about 3–5 nm diameter, as we have observed previously [58].

The best conditions for deposition onto each electrode were determined with respect to the potential applied during each step and the time during which each step is applied. The electrocatalytic activity of Pt/carbon electrodes was investigated by cyclic voltammetry in a 0.1 M CH_3OH + 0.5 M H_2SO_4 solution. The tests were performed from 0.06 to 1.0 V with a scan rate of 10 mV s^{-1}. Figure 16.4 shows the catalytic activity versus particle size for different platinum-supported carbon electrodes. The activity for a Pt/vulcan sample obtained by chemical methods has been also included for comparative purposes [58]. The catalytic activity obtained with the Pt/macroporous carbon disc was almost twice that of the other electrodes. This improved performance of the macroporous carbon support was attributed to its characteristic porosity, which would in principle allow for a better particle distribution over the support. This would in turn lead to smaller particles sizes. The characteristic porosity of macroporous carbon means that the entire 3D surface can be considered a functional electrode. Thus, it seems that the structure of the carbon support has a strong influence on the nature of the platinum particles deposited, which in turn affects the catalytic activity. The influence of particle size on Pt electrocatalytic activity has also been reported by other authors [59].

16.4.2 Influence of Metallic Deposits on Zeolite Membrane Preparation

Although there are still nowadays a few examples on zeolite membrane reactors that are used in reactor assemblies (see, for example, Refs. [51] and [60] and references contained therein), some authors have already shown the potential of using zeolites as an integral part of modern membrane systems, either as a separation layer or as a reaction layer. From the results presented in the previous subsection, we considered that a very interesting prospect could be that of a zeolite layer covering small metallic nanoparticles in order to limit the flow/supply of reactants towards the catalytic active sites, following the example that van der Puil and coworkers reported [61]. Thus, following our work on zeolite membranes, we attempted to grow a continuous silicalite-1 layer on macroporous carbon discs which had been previously coated with small platinum nanoparticles following procedure (1) (see Section 16.3 [62]). The original goal would involve the successful growth of a continuous layer of zeolite on a support surface different from the oxidised carbon disc and the investigation of the effect that the deposited metal nanoparticles would have on the growing zeolite crystals.

From our results, it must be highlighted that the presence of the small metallic nanoparticles in combination with a seeding step by electrophoretic deposition of zeolite nanoparticles prior to the zeolite layer growth, resulted in an unexpected orientation of the crystals forming the zeolite layer depending on the zeolite layer synthesis conditions (Figure 16.5). Thus, it would be possible not only to control access of substances to the surface of the macroporous carbon discs, but also to control their diffusion rates, which constitutes a significant advancement in zeolite membrane preparation.

It might seem unlikely that with an uneven and hydrophobic surface as that of our carbon support such crystal orientation would become possible. Nevertheless, the metallic platinum particles may act as efficient anchoring locations, playing a key role in both crystal growth and orientation. In fact, given the fairly homogeneous coating of Pt particles on the surface of the support (see, for example, Figure 16.3), we may be contemplating a 'metal-like' surface, which from a practical point of view should be deemed as highly convenient. The introduction of these particles with a high density of surface OH groups together with the application of the secondary growth method enhances the crystallisation kinetics. This effect, resulting in oriented zeolite crystals supported on a bed of catalytically active particle would result in systems of potential interest for membrane reactors.

Figure 16.5 *SEM images of two continuous zeolite films grown on Pt-deposited carbon discs showing two different crystalline orientations. (a) Zeolite crystals with a (010) preferential orientation. (b) Zeolite crystals with a (011) preferential orientation. The insets show images of the cross sections of both composites. Note the platinum nanoparticles (bright dots) located between the zeolite layer (top) and the carbon support (bottom) in the inset of (a)*

16.5 Conclusions

Electrochemical deposition proves to be a reliable tool for the preparation of fine tuned metal coatings on a wide variety of substrates (provided that they are electricity conducting). The many parameters that can be controlled with great accuracy such as step potential or time provides the user with a substantially wide array of possibilities for the controlled deposition of different amounts and types of metallic nanoparticles.

Our results show that despite the fact that grown platinum nanoparticles have the appearance of large crystals, closer inspection reveals that they are indeed made up of aggregates of smaller particles of approximately 2–4 nm in size.

The catalytic activity of the prepared nanoparticles highlights not only that the preparation protocol is key to obtaining the desired catalyst, but also the chemical nature and structure of the support, which seems to exert a significant influence on the performance of the prepared systems. The topography of real solid surfaces plays an important role in defining the electronic energy distribution at surface sites, particularly when irregularities at the atomic level are taken into account.

Likewise, surface irregularities at the nanometric level determine the electrocatalytic properties. In fact, the reactivity of small metal clusters has been found to vary by orders of magnitude when the cluster size is changed by only a few atoms. Furthermore, the high surface to volume ratio (as well as the very small particle size) provides nanostructured catalysts with unique properties.

Furthermore, the presence of metallic nanoparticles on the support surface has shown a remarkable effect on the growth of zeolite layers, which would in principle also potentiate their use as continuous membrane reactors (CMRs).

Acknowledgements

The authors would like to thank the Spanish Ministerio de Ciencia e Innovación, FEDER and Generalitat Valenciana (Projects CTQ2006-08958/PPQ, CTQ2009-10813/PPQ, MAT2007-60621, RyC-2009-03813, ACOMP/2009/174 and PROMETEO/2009/047) for financial support. J.A.P. would also like to thank the Spanish Ministry of Science and Innovation for a Juan de la Cierva grant.

References

1. E. Budevski, G. Saikov, W.J. Lorenz, *Electrochim. Acta*, **45**, 2559 (2000).
2. M. Endo, Y.A. Kim, M. Ezaka, K. Osada, T. Yanagiswa, T. Hayashi, M. Terrones, M.S. Dresselhaus, *Nano Lett.*, **3**, 723 (2003).
3. D.L. Boxall, G.A. Deluga, E.A. Kenik, W.D. King, C.M. Lukehart, *Chem. Mater.*, **13**, 891 (2001).
4. J. Solla-Gullón, V. Montiel, A. Aldaz, J. Clavilier, *J. Electroanal. Chem.*, **491**, 69 (2000).
5. W.Z. Li, C.H. Liang, W.J. Zhou, J.S. Qiu, Z.H. Zhou, G.Q. Sun, Q. Xin, *J. Phys. Chem.*, **107**, 6292 (2003).
6. X.H. Xue, T.H. Lu, C.P. Liu, W. Xing, *Chem. Commun.*, **12**, 1601 (2005).
7. V. Tsakova, D. Borissov, S. Ivanov, *Electrochem. Commun.*, **3**, 312 (2001).
8. V. Tsakova, D. Borissov, B. Ranguelov, C. Stromberg, J.W. Schultze, *Electrochim. Acta*, **46**, 4213 (2001).

9. P. He, H. Liu, Z. Li, J. Li, *J. Electrochem. Soc.*, **152**, E146 (2005).
10. H. Tang, J. Chen, L. Nie, D. Liu, W. Deng, Y. Kuang, S. Yao, *J. Colloid Interface Sci.*, **269**, 26 (2004).
11. L.M. Plyasova, I.Y. Molina, A.N. Gavrilov, S.V. Cherepanova, O.V. Cherstiouk, N.A. Rudina, E.R. Savinova, G.A. Tsirlina, *Electrochim. Acta*, **51**, 4477 (2006).
12. J. Ye, H. Cui, Y. Wen, W. Zhang, G. Xu, F. Sheu, *Microchim. Acta*, **152**, 267 (2006).
13. H. Tang, J. Chen, S. Yao, L. Nie, Y. Kuang, Z. Huang, D. Wang, Z. Ren, *Mater. Chem. Phys.*, **92**, 548 (2005).
14. Z. He, J. Chen, D. Liu, H. Tang, W. Deng, Y. Kuang, *Mater. Chem. Phys.*, **85**, 396 (2004).
15. M.M.E. Duarte, A.S. Pilla, J.M. Sieben, C.E. Mayer, *Electrochem. Commun.*, **8**, 159 (2006).
16. M. Ueda, H. Dietz, A. Anders, H. Kneppe, A. Meixner, W. Plieth, *Electrochim. Acta*, **48**, 377 (2002).
17. J.V. Zoval, J. Lee, S. Gorer, R.M. Penner, *J. Phys. Chem. B*, **102**, 1166 (1998).
18. H. Tang, J. Chen, L. Nie, D. Liu, W. Deng, Y. Kuang, S. Yao, *J. Colloid Interface Sci.*, **269**, 26 (2004).
19. R.D. Armstrong, M. Fleischmann, H.R. Thirsk, *J. Electroanal. Chem.*, **11**, 208 (1966).
20. B. Scharifker, G. Hills, *Electrochim. Acta*, **28**, 879 (1983).
21. M.V. Mirkin, A.P. Nilov, *J. Electroanal. Chem.*, **283**, 35 (1990).
22. L. Heerman, A. Tarallo, *J. Electroanal. Chem.*, **470**, 70 (1999).
23. C.A. Foss, M.J. Tierney, C.R. Martin, *J. Phys. Chem.*, **96**, 9001 (1992).
24. M. Platt, R.A.W. Dryfe, E.P.L. Roberts, *Chem. Commun.*, **2002**, 2324 (2002).
25. D. Bera, S.C. Kuiry, S. Patil, S. Seal, *Appl. Phys. Lett.*, **32**, 3089 (2003).
26. L. Xu, L.D. Tung, L. Spinu, A.A. Zakhidov, R.H. Baughman, J.B. Wiley, *Adv. Mater.*, **15**, 1562 (2003).
27. G. Tourillon, F. Garnier, *J. Phys. Chem.*, **88**, 5281 (1984).
28. K. Kost, D. Bartak, B. Kazee, T. Kuwana, *Anal. Chem.*, **60**, 2379 (1988).
29. G.K. Chandler, D. Pletcher, *J. Appl. Electrochem.*, **16**, 62 (1986).
30. J.C. Moutet, Y. Ouennoughi, A. Ourari, S. Hamar-Thibault, *Electrochim. Acta*, **40**, 1827 (1995).
31. D.W. Hatchett, M. Josowicz, J. Janata, D.R. Baer, *Chem. Mater.*, **11**, 2989 (1999).
32. B.J. Hwang, R. Santhanam, Y.L. Lin, *Electronalysis*, **15**, 1667 (2003).
33. J.Y. Lee, T.C. Tan, *J. Electrochem. Soc.*, **137**, 1402 (1990).
34. V. Tsakova, A. Milchev, *Electrochim. Acta*, **36**, 1151 (1991).
35. L. Torsi, M. Pazzuto, P. Siciliano, R. Rella, L. Sabbatini, L. Valli, P.G. Zambonin, *Sens. Actuators B Chem.*, **48**, 362 (1998).
36. R.J. Nichols, D. Schroer, H. Meyer, *Electrochim. Acta*, **40**, 1479 (1995).
37. M. Hepel, Y.M. Chen, R. Stephenson, *J. Electrochem. Soc.*, **143**, 498 (1996).
38. L.M. Abrantes, J.P. Correia, *Electrochim. Acta*, **45**, 4179 (2000).
39. V.D. Jovic, T. Trisovic, B.M. Jovic, M. Vojnovic, *J. Electroanal. Chem.*, **408**, 149 (1996).
40. J.Y. Lee, T.C. Tan, *J. Electrochem. Soc.*, **137**, 1402 (1990).
41. P. Ocon, P. Herrasti, S. Rojas, *Polymer*, **42**, 2439 (2001).
42. L. Martinot, D. Leroy, H. Zhan, C. Licour, C. Jerome, G. Chapelle, C. Calberg, R. Jerome, *J. Mater. Chem.*, **10**, 729 (2000).
43. J.M. Ko, D.Y. Park, N.V. Myung, J.S. Chung, K. Nobe, *Synth. Methods*, **128**, 47 (2002).
44. J.C. Moutet, Y. Ouennoughi, A. Ourari, S. Hamar-Thibault, *Electrochim. Acta*, **40**, 1827 (1995).
45. C. Hable, M.S. Wrighton, *Langmuir*, **9**, 3284 (1993).
46. T. Kessler, A.M. Castro Luna, *J. Appl. Electrochem.*, **32**, 825 (2002).
47. S. Swathirajan, Y.M. Mikahil, *J. Electrochem. Soc.*, **139**, 2105 (1992).
48. A. Kelaidopoulou, E. Abelidou, A. Papoutsis, E.K. Polychroniadis, G. Kokkinidis, *J. Appl. Electrochem.*, **28**, 1101 (1998).

49. M.A. Del Valle, F.R. Diaz, M.E. Bodini, T. Pizarro, R. Cordova, H. Gomez, R. Schreble, *J. Appl. Electrochem.*, **28**, 943 (1998).
50. T. Kessler, A.M. Castro Luna, *J. Solid State Electrochem.*, **7**, 593 (2003).
51. J. Coronas, J. Santamaría, *Topics Catal.*, **29** (1/2) 29 (2004).
52. Y. Morigami, M. Kondo, J. Abe, J.H. Kita, K. Okamoto, *Sep. Purif. Technol.*, **25**, 251 (2001).
53. A.J. Bard (ed.), *Electroanalytical Chemistry* vol. **16**, Marcel Dekker, New York (1989).
54. F. Scholz (ed.), *Electroanalytical Methods. Guide to Experiments and Applications*, Springer-Verlag, Berlin (2002).
55. G.E. Evans, K.V. Kordesch, *Science*, **158** (805), 1148 (1967).
56. F. Montilla, E. Morallon, I. Duo, Ch. Comninellis, J.L. Vazquez, *Electrochim. Acta*, **48** (25/26), 3891 (2003).
57. F. Gloaguen, J.M. Léger, C. Lamy, A. Marmann, U. Stimming, R. Vogel, *Electrochim. Acta*, **44**, 1805 (1999).
58. S. Domínguez-Domínguez, J. Arias-Pardilla, Á. Berenguer-Murcia, E. Morallón, D. Cazorla-Amorós, *J. Appl. Electrochem.*, **38**, 259 (2008).
59. K. Bergamaski, A.L.N. Pinheiro, E. Teixeira-Neto, F.C. Nart, *J. Phys. Chem. B*, **110** (39), 1972 (2006).
60. E.E. McLeary, J.C. Jansen, F. Kapteijn, *Micropor. Mesopor. Mater.*, **90**, 198 (2006).
61. N. van der Puil, E.J. Creyghton, E.C. Rodenburg, S.T. Sie, H. van Bekkum, J.C. Jansen, *J. Chem. Soc. Faraday Trans.*, **92** (22), 4609 (1996).
62. Á. Berenguer-Murcia, E. Morallón, D. Cazorla-Amorós, Á. Linares-Solano, *Micropor. Mesopor. Mater.*, **78** (2/3), 159 (2005).

17

Electrochemical Preparation of Pd Seeds/Inorganic Multilayers on Structured Metallic Fibres

F. Basile, P. Benito, G. Fornasari, M. Monti, E. Scavetta, M. Tonelli and A. Vaccari*

Università di Bologna, Bologna, Italy

17.1 Introduction

There is an increasing interest in H_2 production and use by a coupling catalytic process with dense hydrogen membrane separation in simultaneous or sequential steps. The key point of these technologies is the synthesis of membranes based on a thin Pd (or Pd-alloy) film on a ceramic and metallic support with stable behaviour and tailored features. The applications range from: high temperature ones such as reforming reaction with continuous H_2 separation, or olefin production thorough dehydrogenation reaction and simultaneous H_2 separation; medium temperature applications like H_2 production/separation during water gas shift reaction; and recently low temperature (200–250 °C) distributed feeding of H_2 to control the selectivity. The wide number of applications requires tailored membranes since under different conditions, stability, transport, H_2 absorption or decomposition can be limiting features for the reactions. Therefore from the classical patent, which regards flat and dense membranes made of self supported Pd or Pd alloy dens film and characterised by poor stability and high costs, the research is currently focused on the use of thin films, showing a clear advantage in lower cost and increased flux [1]. In fact hydrogen diffusion through the membrane is usually the rate determining step, therefore hydrogen permeability of Pd-based membranes is inversely proportional to their thickness [2]. Furthermore a thinner membrane means

*Corresponding author: telephone +390512093663, fax +390512093680, e-mail f.basile@unibo.it

Membranes for Membrane Reactors: Preparation, Optimization and Selection, Edited by Angelo Basile and Fausto Gallucci

also a better volumetric efficiency of the membrane separator or reactor. This is the case of the steam methane reforming in a fluidised-bed membrane reactor, in such a case the volume of the reactor has to accommodate enough membrane modules to obtain sufficient hydrogen permeation capacity, therefore, decreasing the membrane thickness would decrease the membrane modules and accordingly the reactor volume [2], resulting in a more compact system.

Thin Pd-based membranes cannot withstand any significant pressure unless a proper support is provided, therefore they need to be supported on ceramic or metallic porous materials with enough mechanical strength. The membrane film can be produced by using coating techniques, such as electroless plating [3–5], sputtering [6,7] or MOVCD [8]. The former is the most widely used one even if the number of cycles necessary to the layer formation has been simplified only recently by using the seeding principle to start the autocatalytic Pd reduction [1,2]. At present, the commonly used substrates for the thin Pd-based membranes are porous ceramics [4,5,9], stainless steel or metallic alloys [2]. Although ceramic substrates are cheap and their surface can be sufficiently smoothed, the interaction between the ceramic substrate and the Pd metallic membrane is difficult and the different thermal expansion coefficients make the adherence very problematic during thermal and operational cycling, furthermore the lack in ductility and the difficulty welding with the reactor parts limit their application [10].

In contrast, the porous stainless steel support can be easily coupled and sealed to the reactor and has similar expansion coefficients than Pd. Those features have led to a wide number of studies concerning thin Pd films coating stainless steel substrates. As drawbacks, the stainless steel support can lead to instability for long time operation at elevated temperature due to the interdiffusion between the Pd-based membrane and the porous stainless steel when they are in direct contact [11,12]. Furthermore the membrane integrity is highly affected by the structure of the porous substrate, such as roughness and pore size and need to be three times larger than the pore size (the pores size is usually in the range of 10 μm) [13]. These two problems can be addressed by using a ceramic layer to avoid contact between Pd and the metal substrate, surface treatment or filling of the pores to reduce the pore size and the pinhole. As a result the overall membrane scheme is a complex composite in a multilayer arrangement with several preparation procedures and equipments that limit the applicability of such a system.

17.2 Brief Review on Preparation Method

The electrochemical deposition or electrodeposition is achieved by making flow an electric current between two or more electrodes separated by an electrolyte. A potentiostat/galvanostat, connected to the electrochemical cell allows to control and maintain the *potential* of the *working electrode* at a controlled level with respect to the *reference electrode* (usually a saturated calomel electrode or an Ag/AgCl electrode, whose potential is stable and well known) by adjusting the *current* at an *auxiliary electrode* (a large area inert metal electrode, such as a platinum gauze). The working electrode can be any conductive substrate and is the material to be coated with the electrodeposited film.

The composition, morphology and texture of the deposited films can be tuned by means of experimental parameters such as potential, current density, deposition time and plating solution composition. The method is well known and widely applied in science and industry to produce metal or alloy coatings using acqueous [14] or non-aqueous [15] solutions containing the salts of the metal(s) to be deposited. A very important feature is that the electrode substrate can be any conductive material of any shape and dimension. The electrochemical synthesis is based on a reduction or oxidation reaction which usually changes the oxidation states of the precipitating species but it can either work by changing the pH and therefore the solubility products or the precipitating counter anion. By changing the applied potential the reaction product can be varied and selected, the film composition can be controlled by varying the bath composition and the film thickness can be controlled by changing the pulse length. Several kind of material can be successfully electrodeposited, by choosing properly the electrode, the deposition solution (pH, composition, concentration), applied potential (or current) and the synthesis pulse length.

Since the electrodeposition process is capable of depositing metals and alloys onto recessed and non-uniform surfaces, it has also found a role in microelectromechanical systems (MEMS), which are the integration of mechanical elements, sensors, actuators, and electronics on a common silicon substrate through microfabrication technology [16]. Yet, cathodic or anodic electrochemical synthesis can be also used to produce oxide and/or hydroxides thin films in conductive materials [17]. Metal hydroxide thin films and powders produced by this technique have been explored extensively for material in energy storage (often using cobalt or nickel precursors). Traditionally, this has meant electrochemical deposition of dense metal hydroxides (cobalt and nickel hydroxides) in the pores of metal electrodes. Currently, the method has been extended to the synthesis of a wide battery of hydroxides/oxides (CeO_2, FeOOH, hydrotalcite-type compounds, ZnO, perovskites, high-temperture oxide superconductors) on different supports several supports (platinum, stainless steel, carbon nanotubes), forming not only monolayers but also stratified compounds.

Finally, it should be remarked that the research in electrochemical deposition is currently focused on the growth of nanostructures with tailored morphology, performed by changing the concentration of the species in the solution or by adding some templates [18], as well as on the synthesis of inorganic/organic thin films can be prepared [19].

The electrodeposition shows advantages in comparison to other coating methods: it is a soft chemical route since the deposition occurs at low temperature, as a consequence the deposition occurs closer to equilibrium than with high-temperature methods, inter-element diffusion is not a problem and the thermal stresses are reduced; the deposition is very fast; the process can be controlled because of its electrical nature; and it can be applied for the coating of miniaturised systems Lastly, simple and inexpensive equipment is required and the scale up is easy. The synthesis takes places, usually at room temperature, at the working electrode/solution film interface and the product is deposited on the electrode surface in the form of a thin film or coating. However, since it is an ambient temperature technique it yields poor crystallised materials.

The preparation technique developed here deals with the electrochemical assisted preparation of inorganic substrates on metallic materials which can be easily coupled both with the electro chemical Pd membrane production and the seeding process of electroless plating. The paper deals with the illustration of the preparation concept and the application aimed to form multilayer coverage of a metallic support.

17.3 Explanation of the Proposed Preparation Method

Usually, metal (e.g., Au, Pt, Rh, Pd) thin films, nanoparticles or nanowires, of different kind, shape and dimension can be successfully prepared by cathodic reduction, starting from the corresponding metal salts; the electrochemical synthesis allows to easily control the features of the metal nanoparticle just by changing the parameters of the deposition (in term of applied current or potential, electrosynthesis length) and the bath solution (pH, concentration, presence or absence of additives). Yet, metal hydroxides or hydrotalcite-type (HT) compounds $[M^{2+}{}_{x}Al^{3+}{}_{1-x}(OH)_2](A^{b-})_{b/x}\cdot H_2O$ ($M = Mg$, Ni) can be succesfully prepared by electrosynthesis, in a single step and in a very short time, inducing a pH modification at the the interface electrode/solution [17]. The method is known as 'base electrogeneration' and it is based on the application of a cathodic potential to a working electrode in a deposition solution containing metals nitrates. Three classes of reactions take place:

1. Reactions which consume H^+ ions such as:
   (a) $H^+ + e^- \rightarrow$ Hads or Habs
   (b) $2H^+ + 2e^- \rightarrow H_2$ ($E^\circ = 0.0$ V)
   (c) $NO_3^- + 2H^+ + 2e^- \rightarrow NO_2^- + H_2O$ ($E^\circ = 0.934$ V)
   (d) $NO_3^- + 10H^+ + 8e^- \rightarrow NH_4^+ + 3H_2O$
2. Electrolysis of water:
   (e) $2H_2O + 2e^- \rightarrow H_2 + 2OH^-$ ($E^\circ = -0.828$ V)
3. Anion (for example, nitrate) reduction reactions:
   (f) $NO_3^- + H_2O + 2e^- \rightarrow NO_2^- + 2OH^-$ ($E^\circ = 0.01$ V)
   (g) $NO_3^- + 7H_2O + 8e^- \rightarrow NH_4^+ + 10OH^-$

   All these reactions cause an increase of pH near to the electrode surface, due both to the consumption of H^+ or to OH^- generation.

   The extent of the pH increase depends from the applied potential and from the nature of the working electrode; in fact each own materials posses a specific conductivity and each redox reaction takes place, at its surface, at a potential value higher than the standard potential, because it is inclusive of an overpotential whose value depends on the electrode material.

   A simple procedure which can be employed to measure the pH generated at the electrode surface at the beginning of the synthesis, consist in performing amperometric tests in the presence of two drops of acid–base indicators, in a 0.3 M KNO_3 solution, with an initial pH similar to that in the metal solution in order to mimic actual synthesis conditions. With the aim of covering the pH range from 6.0 to 12.0, a series of acid–base indicators was used The potential step was applied to observe the change in the colour of the indicator, near to the electrode (Table 17.1).

   The reactions of metal hydroxide precipitation effectively compete with the metal ion reduction reaction:
   (h) $Me^{n+} + e^- \rightarrow M^o$

The reaction with the most positive E° value would be preferred over the others. Reactions of the type (f) and (g) have a more positive E° value compared to most metal ion reduction reactions. As a result metal deposition does not take place in most nitrate baths. Instead, the metal ion deposits in the form of a hydroxide. If the deposition solution contains the nitrates of both a bivalent and trivalent metal of suitable dimension in the opportune ratio and concentration it is possible synthesise thin film of HT compounds.

Reaction (h) could become competitive for easily reducible species (e.g., Rh, Pd, C).

Table 17.1 *Relation between potential applied and pH in the vicinity of the Pt work electrodes, using acid–base indicators*

	Series of acid–base indicators				
Applied potential, *E* (V)	–0.9	–1.0	–1.1	–1.2	–1.3
Pt indicator range	7.5 < pH < 8.7	7.5 < pH < 8.7	8.7 < pH < 9.6	9.7 < pH < 11	9.7 < pH < 11

By using this method, the authors have successfully deposited thin films of controlled thickness and composition of HTs containing Ni and Co as the bivalent metals and Al as the trivalent one, on several different kinds of conductive surfaces (Pt, gold, carbon based electrodes, etc.) [20,21]. The composition of the deposition solution, the deposition potential the pulse length have been optimised, deeply studying the influence of the different parameter on the morphology and composition of the obtained films. The films, obtained with this method result well adherent to the metal support. This procedure has been extended to ferritic alloys, stainless steel containing a small amount of Al, such as FeCrAlloy foams (73% Fe, 20% Cr, 5% Al, 2% Y) [22,23].

The present preparation has been started as preliminary work using the experience achieved in the preparation of sensor and catalyst on structured metallic support based on the synthesis and film electrodeposition on metallic inconel fibres.

The method is applied first to the preparation of metal-Pd nanoparticles to illustrate the effect of the preparation parameters and then applied to the deposition of a ceramic layer on the metal which can be used to support the Pd° seeds for further electrochemical or electroless deposition of the Pd film. The ceramic layer can here be used to inhibit the effect of the Pd migration in the metal structure and to fill the pores of the porous metal support. The potentiality of the method has been here shown using metallic fibre materials with micrometric size. This particular support has been chosen to emphasise the versatility of the method. It is possible to obtain different kinds of coatings just modulating the deposition parameters. As shown in the following scheme thin or dense films can be deposited on the whole support, including interporous spaces, or just on the surface of the single fibre, moreover the present method allow to deposit in this way different coatings, such as metallic nanoparticles or ceramic material, depending on the final application.

Three different preparation procedures will be illustrated as useful tools for the thin Pd membrane preparation: (i) the precipitation of Pd seeds directly on the metal surface using a nitrate containing solution to allow the control of the shape and the homogeneity of the Pd particles synthesis and deposition changing the preparation parameters, (ii) a sequential deposition involving the deposition in a first step of the hydroxide, or HT materials, on the metallic fibres, eventually followed by the Pd seeds precipitation by electrochemical methods, (iii) the simultaneous precipitation of hydroxide or hydrotalcite and the Pd seeds on the metallic fibres (Figure 17.1). The second and third preparation methodologies, inducing the pore filling, were used to mimic the work carried out by Tong *et al.* [10]. Electrochemical measurements were

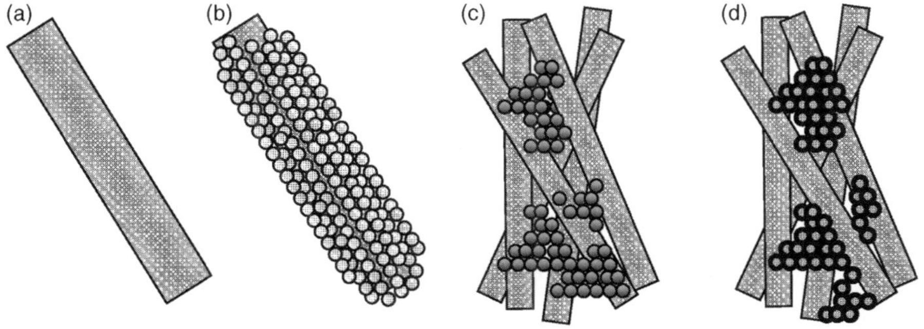

Figure 17.1 *Scheme illustrating the wide range of possibilities using electrochemical synthesis and deposition on fibres: (a) nude fibre, (b) Pd seeds or inorganic coating on fibre, (c) interpore precipitation for pore filling and eventually sequential Pd seed precipitation, (d) interpore simultaneous precipitation of inorganic coating and Pd seeds*

carried out at room temperature using a galvanostat/potentiostat (Autolab PG-STAT 20) in a conventional three electrode cell configuration. A FeCrAlloy metallic microfibre was used as macro support for these preliminary tests due to its high flexibility and to the possibility to show the thickness and the features of the deposited system either in each fibre or in the interconnected fibre materials. Electrode potentials were measured with respect to an aqueous saturated calomel electrode [SCE; i.e., reference electrode (R.E.); Figure 17.1a–d].

A Pt gauze with high surface area was used as the counter electrode (C.E.). The working electrode (W.E.) was the conductive material to be coated with the active layer, i.e., a metallic fibre, or plates.

1. A first set of measurements was performed by depositing, on the inconel fibres a single layer of Pd seeds, by cathodic reduction (at three different potentials, i.e., –0.9, –1.1 and –1.3 V vs SCE) from a 0.02 M $PdNO_3$ solution. For each of the tested potential two different deposition times (600 and 1000 s) were investigated. A 0.1 M KNO_3 solution was used as solvent and supporting electrolyte.
2. The second group of plates was coated with a double layer film by performing a two-step deposition. In a first step a cathodic potential (E = –0.9 V vs SCE) was applied for 600 s in 15 ml of a 0.03 M solution containing $Mg(NO_3)_2 \cdot 6H_2O$ and $Al(NO_3)_3 \cdot 9H_2O$ (molar ratio 2:1) and 0.3 M KNO_3 (or 15 ml of a 0.03 M solution of $Al(NO_3)_3 \cdot 9H_2O$ and 0.3 M KNO_3). This step is aimed to cover the plate with a uniform well adherent layer of Mg/Al HT material. After the first step another cathodic potential (E = –1.3 V) was applied for 600 s to the W.E. in 15 ml of a 0.02 $PdNO_3$ solution (in KNO_3 0.1 M).
3. The third group of plates was coated with a mixed phase by performing a cathodic reduction of a solution containing 0.015 M $Al(NO_3)_3 \cdot 9H_2O$, 0.005 M $PdNO_3$ and 0.1 M KNO_3. The cathodic reduction was carried out in a single step by applying a fixed potential of –1.3 V for 600 s.

After electrosynthesis, all the films were gently rinsed with doubly distilled water and dried under a nitrogen flow which helps also to remove the not well adherent part of the coating.

17.4 Multilayer Preparation on Metal Substrates

A first group of plates has been coated with Pd particles deposited from a 0.02 M solution of $PdNO_3$ by cathodic reduction. In order to find the best experimental conditions to deposit Pd nanoparticles with small dimension homogeneously three different applied potentials and deposition times were carried out. Pd(II) is an easily reducible cation, and therefore even in the presence of NO_3^- the metal precipitation always prevails over the palladium hydroxide formation.

Depending on the synthesis conditions it is possible change the coverage of the supports as well as the shape and size of the metallic nanoparticles. On the whole, two type of palladium particles were obtained: those precipitated on the surface of the support and those obtained among the fibres, that is, inside the pores of the material. Performing the synthesis at –0.9 V for 600 s allowed obtaining small Pd nanoparticles, mainly distributed on the fibre walls, a low coverage of the support were obtained (Figure 17.2a). Increasing the synthesis time did greatly modify neither the shape nor size of the particles, but it did increase the amount of electro-generated Pd particles. If a more catodic potential was applied (–1.1 or –1.3 V for 600 s) the amount of particles increased. At –1.1 V a nonhomogeneous deposition occurred with disordered

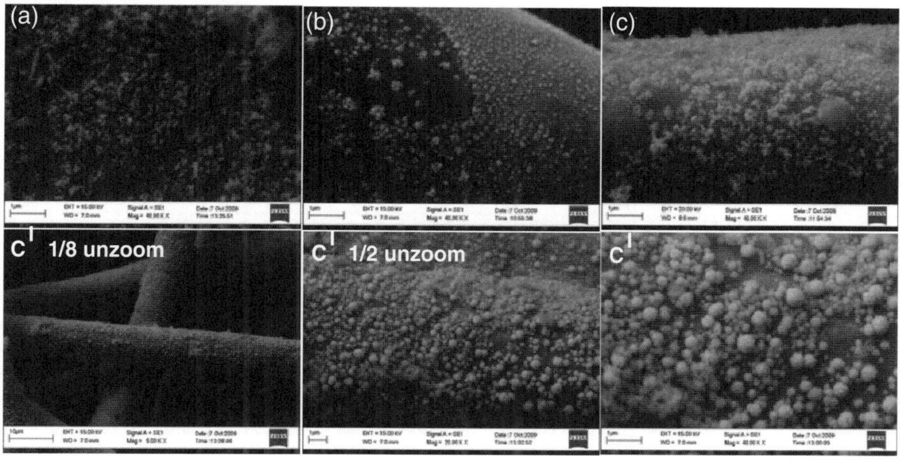

Figure 17.2 *SEM images of a Pd coated metallic inconel fibres prepared at: (a) –0.9 V for 600 s, (b) –1.1 V for 600 s, (c) –1.3 V for 600 s, (C') –1.3 V for 1000 s, shown with three different magnifications (5 k ×, 20 k ×, 40 k ×)*

Pd particles growth, while using the highest potential an homogeneous coverage of the support has been obtained (Figure 17.2c) and round shape particles made of nanocrystallites started to growth. Increasing the synthesis time the round agglomerates of nanocrystallites increased in number and homogeneity and for a the synthesis time of 1000 s, the support was quite homogenously covered. Moreover, depending on the synthesis conditions and especially at high preparation time and potential, some precipitation within the pores occurred. These results suggested that the electrochemical route could be applied for preparing seeds of Pd nanoparticles, useful as precursors in the electroless plating (Figure 17.2).

The second group of fibre plates shows a two-step deposition. This approach allows depositing a Pd nanoparticles layer over a first ceramic layer, to improve the stability due to interdiffusion, the particles adhesion and reduce the film thickness by pore filling.

The ceramic layer can be deposited by using the electrogeneration method starting from a deposition solution containing $Al(NO_3)_3$ with the aim of obtaining a $Al(OH)_3$ coating or from a solution containing $Mg(NO_3)_2$ and $Al(NO_3)_3$, with the aim of coating the plate with a Mg/Al-HT. These inorganic layers were used as a basis for the Pd deposition, as synthesised or after a calcinations treatment in order to form the respective oxides.

Different potentials and synthesis times have been used and the deposits have been characterised with SEM analysis.

The applied potential that ensure the precipitation of an homogeneous layer of Mg/Al HT or $Al(OH)_3$ layer is –1.2 and–0.9 V, respectively, in this way a deposition mainly on the fibre surface occurs, achieving a full fibre coverage. By increasing the deposition time (600–1000 s) or the potential an interpore coverage can also be obtained. The best results for the $Al(OH)_3$ can be obtained by changing the applied potential. A series of SEM images going from –0.9 to –1.2 V are displayed in Figure 17.2. A quite homogeneous interpore filling is shown, which can be a key point for the preparation of membrane layers on metallic supports.

An increase of the deposition time leads to the increase of the thickness of the deposit and to the concomitant pores filling. In fact, a similar coverage has been obtained for Mg and Al nitrates precipitating as HT using –1.2 V and a longer deposition time ($\geq$1000 s). The preparation method

Figure 17.3 SEM images of $Al(OH)_3$ deposited on inconel fibre plates prepared at: (a) −0.9 V and 600 s (1 and 10 k ×), (b) −1.0 V and 600 s (1 and 5 k ×), (c) −1.2 V and 600 s (1 and 5 k ×)

can be used to carry out a Pd membrane synthesis by electroplating or can be used to support Pd seeds for the electroless plating. For this purpose, the best conditions giving a high interpore coverage have been used to prepare a multilayer ceramic/metallic material. Three different synthetic approaches can be used to achieve an homogeneous layer distribution on the fibres surface and within the pores: the Pd can be precipitated directly on the $Al(OH)_3$ deposited on the fibre (sequential precipitation) or on the oxide obtained by calcination of hydroxdes precipitated on metallic fibres. In the former case the calcination will follow the Pd membrane deposition. As described in literature [10], this guarantee a better coverage of the surface during the membrane formation and a significant porosity is generated by the evolution of water in the $Al(OH)_3$ layer or water and anionic interlayer in the HT material (CO_2, NO_x) during calcination, which leads to a shrinkage of the structure. In both cases the Pd seeds deposition reuiqres high potential (–1.3 V) for 600 s. Figure 17.3 shows a SEM image obtained when Pd seeds are deposited over the Al $(OH)_3$ and the calcined sample. EDS analyses performed on different zones of the sample reveal a dense and homogeneous metallic/ceramic double layer (Figure 17.4).

The Pd is present as homogeneously deposited on the Al hydroxide and on the Al_2O_3, in the latter a porous system is also visible as result of the calcination process. No Pd agglomeration is present probably due to the presence of the Al layer, which can decrease and homogenise the potential and increase the support surface for the precipitation of Pd° particles.

The third approach for multilayer preparation is based on cathodic reduction of a solution containing both $Pd(NO_3)_2$ and $Al(NO_3)_3$. The coating was obtained applying a potential of –1.3 V for 600 s. This experimental condition allows the deposition of a mixed layer consisting of $Al(OH)_3$ and Pd with a lower Pd dispersion. The amount of Al precipitated is lower due to the fact that the Pd reduction competes with the NO_3^- reduction by decreasing the OH^- generation rate and moderating the pH increase on the electrodes vicinity (Figure 17.4c). These phenomena decrease the $Al(OH)_3$ precipitation. However, the morphology of the Pd layer is similar to that obtained on clean fibres under similar conditions, showing the formation of spherical particles on a thin $Al(OH)_3$ layer.

Finally, the sequential approach can used HT materials as inorganic layer in the inconel-Mg/Al-Pd multilayer preparation. The coprecipitation approach cannot be used since more severe reaction conditions, to obtain high pH value are required for the hydrotalcite synthesis

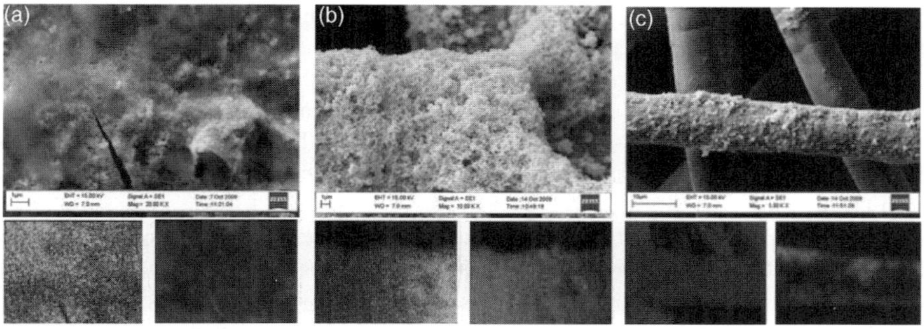

Figure 17.4 SEM-EDAX images of a metallic fibre plate coated with two layers formed by: (a) sequential $Al(OH)_3$ (−1.2 V for 600 s) and Pd° (at −1.3 V for 600 s) synthesis, (b) sequential $Al(OH)_3$ (−1.2 V for 600 s) calcinations and Pd° synthesis (−1.3 V for 600 s), (c) simultaneous coprecipitation of $Al(OH)_3$ and Pd° (−1.3 V and 600 s)

with respect to $Al(OH)_3$ while low pH values are obtained using the coprecipitation method. The inconel-Mg/Al-Pd multilayer synthesis can be performed at −1.2 V and 1000 s for the Mg/Al precipitation and −1.3 V and 600 s for the Pd deposition. Figure 17.5 shows the interesting result of the sample obtained by precipitation of Mg/Al hydrotalcite followed by precipitation of Pd°. An homogeneous coverage of Mg and Al on the fibre and in the interpore demonstrate the precipitation of HT on the whole sample. The Pd precipitation is also homogeneous on the fibres and within the interpores and does not shows any agglomeration of Pd crystallites. The calcination of the HT compounds can give rise to an homogeneous porous system with a sharp distribution of mesoporous size.

17.5 Final Remarks and Conclusion

The potential application of the preparation procedure of electrochemical induced precipitation by base generation is demonstrated. The preparation method allows the precipitation of inorganic

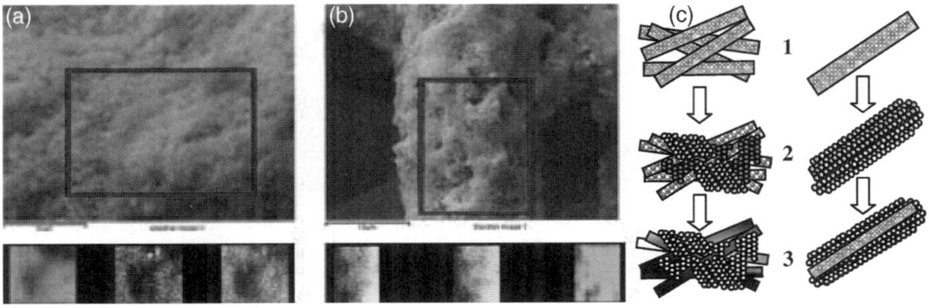

Figure 17.5 Multilayer sequential synthesis of Mg/Al HT (−1.2 V for 1000 s) and Pd° (−1.3 V for 600 s) on a metallic fibre plate, with SEM EDAX of the multilayer (a) within interpore (b) on the fibre. (c) Schematic representation of the preparation steps for the multilayer synthesis on the fibres and within the interpores (among the fibres) showing: (1) inconel support, (2) inconel covered by Mg/Al and (3) Inconel-Mg/Al-Pd seeds

hydroxides and Pd seeds in a controlled way on metallic fibres. By changing the synthesis parameters such as the potential and time a good coverage of the metal or an interpore filling can be obtained. The controlled inorganic layer obtained in the first step of the sequential precipitation approach can be potentially used for the electroplating procedure by adjusting the deposition parameters. The electrochemical depostion will allow to obtain in two steps, a controlled pore filling and substrate coverage with an inorganic layer and a Pd membrane layer deposition.

The second step of the sequential approach proved that the presence of an inorganic interlayer is beneficial for the electrodeposition of Pd seeds producing high Pd dispersion and good coverage, evidencing a large potential to be used as methods for improved electroless synthesis by omitting sensitisation and activation steps. Simultaneous precipitation of inorganic and Pd layer produces a decrease of coverage and a lack of homogeneity.

References

1. S. Abate, G. Centi, S. Perathoner, F. Frusteri, *Catal. Today*, **118**, 189 (2006).
2. A. Li, J.R. Grace, C.G. Lim, *J. Membr. Sci.*, **298**, 175 (2007).
3. A.W. Li, G.X. Xiong, J.H. Gu, L.B. Zheng, *J. Membr. Sci.*, **110**, 257 (1996).
4. A.W. Li, W.Q. Liang, R. Hughes, *Thin Solid Films*, **350**, 106 (1999).
5. S. Uemiya, T. Matsuda, E. Kikuchi, *J. Membr. Sci.*, **56**, 315 (1991).
6. V. Jayaraman, Y.S. Lin, *J. Membr. Sci.*, **104**, 251 (1995).
7. H.D. Tong, J.W.E. Berenschot, M.J. De Boer, J.G.E. Gardeniers, H., Wensink, H.V. Jansen, W. Nijdam, M.C. Elwenspock, E.C. Gielens, C.J.M. van Rijn, *J. Microelectromech. Syst.*, **12**, 622 (2003).
8. G. Xomeritakis, Y.S. Lin, *J. Membr. Sci.*, **133**, 217 (1997).
9. H.Y. Gao, Y.S. Lin, Y. D, Li, B.Q. Zhang, *Ind. Eng. Chem. Res.*, **43**, 6920 (2004).
10. J. Tong, H. Suda, K Haraya, Y. Matsumura, *J. Membr. Sci.*, **260**, 10 (2005).
11. J. Shu, A. Adnot, B.P.A. Grandjean, S. Kaliaguine, *Thin Solid Films*, **286**, 72 (1996).
12. D. Wang, H.H. Tong, H.Y. Xu, Y. Matsumura, *Catal. Today*, **93/95**, 689 (2004).
13. A. Li, J.R. Grace, C.G. Lim, *J. Membr. Sci.*, **306**, 159 (2007).
14. L.P. Bicelli, B. Bozzini, C. Mele, L. D'Urzo, *Int. J. Electrochem. Sci.*, **3**, 356 (2008).
15. W. Simka, D. Puszczyk, G. Nawrat, *Electrochim. Acta*, **54**, 5307 (2009).
16. D. Landolt, *J. Electroanal. Chem.*, **149**, S9 (2002).
17. G.H.A. Therese, P.V. Kamath, *Chem. Mater.*, **12**, 1195 (2000).
18. M. Lai, D. J. Riley, *J. Colloid Interf. Sci.*, **323**, 203 (2008).
19. T. Yoshida, J. Zhang, D. Komatsu, S. Sawatani, H. Minoura, T. Pauporte, D. Lincot, T. Oekermann, D. Schlettwein, H. Tada, D. Wöhrle, K. Funabiki, M. Matsui, H. Miura, H. Yanagi, *Adv. Funct. Mater.*, **19**, 17 (2009).
20. E. Scavetta, A. Mignani, D. Prandstraller, D. Tonelli, *Chem. Mater.*, **19**, 4523 (2007).
21. E. Scavetta, B. Ballarin, M. Gazzano, D. Tonelli, *Electrochim. Acta*, **54**, 1027 (2009).
22. F. Basile, P. Benito, P. Del Gallo, G. Fornasari, D. Gary, V. Rosetti, E. Scavetta, D. Tonelli, A. Vaccari, *Chem. Commun.*, 2008, **2917** (2008).
23. F. Basile, P. Benito, G. Fornasari, V. Rosetti, E. Scavetta, D. Tonelli, A. Vaccari, *Appl. Catal. B*, **91**, 563 (2009).

18

Membranes Prepared Via Spray Pyrolysis

Mingtao Li and Liejin Guo

State Key Laboratory of Multiphase Flow in Power Engineering, Xi'an Jiaotong University, Xi'an, PR China

18.1 Introduction

Spray pyrolysis deposition (SPD) is a cheap and facile method for material preparation, which is widely used for synthesis of powders, films and membranes. This method was first introduced by Aman to prepare MgO in 1956 [1]. Thereafter Chamberlin *et al.* successfully deposited CdS films for solar cell application in 1966 [2]. Since then, spray pyrolysis deposition have been gained a great deal of attention because of the simplicity of apparatus and good productivity. Commonly, in spray pyrolysis a precursor solution or colloidal suspension is atomised, then the atomised precursor droplets are transported into a heated reactor chamber or onto a heated substrate, where in the simplest case the solvent evaporates and the precursor undergoes subsequent reaction to form product powders or films.

The process is very similar to that in chemical vapour deposition (CVD), except that the droplets are still in liquid phase in this process. In many cases SPD instead of CVD is generally used for the fabrication of thin films using metal-inorganic precursors. As an aerosol deposition method spray pyrolysis has several advantages:

1. It is a cheap method and does not need any high vacuum as required in magnetron sputtering and vacuum evaporation.
2. Materials with desired stoichiometric ratio can be easily deposited in film, coating or powder forms.
3. The deposition rate and thickness can be controlled easily by changing spray parameters.

Membranes for Membrane Reactors: Preparation, Optimization and Selection, Edited by Angelo Basile and Fausto Gallucci

4. It is a reproducible method and desired for compact or porous, uniform films.
5. There are virtually no restrictions on substrate material, dimension or its surface profile.
6. By changing composition of the spray solution or spray parameters, it can be used to make membrane layer by layer, so composition gradient membrane or pore gradient membrane throughout the thickness can be made.
7. It is very convenient to scale up to produce large membranes.

In 1993 Li *et al.* successfully applied this method for the deposition of a Pd-Ag alloy membrane by spray pyrolsyis of $Pd(NO_3)_2$ and $AgNO_3$ mixed aqueous solution [3]. Nowadays this method has been adapted for membranes synthesis such as carbon membrane, ceramics membrane and catalytic membrane.

Various reviews concerning spray pyrolysis techniques have been published. As spray pyrolysis for particular material preparation, the reader can refer to these reviews for further information. Albin *et al.* presented a review of the equipments, operation parameters, and optoelectronic materials deposited by spray pyrolysis technique [4]; Patil reviewed different atomisation techniques and thin metal oxide and chalcogenide films deposited by spray pyrolysis [5]; Bohac *et al.* have discussed the mechanism of chemical spray deposition and presented some examples of sprayed yttria-stabilised zirconia (YSZ) films [6]; Perednis *et al.* gave an extensive review of the effects of spray parameters on film quality to demonstrate the importance of the process of optimisation [7].

In this chapter the process of spray pyrolysis is described and some spray apparatus are also introduced as regard to produce qualified membranes. Then some selected membranes prepared by this method were summarised including some organic membranes, inorganic separation membranes, ionic/electronic conductive membranes and catalytic membranes. Finally some remarks and perspectives are given for this technique.

18.2 Spray Pyrolysis Material Preparation Method

The basic process of spray pyrolysis is described above. In order to get pure desired materials, the precursor is commonly prepared carefully such that the products other than the desired compound are volatile.

Generally, all kinds of solution can almost be used as precursor solution such as aqueous solution, organic solution, sol, suspension, and so on. However, aqueous solution may be the best choice because of low cost, safety, easy operation. Also, almost all elements have suitable salts that can be dissolved in water. Solubility is an important factor that has an influence on speed of film deposition or powder production. In order to increase solubility of some materials uneasy dissolving in water, some organic complexant can be used such as critic acid, ethylenediaminetetraacetic acid (EDTA). Also some alcohols can be used to facilitate the formation of films.

Among all steps, the spray process is the most important step and regarded as the heart of the spray pyrolysis. Originally, the precursor is atomised by pressured air pulverising method as shown in Figure 18.1. In such a process, the size and size distribution of the droplets, which affects the morphology and properties of the product, are determined by many factors, such as solution concentration, solution viscosity, surface tension, gas pressure in the nozzle and so on.

Nevertheless, changing property of precursor solution is more difficult rather than changing spray parameters in the nozzle. So generally the mist size is controlled through using different spray pyrolysis parameters or special designed apparatus. In a conventional spray system, the droplets are not uniformed, so it is not easy to get uniform micro grain distribution in the

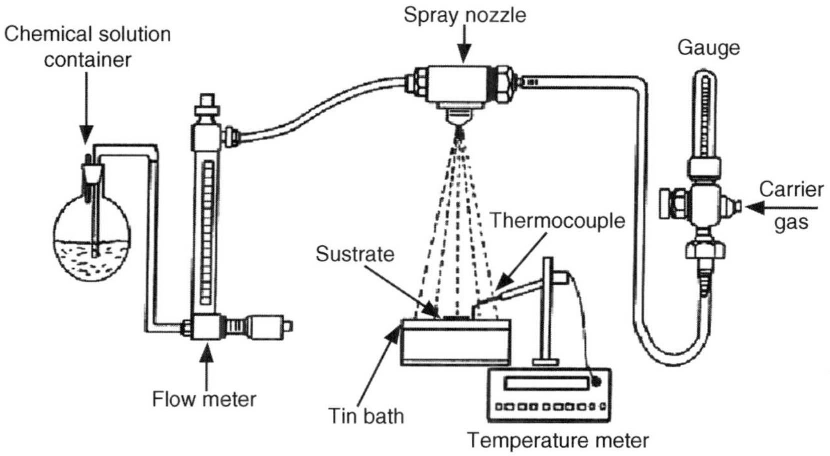

Figure 18.1 *A typical pneumatic nebulising spray pyrolysis system for film deposition. Reprinted from Thin Solid Films, Ortega-Lopez, M., Morales-Acevedo, A., Characterization of $CuInS_2$ thin films for solar cells prepared by spray pyrolysis. Vol. 330(2), 96–101. Copyright (1998) with permission from Elsevier*

deposited layer. Some new techniques are introduced into this method such as ultrasonic pulverisation nozzle and PZT transducer atomisers to control the size of droplets. However, the deposition efficiency is still very low due to large portion of the droplets flush out. In order to get uniform layer and improve the deposition efficiency, some other techniques are also adopted, such as electrostatic spray, and corona spray.

To obtain finer precursor droplets, an ultrasonic pulverising nozzle in place of ordinary pulverising nozzle is used in some experiments [9]. Recently, an ultrasonic nebuliser using piezoelectric transducer (PZT) has been used to simplify the system and to make the precursor droplets smaller and the size distribution more uniform [10]. Figure 18.2 is a typical ultrasonic spray pyrolysis system using PZT transducer. In such case, the droplet size, size distribution, and the rate of atomisation are dominated by the viscosity of the precursor solution, the oscillation frequency and intensity of the PZT and the depth of the precursor solution except the rate of carrier gas. It brings the advantage of control the rate of gas flow rate independently to get film with high

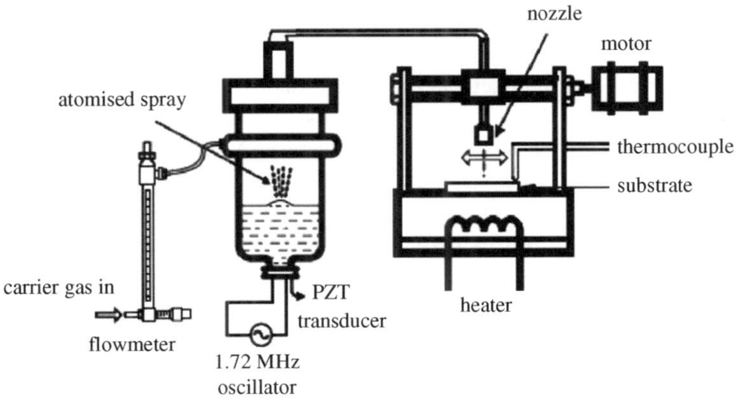

Figure 18.2 *An ultrasonically nebulising spray system with a reciprocating moving nozzle. Adapted from [11]*

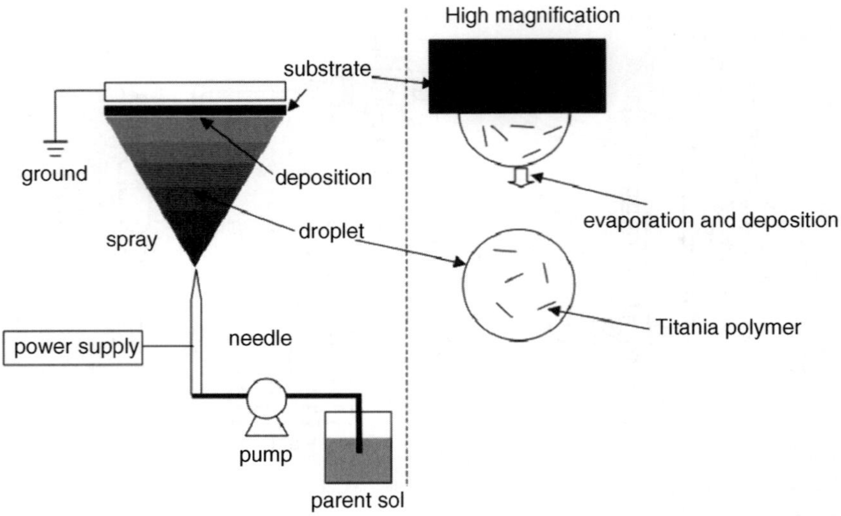

Figure 18.3 *A typical electrostatic spray deposition system. Reprinted from Chemistry of Materials, Nomura, M., et al., Improvement of Thermal Stability of Porous Titania Films Prepared by Electrostatic Sol-Spray Deposition (ESSD). Vol. 15(6), 1283–1288. Copyright (2003) with permission from American Chemical Society*

quality, which is not the case with pneumatic spraying. Also the atomisation chamber functions like a fluidised bed, where the larger droplets cannot be transported by the carrier gas and therefore return to the atomising chamber. This gives rise to good size selectivity for the droplets.

Electrostatic spray pyrolysis is a widely used technique to improve the deposit efficiency. Figure 18.3 is a typical electrostatic spray pyrolysis system. A high positive voltage was applied to the nozzle to disperse liquid as aerosol. The charge also prevents the coalescence of droplets during flight. Furthermore, the electric field allows a high degree of control over the direction of flight and deposition rate over the substrate. Thus charged droplets can be driven by an electric field force and deposited onto a controlled area of the substrate [12]. This can improve the deposition efficiency drastically.

Also a combination of ultrasonic atomising and electrostatic spray has been reported to deposit films with high deposition efficiency as great as 80% [13]. This configuration has been applied for film cathodes preparation in solid oxide fuel cell (SOFC) [14].

In order to get large area products and make the deposited layer uniform, in many cases a robotic stage was used to control the nozzle or the substrate move to and fro in a special manner [11]. Even progressive scanning manner were also applied [16], Such as an typical special designed system for roll to roll deposition of films shown in Figure 18.4.

Except the techniques described above there are some other spray methods widely used in film and/or membrane preparation, such as thermal spray, cool spray, and high velocity oxy fuel spray. Those are not the scope of this chapter.

Before the droplets arriving in the substrate, a series of physical reaction occur: (i) solvent vaporisation, (ii) solute sedimentation, (iii) droplets agglomeration. After the droplets leaving the nozzle, solvent of the droplets begins to vapour and solution is concentrated because of heat and gas flow action. Friedlander *et al.* analysed dynamics of the droplets agglomeration and obtained a relation between the number of droplets and the average dimension of droplets [18]. It is an effective method to hinder droplets agglomeration by decreasing original number of droplets.

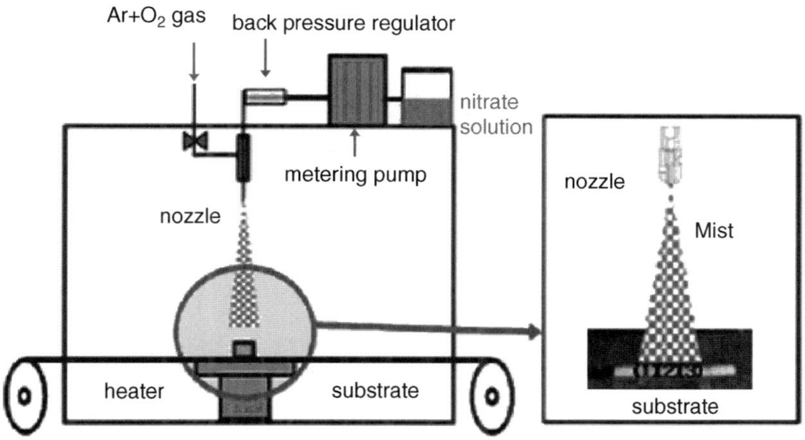

Figure 18.4 *A spray deposition system with a moving substrate for continuous deposition. Adapted from [17]*

In the case of powder preparation, the droplets are heated in the chamber ambient atmosphere and solvent vaporises, the droplets grow into powders. Jayanthi *et al.* proposed a criterion for the formation of hollow or solid particles based on the critical percolation threshold-critical volume fraction concept in 1993 [19]. Jain *et al.* applied this concept of critical volume fraction to different systems and correlated the morphology of the product particles to the material properties and process parameters in a review on spray pyrolysis for powder preparation in 1997 [20].

As for film preparation, a substrate is used as grain receiver, where further chemical reaction occurs to form thin film, dense or porous on the substrate. During the flight of the droplets, the solvent vaporises as the droplets approach the substrate. The reactants diffuse to the substrate and a heterogeneous reaction occurs which leads to the formation of thin solid films. In some case the covered substrate is post-annealed in a furnace and heated as slowly as possible to eliminate some residual undesired stuff and prevent membrane cracking. Obviously the distance between the nozzle and the substrate, the flow rate of the carrier gas, the temperature of the substrate and nature of the precursor solution are critical parameters for film deposition.

18.3 Selected Membranes Prepared Via Spray Pyrolysis Coating Method

Because it is simple and both time and cost-effective to get desired structure with a suitable composition, nowadays spray pyrolysis is widely applied for film and/or membrane preparation.

Spray deposition as a versatile deposition technique can be used for fabricating functional multicomposite organic polymer membrane. For example, an ethylcellulose pseudolatex membranes polymers have been prepared as the coating materials to cover pellets or tablets for better looking, moisture or light protection, taste masking, and controlled drug release [21]. Lin *et al.* prepared a blood plasma separation membrane to fast separate red cell from plasma for biosensor application in medical diagnosis [22].

Nanoporous carbons (NPC) or carbon molecular sieves (CMS) are good candidate materials for selective membrane formation, for example, they can separate nitrogen from oxygen. Carbon ultrafiltration membrane can be made by carbonisation of a pre-existing polymer membrane with identical pore structure [23]. But the pyrolysis of a polymeric microfiltration membrane can enlarge or eliminate the pores of the pre-existing film. Foley *et al.* presented a novel method for producing carbon membranes for ultrafiltration applications using a spray deposition and pyrolysis of poly(furfuryl alcohol)/poly(ethylene glycol) mixtures on macroporous stainless steel supports [24]. They used noncarbonising polymers, such as poly(ethylene glycol) (PEG) to create pores during pyrolysis of deposited polymer layer.

As for asymmetry functional membranes synthesis, Krogman *et al.* reported a powerful and economical spray-assisted layer by layer deposition technique for developing multiple coatings of different morphologies and functions within a single textile membrane [25].

Except organic membranes described above, some inorganic membranes can also be prepared via spray pyrolysis. However, attention should be taken to avoid pinholes and cracks causing membrane leakage, because in most case these membranes present in thin film form.

18.3.1 Pd-Ag Alloy Hydrogen Separation Membrane

The spray pyrolysis was first introduced into membrane preparation by Li *et al.* in 1993 [3]. They successfully deposited a Pd-Ag alloy membrane on the outer surface of a porous aluminum hollow fiber employing flame spray pyrolysis. A mixed solution of $Pd(NO_3)_2$ and $AgNO_3$ was used as precursor solution. The thickness of the resulting alloy membrane was about 1.5–2.0 μm and the separation factor H_2/N_2 was 24. The as-prepared membrane exhibited a relative low hydrogen selectivity compared with the other deposition techniques. It seems to be likely that some pinholes presented in the membrane limited the separation factor. So some improvements that can be employed to avoid pinholes and get dense films will make this technique more practical in membrane preparation.

18.3.2 Porous TiO_2 Membrane

In a membrane reactor the membrane assembly usually serves as a catalyst for catalysis and a separation for product. Nomura *et al.* proposed a membrane reactor, in which a thin porous catalyst layer is combined with a porous metal support to improve the thermal conductivity, as shown in Figure 18.5 [26]. They used an electrostatic sol-spray deposition method to prepare the middle catalyst layer, a porous thin inorganic layer. Titanium tetraisopropoxide (TTIP), ethanol, butyl carbitol, and acetic acid were used to prepare the parent precursor. They obtained a porous

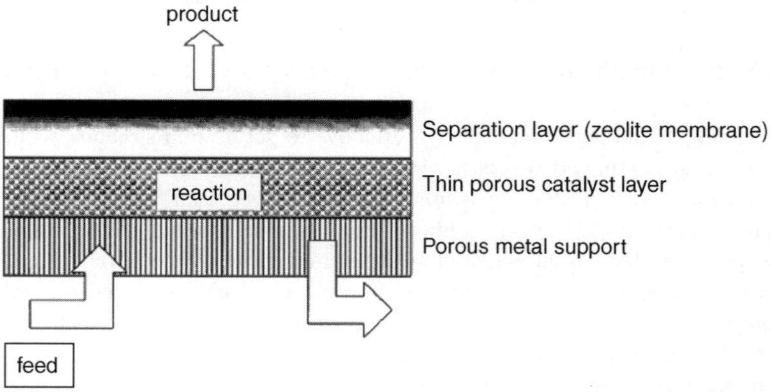

Figure 18.5 *Schematic diagram of composite membrane reactor. Reprinted from Separation and Purification Technology, Nomura, M., et al., Preparation of thin porous titania films on stainless steel substrates for heat exchange (HEX) reactors. Vol. 32(1–3) 387-395. Copyright (2003) with permission from Elsevier*

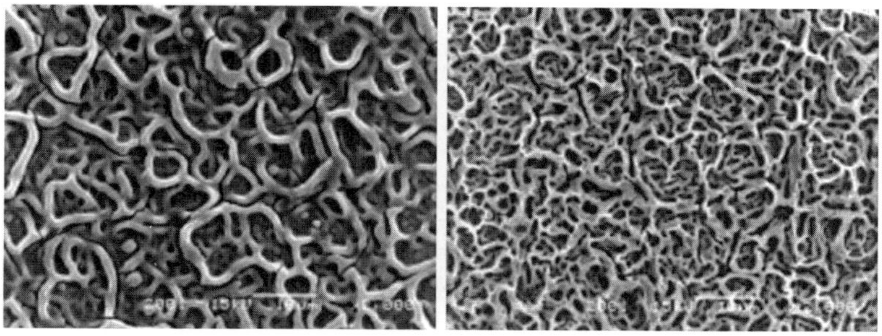

Figure 18.6 *Surface morphological SEM images of spray deposited TiO_2 films. Reprinted from Separation and Purification Technology, Nomura, M., et al., Preparation of thin porous titania films on stainless steel substrates for heat exchange (HEX) reactors. Vol. 32(1–3) 387–395. Copyright (2003) with permission from Elsevier*

titania film with uniform micron sized pores on dense stainless steel disks, and there was no cracks in the film, as shown in Figure 18.6.

Also they improved the thermal stability of porous titania film by using a mixture of aged titania sol and reactive titania sol in different ratios [15]. The porous structures obtained from the mixed sol remained intact by calcination at 600 and 1000 °C, resulting in a macroporous anatase or rutile film, respectively, with uniform pores of a few micrometers in size.

18.3.3 Ionic and Electronic Conductive Membrane in SOFCs

Solid oxide fuel cell (SOFC) is considered as one of the most promising energy conversion devices for its high energy conversion efficiency and low impact to environment. It can be used for distributed power supply and transportation application. Recently it has attracted increasing interest for its several advantages. For the planar SOFC design, electrolyte-supported SOFC has the potential to reduce the electrolyte thickness down to 10–20 μm, which, in turn, reduces the internal resistance of the solid electrolyte and hence the operating temperature of the SOFC. The basic structure of a state of the art electrolyte-supported design consists of a fully dense thin film electrolyte (proton conductor electrolyte or oxide ion conductor) supported on a 1–2 mm thick porous anode or cathode support. The composite membrane was the key component and the electrolyte film in the composite membrane plays an important role in the operation of SOFC.

There are several prerequisites for the electrolyte film: (i) sufficient chemical stability in both reducing and oxidising atmospheres, (ii) good conductivity at the cell operation temperature, (iii) compatibility with anode and cathode, (iv) suitable coefficient of thermal expansion, (v) dense to avoid gas leakage.

Oppositely, the anode and cathode are needed to be porous to facilitate the supply of fuel and elimination of product.

Dense yttria or calcia stabilised zirconia [7,27–31], cerium gadolinium oxide (CGO) [32], barium cerate (BCO) [33–35] electrolyte thin films, interconnector films [36] or porous cathode structures [37], anode structure [38,39] have all been reported prepared by spray pyrolysis. Table 18.1 lists some membrane deposited via spray pyrolysis for SOFC applications.

18.3.3.1 Yttria-Stabilised Zirconia

Yttria-stabilised zirconia (YSZ) is the most commonly used oxygen ionic conductor electrolyte. As described above, decreasing the thickness of the electrolyte can result in the increase of

Table 18.1 *Some spray pyrolysis deposited films for SOFCs application*

Material	Spray solution constitution	Substrate material	Pyrolysis temperature	Preparation method	Remarks	Reference
YSZ	Zr(acac)4, Y(acac)3, methanol	Silicon slices	300–375 °C	Ultrasonic spray pyrolysis via PZT	Proton conductor	[41]
YSZ	Aqueous and EG solution of citric complexes	Optical grade glass, quartz single crystal, β-quartz, ceramic YSZ, ceramic erbia-stabilised zirconia, corundum-708 and corundum-710	250–590 °C	Pneumatic nebulising	Proton conductor	[43]
YSZ	—	Quartz	—	Moving nozzle	Proton conductor	[42]
YSZ	Variety of precursor salts	Inconel	200–300 °C	Electrospray and pressurised spray	Proton conductor	[28]
YSZ	$Zr(acac)_4$, $(ZrO(NO_2)_3$,YCl_3, butylcarbitol	Inconel	196–365 °C	Electrospray and pressurised spray	Proton conductor	[30]
YSZ	$ZrO(NO_3)_2$.xH2O Y $(NO_3)_3.4H_2O$	—	600 °C	Ultrasonic spray pyrolysis	Proton conductor, powder product	[52]
$Gd_{0.1}Ce_{0.9}O_{1.95}$	$Ce(NO_3)_3$, $Gd(NO_3)_3$, water	Glass	300 °C	Air pulverisation	Oxygen ion conductor	[44]
NiO-$Ce_{0.8}Gd_{0.2}O_{1.9-x}$	Nitrate, acetate, chaloride, ethanol, water, polyethylene glycol	Sapphire	250–390 °C	Air pressure	Anode	[38]
NiO-CGO	$Ni(NO_3)_2$,$Ce(NO_3)_3$,$GdCl_3$	Dense polycrystalline 8-YSZ substrates made by tape-casting	600–1200 °C	Air blast spray pyrolysis	anode	[39,45]

$BaCe_{0.8}Sm_{0.2}O_{3-\delta}$	$BaCO_3$, CeO_2 and Sm_2O_3, ethanol, polyvinyl butyral	NiO-BCS anode substrate	1400 °C	Extremely dense, without any obvious holes and adheres	Proton conductor	[34,46]
$BaCe_{0.8}Y_{0.2}O_{3-\delta}$	BCY powder in isopropanol	Porous substrates BCY	1400 °C	Ultrasonic atomising nozzle	Proton conductor	[53]
$Ba(Zr_{0.1}Ce_{0.7})Y_{0.2}O_{3-\delta}$	BZCY7-1, BZCY7-2 with different size suspensions in ethanol	Anode	1400 °C	Suspension spray method	Proton conductor, mixing two BZCY powders with different average size to make dense film	[35]
$La_{0.6}Sr_{0.4}Co_{0.2}Fe_{0.8}O_3$	Nitrates and chlorides, ethanol, diethylene glycol monobutyl ether	Ce0.8Gd0.2O1.9 pellet	270 °C	Air-pressurised spray pyrolysis	Cathode	[47,48]
Lanthanum strontium manganite	Metal acetates, deionised water, metal-organic precursors of La, Sr, and Mn, ethylene glycol dimethyl ether	YSZ	520–580 °C	Ultrasonic spray pyrolysis	Cathode	[49]

conductivity so as to lower operation temperature. However there is a limitation for the thickness, for a gas-tight reason. Spray pyrolysis as a cheap and simple method have been employed to prepare YSZ films by many researchers, including electrostatic, pneumatic, ultrasonic and other techniques. Each technique has advantages and disadvantages in terms of complexity and quality of deposition. Electrostatic spray has high deposit efficiency and produce almost mono-dispersed and fine drops due to effects of applied electric field. However the effect of preferential landing characteristic of the charged droplets can lead to porous and/or cracked films. Pressurised spray deposition (PSD) is reported as a more adequate technique to deposit dense films compared to the ESD technique, however, it requires much more attention to get uniform films because the size distribution of the precursor droplets is more disperse. Ultrasonic spray deposition (USD) produce smaller droplets with a narrow size distribution than electrostatic or pressurised spray, however, it put much limitation on the precursor solution for atomising reason.

Perednis and Wilhelm *et al.* have deposited of thin YSZ films onto inconel applying ESD and PSD [28,30], a compare was carried out with various deposition parameters such as substrate temperature, deposition time, source of precursor, solution flow rate, and so on.

As for the spray parameters, the substrate temperature is the most important spray parameter. In both case of ESD and PSD, the film was cracked at 200 °C and dense at 250 °C. However at too high temperatures (>350 °C) discrete particles are formed on the surface due to slow spreading for ESD, while a dense film are still formed for PSD.

Furthermore, they found adding PEG into the precursor solution can changed the film morphology from cracked to the crack-free. The substrate temperature has to be above the boiling point of the solvent. This guarantees a fast and effective evaporation of the solvent after droplet deposition. They also investigated the performance of SOFCs with spray pyrolysis deposited YSZ electrolytes. Figure 18.7 shows the cross section images of solid oxide fuel cells with sprayed electrolytes [40]. Good performances of cells with more than two layers of different electrolytes were achieved at 700 °C but considerable degradation was observed.

Except pneumatic and electrostatic spray, USD has been also successfully applied by Garcá-Sánchez *et al.* [41]. Also YSZ has been deposited via spray pyrolysis with a moving nozzle on quartz substrate by Rodrigues *et al.* [42], results in production of films more homogeneous, roughness and reduction of morphologic defects. Over and above, Pt loaded YSZ composites membrane has been deposited by spray pyrolysis aqueous and mainly ethylene glycol solutions of citric complexes by Todorovska *et al.* [43].

(a) (b)

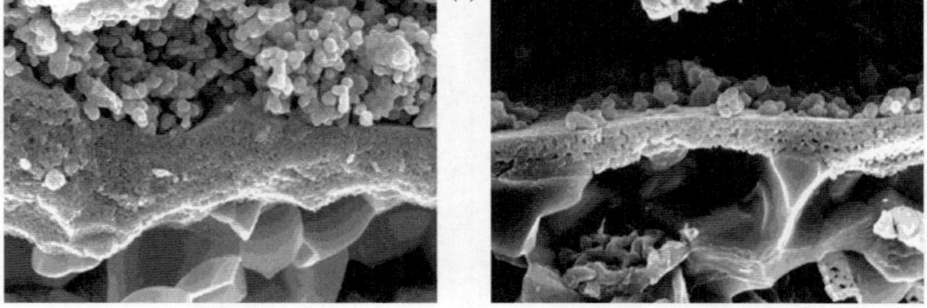

Figure 18.7 *Cross section images of solid oxide fuel cells with sprayed electrolytes. From top to bottom: screen-printed cathode (LSCF), YSZ electrolyte prepared by spray deposition, and anode substrate (Ni–YSZ cermet). (a) Electrostatic spray deposition, (b) air blast atomiser. Reprinted from Solid State Ionics, Perednis D., and Gauckler L. J., Solid oxide fuel cells with electrolytes prepared via spray pyrolysis. Vol. 166(3–4), 229–239. Copyright (2004) with permission from Elsevier*

18.3.3.2 Cerium Gadolinium Oxide

Cerium gadolinium oxide (CGO) is known to be good oxygen ionic conductors and is also widely used as electrolyte in SOFCs, especially in thin film form for lowing SOFCs operating temperature.

Chourashiya *et al.* deposited pure $Gd_{0.1}Ce_{0.9}O_{1.95}$ onto glass substrate via spray pyrolysis an aqueous precursor solution of nitrate [44]. The films were uniform and dense with very small surface roughness. The conductivity of as-prepared films at 350 °C is of the order of 0.5 S cm^{-1}. The film thickness was in the range of 1.5–3.0 μm, and it can be well monitored to a further thinness.

NiO-CGO can also be used as anode in SOFC. Muecke *et al.* successfully deposited nickel oxide (NiO), cerium gadolinium oxide (CGO) and NiO-CGO thin films by air spray pyrolysis on sapphire single crystals [38,39,45]. They found that there exists a limiting substrate surface temperature above which no continuous films can be obtained due to solvent evaporation and Leidenfrost phenomenon. By adding high boiling point solvent such as tetraethylene glycol, they prepared continuous, dense and crack-free films of CGO and NiO-CGO with thicknesses up to 500 and 800 nm, as shown in Figure 18.8.

18.3.3.3 Barium Cerate

Another common used ceramics electrolyte membrane is barium cerate (BCO) and its derivatives, a perovskite-type proton conductor. In contrast to the oxygen ionic conductor fuel cells, water forms at the cathode side, hence the fuel at the anode remains pure and requires no recirculation.

Xie *et al.* successfully fabricated 20 μm thin membranes of $BaCe_{0.9}Nd_{0.1}O_{3-\delta}$ on anode substrates with a suspension spray method [33]. However further reduction of electrolyte thickness will lead to densification problems of membranes. Bi *et al.* prepared an extremely dense $BaCe_{0.8}Sm_{0.2}O_{3-\delta}$ thin membrane via an in situ reaction of sprayed suspension of $BaCO_3$, CeO_2 and Sm_2O_3 [34]. The thickness was about 10 μm on NiO-BCS anode substrate. They also applied this method to improve the adhesion of screen-printed electrolyte membrane to the cathode [46]. Xie *et al.* propose a particle gradation method to produce dense electrolyte membrane through suspension spray [35].

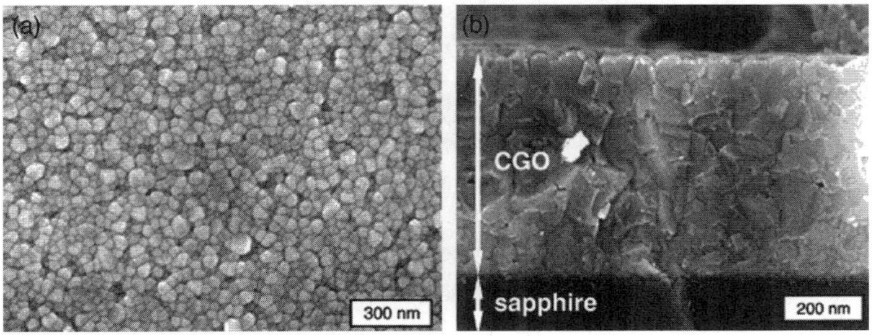

Figure 18.8 *Top and cross section SEM micrographs of a spray deposited CGO film on sapphire. Reprinted from Thin Solid Films, Muecke, U. P., et al., Initial stages of deposition and film formation during spray pyrolysis— Nickel oxide, cerium gadolinium oxide and mixtures thereof. Vol. 517(5), 1522–1529. Copyright (2009) with permission from Elsevier*

18.3.3.4 LSCF

Besides electrolyte and anode, cathode of SOFC can also be prepared via spray pyrolysis. The electrodes should be porous and with good permeability for reactive gas, a little different from that of electrolyte. Beckel deposited LSCF on a $Ce_{0.8}Gd_{0.2}O_{1.9}$ substrate for μ-SOFCs application using nitrates and chlorides mixed with ethanol, diethylene glycol monobutyl ether as precursor solution and systemic investigated properties of the as-prepared films [37,47,48]. The microstructure in combination with the choice of material is the most critical point for performance of the thin film cathode. Smaller grain sizes led to better performance and can be achieved by low annealing temperature. The microstructure achieved by spray pyrolysis proved to be better for the cathode performance than the one achieved by pulsed laser deposition (PLD). It was found the most critical parameter when preparing thin films by spray pyrolysis is the ratio of deposition temperature to solvent boiling point, because it determines the drying and decomposition kinetics of the droplets and the growing film. By keeping this ratio constant, the absolute deposition temperature could be varied by about 100 °C, while still keeping a coherent crack-free film. Figure 18.9 shows some selected SEM cross section images of films deposited by spray pyrolysis in [47].

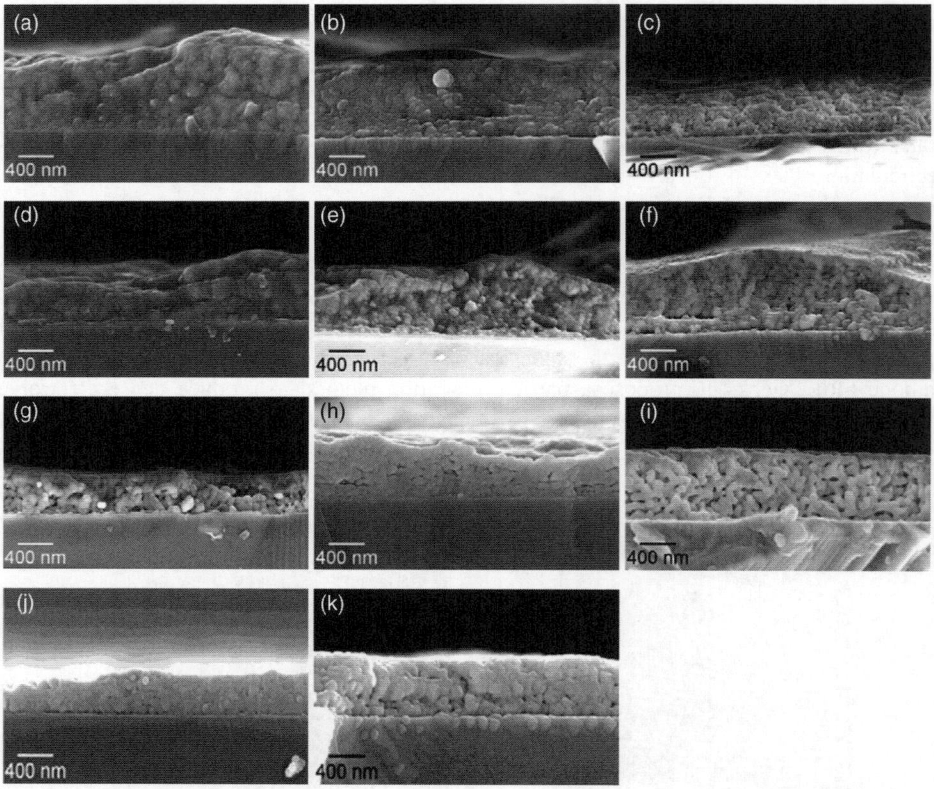

Figure 18.9 *SEM crosssections of films deposited by spray pyrolysis (Films were annealed at different temperatures for varying times. First row, 600 °C: (a) 1 min, (b) 155 min, (c) 178 min. Second row, 700 °C: (d) 1 min, (e) 40 min, (f) 120 min. Third row, 800 °C: (g) 20 min, (h) 60 min, (i) 550 min. Fourth row, 900 °C: (j) 1 min, (k) 35 min. Reprinted from Journal of the European Ceramic Society, Beckel, D., et al., Solid-state dewetting of La0.6Sr0.4Co0.2Fe0.8O3 ± δ thin films during annealing. Vol. 28(1), 49–60. Copyright (2008) with permission from Elsevier*

Hamedani *et al.* successfully deposited a similar cathode film, lanthanum strontium manganite (LSM) on YSZ for intermediate temperature SOFC application [49]. They used a setup similar to CVD equipment. By varying the deposition temperature and the solution flow rate, they get a gradient porous LSM cathode and found metal-organic precursors and organic solvent have proven to be more satisfactory than aqueous solution.

Due to their high electronic and ionic conductivity, these materials can be used as oxygen-semipermeable membranes without electrodes and external electrical circuits. In 1998 Zeng *et al.* prepared perovskite-type ceramic membrane from spray pyrolysis powder [50]. In 2000 Qi *et al.* prepared dense perovskite-structured membranes with desired composition of $La_{0.8}Sr_{0.2}Co_{0.6}Fe_{0.4}O_{3-\delta}$ too by this method [51].

18.4 Catalyst Synthesis and Spread in PEMFC

Proton exchange or polymer electrolyte membrane fuel cell (PEMFC) is another promising energy conversion device for vehicular, portable and small stationary applications. Lowering the Pt loading, improving the mass activity and simplifying the cell manufacturing process are of most importance to make PEMFCs towards commercial reality. Spray pyrolysis can be used in preparation of carbon support Pt catalysts and spread catalysts in both hot-pressed MEA and catalyst-coated membranes (CCM) fabrication.

Xue *et al.* synthesised highly dispersed Pt-Ru/C catalysts by a two-step process by spraying the precursor of mixed carbon black and dissolved H_2PtCl_6 and $RuCl_3$ in poly(ethylene glycol), and post-calculation [54]. The Pt-Ru/C catalysts showed high electrocatalytic activity for methanol oxidation due to the uniform distribution, small average size and high alloying degree of the Pt-Ru particles on the carbon carrier.

Umeda *et al.* prepared membrane electrode assembly by ESD [12]. The dispersion of commercial Pt/C and Nafion solution mixed with methanol, 2-propanol and water was introduced into a glass syringe for ESD ejection. The catalyst layer prepared possessed a fine, smooth structure. Cross sectional SEM photographs of the catalyst-coated membranes are shown in Figure 18.9. A single cell prepared by ESD has the same high performance as a cell prepared by air spraying. They also found this method enabled an extremely high yield of deposition and efficient utilisation of pt because of painting of an electrocatalyst layer over a limited area without the use of any surface mask.

Sun prepared a catalyst-coated membrane by directly spraying coating [55]. Flatted CCMs with total Pt loading of 0.3 mg cm^{-2} has been successfully prepared. The Nafion content at range of 33.3–50.0 wt% in the electrode layer has been found optimal to help improve the fuel cell performance.

18.5 Remarks and Perspectives

Spray pyrolysis is a versatile material preparation technique, which can be used for a variety of materials preparation including organic and inorganic in forms of powders, films and membranes. Unique advantages of spray pyrolysis deposition technique including high throughput manufacturing, high speed deposition capability, minimal heat input into the substrate and improved cost-effectiveness make it very attracting for material preparation. In order to improve the quality of the product and the deposition efficiency, many novel techniques can be ultilised including ultrasonic atomiser, electrostatic spray, and some controlled robotic system. The synthesis process critically determines the intrinsic morphology and the microstructure of the membrane system, and hence the effectiveness of the membrane. The deposition temperature, properties of precursor solution, and some other parameters play important roles in producing desired film for industrial application. As a summary, this technique has many potential applications in preparation of catalysis membrane, separation membrane, and membrane assemblies for SOFC. With carefully optimised parameters high quality of membrane can be obtained to fulfill variety demands.

Acknowledgements

The authors gratefully acknowledge the financial supports from the National Basic Research Program of China under contract 2009CB220000 and the National Natural Science Foundation of China under contract 50821064.

References

1. Aman J, Israeli Patent 8722 (1956).
2. Chamberlin RR and Skarman JS, Chemical spray deposition process for inorganic films. *Journal of the Electrochemical Society*, **113**(1), 86–89 (1966).
3. Li ZY, *et al.*, Preparation of palladium–silver alloy membranes for hydrogen separation by the spray pyrolysis method. *Journal of Membrane Science*, **78**(3), 247–254 (1993).
4. Albin DS and Risbud SH, Spray pyrolysis processing of optoelectronic materials. *Advanced Ceramic Materials*, **2**(3a), 243–252 (1987).
5. Patil PS, Versatility of chemical spray pyrolysis technique. *Materials Chemistry and Physics*, **59**(3), 185–198 (1999).
6. Bohac P and Gauckler L, Chemical spray deposition of ceramic films. *Oxygen Ion and Mixed Conductors and Their Technological Applications*, **368**, 271–294 (2000).
7. Perednis D and Gauckler LJ, Thin film deposition using spray pyrolysis. *Journal of Electroceramics*, **14**(2), 103–111 (2005).
8. Ortega-Lopez M and Morales-Acevedo A, Characterization of $CuInS_2$ thin films for solar cells prepared by spray pyrolysis. *Thin Solid Films*, **330**(2), 96–101 (1998).
9. Baykul MC and Balcioglu A, AFM and SEM studies of CdS thin films produced by an ultrasonic spray pyrolysis method. *Microelectronic Engineering*, **51/52**, 703–713 (2000).
10. Lee JH and Park SJ, Preparation of spherical tin dioxide powders by ultrasonic spray pyrolysis. *Journal of the American Ceramic Society*, **76**(3), 777–780 (1993).
11. Li M, Su J, and Guo L, Preparation and characterization of $ZnIn_2S_4$ thin films deposited by spray pyrolysis for hydrogen production. *International Journal of Hydrogen Energy*, **33**(12), 2891–2896 (2008)
12. Umeda M, Kawaguchi S, and Uchida I, Characterization of membrane electrode assembly for fuel cells prepared by electrostatic spray deposition. *Japanese Journal of Applied Physics, Part 1 Regular Papers, Brief Communications and Review Papers*, **45**(7), 6049–6054 (2006).
13. Siefert W, Corona spray pyrolysis: a new coating technique with an extremely enhanced deposition efficiency. *Thin Solid Films*, **120**(4), 267–274 (1984).
14. Chang C-L and Hwang B-H, Microstructure and electrochemical characterization of $Sm_{0.5}Sr_{0.5}CoO_3$ films as SOFC cathode prepared by the electrostatic-assisted ultrasonic

spray pyrolysis method. *International Journal of Applied Ceramic Technology*, **5**(6), 582–588 (2008).
15. Nomura M, *et al.*, Improvement of thermal stability of porous titania films prepared by electrostatic sol-spray deposition (ESSD). *Chemistry of Materials*, **15**(6), 1283–1288 (2003).
16. Li M, Zhao L, and Guo L, Preparation and photoelectrochemical study of $BiVO_4$ thin films deposited by ultrasonic spray pyrolysis. *International Journal of Hydrogen Energy*, in press (2009).
17. Kim J-G, *et al.*, Reel-to-reel deposition of YBCO films by spray pyrolysis using nitrate precursors. *Superconductor Science and Technology*, **20**(10), 950–954 (2007).
18. Friedlander SK, Smoke, dust, and haze, in *Fundamentals of Aerosol Behavior*, Springer, Heidelberg, 317 pp (1977).
19. Jayanthi GV, Zhang SC, and Messing GL, Modeling of solid particle formation during solution aerosol thermolysis. The evaporation stage. *Aerosol Science and Technology*, **19**(4), 478–490 (1993).
20. Jain S, Skamser DJ, and Kodas TT, Morphology of single-component particles produced by spray pyrolysis. *Aerosol Science and Technology*, **27**(5), 575–590 (1997).
21. Sun Y-M, Huang W-F, and Chang C-C, Spray-coated and solution-cast ethyl cellulose pseudo-latex membranes. *Journal of Membrane Science*, **157**(2), 159–170 (1999).
22. Lin L and Guthrie JT, Preparation and characterisation of novel, blood-plasma-separation membranes for use in biosensors. *Journal of Membrane Science*, **173**(1), 73–85 (2000).
23. Schindler E and Maier F, Manufacture of porous carbon membranes, US Patent 4919860 (1990).
24. Strano MS, *et al.*, Ultrafiltration membrane synthesis by nanoscale templating of porous carbon. *Journal of Membrane Science*, **198**(2), 173–186 (2002).
25. Krogman KC, *et al.*, Spraying asymmetry into functional membranes layer-by-layer. *Nature Materials*, **8**(6), 512–518 (2009).
26. Nomura M, *et al.*, Preparation of thin porous titania films on stainless steel substrates for heat exchange (HEX) reactors. *Separation and Purification Technology*, **32**(1/3) 387–395 (2003).
27. Bohac P and Gauckler L, Chemical spray deposition of YSZ and GCO solid electrolyte films. *Solid State Ionics*, **119**(1/4) 317–321 (1999).
28. Perednis D, *et al.*, Morphology and deposition of thin yttria-stabilized zirconia films using spray pyrolysis. *Thin Solid Films*, **474**(1/2) 84–95 (2005).
29. Matsuzaki Y, Hishinuma M, and Yasuda I, Growth of yttria stabilized zirconia thin films by metallo-organic, ultrasonic spray pyrolysis. *Thin Solid Films*, **340**(1/2) 72–76 (1999).
30. Wilhelm O, *et al.*, Electrospray and pressurized spray deposition of yttria-stabilized zirconia films. *Thin Solid Films*, **479**(1/2) 121–129 (2005).
31. Rupp JLM, Infortuna A, and Gauckler LJ, Microstrain and self-limited grain growth in nanocrystalline ceria ceramics. *Acta Materialia*, **54**(7), 1721–1730 (2006)
32. Setoguchi T, *et al.*, Application of the stabilized zirconia thin film prepared by spray pyrolysis method to SOFC. *Solid State Ionics*, **40/41**(1) 502–505 (1990).
33. Xie K, *et al.*, An ammonia fuelled SOFC with a $BaCe_{0.9}Nd_{0.1}O_{3-\delta}$ thin electrolyte prepared with a suspension spray. *Journal of Power Sources*, **170**(1), 38–41 (2007).
34. Bi L, *et al.*, Preparation of an extremely dense $BaCe_{0.8}Sm_{0.2}O_{3-\delta}$ thin membrane based on an *in situ* reaction. *Electrochemistry Communications*, **10**(7), 1005–1007 (2008)
35. Xie K, *et al.*, A modified suspension spray combined with particle gradation method for preparation of protonic ceramic membrane fuel cells. *Journal of Power Sources*, **179**(2), 576–583 (2008).
36. Furusaki A, Konno H, and Furuichi R, Perovskite-type lanthanum chromium-based oxide films prepared by ultrasonic spray pyrolysis. *Journal of Materials Science*, **30**(11), 2829–2834 (1995)
37. Beckel D, *et al.*, Spray pyrolysis of $La_{0.6}Sr_{0.4}Co_{0.2}Fe_{0.8}O_{3-\delta}$ thin film cathodes. *Journal of Electroceramics*, **16**(3), 221–228 (2006).

38. Muecke UP, *et al.*, Initial stages of deposition and film formation during spray pyrolysis – Nickel oxide, cerium gadolinium oxide and mixtures thereof. *Thin Solid Films*, **517**(5), 1522–1529 (2009).
39. Muecke UP, *et al.*, Microstructure and electrical conductivity of nanocrystalline nickel- and nickel oxide/gadolinia-doped ceria thin films. *Acta Materialia*, **56**(4), 677–687 (2008).
40. Perednis D and Gauckler LJ, Solid oxide fuel cells with electrolytes prepared via spray pyrolysis. *Solid State Ionics*, **166**(3/4) 229–239 (2004).
41. Garcia-Sanchez MF, *et al.*, Nanostructured YSZ thin films for solid oxide fuel cells deposited by ultrasonic spray pyrolysis. *Solid State Ionics*, **179**(7/8) 243–249 (2008).
42. Rodrigues CHM, Toniato M, and Paes HR Jr, Deposition of YSZ 8 mol% films by modified spray pyrolysis system. *Revista Materia*, **13**(3), 533–541 (2008).
43. Todorovska R, Petrova N, and Todorovsky D, Spray pyrolysis deposition of YSZ and YSZ-Pt composite films. *Applied Surface Science*, **252**(5), 1266–1275 (2005).
44. Chourashiya MG, Pawar SH, and Jadhav LD, Synthesis and characterization of $Gd_{0.1}Ce_{0.9}O_{1.95}$ thin films by spray pyrolysis technique. *Applied Surface Science*, **254**(11), 3431–3435 (2008).
45. Muecke UP, Messing GL, and Gauckler LJ, The Leidenfrost effect during spray pyrolysis of nickel oxide–gadolinia doped ceria composite thin films. *Thin Solid Films*, **517**(5), 1515–1521 (2009).
46. Bi L, *et al.*, Screen-printed $BaCe_{0.8}Sm_{0.2}O_{3-\delta}$ thin membrane solid oxide fuel cells with surface modification by spray coating. *Journal of Alloys and Compounds*, **473**(1/2) 48–52 (2009).
47. Beckel D, *et al.*, Solid-state dewetting of $La_{0.6}Sr_{0.4}CO_{0.2}Fe_{0.8}O_{3\pm\delta}$ thin films during annealing. *Journal of the European Ceramic Society*, **28**(1), 49–60 (2008).
48. Beckel D, *et al.*, Electrochemical performance of LSCF based thin film cathodes prepared by spray pyrolysis. *Solid State Ionics*, **178**(5/6) 407–415 (2007).
49. Hamedani HA, *et al.*, Fabrication of gradient porous LSM cathode by optimizing deposition parameters in ultrasonic spray pyrolysis. *Materials Science and Engineering B Advanced Functional Solid-State Materials*, **153**(1/3) 1–9 (2008).
50. Zeng Y, Lin YS, and Swartz SL, Perovskite-type ceramic membrane synthesis, oxygen permeation and membrane reactor performance for oxidative coupling of methane. *Journal of Membrane Science*, **150**(1), 87–98 (1998).
51. Qi X, Lin YS, and Swartz SL, Electric transport and oxygen permeation properties of lanthanum cobaltite membranes synthesized by different methods. *Industrial and Engineering Chemistry Research*, **39**(3), 646–653 (2000).
52. Menzler NH, *et al.*, Materials synthesis and characterization of 8YSZ nanomaterials for the fabrication of electrolyte membranes in solid oxide fuel cells. *Ceramics International*, **29**(6), 619–628 (2003).
53. Lee TH, Dorris SE, and Balachandran U, Thin film preparation and hydrogen pumping characteristics of $BaCe_{0.8}Y_{0.2}O_{3-\delta}$. *Solid State Ionics*, **176**(15/16) 1479–1484 (2005).
54. Xue X, *et al.*, Simple and controllable synthesis of highly dispersed Pt-Ru/C catalysts by a two-step spray pyrolysis process. *Chemical Communications*, **2005**(12), 1601–1603 (2005).
55. Sun L, *et al.*, Fabrication and performance test of a catalyst-coated membrane from direct spray deposition. *Solid State Ionics*, **179**(21/26) 960–965 (2008).

19

Preparation and Characterisation of Nanocrystalline and Quasicrystalline Alloys by Planar Flow Casting for Metal Membranes

J.W. Phair[1] *and M.A. Gibson*[2]

[1]*Division of Fuel Cells and Solid State Chemistry, Risø National Laboratory for Sustainable Energy, The Technical University of Denmark, Frederiksborgvej 399, Roskilde 4000, Denmark*

[2]*CSIRO Process Science and Engineering, Clayton, Victoria 3168, Australia*

19.1 Introduction

Dense metal membranes have seen considerable application in the separation of gases at high temperatures (> 300 °C). Numerous metals have demonstrated selective hydrogen permeation (e.g., Pd) from a mixture of gases making it the most widespread application of a metal membrane. For gaseous separations, the operation of a dense metal membrane in the presence of a potential gradient is largely based on its ability to dissociate diatomic gas molecules (e.g., H_2, O_2, CO, N_2) upon adsorption onto the metal surface. This is followed by absorption of the monatomic molecules into the bulk metal, their diffusion, subsequent re-association and desorption from the permeate side without significant mechanical and chemical degradation of the membrane.

For the separation of hydrogen, metal membranes have been considered a promising candidate from other choices (e.g., nanoporous, polymer, dense ceramics, etc.) due to their ability to attain high-purity hydrogen. Several detailed reviews have assessed the various membrane types for

Membranes for Membrane Reactors: Preparation, Optimization and Selection, Edited by Angelo Basile and Fausto Gallucci

hydrogen separation [1–7]. Traditionally, Pd has been the most widespread metal for hydrogen separation membranes due to its high permeability to hydrogen at high temperatures ($>300\,°C$), its catalytic H_2 dissociation and re-association properties, and its relative good resistance to degradation. However, it remains an expensive material to use and requires alloying (e.g., with Ag or Au) to improve its stability such as the prevention of α and β hydride phase transitions. Pd-based membranes are typically crystalline in structure.

In the search for improved Pd-based membranes or nonPd-based membranes comprising of high-performance alloys with cheaper constituents, considerable attention has focused on controlling the microstructure of the metal and its optimisation. Significant debate has focused on the advantages of crystalline metals compared to amorphous metals. Amorphous metals have been considered as likely replacements given their higher H_2 absorption capacity, lower rate of embrittlement and higher mechanical strength. However, the long-term stability of metallic glasses at elevated temperatures [8,9] and in the presence of hydrogen or other 'toxic' contaminants are major issues that require considerable attention [8,10,11]. Combined with the fact that their reported hydrogen diffusivities are often lower than their crystalline counterparts, attention has turned to redesigning crystalline alloys for hydrogen separation.

A possible route for improving the performance of the membranes is by increasing the amount of nanocrystalline or quasicrystalline phases within the metal. Given the relative recent arrival of nanocrystalline materials, such an advancement is well worth exploring based on initial results. However, the exact effect of nanocrystallinity on the permeation and durability of the membranes remains poorly understood. Extensive research effort is still required to better understand the formation and effects of nanocrystalline and quasicrystalline phases on the performance of metal membranes for hydrogen separation.

The purpose of this chapter is to review the preparation of nanocrystalline and quasicrystalline metal membranes by a selected method (planar flow casting) with examples provided of the properties and performance of metal membranes prepared by this method. The method of metal membrane preparation by planar flow casting is introduced and described in detail, and the major parameters affecting the formation of metals for hydrogen separation membranes are discussed, including their microstructure/nanostructure and performance.

19.2 Properties and Preparation of Nanocrystalline and Quasicrystalline Metals

19.2.1 Properties

Nanocrystalline metals consist of single or multiphase polycrystals with nanoscale grain sizes (1–250 nm). They are characterised by containing a high volume (≥ 50 atom%) of strained crystal lattice regions and defect cores as well as a high density ($10^{19}\,cm^{-3}$) of grain interface boundaries. Nanocrystalline metals may be classified in terms of their dimensionality, that is: 0D = nanocluster, 1D = multilayer, 2D = nanograined layer and 3D = equiaxed bulk solid [12]. They may also be classified in terms of their composition, morphology and distribution of nanograins [13]. A nanocrystalline metal maintains low energy crystal structures while the boundary regions are at a high energy and highly non-equilibrium state [13], with a considerably different form of heterogeneity observed on thermally induced amorphous metals.

Crystallite sizes of less than around 10 nm are required to achieve the highest densities with smaller nanocrystals (<10 nm) having been reported to offer significant advantages over larger nanocrystals. Reducing the crystal size leads to an increase in the number of triple grain junctions in the metal, and an increase in the hardness following the well known Hall–Petch relationship. Other mechanical properties of potential significance for improved membrane properties include ultra-high yield and fracture strengths, and superior wear resistance.

Quasicrystalline metals contain an essentially well ordered crystalline structure but in a nonperiodic and nonrepeating fashion. The typical morphology of an icosahedral quasicrystal has been referred to as a petal-like structure with five branches [14]. Due to this uncommon structure, they display unpredictable and irregular combinations of material properties (e.g., slippery like Teflon or good insulators). The consequences of a quasicrystalline structure for hydrogen separation membranes remains poorly documented. For instance, while the absorption of hydrogen by ZrCuNiAl quasicrystalline alloys is greater than for their amorphous counterparts, at higher hydrogen contents the icosahedral phases begin to decompose due to hydride formation [15]. Furthermore, Ti–Zr–Ni and Ti–Hf–Ni quasicrystals and approximants have been examined for their potential as hydrogen storage alloys with Ti–Hf–Ni being found to be loaded with H at 250 °C, without the formation of a hydrided phase and therefore, a better candidate for hydrogen loading and unloading [16]. Such properties would likely be useful for hydrogen separation membranes but a closer examination of such quasicrystal alloys for hydrogen separation membranes remains to be done.

Defining the effect of the nature and amount of grain boundaries in nanocrystalline metals is critical to enhancing the physical, mechanical and chemical properties of separation membranes. For instance, an increase in the number of grain boundaries facilitate stress–relief mechanisms, serve as a source and sink for dislocations and impurities as well as mediate diffusion processes.

19.2.2 Preparation

Controlling the existence and nature of nanocrystalline or quasicrystalline structure is largely a function of the manufacturing method. The preparation of nanocrystalline metals may be achieved in a variety of ways such as sol-gel deposition, sputtering, electrodeposition, mechanical attrition (high energy milling), laser ablation, inert gas condensation (or gas phase condensation), devitrification, melt-quenching or phase separation of simple alloys. Most of these techniques yield grain sizes <100 nm, however inert gas condensation and electrodeposition have the capability of forming grain sizes of 5–50 nm.

While inert gas phase condensation has problems attaining a fully dense metal structure, electrodeposition has received much interest for its ability to produce a large number of nanocrystalline pure metals and alloys with relatively little capital expenditure and established knowhow. A great advantage of the electrodeposition technique is the ability to produce nanocystalline nickel and other metals with a specific crystal shape and size, in a controlled manner.

In contrast to electrodeposition which appears to be a relatively easy method for controlling the crystal properties of nanocrystalline metals, formation of nanocrystalline or quasicrystalline metals following the devitrification pathway may be more difficult to control but potentially offers considerably better properties. A variety of methods can be used to prepare metallic glasses such as electrodeposition or vapour deposition, planar flow casting, mechanical alloying or splat quenching [17]. To prepare nanocrystalline or quasicrystalline metals from the amorphous state, the crystallisation kinetics must be controlled by heat treatment, irradiation or mechanical attrition such that the amorphous phase crystallises completely. Complete crystallisation of the amorphous phases affords some important advantages such as the ability to form a fully dense nanocrystalline metal alloy or supersaturated solid solution.

Figure 19.1 Metal membrane produced by planar flow casting at CSIRO (courtesy M. Fergus and D. Liang, CSIRO)

Recent advances in the preparation of metal membranes for hydrogen separation by planar flow casting have been achieved by CSIRO, Australia [8] and an image of the ribbons prepared for such applications by planar flow casting is provided in Figure 19.1.

19.3 Preparation of Nanocrystalline and Quasicrystalline Metal Membranes by Planar Flow Casting

Planar flow casting (PFC) is a method of continuously casting an alloy melt into a thin strip [18] (typically $<50\,\mu m$ thick and up to 300 mm wide [19,20]) under conditions of rapid solidification (cooling rates up to $10^6\,K\,s^{-1}$) to generate a range of novel microstructures (e.g., metastable crystalline, quasicrystalline, amorphous phases) with a variety of functional properties direct from the melt or by controlled crystallisation, (see Figure 19.2 [21]).

PFC is highly suited to the fabrication of 100% dense thin foils of nanocrystalline and quasicrystalline alloys useful for hydrogen separation membranes. The basic features of a PFC system are shown schematically in Figure 19.3, which is a compilation of a number of successful designs [22–24]. It depicts all the important process parameters that must be regulated for the most effective operation of the casting facility. Precise control of the cleanliness, superheat and flow rate of the molten metal; the positioning of the casting nozzle and the speed of the wheel are all essential for the production of high quality strip. In practice, PFC employs a rectangular orifice with a high aspect ratio (the ratio of the slot length, N_L, to the slot breadth, N_b, being typically much greater than 10), in the end of the nozzle. The alloy melt is ejected through this slot by applying a gas pressure, P_e, to the top of the melt. The melt impinges on the substrate to form

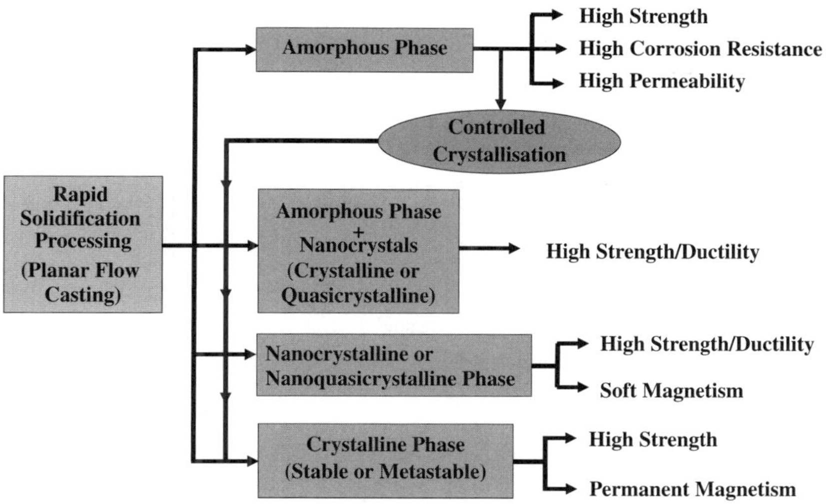

Figure 19.2 *Schematic diagram of the variety of rapid solidification processing routes to obtain a multitude of crystalline, nanocrystalline/quasicrystalline phases and amorphous microstructures for high performance membranes (after [21])*

a puddle and the motion of the substrate drags a thin film of material from the base of the puddle, quenches it into a solid strip and transports it away in a continuous manner (see Figure 19.4).

At present, the PFC process is insufficiently understood to allow prediction of the optimum operating window for the production of uniform, high-quality strips in combination with the desired microstructure. The determination of such an operating window (Figure 19.3) remains largely empirical [25–28], however, progress continues to be made through the application of analytical models for strip formation [29–35] and microstructure selection during rapid solidification [36–39]. The interrelation between the various process parameters is complex and must be considered when optimising the casting conditions, Figure 19.5 [40].

A crucial issue for PFC is that the end of the crucible/nozzle, which feeds the molten metal, must be in close proximity to the surface of the moving substrate (i.e., nozzle/substrate gap, $G_{ns} = 150$–$300\,\mu m$). This gap must be controlled to maintain steady-state casting conditions for long periods. The main advantage of this configuration is that perturbations of the melt puddle menisci are dampened substantially by the mechanical constraint of the melt between the end of the nozzle and the substrate. This results in a cast product with exceptional dimensional uniformity provided the processing conditions have been optimised [41].

Consequently, the melt puddle is the most significant region in the PFC process. Both the overall puddle dynamics and the characteristics of the temperature profile and flow field within the puddle have a pronounced effect on determining the macro- and microstructure of the resulting strip. Experimental investigations into the effects of processing parameters on strip geometry [42–46] invariably find that, under steady-state conditions, the strip width is essentially independent of all variables apart from N_L, for which there is a one to one correspondence. The thickness of PFC strip is controlled substantially by the velocity of the substrate, V_S, with P_e and N_b also having a significant influence.

A minor dependence of strip thickness on G_{ns} has been observed [47], however, G_{ns} has been shown to have a dominant influence on the uniformity of the strip profile across the entire strip width [43,48,49], which is observed to improve considerably with smaller gaps, all other factors

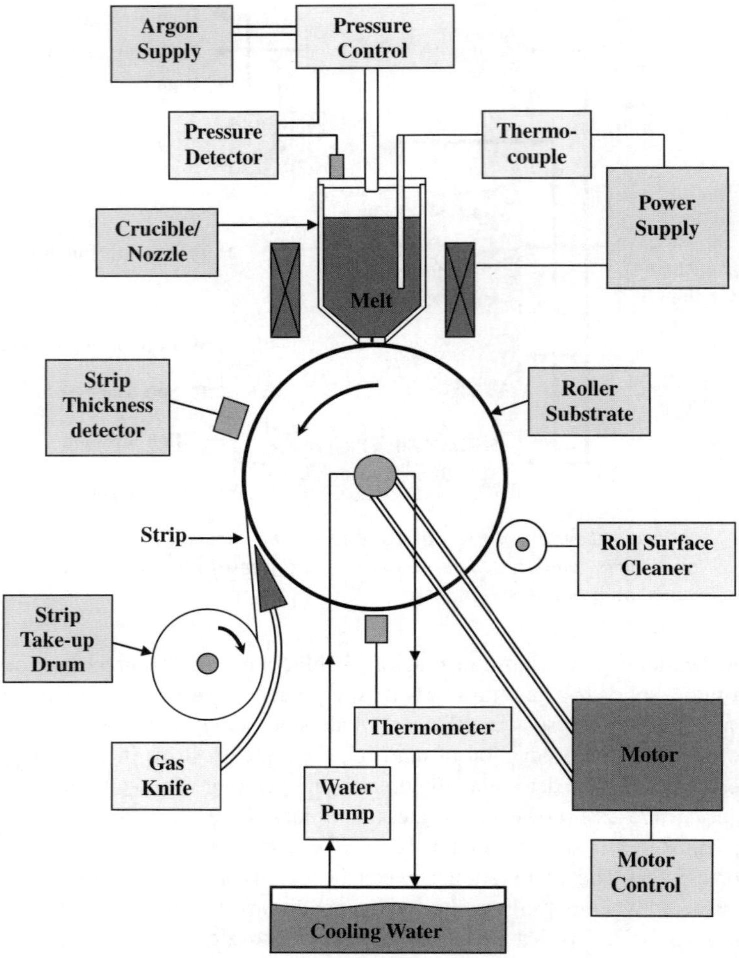

Figure 19.3 *Schematic diagram showing the various elements of a planar flow casting device indicating certain essential features (highlighted in medium grey) characteristic of all such units and other features that are desirable (highlighted in light grey) in a more sophisticated apparatus*

remaining constant. Direct observations of melt puddle dynamics have also confirmed that the casting of non-uniform strip is associated with both large scale oscillations of the melt puddle profile and high-frequency surface waves [47,50]. At best, these instabilities lead to irregularities in the surface, imparting a characteristic pattern as shown in Figure 19.6, and at worst, to the total collapse of the melt puddle, resulting in no strip being formed at all. It is therefore, essential to the production of high quality net shape strip (both crystalline and noncrystalline), that G_{ns} be controlled within the smooth flow regime during the casting process [24,49,51]. Moreover, the greater the width of the strip to be cast, the smaller G_{ns} must be to suppress dimensional non-uniformities and this requires greater precision in process control [51,52].

The heat transfer coefficient between the strip and the substrate is another major process parameter, but difficult to control experimentally. A high heat transfer coefficient facilitates the generation of a large temperature gradient within the strip thickness which, depending on the

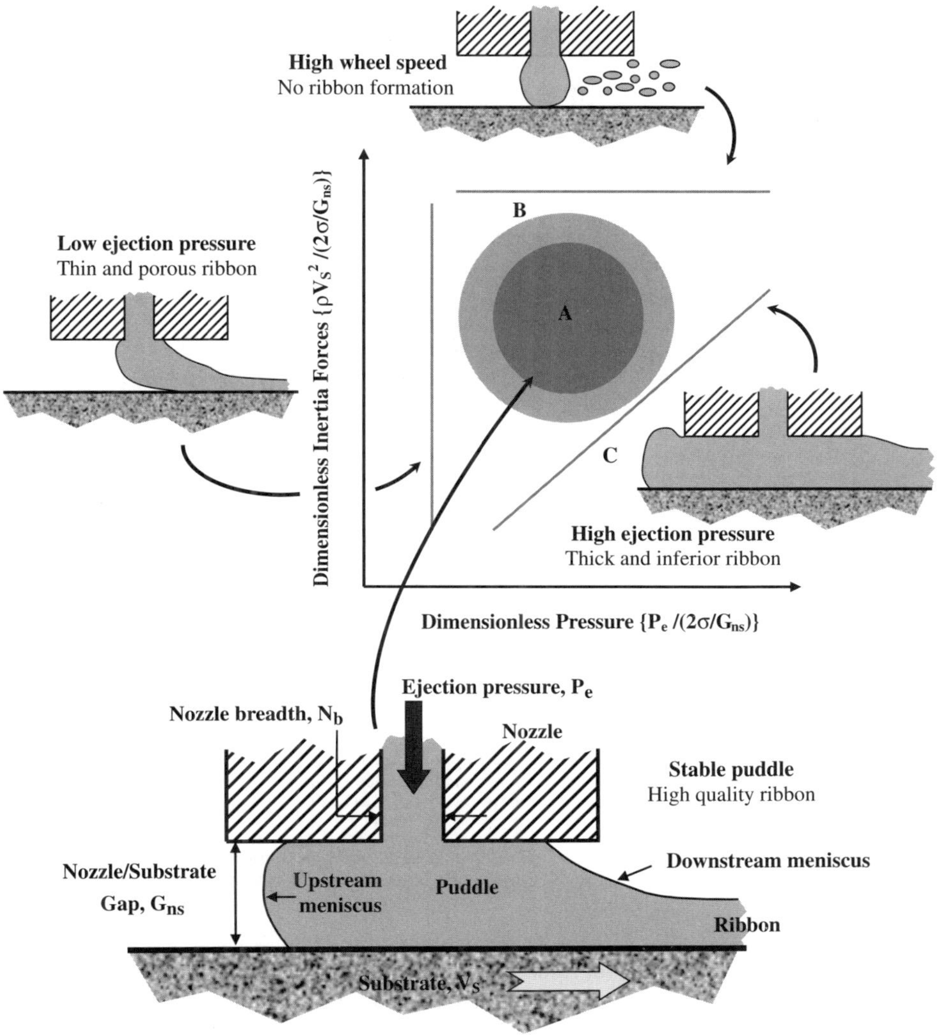

Figure 19.4 Schematic representation of the planar flow casting (PFC) process showing the nozzle/substrate configuration and indicating the main processing variables. The profile of the melt puddle depends on the processing conditions and this, inturn, has a profound influence on the quality of the as-cast product. Schematic illustration of the quality of PFC product as a function of two dimensionless parameters which characterise the casting process (ρ is the alloy density and σ is the liquid alloy surface tension). The dark grey region, A, represents the processing window for high quality strip. The plot is also dependent on the dimensions of the slot (length and breadth) through which the molten metal is ejected (after [26])

alloy composition, can lead to non-equilibrium solidification. Poor contact between the strip and the substrate results in a small temperature gradient and consequently only minor deviations from equilibrium. The wetting behaviour between the substrate surface and the liquid alloy is of fundamental importance in PFC since it is only via this intimate contact that high rates of heat transfer are possible. It is well established that [42,53,54], unless casting is conducted under

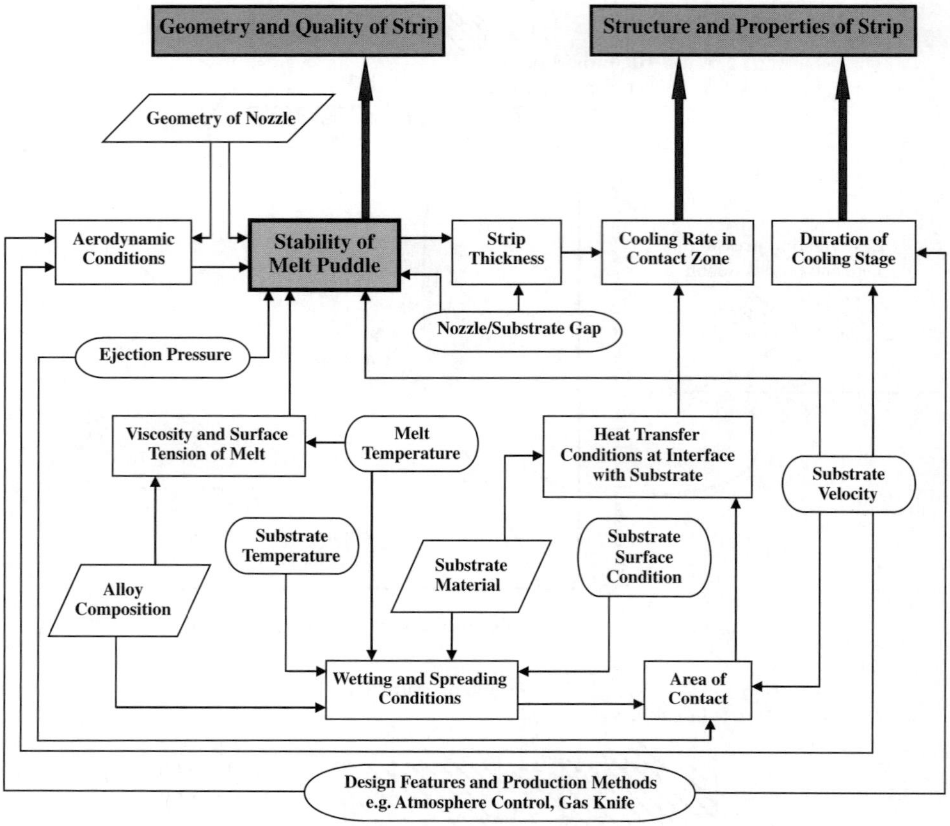

Figure 19.5 *Schematic diagram of the interrelationships between the various process parameters and their influence on the melt puddle stability and the characteristics of planar flow cast strip (after [40])*

vacuum, the gaseous boundary layer associated with the rotating substrate produces a distinctive *wetting pattern* on the surface of the strip in contact with the substrate. This pattern consists of depressions in the surface, produced by the entrapment of gas bubbles under the melt puddle, together with regions of direct contact between the liquid and the substrate. It is essential to maximise the area of metal to metal contact since it is only through this area that rapid heat transfer can be achieved [55].

This contact area depends on numerous factors including the substrate surface condition [53], the alloy/substrate compositional compatibility [56], melt and substrate temperature [57], substrate velocity [58,59] and the composition and pressure of the ambient atmosphere [60,61] (see Figure 19.5). It must also be recognised that rapid heat transfer can only occur during the time that the strip is in intimate contact with the substrate. There is a definite advantage in extending the contact time/distance of the strip on the wheel to obtain better overall quenching conditions, but, more importantly it has been determined that for uniform properties throughout the entire length of strip, the contact time/distance must remain constant [62].

The strip/substrate adhesion distance has been observed to increase during the casting process [54] and a transition can occur to *catastrophic adhesion* where the strip remains on the wheel for a complete revolution and disrupts the casting process [57]. The transition from

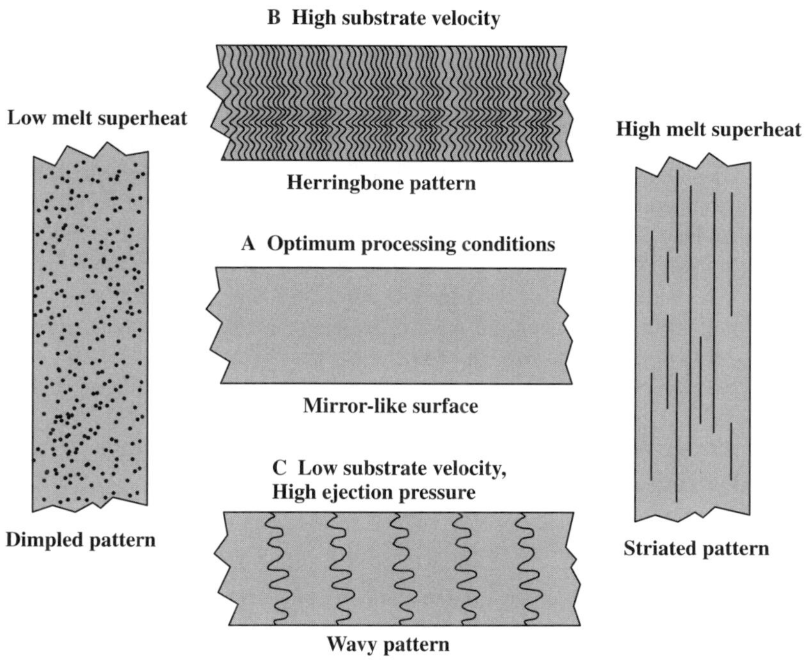

Figure 19.6 *Schematic illustration of a number of characteristic patterns/textures that are often observed on the top surface of planer flow cast strips and the main processing conditions that are responsible for their formation. The letters A, B and C correspond to the positions labelled in the processing parameter map in Figure 19.4 (after [26])*

a freely floating lift-off point to *catastrophic adhesion* is associated with enhanced wetting conditions that result from substrate heating during casting and occurs earlier for smaller wheel diameters, higher substrate velocities, wider strip and greater compositional compatibility between the strip and the substrate. This has a number of practical consequences for the continuous production of strip, necessitating the incorporation of devices to facilitate the controlled detachment of the strip from the substrate. Water cooling of the substrate is also essential for prolonged casting runs.

Apart from ribbon geometry, the scope and scale of the resulting microstructure is also of paramount importance for rapid solidification processing of functional net shape materials. In PFC, the nature of the microstructure is governed primarily by the period of time that the nucleation of the transformation products can be delayed; ranging from microcrystalline for short periods, nanocrystalline if nucleation can be delayed significantly, and to amorphous if nucleation is avoided altogether. Obviously the alloy chemistry has a significant influence on the variety of solidification paths that are available for a particular alloy, but the processing conditions must create the correct environment for the imposed solidification conditions to proceed down a preferred path.

If nucleation occurs easily within the melt puddle (no melt undercooling) then a classical columnar structure consisting of stable crystalline phases is the most likely result. Under the appropriate solidification conditions, this produces a structure where the grains all start near the substrate surface and extend all the way to the 'free' surface. Consequently, the grain boundaries are all parallel and lie across the thickness of the strip. This may be advantageous for hydrogen separation membranes, by providing conduits for hydrogen diffusion across the membrane. The disadvantage is that it creates planes of weakness in the metal. By varying the substrate velocity, one can alter the nature of this microstructure to a limited extent – a transition from columnar to a mixture of columnar and equiaxed grains to a fully equiaxed microstructure with a decrease in casting speed. If nucleation is delayed sufficiently so that it occurs downstream of the melt puddle then significant melt undercooling is possible and, as a result, nonequilibrium solidification paths are more accessible. Under these solidification conditions, homogeneous nucleation throughout the melt is more likely (but not exclusively) and this leads to the formation of a refined, uniform microstructure. In addition, it can also lead to microstructures not accessible under more conventional melt processing routes and this can produce material with enhanced performance such as a large hydrogen storage capacity [63].

19.4 Nanocrystalline and Quasicrystalline Metal Membranes for Hydrogen Separation

19.4.1 General

The manipulation of the various process parameters for PFC to enable the direct production of an optimum crystalline microstructure (including quasicrystalline and nanocrystalline) is difficult, but progress continues to be made [31,64–67]. In general, better control over microstructure generation is obtained by a two-stage process: over-quench to a fully amorphous structure is followed by heat treatment, to nucleate and grow a much finer and more uniform distribution of the target phase through controlled crystallisation (see Figure 19.2). This has been employed to great effect in producing Fe-based nanocrystalline soft magnetic materials [68] and for a range of high strength Al-based nanoquasicrystalline or nanocystalline alloys [69]. This method of production is currently the focus of a number of investigations aimed at the development of a new range of structural amorphous [70,71] and nanocrystalline [72] steels. Careful selection of alloy composition [73] and/or annealing heat treatment [74–76] can facilitate the targeted precipitation of different phases (crystalline or quasicrystailline) which can not be obtained directly from the melt.

It is the behaviour of these nanocrystalline structures that is of great interest to the development of efficient hydrogen separation membranes. Cheng *et al.* [77] determined that for a $Zr_{52.5}Cu_{17.9}$ $Ni_{14.6}Ti_5Al_{10}$ alloy, the nanocrystalline microstructure generated from an amorphous precursor had superior hydrogenation behaviour than the as-cast amorphous structure. Kirchheim *et al.* [78] have shown that the activation energy for hydrogen diffusion decreases dramatically from liquid-quenched, to vapour-quenched to nanocrystalline structures in membrane materials. In addition, Hirscher *et al.* [79] suggest that alloy chemistry and alloy microstructure can have a significant influence on the diffusion kinetics of hydrogen and that there exists some scope to tailor the response of the membrane by adjusting the scale of the microstructure depending on the alloy composition.

While the long term thermal stability of nanocrystalline microstructures is as equally an important an issue as it is for amorphous microstructures [80], huge potential remains for alloy and nanostructure design to generate an extensive range of candidate materials for hydrogen separation. Microstructural stability against both phase transformation and grain growth is crucial to the practical application of nanostructured membrane materials and this can be enhanced through the selective alloying with other metals.

19.4.2 Pd-Based Membrane Materials

Few reports on the utilisation of PFC for the fabrication of alloys where Pd is a major constituent exist – most likely due to the fact that Pd is precious metal and wastage from this method makes it uneconomical. However, since Pd has been traditionally the most popular metal for hydrogen separation, much can be learned from examining the properties of nonPFC generated nanocrystalline Pd relative to conventional crystalline Pd for designing novel alloys.

Conflicting reports exits in the literature on the benefits of nanocrystallinity to improve hydrogen separation properties as a function of the preparative method and composition. For instance, nanocrystalline Pd membranes have demonstrated hydrogen diffusivity up to 10 times greater than conventional Pd [81] with a potentially high resilience to degradation during hydrogen separation [82]. Greater volume fraction of grain boundaries in sputtered nanocrystalline Pd membranes resulted in superior hydrogen permeation compared with conventional Pd foil membranes [83]. However, McCool and Lin [84] observed a decrease in hydrogen permeation with a corresponding decrease in nanocrystalline grain size in sputtered Pd-Ag membranes. Different construction of nanocystallites and grain boundaries are likely to result in different behaviours which must be better understood for a range of fabrication methods before general conclusions can be drawn on the best nanostructure for hydrogen separation.

19.4.3 NonPd-Based Alloy Membrane Materials

NonPd-based alloys for potential candidate membrane materials are typically derived from those elements that have a large intrinsic permeability for hydrogen, namely V, Nb, Zr and Ta [85,86], together with those elements that have a catalytic activity for hydrogen dissociation, namely Co, Ni and Cu. Although, these transition metal membranes are susceptible to severe hydrogen embrittlement. Paglieri *et al.* [87] investigated the use of V-Cu crystalline foils, coated with a thin layer of Pd, for hydrogen separation membrane materials. They determined that increasing the Cu content (below 10 at.% Cu) led to a reduction in the hydrogen permeability of the foil but moreover, also resulted in a greater resistance to hydrogen embrittlement. Komiya *et al.* [88] and Watanabe *et al.* [89] observed an increase in permeability in Nb-Pd alloy membranes together with an enhanced resistance to hydrogen embrittlement. Nishimura and coworkers [90–93] have investigated the hydrogen permeation characteristics of foils fabricated from a number of crystalline V-Ni-based alloys. They determined that a Pd-coated cold rolled $V_{85}Ni_{15}$ alloy membrane had greater hydrogen permeability than a $Pd_{75}Ag_{25}$ alloy and V-Ni-based alloys are highly efficient for hydrogen separation in the temperature range from 200 to 400 °C. However, further testing under membrane operating conditions are required to verify this [94].

The time for the onset of crystallisation can be significantly reduced in Zr-based metallic glasses (specifically $Zr_{55}Cu_{30}Al_{10}Ni_5$) by the incorporation of even moderate amounts of hydrogen into the alloy [95]. However, recent results suggest that hydrogen can also refine the nanocrystalline grain size during subsequent crystallisation of the amorphous phase. Oxidation is also a considerable problem for metal membranes and Köster *et al.* [96] have observed a significant difference in the oxidation behaviour of a Zr-Cu-Ni-Al alloy in the amorphous and nanocrystalline states, with the latter performing much better.

19.4.4 Ni-Ti-Nb-Based Alloy Membrane Materials

Hashi *et al.* have investigated the hydrogen permeability of crystalline ternary Ni-Ti-Nb alloys (several hundred microns thick, 0.20–0.97 MPa upstream pressure, 250–400 °C) [97,98] and the

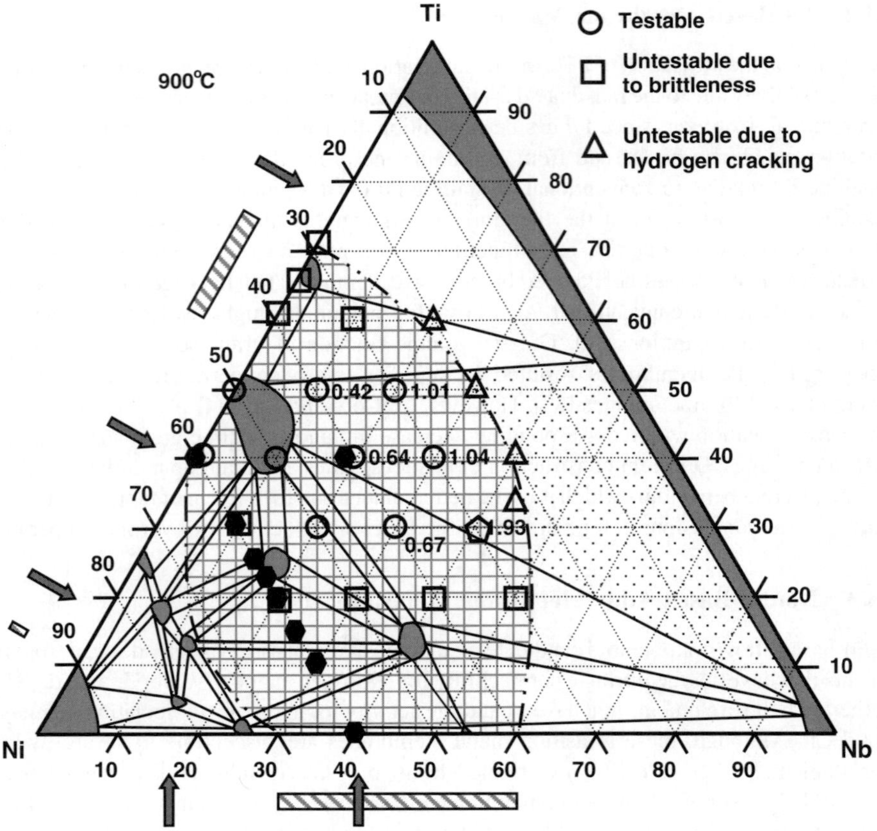

Figure 19.7 Schematic diagram of phase information for the Ni-Nb-Ti ternary system showing an isothermal section at 900 °C [99]. In addition, the positions of the known binary eutectic compositions are indicated by grey arrows. The established glass-forming regions for the Ni-Nb binary [100] and the Ni-Ti binary [101,102] systems are indicated by hatched bars. An estimate of the potential ternary glass-forming region in the ternary section (taking into consideration the three-phase fields and published ternary glass-forming data; solid black hexagons [103]) is indicated by the square hatched area. The compositions of the fully crystalline alloys investigated by Hashi et al. [97,98] are indicated by the open shapes, together with the associated hydrogen permeability ($\times 10^{-8}$ mol m^{-1} s^{-1} Pa$^{-0.5}$)

results are summarised in Figure 19.7 [99–103]. For compositions with a measurable hydrogen permeability, a trend for increasing permeability with increasing Ti and Nb content of the alloy was observed. The highest hydrogen permeability was measured for $Nb_{39}Ti_{31}Ni_{30}$ (indicated by the open pentagon in Figure 19.7), which was explained by a combination of properties associated with particular microstructural elements that are characteristic of the two-phase region in this part of the ternary diagram (dotted shaded region in Figure 19.7). The primary dendrites of Nb(Ti) are considered responsible for hydrogen permeation and the eutectic matrix [Nb(Ti) + NiTi] provides resistance to hydrogen embrittlement. The importance of aligning microstructural constituents was demonstrated by Kishida *et al.* [104]. They showed that the hydrogen permeability for a membrane fabricated from unidirectionally grown $Nb_{19}Ni_{41}Ti_{40}$ aligned perpendicular to the surface of the membrane, was more than double that of a random as-cast microstructure. This aligned microstructure provided an uninterrupted pathway for hydrogen diffusion through the membrane via the Nb-rich bcc phase.

Rolling and annealing both conventional [105] and via planar flow casting [106,107] have been investigated to improve performance, reduce thickness and facilitate more efficient fabrication. It was determined that hydrogen permeation decreased in $Nb_{40}Ni_{30}Ti_{30}$ with increasing rolling reduction [105] due to an increase in the density of lattice defects to trap hydrogen and also the alignment of the bcc phase parallel to the membrane surface. The permeability was restored by annealing at 1100 °C for a week to recrystallise the microstructure in a 120 µm thick membrane. Interestingly, the resistance to hydrogen embrittlement was retained after rolling and annealing although the lamellar nature of the matrix was lost. For greater ease of glass formation, $Nb_{20}Ni_{40}Ti_{40}$ was selected to obtain a 40 µm thick, fully amorphous foil by PFC [107]. The resulting foil was heat treated to produce duplex nanocrystalline or microcrystalline microstructures. For heat treatments between 525 and 650 °C these membranes were very brittle. Only after prolonged heat treatment above 675 °C was the ductility and fracture toughness recovered.

Similarly, Hashi *et al.* also investigated the hydrogen permeability of a number of crystalline membranes in Ni-V-Ti and Ni-Ta-Ti ternary systems and determined $V_{31}Ti_{40}Ni_{29}$ (1.2×10^{-8} mol m^{-1} s^{-1} Pa$^{-0.5}$) and $Ta_{53}Ti_{28}Ni_{19}$ (1.66×10^{-8} mol m^{-1} s^{-1} Pa$^{-0.5}$) to have the highest permeability at 400 °C [108].

Again the best performing alloys were associated with a multicomponent microstructure. Similar findings have been reported for a $V_{53.6}Ti_{28}Ni_{18.4}$ membrane material [109]. In parallel, Kolomytsev and coworkers [110–112] developed a production procedure involving controlled crystallisation through a tailored heat treatment cycle, for the fabrication of multicomponent shape memory materials based on TiNi, with substitutions of Zr and Hf for Ti and Cu, Co, Pd, Ag and Al for Ni, from amorphous precursors. By following a simple set of rules, a range of alloys with good mechanical/functional properties were developed from PFC ribbons. A similar procedure could be used for hydrogen separation membranes.

19.4.5 Ti-Zr-Ni-Based Alloy Membrane Materials

Most work to date on quasicrystalline metals for hydrogen technologies has been conducted in the area of hydrogen storage but useful information may still be learned. A partial isothermal section in the ternary Ni-Ti-Zr phase diagram at 700 °C [113] is shown in Figure 19.8. In addition, the ternary glass-forming region as reported by Rabinkin *et al.* [114] (as indicated by the square hatched area) and the region identified by Molokanov and Chebotnikov [115] to be associated with the formation of a quasicrystalline microstructure (as indicated by the diagonal hatched area), are also shown in Figure 19.8. It was found that there is phase selection competition between a quasicrystalline and an amorphous microstructure in melt-spun ribbons depending on the imposed cooling rate, with the amorphous phase dominating at the highest cooling rates when quenching from the liquid. This was also confirmed by Wang *et al.* [116] in the $Ti_{45}Zr_{35}Ni_{20-x}Cu_x$ ($x = 0$, 1, 3, 5, 7) alloy series.

Yamaura *et al.* [117] produced a single amorphous phase in both $Ni_{63}Ti_7Zr_{30}$ and $Ni_{42}Ti_{28}Zr_{30}$ (see solid black squares in Figure 19.8) via melt spinning and determined the hydrogen permeability at 400 °C to be ~2.95×10^{-9} mol m^{-1} s^{-1} Pa$^{-1/2}$ and 1.05×10^{-8} mol m^{-1} s^{-1} Pa$^{-1/2}$, respectively. Stroud *et al.* [118] demonstrated the feasibility of using a quasicrystalline microstructure in $Ti_{45}Zr_{38}Ni_{17}$ for hydrogen storage (see solid black circles in Figure 19.8). Konstanchuk *et al.* [119] determined that the hydrogen absorption kinetics observed for the icosahedral phase derived from mechanical alloyed $Ti_{45}Zr_{38}Ni_{17}$ were faster than those of a similar

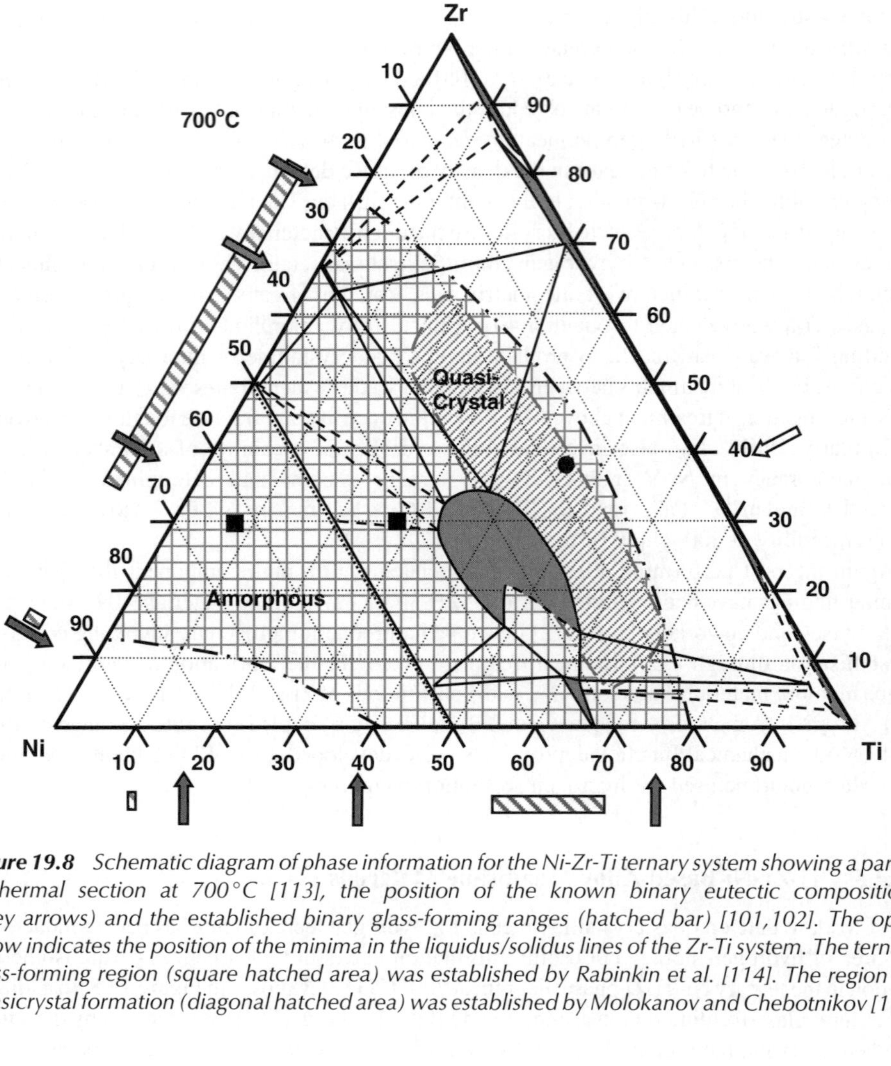

Figure 19.8 *Schematic diagram of phase information for the Ni-Zr-Ti ternary system showing a partial isothermal section at 700 °C [113], the position of the known binary eutectic compositions (grey arrows) and the established binary glass-forming ranges (hatched bar) [101,102]. The open arrow indicates the position of the minima in the liquidus/solidus lines of the Zr-Ti system. The ternary glass-forming region (square hatched area) was established by Rabinkin et al. [114]. The region for quasicrystal formation (diagonal hatched area) was established by Molokanov and Chebotnikov [115]*

phase derived from as-cast melt-spun ribbons. They suggested that this reduction in induction time was a consequence of a more active surface in the former due to the presence of more defects.

The hydrogenation of Ti-Zr/Hf-Ni-based quasicrystals was further investigated by Kelton and coworkers [120,121]. They reported that Ti-Zr-Ni quasicrystals displayed a pressure plateau at 80–200 psi, a storage capacity >4 wt.% and the ability to cycle large quantities of hydrogen. Moreover, they observed the formation of a hydride phase in $Ti_{45}Zr_{38}Ni_{17}$ quasicrystals, which they associated with the formation of a surface barrier that prevented easy, desorption of hydrogen. However, they also noted that $Ti_{40}Hf_{40}Ni_{20}$ (a rational approximant phase) had nearly identical hydrogenation properties as $Ti_{45}Zr_{38}Ni_{17}$ but no evidence of hydride phase formation was observed after loading from the gas phase at 250 °C, which suggests that this composition has better cycling ability. They indicate that the absorption/desorption characteristics in such systems can be improved by the tailoring of the quasicrystalline microstructure. Huang *et al.* [122] noted a strong influence of composition on the hydrogenation properties in the Ti-Zr-Ni system. Jeon *et al.* [123] recently reported that proton beam irradiation was effective in eliminating the oxide barrier to enhance the rates of hydrogen diffusion in Ti-Zr-Ni quasicrystals.

Szajek *et al.* [124] determined that the replacement of Ni in nanocrystalline TiNi-based alloy powders by Mg, Mn or Zr, improved not only the discharge kinetics, but also the cycle life of these materials. Chen *et al.* [125] investigated the influence of Hf additions in the Ni-Zr-Ti system (quaternary section at a constant 20 at% Ni) on the formation ranges for icosahedral, amorphous and crystalline microstructures during rapid solidification. They determined that the formation of a single amorphous phase was favoured for high Hf contents (e.g., $Hf_{40}Ni_{20}Ti_{20}Zr_{20}$). Also, high Ti and Zr contents enhanced quasicrystalline phase formation (e.g., $Ti_{40}Hf_{20}Ni_{20}Zr_{20}$ contained a nanoicosahedral single phase). For compositions close to equi-atomic concentrations of Hf, Ti and Zr the as-solidified microstructure was found to consist of a mixture of a bcc β-(Ti-Zr-Hf-Ni) phase in an amorphous matrix.

Kim *et al.* [126] observed that the Pd-coated icosahedral phase ribbons fractured into a fine powder upon hydrogenation, whereas the as-cast ribbons did not. Kim *et al.* also investigated the response of amorphous ribbons of a similar composition to the icosahedral phase (also containing Si as a glass-forming agent). They found that, even with the surface preparation and Pd coating, the induction times for the amorphous ribbons were longer than those of the icosahedral phase. Liu and coworkers have evaluated the electrochemical performance of amorphous and quasicrystalline $Ti_{45}Zr_{35}Ni_{17}Cu_3$ powders [127], $Ti_{44}Zr_{32}Ni_{22}Cu_2$ and $Ti_{41}Zr_{29}Ni_{28}Cu_2$ melt spun ribbons [128], $(Ti_{45}Zr_{30}Ni_{25})_{100-x}La_x$ ($x = 3, 5, 7$) melt spun alloys [129] and $Ti_{45}Zr_{35}Ni_{13}Pd_7$ melt spun ribbon [130]. It was found that the stability of the amorphous $Ti_{45}Zr_{35}Ni_{17}Cu_3$ powder was not as good as the quasicrystalline counterpart, however, the degradation of the quasicrystalline powder with cycling is a major concern.

The increased glass forming ability in the Ti-Zr-Ni-based alloys with the addition of Cu, La or Pd led to the formation of a mixed microstructure of nanoquasicrystalline and amorphous phases, which improved the high rate dischargeability and the cycling stability of the material, with Pd being the best.

Tanaka *et al.* [131] have studied the hydrogen storage behaviour of both the $Zr_{35}Ni_{65-x}V_x$ ($x = 10$–25) and the $Zr_{25}Ni_{75-x}V_x$ ($x = 10$–50) series of amorphous alloys. Nanocrystalline microstructures, generated from the controlled crystallisation of the amorphous state, displayed hydrogen absorbency and hydriding/dehydriding kinetics superior to those of their amorphous counterparts. Moreover, these properties may well be improved further with a better understanding of the nanocrystallisation process. Zuttel *et al.* [132] investigated the electrochemical hydrogen absorption/desorption behaviour of the $Ni_{42.7}Zr_{36}V_{21.3}$ alloy in both crystalline and amorphous forms. The kinetics of the reaction for the amorphous material were significantly lower than those of the crystalline counterpart due to the presence of a dense and stable oxide film on the surface of the former. However, Zuttel *et al.* [133] also found that the partial substitution of Ti for Zr in this system produced a mixed oxide surface layer that was more readily penetrated by hydrogen and this improved the absorption/desorption kinetics. They do not appear to have applied this technique to the amorphous alloy since it would be expected that the Ti modified alloy would also be glass-forming. To be an effective and viable membrane material, an economical solution to overcome such difficulties associated with tenacious oxide films must be factored into the fabrication process where necessary.

Shimizu *et al.* [134] investigated the hydrogen absorption properties of ZrCo-TiNi pseudo-binary alloys in the amorphous and crystalline condition. Interestingly, the compositions that crystallised into an ordered B2 structure were able to store more hydrogen than those that transformed into a disordered bcc structure. The maximum storage capacity was observed for the

$Zr_{35}Co_{35}Ti_{15}Ni_{15}$ alloy. Recently, Jayalakshimi *et al.* [135] investigated the high temperature properties of hydrogenated melt spun $Ti_{50}Zr_{25}Co_{25}$ ribbons. The ribbon with a hydrogen concentration of 15 at.% displayed an appreciable ductility of 20% at 350 °C, greater than the as-cast ribbon, which was attributed to the formation of a quasicrystalline phase within the amorphous matrix. Huang *et al.* [136] reported on the effect of Co addition in the formation of a quasicrystalline microstructure from the melt in Ti-Zr-(Ni, Co) alloys.

Homma *et al.* (2002) [137] studied the hydrogen absorption of crystalline multiphase alloys with compositions of $V_{77.8}Zr_{7.4}Ti_{7.4}Ni_{7.4}$, $V_{54.5}Zr_{18.25}Ti_{11.25}Ni_{16}$ and $V_{17}Zr_{36}Ti_{17}Ni_{30}$ and determined that the Laves phase in the specimen microstructure acted as the preferred penetration path for hydrogen into the alloy despite being stored in the bcc phase. Hu *et al.* [138] speculated that the addition of V to Ti-Ni alloys, to form a quasicrystalline phase, makes the usual compact Ti oxide film more porous leading to enhanced hydrogen absorption/desorption kinetics. Zhang *et al.* [139] determined that the electrochemical capacity of a nanocrystalline C14 phase in a melt spun $Zr_{23.33}Ti_{10}Mn_{13.33}V_{13.33}Ni_{40}$ alloy was higher than that of the conventional crystalline C14 phase. Moreover, a greater resistance to decrepitation and cracking was observed for the more refined grain microstructure.

Taizhong *et al.* [140] evaluated the hydrogen storage characteristics of a $Ti_{35.71}Cr_{39.29}V_{17.86}$-$Fe_{3.57}Mn_{3.57}$ alloy produced under both conventional and rapid solidification processing conditions. Although, the hydrogen storage performance of the melt-spun material was superior, the full ramifications of the complex behaviour associated with this material and the different transformation paths were not understood completely. Nonetheless, this work indicates that rapid solidification processing can increase the options available to tailor membrane properties.

19.5 Concluding Remarks

There has been a rapid increase in the research and development of nanocrystalline and quasicrystalline metals in recent times. This is due to a number of factors such as improvements to the resolution of scientific instruments, processing methods and modelling capabilities (both hardware and software). Initial results indicate that a great deal of research and development remains to be done in this field to improve the basic understanding of underlying microstructure/property relationships in these materials. In particular, critical technical issues associated with hydrogen embrittlement and both oxide film formation and composition, need to be addressed in order for such alloys to be utilised as practical hydrogen separation membranes. The indications are, however, that with appropriate tailoring of the composition, in conjunction with utilisation of targeted modelling and correct processing, that significant progress can be made. Indeed, few studies are available which investigate the grain boundaries of nanocrystalline and quasicrystalline metals and their effects on hydrogen permeation and resilience to hydrogen degradation. Further studies are required to fully characterise nanocrystalline and quasicrystalline structures as a function of planar flow casting variables and investigate their effects on the properties of the metal as a function as grain size and grain boundaries.

References

1. S.N. Paglieri and J.D. Way, Innovations in palladium membrane research, *Sep. Purif. Methods*, **31**, 1–169 (2002).

2. S. Adhikari and S. Fernando, Hydrogen membrane separation techniques, *Ind. Eng. Chem. Res.*, **45**, 875–881 (2006).
3. J.W. Phair and R. Donelson, Developments and design of novel (non-palladium-based) metal membranes for hydrogen separation, *Ind. Eng. Chem. Res.*, **45**, 5657–5674 (2006).
4. M.D. Dolan, N.C. Dave, A.Y. Ilyushechkin, L.D. Morpeth and K.G. McLennan, Composition and operation of hydrogen-selective amorphous alloy membranes, *J. Membrane Sci.*, **285**, 30–55 (2006).
5. J.W. Phair and S.P.S. Badwal, Materials for separation membranes in hydrogen and oxygen production and future power generation, *Sci. Tech. Adv. Mater.*, **7**, 792–805 (2006).
6. D.S. Sholl and Y.H. Ma, Dense metal membranes for the production of high-purity hydrogen, *MRS Bulletin*, **31**, 770–773 (2006).
7. G.Q. Lu, J.C. Dinz da Costa, M. Duke, S. Giessler, R. Socolow, R.H. Williams and T. Kreutz, Inorganic membranes for hydrogen production and purification: A critical review and perspective, *J. Colloid Interface Sci.*, **314**, 589–603 (2007).
8. M.D. Dolan, S. Hara, N.C. Dave, K. Haraya, M. Ishitsuka, A.Y. Ilyushechkin, K. Kita, K.G. Mclennan, L.D. Morpeth and M. Mukaida, Thermal stability, glass forming ability and hydrogen permeability of amorphous Ni64Zr36-xMx (M = Ti, Nb, Mo, Hf, Ta or W) membranes, *Sep. Purif. Tech.*, **65**, 298–304 (2009).
9. J.B. Qiang, W. Zhang and A. Inoue, Ni-(Zr/Hf)-(Nb/Ta)-Al bulk metallic glasses with high thermal satabilities, *Intermetallics*, **17**, 249–252 (2009).
10. N. Eliaz and D. Eliezer, An overview of hydrogen interaction with amorphous alloys, *Advanced Performance Mater.*, **6**, 5–31 (1999).
11. X.G. Li, T. Otahara, S. Takahashi, T. Shoji, H.M. Kimura and A. Inoue, Hydrogen absorption and corresponding changes in structure and thermal stability of $Zr_{60}Al_{10}Ni_{30}$ amorphous alloy, *J. Alloys Compd.*, **297**, 303–311 (2000).
12. R.W. Siegel, Mechanical properties and deformation behaviour of materials having ultrafine microstructures, in *NATO ASI Series E: Applied Sciences* vol. 233, M. Natasi, D.M. Pakin, H. Gleiter (eds), Kluwer Academic Publishers, Dordrecht (1993).
13. H. Gleiter, Nanocrystalline materials, *Prog. Mater. Sci.*, **33**, 223–315 (1989).
14. D. Wan, G. Yang, S. Chen. M. Zhu and Y. Zhiu, Growth morphology and evolution of quasicrystal in as-solidified Y-rich Mg-Zn-Y ternary alloys, *Rare Metals*, **26**, 435–439 (2007).
15. D. Zander, U. Köster, N. Eliaz and D. Eliezer, Influence of hydrogen on formation and stability of Zr-based quasicrystals, *Mater. Sci. Eng. A.*, **294**, 112–115 (2000).
16. P.C. Gibbons, R.G. Hennig, V.T. Huett and K.F. Kelton, Ti-Zr-Ni and Ti-Hf-Ni quasicrystals and approximants as hydrogen storage alloys, *J. Non-Crys. Solids*, **334/335**, 461–465 (2004).
17. H.H. Leiberman, Amorphous metallic alloys, in *Amorphous Metallic Alloys*, Luborsky, F.E. (ed.), Butterworths, London (1988).
18. M.C. Narasimhan,Continuous casting method for metallic strips, US Patent 4,142,571 (1979).
19. W.A. Hienemann, Rapid solidification, *in Rapidly Quenched Metals V (Conf. Proc.),* S. Steeb *et al.* (eds), Elsevier Science Publishers, Amsterdam (1985).
20. Y. Sato and M. Yamate, New casting technique for a wide ribbon in a single-roller process, *Mater. Sci Eng.*, **A181/A182**, 1310–1316 (1994).
21. M.A. Gibson and A.J. Morton, "Non-equilibrium processing of metals to near net shape", *Mater. Forum*, **20**, 93–130 (1996).
22. M.T. Smith and M. Saletore, Simple, low-cost planar flow casting machine for rapid solidification processing, *Rev. Sci. Instrum.*, **57**, 1647–1653 (1988).
23. M. Motomura, T. Haga and Y. Sakurai, Design and assembling of single roll apparatus for rapid solidification and characteristics of wide Al-Si eutectic alloy foils solidified, *Keikinzoku*, **38**, 528–533 (1988).

24. M.J. Fleetwood, A.G. Law, J.E. Whittle, A.G. Todd and J.H. Vincent, Control of thin strip casting, *in Rapidly Solidified Materials: Properties and Processing*, P.W. Lee *et al.* (eds). ASM, Metal Park, Oh. (1988).
25. J.K. Carpenter and P.H. Steen, Planar-flow spin-casting of molten metals: Process behaviour, *J. Mater. Sci.*, **27**, 215–225 (1992).
26. P.H. Steen, and Ch. Karcher, Fluid Mechanics of spin casting of metals, *Ann. Rev. Fluid Mech.*, **29**, 373–397 (1997).
27. C.J. Byrne, E.A. Theisen, B.L. Reed and P.H. Steen, Capillary puddle vibrations linked to casting-defect formation in planar-flow melt spinning, *Metall. Mater. Trans. B*, **37B**, 445–456 (2006).
28. C.J. Byrne, S.J. Weinstein and P.H. Steen, Capillary stability limits for liquid metal in melt spinning, *Chemical Eng. Sci.*, **61**, 8004–8009 (2006).
29. C. Karcher and P.H. Steen, High-Reynolds-number flow in a narrow gap driven by solidification. I. Theory, *Phys. Fluids*, **13**, 826–833 (2001).
30. C. Karcher and P.H. Steen, High-Reynolds-number flow in a narrow gap driven by solidification. II. Planar flow casting application, *Phys. Fluids*, **13**, 834–840 (2001).
31. G.-X. Wang and E.F. Matthys, Mathematical simulation of melt flow, heat transfer and non-equilibrium solidification in planar flow casting, *Modelling Simul. Mater. Sci. Eng.*, **10**, 35–55 (2002).
32. M. Bussmann, J. Motaghimi, D.W. Kirk and J.W. Graydon, A numerical study of steady flow and temperature fields within a melt spinning puddle, *Int. J. Heat Mass Transfer*, **45**, 3997–4010 (2002).
33. N. Ito, Simulation model for solidification process in planar flow casting using the multi-zone method, *ISIJ Int.*, **48**, 944–953 (2008).
34. A.B. Bichi, W.R. Smith and J.G. Wissink, Solidification and downstream meniscus prediction in the planar-flow spin casting process, *Chemical Eng. Sci.*, **63**, 685–695 (2008).
35. Y.B. Levin, A.N. Shumakov, M.R. Filonov and Yu.A. Anikin, A model of amorphous and nano-crystalline ribbon processing by planar – flow casting, *J. Phys C*, **98**, 072011 (2008).
36. G. Upadhya and D.M. Stefanescu, Modeling of the crystalline-to-amorphous transition in rapidly solidified alloys, *Mater. Sci. Eng.*, **A152**, 215–226 (1992).
37. H. Jones, Formation of phases and microstructures by rapid solidification processing – an update, *Mater. Sci. Eng.*, **A179/A180**, 1–7 (1994).
38. P. Gilgien, A. Zryd and W. Kurz, Metastable phase-diagrams and rapid solidification processing, *ISIJ Int.*, **35**, 566–573 (1995).
39. J. Li, Y.L. Tang, N. Shen and W. Pan, Effects of solidification kinetics on phase selection of Ni-Al alloys, *J. Alloys Compd.*, **329**, 157–161 (2001).
40. G.A. Srebryanski and Y. N. Starodubtsev, Basic principles for improving amorphous strip production technology, *Steel USSR*, **21**, 426–430 (1991).
41. T. Haga and S. Suzuki, Single roll method for foil casting of aluminium alloy, *J. Mater. Processing Tech.*, **137**, 86–91 (2003).
42. H.H. Liebermann, Manufacture of amorphous alloy ribbons, *IEEE Trans. Mag.*, **15**, 1393–1397 (1979).
43. H.C. Fiedler, H. Muhlbach and G. Stephani, The effect of the main processing parameters on the geometry of amorphous metal ribbons during planar flow casting (PFC), *J. Mater. Sci.*, **19**, 3229–3235 (1984).
44. O.P. Pandey, S.N. Ojha, G.M.K. Sarma, E.S. Dwarakadasa and T.R. Anantharaman, High-speed production of aluminium foils by the planar flow casting technique, *Indian J. Tech.*, **29**, 173–178 (1991).
45. S.-M. Lee, N.-H. Myoung and C.-P. Hong, Fabrication of wide tin polycrystalline ribbons by planar flow casting, *J. Korean Inst. Met. Mater.*, **32**, 1180–1186 (1994).

46. J.K. Sung, M.C. Kim, C.G. Park and Y.S. Kim, Theoretical expectation of strip thickness in planar flow casting process, *Mater. Sci. Eng.*, **A181/A182**, 1237–1242 (1994).
47. Z. Gong, P.D. Wilde and E.F. Mattyhs, Numerical modeling of the planar flow melt-spinning process and experimental investigation of its solidification puddle dynamics, *Int. J. Rapid Solidification*, **6**, 1–28 (1991).
48. M. Motomura, T. Haga and Y. Sakurai, Relative position of nozzle to roll as a forming factor in single-roll rapid solidification of eutectic Al-Si alloy foils, *Keikinzoku*, **38**, 534–539 (1988).
49. V.S. Chernov, O.G. Ivan and F.E. Pashchenko, Production and properties of amorphous magnetically soft alloys AMAG, *J. Adv. Mater.*, **1**, 353–358 (1994).
50. P.D. Wilde and E.F. Mattyhs, Experimental investigation of the planar flow casting process: Development and free surface characteristics of the solidification puddle, *Mater. Sci. Eng.*, **A150**, 237–247 (1992).
51. K. Shibuya, Production and magnetic-properties of Fe-B-Si amorphous alloy ribbons – a review, *Mater. Sci. Eng.*, **A181/A182**, 1435–1439 (1994).
52. H. Kemmer and M. Hohmann, Advanced vacuum processes for manufacturing amorphous materials, *in Advanced Materials and Processing (Proc. 2nd Pac. Rim Int. Conf.)*, K.S. Shin *et al.*. (eds), The Korean Inst. Met. Mater., Seoul, Korea, p. 1843–1848 (1995).
53. S.C. Huang and H.C. Fiedler, Effects of wheel surface conditions on the casting of amorphous metal ribbons, *Metall. Trans. A*, **12A**, 1107–1112 (1981).
54. H.H. Liebermann, R.E. Maxwell, R.W. Smashey and J.L. Walter, Process and structural aspects of melt-spun nickel-based superalloy ribbons, *Metall. Trans. A*, **14A**, 1817–1823 (1983).
55. V.I. Tkatch, A.M. Grishin and V.V. Maksimov, Estimation of the heat transfer coefficient in melt spinning process, *J. Phys C*, **144**, 012104 (2009).
56. M.J. Tenwick and H.A. Davies, The structure and properties of rapidly solidified Fe-3 to 9.3wt% Si alloys, *Int. J. Rapid Solidification*, **1**, 143–155 (1985).
57. H.H. Liebermann, Ribbon-substrate adhesion dynamics in chill block melt-spinning processes, *Metall. Trans. B*, **15**, 155–161 (1984).
58. S.C. Huang and H.C. Fiedler, Amorphous ribbon formation and the effects of casting velocity, *Mater. Sci. Eng.*, **51**, 39–46 (1981).
59. A.G. Gillen and B. Cantor, Photocalorimetric cooling rate measurements on a Ni-5wt%Al alloy rapidly solidified by melt-spinning, *Acta Metall.*, **33**, 1813–1825 (1985).
60. W.K. Wang, C.T. Kuo and T.S. Chin, Effect of arc plasma heating on the air pocket formation during the planar flow casting, *Int. J. Rapid Solidification*, **6**, 285–295 (1991).
61. H. Fiedler, L. Illgen, M. Barth and G. Stephani, Influence of CO_2 gas flow around the melt puddle on the surface roughness of Fe-B-Si ribbons, *Mater. Sci. Eng.*, **A133**, 657–661 (1991).
62. H.H. Liebermann, R.J.J. Martis and D.M. Nathasingh, Dependence of some properties on thickness of smooth alloy ribbon, *J. Appl. Phys.*, **55**, 1787–1789 (1984).
63. H.B. Wang, Q. Wang, C. Dong, L. Yuan, F. Xu and L.X. Sun, Composition design for Laves phase-related bcc-V solid solution alloys with large hydrogen storage capacities, *J. Phys C*, **98**, 012018 (2008).
64. A.D. Rud, U. Schmidt, G.M. Zelinska, A.M. Lakhnik, G.Ya. Kolbasov and M.O. Danilov, Atomic structure and hydrogen storage properties of amorphous-quasicrystalline Zr-Cu-Ni-Al melt-spun ribbons, *J. Non-Cryst. Solids*, **353**, 3434–3438 (2007).
65. C.J. Byrne, A.M. Kueck, S.P. Baker and P.H. Steen, In situ manipulation of cooling rates during planar-flow melt spinning processing, *Mater. Sci. Eng. A*, **459**, 172–181 (2007).
66. C. Dong, Q. Wang, J.B. Qiang, Y.M. Wang, G. Han, J. Wu, Y.H. Li, X. Cheng, C.L. Zhu and H. Chen, A cluster line approach for composition rules of quasicrystals and bulk metallic glasses, *J. Phys C*, **98**, 012015 (2008).
67. A.I. Zaitsev, Thermodynamic and physico-chemical study of high temperature melts-the efficient way to construct amorphous and quasicrystalline materials, *J. Phys C*, **98**, 032002 (2008).

68. G. Herzer, Soft magnetic nanocrystalline materials, *Scripta Metall. Mater.*, **33**, 1741–1756 (1995).
69. A. Inoue, Amorphous, nanoquasicrystalline and nanocrystalline alloys in Al-based systems, *Prog. Mater. Sci.*, **43**, 365–520 (1998).
70. Z.P. Lu, C.T. Liu, J. R. Thompson and W.D. Porter, Structural amorphous steels, *Phys. Rev. Lett.*, **92**, 245503-1 to 245503-4 (2004).
71. V. Ponnambalam, S.J. Poon and G.J. Shiflet, Fe-Mn-Cr-Mo-(Y, Ln)-C-B (Ln = Lanthanides) bulk metallic glasses as formable amorphous steel alloys, *J. Mater. Res.*, **19**, 3046–3052 (2004).
72. D.J. Branagan, Enabling factors towards production of nanostructured steel on an industrial scale, *J. Mater. Eng. Performance*, **14**, 5–9 (2005).
73. S. Pauly, J. Das, N. Mattern, D.H. Kim and J. Eckert, Phase formation and thermal stability in Cu-Zr-Ti(Al) metallic glasses, *Intermetallics*, **17**, 453–462, (2009).
74. M.H. Lee, R.T. Ott, M.F. Besser, M. J Kramer and D.J. Sordelet. Compositional dependence on phase selection during devitrification of amorphous Zr-Pt alloys, *Scripta Mater.*, **55**, 505–508, (2006).
75. Q.S. Zhang, W. Zhang, G.Q. Xie and A. Inoue, Eutectic crystallisation behaviour of new $Zr_{48}Cu_{36}Al_8Ag_8$ alloy with high glass-forming ability, *J. Phys. C*, **144**, 012031 (2009).
76. T. Nagase, T. Hosokawa, K. Takizawa and Y. Umakoshi, Electron-irradiation-induced nanocrystallisation in quasicrystal-forming Zr-based metallic glass, *Intermetallics*, **17**, 657–668 (2009).
77. X.-Y. Cheng and F. Wang, Hydrogen absorption and desorption in metallic glass and nanocrystalline $Zr_{52.5}Cu_{17.9}Ni_{14.6}Ti_5Al_{10}$ alloy, *Trans. Nonferrous Met. Soc. China*, **19**, 377–382 (2009).
78. R. Kirchheim, T. Mutschele, W. Kieninger, H. Gleiter, R. Birringer and T.D. Koble, Hydrogen in amorphous and nanocrystalline metals, *Mater. Sci. Eng.*, **99**, 457–462 (1988).
79. M. Hirscher, J. Mossinger and H. Kronmuller, Hydrogen diffusion in nanocrystalline, mesoscopic, and microcrystalline heterogeneous alloys, *J. Alloys Compounds*, **231**, 267–273 (1995).
80. C.C. Koch, Processing-structure-property relationships in ultrafine grain and nanocrystalline materials, *J. Phys C*, **144**, 012081 (2009).
81. R. Kirchheim, T. Mutschele, W. Kieninger, H. Gleiter, R. Birringer and T.D. Koble, Hydrogen in amorphous and nanocrystalline metals, *Mater. Sci. Eng.*, **99**, 457–462 (1988).
82. K.J. Bryden and J.Y. Ying, Nanostructured palladium-iron membranes for hydrogen separation and membrane hydrogenation reactions, *J. Membrane Sci.*, **203**, 29–42 (2002).
83. K.J. Bryden and J.Y. Ying, Nanostructured palladium membrane synthesis by magnetro sputtering, *Mater. Sci. Eng. A*, **204**, 140–145 (1995).
84. B.A. McCool and Y.S. Lin, Nanostructured thin palladium-silver membranes: effects of grain size on gas permeation properties, *J. Mater. Sci.*, **36**, 3221–3227 (2001).
85. N.M. Peachey, R.C. Snow and R.C. Dye, Composite Pd/Ta metal membranes for hydrogen separation, *J. Membrane Sci.*, **111**, 123–133 (1996).
86. T.S. Moss, N.M. Peachey, R. C Snow and R.C. Dye, Multilayer metal membranes for hydrogen separation, *Int. J. Hydrogen Energy*, **23**, 99–106 (1998).
87. S.N. Paglieri, S. A. Birdsell, R.C. Snow, F.M. Smith and C.R. Tewell, Development of palladium composite membranes for hydrogen separation, *in Advanced Materials for Energy Conversion II*, D. Chandra, R.G. Bautisia and L. Schlapbach (eds), TMS, Warrendale, USA (2004).
88. K. Komiya, Y. Shinzato, H. Yukawa, M. Morinaga and I. Yasuda, Alloying effects on hydrogen permeability of niobium, *Mater. Sci. Forum*, **475/479**, 2497–2500 (2005).
89. N. Watanabe, G. Zhang, H. Yukawa, M. Morinaga, T. Nambu, K. Shimizu, S. Sato, K. Matsumoto and I. Yasuda, Hydrogen solubility and resistance to hydrogen embrittlement of Nb-Pd based alloys for hydrogen permeable membrane, *Adv. Mater. Res.*, **26/28**, 873–876 (2007).

90. M. Amano, M. Komaki and C. Nishimura, Hydrogen permeation characteristics of palladium-plated V-Ni alloy membranes, *J. Less Common Met.*, **172/174**, 727–731 (1991).
91. C. Nishimura, M. Komaki, S. Hwang and M. Amano, V-Ni alloy membranes for hydrogen purification, *J. Alloys Compd.*, **330/332**, 902–906 (2002).
92. T. Ozaki, Y. Zhang, M. Komaki and C. Nishimura, Preparation of palladium-coated V and V-15Ni membranes for hydrogen purification by electroless plating technique, *Int. J. Hydrogen Energy*, **28**, 297–302 (2003).
93. T. Ozaki, Y. Zhang, M. Komaki and C. Nishimura, Hydrogen permeation characteristics of V-Ni-Al alloys, *Int. J. Hydrogen Energy*, **28**, 1229–1235 (2003).
94. H. Yukawa, G.X. Zhang, N. Watanabe, M. Moringa, T. Nambu and Y. Matsumoto, Analysis of hydrogen diffusion coefficient during hydrogen permeation through niobium and its alloy, *J. Alloys Compd.*, **476**, 102–106 (2009).
95. N. Ismail, M. Uhlemann, A. Gebert and J. Eckert, Hydrogenation and its effect on the crystallisation behaviour of $Zr_{55}Cu_{30}Al_{10}Ni_5$ metallic glass, *J. Alloys Compounds*, **298**, 146–152 (2000).
96. U. Köster, D. Zander, Triwikantoro, A. Rüdiger and l. Jastrow, Environmental properties of Zr-based metallic glasses and nanocrystalline alloys, *Scripta Mater.*, **44**, 1649–1654 (2001).
97. K. Hashi, K. Ishikawa, T. Matsuda and K. Aoki, Hydrogen permeation characteristics of multi-phase Ni-Ti-Nb alloys, *J. Alloys Compounds*, **368**, 215–220 (2004).
98. K. Hashi, K. Ishikawa, T. Matsuda and K. Aoki, Microstructures and hydrogen permeability of Ni-Ti-Nb alloys with high resistance to hydrogen embrittlement, *Mater. Trans. JIM*, **46**, 1026–1031 (2005).
99. Y. Guanjum and H. Shiming, Study on the phase equilibria of the Ti-Ni-Nb ternary system at 900°C, *J. Alloys Compd.*, **297**, 226–230 (2000).
100. R.C. Ruhl, B.C. Giessen, M. Cohen and N.J. Grant, New microcrystalline phases in the Nb-Ni and Ta-Ni systems, *Acta Metall.*, **15**, 1693–1702 (1967).
101. S.H. Whang, New prediction of glass-forming ability in binary alloys using a temperature-composition map, *Mater. Sci. Eng.*, **57**, 87–95 (1983).
102. D.H. Kim, W.T. Kim and D.H. Kim, Formation and crystallisation of Al-Ni-Ti amorphous alloys, *Mater. Sci. Eng. A*, **385**, 44–53 (2004).
103. W. Zhang and A. Inoue, Efects of Ti on the termal stability and glass-forming ability of Ni-Nb glassy alloy, *Mater. Trans. JIM*, **43**, 2342–2345 (2002).
104. K. Kishida, Y. Yamaguchi, K. Tanaka, H. Inui, S. Tokui, K. Ishikawa and K. Aoki, Microstructures and hydrogen permeability of directionally solidified Nd-Ni-Ti alloys with the Nd-NiTi eutectic microstructure, *Intermetallics*, **16**, 88–95 (2008).
105. K. Ishikawa, S. Tokui and K. Aoki, Microstructure and hydrogen permeation of cold rolled and annealed $Nd_{40}Ti_{30}Ni_{30}$ alloy, *Intermetallics*, **17**, 109–114 (2009).
106. Y. Shimada, M. Matsuda, Y. Kawakami, M. Otsu, K. Takashima, M. Nishida, K. Ishikawa and K. Aoki, Micromechanical characterisation of rapidly-solidified Nb-TiNi hydrogen permeation alloy membranes, *J. Japan Inst. Met.*, **72**, 1015–1020 (2008).
107. M. Nishida, M. Matsuda, Y. Shimada, K. Takashima, K. Ishikawa and K. Aoki, Microstructural and mechanical characterisation of rapidly solidified Nb-TiNi hydrogen permeation alloy, *J. Phys C*, **144**, 012106 (2009).
108. K. Hashi, K. Ishikawa, T. Matsuda and K. Aoki, Hydrogen permeation characteristics of (V, Ta)-Ti-Ni alloys, *J. Alloys Compd.*, **404/406**, 273–278 (2005).
109. T.M. Adams and J. Mickalonis, Hydrogen permeability of multiphase V-Ti-Ni metallic membranes, *Mater. Lett.*, **61**, 817–820 (2007).
110. V. Kolomysev, A. Pasko and A. Sezonenko, Selection rules for the production of bulk metallic glass precursors for shape memory materials, *Archives Metall. Mater.*, **49**, 825–840 (2004).

111. P. Ochin, V. Kolomytsev, A. Pasko, A. Sezonenko, Ph. Vermaut, F. Prima and R. Portier, Amorphous multielementary alloys: A preparation route for shape memory alloys, *J. Alloys Compd.*, **434/435**, 268–271 (2007).
112. A. Pasko, V, Kolomytsev, P. Vermaut, F. Prima, R. Portier, P. Ochin and A. Sezonenko, Crystallisation of amorphous phase and martensitic transformations in multicomponent (Ti,Hf,Zr)(Ni,Cu)-based alloys, *J. Non-Cryst. Solids*, **353**, 3062–3068 (2007).
113. *Handbook of Ternary Alloy Phase Diagrams*, P. Villars, A. Prince and N. Okamoto (Eds), ASM, Materials Park, Ohio (1995).
114. A. Rabinkin, H. Liebermann, S. Pounds, T. Taylor, F. Reidinger and S.-C. Lui, Amorphous Ti-Zr-base METGLAS® brazing filler metals, *Scripta Metall.*, **25**, 399–404 (1991).
115. V.V. Molokanov and V.N. Chebotnikov, Quasicrystals and amorphous alloys in Ti-Zr-Ni system: glassforming ability, structure and properties, *J. Non-Cryst. Solids*, **117/118**, 789–792 (1990).
116. L. Wang, L. Ma, C. Ma and A. Inoue, Formations of amorphous and quasicrystal phases in Ti-Zr-Ni-Cu alloys, *J. Alloys Compd.*, **361**, 234–240 (2003).
117. S. Yamaura, S. Nakata, H. Kimura, Y. Shimpo, M. Nishida and A. Inoue, Hydrogen permeation of the melt-spun Ni-X-Zr amorphous membranes, *Mater. Trans. JIM*, **46**, 1768–1770 (2005).
118. R.M. Stroud, A.M. Viano, P.C. Gibbons and K.F. Kelton, Stable Ti-based quasicrystal offers prospect for improved hydrogen storage, *Appl. Phys. Lett.*, **69**, 2998–3000 (1996).
119. I.G. Konstanchuk, E.Yu. Ivanov, B.B. Bokhonov and V.V. Boldyrev, Hydriding of mechanically alloyed icosahedral phase $Ti_{45}Zr_{38}Ni_{17}$, *J. Alloys Compd.*, **319**, 290–295 (2001).
120. K.F. Kelton, Ti/Zr/Hf-based quasicrystals, *Mater. Sci. Eng. A*, **375/377**, 31–37 (2004).
121. K.F. Kelton, J.J. Hartzell, R.G. Hennig, V.T. Huett and A. Takasaki, Hydrogen storage in Ti-Zr and Ti-Hf-based quasicrystals, *Phil. Mag.*, **86**, 957–964 (2006).
122. H. Huang, R. Li, C. Yin, S. Zheng and P. Zhang, Hydrogenation study of suction-cast Ti40Zr40Ni20 quasicrystal, *Int. J. Hydrogen Energy*, **33**, 4607–4611 (2008).
123. J.K. Jeon, J. Lee, E. Kim, Y.M. Lee, S.H. Lee, H. Shin, S.B. Choi, J.Y. Kim, S.P. Youn and K. Kim, Proton irradiation and hydrogen diffusion in quasicrystals, *J. Korean Phys. Soc.*, **54**, 2046–2049 (2009)
124. A. Szajek, M. Makowiecka, E. Jankowska and M. Jurczyk, Electrochemical and electronic properties of nanocrystalline $TiNi_{1-x}M_x$ (M = Mg, Mn, Zr; x = 0, 0.125, 0.25) ternary alloys, *J. Alloys Compd.*, **403**, 323–328 (2005).
125. N. Chen, D.V. Louzguine, S. Ranganathan and A. Inoue, Formation ranges of icosahedral, amorphous and crystalline phases iin rapidly solidified Ti-Zr-Hf-Ni alloys, *Acta mater.*, **53**, 759–764 (2005).
126. J.Y. Kim, P.C. Gibbons and K.F. Kelton, Hydrogenation of Pd-coated samples of the Ti-Zr-based icosahedral phase and related crystalline phases, *J. Alloys Compd.*, **266**, 311–317 (1998).
127. B. Liu, Y. Wu and L. Wang, Electrochemical properties of amorphous and icosahedral quasicrystalline $Ti_{45}Zr_{35}Ni_{17}Cu_3$ powders, *J. Power Sources*, **159**, 1458–1463 (2006).
128. B. Liu, Y. Wu and L. Wang, Crystallographic and electrochemical characteristics of melt-spun Ti-Zr-Ni-Cu alloys, *Rare Met.*, **26**, 440–444 (2007).
129. B. Liu, J. Liu, G. Mi, Z. Zhang, Y. Wu and L. Wang, Crystallographic and electrochemical characteristics of melt-spun Ti-Zr-Ni-La alloys, *J. Alloys Compd.*, **475**, 881–884 (2009).
130. B. Liu, Q. Li, Z. Zhang, G. Mi, Q. Yuan and L. Wang, Crystallographic and electrochemical characteristics of $Ti_{45}Zr_{35}Ni_{13}Pd_7$ melt-spun alloys, *Int. J. Hydrogen Energy*, **34**, 1890–1895 (2009)
131. K. Tanaka, M. Sowa, Y. Kita, T. Kubota and N. Tanaka, Hydrogen storage properties of amorphous and nanocrystalline Zr-Ni-V alloys, *J. Alloys Compd.*, **330/332**, 732–737 (2002).
132. A. Zuttel, D. Chartouni, C. Nutzenadel, K. Gross and L. Schlapbach, Bulk and surface properties of crystalline and amorphous $Zr_{36}(V_{0.33}Ni_{0.66})_{64}$ alloy as active electrode material, *J. Alloys Compd.*, **266**, 321–326 (1998).

133. A. Zuttel, F. Meli and L. Schlapbach, Surface and bulk properties of Tiy$Zr_{1-y}(V_xNi_{1-x})_2$ alloy system as active electrode material in alkaline electrolyte, *J. Alloys Compd.*, **231**, 645–649 (1995).
134. E. Shimizu, K. Aoki and T. Masumoto, Hydrogen adsorption properties of amorphous and crystalline alloys in the pseudo-binary ZrCo-TiNi system, *J. Alloys Compd.*, **293/295**, 526–530 (1999).
135. S. Jayalakshimi, E. Fleury and D.J. Sordelet, High temperature properties of hydrogenated $Ti_{50}Zr_{25}Co_{25}$ rapidly quenched alloy, *J. Phys C*, **144**, 012120 (2009).
136. H.G. Huang, J.B. Qiang, B. Bai, P. Dong and P.C. Zhang, Effect of Co substitution for Ni on the Ti-Zr-(Ni, Co) pseudo-ternary quasicryatal formation, *J. Non-Crystalline Sol.*, **353**, 1670–1675 (2007).
137. H. Homma, H. Saitoh, T. Misawa and T. Ohnishi, Effects of Laves phases on hydrogen behaviour in V-Zr-Ti-Ni hydrogen absorption alloys, *Mater. Trans. JIM*, **43**, 1110–1114 (2002).
138. W. Hu, J. Wang, L. Wang, Y. Wu and L. Wang, Electrochemical hydrogen storage in $(Ti_{1-x})_2Ni$ (x = 0.05–0.3) alloys comprising icosahedral quasicrystalline phase, *Electrochimica Acta*, **54**, 2770–2773 (2009)
139. S.K. Zhang, Q.D. Wang, Y.Q. Lei, G.L. Lü, L.X. Chen and F. Wu, The phase structure and electrochemical properties of melt-spun alloy $Zr_{0.7}Ti_{0.3}Mn_{0.4}V_{0.4}Ni_{1.2}$, *J. Alloys Compd.*, **330/332**, 855–860 (2002).
140. H. Taizhong, W. Zhu, F. Shanglong, X. Baojia and X. Naixin, Comparison of hydrogen storage characteristics between as-cast and melt-spun $TiCr_{1.1}V_{0.5}Fe_{0.1}Mn_{0.1}$ alloys, *Mater. Sci. Eng. A*, **390**, 362–365 (2005).

20

Preparation and Characterisation of Amorphous Alloy Membranes

Shin-ichi Yamaura and Akihisa Inoue

Institute for Materials Research, Tohoku University, 2-1-1 Katahira, Sendai 980-8577, Japan

20.1 Introduction

Recently, the climate change caused by global warming has done severe damage to our everyday life and the trend toward the suppression of CO_2 emissions and the environmental preservation movement is attracting much attention all over the world. Therefore, it will be important in future to construct a hydrogen-powered society based on fuel cell technologies after the present fossil-fuel consuming society in order to solve the environmental and energy problems.

For the coming hydrogen-powered society, it is also important to establish a mass production system for stable hydrogen supply. Hydrogen production using metallic membranes has been regarded as a promising technique for mass-production of purified hydrogen [1], because it has advantages such as a continuous operation without phase transformation, resulting in cost and energy savings. Pd-based alloys [2,3] such as Pd-Ag and Pd-Cu are now commercially available as hydrogen permeable membrane materials.

However, Pd is one of the most expensive noble metals and is in relatively small supply. Therefore, it is important to develop a new membrane material which can replace the conventional Pd-based alloys. Nowadays, many research groups have made efforts to develop new hydrogen permeable membrane materials without Pd as a base metal. For example, hydrogen permeability of Pd-coated V-Ni alloys [4] and Pd-coated Nb-Ti-Ni alloys [5] have been studied and it was shown that those alloys exhibit excellent hydrogen permeability higher than that of the Pd-based alloys. Since the BCC (body-centred cubic) metals such as V and Nb are well known for their good hydrogen diffusivity [6], they are often adopted as a base metal of new nonPd-based membrane materials.

Membranes for Membrane Reactors: Preparation, Optimization and Selection, Edited by Angelo Basile and Fausto Gallucci

We have studied hydrogen permeation of amorphous alloy membranes and reported that the Pd-coated Ni-Nb-Zr amorphous alloys possess excellent hydrogen permeability, as high as that of Pd metal [7,8]. Amorphous alloys have homogeneous random atomic configurations. There are no grain boundaries or precipitations inside amorphous alloys. This feature has facilitated the appearance of various characteristics such as good mechanical properties [9,10], useful physical properties [11,12] and excellent chemical properties [13,14] which have not been obtained from conventional crystalline alloys.

Hydrogen permeable metallic membranes are generally used at the temperature 573–773 K. Amorphous alloys are thermally less stable than crystalline alloys because amorphous alloys possess a non-equilibrium local structure in the viewpoint of atomic configuration. Amorphous alloy crystallises at a certain temperature inherent to the alloy composition, resulting in disappearance of characteristic properties of amorphous alloy.

The temperature at which an amorphous alloy crystallises is dubbed the crystallisation temperature, T_x. So, it is important to develop an amorphous alloy with the crystallisation temperature T_x as high as possible when the amorphous alloy is applied to practical applications at high temperature, such as hydrogen separation.

20.2 Brief Review of Preparation Methods

Amorphous alloys are usually produced by rapid quenching from the liquid state.

The critical cooling rate for amorphisation is about 105 K s^{-1}. There are many production techniques to synthesise amorphous alloy samples, such as melt-spinning [15], gas atomising [16] and vapor deposition [17]. Single-roller melt-spinning has been widely adopted to produce amorphous alloys in a thin ribbon form with about 30 μm in thickness. A schematic illustration of single-roller melt-spinning apparatus is shown in Figure 20.1. An outer view of the melt-spinning

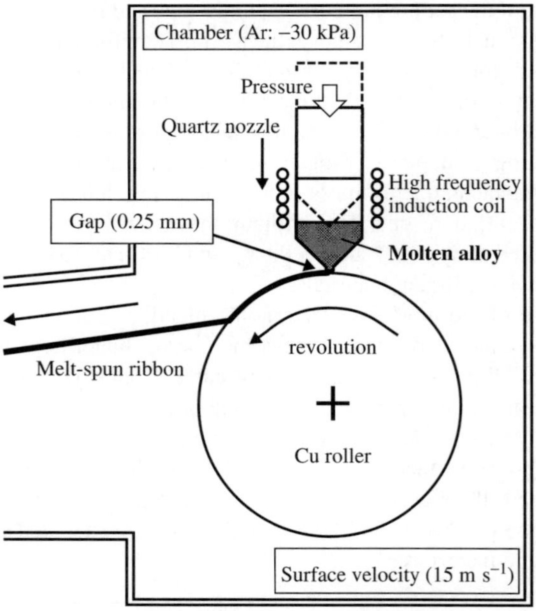

Figure 20.1 *Schematic of apparatus for melt-spinning*

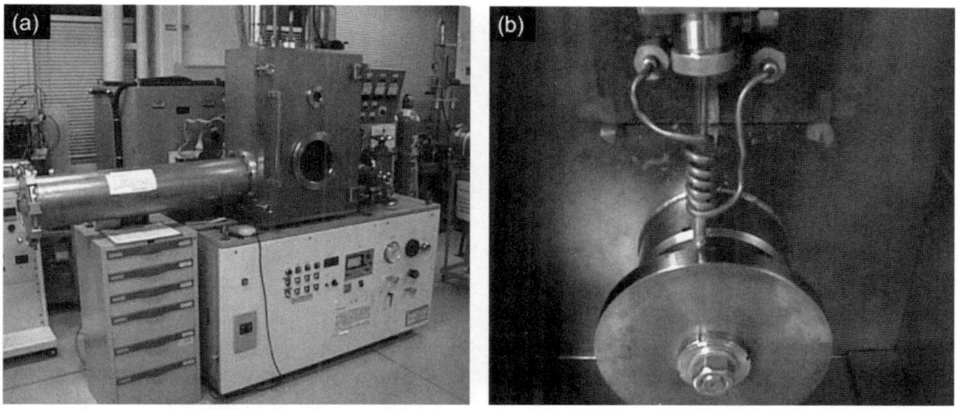

Figure 20.2 *Photographs of (a) melt-spinning apparatus and (b) high-frequency coil and Cu roller inside a chamber*

apparatus and an inner view inside a chamber are also shown in Figure 20.2. The nozzle shown in Figure 20.2b is in the lower position just before injection. The nozzle size and roller width are dependent on the sample width you expected. The principle of this method is simple. First, a crushed alloy ingot is put into a nozzle having a suitable opening. The nozzle is set inside a high-frequency induction coil. The alloy ingot is heated by the coil in a dilute Ar atmosphere inside a chamber or in air until it melts. Then the nozzle moves down to lower position. The gap between the nozzle and the wheel surface is about 0.3 mm. Finally, the molten alloy is injected onto the surface of a revoluting roller (1000–3000 rpm). The diameter of the roller is 20 cm. Thus, melt-spun amorphous samples can be produced. Figure 20.3a shows a photograph of melt-spun $Ni_{60}Nb_{20}Zr_{20}$ amorphous alloy at 10 mm wide. We need rapid quenching to freeze the random atomic arrangement when we produce amorphous alloys. The Cu roller is usually adopted because of its good thermal conductivity. The sample thickness can be controlled by revolution speed. If you want to produce a thinner sample, you should increase the wheel speed and decrease the Ar pressure for injection. Preparation conditions we have to optimise are as follows: (i) gap distance between the quartz nozzle and the surface of the roller, (ii) molten alloy temperature just before injection, (iii) pressure for injection, (iv) surface velocity of the rotating roller, (v) cleanness of the surface of the roller, (vi) material of the roller, (vii) room temperature and humidity, (viii) injection angle of the molten-alloy onto the surface of the roller and (ix) surface temperature of the roller. It is possible to produce amorphous thin samples of more than 150 mm in width by using the nozzle having a wide slit.

Figure 20.3b shows a photograph of the melt-spun $(Ni_{0.6}Nb_{0.4})_{45}Zr_{50}Co_5$ amorphous alloy of 100 mm in width [18]. It is not easy to produce 100 mm wide Ni-Nb-Zr-Co amorphous ribbons by a single-roller melt-spinning because the molten alloy is readily to solidify in the slit of the nozzle and could not be ejected. We finally produced the 100 mm-wide Ni-Nb-Zr-Co ribbons by optimising the conditions of melt-spinning. Before we produced the wide melt-spun ribbons as shown in Figure 20.3b, we had to try to use some kinds of materials for injection nozzles. Previously, the nozzles made of quartz or boron nitride (BN) were adopted. However, it was difficult to produce wide melt-spun ribbons having smooth surface by using quartz and BN nozzles because of the reaction between nozzle material and the melt. So, a carbon nozzle was adopted. Then we produced wide melt-spun ribbons having smoother surface by using carbon nozzles than by quartz/BN nozzles.

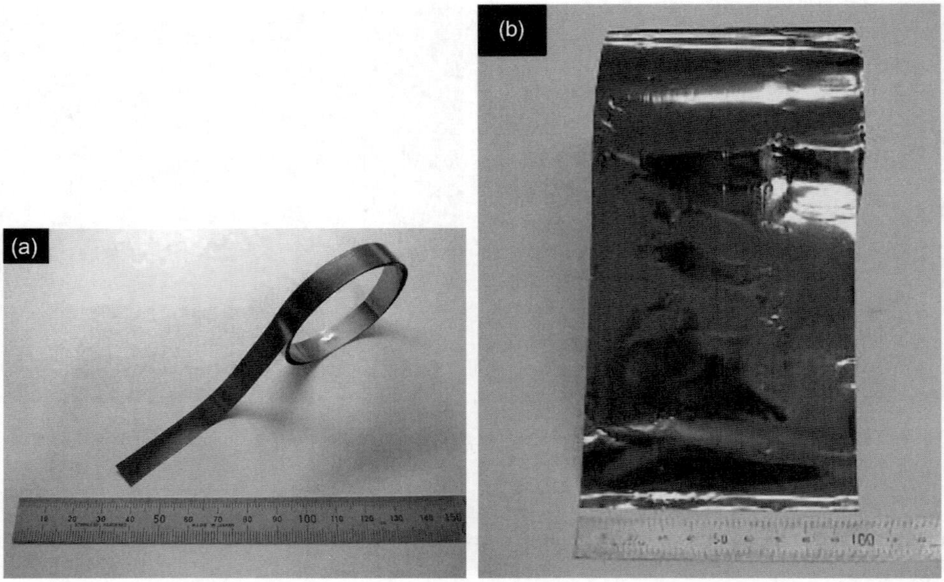

Figure 20.3 *Photographs of melt-spun (a) $Ni_{60}Nb_{20}Zr_{20}$ amorphous alloy (10 mm width) and (b) $(Ni_{0.6}Nb_{0.4})_{45}Zr_{50}Co_5$ amorphous alloy (100 mm width). Reprinted from Y. Shimpo, S. Yamaura, M. Nishida, H.M. Kimura and A. Inoue, Development of melt-spun Ni-Nb-Zr-Co amorphous alloy for high-performance hydrogen separating membrane, J. Membr. Sci., 286, 170–173. Copyright (2006) with permission from Elsevier*

Melt-spinning is a suitable technique to produce amorphous thin membrane directly from a liquid melt, resulting in cost and energy savings. However, it may be difficult to produce amorphous samples having an extremely high melting temperature depending on the alloy compositions and heating capability of the apparatus.

20.3 Experimental Procedure

The detailed sample preparation and experimental procedures in this work are as follows.

20.3.1 Sample Preparation

The mother alloy ingots of the Ni-Nb-Zr-based alloys were produced by arc-melting the raw materials in a dilute Ar atmosphere. The amorphous thin ribbons of the alloys were then produced by a single-roller melt-spinning technique in a dilute Ar atmosphere as mentioned above. The melt-spun ribbon samples of wider than 10 mm and about 40 μm in thickness were obtained.

The amorphicity of melt-spun ribbon specimens was examined by X-ray diffractometry (XRD, Cu-Ka, 40 kV, 40 mA). Amorphous alloy samples show a broad halo pattern indicating random atomic configurations inherent to amorphous alloys. No sharp distinct peaks are seen in the XRD obtained from an amorphous alloy sample. The crystallisation temperature T_x of them was measured by differential scanning calorimetry (DSC) at a heating rate of 0.67 K s^{-1} in an Ar flow.

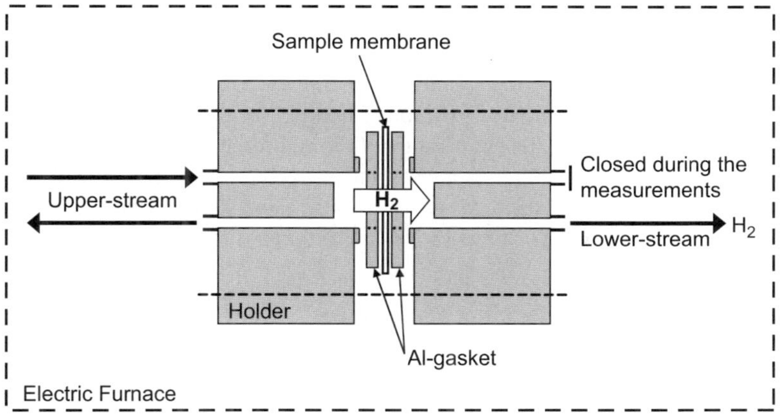

Figure 20.4 *Schematic of sample holder*

Palladium thin layer was deposited on both sides of all the specimens for permeation measurement by RF magnetron sputtering as an active catalyst for promoting hydrogen dissociation and recombination during permeation. Before the deposition, the surface of the specimens was mechanically polished by hand with size 1000 SiC papers to remove a surface oxide layer and cleaned by supersonication in acetone. The thickness of the deposited Pd layer was about 0.1 μm.

20.3.2 Hydrogen Permeability Measurement

A schematic illustration of a cross sectional view of a sample holder used for hydrogen permeation measurements in this work is shown in Figure 20.4. The Al flat rings were used as gaskets because of their low mechanical strength. As clearly shown in the figure, the upper-stream gas was introduced into a holder and then exhausted after passing by a membrane.

Hydrogen permeation measurements were conducted with a conventional gas permeation technique shown in Figure 20.5 (after [19]). A membrane sample was mounted in the gas permeation cell as shown in Figure 20.4 and then heated at the temperature up to the set

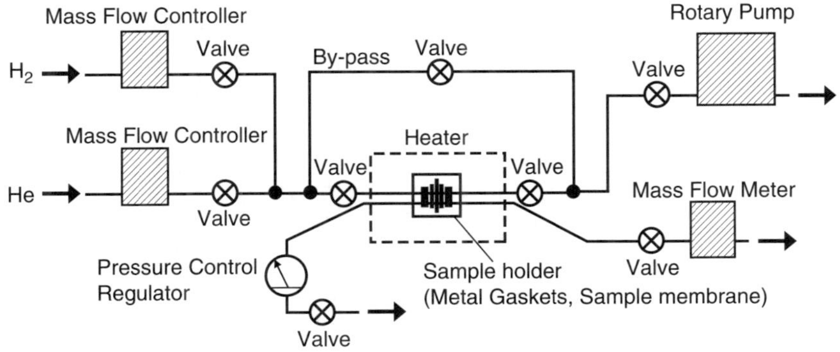

Figure 20.5 *Schematic of apparatus for hydrogen permeation measurement. Reprinted from S. Yamaura, S. Nakata, H.M. Kimura and A. Inoue, Hydrogen permeation of the Zr65Al7.5Ni10Cu12.5Pd5 alloy in three different microstructures, J. Membr. Sci., 291, 126–130. Copyright (2007) with permission from Elsevier*

temperature by using an electric furnace. Pure hydrogen gas was introduced to the upper-stream side of a membrane and then the flux of effluent hydrogen gas from the lower-stream side was measured with a mass flow meter. Before the measurements, pure He gas was introduced to the sample cell and the cell was evacuated by a rotary pump. This procedure was repeated several times before heating. Soon after the temperature of the cell reached to the set temperature, pure H_2 gas was introduced to the cell. It took a few minutes to control the pressure difference and to start the flux measurement. Hydrogen gas pressure in the upper-stream side is controlled with a pressure control regulator. The purity of H_2 gas used in this study was 99.99999% (7Nup).

20.3.3 Methanol Steam Reforming Experiment

This experiment was conducted to show the applicability of amorphous alloys to practical hydrogen production. The methanol solution ($CH_3OH : H_2O = 1 : 1$ mole) was evaporated at about 623 K. The evaporated steam was introduced into the reactor in which there were two chambers divided by a Pd-coated amorphous alloy membrane. Figure 20.6a, b show a schematic illustration and photograph of a reactor for methanol steam reforming used in this study, respectively [20]. The catalyst was filled in the upper-side chamber in the reactor. The catalyst

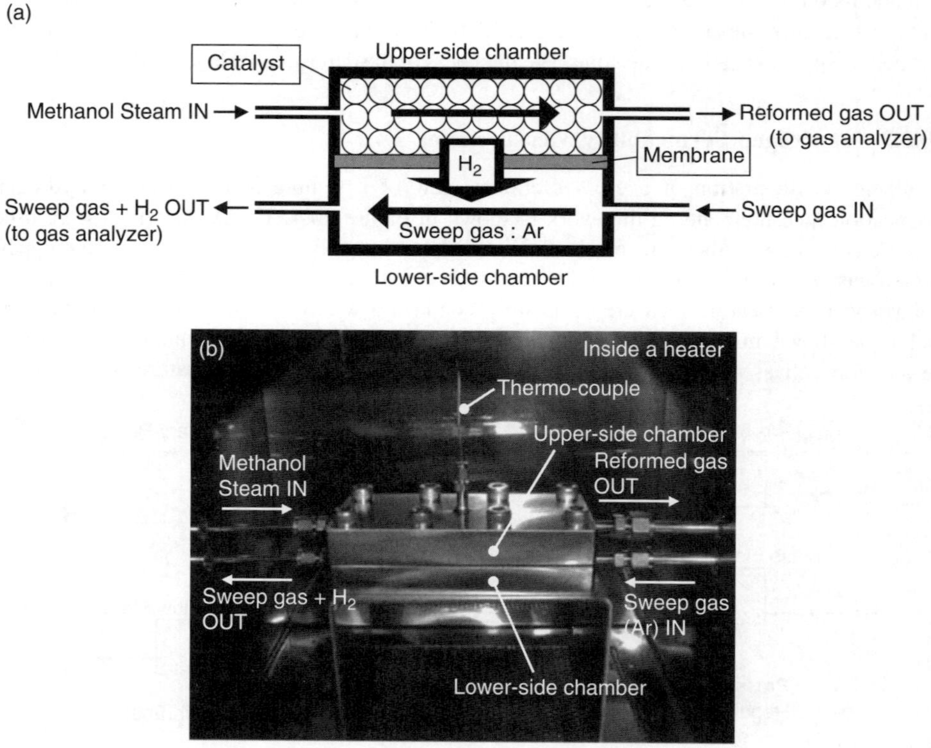

Figure 20.6 *(a) Schematic. Reprinted from S. Yamaura, H.M. Kimura, A. Inoue, Y. Shimpo, M. Nishida and S. Uemiya, Hydroegn Permeability of Melt-Spun Ni-Nb-Ta-Zr-Co Amorphous Alloy Membrane and Its Application to Hydrogen Production by Methanol Steam Reforming, J. Soc. Mater. Scie. Jpn., 57, 1031–1035. Copyright (2008) with permission from The Society of Materials Science, Japan and (b) photograph of reactor for methanol steam reforming*

Table 20.1 Experimental conditions of methanol steam reforming. Reprinted from S. Yamaura, H.M. Kimura, A. Inoue, Y. Shimpo, M. Nishida and S. Uemiya, Hydroegn Permeability of Melt-Spun Ni-Nb-Ta-Zr-Co Amorphous Alloy Membrane and Its Application to Hydrogen Production by Methanol Steam Reforming, J. Soc. Mater. Scie. Jpn., 57, 1031–1035. Copyright (2008) with permission from The Society of Materials Science, Japan

Test condition	Test value
Temperature	573 K
Flow rate of the ($H_2O + CH_3OH$) solution	0.01 $cm^3\,min^{-1}$
Amount of catalyst	8.3 g (Cu-based catalyst) + 8.5 g quartz sand
Carrier gas	Ar
Membrane compositions	$Ni_{40}Nb_{20}Ta_5Zr_{30}Co_5$ amorphous sample
Membrane surface	Pd coating (thickness < 0.1 μm)
Membrane thickness	40 μm
Permeation area	14 × 47 mm

was supplied by Sued-Chemie Catalysts Japan, Inc. (MDC-3, CuO/ZnO type). The oil carbon gaskets were used in this measurement. The reactor was heated at up to 573 K. By using the catalyst, hydrogen can be produced in the following description at around 573 K:

$$CH_3OH + H_2O \rightarrow 3H_2 + CO_2$$

Then generated hydrogen in the upper-side chamber permeates through the amorphous membrane by the hydrogen partial pressure difference and escape from the lower-side surface of the membrane into the lower-side chamber. Ideally the partial pressure of generated H_2 is 0.075 MPa. The carrier gas in the lower-side chamber was Ar. As the flow rate of the carrier Ar gas increased, the hydrogen permeation of the membrane increased.

The compositions of the reformed gas in the upper-side chamber and the permeated gas in the lower-side chamber were analysed by using the TCD-type gas chromatograph. Detailed conditions of methanol steam reforming in this study are summarised in Table 20.1 [20].

20.4 Hydrogen Permeation of Ni-Nb-Zr Amorphous Alloy Membranes

The hydrogen permeability of amorphous alloys has been studied for more than two decades [21]. For example, hydrogen permeability has already been reported for Fe-Ni-P-B alloy ribbons [22], $LaNi_5$ alloy thin films [23,24], Fe-Ti alloy thin films [25], Ni-P thin films [26] and Ni-Zr alloy ribbons [27]. We have also previously reported that the Pd-coated Ni-Nb-Zr amorphous alloys possess excellent hydrogen permeability, as high as that of Pd metal [7,8].

Recently, Inoue *et al.* found that amorphous alloys can be formed in the Ni-Nb-Ti-Zr system [28]. Subsequently, Kimura *et al.* reported that Ni-Nb-Zr amorphous alloys can be produced in a wide composition range by melt-spinning [29]. In this paper, the hydrogen permeation behavior of the $(Ni_{0.6}Nb_{0.4})_{100-x}Zr_x$ ($x = 0$, 20, 30, 40, 50 at%) amorphous alloys produced by melt-spinning was studied [8].

20.4.1 Hydrogen Permeation

In general, the permeation of hydrogen through a membrane is thought to occur through the following three steps [6]: (i) the dissociation of hydrogen gaseous molecules into hydrogen atoms on the surface of the upper side of the membrane, (ii) the diffusion of hydrogen atoms through the membrane, and (iii) the recombination of hydrogen atoms on the surface of the permeate side of the membrane. When the hydrogen behavior is interpreted on the basis of the above mentioned mechanism, the hydrogen permeation rate J [mol s^{-1}] can be estimated by the following equation:

$$J = \frac{P.S}{t}\left(\sqrt{P_{\text{upper}}} - \sqrt{P_{\text{lower}}}\right)$$

Where P is the hydrogen permeability [mol · m^{-1} · s^{-1} · Pa$^{-1/2}$] and t is the membrane thickness [m]. The hydrogen pressures of the upper side and of the permeate side of the membrane are P_{upper} and P_{lower} [Pa], respectively. S is the permeation area of the membrane sample [m^2].

Figure 20.7 shows the XRD patterns of the melt-spun $(Ni_{0.6}Nb_{0.4})_{100-x}Zr_x$ (x = 0, 20, 30, 40, 50 at%) alloys. No distinct diffraction peaks are observed in the range 20–80°. This indicates that all the alloys possess a single amorphous phase. The crystallisation temperature T_x of the $(Ni_{0.6}Nb_{0.4})_{70}Zr_{30}$ amorphous alloy is 794 K. There observed a tendency for T_x to decrease with increasing Zr content.

Figure 20.8 indicates the Arrhenius plots of the hydrogen permeability, P, of the alloys. Plots of the Pd-23 mass%/Ag alloy and Pd metal are included in the figure. The permeation data of the Pd-23 mass%/Ag alloy were also obtained by the present author and those of Pd metal were referred to the previous report [25]. One can see the tendency in the figure that the permeability of the alloys increases with increasing Zr content at 673 K and with increasing temperature. Besides, a

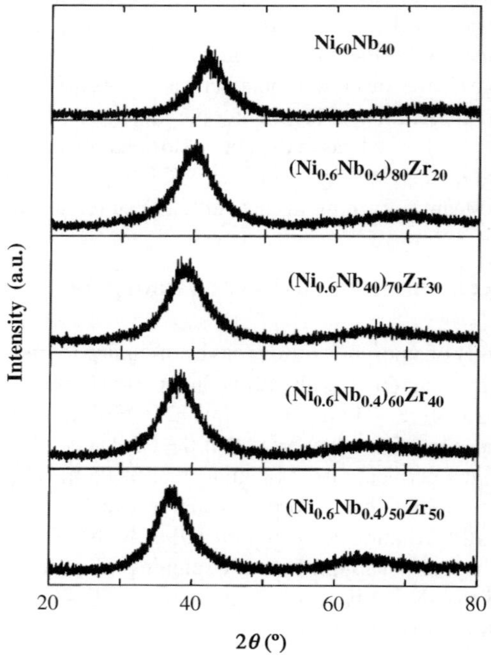

__Figure 20.7__ XRD patterns of Ni-Nb-Zr melt-spun alloys. Reprinted from S. Yamaura, M. Sakurai, M. Hasegawa, K. Wakoh, Y. Shimpo, M. Nishida, H.M. Kimura, E. Matsubara and A. Inoue, Hydrogen permeation and structural features of melt-spun Ni-Nb-Zr amorphous alloys, Acta Mater., 53, 3703–3711. Copyright (2005) with permission from Elsevier

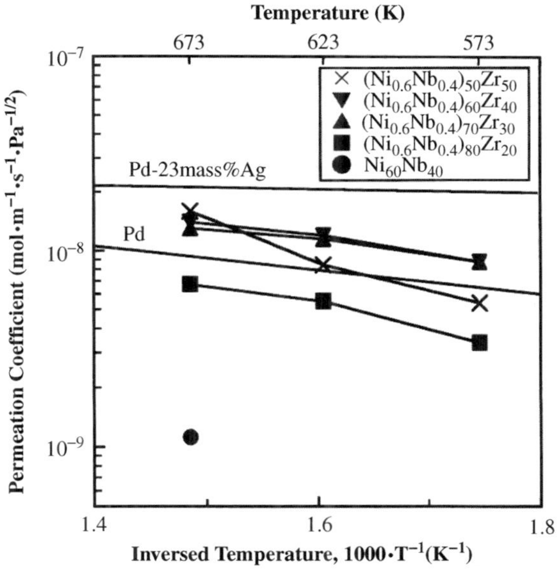

Figure 20.8 *Arrhenius plots of the hydrogen permeability of Ni-Nb-Zr amorphous alloys. Reprinted from S. Yamaura, M. Sakurai, M. Hasegawa, K. Wakoh, Y. Shimpo, M. Nishida, H.M. Kimura, E. Matsubara and A. Inoue, Hydrogen permeation and structural features of melt-spun Ni-Nb-Zr amorphous alloys, Acta Mater., 53, 3703–3711. Copyright (2005) with permission from Elsevier*

halo peak shown in Figure 20.7 shifts to lower angles with increasing Zr content, indicating that the average atomic distance in the amorphous structure increases by the addition of Zr. It can be seen from Figures 20.7, 20.8 that the amorphous alloy having a larger atomic spacing possesses higher hydrogen permeability in this alloy system. In order to investigate the local atomic configuration of the alloys more quantitatively, the RDF analysis was conducted.

20.4.2 Local Atomic Configuration of the Alloys

Figure 20.9 shows the ordinary pair distribution functions before and after hydrogenation, calculated by the Fourier transformation of the ordinary interference functions obtained from the XRD data. The solid and dotted lines correspond to the curves for the as-spun and hydrogenated samples. The $(Ni_{0.6}Nb_{0.4})_{70}Zr_{30}$ and $(Ni_{0.6}Nb_{0.4})_{50}Zr_{50}$ amorphous alloys absorbed hydrogen up to 33.7 at% and 48.9 at%, respectively. As clearly shown in the figures, a distinct peak corresponding to the presence of the Zr-Zr pairs appears drastically by Zr addition and the atomic distance between the Zr atoms increases by hydrogenation. Since the $Ni_{60}Nb_{40}$ amorphous alloy absorbed hydrogen only up to 1.5 at%, there is no difference in the curve profiles between as-spun and hydrogenated samples. Furthermore, the chemical ordering that the number of Zr coordinate is much higher than that of Ni and Nb coordinates was found in the $(Ni_{0.6}Nb_{0.4})_{70}Zr_{30}$ and $(Ni_{0.6}Nb_{0.4})_{50}Zr_{50}$ amorphous alloys from more detailed analysis. These results indicate that hydrogen absorption causes significant structural change in the amorphous alloys.

The results of structural analysis mentioned above lead to conclusion that hydrogen may easily permeate between the Zr-Zr pairs where Zr-Zr bond length is expanded by the hydrogen atoms in those amorphous alloys. As a result, the excellent hydrogen permeation in the $(Ni_{0.6}Nb_{0.4})_{70}Zr_{30}$ and $(Ni_{0.6}Nb_{0.4})_{50}Zr_{50}$ amorphous alloys is achieved.

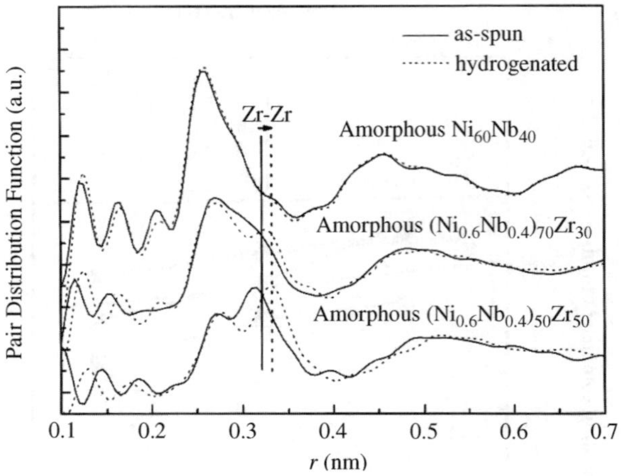

Figure 20.9 *Ordinary radial distribution functions in melt-spun alloys before and after heat-treatment in a hydrogen atmosphere. Reprinted from S. Yamaura, M. Sakurai, M. Hasegawa, K. Wakoh, Y. Shimpo, M. Nishida, H.M. Kimura, E. Matsubara and A. Inoue, Hydrogen permeation and structural features of melt-spun Ni-Nb-Zr amorphous alloys, Acta Mater., 53, 3703–3711. Copyright (2005) with permission from Elsevier*

20.4.3 Long-Term Durability Tests

The requirements which the hydrogen permeable membranes have to satisfy are not only high hydrogen permeability (high hydrogen solubility and high hydrogen diffusivity) but also excellent immunity from hydrogen embrittlement. Hydrogen embrittlement is known to be one of the most important subjects of hydrogen permeable metallic membranes.

Cobalt was added to the Ni-Nb-Zr ternary amorphous alloys expecting the mitigation of hydrogen embrittlement [18]. Moreover, it was found that the intermetal diffusion of Pd was observed at 673 K. The disappearance of Pd layer at 673 K caused the decrease in surface catalyst activity for hydrogen dissociation and recombination. Therefore, hydrogen permeation tests were conducted for about 350 h at 573 K. Figure 20.10 shows the time-dependent change of the hydrogen permeation rates of the melt-spun $(Ni_{0.6}Nb_{0.4})_{55}Zr_{40}Co_5$ and the $(Ni_{0.6}Nb_{0.4})_{45}Zr_{50}Co_5$ amorphous alloys at 573 K. The hydrogen pressures of the upper- and the lower-side surfaces of the $(Ni_{0.6}Nb_{0.4})_{55}Zr_{40}Co_5$ amorphous alloy membrane was 0.35 MPa and 0.1 MPa, respectively. Those for the $(Ni_{0.6}Nb_{0.4})_{45}Zr_{50}Co_5$ amorphous alloy membrane was 0.3 MPa and 0.1 MPa, respectively. The permeation rates decreased very slowly and rapid change is not seen. The samples could permeate hydrogen even after 350 h. The decrease of permeation rate of the $(Ni_{0.6}Nb_{0.4})_{45}Zr_{50}Co_5$ sample is larger than that of the $(Ni_{0.6}Nb_{0.4})_{55}Zr_{50}Co_5$ sample. Figure 20.11 shows the XRD patterns of the samples after the permeation test at 573 K. The XRD pattern obtained from the $(Ni_{0.6}Nb_{0.4})_{55}Zr_{50}Co_5$ sample shows only sharp diffraction peaks of Pd while that obtained from the $(Ni_{0.6}Nb_{0.4})_{45}Zr_{50}Co_5$ sample shows a small diffraction peak identified as Nb. This is because the crystallisation temperature of the Ni-Nb-Zr-Co amorphous alloy tends to decrease with increasing Zr content. The crystallisation temperatures of the $(Ni_{0.6}Nb_{0.4})_{55}Zr_{50}Co_5$ and the $(Ni_{0.6}Nb_{0.4})_{45}Zr_{50}Co_5$ amorphous alloys are 757 and 727 K, respectively. As a result, it is concluded that the slight deterioration at 573 K is caused mainly by the crystallisation in the $(Ni_{0.6}Nb_{0.4})_{45}Zr_{50}Co_5$ amorphous sample.

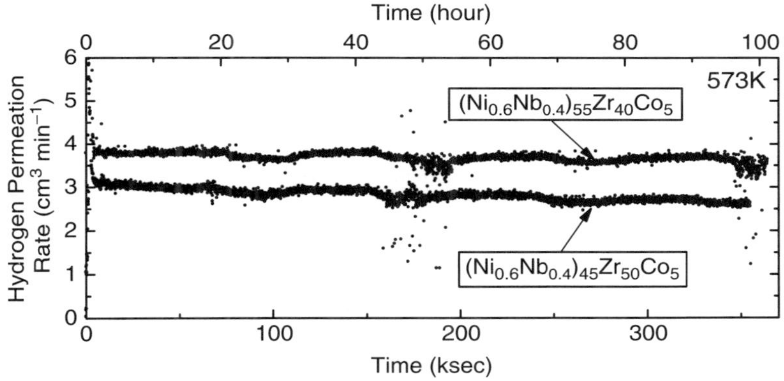

Figure 20.10 *Time-dependent change of the hydrogen permeabilities of melt-spun $(Ni_{0.6}Nb_{0.4})_{55}Zr_{40}Co_5$ and $(Ni_{0.6}Nb_{0.4})_{45}Zr_{50}Co_5$ amorphous alloys at 573 K. Reprinted from Y. Shimpo, S. Yamaura, M. Nishida, H.M. Kimura and A. Inoue, Development of melt-spun Ni-Nb-Zr-Co amorphous alloy for high-performance hydrogen separating membrane, J. Membr. Sci., 286, 170–173. Copyright (2006) with permission from Elsevier*

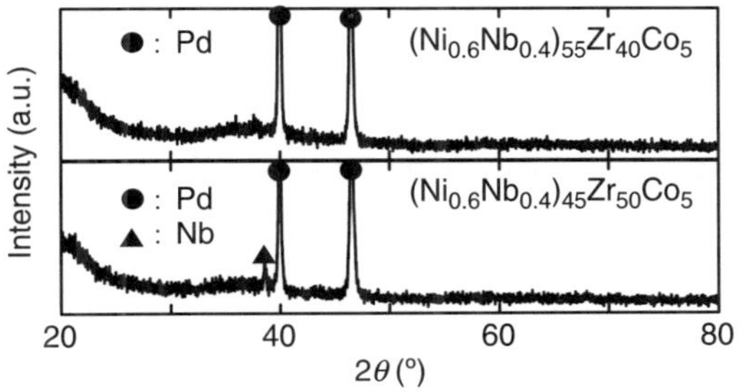

Figure 20.11 *XRD patterns of $(Ni_{0.6}Nb_{0.4})_{55}Zr_{40}Co_5$ and $(Ni_{0.6}Nb_{0.4})_{45}Zr_{50}Co_5$ samples after permeation test at 573 K. Reprinted from Y. Shimpo, S. Yamaura, M. Nishida, H.M. Kimura and A. Inoue, Development of melt-spun Ni-Nb-Zr-Co amorphous alloy for high-performance hydrogen separating membrane, J. Membr. Sci., 286, 170–173. Copyright (2006) with permission from Elsevier*

20.5 Hydrogen Production by Methanol Steam Reforming Using Melt-Spun Ni-Nb-Ta-Zr-Co Amorphous Alloy Membrane

In this work, Ta was added to the Ni-Nb-Zr-Co quaternary alloy to increase its thermal stability. The $Ni_{40}Nb_{20}Ta_5Zr_{30}Co_5$ alloy was used for further investigation [20]. Figure 20.12 shows the DSC curve obtained from an as-spun sample heated in an Ar flow. The crystallisation temperature T_x of the sample is 825 K. The T_x of the previously reported $(Ni_{0.6}Nb_{0.4})_{70}Zr_{30}$ (=$Ni_{42}Nb_{28}Zr_{30}$) amorphous alloy is 794 K, measured at the heating rate of 0.67 K s^{-1}. Therefore, it was found that Ta and Co additions to the Ni-Nb-Zr alloy increased T_x by about 30 K.

Figure 20.13 shows the results of the gas chromatograph analysis of the reformed gas extracted from the upper-side chamber and the permeated gas from the lower-side chamber. As you can see

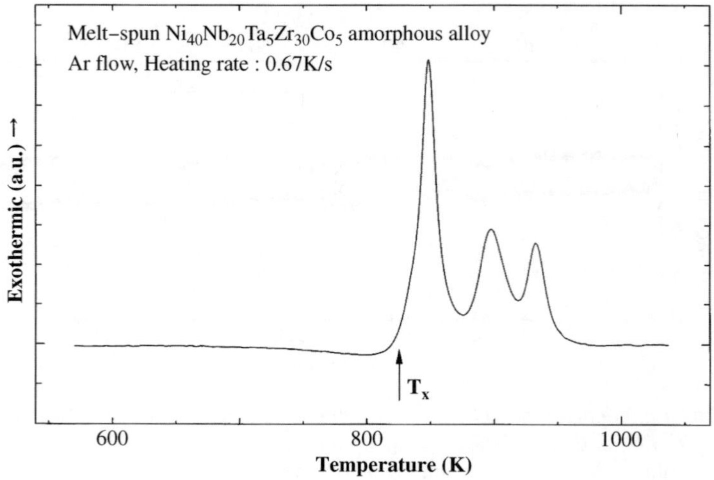

Figure 20.12 *DSC curve of $Ni_{40}Nb_{20}Ta_5Zr_{30}Co_5$ amorphous alloy. Reprinted from S. Yamaura, H.M. Kimura, A. Inoue, Y. Shimpo, M. Nishida and S. Uemiya, Hydroegn Permeability of Melt-Spun Ni-Nb-Ta-Zr-Co Amorphous Alloy Membrane and Its Application to Hydrogen Production by Methanol Steam Reforming, J. Soc. Mater. Scie. Jpn., 57, 1031–1035. Copyright (2008) with permission from The Society of Materials Science, Japan*

in the gas-chromatograms, the reformed gas contains hydrogen, nitrogen (N_2), carbon dioxide (CO_2) and carbon monoxide (CO). As clearly shown in this figure, CO_2 and CO gases were removed from the permeated gas by using the Pd-coated amorphous membrane. Ar gas is not detected in this analysis because Ar gas is a sweep gas in this analysis. N_2 gas was observed in both gases. We did not use a vacuum pump before introducing the evaporated gas in the upper-side chamber and the sweep gas in the lower-side chamber of the reactor. So, the N_2 gas may come from air which slightly remained in the reactor chambers. Moreover, N_2 may be a contamination

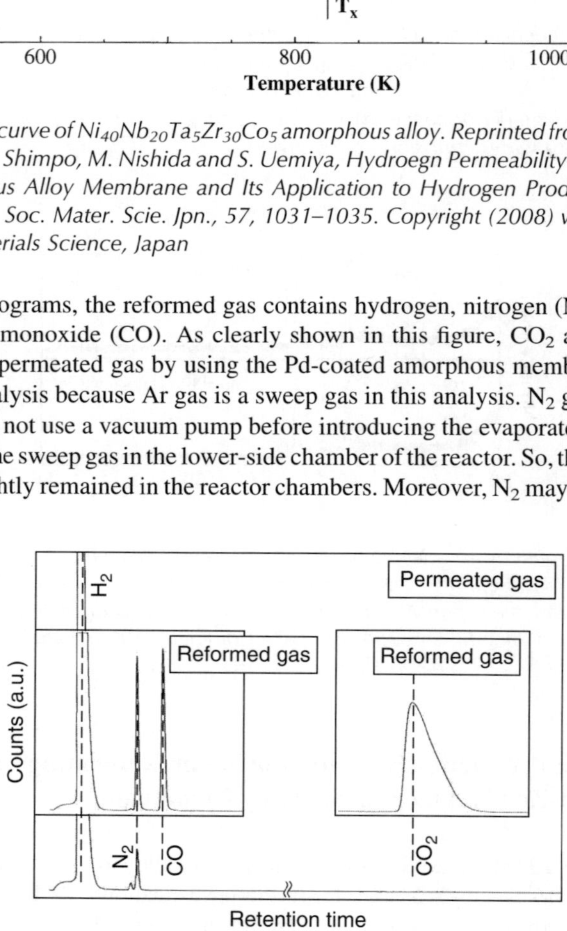

Figure 20.13 *Gas chromatograph analyses of reformed and permeated gases in methanol steam reforming.Reprinted from S. Yamaura, H.M. Kimura, A. Inoue, Y. Shimpo, M. Nishida and S. Uemiya, Hydroegn Permeability of Melt-Spun Ni-Nb-Ta-Zr-Co Amorphous Alloy Membrane and Its Application to Hydrogen Production by Methanol Steam Reforming, J. Soc. Mater. Scie. Jpn., 57, 1031–1035. Copyright (2008) with permission from The Society of Materials Science, Japan*

caused by taking gas samples from the gases by injection-type syringe. A small amount of air is inevitably mixed into the gas sample in the needle of a syringe. So, we can ignore the N_2 peak in the gas-chromatograms. Therefore, it can be concluded that the permeated gas includes only pure hydrogen gas and sweep Ar gas and that any other by-products are not detected in the permeated gas. In this work, it was found that it is possible to produce pure hydrogen by methanol steam reforming using amorphous alloy membrane.

20.6 Final Remarks and Conclusions

In this closing section, it is essential to discuss the challenges to develop a practical nonPd-based metallic membrane for hydrogen separation.

For one thing, it is important to develop an amorphous alloy having the T_x as high as possible by modifying the alloy compositions, resulting in high stability of the alloy at high temperature, in order to suppress the degradation of hydrogen permeation caused by the crystallisation of the alloy.

Another challenge is to disuse the Pd surface layer on the membrane. This requirement is not limited to amorphous alloys but all the nonPd-based alloys. All of the present candidate materials to replace Pd-based alloys require thin Pd layers on both sides of the membrane surface. The Pd layer promotes the dissociation and recombination of $H_2 \leftrightarrow 2H$ as a catalyst on the surface of the membrane, resulting in a larger flux of permeated hydrogen through nonPd-based alloy membranes. To the author's knowledge, there have also been some reports of hydrogen permeability in Ni-coated membranes measured electrochemically in solution [30]. Due to lack of supply, it will be important for the future to develop nonPd-based alloy membranes that do not require Pd coating. Disuse of Pd can also lead to cost savings. Therefore, it is important to understand the roles of surface coating on hydrogen permeable metallic membranes and to study the effect of coating elements on the hydrogen permeation of nonPd-based alloy membranes.

The other challenge is to suppress the hydrogen embrittlement that occurs significantly after hydrogenation of the membrane alloys. Since the alloys can absorb a larger amount of hydrogen at lower temperatures (373~473 K) than at the temperature for the measurements (573~673 K), cyclic thermal shock between room temperature and operating temperature causes membrane cracking. Expansion that weakens the atomic bonds [31] and the formation of hydrides after hydrogenation are important factors to bring hydrogen embrittlement. It is fundamental to understand the embrittlement mechanism in detail with a view to suppress the hydrogen embrittlement.

In this work, hydrogen permeation was examined for the Pd-coated Ni-Nb-Zr amorphous alloys. The long time permeation tests were also conducted with the Pd-coated Ni-Nb-Zr-Co amorphous alloy membranes. Then the Pd-coated Ni-Nb-Ta-Zr-Co amorphous alloy membrane was applied to hydrogen production by methanol steam reforming. The results obtained are summarised as follows:

1. Hydrogen permeability of the amorphous $(Ni_{0.6}Nb_{0.4})_{100-x}Zr_x$ ($x = 0$, 20, 30, 40, 50 at%) alloys was measured and the permeation mechanism was discussed. The permeability of the Ni-Nb-Zr amorphous alloys is strongly dependent on their alloy compositions. The permeability of the alloys increases with increasing Zr content and temperature. The hydrogen permeabilities were 1.3×10^{-8} and 1.59×10^{-8} [$mol \cdot m^{-1} \cdot s^{-1} \cdot Pa^{-1/2}$] at 673 K for the $(Ni_{0.6}Nb_{0.4})_{70}Zr_{30}$ and the $(Ni_{0.6}Nb_{0.4})_{50}Zr_{50}$ amorphous alloys. The bond length between Zr-Zr atoms is expanded by hydrogenation while that of other atomic pairs do not show clear

increase. This may be because hydrogen interacts selectively with Zr atoms. Hydrogen easily diffuses through such Zr-Zr bondings expanded by hydrogenation in those amorphous alloys, possibly resulting in the excellent hydrogen permeation.
2. Long-term permeation tests were conducted with the Ni-Nb-Zr-Co amorphous alloy membranes. The permeability of the $(Ni_{0.6}Nb_{0.4})_{55}Zr_{40}Co_5$ and the $(Ni_{0.6}Nb_{0.4})_{45}Zr_{50}Co_5$ amorphous alloys did not almost decrease at 573 K even after 100 h and decreased gradually in 350 h. Crystallisation occurs in the $(Ni_{0.6}Nb_{0.4})_{45}Zr_{50}Co_5$ amorphous alloy even at 573 K after the long-term duration test.
3. Hydrogen production test by methanol steam reforming was conducted with the $Ni_{40}Nb_{20}$-$Ta_5Zr_{30}Co_5$ amorphous alloy membrane. Comparing the reformed gas with the permeated gas, it was found that the carbon monoxide and carbon dioxide are removed from the reformed gas and the permeated gas contains hydrogen gas and sweep Ar gas only. We have successfully demonstrated the possibility of producing pure hydrogen by methanol steam reforming with the amorphous alloy membrane.

References

1. E. Kikuchi, Membrane reactor application to hydrogen production, *Catal. Today*, **56**, 97–101 (2000).
2. S. Uemiya, N. Sato, H. Ando, Y. Kude, T. Matsuda and E. Kikuchi, Separation of hydrogen through palladium thin film supported on a porous glass tube, *J. Membr. Sci.*, **56**, 303–313 (1991).
3. J.R. Young, Purity of hydrogen permeating through Pd, Pd-25%Ag, and Ni, *Rev. Sci. Instrum.*, **34**, 891–892 (1963).
4. C. Nishimura, M. Komaki, S. Hwang and M. Amano, V-Ni alloy membranes for hydrogen purification, *J. Alloy. Comp.*, **330/332**, 902–906 (2002).
5. K. Hashi, K. Ishikawa, T. Matsuda and K. Aoki, Microstructures and hydrogen permeability of Nb-Ti-Ni alloys with high resistance to hydrogen embrittlement, *Mater. Trans.*, **46**, 1026–1031 (2005).
6. S. Uemiya, Brief review of steam reforming using a metal membrane reactor, *Top. Catal.*, **29**, 79–84 (2004).
7. S. Yamaura, Y. Shimpo, H. Okouchi, M. Nishida, O. Kajita, H.M. Kimura and A. Inoue, Hydrogen permeation characteristics of melt-spun Ni-Nb-Zr amorphous alloy membranes, *Mater. Trans.*, **44**, 1885–1890 (2003).
8. S. Yamaura, M. Sakurai, M. Hasegawa, K. Wakoh, Y. Shimpo, M. Nishida, H.M. Kimura, E. Matsubara and A. Inoue, Hydrogen permeation and structural features of melt-spun Ni-Nb-Zr amorphous alloys, *Acta Mater.*, **53**, 3703–3711 (2005).
9. M. Hagiwara, A. Inoue and T. Masumoto, Mechanical properties of Fe-Si-B amorphous wires produced by in-rotating-water spinning method, *Metall. Trans.*, **13A**, 373–382 (1982).
10. A. Inoue, B.L. Shen, H. Koshiba, H. Kato and A.R. Yavari, Cobalt-based bulk glassy alloy with ultrahigh strength and soft magnetic properties, *Nat. Mater.*, **2**, 661–663 (2003).
11. H.S. Chen, The influence of structure on electrical resistivities of Pd-Au-Si and Au-Ge-Si glass forming alloys, *Sol. Stat. Comm.*, **33**, 915–919 (1978).
12. A. Inoue, Bulk amorphous alloys with soft and hard magnetic properties, *Mater. Sci. Eng.*, **A226/A228**, 357–363 (1997).
13. T.M. Devine, Anodic polarization and localized corrosion behavior of amorphous $Ni_{35}Fe_{30}Cr_{15}P_{14}B_6$ in near-neutral and acidic chloride solutions, *J. Electrochem. Soc.*, **124**, 38–44 (1977).

14. R.B. Diegle, Localized corrosion of amorphous Fe-Ni-Cr-P-B alloys, *Corrosion*, **35**, 250–258 (1979).
15. T. Masumoto and R. Maddin, The mechanical properties of palladium 20 a/o silicon alloy quenched from the liquid state, *Acta Metall.*, **19**, 725–741 (1977).
16. A. Inoue, T. Komura, J. Saida, M. Oguchi, H.M. Kimura and T. Masumoto, Production of flaky amorphous alloy powders in Co-Si-B sysetm by a 2-stage quenching technique of gas atomization and centrifugal spinning, *Int. J. Rapid Solidif.*, **4**, 181–195 (1989).
17. H. Yoshioka, K. Asami, A. Kawashima and K. Hashimoto, Laser-processed corrosion-resistant amorphous Ni-Cr-P-B surface allys on a mild steel, *Corros. Sci.*, **27**, 981–995 (1987).
18. Y. Shimpo, S. Yamaura, M. Nishida, H.M. Kimura and A. Inoue, Development of melt-spun Ni-Nb-Zr-Co amorphous alloy for high-performance hydrogen separating membrane, *J. Membr. Sci.*, **286**, 170–173 (2006).
19. S. Yamaura, S. Nakata, H.M. Kimura and A. Inoue, Hydrogen permeation of the $Zr_{65}Al_{7.5}Ni_{10}Cu_{12.5}Pd_5$ alloy in three different microstructures, *J. Membr. Sci.*, **291**, 126–130 (2007).
20. S. Yamaura, H.M. Kimura, A. Inoue, Y. Shimpo, M. Nishida and S. Uemiya, Hydroegn permeability of melt-spun Ni-Nb-Ta-Zr-Co amorphous alloy membrane and its application to hydrogen production by methanol steam reforming (in Japanese), *J. Soc. Mater. Sci. Jpn*, **57**, 1031–1035 (2008).
21. M.D. Dolan, N.C. Dave, A.Y. Ilyushechkin, L.D. Morpeth and K.G. McLennan, Composition and operation of hydrogen-selective amorphous alloy membranes, *J. Membr. Sci.*, **285**, 30–55 (2006).
22. R.W. Lin and H.H. Johnson, Hydrogen permeation in the metallic glass $Fe_{40}Ni_{40}P_{14}B_6$, *J. Non-Cryst. Solids*, **51**, 45–46 (1982).
23. G. Adachi, H. Nagai and J. Shiokawa, $LaNi_5$ film for hydrogen separation, *J. Less-Common Metals*, **97**, L9–L10 (1984).
24. H. Sakaguchi, H. Seri and G. Adachi, Preparation of amorphous and crystalline lanthanum-nickel ($LaNi_{5.0}$) thin films using a sputtering method and thermodynamics of hydrogen absorption in these films, *J. Phys. Chem.*, **94**, 5313–5316 (1990).
25. M. Amano, Y. Sasaki, K. Nakamura, C. Nishimura, M. Komaki and M. Shibata, Research and development of high-performance membrane alloys for hydrogen separation (in Japanese), *Annu Rep Nat Res Inst Metals*, **11**, 277–287 (1990).
26. B.S. Liu, H. Li, Y. Cao, J.Y. Deng, C. Sheng and S. Zhou, Preparation and characterization of Ni-P amorphous alloy/ceramic composite membrane, *J. Membr. Sci.*, **135**, 33–39 (1997).
27. S. Hara, K. Sakai, N. Itoh, H.M. Kimura, K. Asami and A. Inoue, An amorphous alloy membrane without noble metals for gaseous hydrogen separation, *J. Membr. Sci.*, **164**, 289–294 (2000).
28. A. Inoue, W. Zhang and T. Zhang, Thermal stability and mechanical strength of bulk glassy Ni-Nb-Ti-Zr alloys, *Mater. Trans.*, **43**, 1952–1956 (2002).
29. H.M. Kimura, A. Inoue, S. Yamaura, K. Sasamori, M. Nishida, Y. Shimpo and H. Okouchi, Thermal stability and mechanical properties of glassy and amorphous Ni-Nb-Zr alloys produced by rapid solidification, *Mater. Trans.*, **44**, 1167–1171 (2003).
30. W.C. Chian, W.D. Yeh and J.K. Wu, Hydrogen permeation in $Fe_{40}Ni_{38}B_{18}Mo_4$ and $Fe_{81}B_{13.5}$-$Si_{3.5}C_2$ amorphous alloys, *Mater. Lett.*, **59**, 2542–2544 (2005).
31. S. Jayalakshmi, S.O. Park, K.B. Kim, E. Fleury and D.H. Kim, Studies on hydrogen embrittlement in Zr- and Ni-based amorphous alloys, *Mater. Sci. Eng.*, **A449/A451**, 920–923 (2007).

21

Membranes Prepared Via Phase Inversion

M.G. Buonomenna[1], S.-H. Choi[2,3], F. Galiano[2] and E. Drioli[1,2]

[1]Department of Material and Chemical Engineering, University of Calabria and Consortium INSTM, Rende, CS, Italy
[2]Institute on Membrane Technology, ITM-CNR c/o University of Calabria, Rende, CS, Italy
[3]Green Chemistry and Environmental Biotechnology, University of Science and Technology, Daejeon, Korea

21.1 Introduction

The most important part in membrane reactors is the membrane itself.

The traditional membrane separation processes such as reverse osmosis, micro-, ultra- and nanofiltration, electrodialysis, pervaporation, and so on, already largely used in many different applications, are today combined with membrane systems as catalytic membrane reactors [1]. Typical functions of a membrane in general, and of polymeric membranes, in particular, in a MR are:

1. Selective removal of products from the reaction mixture to circumvent the limitation by the equilibrium or to avoid consecutive reactions (*extractor*).
2. Control the addition of reactants to influence the selectivity of the reaction (*distributor*).
3. Intensify the contact between reactant and catalyst (*contactor*).
4. Recycling of the catalyst (retaining in the reactor of homogeneous catalysts and entrapment of catalyst within membranes).

Inorganic catalytic membranes are typically applied in high temperature reactions, but they suffer from some important drawbacks: the cost of the membranes is very high, their lifetime is

Membranes for Membrane Reactors: Preparation, Optimization and Selection, Edited by Angelo Basile and Fausto Gallucci

limited (poisoning and fouling of the membranes) and high-temperature applications lead to difficulties when materials with different thermal expansion coefficients are combined (delamination of the membrane's top layer from the support).

The use of polymeric membranes in MRs, in addition to the advantage of combining reactions with membrane separation, entails some important new possibilities [2]. The polymeric environment can actively take part in the reaction by its affinity towards reagents and products, with a potential beneficial effect on the catalyst's performance. A much wider choice of polymeric membranes is available as compared with metallic or ceramic membranes, and the costs are generally lower. The relatively low operating temperature of polymeric catalytic membranes is also associated with less stringent demands for the materials needed in the module construction. Although polymeric membranes are less resistant to high temperature and aggressive chemicals than inorganic or metallic membranes, many polymeric materials resistant under rather harsh conditions, such as Teflon® AF, Nafion®, Hyflon®, polyvinylidene fluoride (PVDF), polyimides (PI) are available.

Different methods of polymeric membrane preparation have been reported in literature [3–6]. Membranes either have a symmetric (isotropic) or an asymmetric (anisotropic) structure. Symmetric membranes have a uniform structure throughout the entire membrane thickness, whereas asymmetric membranes have a gradient in structure. In particular this is evident, for integrally skinned with a dense top layer. Compared to dense symmetric membrane, for which the separation properties are determined by the overall structure, asymmetric membranes with a dense skin layer on a porous sublayer have the most impact in membrane market.

In fact, for dense homogeneous membranes the permeant flow across the membrane is quite low, since a minimal thickness is required to guarantee membrane mechanical stability. Asymmetric membranes combine high permeant flow, provided by a very thin selective top layer and a reasonably mechanical stability, resulting from the porous *sublayer* [3].

The most common method used to prepare polymeric membranes (both symmetric and asymmetric) is the phase inversion process. This method covers a range of different techniques such as solvent evaporation, precipitation by controlled evaporation, thermal precipitation, precipitation from the vapour phase and immersion precipitation [7]. In the case of porous membranes as those employed in microfiltration, other methods other than phase inversion are applied as sintering of powders, stretching of films, irradiation and etching of films [7].

21.2 Brief Review

The most studied polymer system for catalytic reactions is polydimethylsiloxane (PDMS) (Table 21.1). This highly permeable elastomer combines a fairly high thermal (up to 250 °C) and mechanical stability with chemical resistance, the latter being of utmost importance under reactive conditions.

The first example of PDMS catalytic membranes was that based on zeolite Y with Fe-phtalocyanine (FePc) immobilised in its cage [8]. In the oxidation of organic substrates by using aqueous hydrogen peroxide, the PDMS based membrane avoids the use of solvent: the elastomer polymer keeps in contact both immiscible reagent phases (Figure 21.1).

Water present in the peroxide phase was excluded from the hydrophilic catalyst, while the organic substrate was sorbed in the PDMS matrix to form a reservoir of reagents. In this way, the polymeric membrane is not an inert support: when the appropriate polymer was selected with

Table 21.1 *Structures of some polymers used for the preparation of polymeric membranes for application in membrane reactors*

Polymer	Structure	References
PDMS	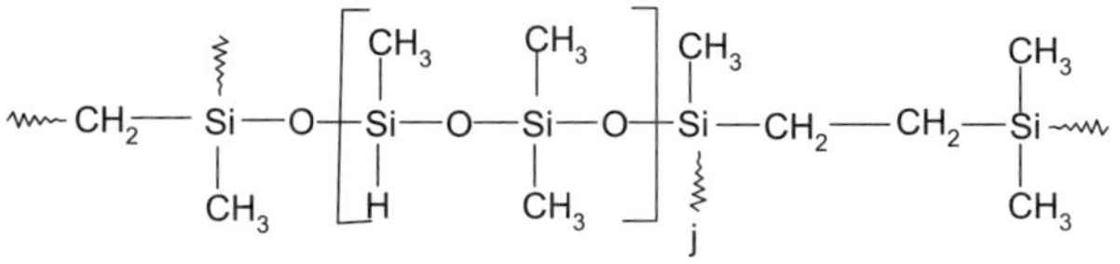	8–14,18
Nafion®	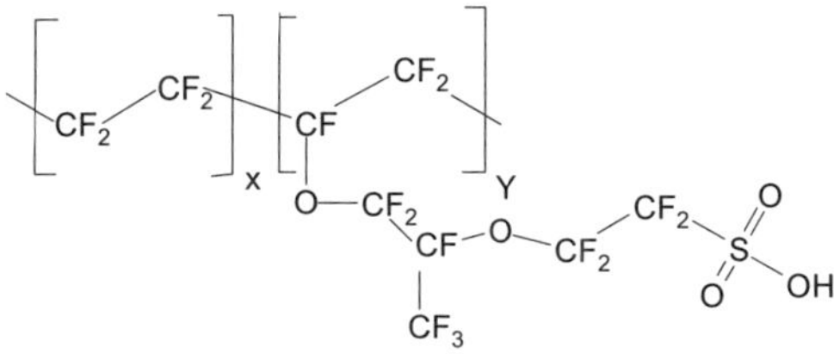	15, 18
Teflon® AF[a]	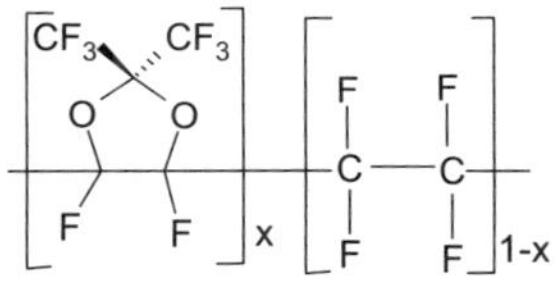	18
Hyflon® AD[b]	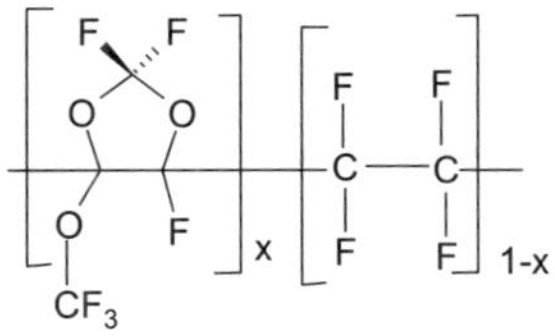	18–22
PVA	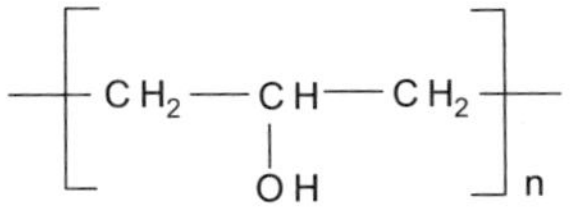	23,24,25
PES	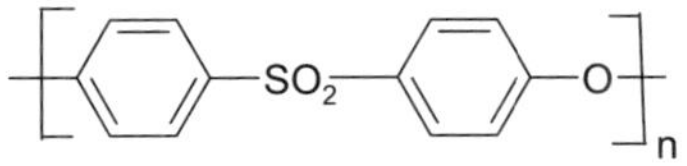	26
PAI	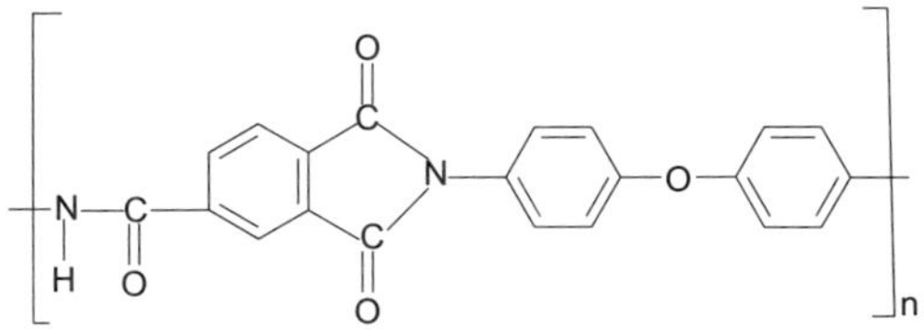	27
PEBAX[c]	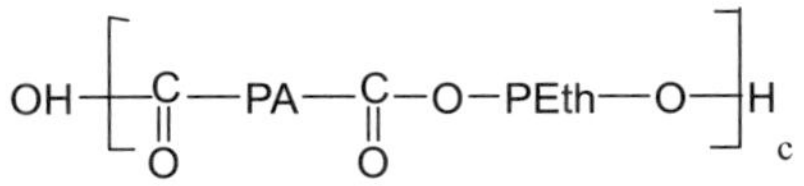	28

(continued)

Table 21.1 *(Continued)*

Polymer	Structure	References
PVDF	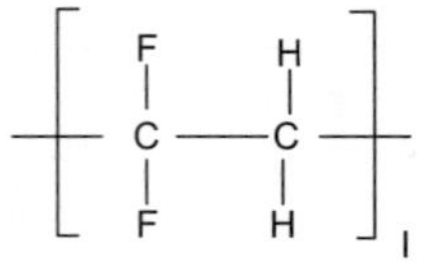	29–35
PEEKWC	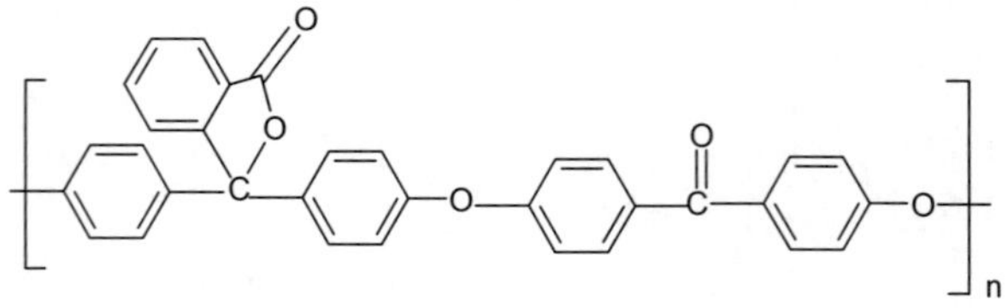	32, 33
PEI	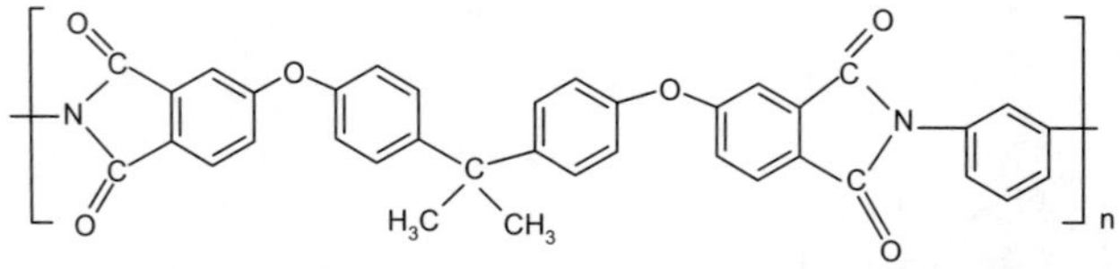	36

[a]Teflon AF2400, $x = 0.87$; another commercial grade offered by DuPont is Teflon AF1600, $x = 0.65$.
[b]Hyflon AD80, $x = 0.80$; another commercial grade offered by Solvay Solexis is Hyflon AD60, $x = 0.60$.
[c]PA represents the polyamide segment; PEth represents the polyether segment.

respect to selective sorption and diffusion of reagents and products, a different reaction path is observed compared to homogeneous system and over-oxidation reactions are avoided [9].

Nafion®, a perfluorinated ionomeric polymer (Table 21.1), is a promising and stable host for catalysts [15]. Lead ruthenate pyrochlore (Py) has been reported as an active catalyst for fuel cells, organic syntheses, charge storage capacitors and chemical sensors [16, 17]. Ke *et al.* [15] reported the preparation and use of a rugged lead ruthenate pyrochlore Nafion 417 membrane catalyst (NPy) in the selective oxidation of benzyl alcohol to benzaldehyde in a triphasic reaction.

Fritsch *et al.* [18] used polymeric composite catalytic membranes for dimerisation of isobutene to isooctane. Catalyst such as silica supported Nafion and supported phosphotungstic acid were mixed in solution with a polymeric binder. Teflon®, Hyflon®, PDMS polymers were

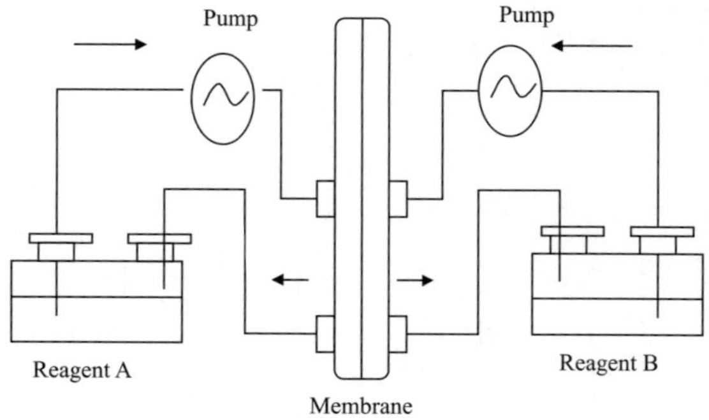

Figure 21.1 *Scheme of a biphasic membrane reactor. The PDMS catalytic membrane prevents the use of a solvent for the catalyst, keeping the two reaction phases in contact. Reprinted from Journal Molecular Catalysis A: Chemical, Parton, R.F., et al., Membrane occluded catalysis: a higher order mimic with improved performance. Vol. 113, 113, 283. Copyright (1996) with permission from Elsevier*

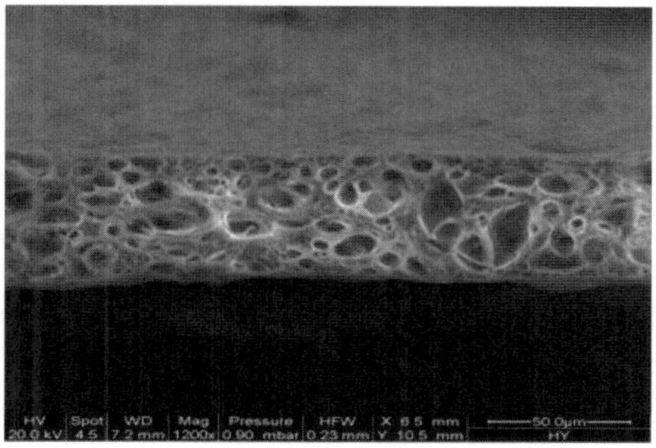

Figure 21.2 *Asymmetric Hyflon AD60X-based membrane prepared by nonsolvent-induced phase inversion, using Galden® HT 55 as solvent and n-hexane as nonsolvent*

used as binders to form a reactive layer on top of a support. The polymeric catalytic membranes were applied in a forced-flow membrane reactor.

The discovery in the past 20 years of amorphous, solvent processable perfluoropolymers such as Teflon® AF, Hyflon® AD has created new opportunities for membrane applications in harsh conditions. Perfluoropolymers can be fabricated into thin, high flux composite or anisotropic membranes (Figure 21.2), retaining the excellent stability typical of fluorinated material.

Gallo *et al.* reported [19, 20] the preparation by means of nonsolvent induced phase inversion of catalytic membranes having ruthenium porphyrin complexes embedded in the perfluoropolymer Hyflon® AD60X and their use in organic solvents as benzene and n-hexane for the aziridination reaction of olefins. Caselli *et al.* used [21] Co(II)-salen complexes, a class of catalysts that has found a remarkable success in the cyclopropanation of olefins with diazoacetates, into polymeric membranes prepared by means of phase inversion.

Vital *et al.* [23–25] evaluated the performance of Nafion® 112, Nafion® 115 and PVA membranes containing sulfonic acids groups as solid acid catalysts in the transesterification of soybean oil with methanol.

Fritsch *et al.* [26, 27] developed catalytically reactive porous membranes for hydrogenation of edible oil in a membrane reactor. PES- and PAI-based membranes with large pores at high porosity have been activated with different Pt contents and about of half of the linoleic acid (the major compound in sunflower oil triglycerides) was hydrogenated. Membrane reactor tests were done in the flow-through mode with permeates recirculation at 100 °C and 4 bar hydrogen pressure (see next paragraph).

PEBAX [28] membranes have been prepared and applied for the hydrogenation of acetophenone in a pervaporative membrane reactor: palladium clusters have been homogeneously distributed inside the polymeric membranes. Acetophenone is reduced by palladium catalyst and molecular hydrogen to the aromatic hydrocarbon ethyl benzene.

Bottino *et al.* [29] prepared two types of catalytic PVDF membranes by the phase inversion process for the hydrogenation of methylenecyclohexane: the first one was obtained from a 20% (w/w) solution of PVDF in N,N-dimethylformamide. The membrane was impregnated with a salt of Pd. The Pd in the membrane was reduced to the metallic form by a solution of sodium boron hydride, at 50 °C. The second type of membrane was prepared by adding a pore former (PVP) in the casting solution. The subsequent impregnation conditions were the same as those reported for the first type of membrane.

PVDF-based asymmetric membranes were prepared, characterised and used in different membrane reactors for oxidation reactions [30–35] due to its excellent chemical resistance. These membranes have been compared with those based on PEEKWC (Table 21.1) for the photooxidation of alcohols in water [33] and for the oxidation of secondary aromatic amines [34].

PEEKWC membranes embedding ruthenium complexes constitute the first example of a polymeric catalytic membrane employed under very forcing conditions (160°C and 5MPa of CO pressure). The reaction investigated was the reduction of nitrobenzene to aniline by CO/H_2O [36].

Molinari and Poerio [30] studied the effects of different solvents on morphology and catalytic properties of PVDF membranes entrapping during the membrane preparation two types of copper (II) oxide (powder and nanopowder).

Fontananova *et al.* [32] investigated the heterogenisation of polyoxometalates (POMs) in PVDF membranes for the aerobic photo-oxidation of phenol in water. Flat sheet membranes at different catalyst loadings by phase inversion technique induced by non solvent have been prepared.

In a different approach compared to the catalyst heterogenisation, inert PVDF membranes characterised by different porosity and morphology have been used as interfaces for biphasic oxidation of industrial interest such as the oxidation of aromatic alcohols and the oxidation of cyclohexene to adipic acid [35]. The membranes have been prepared by means of non solvent induced phase inversion: the use of different preparation conditions (type of solvent, exposure time before coagulation) induced the formation of membranes with well controlled physico-chemical properties.

The reduction of nitrate to nitrogen in aqueous solutions by means of catalytically active membranes has been investigated by LuÈdtke *et al.* [37]: a heterogeneous catalyst (Pd/Cu) has been incorporated in the casting solution of polyetherimide (PEI).

21.3 Explanation of the Phase Inversion Process

During the phase inversion process, a thermodynamically stable polymer solution is subjected to a liquid–liquid demixing during which the cast polymer film separates into a polymer-rich (membrane matrix) and a polymer-lean phase (membrane pores).

Polymeric membranes can be prepared by means of phase inversion from virtually all polymers that are soluble at a certain temperature in an appropriate solvent or solvent mixture and can be precipitated as a solid phase by [7]:

1. Cooling of a polymer solution, which separates at a certain temperature in two phases, that is, a liquid phase forming the membrane pores and a solid phase forming the membrane structure (thermally induced phase inversion).
2. Precipitation by solvent evaporation allowing a dense homogeneous membrane to be obtained.
3. Addition of a nonsolvent or nonsolvent mixture to a homogeneous solution (immersion-precipitation or nonsolvent-induced phase inversion).
4. Precipitation from the vapour phase, consisting of a nonsolvent saturated with a solvent. The high solvent concentration in the vapour phase prevents the evaporation of solvent from the cast film. Membrane formation occurs because of diffusion of nonsolvent into the cast film.
5. Precipitation by controlled evaporation or dry phase inversion: the polymer is dissolved in a mixture of solvent and nonsolvent; since the solvent is more volatile than the nonsolvent, the evaporation step leads to polymer precipitation own to a higher nonsolvent and polymer content.

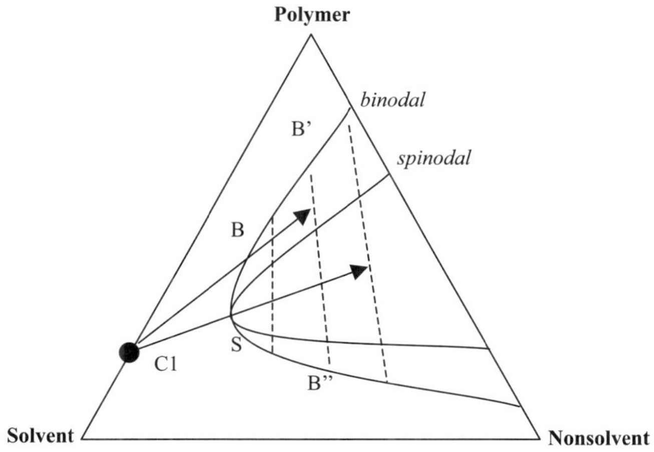

Figure 21.3 *Ternary phase diagram (solvent/polymer/nonsolvent) for membrane formation via immersion–precipitation*

There are different variations to diffusion or nonsolvent induced phase inversion: an evaporation step to the precipitation is used to change the composition in the surface of the cast film (dry–wet phase inversion) [38–40]; an annealing step is applied to change the structure of the precipitated membrane [41].

Immersion–precipitation is the most common phase inversion method. The thermodynamic behaviour of a polymer solution subjected to immersion–precipitation can be represented in polymer/solvent/nonsolvent phase diagram (Figure 21.3).

In this diagram, the initial polymer solution (C1) is situated in the stable region outside the binodal. After demixing polymer solution arrives in the metastable region between the binodal and the spinodal (binodal demixing).

In this region, polymer solutions will 'phase separate' into a polymer-lean and a polymer-rich phase, indicated on the phase diagram by the tieline ends (B′ and B″). Phase separation takes place according to the 'nucleation and growth' mechanism, in which the formed nuclei grow and most often progress towards a phase coalescence. 'Spinodal decomposition', the second and less frequent mechanism, occurs whenever the polymer solution (from C1 to S) directly moves to the thermodynamically unstable zone within the spinodal.

For the phase inversion process the moment at which the developing membrane structure gets solidified (kinetic aspects) makes the difference between the demixing types: (a) in the instantaneous demixing the binodal is crossed already at time t and demixing will start directly (Figure 21.4a), (b) in the delayed demixing, at time t, all positions in the film are still situated within the thermodynamically stable region and demixing will only start when more nonsolvent has diffused into the polymer film in such a way that the binodal can be crossed (Figure 21.4b).

In general, one of the main variables in the immersion–precipitation process is the choice of the solvent/nonsolvent system. The solvent and nonsolvent must be completely miscible. For a given polymer, a very large number of combinations of solvent and nonsolvent are possible all with their own specific thermodynamic behaviour. Where a high mutual affinity exists a porous membrane characterised by macrovoids is obtained, whereas in the case of low mutual affinity an asymmetric membrane with a dense nonporous top layer is obtained. In Figure 21.5, the SEM analysis of the cross section of the PEEKWC based membranes prepared by means of two different solvent/nonsolvent pairs [NMP/water (Figure 21.5a) and THF/water (Figure 21.5b)] is shown.

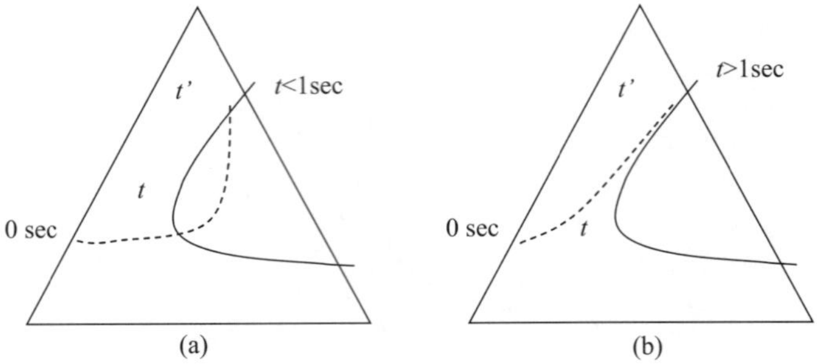

Figure 21.4 *Composition paths of two phase separation processes. (a) Instantaneous demixing. (b) Delayed demixing*

In the case of DMA the skin region of the membrane consists of a thin top layer with a typical structure of closely packed polymeric spheres, so called nodular structure, supported by elongated macrovoids. Using THF as the solvent, a porous sublayer with a closed cell structure has been obtained: the size and the number of the cells gradually decrease, moving from the bottom towards the completely dense skin layer. In Figure 21.6, the schematisation of the mechanism for the pore formation during phase inversion induced by nonsolvent is shown. The formation of macrovoids (as those of PEEKWC membrane in Figure 21.5a) is consequent upon the fast diffusion in the casted film of the nonsolvent coming from the coagulation bath and the immediate mixing with the solvent. For pairs of solvent/nonsolvent characterised by low mutual miscibility the outward solvent diffusion prevails on the nonsolvent inward diffusion inducing the formation of an anisotropic structure with closed pores (Figure 21.5b).

This difference is due to the volatility of THF, which causes an increase of the polymer concentration at the interface. The low THF/water miscibility determines the formation of closed cells. In general, using solvent/nonsolvent pairs with a high mutual affinity (DMF/water, DMA/water, NMP/water), porous membranes were obtained. The combination of a solvent with a nonsolvent with low miscibility allows the formation of membranes with a dense skin layer, suitable for gas separation applications [42].

Although other parameters such as polymer concentration, addition of solvent to the non-solvent bath, addition of nonsolvent to the polymer solution, the temperature of the coagulation

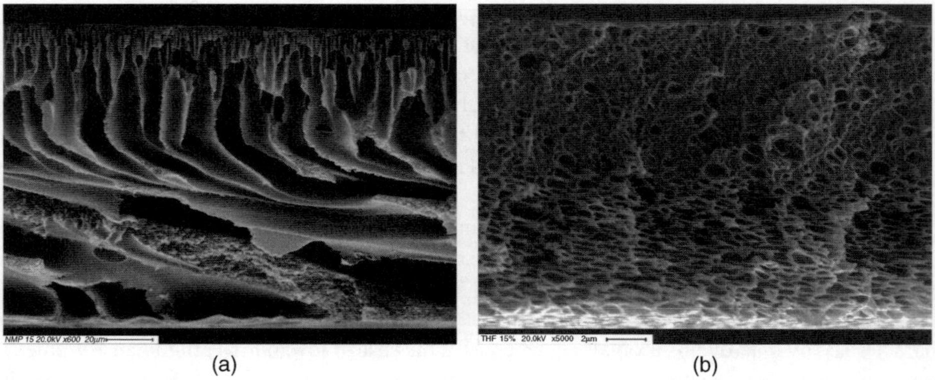

Figure 21.5 *SEM images in the cross section of PEEKWC membranes prepared with two different solvent/nonsolvent pairs: (a) NMP/water, (b) THF/water*

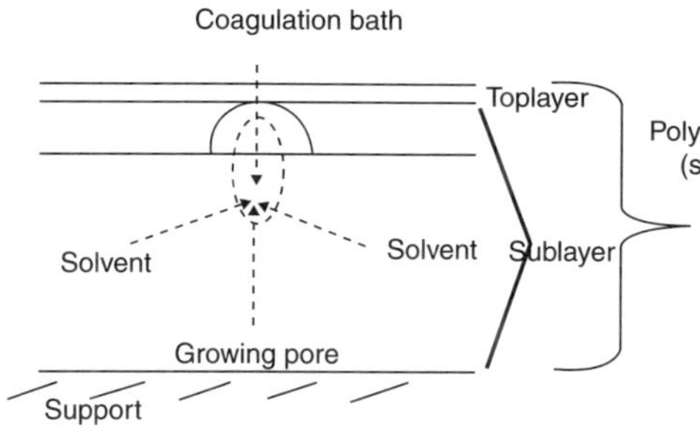

Figure 21.6 *Schematisation of the mechanism for macrovoid formation during phase inversion induced by nonsolvent*

bath and of the polymer solution and the addition of organic and inorganic additives as PVP, salts [29, 43–46] exist which have an influence on the type of membrane structure, the choice of solvent/nonsolvent pair is crucial to affect the final membrane structure.

In Figure 21.7, SEM of the cross section of membranes based on PVDF prepared with different solvent/nonsolvent pair at two different temperatures is shown.

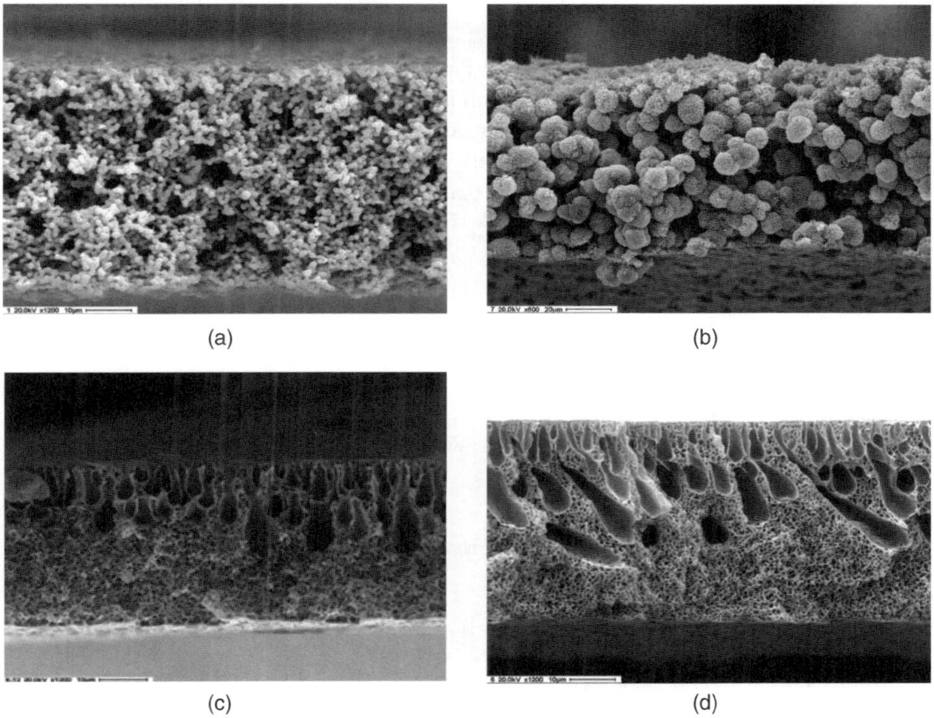

Figure 21.7 *SEM of cross section of PVDF membranes prepared with two solvent/nonsolvent pairs at 25 and 60 °C. (a) DMA/1-octanol, 25 °C. (b) DMA/1-octanol, 60 °C. (c) DMA/water, 25 °C. (d) DMA/water, 60 °C*

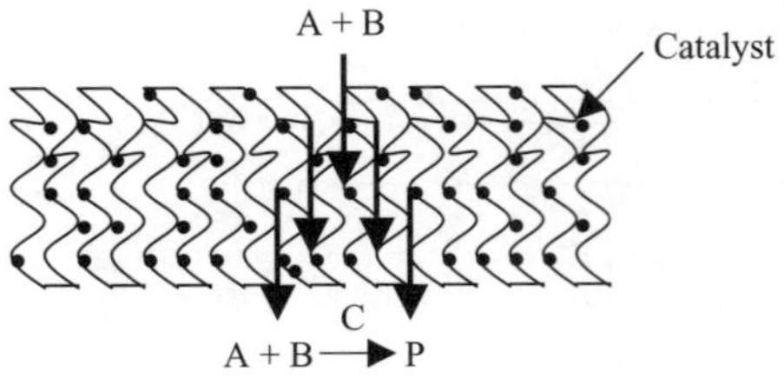

Figure 21.8 Schematisation of forced flow-through membrane reactor

It is evident that the solvent/nonsolvent pair, that is, DMA/water or DMA/1-octanol affects the membrane morphology more than the temperature of the casting solution. The pair DMA/water due to the high miscibility between the components, caused an instantaneous demixing which provokes the formation of macrovoids across the membrane cross section. In the case of DMA/1-octanol, delayed demixing occurred characterised by the presence of well defined polymeric globules, for which the size increases with the temperature. The understanding of the parameters, which determine these different morphologies, was important for the preparation of hydrophobic membranes as contactor/interfaces for biphasic reaction [35].

21.4 Some Applications

In this section some examples of the application of unselective porous catalytic membranes forcing the reactants to flow thorough the membrane, are reported (Figure 21.8). The catalyst is inside the polymeric membrane pores and the reactants flow convectively through the pores: the intensive contact allows for high catalytic activity with negligible mass transport resistances [47].

Polymeric membranes based on PA with pore size of 400 nm [48] are applied to water denitrification in a flow-through operation mode (Figure 21.9): different catalytic membranes are

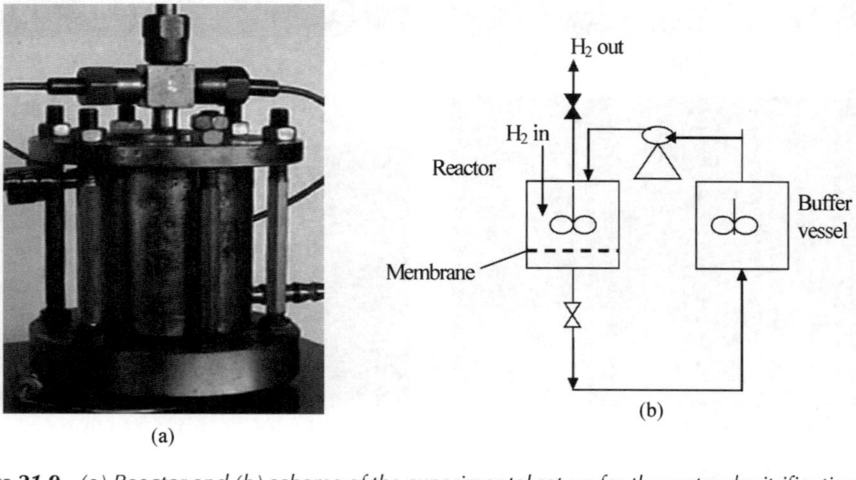

Figure 21.9 (a) Reactor and (b) scheme of the experimental set-up for the water denitrification over catalytic polymeric membranes adapted from [48])

prepared with either copper or palladium or both catalysts. Catalyst is introduced by wet impregnation of the membrane in a catalyst precursor solution followed by chemical reduction.

Both palladium and copper were deposited simultaneously on the isotropic porous polymeric membranes. The catalytically active membranes were applied to investigate the molecular mechanism of the reaction.

The reaction is performed in a stirred reactor module with the catalytic flat sheet membrane and recirculation of the liquid reactants (Figure 21.9). Pd and Cu loaded membranes were poorly active in the reaction, while bimetallic (Pd and Cu)-loaded membranes exhibited high catalytic activity. The study demonstrated that the catalytic activity of the Pd-Cu system in reduction of nitrate ions by hydrogen in water apparently is related to hydrogen spill over that represents an important part of the reaction mechanism.

The same reaction, i.e. water denitrification, has been studied by LuÈdtke *et al.* [37]: different catalytic membranes have been tested by introducing the catalyst in the membrane casting solutions. Flat sheet and hollow fibre membranes were prepared according to the phase inversion process. The polymer solution contained the membrane polymer PEI, the solvents dimethylacetamide (DMAc) or dimethylformamide (DMF), the nonsolvent γ-butyrolactone (GBL) and the catalyst.

In Figure 21.10, the scheme of the experimental set up is shown. Nitrate-containing water was pumped from a reservoir using a centrifugal or piston pump into a hollow fibre module where the nitrate solution was enriched with hydrogen. The hydrogen-enriched water was then fed to the catalytic hollow fibre module. Due to a pressure gradient over the membrane the hydrogen enriched nitrate solution permeated through the membrane where the reaction took place. The volume flow density has been be varied by applying different pressure differences over the membrane. The catalytic membrane was operated in the cross flow mode.

The activities per unit area of membrane are rather low: this fact might be explained by assuming that a certain fraction of the catalyst particles is not accessible being completely surrounded by the polymeric matrix.

Fritsch and Bengtson [27] studied the selective hydrogenation of sunflower oil in a membrane reactor by using high flux polymeric membranes from PES and PAI. Industrially vegetable oil hydrogenation takes place in stirred tank reactors with dispersed catalysts at 170–200 °C and 2–5 bar of H_2 pressures. In this process up to 50% of the undesired trans-isomerised fatty acids are generated. By using a membrane reactor in the flow-through configuration with permeate

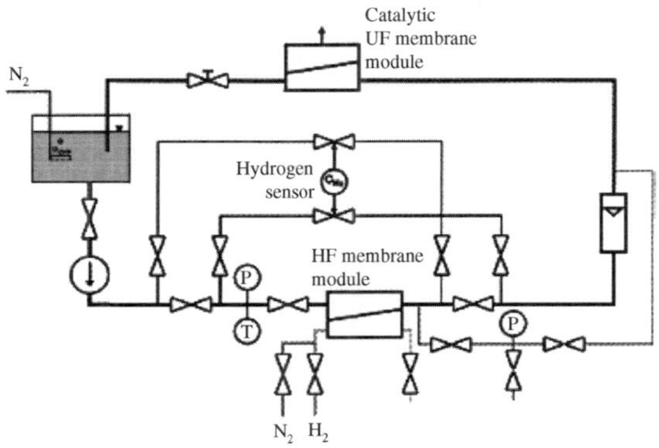

Figure 21.10 *Experimental set-up used for the denitrification of aqueous solution by means of catalytic polymeric membranes Reprinted from Choi, S.H, Scura, F., Barbieri, G., Mazzei, R., Giorno, L., Drioli, E., Kim, J.H. Membrane Journal 19, 72. Copyright (2009) with permission from Korea Institute of Science and Technology Information*

recirculation at 100 °C and 4 bar hydrogen pressure, the percentage of the undesired products can be reduced and at the same time the heterogenisation of the catalyst avoids expensive filtration steps for its recover.

The catalytic membranes were obtained by two different methods: by wet impregnation of the membrane in a catalyst precursor solution (palladium acetate or hexachloroplatinate) or by addition of ready-made supported catalysts to the membrane casting solutions. This last method resulted in membranes with a low catalyst content and low activity for the hydrogenation reaction. However, durable, well fixed catalysts inside the membrane pores have been obtained by wet impregnation and reduction by calcinations in the presence of high amounts of citric acid (Figure 21.11).

PVDF catalytic polymeric membranes filled with two types of copper oxides have been prepared characterised (Figure 21.12) and tested in a membrane reactor (flow-through configuration) for the liquid phase oxidation of benzene to phenol [30].

The catalytic tests were performed in the experimental plant schematised in Figure 21.13. The reaction has been carried out in an ultrafiltration membrane reactor to allow to control the contact time between the feed solution and the catalyst by changing the permeate flow rate operating at different transmembrane pressures checked by a manometer.

The batch reactor was thermostated at 35 °C.

Permeate flow rate through the membrane was affected by the addition of CuO particles and by morphological properties such as membrane thickness and pore size. The addition of the nanosised inorganic filler led to an increase in the PVDF membrane permeability. This trend was not observed when powder particles of copper oxide were used. The advantage to entrap the CuO nanoparticles in the polymer matrix is surely the control of contact time of substrate with the catalyst but also the improvement of some properties of membranes such as membrane permeability. This variation of membrane permeability allowed to change the contact times of the catalytic membrane from high to low values (from 19 to 4 s) by changing the transmembrane pressure thus controlling the successive oxidation of the phenol.

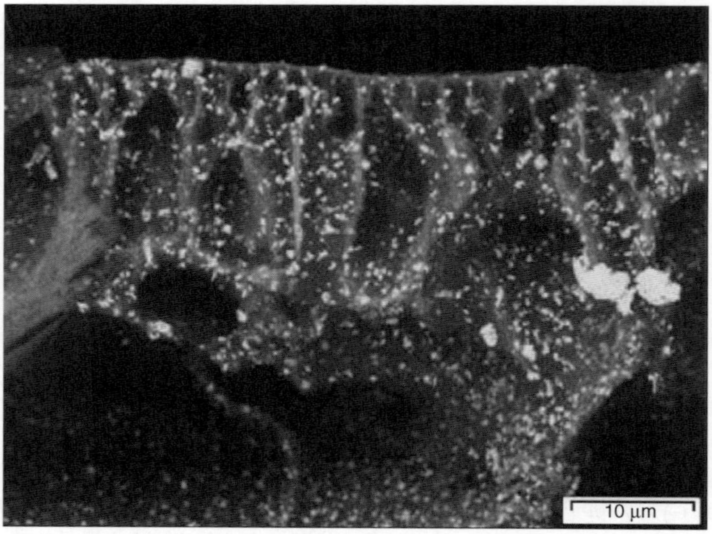

Figure 21.11 BSE analysis of the membrane cross section based on PES: a pre-prepared supported catalyst was added to the casting solution (adapted from [27])

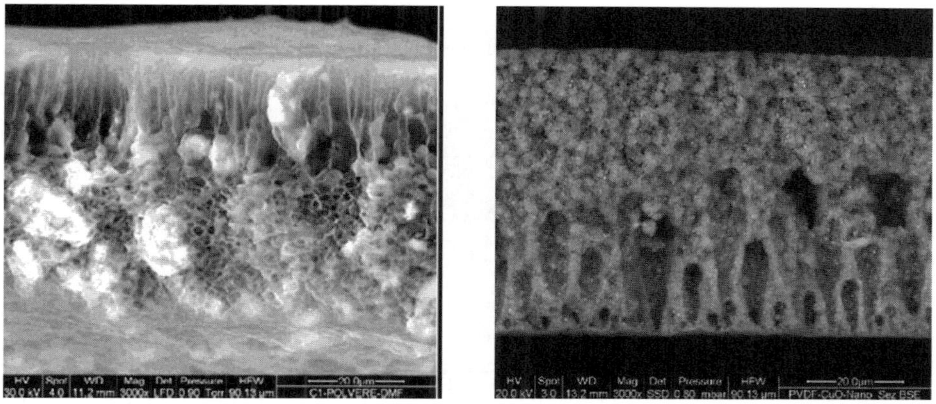

Figure 21.12 *SEM analysis of PVDF catalytic membranes used for the oxidation of benzene to phenol Reprinted from Applied Catalysis A: General, Molinari, R. and Poerio, T., Preparation, characterisation and testing of catalytic polymeric membranes in the oxidation of benzene to phenol. Vol. 358, 119. Copyright (2009) with permission from Elsevier*

Choi *et al.* [49] studied the biodegradation of phenol in wastewater by enzyme-loaded membrane reactor. For more examples of biocatalytic membrane reactors please refer to, for example, [50]. For the specific example reported here, that is, the biodegradation of phenol, an enzyme-loaded capillary membrane (hollow fibre configuration) has been used with polyphenol oxidase homogeneously immobilised in the polymeric asymmetric porous fibres. The feed solution containing water and phenol is fed to the shell side of the membrane module: depending on operating transmembrane pressure difference, a part of the feed flows through membrane pores and herein the degradation reaction of phenol occurs by means of the enzyme immobilised in the porous walls (Figure 21.14).

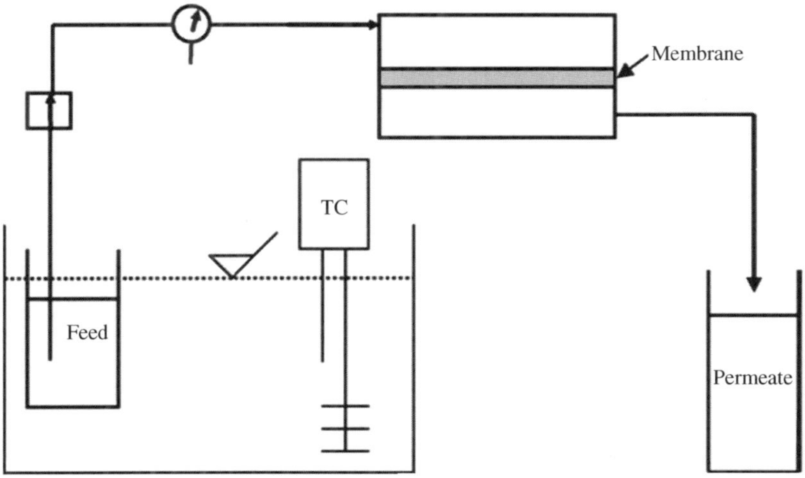

Figure 21.13 *Experimental set-up for the reactor used in the oxidation of benzene to phenol. Reprinted from Applied Catalysis A: General, Molinari, R. and Poerio, T., Preparation, characterisation and testing of catalytic polymeric membranes in the oxidation of benzene to phenol. Vol. 358, 119. Copyright (2009) with permission from Elsevier*

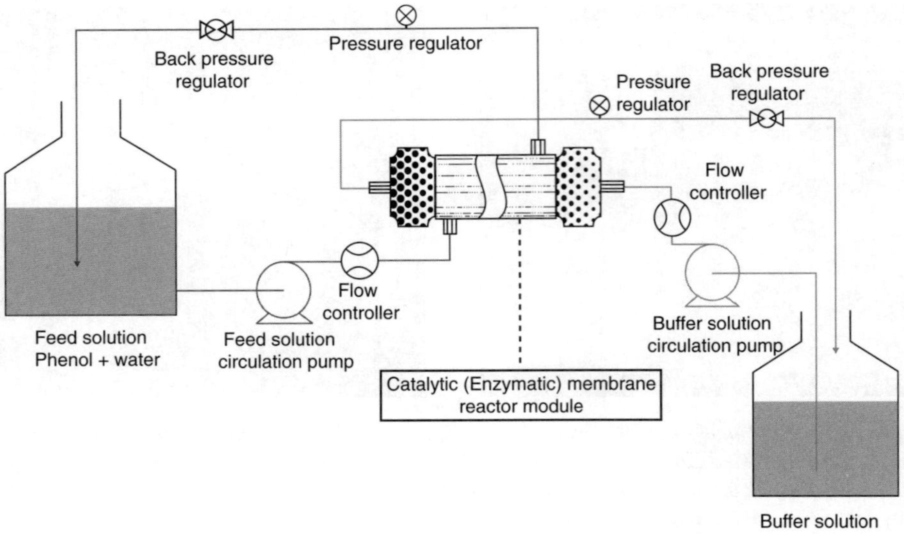

Figure 21.14 *Schematisation of the catalytic membrane reactor used for the phenol wastewater degradation described in [49]*

21.5 Conclusions

The advantage of polymeric membranes compared to inorganic membrane concerns their preparation. In fact, the control of thickness, porosity and pore size, the large scale preparation and the crack-free synthesis are some important characteristics of the membrane preparation by means of phase inversion. However, their application in membrane reactor is limited by the stability at high temperatures. For reactions carried out at temperatures below 200 °C, catalytic polymeric membranes with metallic complexes or metals as catalyst can be prepared easily with well controlled porosity, pore size, hydrophobicity and morphology to tune the transport of the reaction species and then the final conversion and selectivity. In general, the increasing number of works in this field, deals with polymeric catalytic membranes in flow-through mode: in this configuration the possibility to prepare in short time by means of phase inversion reproducible membranes with different permeability to study the effect of the contact time plays a fundamental role to understand the reaction system under investigation. The limitation for this versatile preparation method is that it can be used only with polymers soluble in organic solvents. For insoluble polymers other preparation techniques such as melt extrusion are available.

References

1. Drioli, E. and Fontananova, E. Membrane technology and sustainable growth, *Chemical Engineering Research and Design*, **82**(A12) 1557 (2000).
2. Vankelecom, I.F.J. and Jacobs, P.A. Dense organic catalytic membranes for fine chemical synthesis, *Catalysis Today*, **56**, 147 (2000).
3. Nunes, S.P. and Peinemann, K.V. *Membrane Technology*. John Wiley & Sons, Weinheim (2001).
4. Strathmann, H. Membrane separation processes. *Journal of Membrane Science*, **9**, 121 (1981).

5. Strathmann, H., Giorno, L. and Drioli, E. *An Introduction to Membrane Science and Technology*. Consiglio Nazionale delle Ricerche. Rome (2006).
6. Lonsdale, L.K. The growth of membrane technology. *Journal of Membrane Science*, **10**, 81 (1982).
7. Mulder, M. *Basic Principles of Membrane Technology*. Kluwer, Dordrecht (1984).
8. Parton, R.F., Vankelecom, I.F.J., Tas, D., Casselaman, M., Bezoukhanova, C.P., Uytterhoeven, J.B. and Jacobs, P.A. *Nature*, **370**, 541 (1994).
9. Dioos, B.M.L., Vankelecom, I.F.J. and Jacobs, P. *Adv. Synth. Catal.*, **348**, 1413 (2006).
10. Langhendries, G., Baron, G.V., Vankelecom, I.F. J, Parton, R.F., Jacobs, P.A. *Catalysis Today*, **56**, 131 (2000).
11. Parton, R.F., Vankelecom, I.F.J., Tas, D., Janssen, B.M., Knops-Gerrits, P.P., Jacobs, P.A. *Journal Molecular Catalysis A: Chemical*, **113**, 283 (1996).
12. Janssen, K.B.M., Laquire, I., Dehaen, W. *Tetrahedron: Asymmetry*, **8**, 3481 (1997).
13. Guedes, D.F.C., Leod, T.C.O., Gotardo, M.C.A.F., Schiavon, M.A., Yoshida, I.V.P., Ciuffi, K.J., Assis, M.D., *Applied Catalysis A: General*, **296**, 120 (2005).
14. Vital, J., Ramos, A.M., Silva, I.F., Castanheiro, J.E., *Catalysis Today*, **67**, 217 (2001).
15. Ke, J.H., Kumar, A.S., Sue, J.W., Venkatesan, S., Zen, J.M. *Journal of Molecular Catalysis*, **233**, 111 (2005).
16. Takeda, T., Kanno, R., Kawamoto, Y., Takeda, Y., Yamamoto, O., *J. Electrochem.*, **147**, 1730 (2000).
17. Bang, H.J., Lu, W.C., Fei, P., *J. Electrochem. Comm.*, **2**, 653 (2000).
18. Fritsch, D., Randjelovic, I., Keil, F., *Catal. Today*, **98**, 295 (2004).
19. Gallo, E., Buonomenna, M.G., Viganò, L., Ragaini, F., Caselli, A., Cenini, S. and Drioli, E. *Journal of Molecular Catalysis A: Chemical*, **282**, 85 (2008).
20. Buonomenna, M.G., Gallo, E., Ragaini, F., Caselli, A., Cenini, S., Drioli, E. *Applied Catalysis A: General*, **335**, 37 (2008).
21. Caselli, A., Buonomenna, M.G., de Baldironi, F., Laera, L., Fantauzzi, S. Ragaini, F., Gallo, E., Golemme, G., Cenini, S. and Drioli, E. *Journal of Molecular Catalysis A: General*, **317**, 72 (2010).
22. Carraro, M., Gardan, M., Scorrano, G., Orioli, E., Fontananova, E., Bonchio, M. *Chem. Comm.*, **2006**, 4533 (2006).
23. Guerriero, L., Casthanheiro, J.E., Fonseca, I.M., Martin-Aranda, R.M., Ramos, A.M. and Vital, J., *Catalysis Today*, **118**, 166 (2006).
24. Casthanheiro, J.E., Fonseca, I.M., Ramos, A.M., Oliveira, R., Vital, J., *Catalysis Today*, **104**, 296 (2005).
25. Casthanheiro, J.E., Ramos, A.M., Fonseca, I.M., Vital, J., *Catalysis Today*, **82**, 187 (2003).
26. Fritsch, D. and Bengston, G., *Catalysis Today*, **118**, 121 (2006).
27. Fritsch, D. and Bengston, G., *Advanced Engineering Materials*, **5**, 386 (2006).
28. Bengston, G., Panek, D., Fritsch, D., *Journal of Membrane Science*, **293**, 29 (2007).
29. Bottino, A., Capannelli, G., Comite, A., Di Felice, R., *Desalination*, **144**, 41 (2002).
30. Molinari, R. and Poerio, T., *Applied Catalysis A: General*, **358**, 119 (2009).
31. Molinari, R., Poerio, T. and Augurio, P., *Desalination*, **241**, 22 (2009).
32. Fontananova, E., Drioli, E., Donato, L., Bonchio, M., Carraio, M., Gardan, M., Scorrano, G., *Desalination*, **200**, 705 (2006).
33. Bonchio, M., Carraro, M., Scorrano, G., Fontananova, E., Drioli, E., *Advanced Synthesis and Catalysis*, **345**, 1119 (2003).
34. Buonomenna, M.G., Drioli, E., Bertoncello, R., Milanese, L., Prins, L.J., Scrimin, P. and Licini, G. *Journal of Catalysis*, **238**, 221 (2006).
35. Buonomenna, M.G. and Drioli, E., *Applied Catalysis B: Environmental*, **79**, 35 (2008).

36. Viganò, M., Ragaini, F., Buonomenna, M.G., Lariccia, R., Caselli, A., Gallo, E., Cenini, S., Jansen, J.C., Drioli, E., *Chem. Cat. Chem.*, **2**, 1150 (2010).
37. LuÈdtke, K., Peinemann, K.V., Kascheb, V., Behling, R.D., *Journal of Membrane Science*, **51**, 3 (1998).
38. Pinnau, I. and Koros, W.J. *Journal of Polymer Science, Polymer Physics*, **31**, 419 (1993).
39. Jansen, J.C., Buonomenna, M.G., Figoli A. and Drioli, E. *Journal of Membrane Science*, **272**, 188 (2006).
40. Kawakami, H., Mikawa, M. and Nagaoka, S. *Journal of Membrane Science*, **137**, 241 (1997).
41. Mahendran, R., Malaisamy, R., Mohan, D. *European Polymer Journal*, **40**, 623 (2004).
42. Buonomenna, M.G., Figoli, A., Jansen, J.C. and Drioli, E. *Journal of Applied Polymer Science*, **92**, 576 (2004).
43. Buonomenna, M.G., Macchi, P., Davoli, M. and Drioli, E., *European Polymer Journal*, **43**, 1557 (2007).
44. Bottino, A., Capannelli, G., Ponticelli, O., Piaggio, P. *Journal of Membrane Science*, **166**, 23 (2000).
45. Cheng, L.P., Shaw, H.Y. *Journal of Polymer Science Polymer Physics*, **38**, 747 (2000).
46. Young, T.H., Cheng, L.P., Lin, D.J., Fang L., Chuang W.Y. *Polymer*, **40**, 5315 (1999).
47. Westermann, T. and Melin, T. *Chemical Engineering and Processing*, **48**, 17 (2009).
48. Ilinich, O.M., Gribov, E.N., Simonov, P.A. *Catalysis Today*, **82**, 49 (2003).
49. Choi, S.H., Scura, F., Barbieri, G., Mazzei, R., Giorno, L., Drioli, E., Kim, J.H. *Membrane Journal*, **19**, 72 (2009).
50. Giorno, L., Drioli, E. *Trends Biotechnol.*, **18** (8), 339 (2000).

22

Porous Flat Sheet, Hollow Fibre and Capsule Membranes by Phase Separation of Polymer Solutions

Mathias Ulbricht[1] and Heru Susanto[1,2]

[1]Lehrstuhl für Technische Chemie II, Universität Duisburg-Essen, Essen, Germany
[2]Department of Chemical Engineering, Universitas Diponegoro, Semarang, Indonesia

22.1 Introduction

Membrane technologies have become indispensable for separation engineering in many different areas, and the development in terms of number and size of installed industrial processes is very rapid. With the impressive success in industry, an increasing number of special membrane applications are explored. This includes also membranes contactors and membrane reactors. In any case, the membrane itself is the heart of the process and decides upon its performance, efficiency and reliability.

Inorganic membranes are nowadays increasingly considered for membrane reactor processes. However, the diversity of membranes from synthetic polymers and the membrane area produced from polymers in industrial scale is still much larger. Hence, polymers play also a very important role as materials for membrane reactors. This is due to the facts that, (i) many different types of polymeric materials are commercially available, (ii) different types of selective barriers, i.e., porous, nonporous and charged, can directly be formed from polymers, (iii) production of large membrane area is possible at reasonable cost, and (iv) various shapes and formats including membranes for modules with very high area to volume ratio can easily be obtained [1,2]. Moreover, when a heterogeneous catalyst is incorporated in a polymer matrix or immobilised on its surface, a well chosen polymeric environment can regulate the selective sorption of reagents

Membranes for Membrane Reactors: Preparation, Optimization and Selection, Edited by Angelo Basile and Fausto Gallucci

and products with a beneficial effect on the catalyst's performance. This is another example for the superior flexibility of organic polymers with respect to 'tailored' barrier and matrix properties; in contrast, the variety of chemistries available with inorganic materials is much more limited. Also, the prices for inorganic membranes are still much higher than for typical polymeric membranes, and many inorganic materials are very brittle. Compared to inorganic membranes, polymeric membranes have of course lower resistance towards high temperatures, organic solvents and oxidative conditions. Nevertheless, there are still some polymer membranes which remain stable up to about 350 °C or more and at quite harsh conditions.

In general, polymeric membranes for reactive processes or as parts of membrane reactors can be classified, (i) as tool for facilitating reactive processes (as separation unit and as interface between two phases), and (ii) as compartment for a chemical conversion. Those membranes can be porous or nonporous. In this chapter, the existing methods and challenges in preparation and manufacturing of porous polymeric membranes via phase separation of polymer solutions are described. This is because most of the commercial polymeric membranes are prepared via phase separation methods. Another chapter will focus on alternative preparation methods for porous polymer membranes (see Chapter 23).

22.2 Porous Polymeric Membranes Classification

Most porous polymeric membranes had originally been developed for size-based separations of mixtures in liquid phase, either driven by a pressure difference (microfiltration/MF/; ultrafiltration/UF/; nanofiltration/NF/) or a concentration difference (dialysis) [3]. However, some of these membranes can also be applied as membrane contactor between two different liquid phases (one phase filling the pores of the membrane: membrane extraction; or a gas phase filling the pores of the membrane: membrane distillation), and between a liquid and a gas phase (the gas phase extending in the pores of the membrane: membrane absorption, membrane oxygenation/aeration) [4]. Moreover, macroporous membrane adsorbers, for fast flow-through separations, can also be considered a special type of membrane contactor [1,4]. Furthermore, porous polymeric membranes are also relevant as support for thin-film composite membranes with nonporous barrier or for pore-filling immobilised liquid membranes [1,4].

Porous polymeric membranes can be classified according to their geometry, their cross section structure and their pore size (Figure 22.1). Typical membrane geometries include flat sheet, capillary/hollow fibre and capsule. The simplest geometry is flat sheet, and the largest diversity of membranes from laboratory to industrial scale can be found in this format. Tubular membranes are also available for many industrial MF or UF processes; they have on the one hand a geometry which is analogous to capillary membranes, on the other hand they are more similar to flat sheet membranes because they are cast on a thick macroporous tube as support. Capillary and hollow fibre membranes are both self-supporting; and a rough classification can be made based on the capillary/fibre diameter (>/<0.5 mm). Capsules with perm-selective walls could be considered relatively crude mimics of living cells and are mainly used for the immobilisation of drugs, enzymes or cells [5]; different methods had been reported leading to objects with different size (diameters range from submicrometres to millimetres), made from various polymeric materials via different formation mechanisms. Irrespective of the interesting engineering concepts which are possible with membrane capsules, industrial applications are still rare, mainly because with all the other membrane formats it is much easier to *freely* change the compositions in the phases on *both* sides of the membrane. This is more complicated with the interior of a capsule.

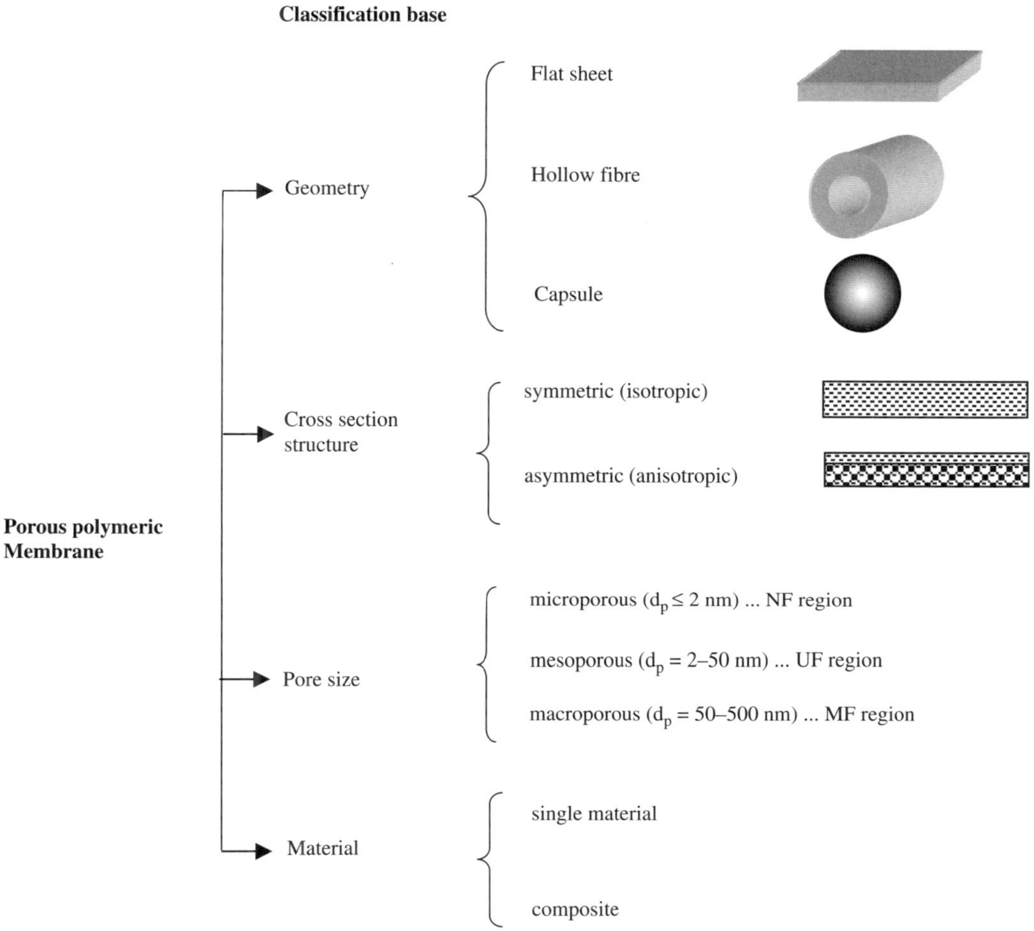

Figure 22.1 *Overview on classifications for porous polymeric membranes*

The cross section structure of porous polymer membranes can be isotropic or anisotropic (cf. Figure 22.1). Isotropic membrane cross sections can be found for macroporous MF membranes (also often used in membrane contactors); because of the relatively large barrier pore diameters, sufficient fluxes can be obtained with overall membrane thicknesses of 100–250 μm. An anisotropic membrane (also called 'asymmetric') has a thin porous selective barrier (thickness often less than 1 μm), supported mechanically by a much thicker porous substructure having larger pore size. In this way, the effective thickness of the selective barrier can be reduced, resulting in high permeate flux irrespective a small barrier pore size.

Based on its barrier pore size, a porous polymeric membrane should be classified according to the IUPAC nomenclature (cf. Figure 22.1). However, one should note that this classification is different from the nomenclature for pressure-driven membrane processes (MF: from 1.0 to 0.05 μm; UF: from 50 to 2 nm; NF: under 2 nm). The pore size distribution in porous polymeric membrane is typically rather broad, resulting in limited size selectivity. Typically, the specification of a commercial UF membrane is not the pore size, but the 'cut off' value, that is, the molar mass of solute for which more than 90% rejection is observed. Pore size (in particular for the largest pore) and wetting properties of the material are the crucial membrane selection criteria for membrane contactor applications.

Anisotropic porous polymer membranes can in principle be made from a single material, or from two or more materials (cf. Figure 22.1). An 'integrally asymmetric' membrane is over the cross section homogeneous with respect to its chemical composition, whereas a thin film composite membrane consists of different materials for the thin selective barrier layer and the support structure. Another alternative are so called mixed matrix membranes, typically based on a membrane polymer and containing another dispersed material (e.g., metal oxides, zeolites, other minerals, carbon etc.). Composite and mixed matrix membranes combine two (or more) materials having different characteristic(s) with the aim to achieve synergistic properties, either with respect to higher membrane separation performance (higher selectivity or/and flux; e.g. by using the pores of zeolites) or higher membrane stability (e.g. improved mechanical properties by inorganic fillers).

22.3 Polymers for Porous Membranes

22.3.1 General Considerations

Nowadays, polymeric membranes are dominated by synthetic polymers, even though some membranes from natural polymers can still be found. Important characteristics of polymers for preparing porous membranes are: (i) state of polymer, (ii) polymer structure, (iii) network structure, and (iv) hydrophilicity/hydrophobicity balance.

The state of polymers can be rubbery, glassy or (semi-)crystalline. This characteristic will largely influence the mechanical strength as well as the thermal and chemical stability [6]. In particular, via the mechanical strength as function of conditions, this is important for the preservation of shape and size of the pores, and, for capillary/hollow fibre membranes, for the integrity of the entire membrane. Polymers with crystalline structure will have relatively high mechanical strength as well as pronounced chemical resistance and thermal stability. However, many membrane polymers are amorphous. Polypropylene (PP), poly(vinylidene fluoride) (PVDF) and cellulose are prominent examples of membrane polymers that are semicrystalline, that is, form a matrix with crystalline domains. The melting temperature (T_m) and the glass transition temperature (T_g) are important properties related to the polymer state and structure. T_m (a transition temperature from crystalline to liquid state) is critical for (semi-)crystalline polymers, whereas T_g [a transition temperature from solid (glass) to super-cooled melt (rubber-like)] is important for amorphous polymers. Chemical crosslinking can, similar to crystalline domains (physical cross-linking) in an amorphous polymer, increase the mechanical strength, relative to the same polymer without crosslinking, even at temperatures above T_g. Moreover, in the development of polymeric membranes for applications in organic solvents, chemical crosslinking is a preferred strategy to ensure the integrity of the membranes and/or to reduce excessive swelling [1].

The polymer structure (homopolymer vs. block or graft copolymer; the latter contain two or more different repeating units within the same polymer chain) also influences the polymeric membrane performance. Block or graft copolymers are often used to obtain synergetic effects between properties of the different components resulting in high performance membranes. In addition, blending of the membrane polymer with another polymer or copolymer is also often done to obtain a better performance or stability. Both polymers should be miscible as indicated by one T_g value between those for the two (co)polymers. A heterogeneous (phase separated) polymer blend will be characterised by two (or more) T_g values for the individual phases.

Hydrophilicity or hydrophobicity is influenced by the functional groups of the membrane polymer. Hydrophilic polymers have a higher affinity to water than to other substances; therefore they are less prone to fouling in aqueous systems than hydrophobic polymers.

22.3.2 Key Characteristics

The barrier property for either educts or products (mainly influenced by the pore structure) is the most important characteristic for porous polymer membranes used for membrane reactors. However, the following characteristics should also be considered during preparation of porous polymeric membranes:

1. *Film forming property.* This property shows the ability of a polymer to be processed into a cohesive film. This is influenced by the macromolecular structure, especially molar mass and attractive interactions between chain segments. Poly(ether sulfones) (PES), polysulfones (PSf), polyamides (PA) or polyimides (PI) are examples for excellent film-forming materials [7].
2. *Mechanical properties.* These properties involve film strength, film flexibility and compaction stability. Mechanical strength is very important for preparing hollow fibre membranes because they are self-supporting. In order to improve their mechanical strength, flat sheet porous polymer membranes are often prepared on a nonwoven support material.
3. *Thermal stability.* Thermal stability is very important for preparing porous polymer membranes to be applied for membrane reactors (except for enzyme membrane reactors). To preserve the integrity of the barrier and membrane pore structure, the operation temperature should be lower than the glass transition temperature (T_g).
4. *Chemical stability.* This property includes the resistance of the polymer at extreme pH values, in presence of oxidation agents and the stability in special solvents (for processes with non-aqueous mixtures).
5. *Hydrophilicity and hydrophobicity.* The wettability of the material is indicated by its hydrophilicity. Hydrophilic materials such as polyacrylonitrile (PAN) are often chosen for preparing membranes as separation tools for aqueous systems, where the pores are utilised (MF, UF). Hydrophobic polymers such as PP or PVDF are chosen for preparing the membrane as contactor between two aqueous phases (in this case, the penetration of liquid into the membrane should be avoided).

The polymers that are frequently used for commercial porous polymer membranes are presented in Table 22.1. It is important to note that contact with common solvents will destroy the respective polymer membrane structure, but these solvents can be used for membrane formation via phase separation of polymer solutions (see Section 22.4). Furthermore, as discussed above, T_g will be the upper temperature limit for operation of polymer membranes. However, it should be also taken into account that T_g can significantly be reduced in presence of common or partially compatible solvents.

22.4 Polymeric Membrane Preparation Via Phase Separation

By far most of polymeric membranes, including MF and UF membranes as well as porous supports for other membrane processes, are produced via phase separation. The method is often also called 'phase inversion'. In principle, a polymer solution containing the membrane polymer

Table 22.1 *Polymers for preparation of porous membranes via phase separation and some of their characteristics*

Membrane polymer	Common solvent	T_g (°C)	pH resistance
Regenerated cellulose	Stable in most organic solvents[a]	High crystalline content	4–9
Cellulose acetate	Acetone, dioxane, dimethylacetamide (DMAc), dimethylformamide (DMF), dimethylsulfoxide (DMSO), tetrahydrofuran (THF)	Around 135[b]	3–7
Cellulose nitrate	Acetic acid, acetone, alcohols, cyclohexane	~50	4–8
Polyacrylonitrile	DMAc, DMF, nitric acid	100	2–10
Polycarbonates	Chloroform, dichloromethane, methylene chloride, ethylene chloride	145–150	4–9
Polyamides (e.g., Nylon-6)	m-Cresol, formic acid, methanol, phenolic or fluorinated solvents	50	4–7
Aromatic polyamide (e.g., Kevlar)	Sulfuric acid, polar aprotic solvent + ~5% $LiCl_2$	>400	4–7
Polyimide (e.g., Kapton, Matrimid, P84)	DMAc, DMF, dioxane, N-methylpyrrolidone (NMP)	300	2–10
Poly(etherimide)	NMP, DMAc, DMF	210–216	1–9
Polysulfone	DMAc, DMF, DMSO, NMP	198	2–13
Poly(ether sulfone)	DMAc, DMF, DMSO, NMP	225	2–13
Polyethylene terephthalate	DMSO (hot), halogenated aliphatic carboxylic acids, nitrobenzene, phenol–tetrachloroethane mixtures	70	2–9
Poly(vinylidene fluoride)	DMAc, DMF, NMP, DMSO	–40, (T_m ~175)	2–11
Polyethylene	Aromatic hydrocarbons (e.g., toluene or xylene), chlorinated solvents (e.g., trichloroethane), trichlorobenzene	–120	1–14
Polypropylene (atactic)	Benzene, chlorinated hydrocarbons, cyclohexane, diethyl ether toluene	–20	1–14
Polypropylene (isotactic)	Above 80 °C: 1,2,4 trichlorobenzene, halogenated hydrocarbons, di-n-amyl ether, vegetable oil	–23 to –14	1–14

[a]Typically prepared from cellulose acetate as precursor.
[b]Depending on degree of acetylation.

(one phase) is transformed by a precipitation/solidification process into two separate phases (a polymer-rich solid and a polymer-lean liquid phase). Before the solidification, a transition of the homogeneous liquid into two liquids (liquid–liquid demixing) occurs. The 'proto-membrane' is made from the solution of the membrane polymer by casting a film on a suited substrate or by spinning through a spinneret together with a bore fluid. The first stage to obtain membrane capsules is forming either a drop of a polymer solution or a liquid drop with a shell from a polymer solution. Based on the way the polymer solution is solidified, phase separation methods can be distinguished into four techniques, namely:

1. *Thermally induced phase separation (TIPS).* A low molecular weight liquid that acts as solvent at high temperature and as nonsolvent at low temperature is the key requirement for this technique. A system of polymer and solvent is used which has an upper critical solution temperature. The solution is cast or spun at high temperature, and cooling leads to demixing/precipitation. The liquid is then removed from the pores after formation of the membrane.
2. *Nonsolvent-induced phase separation (NIPS).* This technique is also called immersion precipitation or wet method. A polymer solution is cast or extruded and is subsequently immersed in a coagulation bath containing a nonsolvent for the polymer (typically water). Demixing and precipitation occur due to the exchange of solvent (from polymer solution) and nonsolvent (from coagulation bath), that is, the solvent and the nonsolvent must be miscible.
3. *Vapor-induced phase separation (VIPS).* This technique is based on the fact that uptake of nonsolvent from vapour can also cause phase separation of a polymer solution. The polymer solution is exposed to an atmosphere containing a nonsolvent (typically water); absorption of nonsolvent causes demixing/precipitation.
4. *Evaporation-induced phase separation (EIPS).* In this technique, phase separation is achieved by solvent evaporation. A polymer solution is made in a volatile solvent or in a mixture of a volatile solvent and a less volatile nonsolvent, and solvent is allowed to evaporate, leading to precipitation or demixing/precipitation.

The TIPS and the NIPS process can be used to prepare all membrane formats, including the *a priori* self-supporting hollow fibres and capillaries because of the compatibility with very short residence times between formation of the liquid proto-membrane and the onset of phase separation. The 'slower' EIPS and VIPS processes are much less frequently used; but they have also relevance in combination with the NIPS process (see below).

22.4.1 TIPS Process

The TIPS process is used to prepare membranes with a relatively isotropic macroporous structure for MF, for gas–liquid or liquid–liquid contactor, or as support for liquid membranes. In particular, for polymers which can not be processed by NIPS because no suited combination of solvent and nonsolvent is available, this method is a very valuable alternative. Most important polymers processed by this method are polyolefines, in particular PP (solvents in technical manufacturing processes are mineral or vegetable oils). The actual mechanism and resulting structures can be more complicated because crystallisation can also occur simultaneously with phase separation and solidification [8]. The range of pore sizes which can be achieved is from 0.05 μm to several micrometres. Figure 22.2 shows an example for a commercial flat-sheet membrane made via TIPS from PP.

22.4.2 NIPS Process

Most of commercial porous polymeric membranes are prepared by the NIPS process and have a more or less pronounced anisotropic cross section structure. The integrally asymmetric porous membrane structure was firstly introduced by Loeb and Sourirajan [9]. This had been a tremendously important achievement to combine the desired high selectivity of a reverse osmosis or UF separation with a high flux. The top layer acts as thin selective barrier and a porous sublayer provides high mechanical strength. Preparation of porous polymer membranes via the NIPS technique basically involves four steps:

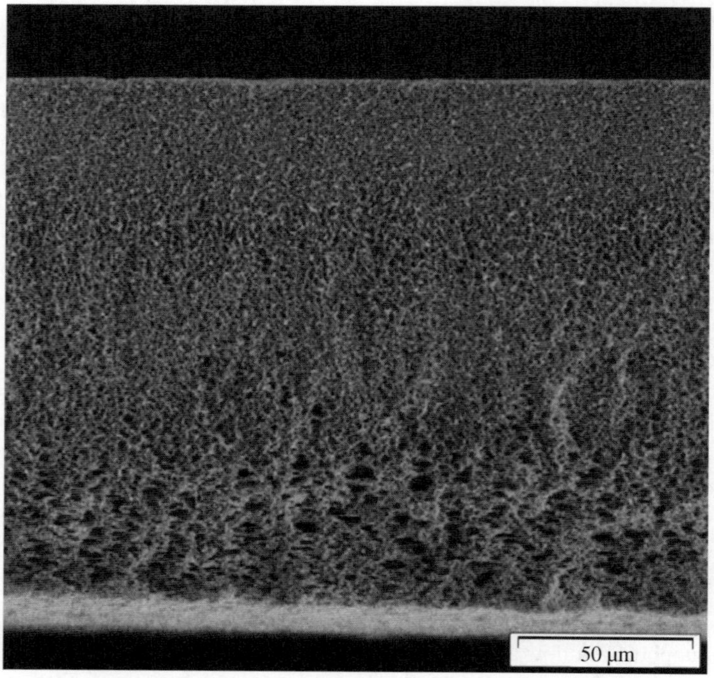

Figure 22.2 *Relatively isotropic macroporous membrane for microfiltration made from polypropylene (type 2EHF, nominal pore diameter 0.2 μm; Membrana GmbH, Wuppertal, Germany)*

1. Polymer dissolution in a single or mixed solvent,
2. Casting the polymer solution as film (proto-membrane) on suited substrate for flat sheet, spinning (extruding) it as free liquid film for hollow fibre/capillary or extruding it as drop for membrane capsules,
3. Precipitation by immersion in nonsolvent coagulation bath,
4. Post-treatments such as rinsing, annealing and drying.

Often the conditions and residence time in the proto-membrane state (2) have a major influence on the final membrane properties because either solvent evaporation or nonsolvent uptake from vapour phase can happen. The membranes resulting from this process have typically a very thin (<1 μm, often even less than 100 nm) top skin layer (selective barrier), which is either nonporous or porous.

The mechanism of phase separation can be described based on ternary phase diagrams with the three main components, namely polymer, solvent and nonsolvent. Besides thermodynamics aspects, the onset and rate of precipitation in the liquid film (both are different depending on the distance to the plane of first contact with the coagulation bath) also determine the resulting pore structure; the mass transfer rate (nonsolvent in-flow and solvent out-flow) can have tremendous influence. In general, two mechanisms are distinguished: (i) instantaneous liquid–liquid demixing, which will result in a porous membrane, (ii) delayed onset of liquid–liquid demixing, which can result in a membrane with less porous or dense barrier [10,11]. However, the rate of precipitation decreases from the top surface (in most cases, this plane of first contact with the coagulation bath will be the barrier in the final membrane) to the bottom surface of the cast film. As precipitation slows down, the resulting pore sizes increase because the two phases have more time to separate.

The formation of 'macrovoids' is related to instabilities within the liquid film causing a rapid convective influx of nonsolvent into the proto-membrane [12,13]. This is often an undesired phenomenon, because the resulting membranes are less stable or are too inhomogeneous (e.g., when considered as compartment for an adsorptive separation or a chemical conversion). More information on phenomena during phase separation and underlying mechanisms can be found in overviews by Mulder [10] or van de Witte *et al.* [11].

In practical applications, most systems for membrane preparation contain more than three components (e.g., polymer blends as membrane materials and solvent mixtures for casting solution and coagulation bath). Consequently, the mechanisms can be very complex. The important variables for control of the membrane characteristics are:

1. *Characteristics of the casting solution.* The casting solution characteristics involve: solvent and polymer concentration. Most important is the selection of a suitable solvent for the polymer, i.e., the strength of mutual interactions is inversely proportional to the ease of precipitation by the nonsolvent. The use of a solvent having higher Hildebrand–Hansen solubility parameter (cf. [14]) will result in a more porous structure. Polymer concentration significantly determines the membrane porosity. Increasing polymer concentration leads to higher fraction of polymer and consequently decreases average membrane porosity and pore size. In addition, it can also increase the thickness of the skin layer. However, increasing polymer concentration could also suppress macrovoid formation and enhance the tendency to form a more stable and homogeneous sponge-like structure. In general, macroporous membranes (for MF) can be obtained within a range of polymer concentrations of 9–12 wt% whereas mesoporous membranes (for UF) are obtained within the range 12–20 wt%.
2. *Solvent/nonsolvent system.* The solvent used during the NIPS process must be miscible with the nonsolvent. An aprotic polar solvent having high Hildebrand–Hansen solubility parameter [14] such as NMP, DMF, DMAc or DMSO should be used for achieving rapid precipitation (instantaneous demixing) upon immersion in the nonsolvent water. Consequently, a high porosity anisotropic membrane can be achieved. For slow precipitation, yielding low porosity or nonporous membrane, solvents having a relatively low solubility parameter, like THF or acetone, should be preferable.
3. *Characteristics of coagulation bath.* The coagulating medium for precipitating the proto-membrane also determines significantly the resulting pore structure. The presence of a fraction of solvent in the coagulation bath can slow down liquid–liquid demixing rate resulting in a less porous barrier structure. However, the opposite effect can also occur because addition of solvent can decrease polymer concentration (in the skin layer of the proto-membrane) leading to a more open porous barrier structure. The amount of the solvent to be added strongly depends on the solvent/nonsolvent interactions. As mutual affinity of solvent and nonsolvent increases, more solvent is required to achieve an effect onto membrane structure. Further, addition of solvent into coagulation bath can reduce the formation of macrovoids and yield sponge-like structure. Nevertheless, this approach is typically too costly because the amount of nonsolvent in the coagulation bath is usually in a very large excess relative to the amount of polymer solution.
4. *Additives to the casting solution.* To increase the membrane performance, additive or modifier is usually added in the casting solution. The additives could be (a) co-solvent to increase flux or rejection, (b) pore forming and hydrophilic agents such as poly(vinyl pyrrolidone) (PVP) or poly(ethylene glycol) (PEG), (c) nonsolvent to promote more porous structure and reduce the macrovoid formation, and (d) crosslinking agents to increase chemical stability as well as to reduce macrovoid formation. Figure 22.3 shows the effect

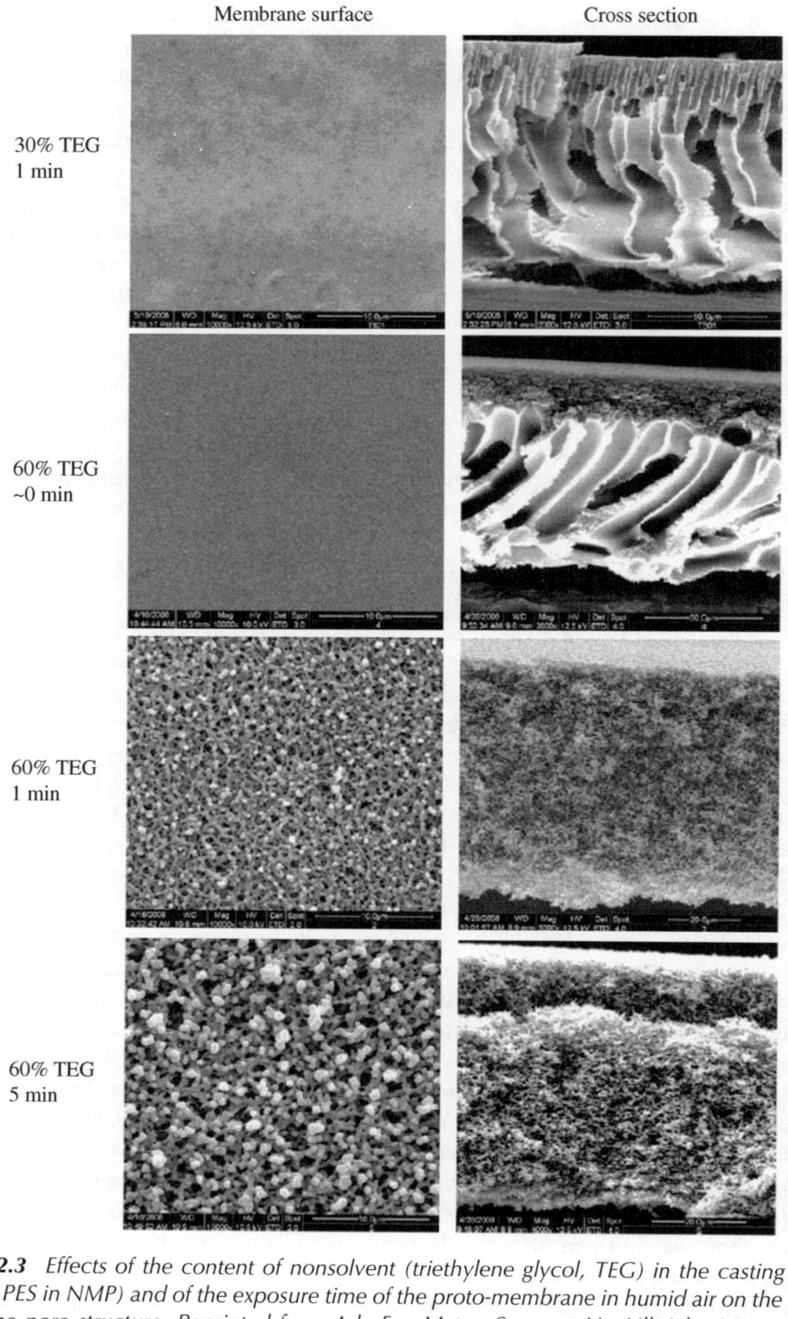

Figure 22.3 *Effects of the content of nonsolvent (triethylene glycol, TEG) in the casting solution (10 wt% PES in NMP) and of the exposure time of the proto-membrane in humid air on the resulting membrane pore structure. Reprinted from Adv Eng Mater, Susanto, H.; Ulbricht, M., et al., High performance polyethersulfone microfiltration membranes having high flux and stable hydrophilic property, 8, 328. Copyright (2006) with permission from Elsevier*

of the addition of a nonsolvent (triethylene glycol, TEG) on the PES membrane structure; in particular the reduction of macrovoids with increasing TEG content is very obvious [15]. Also, amphiphilic block copolymers can have a significantly higher efficiency as surface modifier as compared to homopolymers. This has been demonstrated for a commercial block copolymer from the Pluronic® series (PEG-*b*-PPG-*b*-PEG) versus PEG and PVP with similar molar masses as additives to prepare hydrophilised PES UF membranes [16].

5. *Exposure time of proto-membrane before precipitation.* Exposure time before coagulation is another parameter that can be adjusted to 'fine tune' the membrane pore structure. However, the effect of exposing to atmosphere before immersion will depend on the solvent properties (e.g., volatility, affinity to the nonsolvent water) and atmosphere (e.g., temperature, humidity). This step has significant effects on the characteristics of the skin layer and the degree of anisotropy of the resulting membrane. The resulting process can then be classified as combination of VIPS or EIPS followed by NIPS [17]. Indeed, the membrane pore structure can be 'tuned' by adjusting the exposure time before coagulation (cf. Figure 22.3).

The NIPS process can also be used to prepare membrane capsules from polymer solutions. A straightforward method is forming drops which are then precipitated in a coagulation bath [18]. A more advanced method where the size of the drops and resulting capsules as well as the membrane structure may be better adjusted and controlled is based on membrane emulsification, i.e. the extrusion of droplets through the macropores of a membrane into a liquid which is completely immiscible with the liquid drop; subsequent coagulation takes place upon transition of the drops into a second liquid (a nonsolvent for the polymer but miscible with the polymer solvent). Figure 22.4 shows a capsule obtained by such procedure [19]. The similarity with porous morphologies, including macrovoids, which are typical for flat sheet and capillary/hollow fibre polymeric membranes obtained from NIPS process is obvious.

Significant efforts are made in order to improve the stability of membranes from organic polymers in organic solvents, also because this would enable many more applications of membrane reactor concepts. The focus with porous membranes is on UF and NF types. Two main strategies can be distinguished, that is, the addition of nanoparticles in the casting solution to obtain mixed matrix membranes via phase separation, or the crosslinking of the membrane after preparation. Improvements in mechanical and chemical stability of UF membranes, for example, from polysulfone, have been achieved with ZrO_2 particles [20,21]. Analogous effects have been achieved for UF membranes from poly(vinylidene fluoride) and polyamideimide by incorporation of ZrO_2 particles in the casting solution before phase separation [22]. Recently, properties of solvent-resistant NF membranes from the polyimide P84 have also been modified by addition of TiO_2 nanoparticles [22]. A very promising post-crosslinking strategy is based on a copolymer of polyacrylonitrile (PAN) with a relatively small content of glycidyl methacrylate so that the membrane formation via phase separation is still controlled by the properties of the PAN [23]. Via the reaction with ammonia as bi- or trifunctional crosslinking agent, the membrane pore structure could be stabilised in a three dimensional chemical network. Pore structure, flux and rejection properties were only slightly changed, but the chemical stability was so much increased that the crosslinked polymeric membranes could be even used for UF separations of strongly acidic and alkaline aqueous solutions as well as with most organic solvents. For the preparation of solvent-resistant NF membranes, phase separation of casting solutions of polyimide and subsequent chemical crosslinking of the membranes, for instance with diamines, is now also a common method [24].

In the past decade, block copolymers with controlled structure (two chemically incompatible blocks, both blocks with monodisperse chain lengths) have drawn much interest because such

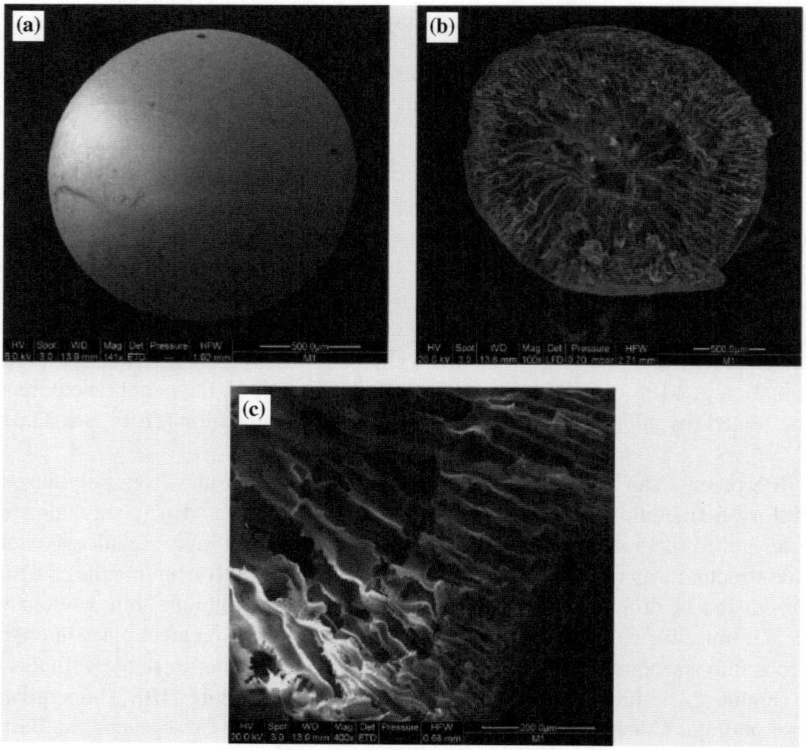

Figure 22.4 Polyetheretherketone (PEEKWC) capsules obtained by emulsification through a membrane with a pore diameter of 550 μm using a solution of PEEKWC in DMF and dodecane as first liquid phase and a water isopropanol as second liquid phase. Reprinted from Figoli, A.; De Luca, G.; Longavita, E.; Drioli, E. PEEKWC Capsules Prepared by Phase Inversion Technique: A Morphological and Dimensional Study, Separ Sci Techn, 42, 2809. Copyright (2007) with permission from Taylor and Francis, http://www.informaworld.com.

materials can form self-assembled structures with characteristic dimensions in the range 2–50 nm. This shows also much promise to yield ultimately membrane barrier structures with controlled porosity [25]. Recently, it had been demonstrated that such 'tailored' block copolymers as membrane material in combination with an EIPS/NIPS process can indeed lead to asymmetric membranes with barrier layers controlled by microphase separated block copolymer morphologies: either an asymmetric membrane with a top layer having very regular pores at high porosity [26], or an UF membrane with barrier pores which are switchable by changes of pH and/or temperature [27].

22.5 Industrial Manufacturing of Porous Polymeric Membranes

The fundamentals and the key steps for manufacturing of polymeric flat sheet and hollow fibre/capillary membranes are the same (cf. Section 22.4), and both processes have been implemented in large industrial scale. Still, the variety of commercial flat-sheet membranes is much wider. This is also because the complexity of the manufacturing processes is typically higher for hollow fibre/capillary membranes. In the following, both processes will be briefly covered, with more emphasis on membrane manufacturing by fibre or capillary spinning.

22.5.1 Flat Sheet Membranes

Flat sheet membranes are very often used in industrial applications, either in plate and frame or in spirally wound modules. The scale of the application dictates the module size, and continuous membrane manufacturing is nowadays done with width of up to 1.5 m and at speeds of up to 10 m min^{-1}. Scale and diversity of production processes are much larger for NIPS than for TIPS. In many cases, the membranes are mechanically supported by casting and phase separation on a nonwoven support (Figure 22.5).

In order to optimise membrane performance, variations of the phase separation conditions are also performed in industrial membrane manufacturing. For instance, advanced MF membranes can also have anisotropic cross section morphologies, for example, an 'hour glass' morphology which can be achieved by choosing strongly anisotropic boundary conditions for the proto-membrane before coagulation (cf. [17]). Moreover, other important innovations had also been introduced in the industrial manufacturing of porous polymer membranes. Examples are the simultaneous casting of two layers of polymer solutions with different composition to yield MF or UF membranes with two layers of different porosity (e.g., from PVDF), and the casting of thin films of polymer solutions on a porous support to achieve UF composite membranes (e.g., with barrier layer from macrovoid-free PES or regenerated cellulose). An overview on this development can be found in the review by Allegrezza *et al.* [29].

As a consequence, structure and separation performance of commercial MF membranes with the same specification can be largely different; this has in detail been studied, for instance, for five different PES MF membranes, all with a nominal pore size of 0.2 μm [30].

22.5.2 Hollow Fibre/Capillary Membranes

The very high packing density offered by hollow fibre/capillary membrane modules leads to numerous applications in various fields. This is of particular interest for membrane contactors

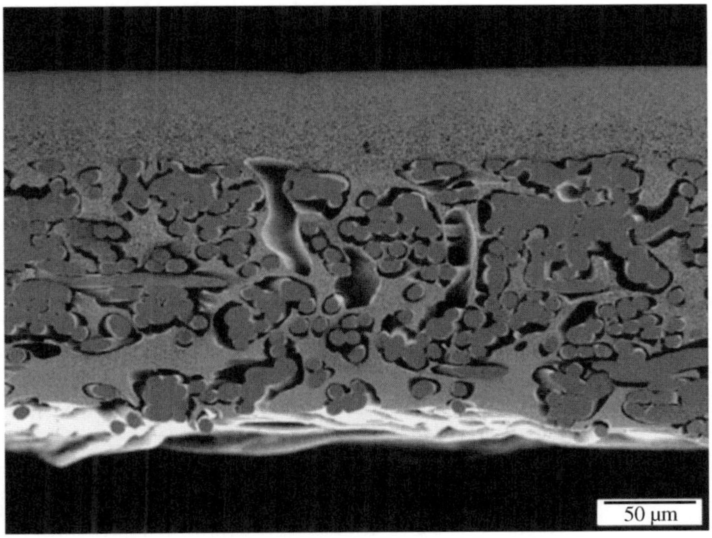

Figure 22.5 *SEM micrograph of a microtome cross section of a porous polymer membrane with an anisotropic structure on a nonwoven as mechanical support. Reprinted from Ind Eng Chem Res, Akamatsu, K.; Yamaguchi, T. Novel Preparation Method for Obtaining pH-Responsive Core-Shell Microcapsule Reactors 46, 124. Copyright (2007) with permission from American Chemical Society*

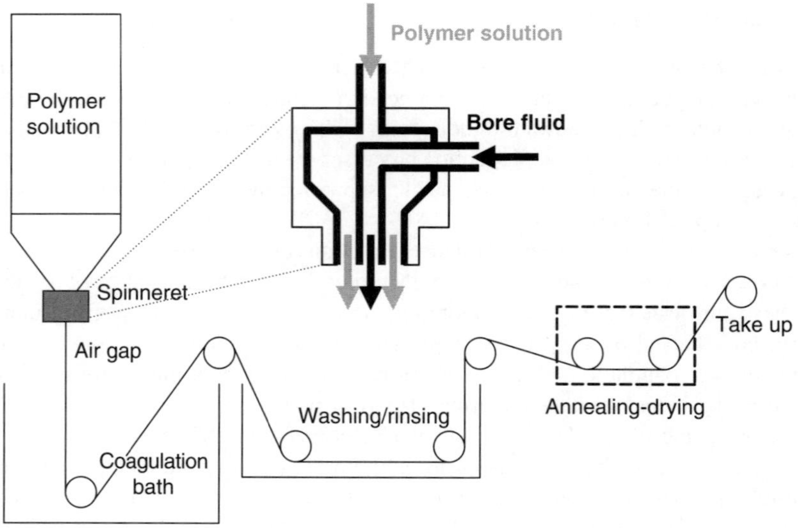

Figure 22.6 Schematic overview on preparation of hollow fibre or capillary membranes via spinning/ NIPS process

and reactors. Figure 22.6 shows the schematic overview of hollow fibre membrane preparation by using the NIPS process (only a few capillary membranes are prepared via the TIPS process; in this case the phase separation is induced by cooling in the air gap, and the first bath is used for cooling purpose). In reality, many fibres are spun in parallel, and the length of the manufacturing line is at least 10 m, often much longer. The technique for preparation of flat sheet can be adapted to prepare hollow fibre membranes in a spinning system. However, phase separation can start at two sides of the liquid film (proto-membrane), that is, the bore or lumen side and the shell or outside side surface (during the making of flat sheet membranes, phase separation generally starts from only one side). Polymer solution is extruded through the outer section of the spinneret, whereas air or a bore liquid is forced through the core of the spinneret yielding the hollow fibre shape. After a very short residence time in the air gap, the proto-membrane undergoes phase separation by immersion into the coagulation bath containing non solvent. Thereafter the fibre is treated (e.g., washing, rinsing, annealing, drying) before taking up as product. The process parameters, which govern the membrane characteristics, are the dimension of spinneret, the extrusion rate of the polymer solution, the injection rate of air/bore liquid and the residence time in the air gap. In addition, the drawing ratio, that is, the ratio between extrusion rate and drawing rate of the already coagulated fibre, can also be used to manipulate the membrane properties. A higher drawing ratio will cause an elongation of the fibre and can also cause an orientation of polymer chains. This can yield changed barrier pore structure and/or improved mechanical stability of the resulting membranes. Before use for practical applications, the fibres are packed into bundles and potted into a tube as module. More details on preparation of fibre membranes can be found in overview articles [31,32].

The porous structure of the resulting membranes will of course also depend on the influences discussed in Section 22.4 (see, e.g., [12]). Figure 22.7 shows a representative example of a commercial UF capillary membrane. The anisotropic pore structure can clearly be seen, but the largest pores are in the middle of the cross section, indicating that phase separation had started from both the outside and the inside surface (cf. above).

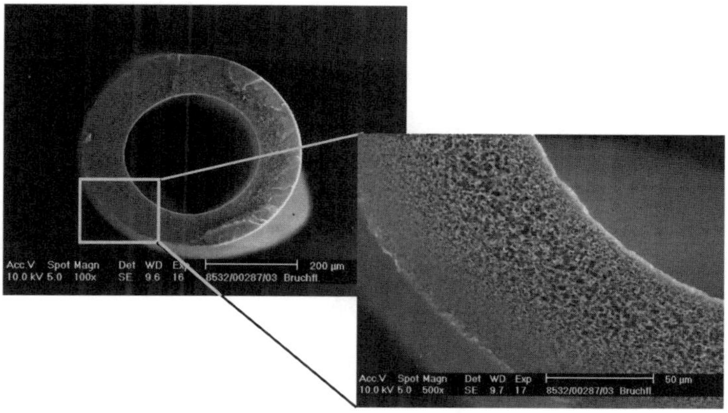

Figure 22.7 *Cross section of a capillary UF membrane made from PES (MicroPES TF10, Membrana GmbH, Wuppertal, Germany). Copyright (2010) with permission from Membrana GmbH*

An interesting extension of the fibre spinning concept is the manufacturing of multibore capillary membranes (Figure 22.8). In this case, the barrier layer of the UF membrane is on the inner surface of each tube in the multibore capillary, and the much larger pore size and porosity of the walls leads to a drastically lower flow resistance of the matrix so that all tubes contribute in a similar way to the UF performance. The advantage of that concept is that a high membrane area to module volume ratio can be achieved with very robust capillaries.

Table 22.2 gives exemplaric information on industrial membranes prepared by phase separation of polymer solutions and respective manufacturers as well as typical processes where such membranes are already widely applied.

22.6 Applications in Membrane Reactor Processes

Polymeric membranes can be used in membrane reactor processes in two different ways:

1. The combination of a reactor for a catalytic reaction with a separate membrane can enable a continuous process also for homogenous catalysis.

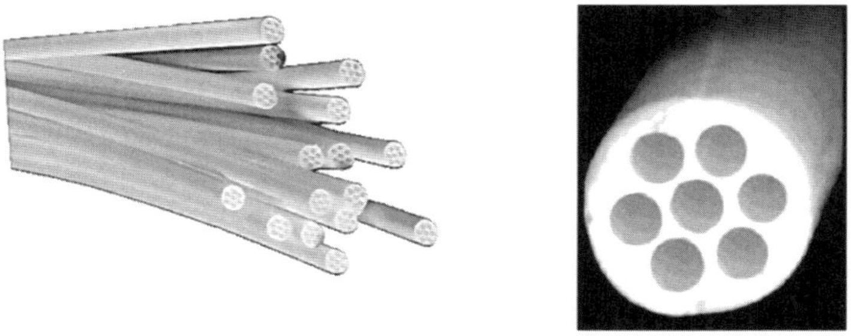

Figure 22.8 *Cross section of a multi-bore capillary UF membrane made from PES and bundle of capillaries (Multibore®, capillary diameter 0.9 mm; Inge AG, Greifenberg, Germany). Reproduced from www.inge.ag, with permission from Dr. Peter Berg, Germany*

Table 22.2 *Overview on typical industrial porous polymer membranes for selected important applications*

Membrane process	Membrane polymer	Membrane shape	Typical process mode, driving force	Application	Typical membrane manufacturers
Microfiltration	PES, PSf, PA, PVDF, PP, PE, cellulose derivatives	Flat sheet	Pressure-driven, up to 3 bar	Clarification, concentration, sterile filtration	Asahi, GVS, Koch, Membrana, Millipore, Pall, Sartorius
	PES, PVDF	Tubular	*As above*	*As above*	Koch, Siemens
	PES, PVDF, PP, PE	Capillary	*As above*	*As above*	Membrana, X-flow (Norit)
Micro-/ ultrafiltration	PES, PSf	Capillary/ hollow fibre	Out–in filtration by vacuum	Membrane bioreactor for waste water	X-Flow (Norit), Koch
Ultrafiltration	PES, PSf, PVDF, PAN, cellulose derivatives	Flat sheet	Pressure-driven, up to 6 bar	Clarification, concentration, fractionation, sterile filtration	AlvaLafal, Koch, Membrana, Millipore, Pall, Sartorius
	PES, PVDF	Tubular	*As above*	*As above*	Koch
	PES	Capillary	Pressure-driven in–out, up to 6 bar	Surface water filtration, filtration prior desalination	Inge, X-flow (Norit), Zenon (GE)
Dialysis	PSf, PSf-PA copolymer, PAN, regenerated cellulose	Hollow fibre	Concentration difference	haemodialysis, fractionation (e. g., ethanol removal from wine or beer)	Asahi, Gambro, Fresenius, Membrana
Gas/liquid contactor	PP	Hollow fibre, capillary	Concentration difference	Blood oxygenation, dehumidification of air	Membrana
Membrane distillation	PP, other polyolefines	Flat sheet, hollow fibre	Temperature difference	Water desalination	Membrana
Membrane adsorber	Regenerated cellulose, PES, PVDF; all with functional layer	Flat sheet, used as stack	convective flow by trans-membrane pressure	Removal of trace impurities (*polishing*), *capturing* of biomolecules or particles	Millipore, Pall, Sartorius

2. The catalyst can be immobilised on/in the membrane, and this may even lead to a synergy between the selectivities and efficiencies of membrane separation and catalysis.

Many specific examples exist for case (1), but with polymeric membranes almost all of those are for aqueous systems. By far most developed is the so-called membrane bioreactor (MBR) for waste water treatment, where polymeric MF of UF membranes are used to retrieve purified water from the reactor and at the same time to increase the efficiency of the microbiological processes [33]. (Bio)catalytic chemical reactions in combination with separation membranes have also been reviewed elsewhere [34]. Some specific examples for the combination of homogenoues catalysis in organic phase with NF can be found in another review which is focussed on solvent-resistant polymeric membranes [35].

Conceptually very interesting and more challenging is the design of catalytic membranes with true synergies between membrane and catalyst, that is, materials and processes where the catalytic membranes are more than just a solid support for catalyst immobilisation. Cases where the solid catalyst is incorporated in or combined with a polymer matrix and the selective sorption or permeation properties of the nonporous polymer have influence on the catalytic process (e.g., the catalytic synthesis/purification of isobutene from C_4 raffinates [36]), are out of the scope of this chapter. Catalytic membranes with a porous barrier can be used as contactor, either between two different phases or in flow-through mode in order to reduce mass transfer resistances [4]. In addition, the selectivity of the porous barrier could be used as well. The preparation of the catalytic membrane can be done either in one step, that is, with the catalyst in the polymer solution for phase separation, or by immobilisation of the catalyst on an already prepared membrane. The first approach is straighforward when the catalyst shall be embedded in a nonporous polymer, but it is not very well feasible for the preparation of porous membranes because both the pore structure and the distribution of the catalyst must be tuned simultaneously. Here, some examples for the second approach will be discussed, and the focus will be on cases where the catalyst is directly immobilised on/in membranes obtained from phase separation. Other examples where the catalyst is immobilised on/in porous membranes after a tailored surface functionalisation of an already prepared membrane is covered in Chapter 23.

All established methods for enzyme immobilisation can be used with porous membranes as support. Infiltration of the porous structure, even in combination with size exclusion by the selective barrier layer, can additionally be used to guide enzyme distribution. Ulbricht and Papra had performed a systematic study with amyloglucosidase about the influences of different immobilisation methods – adsorption, crosslinking and covalent coupling – in polyacrylonitrile membranes with the same pore structure onto enzyme activity and stability; overall best results were achieved by crosslinking of infiltered enzyme using glutaraldehyde [37]. Covalent immobilisation of different enzymes in membranes from various polyacrylonitrile copolymers has also been investigated [38]. Another analogous example was the covalent immobilisation of glucose oxidase in porous membranes prepared from poly(vinylidene fluoride) with grafted poly (acrylic acid) side chains [39]. With lipase, an enzyme known to be active at water/oil interfaces, a hydrophobic (polypropylene) matrix was shown to be a much better carrier for conversions performed in an organic solvent than a hydrophilic (polyamide) membrane [40]. A very comprehensive description of an industrial process implemented in the 1990s has been provided by Lopez and Matson [41]. The enzyme-mediated resolution of a racemic mixture yielding a chiral intermediate for a drug has been implemented in a biphasic hollow fibre membrane reactor where the enzyme lipase had been immobilised in the support pores of an UF membrane. This approach improved the productivity and practicability of the biotransformation by enhancing the enzyme/substrate contact and by efficiently removing an inhibiting reaction product. Ultimately, the process was scaled up to produce 75 t year^{-1} of intermediate.

Much less examples can be found for reactions with polymeric membranes in combination with synthetic catalysts for conversions in organic phase. Fritsch and Bengtson had developed new catalytically active porous membranes, made of either polyethersulfone or polyamideimide, without or with Al_2O_3 filler, for the hydrogenation of edible oils in the flow-through mode [42]. The catalytic activity in the polymer membrane was achieved by two different methods: (i) by wet impregnation of the membrane in a catalyst precursor solution (palladium acetate or hexachloroplatinate), or (ii) by addition of ready-made supported catalysts to the membrane casting solutions. The latter, simultaneous method resulted in membranes with low catalyst content and a corresponding low activity in the hydrogenation reactions. This development was motivated by efforts to reduce the formation of trans isomers during hydrogenation of edible oil which occurs in many conventional reactors. Further, it could be demonstrated that such three-phase chemical reactions can be accomplished very efficiently by using membrane reactor technology. Subsequent work has been conducted for improving the catalytic membrane performance also at higher temperature [43].

Porous polymer membrane flow-through reactors have a great opportunity to replace conventional fixed bed reactors because the mass transfer limitations within particles and in the inter-particle space can be reduced. Both effects can decrease the selectivity for a desired product. The group of Schomäcker [44] had developed a porous catalytic membrane made from crosslinked polyacrylic acid, activated by immobilised Pd particles as catalyst. The reactor was designed as a loop of a saturation vessel and a membrane module for a partial hydrogenation reaction. Because the catalyst was immobilised in a porous membrane, it was in close contact with the substrate flowing through the membrane pores. The results indicated that with a defined pore size, a defined amount of catalyst and convective mass flow through the porous membrane, consecutive reactions can be better controlled than with the same catalyst in a fixed bed reactor. This was because of minimised backmixing and improved adjustment of residence time: A better selectivity for an intermediate product had been achieved.

22.7 Conclusions and Outlook

Separation and integrated processes based on porous polymeric membranes are very well established in many application fields, including very large industrial scale. A wide range of porous membranes have been developed and is industrially produced based on phase separation of polymer solutions. A particular advantage is the option to 'fine tune' pore structure and surface and bulk properties of the membranes by selection of the polymeric material and adapted manufacturing process conditions. Moreover, with the development of the technologies and the markets, the cost of polymeric membranes has reduced very much in the recent years. The existing polymeric membranes are very powerful engineering tools, especially in aqueous system. However, applications in membrane reactors are often limited by solvent, temperature and chemical stability. Therefore, the development of polymeric membranes with improved stability is a very active field. In addition, the preparation of polymeric membranes with a truly well defined pore structure (sharp size distribution and high porosity, at low barrier thickness) is currently intensively investigated; and progress toward that aim will certainly also promote the further development of membrane reactors. Significant development is also done to prepare porous polymeric membranes by other processes than phase separation, and surface functionalisation of already prepared membranes is another important tool to improve the functionality of porous membranes for membrane reactor applications (cf. Chapter 23).

References

1. Ulbricht, M. *Polymer*, **47**, 2217 (2006).
2. Vankelecom, I.F.J. *Chem Rev*, **102**, 3779 (2002).
3. Baker, R.W. *Membrane Technology and Applications*, 2nd edn, John Wiley & Sons, Ltd., Chichester (2004).
4. Sirkar, K.K. *Ind Eng Chem Res*, **47**, 5250 (2008).
5. Chang, T.M.S. *Nat Rev*, **4**, 221 (2005).
6. George, S.C.; Thomas, S. *Prog Polym Sci*, **26**, 985–1017 (2001).
7. Ho, W.S.; Sirkar, K.K. *Membrane Handbook*, Van Nostrand Reinhold, New York (1992).
8. Hanks, P.L.; Lloyd D.R. *J Membr Sci*, **306**, 125 (2007).
9. Loeb, S.; Sourirajan, S. *Adv Chem Ser*, **38**, 117 (1962).
10. Mulder, M. *Basic Principle of Membrane Technology*, 2nd edn, Kluwer Academic, Dordrecht, Chapters 2, 3, 5 (1996).
11. Van de Witte, P.; Dijkstra, P.J.; van den Berg, J. W. A.; Feijen, J. *J Membr Sci*, **117**, 1 (1996).
12. Albrecht, W.; Weigel, T.; Schossig-Tiedemann, M.; Kneifel, K.; Peinemann, K. V.; Paul, D. *J Membr Sci*, **192**, 217 (2001).
13. Prakash, S.S.; Francis, L.F.; Scriven, L.E. *J Membr Sci*, **313**, 135 (2008).
14. Grulke, E.A. In *Polymer Handbook*, 4th edn, Brandrup, J.; Immergut, E.H.; Grulke, E.A.; Abe, A.; Bloch, D. R. (eds), John Wiley & Sons, Inc., New York, Chapter 7, p. VII/675 (1999).
15. Susanto, H.; Ulbricht, M. *J Membr Sci*, **327**, 125 (2009).
16. Khare, V.P.; Greenberg, A.R.; Krantz, W.B. *J Membr Sci*, **258**, 140–156 (2005).
17. Susanto, H.; Stahra, N.; Ulbricht, M. *J Membr Sci*, **342**, 153 (2009).
18. Wang, G.J.; Chu, L.Y.; Zhou, M.Y.; Chen, W.M. *J Membr Sci*, **284**, 301 (2006).
19. Figoli, A.; De Luca, G.; Longavita, E.; Drioli, E. *Separ Sci Technol*, **42**, 2809 (2007).
20. Olson, D.A.; Chen, L.; Hillmyer, M.A. *Chem Mater*, **20**, 869 (2008).
21. Genné, I.; Kuypers, S.; Leysen, R. *J Membr Sci*, **113**, 343 (1996).
22. Ebert, K.; Fritsch, D.; Koll, J.; Tjahjawiguna, C. *J Membr Sci*, **233**, 71 (2004).
23. Soroko, I.; Livingston, A. *J Membr Sci*, **343**, 189 (2009).
24. Hicke, H.G.; Lehmann, I.; Malsch, G.; Ulbricht, M.; Becker, M. *J Membr Sci*, **198**, 187 (2002).
25. See-Toh, Y.H.; Silva, M.; Livingston, A. *J Membr Sci*, **324**, 220 (2008).
26. Peinemann, K. V.; Simon, P.; Abetz, V. *Nat Mater*, **6**, 992 (2007).
27. Schacher, F.; Ulbricht, M.; Müller, A.H.E. *Adv Funct Mater*, **19**, 1040 (2009).
28. Abetz, V.; Brinkmann, T.; Dijkstra, M.; Ebert, K.; Fritsch, D.; Ohlrogge, K.; Paul, D.; Peinemann, K.V.; Nunes, S.P.; Scharnagl, N.; Schossig M. *Adv Eng Mater*, **8**, 328 (2006).
29. Allegrezza, A.; Ireland, T.; Kools, W.; Phillips, M.; Raghunath, B.; Wilkins, R.; Xenopolus, A. *Membranes in the Biopharmaceutical Industry in Membranes for the Life Sciences*, Peinemann K. V.; Nunes S. P. (eds), Wiley-VCH, Weinheim, p. 69 (2008).
30. Ulbricht, M.; Schuster, O.; Ansorge, W.; Ruetering, M.; Steiger, P. *Sep Purif Techn*, **57**, 63 (2007).
31. I. Moch Jr, Hollow fibre membranes, in *Encyclopedia of Chemical Technology*, Vol. 13, 4th edn, John Wiley–InterScience Publishing, New York, p. 312 (1995).
32. McKelvey, S.A.; Clausi D.T.; Koros, W.J. *J Membr Sci*, **124**, 223 (1997).
33. Leiknes, T. Wastewater treatment by membrane bioreactors, in *Membrane Operations. Innovative Separations and Transformations*, Drioli, E.; Giorno, L. (eds), Wiley-VCH, Weinheim, p. 363 (2009).

34. Giorno, L.; Mazzei, R.; Drioli E. Biochemical membrane reactors in industrial processes, in *Membrane Operations. Innovative Separations and Transformations*, Drioli, E.; Giorno, L. (eds), Wiley-VCH, Weinheim, p. 397 (2009).
35. Vandezande, P.; Gevers, L.E.M.; Vankelecom, I.F.J. *Chem Soc Rev*, **37**, 365 (2008).
36. Choi, J.S.; Song, I.K.; Lee, W.Y. *J Membr Sci*, **166**, 159 (2000).
37. Ulbricht, M.; Papra, A. *Enzyme Microb Technol*, **20**, 61 (1997).
38. Godjevargova, T; Gabrovska, K. *J Biotechnol*, **103**, 107 (2003).
39. Ying, L.; Kang, E.T.; Neoh, K.G. *J Membr Sci*, **208**, 361 (2002).
40. Trusek-Holownia, A.; Noworyta, A. *J Biotechnol*, **130**, 47 (2007).
41. Lopez, J.L.; Matson S.L. *J Membr Sci*, **125**, 189 (1997).
42. Fritsch, D.; Bengtson, G. *Catal Today*, **118**, 121 (2006).
43. Fritsch, D.; Bengtson, G. *Adv Eng Mater*, **8**, 386 (2006).
44. Schmidt, A.; Haidar, R.; Schomäcker, R. *Catal Today*, **104**, 305 (2005).

23

Porous Polymer Membranes by Manufacturing Technologies other than Phase Separation of Polymer Solutions

Mathias Ulbricht[1] and Heru Susanto[1,2]

[1]Lehrstuhl für Technische Chemie II, Universität Duisburg-Essen, Essen, Germany
[2]Department of Chemical Engineering, Universitas Diponegoro, Semarang, Indonesia

23.1 Introduction

Organic polymers are the most important materials for synthetic membranes for separation and reaction engineering applications. This is because: (i) many different types of polymeric materials are commercially available, (ii) different types of selective barriers (i.e., porous, nonporous, charged) can directly be formed from polymers, (iii) production of large membrane area is possible at reasonable cost, and (iv) various shapes and formats (flat sheet, capillary/hollow fibre, capsule) can easily be obtained [1,2]. For the classification of porous polymer membranes and selection criteria for membrane polymers, the reader should also refer to Chapter 22.

Most of the porous membranes produced in industrial scale are prepared using one of the various variants of phase separation of polymer solutions (cf. Chapter 22). However, a variety of other preparation or manufacturing methods is also applied, developed or investigated. Traditionally, this is because some of the interesting membrane polymers can not well be processed by phase separation, mainly because no suited solvent is available. Another motivation is to develop methods for membrane preparation with solvent-free processes; this could lead to reduced amounts of organic solvents and water for the actual structure formation as well as the subsequent washing steps. Finally, alternative preparation methods are also relevant when unusual

Membranes for Membrane Reactors: Preparation, Optimization and Selection, Edited by Angelo Basile and Fausto Gallucci

membrane formats are explored, for example in microreactor systems. This chapter will cover these approaches classified into methods based on polymer extrusion, spinning of nanofibres and *in situ* polymerisation/polymer cross-linking. In addition, surface functionalisations of already established membranes are briefly covered; this is of large relevance for membrane industry and applications, and it is also applied to membranes made via phase separation of polymer solutions (cf. Chapter 22) in order to prepare catalytic membranes.

23.2 Technologies Based on Extrusion of Polymer Films

The extrusion of a thermoplastic polymer from its melt is a widely established general way for processing synthetic plastics. Of course, many applications of the resulting nonporous materials, for example, for packing purposes, can be described in the context of membrane (barrier) properties; but this is out of the scope here. Interesting, however, are processes where the film formation by extrusion is followed by or combined with (controlled) formation of a porous (barrier) structure.

23.2.1 Pore Formation by Film Stretching

Porous membranes as flat sheets can be obtained if during or after the solidification upon cooling, stretching of the nonporous precursor film is done in one or two directions. Typically, the resulting materials have to undergo a thermal post-treatment in order to reduce internal tensions in the porous membrane. This process is limited to semi-crystalline polymers because the crystalline domains ensure the integrity of the matrix in between the opened amorphous domains. The stretching (50 to 300%) leads to a deformation of the amorphous phase yielding a slit-like pore structure. Depending on the microstructure of the polymer, a relatively regular pore structure and morphology can be achieved. The resulting pore size can be controlled via adjusting the rate and extend of the stretching. Very high porosity (up to 90%) can be obtained. Two famous commercial membranes made by this method are Celgard® (from polypropylene, PP) and GORE-TEX® [from poly(tetrafluoroethylene), PTFE; Figure 23.1].

This process is used for attractive barrier polymers that are hard or impossible to be processed into a porous structure by phase separation methods (for PTFE) or as a process alternative (for PP;

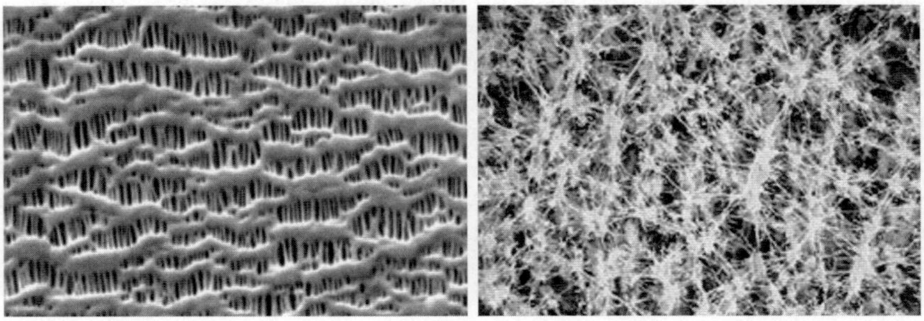

Figure 23.1 *Commercial membranes prepared via polymer film stretching process. Left: Celgard® PP MF membrane (taken from www.celgard.com). Right: GORE-TEX® PTFE membrane (taken from http://www.sterlitech.com; Copyright (2010) with permission from W. L. Gore & Associates)*

membranes from this polymer can also be obtained via thermally induced phase separation; cf. Chapter 22). The latter is relevant because it is an intrinsically solvent-free manufacturing process. However, additives such as softening agents or particles, can be used to promote the formation of desired pore structure via 'tailored' microstructure of the polymer or the function as a 'pore template', respectively [4].

The typical process for the chemically inert PTFE (having a high melting temperature) is more complicated. It involves also elements of sintering polymer particles. PTFE fine particles are homogeneously dispersed in an oil lubricant to form a paste for extrusion. The paste is then compressed at high pressure to produce a void-free preform. Thereafter, extrusion of the preform is carried out at temperatures below 50 °C. During the extrusion process, the lubricated PTFE particles become agglomerated. Finally, the extrudate passes through a rolling and stretching process before drying in an oven to remove the lubricant to yield the pore structure. The influence of process conditions on pore structure had been analysed in detail in a recent publication [5].

The stretching method has originally been designed for producing flat-sheet membranes, but the companies Celgard and Membrana have developed also hollow PP fibres; these membranes are the state of the art material for a special membrane contactor used in large scale for blood oxygenation [6] (cf. Section 23.6).

Overall, depending on the precursor film and its preparation as well as the degree of stretching, membranes with average pore sizes between 0.02 and 10 μm can be produced. For some applications it is attractive that the membranes can be made relatively thin (down to about 10 μm). The resulting membranes are used in some special ultra- and microfiltration applications (e.g., for aggressive streams in the semiconductor industry), as membrane contactor (e.g., blood oxygenation or membrane distillation), as support material for composite membranes (e.g., as battery separator) and as barrier material in textile applications (cf. Section 23.6).

23.2.2 Pore Formation by Track Etching

Membranes with very regular pores of sizes down to around 10 nm can be prepared by track etching [7]. A relatively thin ($<35\,\mu$m) polymer film [typically from poly(ethylene terephthalate), PET, or aromatic polycarbonate, PC] is first bombarded with fission particles from a high-energy source. These particles pass through the film, breaking polymer chains and creating damaged 'tracks'. Thereafter, the film is immersed in an etching bath (strong acid or alkaline aqueous solution), so that the film is preferentially etched along the tracks, thereby forming pores. The pore density is determined by irradiation intensity and exposure time, whereas etching time determines the pore size. However, the microstructure, the degree of crystallinity and the film orientation have also a significant effect on the quality of the pore structure. The advantage of this technique is that uniform and cylindrical pores with very narrow pore size distribution can be achieved (Figure 23.2). Other pore shapes, in particular conical or double-conical, have also been produced by using special etching conditions [8]. In order to avoid the formation of double or multiple pores, produced when two nuclear tracks are too close together, the membrane porosity is usually kept relatively low, that is, typically less than 10%.

Track etched membranes are commercially available with pore sizes from 10 μm to 10 nm and thicknesses between 35 μm and 6 μm. Such membranes with an isoporous cylindrical structure (confirmed by independent characterisation) are also frequently used as model system for fundamental investigations of the interplay between pore size, functionalisation of the pore surface and barrier or transport properties [1,8]. Recently, PET track etched membranes had been functionalised with controlled, surface-initiated graft copolymerisations to obtain advanced stimuli-responsive macroporous materials [9,10] (cf. Section 23.5).

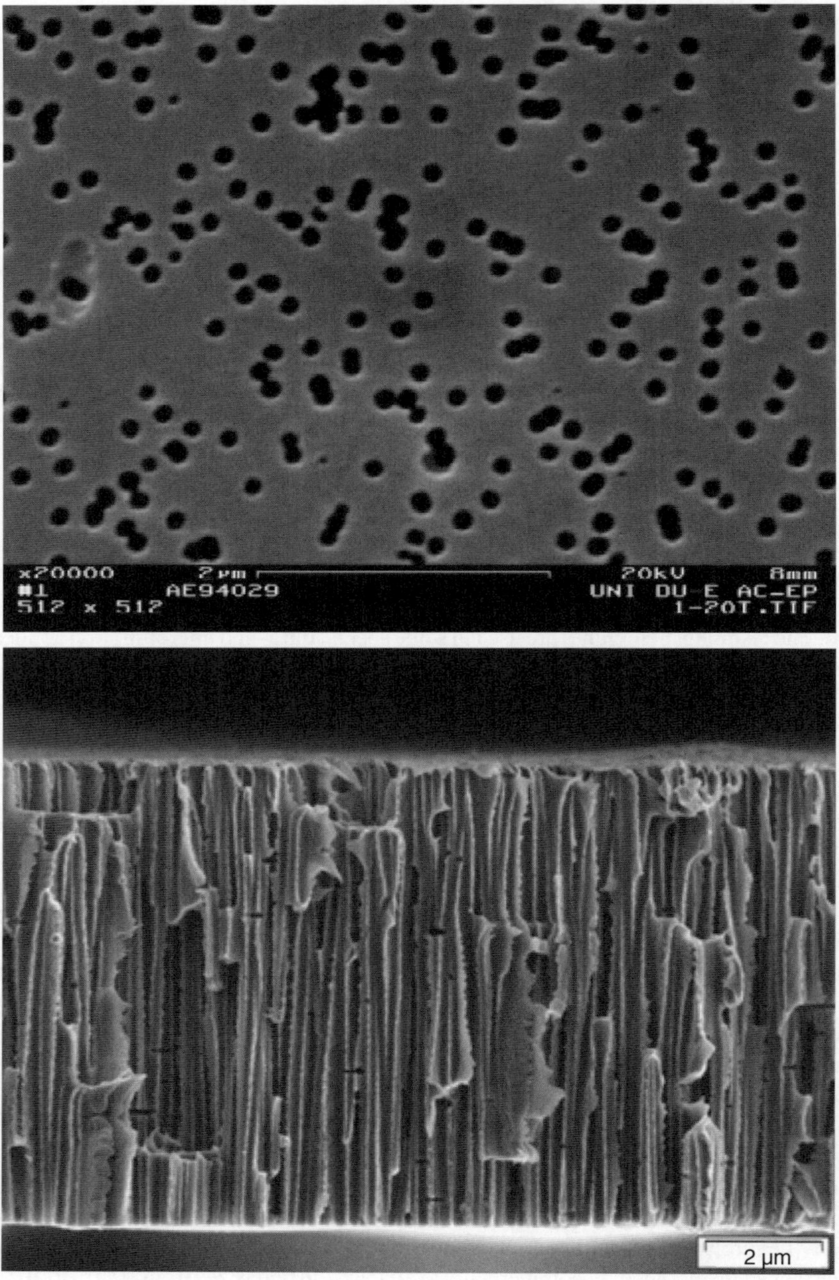

Figure 23.2 *PET track etched membranes. Top view (above) and cross section (below) of a membrane with nominal pore diameter of 100 nm (Rotrac®, Oxyphen GmbH, Dresden, Germany)*

Those membranes are the only commercial polymeric membranes with very narrow pore size distribution. However, technical applications in micro- or ultrafiltration, for example, for the fractionation of macromolecules in solution, are hindered by the low porosity and isotropic pore shape over membrane thickness (low permeability), significant fouling (mainly pore blocking) and the relatively high price of the material. Nevertheless, track etched membranes are used in some special analytical applications (e.g., for size-based fractionation and quantitative analysis of (bio)particles; cf. Section 23.6).

23.2.3 Pore Formation by Foaming

Manufacturing of foams from polymers, mainly for isolation materials, is an important special way of plastic processing. Various physical and chemical 'blowing' agents are used, depending on the structure of the polymers and the final applications. For the preparation of porous membranes, such methods have gained larger interest because solvent-free manufacturing processes could ultimately be developed. The main challenges with respect to useful porous membranes are to achieve high porosities and adjustable pore sizes and, furthermore, to obtain 'foams' with interconnected pores. Important achievements had been made, demonstrating that porous polymeric membranes can indeed be produced by extrusion of the pressurised melt with supercritical carbon dioxide as 'blowing' agent. Hollow fibre membranes, which are in principle suited for microfiltration, had been prepared from polycarbonate in a continuous process [11] (Figure 23.3). Flat sheet membranes from polysulfone with well interconnected cells had been obtained by batch extrusion using carbon dioxide with traces of organic solvents as 'pore opener' [12] (Figure 23.3).

Overall, the feasibility of the technology had been demonstrated for a couple of interesting thermoplastic membrane polymers, and this could also be extended to other foaming agents. However, more work towards theoretical understanding of all effects onto control of resulting pore structures and with respect to the technical implementation has to be done before this method would become truly relevant for industrial membrane manufacturing. The resulting membranes could be interesting for microfiltration and as membrane contactor or support for composite membranes.

23.3 Electrospinning of Porous Polymer Membranes

Technologies for spinning of fibres from synthetic polymers have for many decades been the basis for woven or nonwoven materials for traditional and novel ('technical') textile applications. However, due to the diameter of synthetic polymer fibres, the porosity is only suited for filter applications, that is, the removal or fractionation of particles with diameters much more than 1 μm. Nevertheless, some polymeric nonwovens are also important for polymeric membranes because they are used as support to increase the mechanical strength of reverse osmosis, nanofiltration and ultrafiltration membranes [1].

Electrospinning is a relatively novel technology that has very rapidly been further developed and explored during the last decade. The process can be summarised as follows. An electrical charge is used to draw a fibre from a reservoir containing a polymer melt or solution. As a jet of charged fluid polymer sprays out the bottom of a nozzle, an electric field forces the stream to whip back and forth, stretching the fibres so that their diameter can shrink, from several tens of micrometres by several orders of magnitude. The fibre forms a thin porous membrane as it hits the

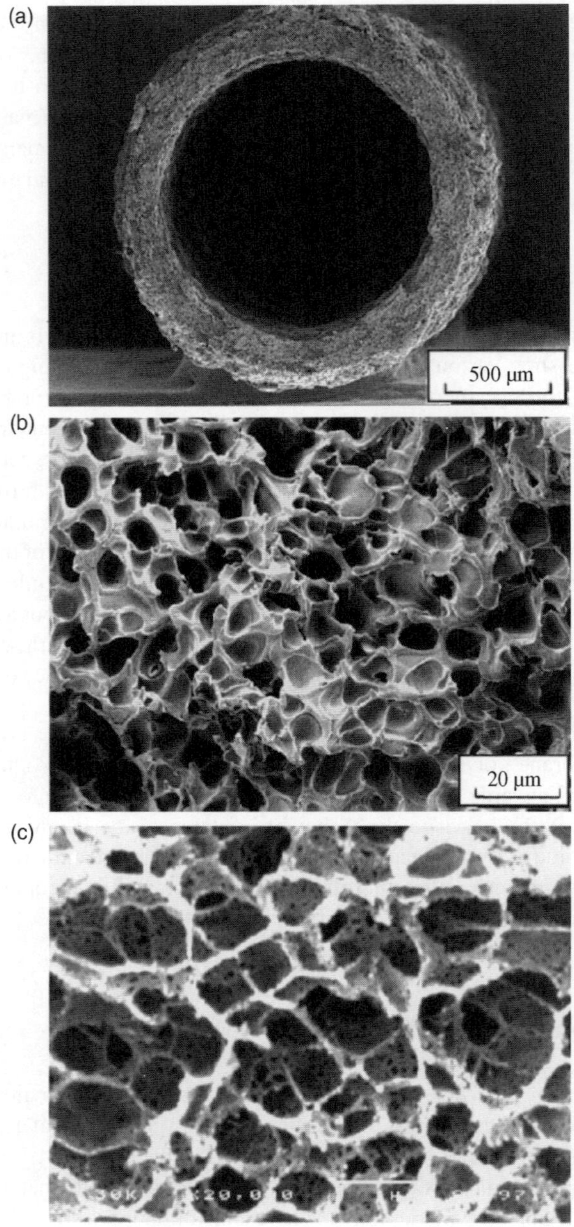

Figure 23.3 *Porous membranes prepared by extrusion from pressurised polymer melt in combination with foaming by dissolved carbon dioxide. Overview (a) and (b) cross section of a polycarbonate hollow fibre. (a) Reprinted from Huang, Q.; Seibig, B.; Paul, D. J Membr Sci, Polycarbonate hollow fiber membranes by melt extrusion. 161, 287. Copyright (1999) with permission from Elsevier. (b) Reprinted from J Membr Sci, Krause, B.; Boerrigter, M.E.; van der Vegt, N.F.A.; Strathmann, H.; Wessling, W. 187, 181. Copyright (2001) with permission from Elsevier. (c) Detail from the cross section of a polysulfone membrane obtained using a small fraction of tetrahydrofuran as pore opener; white scale bar 1 mm.*

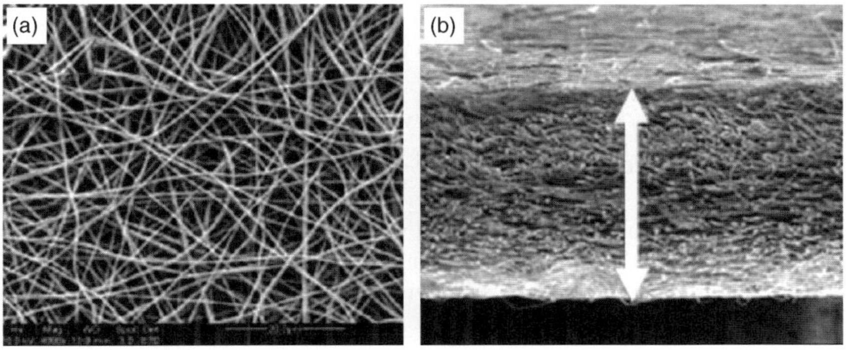

Figure 23.4 SEM micrographs of a electrospun polyvinylidene fluoride (PVDF) membrane: (a) top view, (b) cross section. Reprinted with permission from J Membr Sci, Gopal, R.; Kaur, S.; Feng, C. Y.; Chan. C.; Gopal, R.; Ng, W.G.; Ramakrishna, S.; Tabe, S.; Matsuura, T. 289, 210. Copyright (2007) with permission from Elsevier

surface below the nozzle. Many synthetic or biopolymers can be used. Such electrospun 'membranes' have a unique combination of stretchiness and strength and are easy to handle, making them suitable for a wide range of applications. Due to the small fibre diameter, down to around 100 to 10 nm, the resulting nonwoven-like structures could also have filter properties in the range of microfiltration and, potentially, ultrafiltration. Because such 'membranes' can be prepared with a very high porosity (more than 90%), they are already used as effective particle filters for treatment of gaseous streams. Electrospun polysulfone or polyvinylidene fluoride membranes, with a maximum effective pore diameter of ~5 μm, had been evaluated as prefilters or microfiltration membranes for water treatment [13] (Figure 23.4).

Detailed investigations of the barrier pore structure of electrospun polymeric membranes and, hence, their selectivity in microfiltration, had revealed that pore size and pore size distribution were strongly associated with the 'fibre mass' per area (i.e., the membrane thickness adjusted by the deposition time and conditions), the fibre diameter, and the fibre length [14]. An increase in membrane thickness caused a decrease in pore size and a shift in pore size distribution towards lower values. Larger fibre diameter resulted in the formation of larger pores. The dependence of pore size on fibre length was more complex, but a correlation with the pore size of electrospun membranes could also be established.

Composite membranes based on advanced electrospun nanofibre layer architectures had also been proposed: electrospinning on a nonwoven polymer support lead a highly permeable support for ultrafiltration membranes; however, the selective layer was a thin film of a cross-linked hydrophilic polymer, for example, polyvinyl alcohol [15]. This technology holds even more promise because electrospinning from polymer solutions can be combined with phase separation, leading to hierarchical pore structures with porosity in the nanofibres superimposed on the nonwoven macropore structure.

First commercial products are composites of electrospun nanofibre webs on conventional filters with a separation profile like microfiltration membranes (Ultra-Web® or Spider-Web® from Donaldson; cf. Section 23.6). The more efficient removal of particles (size $\geq 0.3\ \mu$m) from gaseous streams at much higher fluxes than conventional filters is due to a pronounced surface filter effect of the robust nanofibre network. It is expected that many more micro- and ultrafiltration membranes for special applications with gaseous and liquid feed streams will follow. Other emerging membrane-related applications are the use as scaffold for tissue engineering or as affinity membrane adsorber for bioseparations.

23.4 *In Situ* Polymerisation of Porous Membranes

Polymerisation of many monomers in bulk leads to nonporous polymers, while polymerisation in monomer solutions (with the solvent/polymer interactions as ruling factor) can lead to phase separation at certain monomer conversion; aggregation and agglomeration will then yield a porous polymer as product. With simultaneous cross-linking (content of cross-linker monomer as ruling factor), *in situ* polymerisation of monomer/cross-linker/solvent ('porogen') systems can yield porous monolithic polymers [16]. Such monoliths with bimodal pore size distribution and functional internal surface had been developed for various attractive applications, for example, for fast chromatographic separations including those with membrane-like discs [17]. Both general types of polymerisation, chain growth and step growth, can be used to obtain porous polymeric materials. For the preparation of porous membranes, the following *in situ* polymerisation approaches have been used or are currently being explored:

1. Formation of a porous polymer (agglomerate or monolith) by phase separation (induced by solvent or cross-linking).
2. Formation of a porous, molecularly imprinted polymer (MIP) – similar to (1), but a template is added to the polymerisation mixture (porosity is essential for MIPs because the 'imprints' should be accessible for extraction of template and re-binding of solute during application; for more details on this topic see [18]).
3. Formation of a nonporous polymer around (self-assembled) 'templates' for pores; removal of the template yields a porous structure (for an overview see [1]).
4. Fixation of a self-assembled ('nanoporous') structure in a nonporous polymer (for an overview, see [1]).

All the different approaches could, in principle, be performed to yield self-supporting or composite materials (on or in suited porous supports), and all three membrane geometries (flat sheet, capillary/hollow fibre, capsule) are accessible by this means. Similar to well established interfacial polymerisation toward nonporous polyamide barrier layers of composite membranes for reverse osmosis or nanofiltration [1], other *in situ* polymerisation methods can also lead to ultrathin barrier layers (thickness: a few tens of nanometres). One of the advantages of *in situ* preparation methods is that desired membranes can be prepared 'in place', and this is gaining increasing attention in the field of microfluidic reactor systems [19]. The following examples focus on approach 1 (see above), that is, the preparation of porous membranes via *in situ* polymerisation with phase separation.

Typical polymer systems are either based on radical polymerisation of vinyl monomers, for instance, the cross-linking copolymerisation of (meth)acrylate or styrene derivatives, or on polycondensation, for instance, the synthesis of (cross-linked) polyamides. In order to prepare conventional porous self-supporting membranes, this approach is less attractive because: (i) microfiltration membranes from phase separation of engineering polymers are readily available (cf. Chapter 22), (ii) for membranes with smaller pores (in the meso- and micropore range) thin barrier layers are desired (such anisotropic ultrafiltration membranes are also well accessible by phase separation methods; cf. Chapter 22), and (iii) *in situ* preparation of thin porous layers from monomers on porous supports is not straightforward because the reactive mixture can easily penetrate the porous support so that the overall barrier resistance would become too large. The controlled preparation of porous, functional polyacrylate-based monoliths with high specific surface area via *in situ* photo-initiated copolymerisation within the pores of microfiltration membranes (from PET or PP) had been demonstrated; crucial for a defect-free pore filling was a pre-modification of the base membrane with a compatible grafted polyacrylate [20].

The group of Uragami had performed comprehensive work on preparation and characterisation of enzyme membranes; in one recent work the *in situ* polymerisation had been studied: a mixture consisting of vinylised urease, acrylamide, 2-hydroxylethylmethacrylate and a cross-linker monomer yielded urease membranes which had then been evaluated under flow-though conditions [21].

An impressive example for the preparation of a porous separation membrane 'in place' in a microfluidic system was the UV-initiated polymerisation of a mixture of zwitterionic 2-(N-3-sulfopropyl-N,N-dimethylammonium)ethyl methacrylate and cross-linker methylene bisacrylamide, using a focussed 355 nm laser beam [22]. Microdialysis membranes were obtained. By controlling the course of phase separation during cross-linking polymerisation via the ratio between solvent (water) and nonsolvent (2-methoxyethanol), the molecular weight cut-off of the membranes could be engineered for different applications. Recently, the *in situ* formation of various catalytic polymeric membranes at the laminar interface between two different liquids in a microfluidic device had been investigated and used for instantaneous allylic arylation reaction of allylic esters and aryl boron reagents under microflow conditions [23]. One of the membranes had been formed via 'coordinative convolution' of palladium ions and polymeric phosphine ligands exposed by a poly(acrylamide) copolymer with triarylphosphine side groups. The membranes had been used as contactor in the microchannel, and completion of the catalytic reaction had been achieved in a residence time of 1 s.

Interfacial polymerisation reactions in emulsions can conveniently be used to prepare macro- or microcapsules with (porous) membranes as walls. This is relevant to create synthetic mimics of living cells, but such materials also have a large practical potential, for instance for drug delivery systems [24]. Extensive pioneering work on the preparation of such capsules with 'porous walls' had been done with polyamides (e.g., Nylon) by the group of Okahata (for example, see [25]). This research has been continued and extended by other groups. In recent years, membrane emulsification had also been introduced in the preparation in order to obtain monodisperse capsule size distributions [26] (Figure 23.5).

In combination with a post-functionalisation of the pores by grafting of stimuli-responsive polymers (cf. Section 23.5), microcapsules with 'gated' walls had been obtained; the content of the capsule could be released as a function of, for example, changed temperature/pH value or the presence of specific ions in the surrounding solution [27,28]. Another release mechanism was based on the combination of a stimuli-responsive grafted polymer with the enzyme glucose oxidase, and the 'smart' capsules released their content in the presence of glucose [28].

23.5 Surface and Pore Functionalised Membranes

Materials surface properties are of crucial importance in many applications. With porous polymers, a rich diversity of surface properties results from the used material and the chosen membrane preparation method. Nevertheless, post-functionalisation has become a very important field for membrane technology because the requirements with respect to pore structure and

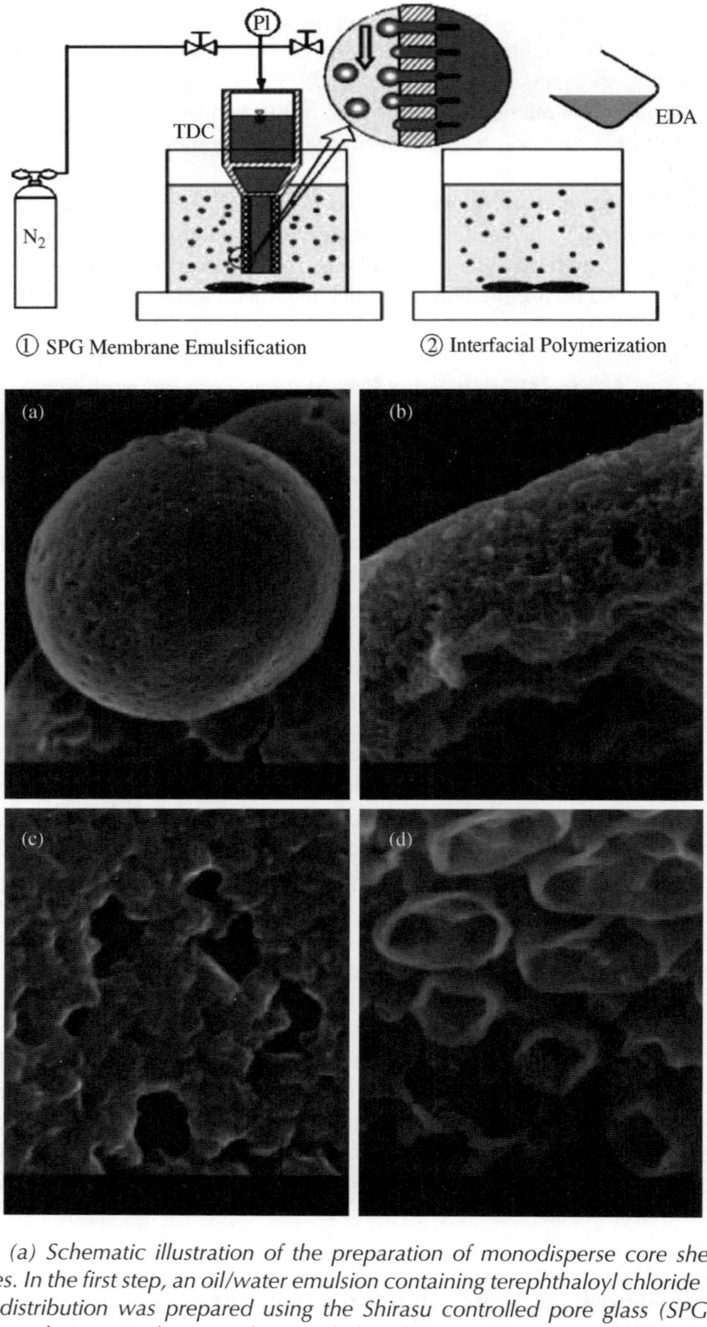

Figure 23.5 *(a) Schematic illustration of the preparation of monodisperse core shell polyamide microcapsules. In the first step, an oil/water emulsion containing terephthaloyl chloride (TDC) with a narrow size distribution was prepared using the Shirasu controlled pore glass (SPG) membrane emulsification technique. In the second step, ethylenediamine (EDA) was added to the emulsion to form a polyamide shell membrane using the interfacial polymerisation technique. (b) SEM micrographs of core shell microcapsules prepared using a two-step SPG membrane emulsification and interfacial polymerisation method: (a) general morphology, (b) cross sectional view, (c) outer surface, (d) inner surface. Reprinted from Akamatsu, K.; Yamaguchi, T. Novel preparation method for obtaining pH-responsive core shell microcapsule reactors, Ind Eng Chem Res 2007, 46, 124. Copyright (2007) with permission of American Chemical Society*

membrane stability can not be combined with all desired surface properties. Consequently, many commercial polymeric membranes are surface modified. The intention of a surface modification of a membrane is either to minimise undesired ('secondary') interactions (adsorption or adhesion) which reduce the performance (membrane fouling), or to introduce additional interactions (affinity, responsiveness, biocompatibility or catalytic properties) for improving the selectivity or creating an entirely novel separation function [1].

Figure 23.6 illustrates with an established porous polymeric membrane as example the various adaptations of interface properties to adjust separation performance in completely different membrane-based processes. The chosen microfiltration membrane from PP had been prepared via the thermally induced phase separation process (cf. Chapter 22); however, membranes from PP can also be prepared via film extrusion and stretching (cf. Section 23.2.1).

Membranes from PP are very versatile, especially because of the chemical stability of the membrane polymer and the adjustable porosity. However, PP is a hydrophobic polymer. This is a problem for microfiltration of aqueous solutions because wetting of the membrane pores may be incomplete leading to lower fluxes. In addition, adsorption and adhesion of macromolecules and colloids on the pore surface can accelerate membrane fouling, i.e. a loss in flux during the microfiltration process [29]. However, for membrane distillation of water, wetting of the pores must completely be prevented (the relatively large porosity and pore diameter is a precondition for high fluxes) [4]. Exactly the same criteria have to be met for membrane aeration or oxygenation [6]. Furthermore, the stability of pore-filling liquid membranes depends also on the wetting of the pore surface, that is, in this case a minimum of the interfacial energy should be achieved [4]. When the porous membrane shall be used as base material for membrane adsorption (or chromatography), the interactions with solutes should typically be reversible, and the relatively low specific surface area of conventional porous membranes should be compensated by introducing a three-dimensional (polymeric) binding layer on the pore surface for higher binding capacity [1].

Depending on the desired application, either the functionalisation of the entire pore surface or only of the outer membrane surface may be sufficient. A key feature of a successful (i.e., 'tailored') surface functionalisation is a *synergy* between the useful properties of the base membrane and the novel functional polymer (layer). This is best achieved by a functionalisation reaction or process which essentially preserves the bulk structure of the porous base membrane.

In some cases, a beneficial effect can already be achieved by a choosing a suited solvent or a modifier for a noncovalent surface modification. For instance, the wetting of all pores in hydrophobic microfiltration membranes can be achieved with alcohols (e.g., isopropanol), and a precoating of the membrane pores with surfactants exposing hydrophilic groups on the surface may also improve the fouling resistance for a certain period of time. Recently, it has been demonstrated with PP microfiltration membranes that the efficiency and long-term stability of such noncovalent hydrophilic surface modification can be further improved by using amphiphilic modifiers in combination with a swelling step for the membrane polymer; the process is named surface modification by 'entrapment' of amphiphilic modifiers [30].

A permanent surface functionalisation can only be obtained by a reactive process. For macroporous membranes, for example, those for microfiltration, a reactive coating via an *in situ* cross-linking polymerisation of hydrophilic acrylate-based monomer mixtures is the most widely applied approach [1]. A chemical attachment of the very thin coating is not essential because the interpenetration of the porous membrane structure with the *in situ* formed chemically cross-linked polymer network ensures long-term stability. For instance, many companies provide 'hydrophilic' PP or PVDF membranes prepared by a variant of that approach (cf. Section 23.6).

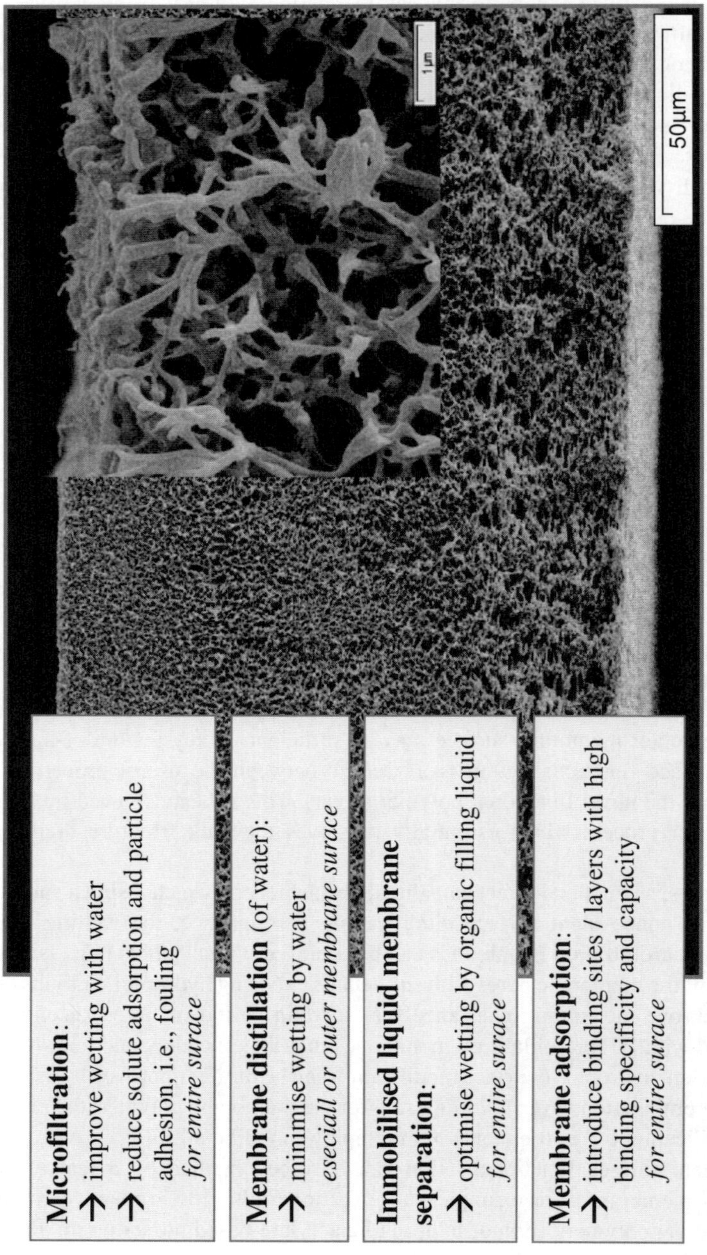

Figure 23.6 *Optimising the interfaces on and within membranes is crucial for effective membrane-based processes. Cross section overview and detail from scanning electron microscopy for a polypropylene (PP) microfiltration membrane (barrier pore diameter 0.4 μm; 2E HF, Membrana GmbH, Wuppertal, Germany), with guide lines for surface functionalisation to meet minimum criteria for dedicated applications*

Besides coating methods, the most common strategies for surface functionalisation of porous membranes can be classified into 'grafting to' (attachment of pre-synthesised functional entities) and 'grafting from' methods (*in situ* formation of functional entities via graft copolymerisation of functional monomers) [1]. Depending on the reactivity/stability of the base polymer, either physical or chemical methods for surface activation or coupling are applied. Overall, the physical activation by plasma, UV irradiation or electron beam has the advantage that conditions can be adapted to a wider range of base polymers than direct chemical modification of the base membrane [1]. More information on surface functionalisation of porous polymer membranes can be found in recent reviews: one is exemplaric for the surface engineering of porous PP membranes [31], another one will provide a broader overview on tailored photo-functionalisations of polymeric membranes [32]. It is beyond the scope of this chapter to give a comprehensive insight into that field. Instead, a couple of relevant examples shall illustrate the diversity and potential of surface functionalisations of porous polymer membranes to enable or facilitate membrane reactor applications.

Stimuli-responsive porous membranes via controlled 'grafting from' of functional polymer layers. Using tailored grafted functional polymer layers on the pore walls of membranes, it is possible to reversibly change the permeability and/or selectivity. The most straightforward mechanism is the alteration of the effective pore diameter by changing the conformation of a grafted polymer via solution conditions as 'stimulus'. For example, reversible switching of permeability had been achieved using photografted pH-responsive poly(acrylic acid) on PP microfiltration membranes [33], or using photografted temperature-responsive poly(N-isopropylacrylamide) on PET track etched membranes [9]. Recently, it has been demonstrated that with controlled, so-called 'living', graft copolymerisation methods, a much higher precision of such pore 'switching' effects can be achieved. Examples are membrane pores that can change from a closed state (either for pressure-driven flux or for diffusion of macromolecules) to an open state, or the 'switching' of effective pore size by two different stimuli which is possible for grafted block copolymers [10,34] (Figure 23.7).

Macroporous membrane adsorbers with grafted polymer layers comprising functional groups for reversible binding. Surface functionalised membrane adsorbers for fast purifications of proteins and other (bio)nanoparticles have been prepared via photo-grafting of two- or three-dimensional layers with suited functional groups on microfiltration membranes, for instance from PP [35–38], or from stabilised regenerated cellulose [39]. Remarkable binding selectivity for certain proteins could be achieved by the choice of special monomers with specific side groups for multipoint molecular recognition [38]. While the grafted layer thickness was important for binding capacity, it had to be adapted to membrane pore size in order to assure sufficient membrane permeability. This is necessary in order to use the main advantage of macroporous membrane adsorbers, that is, the reduction of mass transfer limitations by directional convective flow though the membrane pores [1,4]. Such 'tailored' membranes can also be used for other adsorptive separations (e.g., viruses), and they may also be adapted for the covalent immobilisation of (bio)catalysts (see Section 23.7).

23.6 Overview on Technical Porous Polymeric Membranes

Table 23.1 gives exemplaric information on industrial membranes and respective manufacturers as well as typical processes where such membranes are already widely applied. These established porous polymeric membranes which are not prepared by phase separation of polymer solutions have a macroporous barrier structure which is immediately suited for microfiltration and several membrane contactor applications. Electrospun nanofibre filter materials (Ultra-Web® and

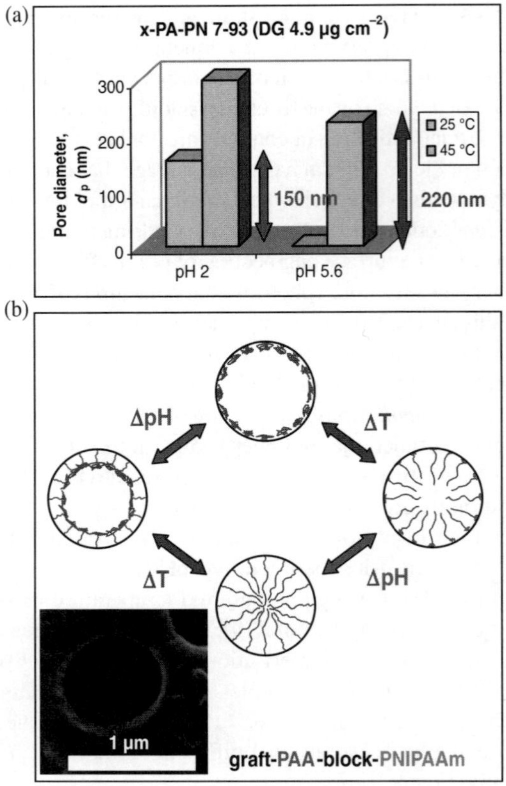

Figure 23.7 *(a) Effective hydrodynamic layer thickness from water flux measurement for a track etched PET membrane (pore diameter 790 nm) which had been sequentially photografted under pseudo-living conditions, first with poly(acrylic acid) (PAA) and thereafter with poly(N-isopropyl-acrylamide) (PNIPAAm). (b) Schematic visualisation of the reversible switching effects with respect to pore size as a function of the conformation changes of the grafted functional polymer blocks. Reprinted from Geismann, C.; Tomicki, F.; Ulbricht, M. Block copolymer grafted poly(ethylene terephthalate) capillary pore membranes distinctly and fast switchable by two different stimuli, Separ Sci Techn 44, 3312. Copyright (2009) with permission from Taylor & Francis*

Spider-Web®, from Donaldson) have been included because their rejection properties in gas filtration place them also in the range of microfiltration (cf. Section 23.3). Except for commercial adsorber membranes, other surface-modified or -functionalised membranes have not been included because of the large diversity of commercial membranes (cf. Section 23.5), many of them based on pore structures from phase separation of polymer solutions. In order to obtain a complete overview on commercial porous polymer membranes, the reader should also refer to the analogous section of Chapter 22.

23.7 Applications in Membrane Reactor Processes

Concepts and examples for porous membranes in catalytic reactors have already been discussed in Chapter 22; the focus here is exclusively on macroporous membranes where a surface

Table 23.1 *Overview on typical technical porous polymeric membranes (not prepared via phase separation) for selected important applications*

Membrane process	Membrane polymer	Membrane shape	Typical process mode, driving force	Application	Membrane manufacturer
Filtration, microfiltration	e.g. polyamide (electrospun)	Flat sheet composite filter	Pressure-driven	Air and gas filtration	Donaldson
Microfiltration	PP, PE, PTFE	Flat sheet	Pressure-driven, up to 3 bar	Clarification, concentration, sterile filtration	Asahi, Donaldson, Gore, GVS, Membrana (Celgard), Millipore
	PP	Capillary	As above	As above	Membrana (Celgard)
Microfiltration	PET (te), PC (te)[a]	Flat sheet	Pressure-driven, up to 3 bar	Sterile filtration, analytical applications	Oxyphen, Whatman
Gas/liquid contactor	PP, PE, PTFE	Hollow fibre, capillary	Concentration difference	Blood oxygenation, dehumidification of air	Membrana (Celgard), GVS
Membrane distillation	PP, other polyolefines	Flat sheet, hollow fibre	Temperature or vapour pressure difference	Water desalination	Membrana (Celgard)
Membrane adsorber	Cellulose, poly-ethersulfone, PVDF with grafted functional layer[b]	Flat sheet, used as stack	Convective flow by trans-membrane pressure	Removal of trace impurities (polishing), capturing biomolecules or particles	Sartorius, Pall, Millipore

[a] te = track etched.
[b] Prepared by post-functionalisation of membranes obtained from phase separation of polymer solutions.

functionalisation had been performed after membrane formation (cf. Section 23.5). Such membranes can be used as membrane contactors in catalysis. With convective flow through the membrane, the impact of mass transfer resistances on the rate of the catalytic conversion can be reduced, and the contact time of substrate with the catalyst can be adjusted [1,4].

Various enzyme membranes have been prepared, typically by covalent protein coupling to functional groups on the pore surface. The ability to bind protein in polymer layers obtained via 'grafting from' (cf. Section 23.5) in multilayers at relatively high density had frequently been used to increase or adjust the catalytic activity; examples are porous polyolefine hollow fibre membranes grafted via radiation initiated polymerisation [40–42]. Exemplaric enzymes in those investigations were ascorbic acid oxidase [40], a glucanotransferase [41], or a lipase [42].

Model studies had been performed with isoporous track etched PET membranes with pore diameters between 100 nm and 3 μm, which had been functionalised via 'grafting from' reactions and covalent enzyme immobilisation in order to prepare enzyme-membranes as convective flow-through microreactors [43,44]. The resulting membranes can be considered an array of tubular plug flow reactors. In these studies it had been demonstrated that with such reactor type, enzymatic polymerisations of sugars (facilitated by glucosyl or fructosyl transferases) could be successfully performed, which are not possible with the enzyme immobilised in porous particles because the product with large molar mass must be removed efficiently from the enzyme in order to avoid its deactivation [43,44].

The surface selective *in situ* preparation of catalytic metal nanoparticles on the prefunctionalised pore walls of track etched membranes has been elaborated as an alternative to enzyme immobilisation [45,46]. With the thus obtained flow-through reactors, systematic studies of the influence of residence time on conversion and selectivity of the catalytic reduction of nitroaromatic compounds have been performed. The same concept of membrane preparation with layer by layer deposition of polyelectrolytes as prefunctionalisation had been transferred to a polyamide microfiltration membrane with less regular pore structure [46].

The combination of reactive coating of the pore surface of a PVDF microfiltration membrane via *in situ* cross-linking polymerisation (cf. Section 23.5) with the controlled reactive deposition of catalytic nanoparticles has lead to very interesting composite membranes (Figure 23.8). These macroporous catalytic membranes were investigated for flow-through decontamination of water from dissolved chlorinated aromatic micropollutants [47].

A new photocatalytic membrane had been prepared by combination of plasma treatment of porous PVDF membranes with the subsequent chemical immobilisation of tungsten-based catalysts [48].

23.8 Conclusions and Outlook

Separation and integrated processes based on porous polymeric membranes are very well established in many application fields, including very large industrial scale. A wide range of porous membranes has been developed and is industrially produced based on phase separation of polymer solutions and alternative methods. The development of alternative methods is driven by the need to intensify membrane manufacturing itself, for instance by using less solvents, or the need for higher preparation flexibility, for instance in the context of miniaturisation. It should be mentioned that other emerging techniques based on self-assembly of functional macromolecular architectures will certainly in the future widen the scope of membrane preparation. For instance, the layer by layer deposition of polyelectrolytes which has mainly

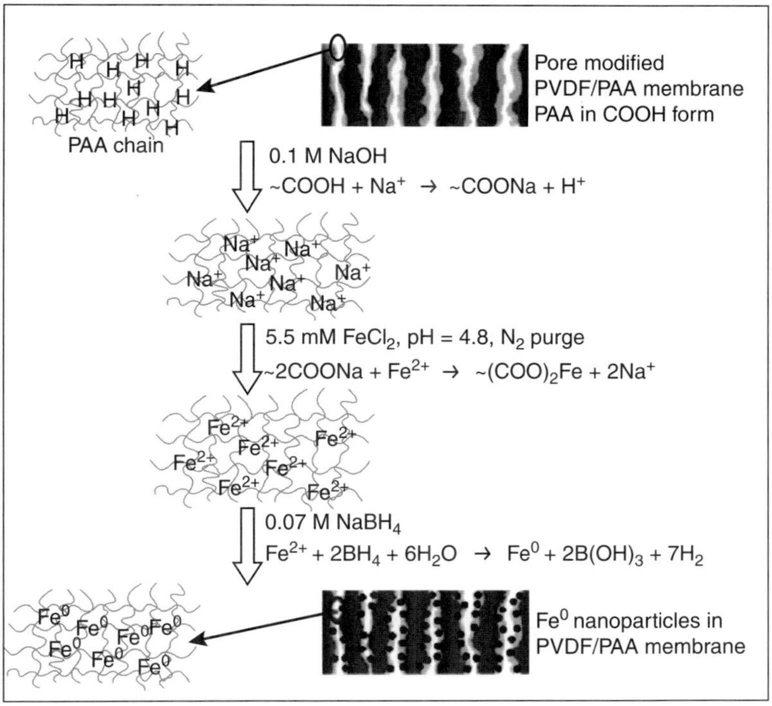

Figure 23.8 *Schematic overview on the preparation of catalytic iron nanoparticles on the pore surface of PVDF membranes which had been pre-functionalised by in situ cross-linking polymerisation of PAA. Reprinted from Xu, J., Bhattacharyya, D., Fe/Pd Nanoparticle immobilization in microfiltration membrane pores: Synthesis, characterization, and application in the dechlorination of polychlorinated biphenyls, Ind Eng Chem Res 46, 2348. Copyright (2007) with permission from American Chemical Society. The resulting catalytic membranes can be used for flow-through decomposition of micropollutants in water*

been explored to prepare nonporous membrane barriers (cf. [1]) could also be used to prepare porous barrier structures in all membrane shapes, including hollow capsules (cf. [22]). Furthermore, by 'down scaling' the particle sintering technique which is currently used to prepare coarse macroporous polymeric filters or supports for membranes [49], it can be imagined that 'well packed' nanoparticles could also directly be used as building blocks for meso- or microporous membrane barriers (cf. [50]). Further development of advanced polymeric materials with properties (chemical stability and other) dedicated for demanding membrane applications will also increase the repertoire of nonconventional manufacturing technologies. Finally, surface functionalisation of porous membranes can add another dimension to their application potential. All these current and future achievements will certainly also promote the further development of membrane reactors.

Acknowledgements

Parts of this chapter (Sections 23.2, 23.3) have been taken in modified form from the chapter 'Polymer membranes' by Mathias Ulbricht, to appear in the book *Porous polymers*, edited by Michael S. Silverstein *et al.*, published by Wiley-VCH.

References

1. Ulbricht, M. *Polymer*, **47**, 2217 (2006).
2. Vankelecom, I.F.J. *Chem. Rev.*, **102**, 3779 (2002).
3. http://www.gore.com/en_xx/products/filtration/micro/gore_microfiltration_media.html.
4. Sirkar, K.K. *Ind Eng Chem Res*, **47**, 5250 (2008).
5. Huang, R.; Hsu, P.S.; Kuo, C.Y.; Chen, S.C.; Lai, J.Y.; Lee, L.J. *Adv Polym Techn*, **26**, 163 (2007).
6. Wiese, F. *Membranes for Artificial Lung in Membranes for the Life Sciences*, Peinemann K.V., Nunes S.P. (eds), Wiley-VCH, Weinheim, p. 49 (2008).
7. Fleischer, R.L.; Alter, H.W.; Furman, S.C.; Price, P.B.; Walker, R.M. *Science*, **172**, 255 (1972).
8. Baker L.A.; Jin, P.; Martin, C.R. *Critical Reviews in Solid State and Materials Sciences*, **30**, 183 (2005).
9. Geismann, C.; Yaroshchuk, A.; Ulbricht, M. *Langmuir*, **23**, 76 (2007).
10. Friebe, A.; Ulbricht, M. *Macromolecules*, **42**, 1838 (2009).
11. Huang, Q.; Seibig, B.; Paul, D. *J Membr Sci*, **161**, 287 (1999).
12. Krause, B.; Boerrigter, M.E.; van der Vegt, N.F.A.; Strathmann, H.; Wessling, W. *J Membr Sci*, **187**, 181 (2001).
13. Gopal, R.; Kaur, S.; Feng, C. Y.; Chan. C.; Ramakrishna, S.; Tabe, S.; Matsuura, T. *J Membr Sci*, **289**, 210 (2007).
14. Li, D.; Frey, M.W.; Joo, Y.L. *J Membr Sci*, **286**, 104 (2006).
15. Yoon, K.; Hsiao, B. S.; Chu, B.; *J Membr Sci*, **338**, 145 (2009).
16. Svec F. *J Separ Sci*, **27**, 1419 (2004).
17. www.monoliths.com.
18. Ulbricht, M. Molecularly imprinted polymer films and membranes in molecularly imprinted materials, in: *Science and Technology*, Yan, M., Ramström, O. (eds), Marcel Dekker, New York, p. 455 (2005).
19. de Jong, J.; Lammertink, R.G.H.; Wessling, M. *Lab Chip*, **6**, 1125 (2006).
20. Salam, A.; Ulbricht, M. *Macromol Mater Eng*, **292**, 310 (2007).
21. Uragami, T.; Ueguchi, K.; Watanabe, M.; Miyata, T. *Catalysis Today*, **118**, 158 (2006).
22. Song, S.; Singh, A. K.; Shepodd, T. J.; Kirby, B. *J. Anal Chem*, **76**, 2367 (2004).
23. Yamada, Y.M.A.; Watanabe, T.; Torii, K.; Uozumi, Y. *Chem Commun*, **2009**, 5594 (2009).
24. Sukhorukov, G.B.; Moehwald, H. *Trends in Biotechnology*, **25**, 93 (2006).
25. Okahata, Y.; Noguchi, H.; Seki, T. *Macromolecules*, **19**, 493 (1986).
26. Akamatsu, K.; Yamaguchi, T. *Ind Eng Chem Res*, **46**, 124 (2007).
27. Chu, L.Y.; Yamaguchi, T.; Nakao, A. *Adv Mater*, **14**, 386 (2002).
28. Chu, L. Y.; Liang, Y. J.; Chen, W. M.; Ju, X. J.; Wang H. D. *Colloids and Surfaces B, Biointerfaces*, **37**, 9 (2004).
29. Belfort, G.; Davis, R.H.; Zydney, A.L. *J Membr Sci*, **96**, 1 (1994).
30. Guo, H.; Ulbricht, M. *J Membr Sci*, **349**, 312 (2010).
31. Wan L.S.; Liu Z.M.; Xu Z.K. *Soft Matter*, **5**, 1775 (2009).
32. He, D. M.; Susanto, H.; Ulbricht, M. *Prog Polym Sci*, **34**, 62 (2009).
33. Ulbricht M., *React Funct Polym*, **31**, 165 (1996).
34. Geismann, C.; Tomicki, F.; Ulbricht, M. *Separ Sci Technol*, **44**, 3312 (2009).
35. Borcherding, H.; Hicke, H.G.; Jorcke, D.; Ulbricht, M. *Ann NY Acad Sci*, **984**, 470 (2003).
36. Ulbricht, M.; Yang, H. *Chem Mater*, **17**, 2622 (2005).
37. He, D. M.; Ulbricht, M. *J Membr Sci*, **315**, 155 (2008).
38. He, D.M.; Sun, W.; Schrader, T.; Ulbricht, M. *J Mater Chem*, **19**, 253 (2009).
39. Wang, J.; Faber, R.; Ulbricht, M. *J Chromatogr A*, **1216**, 6490 (2009).

40. Kawai, T.; Saito, K.; Sugita, K.; Sugo, T.; Misaki, H. *J Membr Sci*, **191**, 207 (2001).
41. Kawakita, H.; Sugita, K.; Saito, K.; Tamada, M.; Sugo, T.; Kawamoto, H. *J Membr Sci*, **205**, 175 (2002).
42. Abrol, K.; Qazi, G.N.; Ghosh, A.K. *J Biotechnol*, **128**, 838 (2007).
43. Becker, M.; Provart, N.; Lehmann, I.; Ulbricht, M.; Hicke, H.G. *Biotechnol Progr*, **18**, 964 (2002).
44. Hicke, H. G.; Becker, M., Paulke, B. R.; Ulbricht, M. *J Membr Sci*, **282**, 413 (2006).
45. Dotzauer, D.M.; Dai, J. H.; Sun, L.; Bruening M. L. *Nano Letters*, **6**, 2268 (2006).
46. Dotzauer, D.M.; Bhattacharjee, S.; Wen, Y.; Bruening, M.L. *Langmuir*, **25**, 1865 (2009).
47. Xu, J.; Bhattacharyya, D. *Ind Eng Chem Res*, **46**, 2348 (2007).
48. Lopez, L.C.; Buonomenna, M.G.; Fontananova, E.; Iacoviello. G.; Drioli, E.; d'Agostino, R.; Favia, P. *Adv Funct Mater*, **16**, 1417 (2006).
49. www.porexfiltration.com.
50. Lin, Y.; Skaff, H.; Böker, A.; Dinsmore, A.D.; Emrick, T.; Russell, T.P. *J Am Chem Soc*, **125**, 12690 (2003).

24

Palladium-Loaded Polymeric Membranes for Hydrogenation in Catalytic Membrane Reactors

V.V. Volkov[1], I.V. Petrova[1], V.I. Lebedeva[1], V.I. Roldughin[2] and G.F. Tereshchenko[1]

[1]*Topchiev Institute of Petrochemical Synthesis, Russian Academy of Sciences, Moscow, Russia*
[2]*Frumkin Institute of Physical Chemistry and Electrochemistry, Russian Academy of Sciences, Moscow, Russia*

24.1 Introduction

Catalytic reactions of organic compounds hydrogenation play their important role in many areas of technology and science. Among known catalysts for hydrogenation, metals of platinum family (Pt, Pd, Ni) are known to be most efficient. Palladium catalysts for selective hydrogenation hold the most unique position. Palladium and its alloys show a unique potency to dissolve huge amounts of hydrogen and, as a membrane materials, they also possess a selective permeability towards hydrogen.

Thus, fundamental concepts on membrane catalysis and catalytic membrane reactors (CMRs) have been advanced in the 1960s by Gryaznov [1]; in 1964, Gryaznov *et al.* [2] has discovered the phenomenon of conjugating of chemical reactions with the evolution (dehydrogenation reaction) and consumption (hydrogenation reaction) of hydrogen on palladium membranes.

Early studies in this direction have been focused on the use of relatively thick nonporous metallic membranes (50–100 μm). Even though the proposed concept seemed to be fascinating, the use of thick palladium membranes (foils, tubes) appeared to be limited. In addition to their low permeability, the cost of the above membranes appeared to be high for their commercial

Membranes for Membrane Reactors: Preparation, Optimization and Selection, Edited by Angelo Basile and Fausto Gallucci

application. This cost can be reduced by decreasing thickness of palladium layer, and this effect can be achieved by the development of composite membranes.

In this connection, the use of polymer supports for the preparation of the composite membranes with thin selective layers based on Pd and its alloys were proposed for the first time in 1979 [3]. The most highly permeable rubbery polymer, poly(dimethylsiloxane) (PDMS), was used as a polymer support. For these catalytic membranes (CMs), conversion of cyclopentadiene at 151 K was equal to 89%, and selectivity towards cyclopentane was 93%. For the preparation of the above membranes, the consumption of Pd was 100 times lower as compared with the case of the Pd foil [3].

The logical follow-up step providing cost reduction and reduced consumption of expensive noble metals was concerned with the development of CMs based on the polymer–metal composites [4]. The polymer matrix was also based on PDMS and other highly permeable poly(organosiloxane)s containing palladium nanoparticles, which were preliminary immobilised on silica gel. The polymeric catalytic membranes (pCMs) were tested on a flow-type reactor for the gas-phase hydrogenation of cyclopentadiene. The maximum conversion of cyclopentadiene of 98.9% was attained, and selectivity towards cyclopentene was 79% [4].

Thus, Gryaznov and coworkers were the first who proposed to use the advantages of polymeric membranes and catalytic performance of Pd and other noble metals. Follow-up studies on the development of pCMs were performed in the two following directions: (i) dense catalytic membranes and (ii) porous catalytic membranes. Nowadays, loads of experimental results have been accumulated for Pd-containing pCMs, and some results have been discussed in the comprehensive reviews on membrane catalysis by Vankelecom [5], Dioos, Vankelecom and Jacobs [6], and Ozdemir, Buonomenna and Drioli [7]. To our knowledge, this paper is the first survey devoted to the preparation of Pd-loaded pCMs for hydrogenation reactions in CMRs.

24.2 Synthesis and Hydrogenation Studies

24.2.1 Dense Catalytic Membranes

The efficient selection of polymers for dense catalytic membranes should take into account the fact that polymer matrix should be loaded with sufficient amounts of catalytic particles without any penalties in the plasticity of composites. In this case, high adhesion between polymer and surface of loaded particles is required. In addition to its high strength, polymeric material should be highly permeable towards reagents and reaction products. Transfer rate of components should be sufficiently high to impose no limitations on the catalysed reaction.

Numerous works by Fritsch and coworkers [8–13] have been devoted to the preparation of dense catalytic membranes. In these works, highly permeable membrane materials based on poly (amide imide)s (PAI) have been worked out, and new methods for the preparation of Pd-loaded pCMs have been advanced. At the first stage, a polymeric membrane is formed as a homogeneous membrane containing palladium acetate as a precursor. For these purposes, N-methyl-2-pyrrolidone (NMP) as a common solvent for PAI and Pd-acetate were used. Then, Pd(II) was reduced to Pd(0) by immersing the Pd-salt-filled membrane in the solution of $NaBH_4$/MeOH. When metal content is 30 wt%, it is possible to preserve the mechanical characteristics of PAI; when CMs contain 15 wt%, metal clusters are no longer soluble in NMP and THF [9]. CMs were studied for the reaction of decomposition of nitrous oxide by hydrogen to nitrogen and water catalysed Pd/Ag. The corresponding structural studies have been performed in [14,15] and are discussed in Section 24.2. In addition to homogeneous membranes with a thickness of 30–80 μm, thin-film composite membranes containing 5 wt% of Pd were prepared on the PVDF porous supports [8].

The proposed synthesis procedure has been later adapted for the preparation of dense pCMs based on elastomer, poly(ether-*b*-amide) (PEBA) [10–12]. PEBA is known as a polymer, which is able to concentrate slightly polar chemicals such as phenols from water; these membranes were used to concentrate and react organics in a pervaporative CMR. As a model for hydrodehalogenation reaction, hydrogenation of 4-chlorophenol in aqueous solution at 30–50 °C using dissolved hydrogen was studied. In the pervaporative CMR, only 80% conversion of 4-chlorophenol was achieved [10]. This consequence is related to the fact that PEBA is well permeable both for 4-chlorophenol and reaction products; as a result, appreciable amounts of reaction products diffuse back from membrane to feed solution, and marked amounts of unreacted 4-chlorophenol pass through the pCM [10]. Moreover, this process is aggravated by a marked catalyst poisoning by the formed HCl [11]. To prevent undesirable effects, hydrogenation reaction of nonchlorinated compound was studied, namely, hydrogenation of acetophenon using PEBA and PDMS as basic membrane materials [12]. The supported catalyst was prepared by suspending silica in ethanol, adding polyvinylpirrolidone (PVP) and Pd-acetate and refluxing the suspension to reduce Pd salt and forming metal nanoclusters. The reaction depends strongly on the H_2 solubility in the membrane. The acetophenon conversion achieves 83% at a hydrogen pressure of 4 bar and at 50 °C. For the gas-phase hydrogenation reaction of propylene into propane in CMR, PDMS-based composite membranes containing Pd nanoclusters were obtained (PDMS layer thickness was about 4 μm) [13]. Feed H_2/C_3H_6 mixture was supplied to CMR at a pressure of 1.2 bar; at the permeate side, the pressure was maintained at 100 mbar to provide the driving force for directed mass transfer through pCM. The corresponding model has been advanced (this model is discussed in Section 24.3). Noteworthy is that Brandão *et al* [16] reported experimental and modeling results on sorption, diffusion and permeability of Ar, H_2, propane, propylene and propyne in dense PDMS membranes with Pd nanoclusters; this material is similar to that used in the catalytic PDMS composite membranes [13].

Gao and coworkers [17] prepared dense catalytic membranes (CMs) based on several kinds of modified or unmodified poly(2,6-dimethyl-1,4phenylene oxide) (PPO) and polysulfone (PSF) by refluxing a Pd-chloride precursor in ethanol solution. It was shown that, in the gas phase hydrogenation of diene to monoene, selectivity of polymer-anchored Pd catalysts is strongly controlled by the hydrogen partial pressure.

The above versions for the preparation of dense pCMs are based on the use of such polymeric materials as PAI, PDMS, PEBA, PSF and PPO, for which commercial technological procedures allowing preparation of high-performance composite or asymmetric membranes have been worked out. Moreover, some publications have been devoted to the preparation and characterisation of dense pCMs (strictly speaking, free-standing films) based on unconventional membrane materials (e.g., [18–20]). Hydrogenation experiments (hydrogenation of propene [18], ethylene and propylene [19] and 1,3-butadiene [20]) were performed as batch reactions. A detailed analysis of these works is beyond the scope of this review.

24.2.2 Pd-Loaded Gas Separation Membranes

Interesting approach for the *in situ* preparation of catalytic membrane reactors has been advanced by Xu and coworkers [21–25]. Asymmetric gas separation membranes used as supports for palladium catalysts were hollow fibres based on cellulose acetate (CA) [21–25], PSF [22,24], and polyacrylonitrile (PAN) [22]. All of CA, PSF, and PAN fibres have a thin dense outer skin layer on a microporous sponge layer.

Approximately 50–70 hollow fibres were collected and clued (by epoxy resin) into a membrane modulus; later, the membranes were catalytically activated (*in situ* CMR preparation). The precursor was prepared in the following way: palladium was anchored first on poly (vinylpyrrolidone) (PVP) [21–25], ethyl cellulose (EC), or melamine–formaldehyde resin (AR) [22,24,25] via corresponding reactions of $PdCl_2$ and PVP, EC or AR. The Pd-loaded catalytic membranes were prepared by pumping an aqueous solution of these metalated polymer precursors through inside of CA, PSF, or PAN hollow fibres in order to retain the precursors in the membrane porous support. NH_2NH_2 or $NaBH_4$ were then used to reduce Pd(II) to its low valence state. The resultant CMRs were studied in the hydrogenation reaction with various conjugated dienes (cyclopentadiene [21,22,25], isoprene [22], butadiene [22,24], propadiene and propyne [23]). The diene and hydrogen were supplied, respectively, inside and outside of hollow fibres. All hydrogenation reactions were conducted in gas phase.

Initial and catalytic membranes were tested with respect to the permeability towards H_2 and N_2 (Table 24.1). Solubility of H_2 in palladium is high; hence, in our opinion, interpretation of the experimental evidence on H_2 permeability seems to be more complicated as compared with N_2. As follows from Table 24.1, N_2 permeability through catalytic membranes appears to be by nearly two orders of magnitude lower than that of virgin membranes. This marked decrease in permeability was explained by the additional resistance provided by the anchored palladium complexes within the porous membrane layer.

Under mild hydrogenation conditions (atmospheric pressure and temperature 40 °C), the prepared pCMs appear to be stable, active and selective in the hydrogenation of dienes and alkynes. Both activity and selectivity of catalytic hollow fibres strongly depend on the polymeric material of hollow fibres as well as on type of polymers used for anchoring Pd complex. The best results were obtained for PVP-Pd and CA or PAN hollow fibres. By varying experimental conditions, conversion of cyclopentadiene can be increased up to 91% or even higher [22]. By using bimetallic (Co/Pd) catalytic membranes, the isomerisation of 1-butene was inhibited and the synergic effect of bimetallic catalyst was significant. The synergetic effect was only observed when reduction was done with $NaBH_4$ instead of NH_2NH_2. It was assumed that small palladium clusters were deposited on superfine cobalt boride particles when $NaBH_4$ was used as a reducing agent [24].

In the conclusion, we would like to mention that H_2 is a fast permeating component of gaseous mixtures. Hence, all CMRs based on gas separation hollow fibre membranes discussed in this section can appear to be useful for the hydrogenation reactions by using not only pure H_2 but also H_2-containing mixtures. In this case, selective layer of gas separation membrane is responsible

Table 24.1 *Permeability of the original and catalytic hollow fibres (driving pressure 5 atm, room temperature). Adapted from [22]*

Fibre	Permeability ($m^{-3}\,m^{-2}\,h^{-1}\,bar^{-1}$)		Permselectivity (α_{H_2/N_2})
	H_2	N_2	
CA	$3.2 \cdot 10^{-1}$	$8.0 \cdot 10^{-2}$	4.0
PVP-Pd/CA	$3.0 \cdot 10^{-2}$	$1.3 \cdot 10^{-3}$	23.9
EC-Pd/CA	$2.4 \cdot 10^{-2}$	$5.9 \cdot 10^{-4}$	40.0
AR-Pd/CA	$1.8 \cdot 10^{-2}$	$5.6 \cdot 10^{-4}$	32.1
PSF	$1.3 \cdot 10^{-1}$	$2.0 \cdot 10^{-3}$	65.8
PVP-Pd/PSF	$1.9 \cdot 10^{-3}$	$2.1 \cdot 10^{-5}$	91.4

for H_2 purification prior to its supply onto the catalyst, which is immobilised within the porous membrane support.

24.2.3 Porous Catalytic Membranes

In the case of porous membranes compared to dense ones, the requirements on polymeric materials are less severe because all components are transferred through pores rather than through polymer matrix. In this case, the requirements are reduced to thermal and chemical stability of polymers and their surface characteristics such as hydrophilicity or hydrophobicity.

24.2.3.1 Gas Phase Hydrogenation

Fritsch and coworkers [26] have studied the advantages of the combination of polymeric and inorganic membranes for the preparation of hybrid organic–inorganic porous CMs for gas-phase hydrogenation. This approach can be described as follows. Low-cost ultrafiltration membranes are used, and their inner space is modified with TiO_2 particles with the surface-deposited palladium catalyst. In this case, the diameter of pores decreases down to 10 nm, and the Knudsen flow regime is provided. Moreover, palladium catalytic function appears to be decoupled with the polymer matrix. The membranes were studied in the catalytic membrane reactor under 'flow-through with retentate' regime with respect to the hydrogenation of propyne to propene or propyne to propane. In this connection, the key aspect is concerned with the proper selection of pore sizes and gas permeability of the catalytic membrane as this selection is responsible for the best combination of high conversion and performance of the whole process.

Table 24.2 presents the data on changes nitrogen flow rate through the porous membranes based on PAN, polyetherimide (PEI), and PAI after their modification with TiO_2 and subsequent deposition of palladium. Flat sheet membranes were prepared as supported membranes according to the phase inversion process. After each stage of modification, permeability of membranes is seen to be decreased. A broad distribution in flow rates indicates a poor reproducibility of the above membranes. According to [26], 'PAI+' membranes seem to be best candidates for their practical application because preparation of initial membranes with high performance and without any collapse of small pores does not require any stage of solvent exchange. Moreover, due to the introduction of TiO_2, the degree of swelling of membranes in

Table 24.2 *Nitrogen permeability of various membrane materials before and after pore modification with titanium dioxide and palladium (adapted from [26])*

Membrane material	Number of stamps	Permeability of original ($m^3\,m^{-2}\,h^{-1}\,bar^{-1}$)	Permeability of TiO_2 modified ($m^3\,m^{-2}\,h^{-1}\,bar^{-1}$)	Permeability of TiO_2 and Pd modified ($m^3\,m^{-2}\,h^{-1}\,bar^{-1}$)
PAN	10	50–60	20–30	0.3–7.0
PAN filled by TiO_2 at membrane casting (PAN+)	5	100	14	1.0–2.0
PAI	10	20	9–12	0.2–2.6
PAI filled by TiO_2 at membrane casting (PAI+)	15	70–85	10–15	0.2–1.5
PEI	3	120–140	40	3.0–17.0

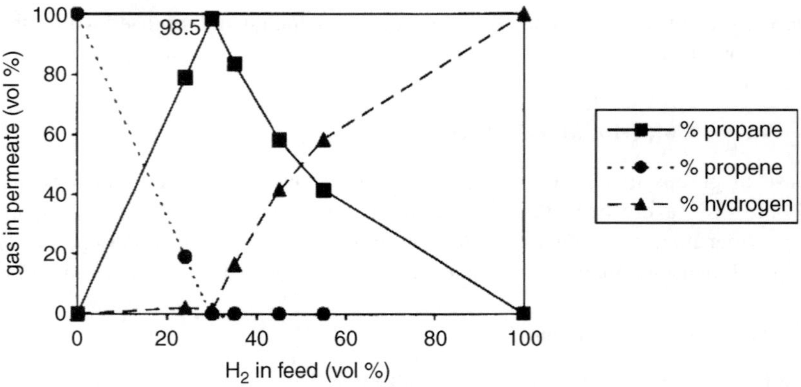

Figure 24.1 *Hydrogenation of propene at different H_2 concentration in feed (membrane PAI+; 0.29 wt% TiO_2; 1.31 wt% Pd). Reprinted from Ziegler S., Theis J., Fritsch D., Palladium modified porous polymeric membranes and their performance in selective hydrogenation of propyne // J. Membr. Sci.; 187: 71. Copyright (2001) with permission from Elsevier*

organic solvents is reduced, and thermal stability is improved (up to 200 °C). Comparing the data summarised in Tables 24.1 and 24.2, one can conclude that high gas transport characteristics can be achieved in the pCMs based on ultrafilrtration membranes.

Figure 24.1 presents the results on the hydrogenation of propyne using catalytic membrane based on 'PAI+'. As follows from Figure 24.1, the maximum yield of 98.5% of the product propane is observed when hydrogen content is 30 vol% and permeate flux is 0.26 $m^3 m^{-2} h^{-1} bar^{-1}$. In addition, selective hydrogenation of propyne in propene was studied where conversion only to propene was required. Depending on the type of membrane, selectivity towards target product, propene, was equal to 80–99%. Noteworthy is that this reaction is the key reaction for the petrochemical industry in the production of polypropylene.

Hydrogenation of propyne and cyclohexane was also studied by Schomacker and coworkers [27,28] when porous membranes based on polyacrilic acid (PAA) were used. Pd nanoparticles (Pd-acetate as a precursor and ethanol or $NaBH_4$ as a reducing agents), stabilised by block copolymer polysterene-co-polyethyleneoxide were immobilised within the PAA network, and flat sheet membranes of diverse porosity were obtained.

Bhattacharyya and coworkers [29–31] have proposed an alternative approach for the hydrogenation reaction in the presence of water but without any supply of gaseous hydrogen. In this case, bimetallic catalysts, including Fe-Pd particles, are used. The role of Fe is the generation of hydrogen by corrosion reaction and Pd serves as a catalyst. Nanoscale Fe-Pd particles in PAA/PES [29] and PAA/PVDF porous membranes [30] were synthesised. Catalytic activity of membranes was tested in the reaction of catalytic hydrodechlorination of trichloroethylene and polychlorinated biphenyls. High catalytic activity of Pd was confirmed by low activation energy at 25–65 °C. However, deposition of iron hydroxide on the surface of Fe/Pd nanoparticles entails deactivation of the catalyst.

24.2.3.2 Three-Phase Catalytic Membrane Contactors

All the above-discussed studies on porous pCMs were devoted to the hydrogenation in the gas phase. Below, we will consider the results on three-phase catalytic membrane contactors (CMCs) for the hydrogenation in liquid phase. Most works have been focused on the hydrogenation of sunflower oil [32–35]. The above works solved the challenging problem on the reducing of the formation of trans-isomers during hydrogenation. Some publications reported the results on the hydrogenation of methylenecyclohexane [36] and water-dissolved oxygen [37–43]. According to the classification advanced by Dalmon [44], depending on the mode of supply of reagents, two types of contractors exist: interfacial contactor and flow-through contactor. Noteworthy is that the correct design of pCM makes it possible to achieve both interfacial [37–43], and flow-through [32–36] contactor modes.

Flow-Through Mode. Ilinitch *et al.* [32] prepared pCM by loading a nanosised Pd catalyst into the pores of polymeric Nylon-6-based membrane, and hydrogenation of sunflower oil was studied with respect to cis/trans selectivity of the resultant membrane. The results on the membrane reactor were compared with the results on a conventional slurry reactor operating with palladium on charcoal catalyst. In the case of the CMR, the formation of trans isomers was shown to be lower at comparable hydrogenation levels.

Fritsch and coworkers [33–35] proposed new high oil flux pCMs based on PES and PAI with and without inorganic alumina filler. Casting solution contains pore-forming agent, pluronic (a water soluble ethyleneoxide-propyleneoxide-b-copolymer with a molecular mass of 12 kDa). Membrane casting only at temperatures below 10 °C and subsequent removal of pluronic by water provide the formation of large pores and development of high porosity. In this case, the formation of water soluble agglomerates of pluronic at temperatures below 10 °C is the reason behind the formation of large pores and related high oil flux (900–2000 $l\,m^{-2}\,h^{-1}\,bar^{-1}$ at 60 °C). The alumina particles are arranged at the interface of inner membrane pores and, therefore, are highly accessible to the reactants in the flow-through contactor mode. Addition of citric acid to the impregnating solution is important to control the dispersion of nanoparticles and to reduce calcination temperature from 350 down to 175 °C. Pd- and Pt-containing pCMs are characterised by similar catalytic activity. However, as compared with Pt catalyst, undesirable process of isomerisation to trans-fatty acids appears to be more pronounced for the Pd catalyst.

Hydrogenation of methylenecyclohexane to methylcyclohexane was studied for the two types of porous catalytic PVDF membranes prepared by the phase inversion process [36]. Both membranes were cast from the DMF solutions of PVDF; in one case, PVP was added. The prepared porous asymmetric membranes were filled with the acidic solution of $PdCl_2$, and Pd(II) was reduced to Pd(0) by $NaBH_4$. Average dimensions of Pd particles are equal to 4.5 and 3.5 nm for the membranes with and without addition of PVP, respectively. Methylenecyclohexane was presaturated with hydrogen and supplied to CMR, where liquid-phase hydrogenation took place. Comparing catalytic activity of both membranes, one can conclude that the *turn over number* (TON) is lower for the PVDF-PVP-Pd membrane, and this fact can be explained by the smaller dimensions of Pd nanoparticles.

Interfacial Mode. In recent years, marked contribution to the development and study of interfacial catalytic membrane contactors has been provided by the Russian–Dutch joint team [37–43]. As a result, Pd-loaded CMs based on porous polypropylene (PP) hollow fibre membranes have been worked out (Accurel Q3/2, Accurel S6/2, and Celgard X50); laboratory-scale and pilot interfacial CMCs have been designed; and removal of dissolved oxygen from water has been studied [43]. To our knowledge, this is the only example for the use of the interfacial contactor mode based on pCMs.

Noteworthy is that the process of catalytic removal of dissolved oxygen (DO) from water presents an encouraging example for the industrial application of Pd catalysts supported on ion-exchange resins [45,46]. The process of DO removal is carried out in two stages: (i) absorption of hydrogen and (ii) passage through a fixed-bed catalytic reactor.

The interfacial catalytic contactor approach provides opportunity for one stage process. Hydrophobic porous pCM should perform three principal functions: (i) well defined and easily controlled position of gas–liquid interface, (ii) accessibility of a catalyst for reagents (hydrogen and oxygen), (iii) high parameters of hydrogen mass transfer.

Prior to the deposition of Pd onto the membrane surface, hollow fibres were cleaned with surfactants and organic solvents. Subsequently, their outer surface was etched either by strong inorganic acid or by strong inorganic base. In the next step, one of the two methods of chemical (electroless) deposition of Pd was used: (i) reduction of palladium tetra-aminochloride by NH_2NH_2 (Accurel Q3/2, Accurel S6/2) [40,42,43] and (ii) reduction of $PdCl_2$ or $PdAc_2$ by aliphatic alcohol (Accurel Q3/2 [40,42], Celgard X50 [41,43]).

Later, after the development of the method for the coating of single membrane fibres (*ex situ* coating), this method has been adapted for the coating of a commercially available membrane module (commercial Liqui-Cel™ Extra-Flow membrane contactor, 1.4 m^2 membrane surface area) without its disassembly (*in situ* coating). As an intermediate step, laboratory-scale modules were coated. Both commercial and laboratory-scale modules were tested in a continuous flow system in both once-through and recirculation mode.

Figure 24.2 shows the oxygen removal rate for *ex situ* and *in situ* coated modules. It should be noted that gas phase is stagnant; therefore, physical stripping of oxygen is prevented. Within 80 min at room temperature concentration of DO in water decreases down to 110 and 130 mg/l for *ex situ* and *in situ* modules, respectively. In both cases, rate constants well agree within experimental error (0.189 ± 0.020 and $0.188 \pm 0.020\,mg\,l^{-1}\,min^{-1}$). This fact indicates that the procedure for the deposition of catalyst onto individual hollow fibres can be used for the deposition of Pd onto the membrane surface directly in the membrane module (without its disassembly).

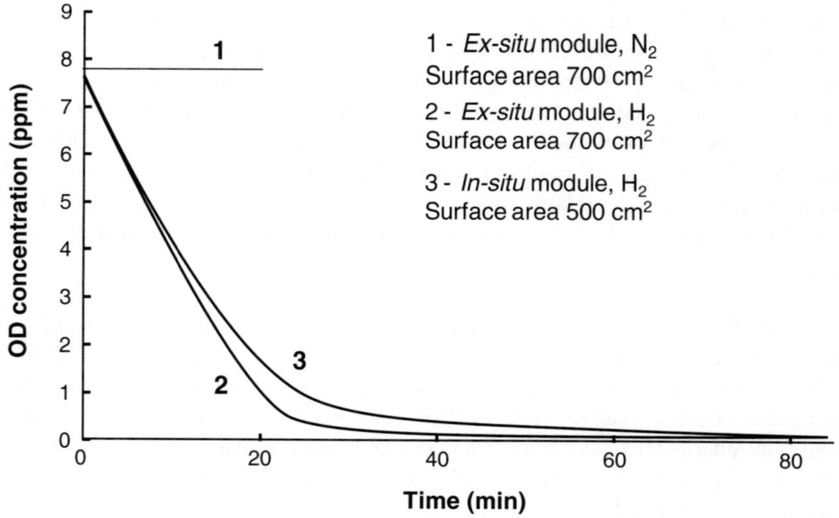

Figure 24.2 *Oxygen removal by in situ and ex situ coated hollow fibre modules in recirculation mode. Reprinted from Journal of Membrane Science, van der Vaart, R., Preparation and characterisation of palladium-loaded polypropylene porous hollow fibre membranes for hydrogenation of dissolved oxygen in water. Vol. 299: 38. Copyright (2007) with permission from Elsevier*

In situ method can be also used for the commercially available Liqui-Cel modules. For this contactor in the water once-through mode (initial concentration of the DO is 8 mg l^{-1}), the DO conversion is increased from 80 to 96% when contact time increases from 0.5 to 3.0 min. As to the water recirculation mode, concentration of DO in water decreases down to sub-ppb level [43].

24.3 Characterisation of Palladium Nanoparticles in Catalytic Membranes

Characterisation of nanoparticles presents the important part of studies on catalytic membranes. For example, characteristics of nanoparticles control both catalytic activity and permeability of membranes. In this section, we will consider characteristics of palladium particles incorporated or synthesised in polymer membranes.

As expected, the most traditional methods for the characterisation of Pd nanoparticles in pCMs are the methods of X-ray diffraction (XRD) and electron spectroscopy [10,11,13,42,47,48]. For example, the average dimensions of particles d_p can be estimated by the XRD [49]. For crystalline nanoparticles, these dimensions coincide with the coherent scattering region (CSR) and can be estimated at the halfwidth w of diffraction peaks. Correlation between d_p and w is described by the Sherrer formula:

$$d_p = \frac{\lambda}{w \cos \theta}, \tag{24.1}$$

where λ is the wavelength, and θ is the angular position of the X-ray reflection

Figure 24.3 presents the XRD spectra for Pd nanoparticles with dimensions of 3.7 and 1.5 nm. Spectra corresponding to the particles with different dimensions are seen to be appreciably different. Note that, during catalytic membrane experiments, nanoparticles change (reduce) their sizes. Even though dissolution of particles seems to be very unusual process, this process should be taken into account for the analysis of the characteristics of the metal–polymer catalytic systems [45].

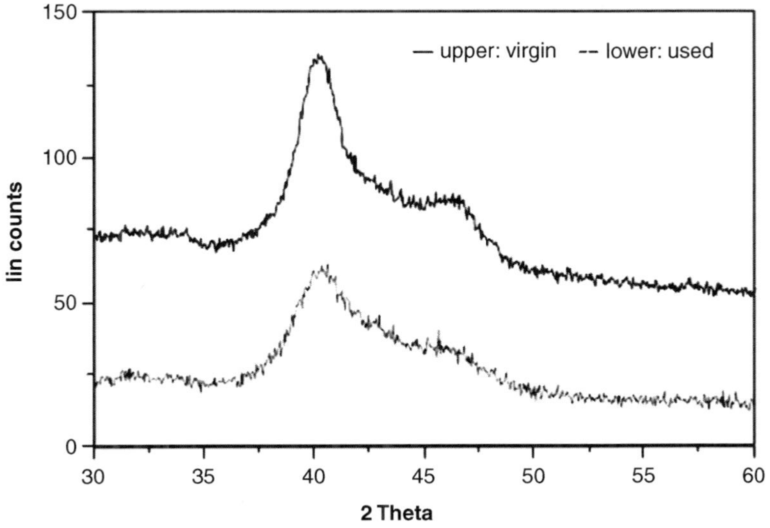

Figure 24.3 *XRD spectra of membrane before and after use in hydrodechlorination of 4-chlorphenol. Reprinted from Chemical Engineering and Processing*

Table 24.3 *EXAFS and XRD results (adapted from [42])*

Sample	PP membrane	Pd content (% w/w)	Pd content ($\mu g\, cm^{-2}$)	Amplitude	Inter-atomic distance (Å)	Effective coordination number	D_{111} (nm)
Pd foil	—	100		3.028	2.72	12	—
M-1	Accurel S6/2	48	9790	2.734	2.72	10	37
M-2	Accurel Q3/2	24	960	2.590	2.74	10	31
M-3	Accurel Q3/2	0.73	30	2.170	2.73	8	34
M-4	Accurel Q3/2	0.68	28	2.007	2.73	8	25
M-5	Accurel Q3/2	0.36	25	1.996	2.73	8	26
M-6	Accurel Q3/2	0.3	12	1.445	2.74	6	18

For larger particles, XRD spectrum coincides with the calculated spectrum [42]. This coincidence of XRD patterns makes it possible to gain additional information concerning the state of Pd nanoparticles in pCMs: particles are only metallic Pd^0. Note that the recorded EXAFS spectra also indicate the existence of only Pd-Pd bonds in the Pd-loaded PP hollow fibre membranes [42]. As follows from Table 24.3, the first interatomic distance is virtually the same for all samples and its value is equal to the similar values measured for the metallic Pd foil. Moreover, even though the first coordination number for the palladium foil is equal to 12, this parameter appears to be equal to 10, 8, and 6 for the pCMs with Pd-content of 9790, 15–30, and 12 $\mu g\, cm^{-2}$, respectively. In our opinion, this evidence can be explained by the following reasons: (i) small dimensions of Pd clusters (decreased dimensions of CSR, D_{111}, Table 24.3), (ii) numerous pointlike lattice defects, (iii) the presence of small-sized X-ray amorphous particles, which cannot be detected by the XRD method.

In our opinion, it seems reasonable to add some comment concerning the chemical structure of Pd nanoparticles. Even though conclusions, drawn in the above-cited works that Pd exists on the surface as Pd^0 nanoparticles, are valid, the direct method for the chemical analysis is the ESXA method [49]. It is astonishing that, to our knowledge, this method has not been yet used for the analysis of the surface of Pd-containing pCMs. Also, the method of energy dispersive X-ray spectroscopy (EDS) is often included in the traditional SEM observations [49]. This method makes it possible to estimate the distribution of elements in the scanned surface plane (see, e.g., [43,50]) and to identify the distribution of elements in individual particles. As an example, Figure 24.4 presents the data on the EDS analysis for the distribution of Pd particles on the surface of the PP membrane [43]. The analysis of dark region (point 2 in Figure 24.4) detected races of metallic Pd presented in this region (Figure 24.4c).

Demir *et al.* [50] synthesised Pd nanoparticles via reduction of $PdCl_2$ with NH_2NH_2 in the polymer matrix based on (acrylonitrile)–(acrylic acid) copolymer fibre mats. Fibres were prepared by electrospinning of solutions of PAN-AA and $PdCl_2$ in DMF. An important point is that high-resolution electron microscope images demonstrate that the Pd particles are located on the surface of the fibres. It is assumed that NH_2NH_2 penetrates into the fibres, reduce Pd(II) to Pd(0), and then Pd atoms diffuse to the fibres surface and agglomerate. According EDS analysis, Cl ions are not detected. The average Pd-crystallite sizes, estimated from the peak broadening of XRD peaks, are about 10 nm.

The typical example on the study of Pd nanoparticles is the work by Tröger *et al.* [14]. Characteristics of dense pCMs based on PAI were studied. The above membranes were loaded with Pd particles and by bimetallic Pd/Ag, Pd/Cu, Pd/Co and Pd/Pb particles. The pCMs were studied by the methods of X-ray absorption fine structure (XAFS), small-angle X-ray scattering

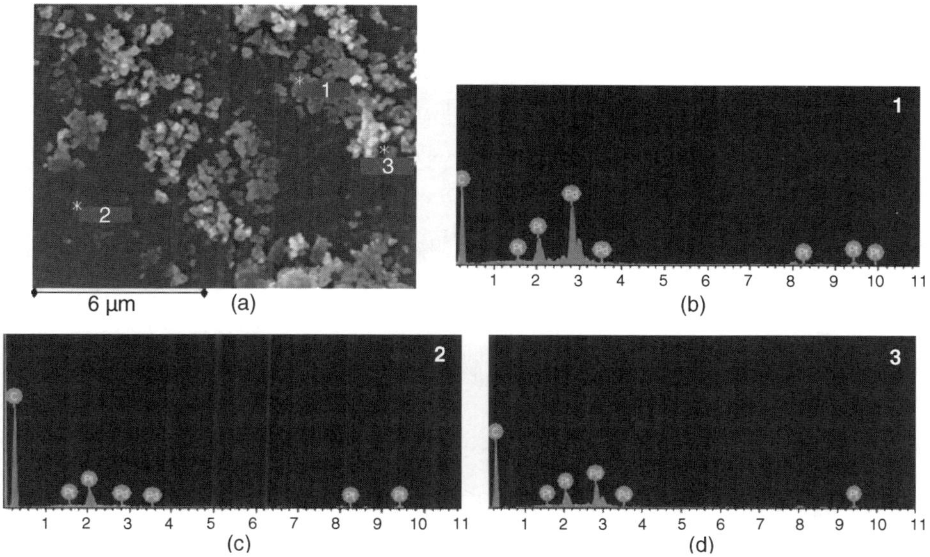

Figure 24.4 *Scanning electron micrograph image of MX/1.94 [1.94 wt.% of Pd] (a) and energy dispersive X-ray analysis: (b) point 1, (c) point 2, (d) point 3. Reprinted from Volkov V.V., Petrova I.V., Lebedeva V.I., Plyasova L.M., Rudina N.A., van Erkel J., van der Vaart R., Tereshchenko G.F., Chapter 40, Starov V.M. (Ed.) Copyright (2009) with permission from Taylor & Francis Group*

(SAXS), XRD, and transmission electron microscopy (TEM). The TEM observations show a homogeneous distribution of metallic nanoclusters of 1–3 nm. Besides, a smaller content of larger aggregates up to 30 nm is also observed.

Information concerning the size distribution of particles can be extracted from the SAXS data [14,20]. For example, Figure 24.5 presents scattering intensity plotted against scattering vector for the Pd particles and Pd/Ag system (3:1), and the curves indicate that, in both systems, the size distribution of nanoparticles shows the bimodal character.

Table 24.4 presents the data on the dimensions of palladium particles obtained by different methods under different conditions [14]. For relatively small particles, all results fairly agree. For large particles, a marked difference is observed, and this difference can be related to the fact that, for example, TEM observations reveal not primary particles (which are detected by the XRD) but only their aggregates.

In this connection, the most efficient procedure for the analysis of the aggregates of Pd nanoparticles on the membrane surface is the combination of various methods such as SEM, XRD, AFM, and optical microscopy [51]. In this work, two modes of the deposition of Pd onto the outer surface of porous PP membranes were used: consecutive and continuous regimes (catalytic MC-CS and MC-CN membranes, respectively). The value of CSR for Pd nanocrystals is equal to 10 and 40 nm for consecutive and continuous regimes, respectively. Figure 24.6 presents the SEM images of the MC-CS and MC-CN membranes. The continuous method allows the formation of adlayers composed of clusters with dimensions below 100–200 nm as well as the development of relatively dense agglomerates of such clusters. In the case of consecutive method, less dense adlayers are formed on the surface of membranes, and these adlayers are composed of clusters, whose dimensions are comparable to the dimensions of CSR, and agglomerates of these clusters which, in the optical region, manifest themselves as black spots, and this pattern is indicative of their high porosity.

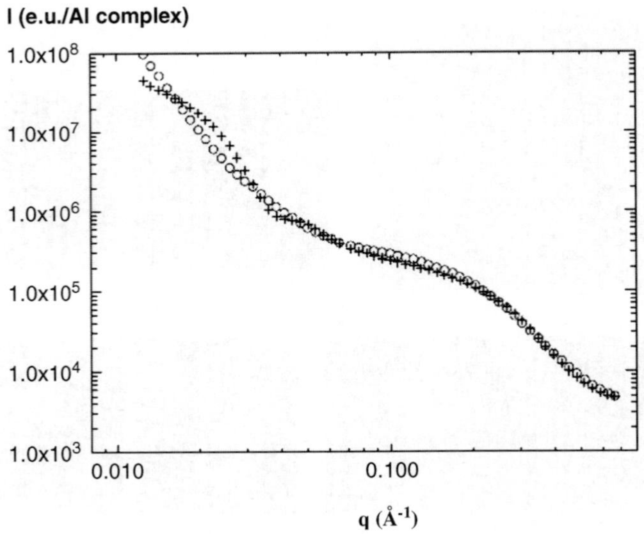

Figure 24.5 *SAXS data of 15 wt% metal-loaded membrane: Pd (plus signs) and Pd/Ag (circles). With kind permission from Springer Science+Business Media: Z. Phys. D, Structural characterization of catalytically active metal clusters in polymer membranes, 1997; 40: 81, Tröger L., et al.*

Table 24.4 *Pd cluster sizes (Å) observed in selected membranes by different techniques (adapted from [14])*

Loading	Preparation	TEM	XAFS	XRD
Pd	NMP, 3 h	10	13	9
Pd	NMP, 5 h	10	13	10
Pd	THF, 1 h	50–100	39	—
Pd	THF, 16 h	10–30	52	23

24.4 Kinetic Studies

The kinetics of reactions in polymer membrane reactors containing Pd catalyst has been described in detail in [13,21–23,28–30]. Dechlorination reaction and hydrogenation reaction of organic compounds have been revisited. Initially, for dechlorination reactions, catalysts based on Fe nanoparticles were used. As was shown in later works, bimetallic catalysts appeared to be far more efficient (see [29]); among them, Fe/Pd catalysts were selected as best candidates.

To compare characteristics of catalysts, the rate of kinetic process should be normalised by the surface area of catalyst. According to [52], dechlorination reaction of chlorine-containing organic compounds on metallic catalyst proceeds via the pseudo-first-order reaction. In the batch system, dechlorination rate can be described by:

$$dc/dt = -k_{SA} s_s \rho_m c \qquad (24.2)$$

where c is the overall concentration of reagents, including aqueous and membrane phase (mg l^{-1}), k_{SA} is the surface area normalised reaction rate coefficient (l h^{-1} m^{-2}), s_s is the specific surface area of metal, ρ_m is the weight concentration of metals in the solution (g l^{-1}), and t is time (h).

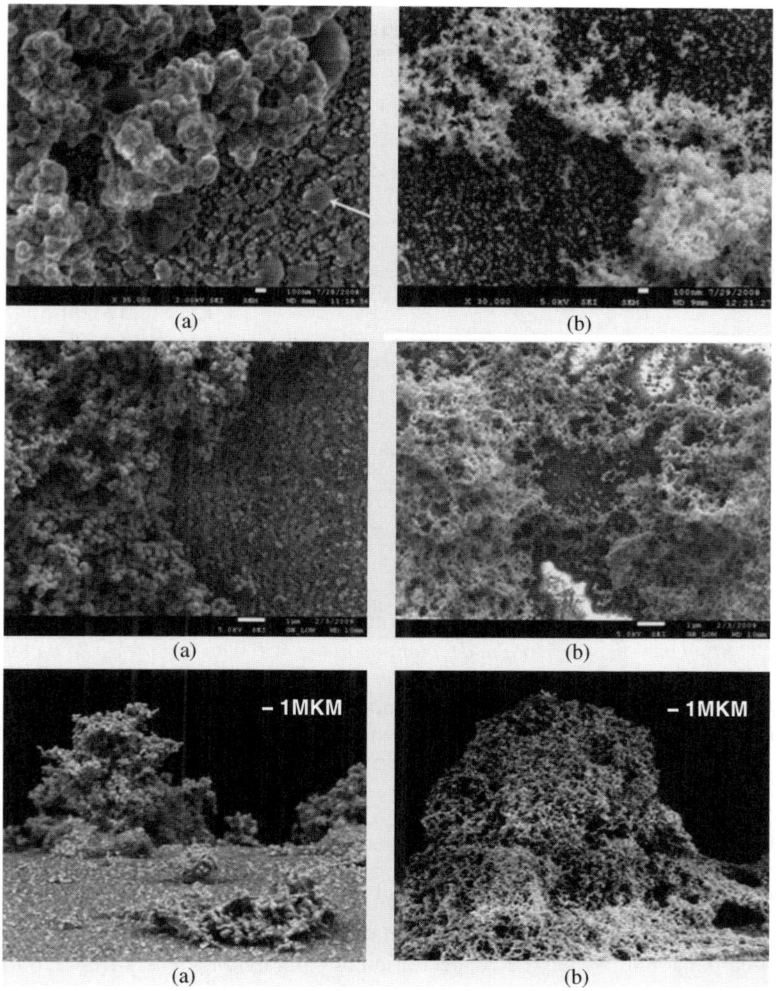

Figure 24.6 *SEM images for (a) MC–CN (left side) and (b) MC–CS (right side). The dimensions of CSR are ≈40 and ≈10 nm, respectively. Scale in the images is 100 nm or 1 μm. The two upper rows of images recorded perpendicular to the surface show top views in two scales; the bottom images show the view along the surface [51]*

For various catalysts, experimental results [29] show that when the kinetics of dechlorination of trichloroethylene is described in terms of Equation (24.2) and dimensions of particles (or, in other words, surface of catalytic particles) are estimated by the electron microscopic observations, reaction constant for the Fe/Pd catalyst ($k_{SA} = 0.948 \pm 0.050\, l\, h^{-1}\, m^{-2}$) appears to be by more than one order of magnitude higher as compared with that for the Fe/Ni catalyst.

The rates of catalytic reactions strongly depend on the dimensions of particles. In [30], hydrodechlorination of 2,2′-dichlorobiphenyl was studied for the catalytic particles of various dimensions. The experiments in CMR were performed for the Fe/Pd nanoparticles inside PAA/PVDF membranes. The average particle size was 30 ± 5.7 nm (TEM measurements), as

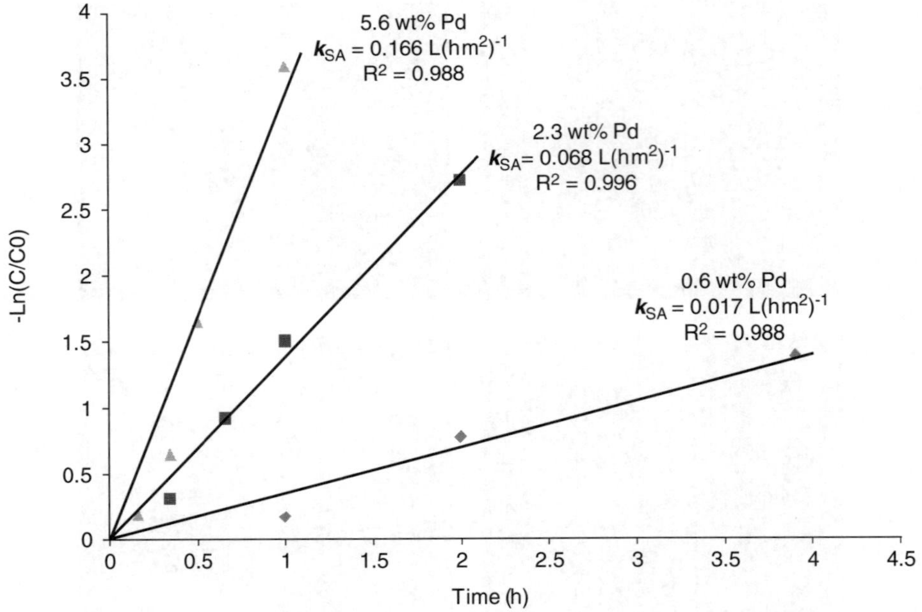

Figure 24.7 *Best linear fit of k_{SA} for the dechlorination of DiCB with various Fe/Pd nanoparticles in PAA/PVDF membranes. Metal loading = 16 mg in 20 ml. Reprinted from Industrial & Engineering Chemistry Research, Xu, J., et al., Fe/Pd Nanoparticle Immobilization in Microfiltration Membrane Pores: Synthesis, Characterization, and Application in the Dechlorination of Polychlorinated Biphenyls. Vol. 46, 2348. Copyright (2007) with permission from American Chemical Society*

a result, the external surface area for the nanoparticles was calculated to be $\sim 25\,m^2\,g^{-1}$. In the case of Pd-coated iron particles with dimensions of $\approx 120\,\mu m$, the rate constant is equal only to $0.00011\,l\,h^{-1}\,m^{-2}$, and this parameter is 600 times lower than that for the nanoscale Fe/Pd particles. According to this work, reaction rate constant also depends on the concentration of the deposited palladium. When the concentration of palladium increases from 0.6 to 5.6 wt%, the reaction rate constant increases from 0.017 to $0.166\,l\,h^{-1}\,m^{-2}$ (Fig 24.7) [30]. Calculations based on reasonable assumptions concerning the structure of catalytic particles make it possible to conclude [30] that the reaction rate is primarily controlled by the sites linked to palladium atoms: reaction rate appears to be directly proportional to the number of Pd surface sites. Note that, in the case of membranes containing incorporated Fe/Pd particles, activation energy of hydrodechlorination of 2,2′-dichlorobiphenyl is equal to $24.5\,kJ\,mol^{-1}$, and this value is 4–5 times lower than that of noncatalysed process or for the systems containing Ni-Mo catalyst [30]. The reaction hydrodechlorination of chlorobenzene on pCM with Pd catalyst is the first-order reaction [53].

Hydrogenation of propyne was also shown to be the first-order reaction [28]. This reaction was analysed in terms of the Langmuir–Hinshelwood model, and this analysis shows that this reaction with the Pt catalyst follows the first order with respect to H_2 and zero order with respect to propyne and propene.

In [13], the kinetics of propylene hydrogenation on PDMS membranes was compared with the data reported in [54], where this reaction was performed on free Pd clusters coated with the stabilising surfactant layer [n-$(C_{18}H_{37})_4N^+Br^-$]. This stabilising agent does not slow down mass

Table 24.5 *Kinetic parameters of propylene hydrogenation over palladium nanoclasters stabilised on different matrices (T = 308 K) (adapted from [13])*

Parameter	Matrix	
	Surfactant	PDMS
k (mol g_{Pd}^{-1} s^{-1})	5.569	18.3
$K^g_{C_3H_6}$ (m^3 mol^{-1})	$2.55 \cdot 10^{-2}$	$1.31 \cdot 10^{-1}$
$K^g_{H_2}$ (m^3 mol^{-1})	$9.7 \cdot 10^{-4}$	$9.7 \cdot 10^{-4}$
Average size of Pd clusters (nm)	7.3	8.8

transfer but blocks nearly half of the surface of nanoparticles. In both cases, local reaction rate can be described as:

$$r_{C_3H_8} = \frac{kK^g_{C_3H_6}K^g_{H_2}c^g_{H_2}c^g_{C_3H_6}}{\left(1+\sqrt{K^g_{H_2}c^g_{H_2}}+K^g_{C_3H_6}c^g_{C_3H_6}\right)^3} \tag{24.3}$$

where k is the kinetic rate constant, $c_i{}^g$ is the gas-phase concentration of species i, and $K_i{}^g$ is the adsorption equilibrium constant on catalyst surface.

Comparative data on this reaction for the two above cases are summarised in Table 24.5.

Kinetic parameters are listed in Table 24.5 for the surfactant-stabilised Pd and PDMS-supported Pd, which was prepared by the above-mentioned fitting process. A fivefold increase in the adsorption constant for propylene was obtained for the Pd/PDMS system under the assumption that certain surface sites on the surfactant-stabilised Pd are blocked off. Noteworthy is that a threefold in the kinetic constant k indicates that the Pd nanoclusters in the PDMS membrane appear to be more efficient as compared with surfactant-stabilised ones.

24.5 Conclusions

Membrane catalysis and catalytic membrane reactors dated back to palladium metallic membranes and hydrogenation in gaseous and liquid phases. Pioneering publications on the preparation of Pd-containing pCMs appeared in the early 1980s and, since the mid-1990s, this direction is known to include two subdirections: (i) dense pCMs and CMRs, and (ii) porous pCMs and CMCs in flow-through and interfacial modes.

In pCMs, size distribution of Pd nanoparticles in polymer matrices can be controlled by the conditions of synthesis, and this approach offers fascinating advantages for the control over the whole catalytic process. This approach makes it possible to obtain the ensemble of palladium particles in CMs with either polymodal or unimodal size distribution as well as to prepare pCMs based on hybrid polymer/metal oxides/Pd catalytic systems.

Our analysis shows that parameters of catalytic process are strongly controlled by the size and composition of particles (in the case of the composite or mixed nanoparticles), by the nature of polymer matrix, and parameters of the catalytic process. The use of mixed particles can substantially improve characteristics of catalytic processes.

Acknowledgement

The authors gratefully acknowledge the support of the Ministry of Education and Science of the Russian Federation (GK No. 02.740.11.0818).

References

1. Gryaznov V.M., Catalysis by selective permeable membranes, *Dokl. Akad. Nauk. USSR*, **189**, 794 (1969).
2. Gryaznov V.M., *A method for simultaneous carrying out catalytic reactions involving hydrogen evolution and consumption,* USSR Patent 274092 (1964).
3. Gryaznov V.M., Smirnov V.S., Vdovin V.M., Ermilova M.M., Gogua L.D., Pritula N.A., Litvinov I.A., *Method of preparing a hydrogen-permeable membrane catalyst on a base of palladium or its alloys for the hydrogenation of unsaturated organic compounds,* US Patent 4132668 (1979).
4. Gryaznov V.M., Smirnov V.S., Vdovin V.M., Ermilova M.M., Gogua L.D., Pritula N.A., Fedorova G.K., *Membrane catalyst for hydrogenation of organic compounds and method for preparing same,* US Patent 4394294 (1983).
5. Vankelecom I.F.J., Polymeric membranes in catalytic reactors, *Chem. Rev.*, **102**, 3779 (2002).
6. Dioos B.M.L., Vankelecom I.F.J., Jacobs P.A., Aspects of immobilisation of catalysts on polymeric supports, *Adv. Synth. Catal.*, **348**, 1413 (2006).
7. Ozdemir S.S., Buonomenna M.G., Drioli E., Catalytic polymeric membranes: Preparation and application, *Appl. Catal. A: General*, **307**, 167 (2006).
8. Fritsch D., Peinemann K.V., Novel highly permselective 6F-poly(amide-imide)s as membrane host for nano-sized catalysts, *J. Membr. Sci.*, **99**, 29 (1995).
9. Fritsch D., Peinemann K.V., Catalysis with homogeneous membranes loaded with nanoscale metallic clusters and their preparation, *Catal. Today*, **25**, 277 (1995).
10. Bengtson G., Scheel H., Theis J., Fritsch D., Catalytic membrane reactor to simultaneously concentrate and react organics, *Chem. Eng. J.*, **85**, 303 (2002).
11. Bengtson G., Oehring M., Fritsch D., Improved dense catalytically active polymer membranes of different configuration to separate and react organics simultaneously by pervaporation, *Chem. Eng. Process.* **43**, 1159 (2004).
12. Bengtson G., Panek D., Fritsch D., Hydrogenation of acetophenone in a pervaporative catalytic membrane reactor with online mass spectrometric monitoring, *J. Membr. Sci.*, **293**, 29 (2007).
13. Brando L., Fritsch D., Mendes A.M., Madeira L.M., Propylene hydrogenation in a continuous polymeric catalytic membrane reactor, *Ind. Eng. Chem. Res.*, **46**(16), 5278 (2007).
14. Tröger L., Hunnefeld H., Nunes S., Oehring M., Fritsch D., Structural characterization of catalytically active metal clusters in polymer membranes, *Z. Phys. D*, **40**, 81 (1997).
15. Tröger L., Hunnefeld H., Nunes S., Oehring M., Fritsch D., Structural characterization of catalytically active metal nanoclusters in poly(amide imide) films with high metal loading, *J. Phys. Chem. B*, **101**, 1279 (1997).
16. Brandão L., Madeira L.M., Mendes A.M., Mass transport on composite dense PDMS membranes with palladium nanoclusters, *J. Membr. Sci.*, **288**, 112 (2007).
17. H. Gao, Y. Xu, Sh. Liao, D. Yu, Catalytic hydrogenation and gas permeation properties of metal-containing poly(phenylene oxide) and polysulfone, *J. Appl. Polymer Sci.*, **50**, 1035 (1993).
18. Carlin R.T., Fuller J., Ionic liquid–polymer gel catalytic membrane, *Chem. Commun.*, **1997**, 1345 (1997).

19. Ciebien J.F., Cohen R.E., Duran A., Catalytic properties of palladium nanoclusters synthesized within diblock copolymer films: hydrogenation of ethylene and propylene, *Supramol. Sci.*, **5**, 31 (1998).
20. Ciebien J.F., Cohen R.E., Duran A., Membrane catalysts for partial hydrogenation of 1,3-butadiene: Catalytic properties of palladium nanoclusters synthesized within diblock copolymer films *Mater. Sci. Eng. C*, **7**, 45 (1999).
21. Gao H., Liao S., Xu Y., Liu R., Liu J., Li D., Selective hydrogenation of cyclopentadiene in a catalytic cellulose acetate hollow-fiber reactor *Catal. Lett.*, **27**, 297 (1994).
22. Gao H., Xu Y., Liao S., Liu R., Liu J., Li D., Yu D., Zhao Y., Fan Y., Catalytic polymeric hollow-fiber reactors for the selective hydrogenation of conjugated dienes, *J. Membr. Sci.*, **106**, 213 (1995).
23. Liu Ch., Xu Y., Liao Sh., Yu D., Zhao Y., Fan Y., Selective hydrogenation of propadiene and propyne in propene with catalytic polymeric hollow-fiber reactor *J. Membr. Sci.*, **137**, 139 (1997).
24. Liu Ch., Xu Y., Liao S., Yu D., Mono- and bimetallic catalytic hollow-fiber reactors for the selective hydrogenation of butadiene in 1-butene *Appl. Catal. A: General*, **172**, 23 (1998).
25. Liu Ch., Xu Y., Liao S., Yu D., Selective hydrogenation of cyclopentadiene in mono- and bimetallic catalytic hollow-fiber reactors, *J. Mol. Catal. A*, **157**, 253 (2000).
26. Ziegler S., Theis J., Fritsch D., Palladium modified porous polymeric membranes and their performance in selective hydrogenation of propyne *J. Membr. Sci.*, **187**, 71 (2001).
27. Beyer A., Schomacker R., Reichert K.H., Synthesis and characterization of palladium containing membranes based upon polyacrylic acid *Colloid Polym. Sci.*, **281**, 862 (2003).
28. Groschel L., Haidar R., Beyer A., Colfen H., Frank B., Schomacker R., Hydrogenation of propyne in palladium-containing polyacrylic acid membranes and its characterization *Ind. Eng. Chem. Res.*, **44**, 9064 (2005).
29. Xu J., Dozier A., Bhattacharyya D., Synthesis of nanoscale bimetallic particles in polyelectrolyte membrane matrix for reductive transformation of halogenated organic compounds *J. Nanopart. Res.*, **7**, 449 (2005).
30. Xu J., Bhattacharyya D., Fe/Pd nanoparticle immobilization in microfiltration membrane pores: synthesis, characterization, and application in the dechlorination of polychlorinated biphenyls *Ind. Eng. Chem. Res.*, **46**, 2348 (2007).
31. Xu J.; Bhattacharyya D., Membrane-based bimetallic nanoparticles for environmental remediation: synthesis and reactive properties, *Environ. Prog.*, **24**, 358 (2005).
32. Ilinitch O.M., Simonov P.A., Cuperus F.P., Nanosize palladium loaded catalytic membrane: Preparation and cis-trans selectivity in hydrogenation of sunflower oil, *Stud. Surf. Sci. Catal.*, **118**, 55 (1998).
33. Fritsch D., Bengtson G., Development of catalytically reactive porous membranes for the selective hydrogenation of sunflower oil, *Catal. Today*, **118**, 121 (2006).
34. Bengtson G., Fritsch D., Catalytic membrane reactor for the selective hydrogenation of edible oil: platinum versus palladium catalyst, *Desalination*, **200**, 666 (2006).
35. Fritsch D, Bengtson G., Catalytic polymer membranes for high temperature hydrogenation of viscous liquids, *Adv. Eng. Mater.*, **8**(5), 386 (2006).
36. Bottino A., Capannelli G., Comite A., Di Felice R., Polymeric and ceramic membranes in three-phase catalytic membrane reactors for the hydrogenation of methylenecyclohexane, *Desalination*, **144**, 411 (2002).
37. Vaart R., Hafkamp B., Koele P.J., Querreveld M., Jansen A.E., Volkov V.V., Lebedeva V.I., Gryaznov V.M., Oxygen removal from water by two innovative membrane techniques, *Proc. Int. Conf. Euromembrane 2000*, Jerusalem, Israel, 24–27 September, p. 359 (2000).
38. Van der Vaart R., Hafkamp B., Koele P. J., Querreveld M., Jansen A. E., Volkov V. V., Lebedeva V. I., Gryaznov V. M., Oxygen removal from water by two innovative membrane techniques, *Ultrapure Water*, **18**, 27 (2001).

39. van der Vaart R., Elizarova A.V., Volkov V.V., Lebedeva V.I., Gryaznov V.M., *Polymeren voorzien van metal,* Patent NL1023364 (2004).
40. Lebedeva V.I., Gryaznov V.M., Petrova I.V., Volkov V.V., Tereshchenko G.F., Shkol'nikov E.I., Plyasova L.M., Kochubey D.I., van der Vaart R., van Soest-Verecammen E.L.J., Porous Pd-containing polypropylene membranes for catalytic water deoxygenation, *Kinet. Catal.*, **47**, 867 (2006).
41. van der Vaart R., Petrova I., Lebedeva V., Volkov V., Kochubey D., Tereshchenko G., In-situ application of catalytic phase to commercial membrane contactor for removal of dissolved oxygen from water, *Desalination* **199**, 424 (2006).
42. van der Vaart R., Lebedeva V.I., Petrova I.V., Plyasova L.M., Rudina N.A., Kochubey D.I., Tereshchenko G.F., Volkov V.V., van Erkel J., Preparation and characterisation of palladium-loaded polypropylene porous hollow fibre membranes for hydrogenation of dissolved oxygen in water, *J. Membr. Sci.*, **299**, 38 (2007).
43. Volkov V.V., Petrova I.V., Lebedeva V.I., Plyasova L.M., Rudina N.A., van Erkel J., van der Vaart R, Tereshchenko G.F., Catalytic nanoclusters of palladium on the surface of polypropylene hollow fiber membranes: removal of dissolved oxygen from water, in *Nanoscience: Colloidal and Interfacial Aspects*, Starov V.M. (ed.), Taylor and Francis Group, Chapter 40 (2009).
44. Miachon S., Dalmon J.-A., Catalysis in membrane reactors: what about the catalyst, *Top. Catal.*, **29**, 59 (2004).
45. Kralik M., Biffis A., Catalysis by metal nanoparticles supported on functional organic polymers, *J. Mol. Catal. A*, **177**, 113 (2001).
46. Gross M.S., Pisarello M.L., Pierpauli K.A., Querini C.A., Catalytic deoxygenation of water: Preparation, deactivation, and regeneration of palladium on a resin catalyst, *Ind. Eng. Chem. Res.*, **49**, 81 (2010).
47. Sablong R, Schlotterbeck U., Vogt D., Mecking S., Catalysis with soluble hybrids of highly branched macromolecules with palladium nanoparticles in a continuously operated membrane reactor, *Adv. Synth. Catal.*, **345**, 333 (2003).
48. Turkenburg D.H., Antipov A.A., Thathagar M.B., Rothenberg G., Sukhorukov G.B., Eiser E., Palladium nanoclusters in microcapsule membranes: From synthetic shells to synthetic cells, *Phys. Chem. Chem. Phys.*, **7**, 2237 (2005).
49. Chorkendorf I., Niemantsverdriet J.W., *Concepts of Modern Catalysis and Kinetics*, Wiley-VCH, Weinheim (2007).
50. Demir M.M., Gulgun M.A., Menceloglu Y.Z., Erman B., Abramchuk S.S., Makhaeva E.E., Khokhlov A.R., Matveeva V.G., Sulman M.G., Palladium nanoparticles by electrospinning from poly(acrylonitrile-*co*-acrylic acid)-$PdCl_2$ solutions. Relations between preparation conditions, particle size, and catalytic activity, *Macromolecules*, **37**, 1787 (2004).
51. Volkov V.V., Lebedeva V.I., Petrova I.V., Bobyl A.V., Konnikov S.G., Ulin V.P., Roldughin V.I., Tereshchenko G.F., Adlayers of palladium particles and their aggregates on porous polypropylene hollow fiber membranes as hydrogenization contractors/reactors, *Adv. Coll. Inter. Sci.*, in press (2010).
52. Jonson T.L., Scherer M.M., Tratnyck P.G., Kinetics of halogenated organic compounds degradation by iron metal, *Environ. Sci. Technol.*, **32**, 2634 (1996).
53. Fritsch D., Kuhr K., Mackenzie K., Kopinke F.-D., Hydrodechlorination of chloroorganic compounds in ground water by palladium catalysts. Part 1. Development of polymer-based catalysts and membrane reactor tests *Catal. Today.*, **82**, 105 (2003).
54. Brando L., Fritsch D., Madeira L.M., Mendes A.M., Kinetics of propylene hydrogenation on nanostructured palladium clusters, *Chem. Eng. J.*, **103**, 89 (2005).

25

Membrane Prepared via Plasma Modification

Marek Bryjak and Irena Gancarz

Department of Polymer and Carbon Materials, Wroclaw University of Technology, Wroclaw, Poland

25.1 Introduction

Various preparation techniques are used to obtain polymer membranes useful in construction of bioreactors. Some of them, presented in the previous chapters, show how important is selection of a proper membrane materials. However, in the case of polymer membranes the problem seems to grow faster due to the shortage of new materials useful for membrane preparation. As was predicted 15 years ago [1], the world production of polymers reached its plateau. It is estimated today at the level of 220–250 × 10^6 t $year^{-1}$. What is more, the production of brand new polymers has not been commercialised during the past decade – and no launch is predicted for the very near future.

The limits of production have forced material scientists to search for some alternative methods that increase the number and variety of offered materials. In the case of polymer materials such approach seems to have significant importance as organic materials are easily modifiable. There are two ways to do that: (i) to modify a polymer first and prepare membrane from its derivative, and (ii) to prepare membrane first and them modify it. As the first way needs the optimisation of membrane manufacturing process for each derivative, the second one seems to be easy to carried out. There are many of membranes available on the market and the effort of membrane modification can be focused on the alteration of their properties only. This approach is

Membranes for Membrane Reactors: Preparation, Optimization and Selection, Edited by Angelo Basile and Fausto Gallucci

faster, less complicated and, what is also important, it offers possibility to obtain whole bunch of new membranes based on one starting matrix. Plasma modification of commercially available membranes seems to be the best method for increasing the quality and diversity of membranes, while the careful control of the modification process allows tailoring membranes for any end-user's request.

Plasma has been discovered by Crookes while Langmuir was the fist person who introduced its name. Today plasma application covers a variety of scientific and technical disciplines ranging from physics to some aspects of chemistry and material science. Different types of plasmas are related to various applications and wide range of *plasma science* makes the subject difficult to classify. Plasma generally consists of negatively charged electrons, positively charged ions, and neutral atoms or molecules or both. It can be obtained when gases are excited into energetic states by radio frequency, microwave, or electrons from a hot filament discharge. Plasma forms highly reactive chemical environment in which many surface reactions can take place. Modification can occur rapidly with the formation of various active species on the surface. In practice, plasma assisted modification has been intensively studied since 1960s.

There are two important issues of plasma use for polymer membrane modification: the technique meets ecological limits for clean technology and it is very fast. The modification lasts usually 1–2 min. For these reasons, its popularity has attracted attention for the last four decades and it became a useful tool for alteration of membrane properties and tailoring them according to a particular need. When plasma acts on polymer membrane two opposite processes take place: ablation and deposition. The first one is caused by etching of polymer chains while the second describes deposition of plasma polymer on the surface. Their balance depends mostly on the form of plasma used and process parameters. What results from action them is alteration of surface chemistry. New functional groups are created on/in the surface layer that are able to change the character of the membrane and turn it to a new item. Having available such powerful tools each material scientist is able to prepare membrane with the properties that are requested. Description of the applied methods is presented below.

25.2 Membrane Treatment with Microwave Plasma

A laboratory microwave plasma device, assembled with 2.45 GHz frequency generator, a glass reaction chamber and a vacuum line is commonly used. Remote plasma is generated in a quartz tube at the top of a reaction chamber. Pieces of membranes are fixed at various distances from the edge of plasma. Pulse frequency, duty time, plasma power, pressure into the chamber, flow rate of gas are adjustable parameters. For same kinds of plasma, argon is used as the carrier gas. Figure 25.1 shows one of the example of the laboratory microwave plasma device.

25.2.1 Membrane Treated by Dielectric Barrier Discharge

Dielectric barrier discharge (DBD) is the electrical discharge between two electrodes separated by an insulating dielectric barrier. DBD is sometimes called silent discharge, ozone production discharge or partial discharge. The process uses high voltage of alternate current with lower frequency. Usually, DBD devices are made in planar configuration with parallel plate electrodes of 0.1 mm distance separated by a dielectric layer. Due to the atmospheric pressure level, such processes require high energy levels to sustain. The critical points of dielectric barrier discharge

Figure 25.1 Remote microwave chamber with ignited plasma

plasma is the tight fitting of polymer membrane to electrode and smooth flow of the gas. The example of the barrier plasma device is shown in Figure 25.2.

25.3 Modes of Plasma Use

Plasma treatment of polymer membranes can be carried out in three different modes: (i) when the nonpolymerisable gas molecules are used, (ii) when the polymerisable vapours are used, (iii) when plasma activates grafting of polymer chains to the membrane surface.

It is commonly assumed that the first mode is mostly related to the ablation process and the second one to deposition of plasma polymers. The third one serves for coating of membranes with thin layer of different material. In the practice, however, there are some exemptions of this simple classification.

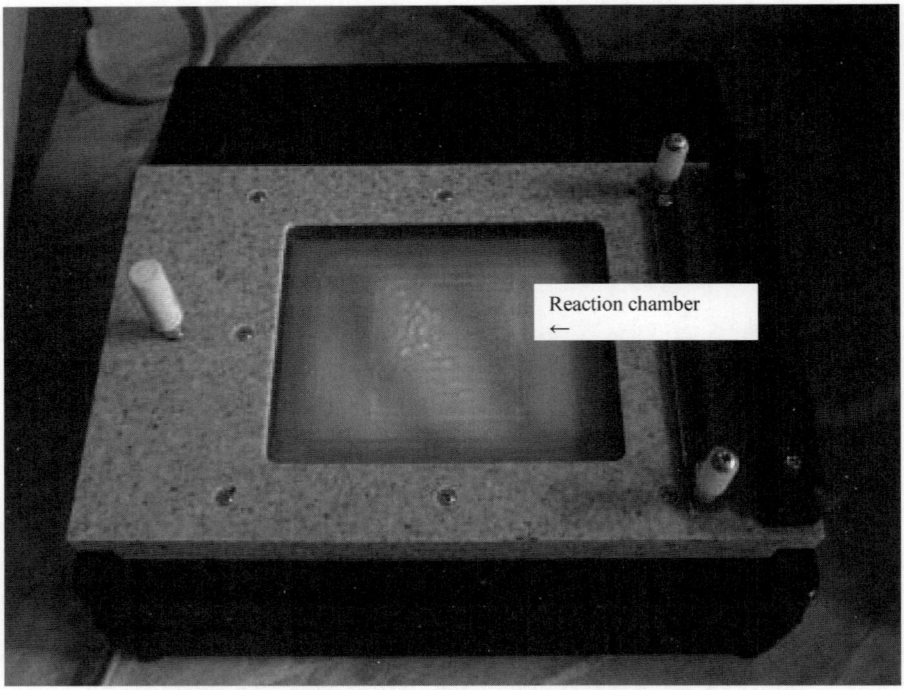

Figure 25.2 DBD reactor with ignited plasma

25.4 Plasma of Nonpolymerisable Gas

The plasma of nonpolymerisable gases causes the significant alteration of surface chemistry and membrane etching. Commonly, it is used to improve membrane permeability and reduce fouling phenomenon. Always, after the plasma treatment the improvement of filtration properties are observed – permeate flux gets larger, adsorption of permeant drops down and flux recovery is high. Cleaning of the plasma treated membranes is easier and in many cases fouling is completely reversible.

25.4.1 Carbon Dioxide Plasma

When microwave plasma is used with oxidative gases the treated polymer is subjected to degradation process [2,3]. Plasma of such gases as air, oxygen or carbon dioxide is known to etch polymer surfaces to a great extent [4,5]. Polysulfone [6] and poly(ether sulfone) [7] ultrafiltration membranes were the most frequently CO_2-plasma treated supports [4,8]. Strong etching character of CO_2 plasma was used to tailor the properties of asymmetric cellulose acetate membranes by simultaneous modification chemistry and structure of the top layer [10]. When plasma is used to porous membranes, one may achieve an enlargement of pore diameter. Hence, the ordinary ultrafiltration polysulfone membrane exposed to CO_2 plasma increases pores dimension with the extent of plasma action [4].

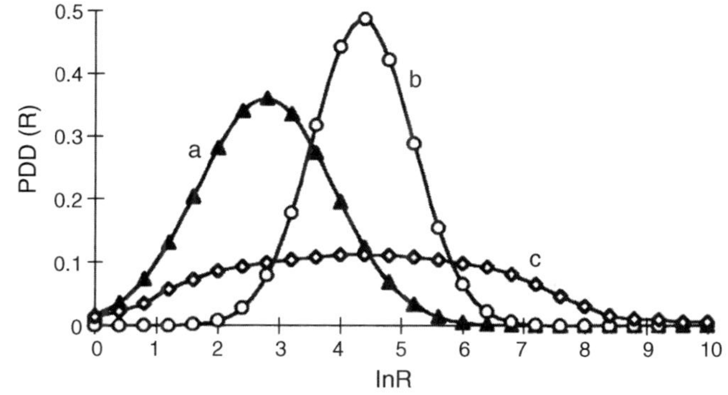

Figure 25.3 *Pore size distribution for: (a) untreated membrane, (b) membrane exposed to CO_2 plasma for 2 min, (c) membrane exposed for 10 min. Reprinted from I. Gancarz, M. Bryjak, G. Pozniak, Modification of polysulfone membrane. 1.CO_2 plasma treatment, Eur. Polym. J., 35 1419–1428. Copyright (1999) with permission from Elsevier*

25.4.2 Case Study on CO_2 Plasma Action

A similar phenomenon was observed by authors of this chapter. Prolonged exposition to the carbon dioxide plasma has resulted in material ablation and an increase in pore diameter [9], see Figures 25.3, 25.4.

Plasma does not alter the membrane morphology but also changes the surface character. Some polar groups created on the membrane can significantly increase the polar component of the surface tension.

In this case, the chemical reconstruction of membrane surface took place at very beginning of the process – during first 60 s. After that time surface was polar at almost 50% level. However, an unusual phenomenon was observed when the flux of distilled water was compared (Figure 25.5).

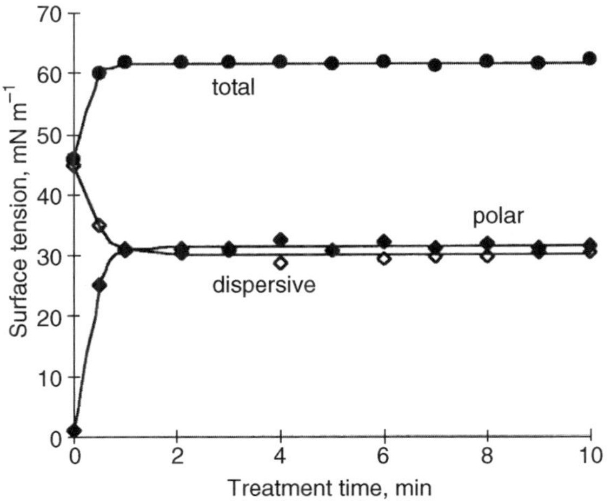

Figure 25.4 *Change of surface tension for polysulfone membranes treated with CO_2 plasma. Reprinted from I. Gancarz, M. Bryjak, G. Pozniak, Modification of polysulfone membrane. 1.CO_2 plasma treatment, Eur. Polym. J., 35 1419–1428. Copyright (1999) with permission from Elsevier*

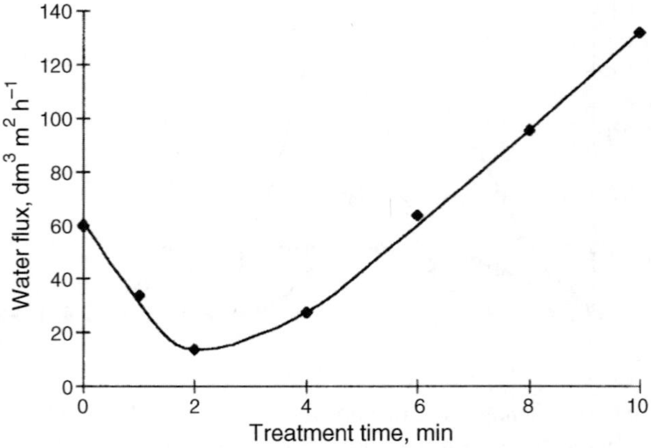

Figure 25.5 *Water flux through plasma modified membrane. Pressure 0.1 MPa. Reprinted from I. Gancarz, M. Bryjak, G. Pozniak, Modification of polysulfone membrane. 1.CO_2 plasma treatment, Eur. Polym. J., 35 1419–1428. Copyright (1999) with permission from Elsevier*

As ablation and deposition took place simultaneously one could expect that deposition might dominate at the beginning of the process while much longer exposition to plasma activated the etching processes.

25.4.3 Nitrogen Plasma Action

Nitrogen is the reactive plasma medium that incorporates various chemical functionalities onto polymer surface, making it more hydrophilic and prone for further reactions. After exposition of membrane to nitrogen plasma, such functional groups as amine, imine, amide or nitrile could be created [11,12]. In some cases, polymer degradation and etching processes are observed also [13–15].

25.4.4 Case Study on Nitrogen Plasma Action

When one compares the ablation effects caused by N_2 plasma, one realises its low etching power – membrane is not so intensively degraded [14]. Pore size distribution for polysulfone membrane exposed to nitrogen plasma is shown in Figure 25.6.

Action of less aggressive plasma on polysulfone membrane is also manifested in the slower changes of surface tension of modified membranes. When for CO_2 plasma the alteration of surface character was observed after 1 min, in the case of nitrogen plasma that effect occured after 5 min of plasma treatment (Figure 25.7).

The fact of not so strong etching of membrane was well manifested in water fluxes. They changed at the beginning of the plasma treatment and then they kept constant. Some polymer fragments fried at the first stage of plasma action were able to deposit on the surface and reduced the number of pores to be active in water transport (Figure 25.8).

According to the preliminary assumptions, the surface of plasma modified polymer should have basic character mostly. However, when someone carried out surface titration, its procedure is described elsewhere [16], one would get the surprising result – surface was partly amphoteric. The titrated surface had 0.28 $\mu m/m^2$ of basic and 0.06 $\mu m/m^2$ of acidic sites [16]. It was the effect of post reactions when plasma activated membrane was exposed to the air.

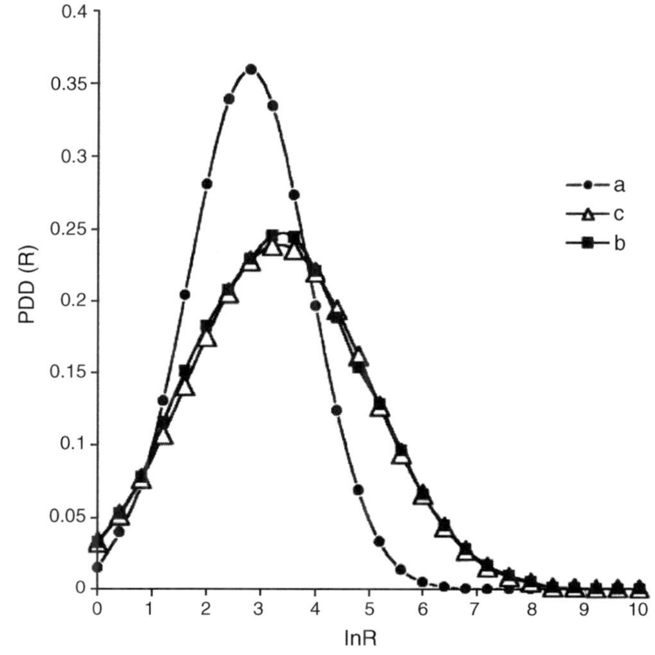

Figure 25.6 Effect of nitrogen plasma action on pore size distribution: a = untreated membrane, b = membrane after 2 min, c = 10 min of plasma action. Reprinted from I. Gancarz, G. Pozniak, M. Bryjak, Modication of polysulfone membranes 3. Effect of nitrogen plasma, Eur. Polym. J., 36, 1563–1569. Copyright (2000) with permission from Elsevier

25.4.5 Ammonia Plasma

It has been shown that ammonia plasma treatment turn surface character to more basic due to appearance of nitrogen containing moieties [17–20]. However, it was difficult to generate plasma

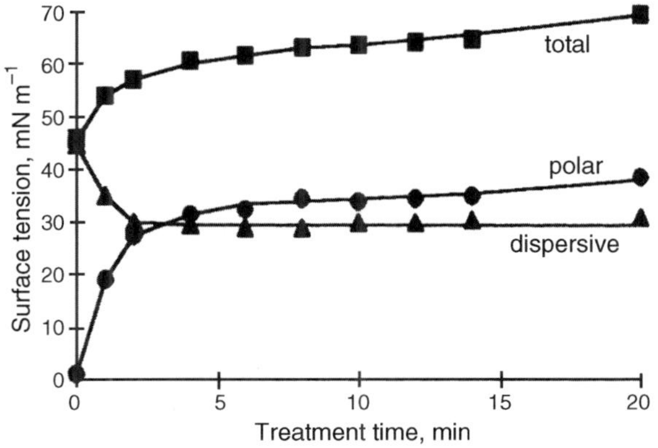

Figure 25.7 Change of surface tension for polysulfone membranes treated with N_2 plasma. Reprinted from I. Gancarz, G. Pozniak, M. Bryjak, Modication of polysulfone membranes 3. Effect of nitrogen plasma, Eur. Polym. J., 36, 1563–1569. Copyright (2000) with permission from Elsevier

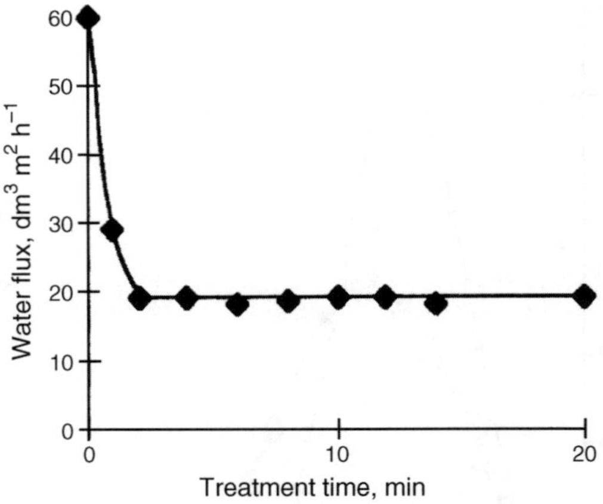

Figure 25.8 *Water flux through N_2 plasma-modified membrane. Pressure 0.1 MPa. Reprinted from I. Gancarz, G. Pozniak, M. Bryjak, Modication of polysulfone membranes 3. Effect of nitrogen plasma, Eur. Polym. J., 36, 1563–1569. Copyright (2000) with permission from Elsevier*

when only ammonia was applied. As it happened for other large molecules, presence of argon in the reactant mixture improved plasma stability [21].

25.4.6 Case Study on Ammonia Plasma Action

In the case of ammonia treatment two kinds of plasma should be evaluated: neat ammonia plasma and plasma of mixture of ammonia with argon. Below, are shown typical results obtained by the use of these two plasmas (Figure 25.9).

When one compares these figures one gets the impression that NH_3 and NH_3/Ar plasmas do not degrade the membrane material in the time of their action. Hence no morphology changes appeared during the process. Both plasmas modified however the surface character and made them more hydrophilic (see Figure 25.10).

The instability of ammonia plasma caused the scattering of measured points in graph (a). As was observed for the previously described case, membrane became 50% hydrophilic after longer exposition to both plasmas. However, the careful study of XPS spectra revealed higher concentration of nitrogen on the surface for NH_3/Ar plasma [21]. The N/C ratio was zero for polysulfone, 0.113 for ammonia and 0.223 for ammonia/argon plasma. The difference appeared also in the forms of functional groups. For NH_3 plasma the relative fraction of C-N functionalities was 9.2% while for NH_3/Ar plasma that value increased to 22.7% [21]. It shows that presence of argon in the reactant mixture is highly profitable – more functional groups are generated on the polymer surface. Besides, as mentioned before, argon stabilises plasma.

25.4.7 Plasmas of Other Gases

Many other gases were used for membrane modification. **Oxygen plasma** was used to improve filtration properties of membranes [22]. The surface of silicone membrane under the influence of oxygen plasma changed into silica; such membranes could be applied in gas/vapour separation and in reverse osmosis [23]. **Argon plasma** does not introduce any functional groups to the surface. It breaks the chemical bonds and removes hydrogen atoms creating the free radicals. For this reason argon plasma is applied in plasma-induced polymerisation. For

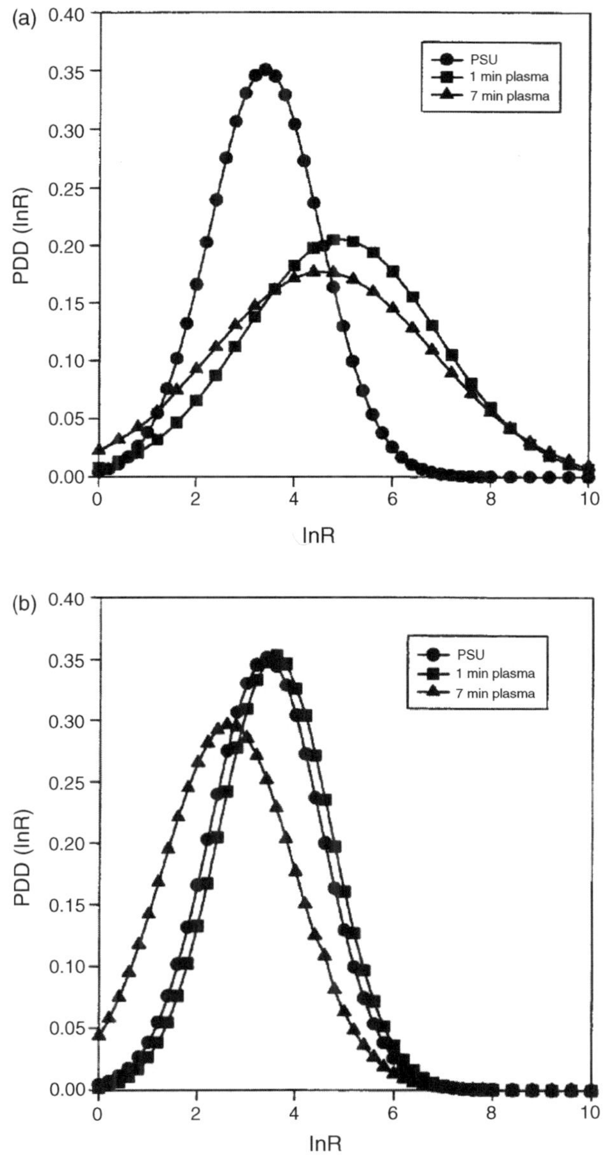

Figure 25.9 *Pore size distribution of polysulfone membrane modified with plasma. (a) ammonia plasma, (b) ammonia argon plasma. Reprinted from M. Bryjak, I. Gancarz, G. Pozniak, W. Tylus, Modification of polysulfone membranes 4. Ammonia plasma treatment, Eur. Polym. J., 38, 717–726. Copyright (2002) with permission from Elsevier*

membrane modification it is used rather rarely; for example PMMA membrane treated in argon plasma showed a decrease of CO_2 sorption and increase of He/CO_2 selectivity in high pressure gas separation [24]. Argon, as a plasma stabilising agent is often added to the organic compounds that are plasma polymerised. *Water vapour plasma* [25] has similar effect on filtration membranes as most other gases – it increases water flux, improved flux recovery and

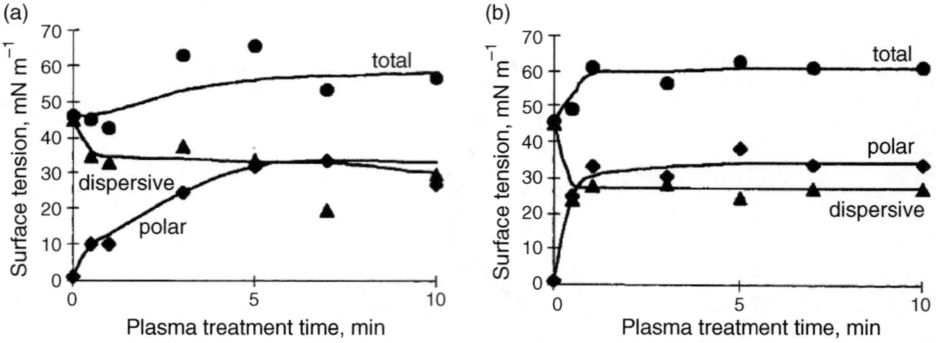

Figure 25.10 *Pore size distribution of polysulfone membrane modified with plasma: (a) ammonia plasma, (b) ammonia argon plasma. Reprinted from M. Bryjak, I. Gancarz, G. Pozniak, W. Tylus, Modification of polysulfone membranes 4. Ammonia plasma treatment, Eur. Polym. J., 38, 717–726. Copyright (2002) with permission from Elsevier*

makes easier membrane cleaning. *CCl_4 plasma* remarkably improves selectivity of poly(4-methyl-2-pentyne) membrane in gas separation process [26] while CF_4 plasma causes hydrophobisation of cellulose membranes [10].

The plasma of gas mixture sometimes acts better that plasma of single gas. $N_2 + H_2$ plasma was found to generate many amino groups on the surface of membranes (the case of PVDF [27] or PP membranes [28]). Plasma of nitrogen in mixture with NH_3 and CO_2 was applied to create amino and carboxyl functionalities, respectively [29]. The other gas mixtures used as a plasma medium for membrane modification were O_2-N_2 [22], SO_2-H_2 [30] or acetylene-N_2 [32].

25.4.8 Plasma of Nonpolymerisable Species: Summary

Such neat gases as Ar, N_2, O_2, CO_2 and NH_3 as well as gas mixture are used commonly. In many cases, an increase of pore diameter and widening of pore size distribution show that polymer ablation is the main process. In the case of larger molecules, argon presence in the reactant mixture is helpful. It allows stabilisation of plasma and makes it homogenous. However, in same cases deposition of fried fragments of etched polymer chain can interfere to the process. After preliminary improvement of surface hydropilicity the recovery of hydrophobicity is observed for long time plasma treatment. In all cases, the hydrophilic character of membrane disappeared with time of storage for treated membranes. The obtained membranes are less prone for membrane fouling and easy for regeneration. In all investigated cases, the flux recovery after membrane cleaning was almost 100% while for not treated membranes this value varies from 70 to 90%.

25.5 Plasma of Polymerisable Species

Plasma polymerisation offers the interesting technique for modification of polymer surfaces and deposition of thin polymer films of several nanometers to one micrometer thickness. Such films are made of highly crosslinked material and show good adhesion to the substrate. Plasma polymerisation is the complex process that depends on many parameters: reactor geometry, temperature, plasma parameters (power, frequency, due time), pressure and flow rate. Two processes take place during the plasma polymerisation: plasma-induced polymerisation and polymer state polymerisation. In the case of the first one polymerisation appears on the surface with easily polymerisable monomers. In the case of the second mechanism process appear in bulk plasma state when some bonds break to form radicals or ions. Hence, any kind of molecule, even that without multiple bonds, can be used in polymer-state polymerisation.

25.5.1 Allyl Alcohol Plasma

Plasma polymerisation of allyl alcohol was intensively investigated [32–36] and high retention of hydroxyl group (50–70%) was confirmed experimentally. This monomer has a sufficiently high vapour pressure to keep a constant flow rate into a reactor. Due to the presence of a double bond in the molecule, its plasma polymerisation occurs more readily and faster than deposition of any saturated alcohol. This monomer, however, has rarely been used for membrane modification. To the best of knowledge of the authors, allyl alcohol has been applied to porous ceramic filters used for gas separation [37], to poly(vinyl alcohol) membrane applied in pervaporation process [38] and to polysulfone for enzyme immobilisation [39].

25.5.2 Case Study on Plasma Polymerisation of Allyl Alcohol

Plasma of allyl alcohol as well as mixture of allyl alcohol with argon were applied for membrane modification [39]. The alteration of surface energy of samples exposed to both plasmas is shown in Table 25.1.

The scattering of data for AllOH plasma shows its unstabile character. After addition of argon, plasma became more stable and reconstruction of suface chemistry went more smoothly. The detailed XPS studies revieled that in the case of AllOH plasma, the deposited plasma polymer retained almost twice more C-OH functionalities than polymer deposited in presence of argon [39].

25.5.3 Amine Plasma

Such amines as allylamine [40–43], vinylpyridine [44,45], ethylene diamine [43,46,47], n-propylamine [43], n-butylamine [40], aniline [48] and N,N-dimethylaniline [49,50] have been used for plasma modification of membranes. As the porous substrate for amine plasma

Table 25.1 *Both components of suface energy and surface polarity in course of plasma action. Reprinted from I. Gancarz, J. Bryjak, M. Bryjak, G. Pozniak, Eur. Polym. J., 39, 1615–1622. Copyright (2003) with permission from Elsevier*

Plasma	Time of action (s)	Surface tension (mN m^{-1})		Polarity (%)
		Polar comp	Dispersive comp	
w/o	0	0.9	45.0	2.0
AllOH	30	18.9	26.2	44.4
	60	17.9	20.7	46.3
	90	13.9	25.0	35.7
	120	19.4	25.1	43.4
	180	35.0	27.1	57.4
AllOH/Ar	30	13.6	29.7	31.4
	60	21.8	25.6	46.0
	90	31.5	26.5	54.3
	120	30.3	28.8	51.2
	180	36.6	27.6	57.0

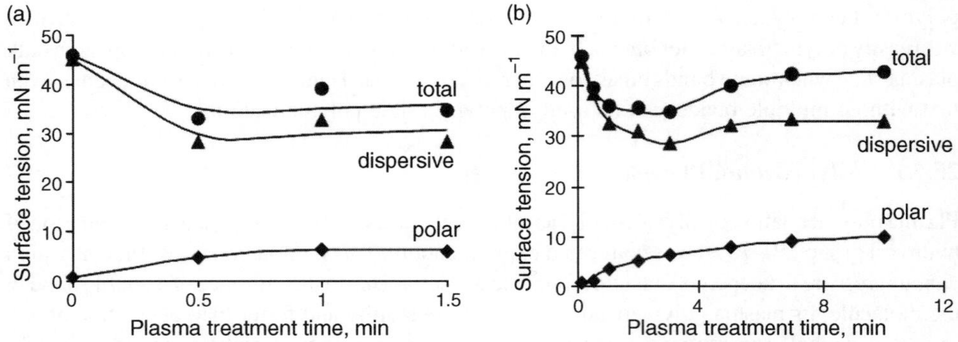

Figure 25.11 *The case of butylamine plasma: (a) ButNH$_2$ plasma, (b) ButNH$_2$/Ar plasma. Reprinted from I. Gancarz, G. Poźniak, M. Bryjak, W. Tylus, Modification of polysulfone membranes 5. Effect of n-butylamine and allylamine plasma, Eur. Polym. J. 38, 1937–1946. Copyright (2002) with permission from Elsevier*

treatment the following membranes were taken: PET track-etched membranes [48–50], polysulfone membranes [40,43], polypropylene membrane [43] and many others. The purposes of plasma treatment were different: from improving the filtration properties [42,49,50], altering the pervaporation selectivity [46], creation of new reverse osmosis membranes [41,43] to immobilisation of biomolecules [47].

25.5.4 Case Study on Butylamine and Allyloamine Plasma Polymerisation

The effect of the modification time on surface character can be monitored by evaluation of surface tension. Figures 25.11 and 25.12 show the alteration of both components of surface tension for pure and mixed gas plasmas, for both butylamine and allylamine vapours.

In the case of allylamine plasma, it was impossible to carried modification longer than 1 min as the plasma went down. The comparison of the effect of use of both amines showed that plasma polymer deposition resulted in creation of two different membranes. The evaluation of pore size distribution confirmed this assumption (see Figure 25.13).

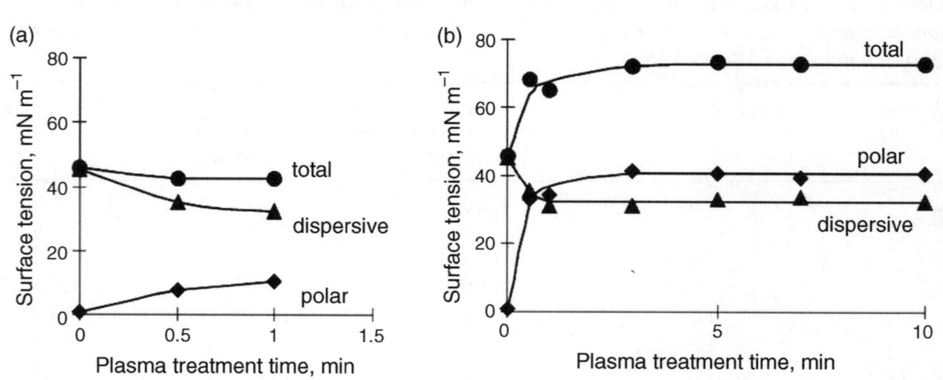

Figure 25.12 *The case of allylamine plasma: (a) AllNH$_2$ plasma, (b) AllNH$_2$/Ar plasma. Reprinted from I. Gancarz, G. Poźniak, M. Bryjak, W. Tylus, Modification of polysulfone membranes 5. Effect of n-butylamine and allylamine plasma, Eur. Polym. J. 38, 1937–1946. Copyright (2002) with permission from Elsevier*

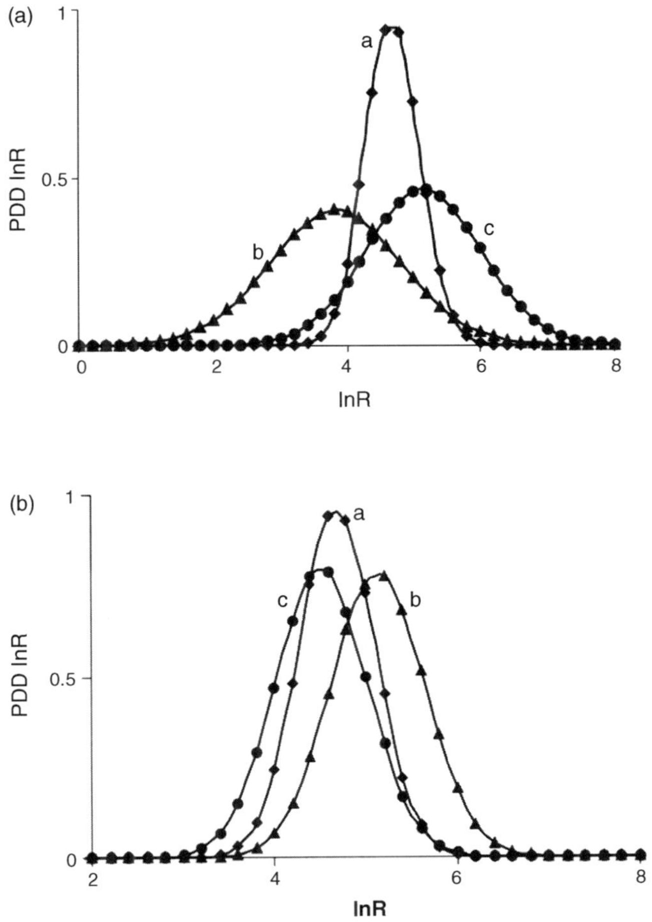

Figure 25.13 *Pore size distributions for* $ButNH_2$ *(a) and* $AllNH_2$ *(b) plasma-treated membranes: a = untreated membrane, b = neat vapour plasma, c = vapour/Ar plasma. Reprinted from I. Gancarz, G. Poźniak, M. Bryjak, W. Tylus, Modification of polysulfone membranes 5. Effect of n-butylamine and allylamine plasma, Eur. Polym. J. 38, 1937–1946. Copyright (2002) with permission from Elsevier*

It seems that allylamine plasma did not deposited polymer on the polysulfone membrane – pore size was not altered significantly, independently of the presence of argon in the reactive mixture. In the case of butylamine plasma presence of argon activated the membrane etching process. The pure $ButNH_2$ plasma deposited polymer layer that reduced the pore diameter. XPS study shed new lights on the vapour polymerisation process. In the case of butylamine the N/C ratio was 0.06 for both kind of plasma, while for allylamine plasma it reached the value of 0.21–0.27. It means that $AllNH_2$ plasma was able to introduce much more nitrogen caring functional groups than $BuTNH_2$ plasma did [40].

25.5.5 Acid Plasma

Among various acids used for plasma polymerisation, acrylic acid is the leader. It easily forms layers of plasma polymer and retains many carboxylic groups. This process was applied to polyurethane membranes used in pervaporation [51,52], polypropylene [43] and polyacrylnitrile [53] membranes for reverse osmosis and to improve electrochemical properties of polypropylene membrane [54]. From other acidic reagents, vinyl acetic acid was applied in plasma copolymerisation process only [55].

25.5.6 Other Kinds of Plasma

The other reagents applied for plasma-polymerisation on the membrane surface are: fluorine bearing compounds (like perfluorohexane [55], perfluoroheptane [56], perfluoroctane [57], hexafluoropropylen [58]), or silicon bearing components (hexamethyldisiloxane [59]). They make the surface hydrophobic and membranes are prepared for a special request.

25.5.7 Plasmas of Polymerisable Species: Summary

Such vapours as AllOH, $AllNH_2$, $ButNH_2$ and AAc were mixed with Ar and used for membrane modification. In same cases membrane ablation dominated but the most frequently observed process was polymer deposition. Depending on the geometry of plasma chamber, concentration of polymerisable component, time of deposition or selection of same plasma parameters different layers were prepared. The spectroscopic studies (XPS and ATR-FTIR) allowed to qualify and quantify surface functionalities. The obtained membranes served for several purposes: as the supports for enzymes immobilisation, as multilayer NF membranes with good rejection of multivalent ions, as brush-like membranes that behaved as membrane valves.

25.6 Plasma-Induced Grafting

Plasma induces radical on the polymer surface that are stable in vacuum but can react rapidly when one exposures them to a reactive gas. After meeting of monomers the radicals initiate their polymerisation (Figure 25.14).

In the presence of oxygen or air, peroxides and hydroperoxides are created. They may initiate polymerisation of the desired monomer in solution after UV irradiation or increasing temperature (Figure 25.15).

This process is named plasma-induced graft copolymerisation or plasma grafting. To some extent, grafting density and length of grafted brush can be controlled by selection of plasma parameters and grafting process. The literature dealing with plasma grafting of various membranes with different monomers is enormously large. Among porous membranes chosen as a support, one can find poly (vinylidene fluoride) [60,61], polytetrafluoroethylene and polyamide [62–65], polypropylene [66–71], polyacrylonitrile [72,73], polysulfone [74,75], PET [76], poly(phenylene oxide) [77]

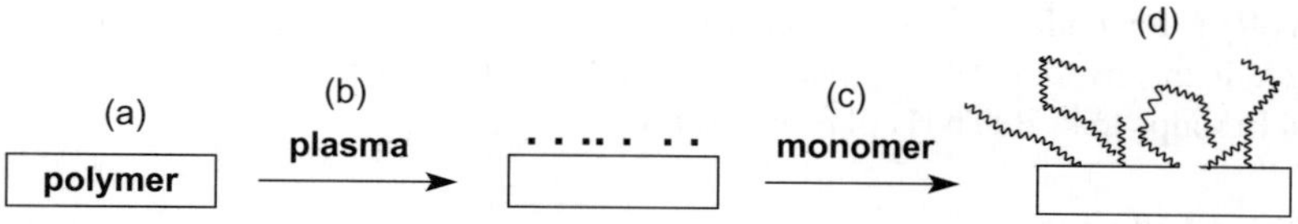

Figure 25.14 Step 1 in plasma-induced grafting. Reprinted from I. Gancarz, J. Kunicki, A. Ciszewski, Application of plasma-induced grafting for modification of alkaline battery separator Chem. Listy, 102, 1467–1472. Copyright (2008) with permission from Springer-Verlag

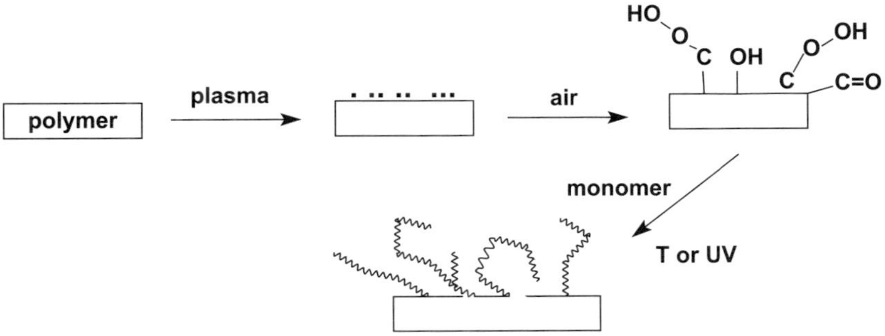

Figure 25.15 Step 2 in plasma-induced grafting

and others. The list of monomers being grafted is also very long, from acrylic and methacrylic acid, acrylamide and N-isopropylacrylamide, glycidyl methacrylate and its derivatives, N-vinylpyrrolidone, styrene to derivatives of styrenesulfonic acid and vinylsulfonic acid. Plasma grafting was applied to get a cation exchange [60] or bipolar [61] membrane, to manufcture thermo-responsive [63] membranes, to improve cell adhesion [78] or to lower membrane electrical resistance [67]. Thanks to increased wettability the filtration properties of membranes get significant improvement – they are less susceptible to protein fouling and easier to clean [73–75,77].

25.6.1 Case Study on Grafting of Acrylic Acid

The good example is grafting of acrylic acid onto the porous polypropylene membrane. Polypropylene is very hydrophobic polymer with surface tension about 30 mN m^{-1} and very low polar component. Acrylic acid is soluble in water, cheap and easy polymerisable and introduces many carboxyl polar functionalities that are capable for further chemical reactions. After plasma grafting, surface polarity reaches the value of 20% [67].

The grafting yield varies with plasma parametres (power, pressure, time, distance from plasma edge in the remote plasma, sample arrangement), polymerisation parameters (solvent, monomer concentration, grafting time) and can reach values up to 20 mmol of acrylic acid per 1 g of polypropylene. When the grafting degree increases, the layer of grafted polymer covers almost the whole surface (Figure 25.16) [69].

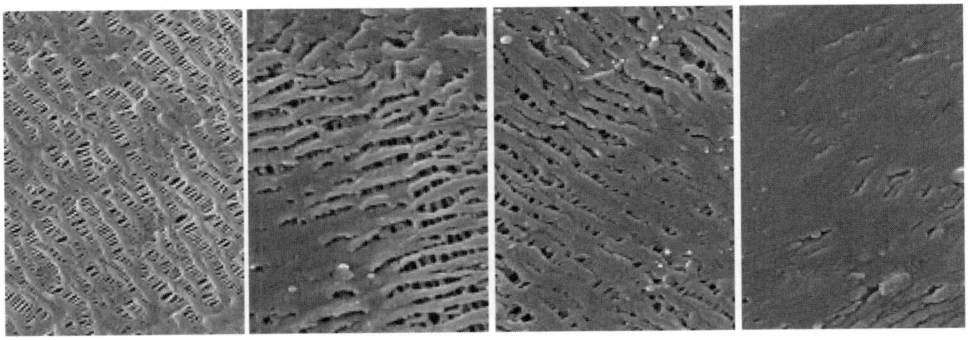

Figure 25.16 SEM view of virgin (a) and grafted (b) 2.3 mmol g^{-1}, (c) 9.7 mmol g^{-1} and (d) 16.2 mmol/g surfaces. Magnification – 6000× [from 69]

Membrane becomes also thicker – estimated thickness increases from 25 μm for unmodified membrane to 29, 32 and 35.5 μm for membrane grafted with 2.3, 9.7 and 16.2 mmol/g, respectively [69].

The plugging of the membrane pores makes it less permeable for water. Those of grafting degree higher than 2 mmol/g are practically impermeable. Despite of that they have good ionic conductivity; the electrolytic area resistance drops from 130 000 mΩ cm^2 for PP to as small a value as 20–50 mΩ cm^2 for modified membranes.

25.6.2 Plasma Modification of Polymer Membranes: Summary

As noted in this chapter, the plasma modified membranes show the unique character. Their bulk properties stay unchanged while surface differs each to the other – depending to the used plasma. Plasma procedure allows creating whole bunch of new membranes having on stock only one type of porous film. Figure 25.17 shows some possibilities to prepare new membranes.

The next benefit of plasma modification is the use of environment friendly technique. The process takes place in the gaseous phase with minimal amount of produced wastes. The very short time of membrane modification is the next profit of this technology. The flexibility and versatility of the plasma system allows to adapt it for production of new membranes very fast. Nevertheless, there are some drawbacks of plasma treatment. Despite of years of study, this method is still not well recognised. Hence, each one plasma device behaves as a new one and its optimisation needs to be carried out. The second one is related to the use of typical conditions for cold plasma. The process runs in vacuum. That request makes some technical problems when one wants to realise the continuous process. To bypass it one can use normal pressure plasmas – like corona or dielectric barrier discharge one. It seems that the methods for plasma modification will be developed in the nearest future as they offer the simplest way for increasing the production of new types of membranes.

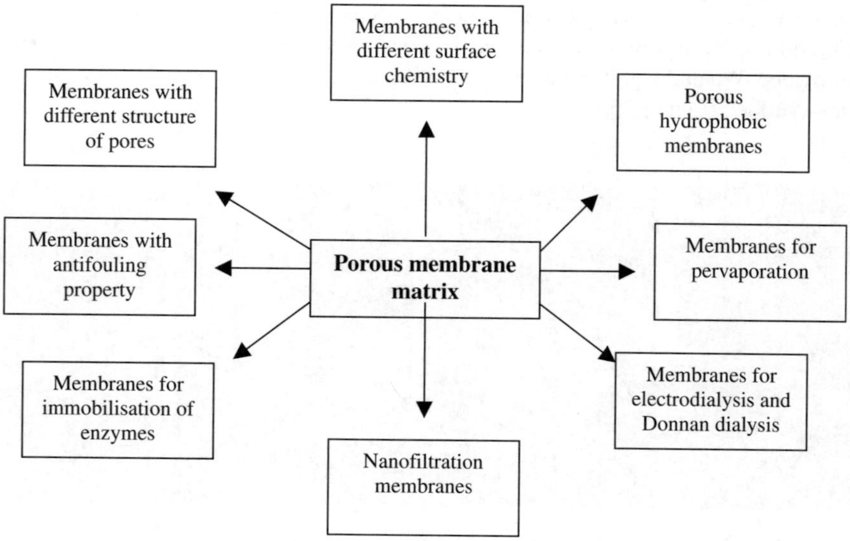

Figure 25.17 Possible routes to prepare new membranes

References

1. L. A. Ultracki, History of commercial polymer alloys and blends, *Polym. Eng. Sci.*, **35**, 2–12 (1995).
2. H. Yasuda, *Plasma Polymerization*. Academic Press, Orlando (1985).
3. S. Pal, S. K. Ghatak, S. De, S. DasGupta, Characterization of CO_2 plasma treated polymeric membranes and quantification of flux enhancement, *J. Membr. Sci.*, **323**, 1–10 (2008).
4. H. Y. Yu, Y. J. Xie, M. X. Hu, J.-L. Wang, Surface modification of polypropylene microporous membrane to improve its antifouling property in MBR: CO_2 plasma treatment, *J. Membr. Sci.*, **254**, 219–227 (2005).
5. D. S. Wavhal, E. R. Fisher, Modification of polysulfone ultrafiltration membranes by CO_2 plasma treatment, *Desalination*, **172**, 189–205 (2005).
6. D. S. Wavhal, E. R. Fisher, Modification of polysulfone ultrafiltration membranes by CO_2 plasma treatment, *Desalination*, **172**, 189–205 (2005).
7. S. Pal, S. K. Glhatak, S. De, S. DasGupta, Evaluation of surface roughness of a plasma treated polymeric membrane by wavelet analysis and quantification of its enhanced performance, *Appl. Surf. Sci.*, **255**, 250 (2008).
8. H. Y. Yu, L. Q. Liu, Z. Q. Tang, M. G. Yan, J. S. Gu, X. W. Wei, Mitigated membrane fouling in an SMBR by surface modification, *J. Membr. Sci.*, **310**, 409–418 (2008).
9. I. Gancarz, M. Bryjak, G. Pozniak, Modification of polysulfone membrane. 1. CO_2 plasma treatment *Eur. Polym. J.*, **35**, 1419–1428 (1999).
10. M. B. O. Riekerink, G. H. M. Engbers, M. Wessling, J. Feijen, Tailoring the properties of asymmetric cellulose acetate membranes by gas plasma etching, *J. Coll. Interf. Sci.*, **245**, 338–348 (2002).
11. K. R. Kull, M. L. Steen, E. R. Fisher, Surface modification with nitrogen-containing plasmas to produce hydrophilic, low-fouling membranes, *J. Membr. Sci.*, **246**, 203–214, (2005).
12. A. J. Wagner, D. H. Fairbrother, F. Reniers, A comparison of PE surfaces modified by plasma generated neutral nitrogen species and nitrogen ions, *Plasmas and Polymers*, **8**, 119–134 (2003).
13. H. Y. Yu, X. C. He, L. Q. Liu, J. S. Gu, X. W. Wei, Surface modification of polypropylene microporous membrane to improve its antifouling characteristics in an SMBR: N-2 plasma treatment, *Water Res.*, **41**, 4703–4709 (2007).
14. I. Gancarz, G. Pozniak, M. Bryjak, Modication of polysulfone membranes 3. Effect of nitrogen plasma *Eur. Polym. J.*, **36**, 1563–1569 (2000).
15. S. Villeger, M. Sixou, J. Durand, A. Ricard, Interaction of N atoms through a nylon membrane in nitrogen flowing post discharges, *J. Phys. D - Appl. Phys.*, **39**, 3826–3838 (2006).
16. M. Bryjak, I. Gancarz, G. Pozniak, Surface evaluation of plasma-modified polysulfone (Udel P-1700). films, *Langmuir*, **15**, 6400–6404 (1999).
17. M. Muller, C. Oehr, Plasma aminofunctionalisation of PVDF microfiltration membranes: comparison of the in plasma modifications with a grafting method using ESCA and an amino-selective fluorescent probe, *Surf. Coat. Technol.*, **116/119**, 802–807 (1999).
18. H. Y. Yu, M. X. Hu, Z. K. Xu, J. L. Wang, S. Y. Wang, Surface modification of polypropylene microporous membranes to improve their antifouling property in MBR: NH3 plasma treatment, *Sep. Pur. Technol.*, **45**, 8–15 (2005).
19. K. R. Kull, M. L. Steen, E. R. Fisher, Surface modification with nitrogen-containing plasmas to produce hydrophilic, low-fouling membranes *J. Membr. Sci.*, **246**, 203–215 (2005).
20. Benrejeb S, Tatoulian M, Khonsari F, Durand NF, Martel A, Lawrence JF, Amouroux J, Legoffic F., Functionalization of nitrocellulose membranes using ammonia plasma for the covalent attachment of antibodies for use in membrane-based immunoassays, *Anal. Chim. Acta*, **376**, 133–138 (1998).
21. M. Bryjak, I. Gancarz, G. Pozniak, W. Tylus, Modification of polysulfone membranes 4. Ammonia plasma treatment *Eur. Polym. J.*, **38**, 717–726 (2002).

22. J. Johansson, H. K. Yasuda, R. K. Bajpai, Fouling and protein adsorption – Effect of low-temperature plasma treatment of membrane surfaces *Appl. Biochem. Biotechnol.*, **70**, 747–758 (1998).
23. K. Beltsios, G. Charalambopoulou, G. Romanos, N. Kanellopoulos, A. Vycor, Membrane with reduced size surface pores. I. Preparation and characterization *J. Porous Mat.*, **6**, 25–32 (1999).
24. C. C. Hu, C. Y. Tu, Y. C. Wang, C. L. Li, K. R. Lee, J. Y. Lai, Effects of plasma treatment on CO_2 plasticization of poly(methyl methacrylate) gas-separation membranes, *J. Appl. Polym. Sci.*, **93**, 395–412 (2004).
25. H. Y. Yu, Z. Q. Tang, L. Huang, G. Cheng, W. Li, J. Zhou, M. G. Yan, J. S. Gu, X. W. Wei, Surface modification of polypropylene macroporous membrane to improve its antifouling characteristics in a submerged membrane-bioreactor: H_2O plasma treatment, *Water Res.*, **42**, 4341–4352 (2008).
26. L. Shao, J. Samseth, M. B. Hagg, Effect of plasma treatment on the gas permeability of poly(4-methyl-2-pentyne). membranes, *Plasma Processes Polym.*, **4**, 823–831 (2007).
27. M. Muller, C. Oehr, Plasma aminofunctionalisation of PVDF microfiltration membranes: comparison of the in plasma modifications with a grafting method using ESCA and an amino-selective fluorescent probe, *Surf. Coat. Technol.*, **116**, 802–809 (1999).
28. J. X. Tang, N. Y. He, N. J. Tan, Q. G. He, H. Chen, A novel substrate for in situ synthesis of oligonucleotide: plasma-treated polypropylene microporous membrane, *Colloids Surf A – Physicochem. Eng. Asp.*, **242**, 53–62 (2004).
29. J. Vartiainen, M. Ratto, S. Paulussen, Antimicrobial activity of glucose oxidase-immobilized plasma-activated polypropylene films *Packaging Technol. Sci.*, **18**, 243–251 (2005).
30. Z. Gugala, S. Gogolewski, Attachment, growth, and activity of rat osteoblasts on polylactide membranes treated with various low temperature radio frequency plasmas *J. Biomed. Mater. Res. A*, **76A**, 288–296 (2006).
31. C. Y. Tu, Y. C. Wang, C. L. Li, K. R. Lee, J. Huang, J. Y. Lai, Expanded poly(tetrafluoroethylene) membrane surface modification using acetylene/nitrogen plasma treatment, *Eur. Polym. J.*, **41**, 2343–2353 (2005).
32. L. Watkins, A. Bismarck, A. F. Lee, D. Wilson, K. Wilson, An XPS study of pulsed plasma polymerised allyl alcohol film growth on polyurethane, *Appl. Surf. Sci.*, **252**, 8203–8211 (2006).
33. A. P. Ameen, R. D. Short, R. J. Ward, The formation of high surface concentrations of hydroxyl groups in the plasma polymerization of allyl alcohol, *Polymer*, **35**, 4382–4391 (1994).
34. C. L. Rinsch, X. Chen, V. Panchalingam, C. R. Savage, Y. H. Wang, R. E. Eberhart, Pulsed radio frequency plasma polymerization of allyl alcohol: controlled deposition of surface hydroxyl groups, *Langmuir*, **12**, 2995–3002 (1996).
35. K. Yoshimura, K. Hozumi, T. Tatsuta, M. Sawai, O. Tsuji, Preparation of hydrophilic plasma-polymers derived from oxygen-containing organic monomers, *J. Appl. Polym. Sci.*, **9**, 1033–1042 (1996).
36. S. Kurosawa, T. Hirokawa, K. Kashima, H. Aizawa, D. S. Han, Y. Yoshimi, Detection of deposition rate of plasma-polymerized films by quartz crystal microbalance, *Thin Solid Films*, **374**, 262–267 (2000).
37. V. Ley, A. P. Kruzic, R. B. Timmons, Permeation rates of low molecular weight gases through a plasma synthesized allyl alcohol membrane, *J. Membr. Sci.*, **226**, 213–226 (2003).
38. M. Rafik, A. Mas, C. Guimon, A. Elharfi, F. Schue, Plasma-modified poly(vinyl alcohol). membranes for the dehydration of ethanol, *Polymer Int.*, **52**, 1222–1229 (2003).
39. I. Gancarz, J. Bryjak, M. Bryjak, G. Pozniak, Plasma modified polymers as a support for enzyme immobilization 1. Allyl alcohol plasma *Eur. Polym. J.*, **39**, 1615–1622 (2003).
40. I. Gancarz, G. Poźniak, M. Bryjak, W. Tylus, Modification of polysulfone membranes 5. Effect of n-butylamine and allylamine plasma *Eur. Polym. J.*, **38**, 1937–1946 (2002).

41. D. T. Tran, S. Mori, D. Tsuboi, M. Suzuki, Formation of plasma-polymerized top layers an composite membranes: influence on separation efficiency, *Plasma Process. Polym.*, **6**, 110–121 (2009).
42. D. T. Tran, S. Mori, M. Suzuki, Characteristics of polyimide-based composite membranes fabricated by low-temperature plasma polymerization, *Thin Solid Films*, **516**, 4384–4391 (2008).
43. I. Kim, H. Kim, S. Soo, Plasma treatment of polypropylene and polysulfone supports for thin film composite reverse osmosis membrane, *J. Membr. Sci.*, **286**, 193–204 (2006).
44. J. Z. Meng, Y. Jiang, Z. Jie, S. Yicai, preparation of highly sulfonated ultra-thin proton-exchange polymer membranes for proton exchange membrane fuel cells, *Surf. Rev. Let.* **16**, 297–309 (2009).
45. M. Koji, C. Satoshi, I. Yasutoshi, A. Takeshi, M. Masao, K. Kenji, O. Zempachi, Preparation of anion-exchange membrane by plasma polymerization and its use in alkaline fuel cells, *Thin Solid Films*, **516**, 3309–3314 (2008).
46. C. H. Yu, I. Kusumawardhana, J. Y. Lai, Y. L. Liu, PTFE/polyamide thin-film composite membranes using PTFE films modified with ethylene diamine polymer and interfacial polymerization: Preparation and pervaporation application, *J Coll. Interf. Sci.*, **336**, 260–269 (2009).
47. B. Liu, Y. H. Yang, Z. Y. Wu, H. Wang, G. L. Shen, R. Q. Yu, A potentiometric acetylcholinesterase biosensor based on plasma-polymerized film, *Sensors Actuators B – Chem.*, **104**, 186–193 (2005).
48. E. N. Demidova, A. Drachev, G. A. Grigoreva, Investigation of electrotransport properties of poly(ethylene terephtalate) track membranes modified by plasma of aniline *Russ. Electrochem.*, **45**, 533–539 (2009).
49. L. Kravets, S. Dmitriev, A. Gilman, A. Drachev, G. Dinescu, Water permeability of poly(ethylene terephthalate) track membranes modified by DC discharge plasma polymerization of dimethylaniline, *J. Membr. Sci.*, **263**, 127–138 (2005).
50. L. Kravets, A. Gilman, A. Drachev, A study on the water permeability of poly(ethylene terephthalate) track membranes modified by DC discharge polymerization of N,N-dimethylaniline *High Energy Chem.*, **39**, 114–120 (2005).
51. D. E. Weibel, C. Vilani, A. C. Habert, C. A. Achete, Surface modification of polyurethane membranes using acrylic acid vapour plasma and its effects on the pervaporation processes, *J. Membr. Sci.*, **293**, 124–135 (2007).
52. D. E. Weibel, C. Vilani, A. C. Habert, C. A. Achete, Surface modification of polyurethane membranes using RF-plasma treatment with polymerizable and nonpolymerizable gases, *Surf. Coat. Technol.*, **201**, 4190–4200 (2006).
53. T. D. Tran, S. Mori, M. Suzuki, Plasma modification of polyacrylonitrile ultrafiltration membrane, *Thin Solid Films*, **515**, 4148–4154 (2007).
54. F. Basarir, E. Y. Choi, S. H. Moon, K. C. Song, T. J. Yoon, Electrochemical properties of PP membranes with plasma polymer coatings of acrylic acid, *J. Membr. Sci.*, **260**, 66–78 (2005).
55. C. Chapman, D. Bhattacharyya, R. C. Eberhart, R. B. Timmons, C. J. Chuong, Plasma polymer thin film depositions to regulate gas permeability through nanoporous track etched membranes, *J. Membr. Sci.*, **318**, 137–149 (2008).
56. S. J. Lue, S. Y. Hsiaw, T. C. Wei, Surface modification of perfluorosulfonic acid membranes with perfluoroheptane (C7F16)/argon plasma, *J. Membr. Sci.*, **305**, 226–237 (2007).
57. D. A. Trofimov, V. M. Shkinev, B. Y. Spivakov, F. Schue, Improvement of pore geometry and performances of poly(ethylene terephthalate) track membranes by a protective layer method using plasma-induced graft polymerization of 1H,1H,2H-perfluoro-1-octene monomer *J. Membr. Sci.*, **326**, 265–276 (2009).
58. A. Taniguchi, K. Yasuda, Waterproofing of carbon paper by plasma polymerization, *J. Appl. Polym. Sci.*, **100**, 1748–1753 (2006).

59. C. H. Lo, J. K. Huang, W. S. Hung, S. H. Huang, M. De Guzman, V. Rouessac, C. L. Li, C. C. Hu, K. R. Lee, J. Y. Lai, Investigation on the variation in the fine structure of plasma-polymerized composite membrane by positron annihilation spectroscopy, *J. Membr. Sci.*, **337**, 297–311 (2009).
60. C. Jian, L. Jiding, C. Cuixian, Surface modification of polyvinylidene fluoride (PVDF) membranes by low-temperature plasma with grafting styrene, *Plasma Sci. Technol.*, **11**, 42–51 (2009).
61. S. D. Li, C. C. Wang, Preparation and characterization of a novel bipolar membrane by plasma-induced polymerization, *J. Membr. Sci.*, **318**, 429–434 (2008).
62. C. L. Hsueh, Y. J. Peng, C. C. Wang, C. Y. Chen, Bipolar membrane prepared by grafting and plasma polymerization, *J. Membr. Sci.*, **219**, 1–13 (2003).
63. Y. Chang, T. Y. Cheng, Y. J. Shih, K. R. Lee, J. Y. Lai, Biofouling-resistance expanded poly(tetrafluoroethylene) membrane with a hydrogel-like layer of surface-immobilized poly(ethylene glycol) methacrylate for human plasma protein repulsions, *J. Membr. Sci.*, **323**, 77–84 (2008).
64. C. Y. Tu, Y. L. Liu, K. R. Lee, J. Y. Lai, Hydrophilic surface-grafted poly(tetrafluoro-ethylene) membranes using in pervaporation dehydration processes, *J. Membr. Sci.*, **274**, 47–55 (2006).
65. M. Yang, L. Y. Chu, Y. Li, X. J. Zhao, H. Song, W. M. Chen, Thermo-responsive gating characteristics of poly(N-isopropyl-acrylamide)-grafted membranes, *Chem. Eng. Technol.*, **29**, 631–636 (2006).
66. E. Y. Choi, B. Bae, S. H. Moon, Characterization of acrylic acid-grafted PP membranes prepared by plasma-induced graft polymerization, *J. Appl. Polym. Sci.*, **105**, 2314–2320 (2007).
67. A. Ciszewski, J. Kunicki, I. Gancarz, Usefulness of microporous hydrophobic polypropylene membranes after plasma-induced graft polymerization of acrylic acid for high-power nickel–cadmium batteries, *Electrochim. Acta*, **52**, 5207–5212 (2007).
68. A. Ciszewski, I. Gancarz, J. Kunicki, M. Bryjak, Plasma-modified polypropylene membranes as separators in high-power alkaline batteries, *Surf. Coat. Technol.*, **201**, 3676–3684 (2006).
69. I. Gancarz, J. Kunicki, A. Ciszewski, Application of plasma-induced grafting for modification of alkaline battery separator, *Chem. Listy*, **102**, 1467–1472 (2008).
70. E. Y. Choi, H. Strathmann, J. M. Park, S. H. Moon, Characterization of non-uniformly charged ion-exchange membranes prepared by plasma-induced graft polymerization, *J. Membr. Sci.*, **268**, 165–174 (2006).
71. Z. Xu, J. Wang, L. Shen, D. Men, Y. Xu, Microporous polypropylene hollow fiber membrane. Part I. Surface modification by the graft polymerization of acrylic acid *J. Membr. Sci.*, **196**, 221–229 (2002).
72. J. Chen, J. Li, Z. P. Zhao, D. Wang, C. X. Chen, Nanofiltration membrane prepared from polyacrylonitrile ultrafiltration membrane by low-temperature plasma. 5. Grafting of styrene in vapor phase and its application *Surf. Coat. Technol.*, **201**, 6789–6792 (2007).
73. Z. P. Zhao, J. Li, D. Wang, C. X. Chen, Nanofiltration membrane prepared from polyacrylonitrile ultrafiltration membrane by low-temperature plasma. 4. grafting of N-vinylpyrrolidone in aqueous solution *Desalination*, **184**, 37–44 (2005).
74. D. S. Wavhal, E.R. Fisher, Hydrophilic modification of polyethersulfone membranes by low temperature plasma-induced graft polymerization, *J. Membr. Sci.*, **209**, 255–269 (2002).
75. I. Gancarz, G. Poźniak, M. Bryjak, A. Frankiewicz, Modification of polysulfone membranes. 2. Plasma grafting and plasma polymerization of acrylic acid *Acta Polym.*, **50**, 317–326 (1999).
76. R. Xie, S. B. Zhang, H. D. Wang, M. Yang, P. F. Li, X. L. Zhu, L. Y. Chu, Temperature-dependent molecular-recognizable membranes based on poly(N-isopropylacrylamide). and β-cyclodextrin *J. Membr. Sci.*, **326**, 618–626 (2009).
77. G. Pozniak, I. Gancarz, W. Tylus, Modified poly(phenylene oxide) membranes in ultrafiltration and micellar-enhanced ultrafiltration of organic compounds, *Desalination*, **198**, 215–224 (2006).
78. P. Lopez-Perez, A. P. Marques, R. M. P. da Silva, I. Pashkuleva, R. L. J. Reis, Effect of chitosan membrane surface modification via plasma-induced polymerization on the adhesion of osteoblast-like cells, *J. Mater. Chem.*, **17**, 4064–4071 (2007).

26

Enzyme-Immobilised Polymer Membranes for Chemical Reactions

Tadashi Uragami

Faculty of Chemistry, Materials and Bioengineering Kansai University Suita, Osaka, Japan

26.1 Introduction

Enzymes are protein catalysts in a living body that participate in a great number of chemical reactions in vivo. The characteristics are the ability to catalyse various reactions, to show a catalytic activity under mild conditions such as normal temperature and normal pressure and to have substance, position, stereo, and reaction specificity. These characteristics are different from catalysts employed in general chemical reactions.

Enzymatic function has already been exploited in a wide variety of fields such as food, fibre, leather, medicine and medical industries. Their catalytic function is very useful for the assembly of reaction processes to conform to the social demands of today such as energy saving, resource saving and environmental conservation. On the other hand, with the remarkable progress in biochemistry and genetic engineering, the mechanisms of enzyme action have been elucidated, new enzymes have been discovered, new manufacturing and purification method have been improved. Furthermore new methods of enzyme utilisation are being investigated.

However, enzymatic activity is remarkably lowered in the presence of heat, strong bases, strong acids and organic solvents. Additionally, the recovery of enzymes from reaction mixtures is very difficult because of their water solubility. Thus, for the purpose of the continuous regeneration of enzymes, they have been immobilised by various methods such as covalent bonding and adsorption onto carriers, crosslinking, gel entrapment and microcapsulation [1–4]. In particular, the immobilisation of enzymes in polymeric membranes separates the dual problems associated with immobilisation–enzymatic catalysis and spatial separation from the membrane.

Membranes for Membrane Reactors: Preparation, Optimization and Selection, Edited by Angelo Basile and Fausto Gallucci

In this chapter, enzyme-immobilised polymer membranes prepared by various immobilisation methods are introduced and the applications of these enzyme-immobilised polymer membranes in membrane reactor processed are discussed.

26.2 Brief Review of the Preparation Method of Enzyme-Immobilised Polymer Membranes

The preparation of enzyme-immobilised polymer membranes is carried out by following essentially the same process as the preparation of immobilised enzyme. To retain sufficient catalytic activity in the immobilised state, the enzyme has to be immobilised under conditions in which the residual groups of the amino acids that comprise the active center of the protein and their higher order structure are not changed.

A number of methods immobilising enzymes have been reported and they can be divided into covalent binding onto supports crosslinking and entrapment. Enzyme immobilised polymer membranes are prepared by methods similar to the immobilised enzyme and are summarised in Figure 26.1, that is: (a) physical adsorption of the enzyme onto a polymer support membrane, (b) crosslinking between enzymes on (a), (c) covalent binding between the enzyme and the

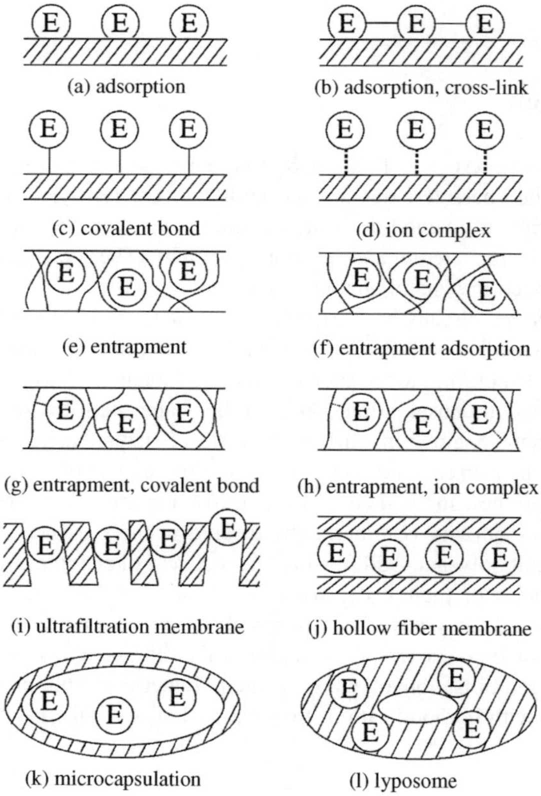

Figure 26.1 *Illustrations of enzyme-immobilised polymer membranes*

polymer membrane, (d) ion complex formation between the enzyme and the polymer membrane, (e) entrapment of the enzyme in a polymer gel membrane, (f) entrapment and adsorption of enzyme in polymer gel, membrane, (g) entrapment and covalent binding between the enzyme and the polymer, (h) entrapment and ion complex formation between the enzyme and the polymer membrane, (i) entrapment of the enzyme in a pore of an ultrafiltration membrane, (j) entrapment of the enzyme in a hollow fibre membrane, (k) entrapment of enzyme in microcapsule, and (l) entrapment of the enzyme in a liposome.

26.3 Preparation of Enzyme-Immobilised Polymer Membranes

26.3.1 Immobilisation of Enzymes on Polymer Membranes by Adsorption

Untreated membranes were employed in these adsorption trials. The membrane was first flushed with deionised water, washed with 0.1 M phosphate buffer (pH 6.5) and finally stored in a buffer solution until used for the immobilisation experiments. Laccase solutions of different concentrations (0.1, 0.2, 0.3, 0.4, 0.5, 0.6 mg m^{-1}) were prepared by dissolving the enzyme into 250 ml of 0.1 M phosphate buffer (pH 6.5). For immobilisation, the laccase solution was re-circulated through a polyether sulphonic spiral membrane at a flow rate of 60 ml min^{-1} and under a pressure of 0.8–1.0 bar for 60 min at 25 °C. The membrane-bound enzyme was then washed with the same buffer until no protein or enzyme activity could be detected in the washing buffer. The loading level of laccase units and protein were determined both by integrating the breakthrough curve and directly by measuring the amount eluted [5].

A membrane sample 50 mm diameter was fixed in a special poly(methylmethacrylate) frame so that only the active membrane layer was in contact with the lipase solution while the membrane supporting layer was prevented from contact by the enzyme solution. The membrane samples were then placed in Petri dishes with 20 ml of 1–10 mg ml^{-1} lipase solution in phosphate buffer (pH 7.0). and were shaken using a ABU-2 device (Mashprom, Tambov) for 1–12 h at 20 °C. After sorption, the membranes were removed from the Petri dishes and rinsed twice with 50 mM phosphate buffer (pH 7). The amount of lipase adsorbed onto the membrane surface was determined as the total protein quantity from the difference between the amount of protein in solution before and after adsorption and in the washing eluant using a QuantiPro™ BCA Assay kit (Sigma-Aldrich, Dorset, UK) [6].

The enzyme was immobilised from a water solution. Before adding any proteins, the membranes were eluted several times with distilled water under pressure and next with 0.05 M Tris-HCl buffer at pH 7.6 (in the case of a polypropylene membrane the procedure was preceded by filtration of with a 50% ethanol solution at the a pressure of 0.06 MPa to make hydrophilic membrane). An aqueous of the protein solution in 0.05 M Tris-HCl buffer at pH 7.6 (6 ml) was filtered at the pressure of 0.06 MPa through the membranes eluted with the buffer. The permeate was circulated for 1 h at 24 °C, and afterward the membranes were eluted several times with the buffer and the Lowry test was used to determine the quantity of bound protein [7].

26.3.2 Immobilisation of Enzymes in Polymer Membranes by Covalent Binding

As shown in Figure 26.2, in order to immobilise ascorbic acid oxidase (AsOM) onto a fibre containing diethylamino (DEA) and ethanolamine (EA) groups, the following solutions were permeated through the pores of a 2 cm fibre using a syringe pump at a constant permeation rate of 1 ml min^{-1} at ambient temperature: (1) 14 mM Tris-HCl buffer (pH 8.0) for equilibration,

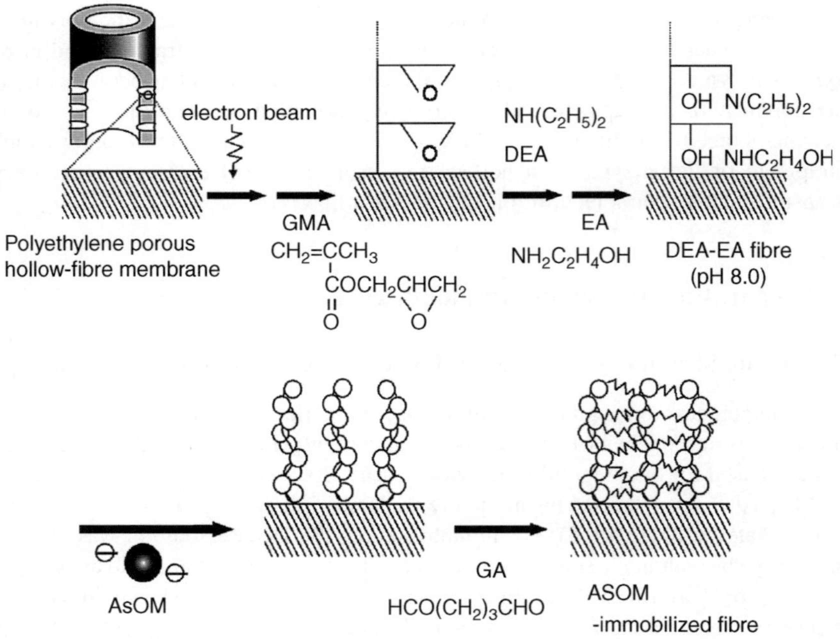

Figure 26.2 Preparation scheme of a matrix for the immobilisation of ascorbic acid oxidase. Reprinted from Journal of Membrane Science, Kawai, T., et al., Immobilization of ascorbic oxydase in multilayers onto porous hollow-fiber membrane. Vol. 191, 207–213. Copyright (2001) with permission from Elsevier

(2) 0.50 g of enzyme per litre of buffer to bind the enzyme onto the diethylamino group containing polymer chains grafted onto the pores of the fibre, (3) a buffer to wash the pores, (4) 0.50 wt% glutaraldehyde aqueous solution to crosslink the enzymes captured by the polymer chains, and (5) 0.50 M NaCl to elute the uncrosslinked enzyme. Through a series of the above procedures, the enzyme concentration in the effluent penetrating the outside surface of the hollow fibre was determined by measuring the UV absorbance at 235 nm. The amount of enzyme immobilised via ion-exchange adsorption and subsequent crosslinking, q, was calculated by the following equation [8].

$$q(\mathrm{mg\,g^{-1}}) = \frac{(\text{amount adsorbed})-(\text{amount washed})/(\text{amount uncrosslinked})}{\text{mass of membrane in the dry state}}$$

Enzyme immobilisation was achieved by activating the hollow fibre with hexamethylene diamine (HMDA). To this accomplish an aqueous solution of 50% (v/v) HMDA was recirculated through the external and internal surface of the hollow fibre for 4 h at 40 °C. Since polypropylene (PP) does not interact with HMDA, the latter was attached to the membrane through the reactive groups of some additives present on the PP matrix. After washing with distilled water, the hollow fibre was treated for 1 h at 30 °C with an aqueous solution of 2.5% (v/v) glutaraldehyde. To immobilise the enzyme, a solution of 10 mg ml^{-1} of β-galactosidase in 0.1 M phosphate buffer (pH 6.5) was re-circulated externally and internally through the hollow fibre for 16 h at 4 °C. After further washings with the same buffer solution to remove the unbound enzymes, the catalytic hollow fibre was ready for use and was assembled into the external housing. Under these conditions, 8.32 mg of β-galactosidase was immobilised onto the two surfaces of the PP membrane. The amount of immobilised enzyme was calculated by measuring the enzyme activity in the initial solution used for the immobilisation, the residual activity of the solution after the immobilisation process, and the activity in all of the solutions used to wash the membrane [9].

The outer surface of the poly(acrylnitrile) (PAN) hollow fibres was hydrolysed and amidated in an aqueous solution containing 1 N NaOH and 1 N 1,6-hexanediamine at 60 °C for 20 min. The resulting fibres were thoroughly rinsed with pH 7 buffer (0.05 M KH_2PO_4 + 0.0291 M NaOH). Unless stated otherwise, this buffer was used throughout the experiments. Afterward, the fibres were reacted with an aqueous solution of 10 wt% gultaraldehyde at 20 °C for 1.5 h, and then the fibres were thenrinsed thoroughly again. Finally, the glutaraldehyde-treated hollow fibres were immersed in an urease aqueous solution of 1 mg ml^{-1} (pH 7) and 4 °C for 24 h [10].

$$\text{PAN-CN} + \text{H}_2\text{N(CH}_2)_6\text{NH}_2 + \text{H}_2\text{O} \xrightarrow[60^\circ\text{C, 20 min}]{} \text{PAN-CONH-(CH}_2)_6\text{-NH}_2 + \text{NH}_3 \quad (26.1)$$

$$\text{PAN-CONH-(CH}_2)_6\text{-NH}_2 + \text{HOC-(CH}_2)_3\text{-COH} \xrightarrow[30^\circ\text{C, 1.5 h}]{}$$
$$\text{PAN-CONH-(CH}_2)_6\text{-N} = \text{CH-(CH}_2)_3\text{-COH} + \text{H}_2\text{O} \quad (26.2)$$

A blend of poly(urethane methacrylate) (PUA), glycidyl methacrylate (GMA) (20%), cross-linker (pentaerythritol tetraacrylate, 5 wt%) and Irgacure® 184 (1 mol%) was prepared in a beaker in a clean room under yellow light. This formulation was coated on a (PP) membrane (0.2 μm pore size) with the help of a bar coater (40 μm). This construct will be termed PUA-D. Prior to coating, a few PP membranes were pre-wetted by isopropanol/methanol (hereafter termed as PUA-I and PUA-M, respectively). After ensuring uniform coating, the disc was exposed to UV radiation (low-pressure mercury lamp, 50 W cm^{-1}) for 10 min under nitrogen flow at ambient temperature. The method was standardised by differential photocalorimetric and IR studies. The membranes were washed, heated at 50 °C under vacuum, dried to ensure the absence of unreacted monomers, and finally cut into flat circular discs (47 mm diameter) and small squares (1 cm^2). Membrane was suspended in 20 ml of 0.5 M of sodium phosphate buffer (pH 7.0) containing different concentrations of Candida rugosa lipase (CRL) and incubated in a rotary shaker at 30 °C (100–120 rpm). After 12 h, the membrane was separated from the enzyme solution and treated with 5% aqueous glutaraldehyde solution at 40 °C for 3 h. The membrane was then washed [11].

26.3.3 Immobilisation of Enzymes in Polymer Membranes by Entrapment

Poly(*N*-isopropylacrylamide-*co*-*N*-acryloxysuccinimide-*co*-2-hydroxyethylmethacrylate) [p(NIPAAm-NAS-HEMA)] hydrogel composite membrane onto a polyester nonwoven substrate was placed in a 0.1 M phosphate buffer (pH 7 unless otherwise specified) solution containing a fixed amount (4 mg unless otherwise specified) of urease. The solution was mixed by stirring at low speed at 4 °C for 24 h, after which the membrane was washed twice with pH 7.5 buffer and stored in the same buffer at 4 °C. The amount of protein immobilised onto the membrane and the activity of the immobilised urease were then determined. To optimise the preparation of the membrane-immobilised urease, the effects of the following factors during the immobilisation step were studied: the molar ratio of NIPAAm to NAS (9, 19, 39), the pH of the coupling buffer solution (5–9), and the concentration of the urease solution (1–10 mg) [12].

The electropolymerisation of pyrrole (1 mM) was carried out on a steel mesh in acetonitrile solution with 0.1 M $NaClO_4$ in a three-electrode cell. The copolymer pyrrole–pyrrole biotin was prepared under the same experimental conditions: a solution containing the monomer (**1**) (2 mM) and pyrrole (1 mM). These textiles coated by the polypyrrole films were characterised by scanning electronic microscopy. The enzymatic membrane was prepared as follows: a 2 mM solution of monomer (**1**) in 0.1 M $NaClO_4$ acetonitrile was electropolymerised by controlled potential oxidation at +0.8 V versus Ag/AgCl electrode on carbon felt in a three-electrode cell. The polypyrrole–biotin modified support was rinsed with distilled water and immersed into an avidin aqueous solution (1 mg ml^{-1}) for 1 h. The resulting carbon porous membrane was washed with distilled water and immersed into a biotinylated enzyme solution (2 mg ml^{-1}) for 1 h [13].

The desired amounts of urease, stylite, and activated charcoal were added to the desired mixtures containing cellulose nitrate, n-propyl alcohol, and *N,N*-dimethyl formamide. These casting solutions were kept overnight at casting temperature (6–7 °C). The membranes were then made by pouring the casting solution onto an applicator for thin layer chromatography, drawing the blade across the glass plate, allowing the solvent to evaporate at 6–7 °C for the appropriate period of time, and immersing the glass plate together with the solution into cold water (6–7 °C). After standing for 24 h in cold water, the urease-, stylite-, and activated charcoal-immobilised cellulose nitrate membranes were removed from the glass plate [14].

26.3.4 Immobilisation of Enzymes in Polyion Complex Membranes with Entrapment and the Formation of Ion Complexes

Sodium polyacrylate (PAANa; 0.4 g) and q-chitosan (q-chito; 1.0 g) were dissolved separately in an aqueous solution of NaBr (20 wt%). The same volume of these two aqueous solutions was mixed, since the ion exchange capacity between PAANa and q-chito was equivalent. The casting solution was prepared by diluting this mixed solution to 0.2 wt% in total polymer concentration with an aqueous solution of NaBr (20 wt%).

A sketch of the membrane preparation is shown in Figure 26.3. The casting solution (2 g) was poured onto a microporous poly(propylene) film mounted on the porous support in an

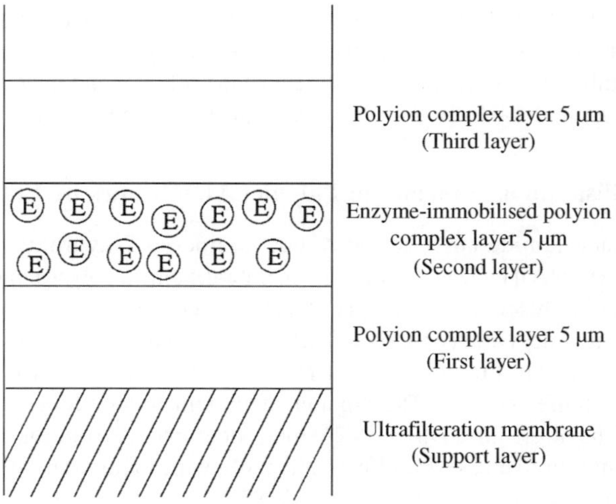

Figure 26.3 Sketch of the enzyme-immobilised polyion complex membrane. Reprinted from Polymer Bulletin, Uragami, T., et al., New method for enzymr immobilization by a polyion complex membrane. Vol. 15, 101–106. Copyright (1986) with permission from Springer

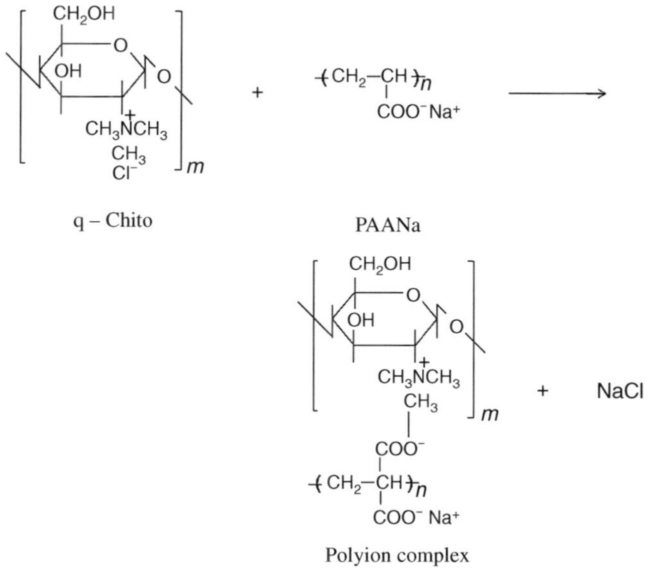

Scheme 26.1 *Formation of polyion complex between q-Chito and PAANa. Reprinted from Membrane, Uragami, T., et al., Hydrolysis of urea through urease immobilizing polyion complex membrane. Vol. 14, 211–216. Copyright (1989) with permission from Membrane Society of Japan*

ultrafiltration cell, and ultrafiltrated at 0 °C and 2 kg cm^{-2} under nitrogen gas supplied from a gas cylinder. Since this ultrafiltration process removed NaBr from the casting solution mixture, the formation of the polyion complex between q-chito and PAANa as shown in Scheme 26.1 was promoted and the first polyion complex layer was formed on the poly(propylene) film. A desired amount of buffer solution containing urease was added to the casting solution (2 g). This mixed solution was put on the first polyion complex layer and then ultrafiltrated at 0 °C and 2 kg cm^{-2} to form a second layer in which invertase was immobilised. Finally, a third polyion complex layer, the same as the first layer, was laminated onto the second layer containing urease. The enzyme immobilised polyion complex membrane constructed in three layers was washed by permeating 0.2 M phosphate buffer solution at 50 °C and 3 kg cm^{-2} to completely revove any excess NaBr remaining in the composite membrane [15–17, 25].

26.3.5 Immobilisation of Enzymes in Ultrafiltration Membranes, Microfiltration Membranes, and Hollow Fibre Membranes

Crude lipase of (400 mg) or pure lipase powder of (49.5 mg) was dissolved in 400 or 200 ml of 50 mM phosphate buffer (pH 7.0), and the solution was gently stirred by a magnetic stirrer for 2 h. Therefore, an initial crude lipase solution of 1 or 2 mg ml^{-1} was used. The solution was then centrifuged at 3000 rpm for 15 min to remove any insoluble substances. Some amount of the solution was stored in a refrigerator for the measurement of the protein concentration. An enzyme solution of 350 ml was circulated in the shell side of the prepared module at a pressure of 50 kPa. Lipase was immobilised into the spongy layer by crossflow filtration of the lipase solution. When the permeate volume amounted to about 120 ml, the circulation was stopped, and the retentate and permeate were removed from the module. The capillary membranes in the module were washed three times using 100 ml of phosphate buffer (50 mM, pH 7.0). The protein concentrations of the original lipase solution, the permeate, the retentate and the washings were then measured by a BCA test kit (Sigma). The amount of protein retained in the membrane was calculated from the protein mass balance among these solutions [18].

A fruit extract solution (5 ml) was recirculated along the shell side at a flow rate of 1.45 ml min^{-1} (axial velocity of 0.028 m s^{-1}) and a transmembrane pressure of 0.10 bar. The enzyme solution was permeated from the shell to thelumen. In this way the β-glucosidase present in the extract (6 kDa monomers or higher molecular weight aggregates), entered the spongy layer, but could not pass through the thin layer (cut-off 30 kDa). After the crossflow ultrafiltration was completed the membrane was rinsed in order to remove any enzyme reversibly adsorbed onto the membrane. Rinsing cycles of 5 min each with 0.1 M borate buffer pH 9 were performed at a flow rate of 2.38 ml min^{-1}, axial velocity of 0.046 m s^{-1} and a transmembrane pressure of 0.15 bar. The removal of non-immobilised enzyme was monitored by spectrophotometer analysis in the UV-vis range [19].

The immobilisation procedure was based on the methods of lipase CLEC/CLEA preparation Lipase precipitation has been was performed from the solution in a phosphate buffer (pH 6.8, 0.05 M) impregnating the microfiltration membranes. Before saturation of the membranes pores with the lipase solution, the membrane samples were pretreated to provide complete wetting with aqueous media. For this purpose, the membrane samples were kept in methanol for 30 min followed by methanol exchange with deionised water over three successive water portions for 30 min each. To saturate the membrane sample with the lipase solution, a portion of lipase solution prepared by dissolution of 250 mg lipase in 3 ml of phosphate buffer was filtered through the membrane in a dead-end filtration cell under a pressure of 2 kPa without stirring. After that, the membrane sample was carefully transferred to a glass vessel containing 20 ml of glutaraldehyde solution in acetone. The glass vessel was tightly sealed and kept for 15 h at 4 °C. After that, acetone and excessive glutaraldehyde were washed out by several portions of cold water (4–8 °C) andthen stored in water overnight at 4 °C. To study the influence of the crosslinker concentration on the activity of the immobilised enzyme, the glutaraldehyde content in the acetone solution was varied in the range from 3 to 20 wt% [20].

The lipase solution was prepared by dissolving crude lipase with an initial concentration of 2.0 g l^{-1} in phosphate buffer and gently stirring for about 2 h at room temperature. The solution was then centrifuged at 3000 rpm for 15 min to remove any insoluble impurities and heated at 60 °C for 7 min. Afterwards, the deactivated lipase was immobilised within the spongy layer of the membrane module by crossflow filtration from shell to lumen. The immobilisation process was carried out at 0.5–0.7 bar and room temperature. In some experiments, an oil in water emulsion was added to the enzyme solution and gently stirred to make it become a homogenous phase. After immobilisation, the rinsing process was carried out using a phosphate buffer solution to remove any reversibly adsorbed enzyme. The rinsing process was repeated until no protein was detected by scanning the optical density of the retentate in the range 800–200 nm. The protein concentrations of the initial feeds, retentates and permeates were used for calculating the amount of immobilised enzyme by mass balance. Next, the pure water permeability through the enzyme-loaded membrane was measured. This allowed us to evaluate the hydraulic resistance due to the irreversibly loaded enzyme [21].

The lipase solutions were prepared by adding the appropriate amounts of lipase powder to a phosphate buffer solution (0.05 M, pH 7.5). The insoluble portion of the enzyme solution was removed by filtration with a filter membrane (pore size: 0.44 μm). A cellulose acetate (CA)/poly(tetrafluoroethylene) (PTFE) composite flat membrane was placed into a dead-end filtration cell with a magnetic stirring apparatus. The PTFE layer was placed upward and the enzyme aqueous solution was filtrated under pressure by the compressed nitrogen of 0.6 MPa. After filtration, the membrane was taken out and rinsed with the phosphate buffer solution for three times. The original enzyme protein concentration and the concentration after filtration with the composite membrane were measured by the Lowry method with an UV-vis spectrophotometer (Agilent 8453 UV-vis spectrophotometer). Different volumes of the enzyme aqueous solution were filtrated to vary the amount of enzyme loading on the membranes [22].

The poly(tetrafluoroethylene) (PTFE) layer was placed upward and the lipase solution was placed in the upper cavity of the filter. The lipase solution was filtrated through the membrane under 0.6 MPa of N_2. In this way, the lipase molecules were trapped in the interface of the layers. The membrane and the filter were then washed with distilled water. The water and the filtrate were carefully collected and the concentration of enzyme was analysed. The amount of lipase entrapped was calculated according the concentration and volume of lipase solution before and after filtration and the concentration and volume of the wash water. Therefore, enzyme loading could be calculated by the amount of enzyme entrapped divided by the area of the membrane [23].

The lipase solutions were prepared by adding appropriate amounts of lipase powder to a phosphate buffer solution(0.05 M, pH 7.5). The insoluble material from the enzyme solution was removed by filtration with a filter membrane. The concentration of the enzyme solution was about 90 mg l^{-1}. A polysulfone flat membrane was placed into a dead-end filtration cell with a magnetic stirring apparatus, and the lipase solution was filtrated under pressure by compressed nitrogen gas at a pressure of 0.6 MPa. The original enzyme protein concentration and the concentration after ultrafiltration were measured by the Lowry method using an UV-vis spectrophotometry (Agilent8453 UV-vis spectrophotometer). Different volumes of the enzyme solution were filtered to vary the amount of enzyme loading on the membranes. To investigate the effects of the location of the enzyme immobilisation, the enzyme solution was ultrafiltered either with the spongy layer up or with the dense layer up. A 20 ml aliquot of glutaraldehyde solution was filtered slowly through the membrane over 2 h under a pressure of 0.4 MPa to crosslink the enzymes. Afterward, the membrane was rinsed with the phosphate buffer solution for three times [24].

26.3.6 Immobilisation of Enzymes in Polymer Membranes by Copolymerisation

In order to immobilise enzymes in the polymer matrix by copolymerisation, vinylised urease (VU) was synthesised using the method reported by Hoffman and coworkers [25], that is, *N*-succinimidylacrylate (NSA; 10 mg) was added to a phosphate buffer solution of 20 mM (pH 7.4) containing urease (120 mg), and the reaction was performed at 36 °C for 1 h to introduce vinyl groups into the urease molecule. Afterwards the resulting vinylised urease (VU) (I) was purified by gel filtration. The degree of vinyl group introduction was 15.9% which was determined by the TNBS method [26]:

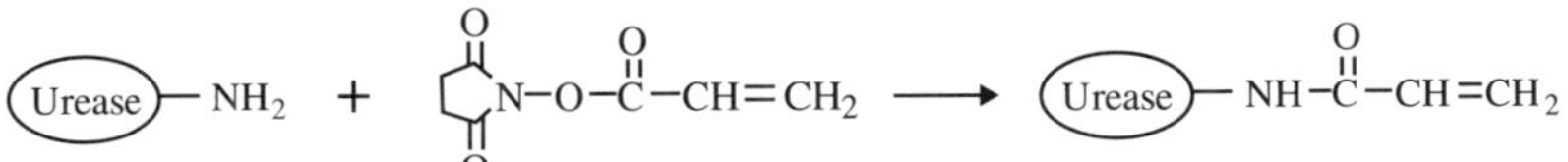

Urease NSA VU (1)

Urease-immobilised poly(VU-AAm-HEMA) membranes were prepared by the copolymerisation of VU with acrylamide (AAm) and 2-hydroxethylmethacrylate (HEMA) using *N,N′*-methylenebiacrylamide (MBAA) as a crosslinker, plus aqueous solutions of ammonium persulphate (APS) and *N,N,N′,N*-tetramethylenediamine (TEMED) as redox initiators, in a molded plate between two glass plates at 25 °C for 3 h, as shown in Figure 26.4 [27].

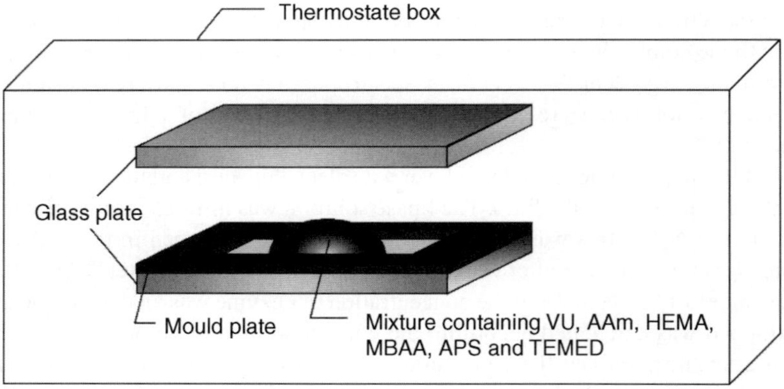

Figure 26.4 *Schema for the preparation of urease-immobilised poly (VU-AAm-HEMA) membranes. Reprinted from Catalysis Today, Uragami, T., et al., Preparation of urease immobilized polymeric membranes and their function, Vol. 118, 158–165. Copyright (2006) with permission from Elsevier*

26.4 Applications of Enzyme-Immobilised Polymer Membranes as Membrane Reactors

26.4.1 Polymer Membranes with Enzymes Immobilised by Adsorption

The immobilisation of a commercial laccase onto a spiral-wound asymmetric polyethersulphone membrane was studied. The immobilised enzyme system displayed a promising half-life of more than 150 h for the oxidation of syringaldazine. The laccase membrane reactor was applied to the biodegradation of a model phenol solution containing 18 phenolic substrates, including chlorophenols, cresols and methoxyphenols. The type and/or the position of the substituent group significantly affected the amount of substrate oxidation [5].

Lipase-immobilised membranes were prepared by both non-covalent and covalent immobilisation methods using: (i) lipase adsorption onto the membranes, (ii) inclusion of the enzyme into the membrane structure by filtration and (iii) covalent attachment of lipase to the membrane. The catalytic properties of these membranes have been studied for the reaction of butyloleate synthesis through the esterification of oleic acid with n-butanol in isooctane. Ultrafiltration membranes composed of regenerated cellulose were used for lipase immobilisation. Lipase inclusion into the wide porous supporting layer of the membrane was the most efficient method for preparing highly effective biocatalytic membranes. The percentage of oleic acid conversion using these membranes was about 70.0–72.7% (with a reaction time of 8 h). It was shown that the distribution profile of the lipase in the membrane was important for the effective enzyme utilisation. The profile imaging atomic force microscopy (AFM) technique was used to visualise the surfaces of the lipase-immobilised biocatalytic membranes. AFM has also been used to directly quantify interactions between the lipase-coated till and the membrane surface. It was concluded that the direct measurements of the interaction force between the enzyme-coated tip and the membrane surface would be a useful and practical approach for the choice of membranes as porous polymeric support for lipase immobilisation through adsorption [6].

The integration of the process of synthesis and dehydration of a reaction phase, in which a biocatalyst is suspended but not dissolved as in water solutions, requires placement of the catalyst in the reactor before directing the stream of the reaction mixture to dehydration process. This placement and the possibility of multiple use of the catalyst may be accomplished by using a separating barrier, for example, an ultrafiltration membrane or by permanently fixing of the catalyst to the matrix, for example, a polymeric membrane. The efficiency and activity of a biocatalyst (lipase CAL-B) immobilised on a polymer membrane by sorption and chemical binding were then determined. A subject of study was the synthesis of geranyl acetate, one of the most well known aromatic compounds. A hydrophobic (polypropylene) matrix was shown to be a much better carrier in the reactions performed in an organic solvent than the hydrophilic (polyamide) membrane being tested. With respect to the reaction kinetics and stability of the native enzyme versus the immobilised preparation, the more attractive type of membrane bioreactor more attractive appears to be the one in which the membrane is used not as a catalyst layer, but only as a barrier that immobilises the native enzyme within the bioreactor volume. When the integrated process proceeds, the method used to collect water in the sorption column during the process, appeared to work very well. The reaction proceeded with a very high efficiency (98.2% for native enzyme and 83.2% for immobilised enzyme, after 120 h) and due to low water concentration in the system (approximately 0.000% v/v) a second phase was not created [7].

26.4.2 Polymer Membranes with Enzymes Immobilised by Covalent Binding

A porous membrane containing an anion exchange group in hollow fibre form was prepared by the radiation-induced graft polymerisation of glycidyl methacrylate, and subsequent conversion of the produced epoxy group into a diethylamino group. Ascorbic acid oxidase (AsOM) was bound to the ionisable polymer chains grafted onto the pore surface during the permeation of the ascorbic acid (AsA) solution through the pores. After crosslinking of the enzyme with glutaraldehyde, AsOM was immobilised at a resulting density of $130\,mg\,g^{-1}$ of fibre, which amounted to a degree of enzyme multilayer binding of 12. The substrate solution was forced to permeate across an AsOM-multilayered porous hollow fibre membrane of 1 mm thickness at a constant permeation rate ranging from 30 to $150\,ml\,h^{-1}$. Regardless of the permeation ratebeing tested, quantitative conversion of AsA into dehydroascorbic acid was observed because the substrate was transported to the highly active enzyme immobilised in the multilayers by convective flow through the pores [8].

A hollow fibre enzyme reactor, operating under isothermal and nonisothermal conditions, was built using a polypropylene hollow fibre onto which β-galactosidase was immobilised. Hexamethylenediamine and glutaraldehyde were used as a spacer and coupling agent, respectively. Glucose production was studied as a function of temperature, substrate concentration, and size of the transmembrane temperature gradient. The actual average temperature differences across the polypropylene fibre, to which a pilot study was done to evaluate the effect of the nonisothermal conditions, were calculated by means of a mathematical approach, which made it possible to determine, using computer simulation, the radial and axial temperature profiles inside the bioreactor and across the membrane. Percentage activity increases, proportional to the size of the temperature gradients, were found when the enzyme activities under nonisothermal conditions were compared to those measured under comparable isothermal conditions. The percent reductions of the production times, proportional to the applied temperature gradients, were also calculated. The advantages of employing nonisothermal bioreactors in biotechnological industrial processes were also discussed [9].

The surface of the polyacrylonitrile hollow fibres was hydrolysed and covalently bonded with urease via glutaraldehyde. The immobilised urease retained higher relative activity than native urease when stored at various pH values. The stabilities of the immobilised urease against pH fluctuations were stronger than those of the native enzyme. The immobilised urease retained 86%

of its initial activity after reusing it 15 times at pH 7. After storing for 42 days at 4 °C and pH 7, the immobilised urease could hydrolyse 15% of the initial concentration of urea at pH 7 and 37 °C after 4 h, whereas the native urease lost almost all of its catalytic ability. The removal of urea using urease-immobilised dialyser was demonstrated by *in vitro* dialysis, and showed a two-fold faster removal rate for urea than a regular dialyser. Furthermore, the improvement in the urea clearance rate by the urease immobilisation to the dialyser increased with the dialysate velocity [10].

Polypropylene was hydrophilised by coating followed by the UV curing of a blend of 2-hydroxyethyl methacrylate (HEMA) terminated polyurethane prepolymer and glycidyl methacrylate (GMA). This enabled the formation of a hydrophobic membrane with increased surface hydrophilicity, biocompatibility and stability. Candida rugosa lipase (CRL) was covalently immobilised on this membrane using 5% glutaraldehyde as a crosslinking agent for post immobilisation stabilisation of the enzyme onto the membrane. The membrane obtained was placed in a batch membrane reactor where the model esterification of oleic acid with octanol was studied. Under optimum conditions, the biocatalytic membrane gave a specific activity of 796.27 units mg^{-1} and 90.26% activity yield. Moreover, there was 85.10% retention of the specific activity. The biocatalytic membrane was observed to retain about 84.23% of its synthetic activity after six cycles [11].

Enzyme–membrane reactors (EMR)EMR had been prepared by covalent immobilisation of fructosyltransferase (FTF) onto amino-functionalised track-etched poly(ethylene terephthalate) membranes with nominal pore diameters of 400, 1000 or 3000 nm, and in all cases the loss of enzyme activity was fast, and the extent of pore blocking was increasing with decreasing pore diameter. Experiments at varied crossflow velocity with the FTF covalently immobilised on the surface of an epoxy-reactive non-porous film suggested that the rate of FTF deactivation could be much reduced by increasing mass transfer efficiency. Consequently, FTF-EMR with pore diameters of 1000 nm had been operated using regular backpulsing during flow-through or with dispersed inert nanoparticles (diameter 110 nm) in the feed solution flowing through the pores. For both variants, a significant performance improvement had been obtained. Finally, a novel nanoparticle composite membrane had been prepared via the covalent immobilisation of epoxy-reactive nanoparticles (diameter 200–230 nm) on the pore walls of the track-etched membranes. Spacing between the particles had been achieved by using a mixture of reactive and inert nanoparticles and suited reaction and washing conditions. The FTF–EMR based on the novel nanoparticle composite membranes had a much higher productivity (prolongation of EMR operation time and more formation of inulin) as compared with all the other membranes or variants investigated in this study. Also a much lower pore-blocking tendency had been observed. This improved performance had been explained by the two functions of the nanoparticles in the membrane – turbulence promoter and support providing increased surface area for covalent enzyme immobilisation [28].

26.4.3 Polymer Membranes with Enzymes Immobilised by Entrapment

A composite membrane made of crosslinked poly(*N*-isopropylacrylamide-*co*-*N*-acryloxysuccinimide-*co*-2-hydroxyethylmethacrylate) [p(NIPAAm-NAS-HEMA)] hydrogel on polyester nonwoven support was synthesised. The composite membrane shows temperature-responsive properties similar to conventional PNIPAAm hydrogels beads, which reversibly swells below and de-swells above their lower critical solution temperature of PNIPAAm (around 32–33 °C). The diffusion of urea through the membrane was temperature-dependent with an effective diffusion coefficient at 20 °C being 18-fold greater than that at 60 °C. Urease was immobilised directly to the membrane by forming covalent bonds between its amino groups and the succinimide ester groups of the membrane. A membrane was prepared with a NIPAAm to NAS molar ratio of 9, and then reacted in pH 7 buffer with 6 mg of urease. This generated the best immobilised enzyme, which gave 0.102 mg protein, 5.71 units of activity per cm^2 of membrane, and 55% relative specific activity. There was negligible internal mass transfer resistance for this preparation judging from the calculated effectiveness factor. Urease showed enhanced thermal stability after immobilisation with a first-order inactivation rate constant at 70 °C which decreased to 12.5% of that of free urease, Membrane-immobilised urease could be utilised in a two-compartment membrane reactor with a temperature swing to substantially enhance the urea hydrolysis rate. The best operating conditions of the membrane reactor was with a temperature cycling between 60 and 20 °C and with a temperature change every 10 min, where the concentration of the product ammonia after 3 h of reaction increased 3.8-fold when compared with isothermal operation at 60 °C [12].

Glucose oxidase (GOD) and peroxidase (POD) have successfully been immobilised into a polypyrrole matrix by an avidin–biotin molecular recognition process. A biotin-labeled pyrrole was the precursor of the electropolymerised polymer. The amount of enzyme entrapped in the polymeric matrix was 60% of the initial amount for GOD and 83% for POD; these values are higher than those previously reported for enzyme entrapment in polymer matrices. These reactive membranes performed a chemical transformation during the permeation of the substrates; the GOD membrane catalysed the oxidation of glucose in the presence of oxygen, whereas the peroxidase membrane catalysed the reduction of hydrogen peroxide in the presence of the oxidant pyrrogallol. The catalytic activity of these membranes has been assessed. A turnover number of 900 in 1 min was determined for the POD membrane. The loss of activity of the enzyme during the membrane formation (42%) by the avidin–biotin method of immobilisation was weak. Moreover the enzymatic POD membrane was remarkably stable after 80 days of storage (less than 6% decrease in activity). These findings point to the interest of using such enzymatic membranes for the treatment of aqueous media [13].

For applications as artificial kidneys the ultrafiltration, hydrolysis, and adsorption characteristics of semipermeable cellulose nitrate membranes containing urease, stylite, and activated charcoal, which were prepared by the use of mixed solvents of n-propyl alcohol and *N,N*-dimethyl formamide, were investigated under various conditions. The optimum composition of the casting solution was found to be a ratio of cellulose nitrate: n-propyl alcohol: *N,N*-dimethyl formamide, of 13.00: 21.75: 65.25 (wt%), containing 3 g of urease, 1.5 g of stylite, and 5 g of activated charcoal per 100 g of the above casting solution. It was observed that the ultrafiltration rate, the hydrolysis, and the adsorption characteristics using aqueous solutions of urea and disodium phenoltetrabromophthalein sulfonate (BSP) as a feed solution were significantly influenced by the solvent composition in the casting solution, the evaporation period, the additional amount of activated charcoal, and the operating temperature and pressure. When a mixture of albumin, sodium chloride, urea, creatinine, BSP, and vitamin B_{12} was used as a model for blood, the ultrafiltration rate was high $2.5 \times 10^{-5}\, g\, cm^{-2}\, s^{-1}$ and urea, creatinine, BSP, and vitamin B_{12} were removed, and also albumin was completely rejected [14].

Figure 26.5 shows the effects of the ultrafiltration period on the ultrafiltration rate, the hydrolysis of urea, the ion exchange capability, the adsorption, and the rejection fraction.

As can be seen from Figure 26.5, the ultrafiltration rate decreased. This result can be attributed to the combined factors of the blockage of pores in the membrane, namely the compaction of membrane, and the concentration polarisation of albumin molecules. The hydrolysis fraction of urea was always 100% and the ammonium ion was completely adsorbed throughout this ultrafiltration period. The hydrolysis fraction of the adsorbed sodium ion eventually approaches

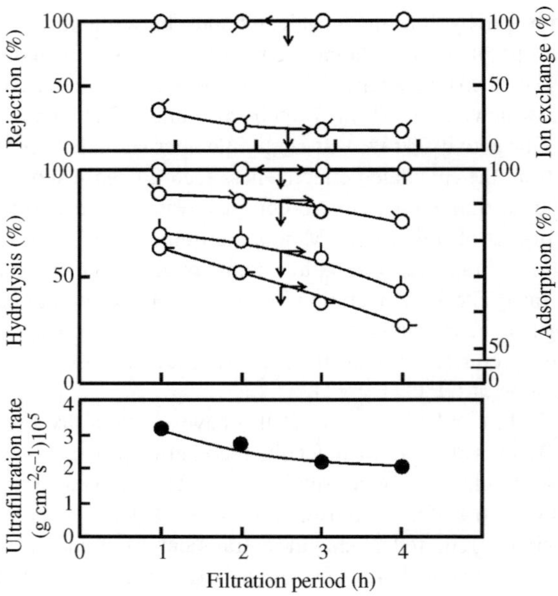

Figure 26.5 Effects of filtration period on the characteristics of the system membrane: cellulose nitrate/urease/stylite/activated charcoal. Operating conditions: 40 °C, 0.5 kg cm^{-2}, pH 7.2. Feed mixture: sodium chloride 900 mg dl^{-1}, urea 60 mg dl^{-1}, BSP 10 mg dl^{-1}, vitamin B_{12} 10 mg dl^{-1}, albumin 1 g dl^{-1}, adsorption fraction of ammonium ion. Reprinted from Angew. Makromol. Chem., Tamura, M., et al., Ultrafiltration, hydrolysis, and adsorption characteristics of membranes from cellulose nitrate, urease, stylite, and activated charcoal, Vol. 79, 67–77. Copyright (2006) with permission from Wiley InterScience

to a constant value, and the albumin was perfectly rejected. The above results can be interpreted as the membrane being able to function as an artificial kidney [14].

26.4.4 Polymer Membrane with Enzymes Immobilised by Entrapment and Ion Complex

A new method of enzyme immobilisation in a polyion complex has been proposed. An enzyme-immobilised polyion complex membrane was prepared by ultrafiltrating a mixture consisting of quarternised chitosan, sodium polyacrylate and invertase in an aqueous NaBr solution. The permeation and hydrolysis characteristics of aqueous sucrose solutions through this invertase-immobilised membrane was studied under various conditions. A hydrotic Michaelis–Menten type reaction was postulated [15].

Figure 26.6 is a Lineweaver–Burke plot of the hydrolysis reaction of urea through this invertase immobilising membrane. This relationship shows a good linearity.

Therefore, this supports the idea that the hydrolysis reaction of urea through the invertase-immobilised polyion complex membrane can be interpreted as the Michaelis–Menten type of reaction.

The kinetic data determined from the intercepts of the axis in Figure 26.6 are summarised in Table 26.1, which also includes those for the batch reaction of the hydrolysis of urea by a native invertase. The value of maximum decomposition rate, V_{max}, of sucrose for the invertase immobilised polyion complex membrane was about 200-fold of that of the native enzyme.

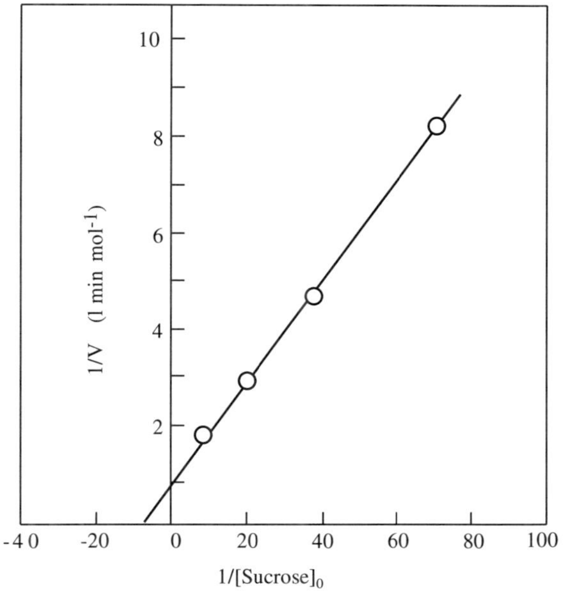

Figure 26.6 *Lineweaver–Burke plot for the decomposition of sucrose by the invertase-immobilised polyion complex membrane at 50 °C and 3 kg cm^{-2}. Reprinted from Polymer Bulletin, Uragami, T., et al., New method for enzymr immobilization by a polyion complex membrane. Vol. 15, 101–106. Copyright (1986) with permission from Springer*

This is mainly caused by the fact that the existence amount of the enzyme per unit volume is higher in the enzyme-immobilised polymer membrane. The reciprocal of the Michaelis constant, $1/K_m$, which is used to evaluate as the affinity constant between the substrate and the enzyme molecules, was slightly larger in the system of enzyme immobilised polymer membrane. A polymer matrix constructed from a polyon complex membrane becomes a barrier for the approach of a substrate molecule to the immobilised enzyme molecule. But once the substrate molecule has contacted the immobilised enzyme molecule, these molecules can interact in the limited space in the polymer matrix.

If some enzymes can be immobilised by the method reported in this study, a composite enzyme-immobilised polymer membrane of the laminate type can be easily prepared as shown in Figure 26.7, in which E_1, E_2, E_3 and E_4 represent different enzymes. When a such composite enzyme-immobilised polymer membrane is applied to a chemical reactions and the resulting product is permeated through E_1 immobilised layer and becomes a substrate for the next enzyme

Table 26.1 *Kinetic data for the hydrolysis of sucrose by native invertase and an invertase immobilised polymer membrane*

Enzyme	K_m (M) × 10	V_{max} (MS^{-1})	$1/K_m$ (M^{-1})
Native[a]	3.27	1.10×10^{-4}	3.06
Immobilised[b]	6.41	2.29×10^{-2}	1.56

[a]Invertase (20 ml) was used in aqueous sucrose solution (50 ml). The reaction conditions were pH 5.3 and 50 °C for 3 h.
[b]Invertase (20 ml) was immobilised in a polyion complex membrane. The permeation conditions were pH 5.3, 50 °C and 3 kg cm^{-2} for 3 h.

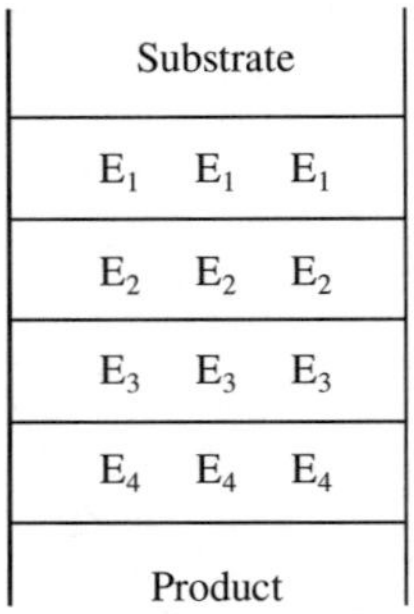

Figure 26.7 *Model of some enzyme-immobilised polyion complex membranes. Reprinted from Membrane, Uragami, T., et al., Hydrolysis of urea through urease immobilizing polyion complex membrane. Vol. 14, 211–216. Copyright (1989) with permission from Membrane Society of Japan*

(E_2) immobilised layer, then the synthesis of materials due to multistage reactions will become possible [15, 17].

26.4.5 Polymer Membranes with Immobilised Enzymes for Ultrafiltration Membranes, Microfiltration Membranes, and Hollow Fibre Membranes

A lipase-immobilised membrane reactor was applied for the optical resolution of racemic naproxen, and the effect of various operational conditions on the reaction rate and enantioselectivity was examined. The membrane reactor consisted of an organic phase dissolving the naproxen ester, a lipase-immobilised polyamide membrane, and an aqueous phase to recover the reaction products. The lipase immobilised into the membrane reactor showed a stable activity for more than 200 h of continuous operation, whereas the native lipase in an emulsion-stirred tank reactor quickly lost its activity showing a half-life time of about 2 h. When crude lipase was used, the biphasic enzyme membrane reactor gave less enantioselectivity compared to the native lipase in the emulsioned tank reactor, wheeas the use of pure lipase gave similar results to the native lipase. This paper discusses the fact that other hydrolases present in the crude enzyme powder caused a decrease in enantioselectivity. The enantioselectivity depended on the substrate concentration, the amount of enzyme loaded into the membrane and the immobilisation site. In fact, these parameters affected the organic/aqueous interface that plays an important role in the enhancement of enantioselectivity [18].

A new combined method has been reported to localise the sites of enzyme immobilisation and to determine its catalytic activity on a polymeric capillary membrane reactor. The useful new method resulted from the merging of the classic in situ enzyme activity assay and Western bloting technique and both results are easily detectable by either at low or at high magnification in light microscopy. β-Glucosidase from olive fruit was selected as enzyme model because of its suitability relevance in the industrial processing of foods, in biotechnology and in pharmaceuticals and for its activity against the synthetic substrate 5-brome-4-chloro-3-indolyl-β-D-glucopyranosyde which develops an insoluble dyed product. The enzyme was physically immobilised within 30 kDa cut-off capillary polysulphone membranes and the results were obtained by means of a polyclonal antibody against β-glucosidase and the synthetic substrate clearly showed the coherent localisation of the immobilisation enzyme sites and its activity [19].

A new multiphase enzyme membrane reactor has been developed for immobilising the lipase from *Candida rugosa* in a polymeric membrane in presence of a stable and uniform oil in water emulsion prepared by membrane emulsification. The reactor has the configuration of a two separate phase membrane reactor comprised by an emulsion + enzyme-loaded membrane and an organic and an aqueous phase recycled along the two separate sides of the membrane. This reactor was tested with different naproxen esters and with triglycerides as reagents and its performance was compared with the traditional two separate phase enzyme membrane reactor where the lipase was immobilised without any emulsion. These results showed that the presence of the emulsion within the membrane improved the catalytic activity and the enantioselectivity of the immobilised enzyme as well as the transport rate of the hydrophobic reagent through the hydrophilic membrane. This work confirmed that immobilisation can either improve enzyme stability or preserve its native selectivity [20].

The transport properties of the phase transfer reaction components through a multiphase enzyme-loaded membrane system were then identified. The multiphase system is of interest for the preparation of a two-separate-phase enzyme membrane reactor used for the enantioselective conversion of the racemic ester of naproxen into the corresponding S-naproxen acid. The transport properties of the system have been measured in absence of biochemical reaction. Therefore, the overall mass transfer coefficients of the reactant (naproxen ester) and product (S-naproxen acid) through the two-separate-phase membrane system have been measured using a deactivated enzyme-loaded membrane. The multiphase system of interest is composed of an organic and an aqueous phase separated by a polymeric membrane containing the immobilised enzyme. The enzyme was immobilised alone or in presence of an oil/water emulsion, in order to either the improve enzyme distribution at the o/w interface or to enhance the transport of reaction components through the membrane. This mass transfer rate through the membrane as a function of the axial velocity was evaluated. The transport performance through the membrane itself, the membrane loaded with enzyme, and the membrane loaded with the enzyme in the presence of emulsion were studied [21].

A specially designed microstructure in the composite membrane with a porous hydrophobic layer and a relatively dense hydrophilic layer is proposed for lipase immobilisation. Firstly, the composite membrane was prepared by coating the cellulose acetate layer with an average pore size of 1.40 nm on the hydrophobic PTFE layer with an average pore size of 76.3 nm. Next, the enzymes were absorbed into the pores of the PTFE layer, and deposited on the interface between the two layers by the filtration process. The relatively dense CA layer was used to reject the enzymes controlling the enzyme loading which prevented enzymes from being dissolved into the aqueous phase. The porous poly(tetrafluoroethylene (PTFE) layer supplied a hydrophobic environment and a large specific surface area for the immobilisation of lipases which were propitious to the activation of lipase. The activity of the immobilised lipase membrane based on the hydrolysis of olive oil was assayed in the biphasic membrane reactor, and a maximum specific activity (1.20 ± 0.04 mm mol^{-1} FFA min^{-1} cm^{-2}) was observed, which was higher than the value reported in some literature. The kinetic parameters of the immobilised lipases, K_m and K_{max} were fitted to the Michaelis–Menten equation. The optimum enzyme loading (0.020 ± 0.002 mg protein cm^{-2}) was obtained with the highest activity and without any diffusion-based limitations. Furthermore, the immobilised lipases retained 80% residual activity after ten hydrolysis cycles. The composite membrane was easily regenerated and the lipases immobilised in the regenerated membrane retained a high activity [22].

The asymmetric hydrolysis of racemic ibuprofen ester catalysed by lipase is one of the most important methods for the chiral separation of ibuprofen. A special microstructure in the composite hydrophilic cellulose acetate (CA)/hydrophobic polytetrafluoroethylene (PTFE) membrane was designed for lipase immobilisation by ultrafiltration. A biphasic enzymatic membrane reactor (EMR) and an emulsion reaction system with free lipase were both used, and the activity, enantioselectivity and half-life of the immobilised versus free enzymes were then compared. The morphology of the composite membrane and the position of the entrapment in the microstructure of the composite membrane where lipase was immobilised were observed by scanning electron microscopy. The effects of the substrate concentration, enzyme loading, reaction temperature and pH on this separation were investigated. The experimental results showed that the lipase was entrapped at the interface of the composite membrane, which is consistent with the interface of the aqueous phase and organic phases. Since the enzyme was immobilised by a new method that did not destroy the enzyme structure, it retained higher activity; the enzyme activity was more than 60% as compared to the free lipase when the enzyme loading was under 1 g protein m^{-2}. Furthermore, the immobilised enzyme provided better chiral selectivity and a longer half-life. High enzyme activity and chiral selectivity were obtained with a substrate concentration of 0.35 M, enzyme loading of 0.7–1.0 g protein m^{-2} in an aqueous phase of pH 8.0 and a temperature of 40 °C [23].

A biphasic enzymatic membrane reactor was made by immobilising Candida Rugosa lipase onto the dense surface of polysulfone ultrafiltration membrane by filtration and then crosslinking it with a glutaraldehyde solution. The reactor was further applied to the hydrolysis of olive oil, the performance of which was evaluated with respect of apparent reaction rate based on the amount of fatty acids extracted into the aqueous phase per minute and per membrane surface. The ultrafiltration and crosslinking processes greatly improved the reaction rate per unit membrane area and the enzyme lifetime. The highest reaction rate reached 0.089 μmol FFA min^{-1} cm^{-2} when the enzyme loading density was 0.098 mg cm^{-2}. These results also indicated that the performance of lipase immobilised onto a membrane surface was superior to that immobilised in the pores, and the apparent reaction rate and stability of the immobilised lipases were greatly improved greatly after crosslinking. It is suggested that the immobilisation of the enzymes by filtration and then crosslinking the enzymes onto the membrane surface is a simple and convenient method to prepare a high-activity immobilised enzyme membrane [24].

26.4.6 Polymer Membranes with Enzymes Immobilised by Copolymerisation

Two kinds of urease-immobilised polymer membranes were prepared. One was prepared by bulk-copolymerising a mixture consisting of vinylised-urease, acrylamide (AAm), 2-hydroxylethylmethacrylate (HEMA), and a crosslinker (urease-immobilised poly(VU-AAm-HEMA) membrane), and the other was prepared by ultrafiltrating a mixture composed of urease, quaternised chitosan, and sodium carboxymethylcellulose in an aqueous sodium bromide solution (urease-immobilised polyion complex membrane). The permeation and hydrolytic characteristics of the aqueous urea solutions were kinetically investigated under various conditions using urease-immobilised membranes. The hydrolysis of urea through the urease immobilised membranes followed Michaelis–Menten kinetics and is discussed herein.

When comparing kinetic data for the hydrolysis of urea through these membranes, the enzymatic activities of the urease-immobilised poly(VU-AAm-HEMA) membranes were lower than those of urease-immobilised polyion complex membranes, but the stabilities of the former were higher than those of the latter. An illustration of these membrane performances is shown in Figure 26.8. These differences could be significantly attributed to differences in the higher-order structure of urease molecules obtained during immobilisation. Although the enzyme activity in conventional and membrane reactors cannot be discussed in a similar way, the membrane reactor has advantages such as: (i) maintaining enzyme stability, (ii) easing the separation of enzyme and product and (iii) enzyme recovery. Therefore, in the near future, if the immobilisation amount of enzyme immobilised in polymer membranes can be increased, then the performance of these enzyme immobilised membrane reactor will be improved [27].

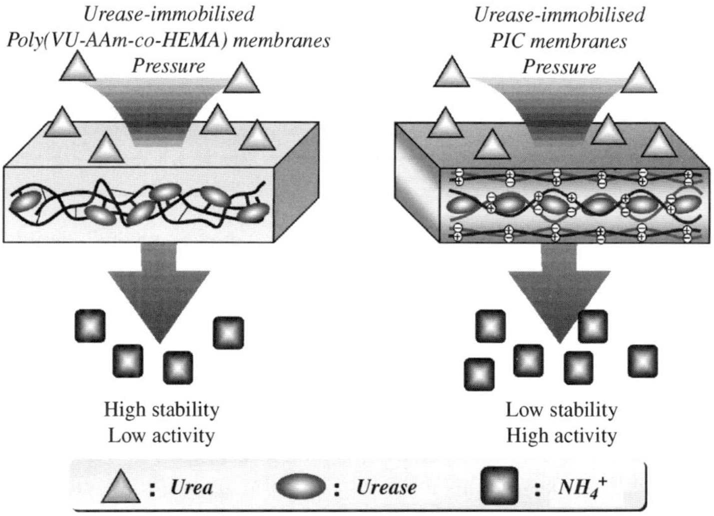

Figure 26.8 *Illustration of the performance of the urea-immobilised poly(VU-AAm-HEMA) membranes and urease-immobilised polyion complex membranes. Reprinted from Catalysis Today, Uragami, T., et al., Preparation of urease-immobilized polymeric membranes and their function, Vol. 118, 158–165. Copyright (2006) with permission from Elsevier*

26.4.7 Industrial Applications

A multiphase/extractive enzyme membrane reactor for production of diltiazem chiral intermediate ha been reported by Matson *et al.* [29]. Enzyme membrane reactors can be used to enhance the productivity and practicality of certain biotransformations by improving substrate/enzyme contact, by providing a simple and reversible means of enzyme 'immobilisation', and by effecting the removal of inhibitory reaction products. This paper describes the development and eventual scale-up of a multiphase/extractive membrane reactor designed to manage reaction problems encountered in a biphasic reaction system characterised by the formation of an inhibitory reaction product.

In particular, the application of this membrane reactor to an enzyme-mediated resolution of a racemic mixture is described – namely, the optical resolution of a chiral intermediate used in the production of diltiazem, a drug used in the treatment of hypertension and angina. The development process is traced from bench-scale studies of process feasibility through optimisation, process reliability, and pilot-plant studies – a process that ultimately culminated in the operation of a commercial-scale membrane reactor facility that currently produces over 75 t $year^{-1}$ of diltiazem intermediate [29].

26.5 Final Remarks and Conclusions

Many enzymes have been immobilised and their immobilised enzymes have been studied fundamentally. However, only a few immobilised enzymes have practical applications in the field. In particular, one application of enzyme-immobilised polymer membranes is as a biosensor. These results can be attributed to the fact that immobilised enzymes can not endure repeated use for a long time. When the problem of an enzyme-immobilised polymer membrane is

Table 26.2 *Some problems in the engineering of enzyme-immobilised polymer membranes*

Area	Problems
Enzyme engineering	Stability, heat resistance, solubilisation in organic solution
Membrane engineering	Selectivity, thin membrane form

reconsidered from the viewpoints of both the enzyme and membrane engineering, they can be summarised in Table 26.2. In the enzyme engineering, the stability of the immobilised enzyme is not sufficient. The development of an enzyme with heat resistance is strongly expected, because a high reaction velocity is demanded from the view of industrial economics.

The development of a soluble enzyme in organic solvents is required because the preparation of enzyme-immobilised polymer membrane then becomes much easier. Further, the membrane engineering field expects the development of an enzyme-immobilised membrane with high selectivity for the separation of products, formed by enzymatic reactions in the enzyme-immobilised polymer membrane. The formation of thin enzyme-immobilised polymer membrane is necessary to obtain reactions with high efficiencies and membrane performance with high permeation and separation. The production of tubular and hollow fibres from a polymer casting solution containing of the enzyme is very important for practical use.

The technology of enzyme-immobilised polymer membrane in which the reaction and separation are carried out simultaneously can be applied to a wide variety of fields which were impossible previously. In the future, this technology will be expanded to other fields.

To achieve this objective, a joint research project is very important, because the science and technology of enzyme-immobilised separation membranes represent a boundary in the academic and industrial fields.

References

1. I. Chihata (ed.), *Immobilized Enzymes*, Kodannsha, Tokyo (1975).
2. K. Mosbach (ed.), *Methods in Enzymolozy, Vol. 44*, *Immobilized Enzymes*, Academic Press, London (1976).
3. S. Fukui, I. Chihata, S. Suzuki (eds), *Enzyme Technology*, Kagaku Dijinn, Tokyo (1981).
4. T. Uragami, Possibility to applications of enzyme-immobilized sepasration membranes, (I) Preparation method and estmation of performance, *Biol. Ind.*, **3**, 638–648 (1986). (II) Applications and future prospects, *Biol. Ind.*, **3**, 743–746 (1986).
5. A. Lante, A. Crapis, A. Krastanov, *et al.*, Biodegradation of phenols by laccase immobilised in a membrane reactor, *Proc. Biochem.*, **36**, 51–58 (2000).
6. N. Hilal, V. Kochkodan, R. Nigmatullin, *et al.*, Lipase-immobilized biocatalytic membranes for enzymatic esterification: comparison of various approaches to membrane preparation, *J. Membr. Sci.*, **268**, 198–207 (2006).
7. A. Trusek-Holownia, A. Noworyta, An integrated process: ester synthesis in an enzymatic membrane reactor and water sorption *J. Biotechnol.*, **130**, 47–56 (2007).
8. T. Kawai, K. Saito, T. Sigo, H. Misaka, Immobilization of ascorbic oxydase in multilayers onto porous hollow fibre membrane, *J. Membr. Sci.*, **191**, 207–213 (2001).
9. N. Diano, V. Grano, S. Rossi, *et al.*, Hollow fibre enzyme reactor operating under nonisothermal conditions, *Biotechnol. Progr.*, **20**, 457–466 (2004).

10. CC. Lin, MC. Yang, Urea permeation and hydrolysis through hollow fiber dialyzer immobilized with urease: storage and operation properties, *Biomaterials*, **24**, 1089–1994 (2003).
11. NS. Pujari, B. Vaidya, S. Bagalko, *et al.*, Poly(urethane methacrylate-co-glycidyl methacrylate)-supported-polypropylene biphasic membrane for lipase immobilization, *J. Membr. Sci.*, **285**, 395–403 (2006).
12. JP. Chen, SH Chiu, A poly(N-isopropylacrylamide-*co*-*N*-acryloxysuccinimide-*co*-2-hydroxyethyl methacrylate) composite hydrogel membrane for urease immobilization to enhance urea hydrolysis rate by temperature swing, *Enzyme Microbiol. Technol.*, **26**, 359–367 (2000).
13. M. Amounas, C. Innocent, S. Cosnier, *et al.*, A membrane based reactor with an enzyme immobilized by an avidin–biotin molecular recognition in a polymer matrix, *J. Membr. Sci.*, **176**, 169–176 (2000).
14. M. Tamura, T. Uragami, M. Sugihara, Ultrafiltration, hydrolysis, and adsorption characteristics of mrmbranes from cellulose nitrate, urease, stylite, and activated charcoal, *Angew, Makromol. Chem.*, **79**, 67–77 (1979).
15. T. Uragami, T. Aketa, S. Gobodani, M. Sugihra, New method for enzymr immobilization by a polyion complex membrane, *Polymer Bull.*, **15**, 101–106 (1986).
16. T. Uragami, T. Aketa, Hydrolisis of urea through urease immobilizing polyion complex membrane, *Membrane*, **14**, 211–216 (1989).
17. T. Uragami, Structures and properties of membranes from polysaccharide derivatives, in *Polysaccharides, Structural Diversity and Functional Versatility*, S. Dumitrium (ed.), pp 887–924, Marcel Dekker, New York (1998).
18. K. Sakaki, L. Giorno, E. Drioli, Lipase-catalyzed optical resolution of racemic naproxen in biphasic enzyme membrane reactors, *J. Membr. Sci.*, **184**, 27–38 (2001).
19. S. Mazzuca, L. Giorno, A. Spadafora, *et al.*, immunolocalization of beta-glucose immobilized within polysulfone capillary membrane and evaluation of its activity in situ, *J. Membr. Sci.*, **285**, 152–158 (2006).
20. N. Hilal, R. Nigmatullin, A. Alpatova, Immobilization of crosslinked lipase aggregates within microporous polymeric membranes, *J. Membr. Sci.*, **238**, 131–141 (2004).
21. L. Giorno, JC. Zhang, E. Drioli, Study of mass transfer performance of naproxen acid and ester through a multiphase enzyme-loaded membrane system, *J. Membr. Sci.*, **276**, 59–67 (2006).
22. J. Xu, YJ. Wang, Y. Hu, *et al.*, Immobilization of lipase by filtration into a specially designed microstructure in the CA/PTFE composite membrane, *J. Mol. Cat. B Enzyme*, **42**, 55–63 (2006).
23. YJ. Wang, Y. Hu, H. Xu, *et al.*, Immobilization of lipase with a special microstructure in composite hydrophilic CA/hydrophobic PTFE membrane for the chiral separation of racemic ibuprofen, *J. Membr. Sci.*, **293**, 133–141 (2007).
24. Y. Wang, X. Jian, G. Luo, *et al.*, Immobilization of lipase by ultrafiltration and crosslinking onto the polysulfone membrane surface, *Bioresour. Technol.*, **99**, 2299–2303 (2008).
25. S. G. Shoemaker, A. S. Hoffman, J. H. Priest, Synthesis and properties of vinyl monomer enzyme conjugates-conjugation of *L*-asparaginase with *N*-succinimidyl acrylate, *Appl. Biochem. Biotechnol.*, **15** 11–24 (1987).
26. A. K. Hazra, S. P. Chock, R. W. Albers, Protein determination with trinitrobenzene sulfonate – A method relatively independent of amino-acid-compound, *Anal. Biochem.*, **137**, 437–443 (1984).
27. T. Uragami, K. Ueguchi, M. Watanabe, T. Miyata, Preparation of urea-immobilized polymeric membranes and their function, *Catal. Today*, **118**, 158–165 (2006).
28. H.-G. Hicke, M. Becker, B.-R. Paulke, M. Ulbricht, Covalently coupled nanoparticles in capillary pores as enzyme carrier and as turbulence promoter to facilitate enzymatic polymerisations in flow-through enzyme–membrane reactors, *J. Membr. Sci.*, **282**, 413–422 (2006).
29. J. L. Lopez, St. L. Matson, A multiphase/extractive enzyme membrane reactor for production of diltiazem chiral intermediate, *J. Membr. Sci.*, **125**, 189–211 (1997).

Final Remarks

Angelo Basile[1] and Fausto Gallucci[2]

[1]*Institute of Membrane Technology, CNR, c/o University of Calabria, Rende, CS, Italy*
[2]*Chemical Process Intensification, Faculty of Chemical Engineering and Chemistry, Eindhoven University of Technology, Eindhoven, The Netherlands*

1 Introduction

In the past two decades, chemical research has been more and more oriented towards process intensification (PI). PI is a revolutionary approach in process and plant design with the ultimate goal of achieving a safer, cleaner, smaller – and cheaper process [1].

According to Stankiewicz and Moulijn [2], a method to achieve process intensification is by using multifunctional reactors. Membrane reactors are a great example of multifunctional reactors.

In fact, as an IUPAC definition, a membrane reactor is a device for simultaneously performing a reaction and a membrane-based separation in the same physical device [3].

In this book the application of membrane reactors for different reaction systems (including membrane bioreactors) has been introduced and discussed in the first review chapter.

The membrane being the core and most important part of a membrane reactor (which indeed determines also the design of the reactor itself), the other chapters have been devoted to the preparation methods of the different membranes (inorganic in the first part of the book and organic in the last part). Examples of membrane reactors for each type of membrane have been shown.

2 Membranes for Membrane Reactors

Generally, membranes can be classified into homogenous or heterotgeneous, symmetric or asymmetric in structure, solid or liquid; they can possess a positive or negative charge as well as

Membranes for Membrane Reactors: Preparation, Optimization and Selection, Edited by Angelo Basile and Fausto Gallucci

being neutral or bipolar. In all cases, a driving force as a gradient of pressure, concentration, and so forth, is applied in order to induce permeation through the membrane.

Thus, the membranes can be categorised according to their nature, geometry and separation regime. Although membranes can be divided per nature into biological and synthetic membranes, membrane reactors almost always use synthetic membranes. *Synthetic membranes* can be subdivided into organic (polymeric) and inorganic (ceramic, metal). *Polymeric membranes* commonly operate between 100–300 °C and are used in low temperature membrane reactors. A great example of application of polymeric membranes is the membrane bioreactor. On the contrary, *inorganic membranes* often have a working temperature above 250 °C, and for this reason are often used in high temperature membrane reactor gas phase reactions.

2.1 Inorganic Membranes

The first part of this book explains how to produce inorganic membranes for membrane reactors.

In particular, a way to use polymeric precursors to prepare membranes able to work at high temperature is described by Yoshimune and Haraya (Chapter 1). These authors describe how a carbon membrane can be produced in microporous structure. There are many studies illustrating how to produce interesting microporous ceramic membranes. However, the application of these kinds of membrane to membrane reactors is very limited so far.

Yoshimune and Haraya have listed some applications of carbon membranes in membrane reactors, such as for H_2-related reactions (e.g., dehydrogenation of cyclohexane on Pt-based catalysts in a hollow fibre carbon-based membrane reactor). Those membranes also affect other reactions involving gases such as H_2O, CO_2, and NH_3, because of the high selectivities and superior chemical resistances of carbon membranes.

Carbon-based membranes can be used as H_2O-selective membranes for the separation of H_2O–alcohol mixtures by pervaporation. The authors suggested that it would be interesting to apply carbon membranes in such reactions where water removal can increase the conversion/selectivity degree.

Producing such a membrane at laboratory scale seems to be a common practice at the moment, however, to foresee a wide application of carbon-based membrane reactors the scaling up challenge should be tackled.

As a matter of fact, it is needed the production of high surface area membranes in a reproducible fashion and in a cost effective manner. Although this challenge seems to be hard to overcome, Yoshimune and Haraya pointed out that some hope can be placed in the availability of new polymer precursors, new production methods, and new module fabrication techniques, which are being exploited at the moment.

Last but not least is the problem/challenge in common with all the inorganic membranes for high temperature applications, namely the membrane/module seals which is a research area itself.

Along with carbon membranes, which seem to be a bridge between polymeric and inorganic membranes, the high temperature membrane reactors make use of ceramic or metallic membranes. These membranes can be produced by various techniques each of which has its own advantage (simplicity, cost effective) and disadvantages (mostly related to the quality of the resulting membranes).

Madaeni and Daraei (Chapter 2) reported how to produce metallic membranes by wire arc spraying. Electric arc spraying is a thermal spray process in which an arc is struck between two consumable electrodes of a coating material. A compressed gas is used to atomise and propel the material to the substrate. The resulting porous metallic layer can act as a membrane. An example of porous stainless steel membrane prepared by arc spraying has been reported.

The technique is quite interesting if a porous membrane is the target. In this case various advantages can be enumerated and, according to Madaeni and Daraei, the main ones are: (i) low cost, (ii) rapid fabrication, (iii) ease of scale up, and (iv) simplicity of producing a membrane with various thicknesses and appropriate mechanical strength without the need for an additional support.

The disadvantage is mainly due to the porosity of the membranes obtained, which should be optimised by opportunely tuning all the spaying parameters such as gun distance and atomiser air pressure. However, the metallic membranes obtained are interesting for water related reactors.

As already said, when working with membranes an important parameter to take into account is the membrane surface area available per unit of reactor volume. A way to increase this ratio is of course by selecting an appropriate membrane geometry (such as hollow fibre). Li and coworkers, at Imperial College London, are pioneers in producing ceramic hollow fibre membranes and this technique is extensively described in their chapter. In particular, Li and coworkers (Chapter 3) tackled the challenges in preparing ceramic hollow fibre membranes which are essentially the cost effective technique (often various steps are required) and the possibility of designing the morphology of the membrane for a specific process.

According to Li *et al.*, by using a combined phase inversion and sintering technique, it is possible to prepare asymmetric ceramic hollow fibres in one go, and also able to design the membrane through a wide range of morphologies.

This latter possibility makes the technique available for producing membranes for a wide range of applications. Li and coworkers pointed out an interesting example in which asymmetric alumina hollow fibre membranes can be used for catalytic membrane reactors. In particular, finger-like voids generated during the fibre preparation process can be impregnated with catalyst while the sponge-like region of the fibre can serve as a separation layer (for instance for hydrogen).

This work points out the versatility of such hollow fibres that can be used as membranes, substrates for catalyst deposition, for further membrane layer deposition or for catalytic membranes.

The cost reduction for the production method combined with the versatility of the combined phase inversion and sintering method for the production asymmetric ceramic hollow fibre membranes is a driving force for wide application and exploitation of membrane reactors in the near future.

Among the applications of membranes in membrane reactors, the two big players are the application of Pd-based membranes for selective hydrogen removal and the application of perovskite-like membranes for selective oxygen addition to the reactor. This is why many chapters in the book are devoted to the different production techniques of such membranes.

For example, Tosti (Chapter 4) has shown in his chapter, how to produce dense self-supporting membranes by rolling technique. In particular, applying the cold rolling, commercial Pd-alloy foils can be easily used for preparing membrane tubes of wall thickness 50–60 μm. Although the thickness of the membranes is higher than membranes prepared by deposition techniques, these membranes are self-supported and the mass transfer resistances are concentrated in the Pd layer, whereas for supported membranes the support itself plays a significant role in the mass transfer resistance and in the membrane cost as well.

An alternative method which can lead to dense membrane foils thinner than 50 micron has been described by Phair and Gibson (Chapter 19), who introduced the planar flow casting method. According to the authors, planar flow casting is a method of continuously casting a molten alloy into a thin strip or generating a range of novel microstructures with a variety of functional properties. It is highly suited to the fabrication of 100% dense thin foils of nanocrystalline and quasicrystalline alloys useful for hydrogen separation membranes.

Self-supported membranes have been tested for pure hydrogen production via reforming of various hydrocarbons and proved to be reliable (no formation of cracks) also for experimental campaigns longer than 1 year. This technique deserves to be further developed in order to reduce the final membrane thickness (and thus the costs) and to prepare dense membranes out of different metal alloys.

As far as oxygen selective membranes are concerned, Tong and O'Hayre (Chapter 5) have provided an extensive description of the preparation methods for mixed ionic and electronic conducting (MIEC) membranes for oxygen separation. Oxygen permeates through MIEC ceramic membranes via a combined electrochemical surface reaction and solid-state ambipolar diffusion process involving both oxygen vacancies and electrons.

The membranes are prepared out of MIEC powder which can be optimised in concentration in order to enhance oxygen transport and stability. Of course, MIEC membranes can be applied to all those high temperature (required for a sufficient membrane flux) reactions involving oxygen. Typical examples are partial oxidation of methane and oxidative coupling of methane, among others. The authors stressed the main concern about these membranes which is related to their stability and mechanical strength. Often, supported membranes are produced to enhance the mechanical strength, but thermal expansion at the high temperatures considered makes the solution challenging. It is clear that there is room for further improvements of these membranes which prospect a big impact on processes, the costly air separation unit not being required if oxygen membrane reactors are employed.

Related to the chapter of Tong and O'Hayre, Zyryanov and coworkers (Chapter 6) concentrated their attention on the microfabrication method behind the production of thin supported oxygen permeable membranes. The selection of the most appropriate MIEC constituents, the support and inter-diffusion layer and the selection of the appropriate microfabricated catalyst for building up a methane reforming membrane reactor, have been discussed. The authors show that each component of the membrane reactor should be optimised and deserves appropriate selection in order to produce an efficient fuel reformer.

An interesting application of membranes in reaction systems is represented by Catalytic Membrane Reactors, in which the membrane has a catalytic and separation function. The membrane can be either self-catalytic, or can be made catalytic by coating the surface of a dense membrane, or by depositing the catalyst material inside the pores of the membrane, or by casting a solution containing the polymeric material and the catalytic material [4–7]. In this respect, Kurungot and Yamaguchi (Chapter 7) described the synthesis and application of a catalytic membrane reactor where the catalytic membrane is produced as a sandwich (layer by layer support/catalyst/selective layer).

As perm-selective layer, a silica film was adopted demonstrating how silica membranes can be considered good candidates for hydrogen production in membrane reactors.

Malygin *et al.* (Chapter 13) illustrated the production of catalytic membranes prepared via molecular layering method. This method consists in the sequential growth of monolayers of structural units of the preassigned chemical composition and texture on the surface of a solid matrix, by carrying out the chemical reactions between the functional groups of a solid and the reagents brought to it under conditions far from equilibrium. The authors showed how to apply the method to synthesise V-P-O-based catalytic layer on a tubular porous aluminum support to produce a catalytic membrane reactor.

Catalytic membranes can be also produced via metal vapour synthesis, as indicated by Pitzalis *et al.* (Chapter 14). In their chapter, the authors demonstrated how to optimise the method for depositing mono- and bimetallic catalysts, onto γ-Al_2O_3 and SiO_2 tubular inorganic membranes with a good control of the morphology, catalyst stability, and so on.

Another family of membranes suitable for membrane reactor applications are zeolite membranes. Zeolites are crystalline microporous aluminosilicates which present quite well defined pore structure and distribution, ion exchange capability as well as catalytic activity for some reactions. Zeolites have been produced as catalysts for a long time, but they can also be produced as membranes (often supported) as described by Téllez and Menéndez (Chapter 8).

There is a large variety of different zeolites that can be selected to produce a membrane for membrane reactors. Each zeolite has its own well defined structure and pore dimension, which makes it easy to select a right (or more appropriate) zeolite for each reaction system.

Research is now trying to tackle the challenge of producing thin membranes without (or with minimal) inter-crystal defects and in a reproducible manner.

These families of inorganic membranes (metal, silica, zeolite, oxygen transport membranes) can be produced with a variety of different methods: all with their pros and cons.

Metal supported membranes can be prepared by diffusion welding as reported by Tosti *et al.* (Chapter 9). A metal layer can be deposited on a porous support (ceramic or metallic) via physical vapour deposition (PVD) as reported of Checchetto *et al.* (Chapter 10) or via electroless plating, which is the most used technique for metal supported membranes at the moment (see Chapter 11, by Broglia *et al.*).

However, silica membranes have been successfully produced via chemical vapour deposition (CVD) as reported by Galuszka and Giddings (Chapter 12). With this technique, microporous silica membranes for hydrogen production have been produced that present good stability at high temperatures (600 °C) and good perm-selectivities (although well below the selectivities of Pd-based hydrogen membranes). According to these authors, although progress in CVD membranes has been achieved so far, a clear correlation between the final property of a membrane and the experimental parameters during the CVD process is not yet available.

Cazorla-Amorós and coworkers (Chapters 15, 16) described electrochemical and electroforetic deposition for membrane (zeolite type) preparation. The electroforetic deposition method is quite interesting for zeolites because it allows the continuous preparation of high quality supported membranes to be used in membrane reactors. The method is anyway applicable to a variety of possible deposits.

Basile *et al.* (Chapter 17) also used an electrochemical method for deposition of Pd (and other metals) on different supports, confirming the versatility of such a method.

Another very versatile method is the spy pyrolysis deposition described by Li and Guo (Chapter 18). According to the literature presented by the authors, spry pyrolysis has been successfully applied for deposition of organic carbon membranes, composite functional membranes, inorganic Pd-Ag alloy hydrogen separation membrane, porous TiO_2 membranes and perovskite conductive oxide membranes. The success of this method is basically due to the simplicity of experimental apparatus and good productivity of this technique.

It should be also reported that many scientists are working with Pd-based membranes because these membranes already showed their potential in improving the performances of reactors. However, if one looks at industrial scale applications, Pd seems to have some limitations because of both its cost and low availability. For this reasons an increasing attention is devoted to research on membrane materials for hydrogen separation different than Pd. An example is given by Yamaura and Inoue (Chapter 20), who demonstrated the feasibility of hydrogen separation through amorphous Ni-Nb-Zr alloys. This is an interesting way to produce stable membranes without Pd (or with limited amount of Pd used as protecting layer).

2.2 Organic Membranes

Organic membrane preparation methods are described in the second part of the book. Indeed, polymeric membranes have been studied for a long time and are already applied in a great variety of processes, ranging from water treatment and desalination to gas separation. The state of the art of these membranes is thus well advanced compared with inorganic membranes.

Polymeric membranes are also applied in membrane reactors. The most important applications are in the field of membrane bioreactors and membrane reactors for fine chemical and pharmaceutical synthesis.

Last but not least, polymeric membranes are interesting for membrane reactors because the matrix of the membrane (the polymer) can actively take part in the reaction by its affinity towards reagents and products, while a much wider choice of polymeric membranes is available as compared with metallic or ceramic membranes, and the costs are generally lower.

The preparation methods described in the book range from phase inversion to plasma membrane and to impregnation membranes (for inorganic films/catalysts onto polymeric membranes).

Buonomenna *et al.* (Chapter 21) described (with a review-like chapter) the phase inversion technique to obtain suitable membranes for membrane reactors application. According to the authors, the phase inversion technique is a simple technique by which a stable polymer solution is subjected to a liquid–liquid demixing during which the cast polymer film separates into a polymer-rich (membrane matrix) and a polymer-lean phase (membrane pores).

Basically, a great variety of polymers are suitable for phase inversion membrane preparation, while probably the most used polymer has been the polydimethylsiloxane. The technique is of course well suited for producing membranes with catalyst incorporated in the matrix, in order to obtain the catalytic polymeric membrane for membrane reactors.

Phase inversion (or phase separation) method is at present the most extensively used in industry for polymeric membrane production. As far as phase separation is concerned, Ulbricht and Susanto (Chapter 22) showed the different kind of membranes that can be obtained via this technique. The technique has been used for producing flat, tubular (hollow fibre) and capsule membranes. Each type of membrane can be used in different reactors.

As pointed out by Ulbricht and Susanto (Chapter 23), significant effort and development has also been made to prepare porous polymeric membranes by other processes than phase separation. The authors show the application of extrusion methods, film stretching, foaming, spinning and electrospinning and so on.

Most of the time, polymeric structures produced by one of the above-mentioned techniques, is used to support a suitable catalyst for a low temperature reaction. Buonomenna *et al.* (Chapter 21) gave some examples, while Volkov *et al.* (Chapter 24) described an interesting application of polymeric membranes as support for palladium catalyst for a variety of reactions. An interesting application of these membranes is the two-phase or three-phase hydrogenation reactions (like de-oxygenation of water).

Since the polymeric matrix can be selected from a great variety of polymers, the selection of the more appropriate structure to host a particular catalyst is much simpler than compared with inorganic membranes. Volkov *et al.* (Chapter 24) described the production of different catalytic membranes for membrane reactors, also with mixed catalysts.

Although the polymeric structure available today is quite wide, the production of new structure reached a plateau. This means that scientists are seeking new techniques for producing novel membrane structures. All the methods described by Buonomenna, Ulbricht and Susanto are essentially based on a selection and modification of a certain polymer with successive production of the desired membrane. In contrast, the approach described by Bryjak and Gancarz

(Chapter 25) forsees the preparation of a certain membrane and the following modification of the surface. In particular, the modification of the membrane structure by plasma treatment is reported in detail. What is interesting in plasma modification is that different plasma techniques can be used; however, plasma modified membranes show a unique character. Their bulk properties stay unchanged, while the surface differs from each to the other – depending on the plasma used. Thus, depending on the membrane reactor application, an adequate plasma modification method can be adopted.

Finally, an expanding example of polymeric membrane reactors is the application to reactions taking place *in vivo*. In this case the catalyst adopted should be able to catalyse these reactions. These catalysts are the enzymes and for this reason a wide literature on enzyme supported membranes is available. Uragami (Chapter 26) described enzyme-immobilised polymer membranes and their applications for in vivo reactions.

According to the author, although many enzymes can be immobilised in membranes, only a few immobilised enzymes have practical applications in the field. This means that research on this very challenging field needs more effort. In particular, one application of enzyme-immobilised polymer membranes can be as a biosensor.

3 Epilogue

Membrane reactors can be use in a wide range of application, ranging from *in vivo* reactions, to high temperature gas phase reactions. Each application needs a specific membrane (type, geometry) and each membrane needs an appropriate preparation method.

This book reports all the up to date preparation methods for inorganic, organic and enzymatic membranes.

References

1. Process Intensification Network, http://www.pinetwork.org/.
2. A.I. Stankiewicz, J.A. Moulijn, Process intensification: transforming chemical engineering, *Chem. Eng. Progr.*, **96** (1), 22–33 (2000).
3. W.J. Koros, Y.H. Ma, T. Shimidzu, Terminology for membranes and membrane processes, *J. Membrane Sci.*, **120**, 149–159 (1996).
4. X. Dong, C. Zhang, X. Chang, W. Jin, N. Xu, A self-catalytic membrane reactor based on a supported mixed-conducting membrane, *AIChE J.*, **54** (6), 1678–1680 (2008).
5. S. Bhatia, C.Y. Thien, A. R. Mohamed, Oxidative coupling of methane (OCM) in a catalytic membrane reactor and comparison of its performance with other catalytic reactors, *Chem. Eng. J.*, **148**, 525–532 (2009).
6. D. Fritsch, G. Bengtson, Development of catalytically reactive porous membranes for the selective hydrogenation of sunflower oil, *Catal. Today*, **118**, 121–127 (2006).
7. K.C. de Souza Figueiredo, V.M. Martins Salim, C.P. Borges, Synthesis and characterization of a catalytic membrane for pervaporation-assisted esterification reactors, *Catal. Today*, **133/135** 809–814 (2008).

Index